AFRICAN BIOGEOGRAPHY, CLIMATE CHANGE, & HUMAN EVOLUTION

THE HUMAN EVOLUTION SERIES

edited by
Russell Ciochon, University of Iowa
Bernard Wood, George Washington University

African Biogeography, Climate Change, and Human Evolution
edited by Timothy G. Bromage and Friedemann Schrenk

AFRICAN BIOGEOGRAPHY, CLIMATE CHANGE, & HUMAN EVOLUTION

EDITED BY Timothy G. Bromage & Friedemann Schrenk

New York Oxford

Oxford University Press

1999

Oxford University Press

Oxford New York
Athens Auckland Bangkok Bogotá Buenos Aires Calcutta
Cape Town Chennai Dar es Salaam Delhi Florence Hong Kong Istanbul
Karachi Kuala Lumpur Madrid Melbourne Mexico City Mumbai
Nairobi Paris São Paulo Singapore Taipei Tokyo Toronto Warsaw

and associated companies in
Berlin Ibadan

Published by Oxford University Press, Inc.
198 Madison Avenue, New York, New York 10016

Library of Congress Cataloging-in-Publication Data
African biogeography, climate change, and human evolution / edited by
Timothy G. Bromage and Friedemann Schrenk.
p. cm. — (the human evolution series)
Revision of papers originally presented at the Wenner-Gren
Foundation International Symposium No. 119, held at
Salima, Malawi, Oct. 25–Nov. 4, 1995.
Includes bibliographical references and indexes.
ISBN 0-19-511437-X
1. Human evolution—Congresses. 2. Paleoecology—Africa—
Congresses. 3. Fossil hominids—Africa—Congresses. 4. Evolution
(Biology)—Africa—Congresses. I. Bromage, Timothy G.
II. Schrenk, Friedemann. III. Wenner-Gren Foundation International
Symposium (119th : 1995 : Salima, Malawi) IV. Series.
GN281.A3 1999
599.93′8—dc21 98-21953

9 8 7 6 5 4 3 2 1

Printed in the United States of America
on acid-free paper

Contents

Part III Fossil Faunas

Part IV Hominid Evolution

List of Contributors

Peter Andrews
Department of Palaeontology
Natural History Museum
Cromwell Road
London SW7 5BD
England

Miranda Armour-Chelu
Virginia Museum of Natural History
1001 Douglas Avenue
Martinsville, VA 24112
USA

Brenda R. Benefit
Department of Anthropology
Southern Illinois University
Carbondale, Illinois 62901
USA

Ray Bernor
College of Medicine
Department of Anatomy
Lab of Evolutionary Biology
Howard University
520 W Street, NW
Washington, DC 20055
USA

Laura C. Bishop
School of Biological and Earth Sciences
Liverpool John Moores University
Byrom Street
Liverpool L3 3AF
England

Timothy G. Bromage
Department of Anthropology
Hunter College, C.U.N.Y.
695 Park Avenue
New York, New York 10021
USA

Mark Collard
Department of Anthropology
University College London
London WC1E 6BT
England

Yves Coppens
Collège de France
11, Place Marcelin Berthelot
75231 Paris Cedex 05
France

George H. Denton
Institute for Quaternary Studies
Boardman Hall
University of Maine
Orono, Maine 04469
USA

Christiane Denys
Laboratoire Mammifères and Oiseaux
M.N.H.N. 55, rue Buffon
75005 Paris Cedex 05
France

Craig Feibel
Department of Anthropology
Rutgers University
131 George Street
New Brunswick, New Jersey 08901
USA

Robert A. Foley
Department of Biological Anthropology
University of Cambridge
Downing Street
Cambridge CB2 3DZ
England

Peter Grubb
35 Downhills Park Road
London N17 6PE
England

Clark Howell
Laboratory for Human Evolutionary Studies
Museum of Vertebrate Zoology
University of California, Berkeley
Berkeley, California 94720
USA

Louise Humphrey
Department of Palaeontology
Natural History Museum
Cromwell Road
London SW7 5BD
England

Thomas M. Kaiser
Institute and Museum of Zoology
Johann-Sebastian-Bach Str. 11-12
D-17489 Greifswald
Germay

Jonathan Kingdon
Department of Zoology
University of Oxford
Oxford 0X1 3PS

Ottmar Kullmer
Department of Geology & Paleontology
Hessisches Landesmuseum Darmstadt
Friedensplatz 1
D-64283 Darmstadt
Germany

Meave Leakey
Department of Palaeontology
National Museums of Kenya
P.O. Box 40658
Nairobi
Kenya

Jeffrey K. McKee
Department of Anthropology
The Ohio State University, Lord Hall
124 W. 17th Avenue
Columbus, Ohio 43210
USA

Eileen M. O'Brien
Institute of Ecology
University of Georgia
Athens, Georgia 30606
USA

R. Norman Owen-Smith
Department of Zoology
University of the Witwatersrand
Wits 2050
South Africa

Charles R. Peters
Department of Anthropology
Institute of Ecology
Baldwin Hall
University of Georgia
Athens, Georgia 30602
USA

Fernando Ramirez Rozzi
Laboratoire Paléoanthropologie et Préhistoire
Collège de France
Statíon Berthelot
Avenue Marcelin Berthelot
92360 Meudon-La-Forèt
France

Michael L. Rosenzweig
Department of Ecology & Evolutionary Biology
University of Arizona
P.O. Box 210088
Tucson, Arizona 85721
USA

Oliver Sandrock
Department of Geology & Paleontology
Hessisches Landesmuseum Darmstadt
Friedensplatz 1
D-64283 Darmstadt
Germany

Friedemann Schrenk
Department of Geology & Paleontology
Hessisches Landesmuseum Darmstadt
Friedensplatz 1
D-64283 Darmstadt
Germany

Nancy E. Sikes
Department of Anthropology
Smithsonian Institution
Museum Support Center MRC-534
Washington, DC 20560
USA

Fred Szalay
Department of Anthropology
Hunter College, C.U.N.Y.
695 Park Avenue
New York, New York 10021
USA

Alan Turner
School of Biological and Earth Sciences
Liverpool John Moores University
Byrom Street
Liverpool L3 3AF
England

Elisabeth S. Vrba
Department of Geology and Geophysics
Yale University, Box 208109
New Haven, Connecticut 06520
USA

Christopher Walker
Department of Anthropology
State University of New York
Stony Brook, New York 11794
USA

Bernard Wood
Department of Anthropology
George Washington University
2110 G Street, NW
Washington, DC 20052
and
Human Origins Program
National Museum of History
Smithsonian Institute
USA

AFRICAN BIOGEOGRAPHY, CLIMATE CHANGE, & HUMAN EVOLUTION

Searching for an Interdisciplinary Convergence in Paleoanthropology

Timothy G. Bromage
Friedemann Schrenk

This volume presents a distillation of ideas and much focused information discussed at the Wenner-Gren Foundation International Symposium No. 119 on African Biogeography, Climate Change, and Early Hominid Evolution, held at the Livingstonia Beach Hotel, Salima, Malawi, between October 25 and November 4, 1995.

The primary aim of the symposium was to explore the means by which a perspective that interprets early humans ecologically and adaptively in the larger context of their habitat specificities might help us to better chronicle human evolution. To help achieve this aim, an interdisciplinary group of scholars with specialties in geology, evolutionary theory, modern savanna ecology, paleoecology, biogeography, vertebrate paleontology and early hominid systematics, behavior, and morphology was brought together. The participants prepared papers in advance of the meeting so that a structured discussion could focus on issues related more to the complementary nature of the various specialties than on any one individual contribution. In light of this discussion, the papers were subsequently changed (or hardened) by the participants following the symposium. Each paper was thereafter peer reviewed and included in the symposium proceedings here.

There are several things we wish to accomplish in this introduction. First, we abstract the volume. Next, we describe the background of events that led us to this symposium and, thus, the reason for convening in Malawi. We then highlight several questions used to navigate contributors toward a level of interdisciplinary convergence pertaining to climate change and human evolution. We end with a statement about the progress that we believe has been made toward our primary aim.

Overview

Part I. Theory

Coppens opens the volume with a lively introduction, detailing historical aspects of the International Omo Research Expedition, describing how this interdisciplinary field venue established his long-term interest in climate change and human evolution, and then providing a grand scenario of human evolution.

Vrba characterizes habitat theory as emphasizing the inherited properties of lineages and the physical and biotic contexts. Hypotheses are generated that underscore the primary interests of the symposium: the relationships between taxonomic patterns and biogeography in the context of environmental change.

Szalay explains that habitat theory can only be as valid as the evolutionary theory it depends upon, and habitat theory, being allied with a hierarchic punctuationist paradigm that is essentially non-Darwinian, fails to stand up to theoretical scrutiny. Although the environment is acknowledged to be a causal factor, it is unlikely to be *the* factor that accounts for the underlying taxonomic patterns exhibited in the fossil record. Darwinian competition appears to drive taxic turnover.

McKee points out that species-level evolutionary patterns come about when, by chance, the variant phenotypes of individuals arise in favorable environments. Changing environments themselves, therefore, do not catalyze evolution.

Rosenzweig characterizes functional groups of organisms as guilds that enable a "community" of fitness functions (the aggregate functions of all competing species) to coexist within a specified habitat. A punctuationist scenario of human evolution is advanced, as example, of how climate change (or any other perturbation to competitive relationships mediated by ecological change) changed the fitness function of the *Homo* lineage.

Turner takes the questions posed to the participants (see below) as his starting point for a discussion that resolutely absorbs habitat theory to explain biogeographic patterns and evolutionary transformations, especially of Carnivora. The theoretical backdrop to the establishment of these patterns is the specific-mate recognition system.

Part II. Geology, Ecology, and Biogeography

Kingdon focuses on the larger quest of finding ever more accurate and ever less biased interpretations of physical and biotic patterns, both global and African. He reminds us that the facts of climatic history and the surviving relics of Africa's biota should be treated as sources of "hard" evidence in the same way as fossils and artifacts.

Denton outlines major Cenozoic climate changes and provides background for the causes of climatic change since the middle Eocene. He proposes that changed Atlantic water circulation may have been the trigger for global cooling, resulting in human adaptation between 2.95 and 2.52 m.y.

O'Brien and Peters discuss the usefulness of a knowledge of African landscapes for studies of early hominid biodiversity. They present biomes as the generators of hominid evolution and describe the Zambezian Regional Centre of Endemism as having conditions particularly favorable to early hominid biodiversity.

Owen-Smith discusses the relevance of climatic change to evolutionary change through its effects on vegetation patterns, habitat conditions, and resource supplies. Special features of African savanna environments are presented that foster cladogenesis among ungulates and hominoids in the Mio-Pliocene.

Grubb provides an examination of faunal evolution as inferred from modern biogeographic distributions. His discussion of mammalian biogeography provides both a rationale and some evidence for major dispersions in their past history, concluding that cycles of disruption and coalescence among biomes, not climate change, have provided the motor for cladogenesis of African mammals during the Quaternary.

Part III. Fossil Faunas

Howell puts the aims of this symposium volume into perspective. In contrast with past decades, we now have in-depth data across a broad front of the natural sciences relevant to the African late Cenozoic and thus to paleoanthropology. Much prior speculation has been succeeded by an expanded body of empirical documentation. His analysis of diversity in extinct and extant African bovids shows remarkable turnover in this group as well as frequent immigration into and emigration out of Africa.

Benefit examines the evolution of Plio-Pleistocene cercopithecoids primarily from the perspective of their dietary adaptations. Because larger bodied monkeys and early hominids shared dietary resources, they experienced the same effects of changes in climate and vegetation.

Bernor and Armour-Chelu show that African hipparions had a complex history of migration, dispersion, adaptation, and evolutionary radiation. These authors demonstrate that the systematic record is particularly valuable when it includes the ecomorphologic reconstruction of chronologically long and biogeographically extensive evolutionary histories.

Bishop reviews the evolution, paleobiogeography, and paleoecology of African suids. Pigs have particular relevance to problems of human evolution and paleoecology. Bishop shows that of the two principal genera, at least one species from each genus underwent postcranial adaptations to mesic habitats.

Denys provides an analysis of late Cenozoic, sub-Saharan rodent taxa and reveals that affinities with vegetation zones are demonstrable. Past biotic zones may have had different latitudinal representation, however. Higher speciation rates are found in the contact zones between regions than within regions.

Grubb et al. examine the relationship between eastern and southern African modern and Pliocene mammal faunas with special reference to the faunas from the Chiwondo Beds in northern Malawi. Hominid cladogenesis probably took place among analogues of todays biotic zones and was affected by similar factors to those facilitating the process in other mammals.

Part IV. Hominid Evolution

Leakey, in her introduction to Part 4, reminds us of the comparatively limited fraction of the ancient landscape surveyed for interpretations of early hominid environments. Her critique of the papers is a useful portrayal of the innovative character of the field, revealing the influences that preconceptions could have on the ways in which the fossil evidence is interpreted.

Feibel introduces the geological context relevant to the reconstruction of hominid ecosystems, identifying the basin and its source area as a functioning ecosystem. The effects of climate change on hominid habitats, as described from individual sites, must be considered in the larger perspective of this system.

Andrews and Humphrey outline factors that underscore the relationship between mammals and their community within the ecosystem. They go on to analyze the spatial and dietary niches of Miocene hominoids and Plio-Pleistocene hominids.

Sikes reviews stable isotopic carbon and oxygen data which do not give evidence of climate change resulting in significantly more grassland habitats in eastern Africa until after 1.7 m.y. *Homo* and *Paranthropus* species are suggested to have habitat specificities to woodland, maintaining sympatry in a manner like that of extant lowland gorillas and chimpanzees.

Collard and Wood consider features of locomotion, diet, brain size, and body shape in their identification of two early hominid grades. One grade, characterized by *Homo ergaster*, is said to have emerged as a result of Late Pliocene climate change.

Foley finds a stronger association between climate change and extinction than with speciation in the hominid record. By considering local events—isolation, dispersion, and contraction—microevolutionary forces come into focus to help explain macroevolutionary patterns over the geographical landscape.

Rozzi et al. examine several enamel microanatomical features and tooth size in the dental sample from members of the Shungura Formation (Omo Basin, Ethiopia). The results parallel those of other mammals, indicating a greater reliance on masticatory demands imposed by climate change during the Late Pliocene.

Bromage advances the need to describe early hominid morphology in a manner complementing the paleoecology. *Australopithecus* (=*Ardipithecus*) *ramidus* is used as a case study, demonstrating that while conventional morphological approaches may provide some indications of the dietary niche, an ecomorphological approach would be rewarding.

Background

Why did we organize this symposium, and why come together in Malawi? By the vagaries of nature, most early hominid discoveries have been made in either eastern or southern Africa, but not in between. The geographic center between them, along Lake Malawi, harbored Late Pliocene sediments called the Chiwondo Beds. We surmised that if one went to a place between the known early hominid localities, then it should be possible to investigate the evolutionary connections between the two regions. Research about such a place could then say something about continental Africa instead of only one site or one region.

We hypothesized that the ecological zone of southeast Africa, called the Zambesian ecozone, was most probably the meeting point for faunas from the north and south since the continent has been in its current position. The Zambesian ecozone comprises the woodlands and grasslands of southern Zaire and Tanzania, Angola, Zambia, Malawi, and northern Mozambique and Zimbabwe. It is situated between the subtropical to temperate ecological domain of southern Africa to the south and the tropical ecological domain of eastern Africa to the north. We reasoned that while animals able to acquire their resources in more than one domain could have flourished in both eastern and southern Africa, those restricted to more specific habitats might have crossed over this region only when their appropriate vegetation zones shifted location. According to a general theory proposed by Vrba (this volume), we figured that these vegetation zones must have shifted their locations about 2.8 m.y., at which time cooler and dryer conditions prevailed around the world (Denton, this volume). If Vrba's theory is correct, there should have been a shift of grassland and woodland vegetation zones northward toward the equator.

To test these ideas we undertook fieldwork in Malawi. It took us 10 years to collect 500 fossils. Most were isolated teeth, not the whole skulls or partial skeletons of other paleoanthropological projects. But one only needs a tooth to record the passing of the animal it once belonged to, and so after our 10 years of surveying in Malawi, we felt that our 500 specimens, many dated by faunal correlation to the Late Pliocene between 3 and 2 m.y., were sufficient to begin testing our ideas.

When we compared the Chiwondo Bed's fauna with those of eastern and southern Africa, it was clear that the Malawi Rift belonged largely to the paleoecological domain of Pliocene eastern Africa. That is, of the animals normally belonging only to eastern or only to southern Africa, the Malawi Rift was dominated by the animals characteristic of eastern Africa. The faunal sample included only a few endemic to southern Africa. But the species shared between eastern and southern Africa still had to be explained. Were they shared because their habitats were the same, because they could live in many different habitats, or could it be (as we thought) that many animals were recorded as shared because some populations had dispersed from their origins in southern Africa to eastern Africa 2.5 m.y. (consistent with Vrba's theory)? Sure enough, when we looked at the whole African record, it was clear that most of these animals had dispersed from southern to eastern Africa about 2.5 m.y., suggesting that the vegetation zones the animals depended on had drifted northward toward the equator during Late Pliocene climate change.

Fortunately, we found both an early hominid mandible we attributed to *Homo rudolfensis* and a maxillary fragment of *Paranthropus boisei* known only from eastern Africa. These specimens represent the southernmost known distribution of these taxa. The biogeographic significance lay in their association with the eastern African animal group. During the drying and cooling of global climates about 2.5 m.y., the southern and more temperate African faunas followed their northward-drifting vegetation zones, but where should the tropical equatorial animals, including the hominids, of eastern Africa go? Our evidence suggests that they stayed in the tropical African ecological domain. Thus *Homo* and *Paranthropus* may have emerged in tropical Africa as a result of the about 2.5 m.y. climatic cooling event and remained there while other animals were dispersing northward toward the equator.

Furthermore, several anatomical characters of the Malawi mandible suggested to us an affinity with *Australopithecus afarensis,* the best known early hominid species from eastern Africa. We thus had the basis for a scenario of hominid evolution rooted with *Australopithecus afarensis* and compliant with both Vrba's habitat theory and the known distributions of early hominids. This scenario depends largely on what we know about the ecology in combination with studies of the anatomical characters (Bromage & Schrenk, 1995).

Our perspective implies a certain "ecocentric" perspective (with its full implications of adaptive shifts in phylogeny) as to the nature of hominid lineages and the way we define their taxonomic segments. We believe that such a perspective is needed at this time to account for the relationship between eastern and southern Africa in hominid evolution, as the more traditional morphological–taxonomic approaches have neglected to account for the fact that these areas belong to seemingly different ecological domains: tropical and temperate zones. Why is it that apparently no Pliocene hominid species are shared between eastern and southern Africa (according to most authorities)? Have the morphologists accurately discriminated paleospecies differences, or could they be attributed to geographic variation? To answer such questions, the habitat specificities of these differences and comparisons with other mammalian faunas traversing Africa's tropical and temperate latitudes need to be examined more closely.

Questions

The symposium aimed to stress the relationship between physical environmental change, lineage and specific differences, and faunal evolution and biogeography. This relationship plunges deep into many areas of paleoanthropology and vertebrate paleontology that conerned all of the areas represented at the symposium, and more. To draw out discussion, some specific questions were posed at the beginning of the symposium. It was not intended that these questions be answered, rather, we hoped that the questions be used to reach the larger issues surrounding our habitat- and ecology-oriented perspective. The questions were:

- What influences might paradigm shifts in evolutionary theory have on the practice of taxonomy, and is such interdependence desirable?
- What constitutes climate change, and how do we define the "turnover pulse" of extinctions and speciations resulting from this change? How might evolutionary theory help us to answer these questions?
- Can the geographic dispersion of fossil taxa be accurately inferred, and, to what extent can we say that taxa either remained endemic or exhibited vicariance of their distribution during periods of habitat change?
- How do we determine the heritability of habitat specificities in modern mammals, and how reliable might our comparisons be to fossil taxa?
- What techniques can be applied to indicate the degree to which early hominids were ecosensitive? Is this sensitivity affected by a material culture?
- What role, if any, did climate change have in the origins of Hominidae and *Australopithecus*, megadontia in robust australopithecines, and encephalization and tool making in *Homo*?
- Can the differences between eastern and southern African hominids be attributed to either geographic variations or lineage differences?

Interdisciplinary Convergence

Paleoanthropology is a venue for interdisciplinary convergence. It is interdisciplinary in the sense that a variety of disciplines focus their research attention on topics of paleoanthropological significance. It is potentially convergent if its participants acquire and use mutually accessible data, intentionally collected for its accessibility, to enlarge our understanding of human evolutionary phenomena. This leads to insights about the connectedness of things; the boundaries of our investigations are limited only by the dimension of our advancing sphere of interests.

Consider the following example. Solar flux conditions the earth's temperatures. The earth's eccentric orbit around the sun, the obliquity of its spin axis relative to the sun, and its distance from the sun (measured at the summer solstice and termed "precession") determine the amount of solar radiation received by the atmosphere and, thus, long-term temperature variation. These orbital peculiarities occur at regular, so-called Milankovitch frequencies, over geological time frames and are theorized to result in the astronomical forcing of climate change cycles (Milankovitch, 1941).

Meanwhile, the earth's surface is in motion. Driven by events deep within the earth's core, the crust is stirred about and periodically plunged into the mantle. This rock cycle was responsible for the coming together of the North and South American continents, the subsequent disruption of tropical Pacific and Atlantic oceanic continuity, and new global atmospheric circulation patterns. Together, the Milankovitch frequencies and the rock cycle, through the medium of altered circulating cloud, rainfall patterns, land surface morphology, and vegetation patterns combined with, perhaps, other temperature-regulating phenomena such as sunspot cycles, established the cascade of events that brought about significant global climate change during the Late Pliocene (see Vrba, this volume; Denton, this volume).

Ensuing cooler and drier conditions favored a tougher savanna vegetation composed of plant species better able to retain their moisture under such conditions (see O'Brien & Peters, this volume; Owen-Smith, this volume). Selection favored more facially robust and large molar-toothed mammals, including early hominids, capable of efficiently processing the tougher, more durable vegetation of the savanna (see Turner, this volume; Rozzi et al., this volume). Selection pressures were sufficient to result in the evolutionary splitting of *Australopithecus* into *Paranthropus* and *Homo* lineages by about 2.5 m.y. (Bromage & Schrenk, 1995).

Larger bones and teeth, with their attendant morphological adaptations, resulted from cellular changes in bone and tooth development. Favored rates and patterns of bone-forming (osteoblast), bone-removing (osteoclast), and tooth-forming (ameloblast) cells were the consequence of selection on the functional phenotypes.

What we want to emphasize in this illustration of interdisciplinary convergence is how the behaviors of individual enamel-forming cells, for instance, are directly and processually linked to the astronomical behavior of the earth's orbit around the sun. These events are occurring over evolutionary and geological time frames through the medium of climate change. There is no apparent design, there is but the connectedness of events in history that provide conditions for the evolutionary changes we observe.

What Have We Achieved?

This symposium followed on the heels of the publication of the proceedings of a conference focusing on the link between climate and evolution, published under the title *Paleoclimate and Evolution, with Emphasis on Human Origins* (edited by Vrba et al., 1995). Four of the authors in this volume (Bernor, Denton, Vrba, and Wood) contributed to *Paleoclimate and Evolution,* and their input provided a link to this symposium.

The symposium participants responded to our questions and our call for the establishment of a new paradigm with zeal. Above all, the discussion was friendly and interesting. The resulting papers, taken on their own, have shown that our attempt to draw out a perspective based on interdisciplinary convergence has been met with measured progress. However, like its predecessor (Vrba et al., 1995), this progress is likely to be realized even more by the reader, comparing the chapters, making connections, getting ideas, and following up problems and issues.

The Wenner-Gren Foundation (WG F) format involves pitching an interdisciplinary group of scientists into a room for about a week (under extremely pleasant circumstances, we might add) to discuss a topic of concern to the lot of them. The format relies heavily on a group dynamic of less than 25 people or so. Each participant brought his or her specialty to the table (the only participant not attending the Malawi symposium had the responsibility of discussing the advent of material culture and any relationship this may have had to climate change). The advantage of this format is that, while being relatively up close and personal, participants have the responsibility of making themselves understood rather than expecting the listener to have been better informed. We all learned a great deal.

The degree to which participants could (or would) subsume the work of others into the conduct of their own field is a measure of how close we are to obtaining interdisciplinary convergence. It is clear that in some respects beliefs in concepts endure despite the flurry of contradictory opinions in the face of, to many outside observers anyway, what appears to be good science. Despite this, the symposium has clearly indicated that sufficient room exists for the development of the ideas, the paradigm, and the interdisciplinary convergence the editors are seeking. One can be sure that many of the participants will build on the experience obtained at the Malawi symposium and consider more deeply its primary aim in their future publications.

The symposium made us aware of deficiencies in the state of the art. There were moments where it seemed that the symposium's hands were tied behind its proverbial back. Sometimes these deficiencies were topics of conversation during coffee and lunch breaks (another thing about the WG F format is that you cannot leave the venue). Besides the often-said need for more data, there were a few areas that we believed could use significant future consideration. One point of concern was the need for a deeper knowledge of the habitat specificities of numerous organisms important to us and the means by which we would project that knowledge into the fossil record. Environmental tolerances vary between organisms, and their responses to climate change are expected to be measured against their habitat specificities. Vertebrate paleontologists need tools by which these specificities can be finessed from the bones and teeth at a much finer level. Benefit, Bernor and Armour-Chelu, Bishop, and Denys (all of this volume) are currently working on the tip of this iceberg. Developmental perspectives, particularly those that examine closely perturbations of growth, would be fertile ground, perhaps, as this would provide quantitative measures of the vulnerability of organisms to specific features of their environment (e.g., Macho, 1996).

Another deficiency, or, perhaps, omission on our part, concerns the boundaries between biomes. A great deal has been researched and said about extinction and speciation pulses resulting from climate forcing of latitudinal biome shifts and habitat expansion and contraction dynamics to help explain biogeographic patterns. We spoke little about biome edges, fuzzy as they are, and the dynamics of faunal interchange at these edges (Denys, this volume, broaches this issue). If we could identify ecotones in the paleobiome, then that would be one thing, but as our discussions tend to be based on more gradual changes in vegetational physiognomy, we felt the desire to know more about the conditions that motivate faunal interchange within a biome and between similar habitats at different latitudes. For instance, we know from studies of modern faunas that biogeographic flux is frequently asymmetrical (Vermeij, 1991). Extinction pulses are followed by immigration pulses often in the temperate-to-tropical direction. What important features of these particular immigrant species predispose *them* to occupy the new region, and are these features visible in the fossil record? If speciation rates are higher among the immigrants, then why would the turnover pulse not be borne out more clearly in the fossil record for all taxa (e.g., Foley, this volume)?

Another issue concerns our tacit claim that larger tooth sizes and increases in masticatory robusticity were common among African megafaunas and early hominids because of their response to a tougher savanna vegetation engendered by increasing aridity during the 2.5 m.y. cooling event (Turner & Wood, 1993a; Turner, this volume; Rozzi et al., this volume). We have reason to believe that such conditions should increase the proportion of grassland (see, however, Sikes, this volume), and consuming proportionately more grass represents increased fiber (and abrasive phytoliths) and selection for comparatively robust masticatory adaptations. But can the broad faunal response be due only, or at all in some cases, to increased grass in the diet? Browsers are known to supplement their diets with grass during seasonal deficiencies of their usual fare. (This had hardly occurred to us until one day in the vicinity of Lake Manyara, Tanzania, we saw giraffes, legs splayed, head down, in the grass. Elephants were curling their tusks around great clumps of grass, tugging them from the ground and moving them to their mouths). Might early hominids have supplemented their diet with grasses as well? A number of food resources are expected to have been exploited by early hominids (O'Brien & Peters, this volume), but for virtually none of them have the mechanical and/or abrasive properties been determined. This important topic deserves significantly more attention if we are to understand the conditions under which numerous organisms adapted to their changing dietary niche.

To sum up the contribution that the Malawi symposium has made to the field, we reiterate Howell's (this volume) assessment of the state of the art:

> The contributions here clearly reflect how rapidly and how far paleobiological and associated biogeographic and paleoenvironmental studies have progressed in recent years. Similarly, they provide a guide to the immediate future in which multidisciplinary researches are mandatory, and cross-disciplinary communication and exchange of perspectives and goals are absolutely essential.

Acknowledgments We are grateful to the Wenner-Gren Foundation for Anthropological Research and its president, Sydel Silverman, for the generous support of the Malawi symposium and the publication of its proceedings. We thank the government of Malawi and Yusuf Juwayeyi for their sustained support of our research and for the warm welcome extended to us at the conference in Salima. We thank Laurie Obbink, together with Anne Singer, for their thoughtful preparation and on-site organization of the symposium.

We owe thanks to all conference participants, who endured a long journey to the warm heart of Africa and who created a truly dedicated atmosphere at the conference table, which is reflected by their essays in this volume. We are grateful to Yves Coppens, Jonathan Kingdon, F. Clark Howell, and Meave Leakey for contributing introductions to parts I–IV in this book.

The preparation of the conference volume would not have been possible without the support of many of our colleagues and students in New York, especially Haviva Goldman and James McMahon, and in Darmstadt, especially Oliver Sandrock, Ottmar Kullmer, Silke Keim, and Kristine Vaaler. We also thank Oxford University Press and its staff for production of this volume and the series editors for helpful advice.

Fond appreciation goes also to the people of Karonga, northern Malawi, who, over the last 15 years, have invited us into their homes. Most important of all, our love and appreciation go to Hediye and Gesine, who tirelessly keep us in their hearts and who keep our families healthy and safe while we're away in the field and the lab.

I

THEORY

1

Introduction

Yves Coppens

In 1967 the International Omo Research Expedition was founded by three "big names" of our scientific community—Camille Arambourg, honorary professor of Palaeontology at the Paris National Museum of Natural History, Louis Leakey, honorary director of the National Museums of Kenya, and Francis Clark Howell, professor of Anthropology at the University of Chicago. A document dividing the exposures of the lower Omo basin into three parts was drafted according to aerial photographs. The French in the south took the exposures surveyed by Camille Arambourg in the early 1930s. The Americans and the Kenyans took the north—the American on the right bank of the Omo River, the Kenyan on the left. This document was then signed in Nairobi by the three co-directors of the expedition; it was also formally decided that the leaders in the field would be Francis Clark Howell as far as the Chicago party was concerned, Richard Leakey for the Kenyan one, and myself for the French mission.

We all left Nairobi in June 1967 and were at work by the beginning of July. The American and the Kenyan parties were a bit disappointed by the northern exposures, these deposits being much younger than expected (around 100,000 years old or so; they were called later the Kibish Formation). It was the south (which would become the Shungura Formation) that was the main area of Pliocene and Early Pleistocene deposits, a thick sequence of sands, silts, and tuffs dated between more than 3 m.y. at the bottom and less than 1 m.y. at the top.

When we arrived in this southern area, looking geomorphologically like a succession of north–south parallel cuestas, climbing from the Omo River on the east to a sort of plateau on the west, Camille Arambourg told me that, for him, the structure of the deposit was a repetition of the same compartment, broken by a series of north–south faults, and containing, consequently, the same homogenous fauna, the one that he had published in the 1940s (Arambourg, 1947).

A new survey, geological and paleontological, quickly demonstrated that each compartment was actually supported by its eastern neighbor and supported its western one. It was, in other words, not one modest repeated unit of sediments, but an impressive true stratigraphic accumulation, artificially divided into steps by an erosion more active on sands and silts than on hardened tuffs (building the cuestas). We had, as soon as the first field season, the paleontological demonstration of a geological history of several million years (more than 2) along a sequence of several hundred meters of sediments (all together, more than 1000 m) (Arambourg et al., 1967; de Heinzelin, 1983).

For the second field season, 1968, the international division of the basin had to be revised. Richard Leakey wanted to give up, to be able to start a survey in Kenya, on the eastern side of the Lake Turkana where he had observed promising exposures. Francis Clark Howell wanted to come to the south to work on the Plio-Pleistocene. The south area was then cut into two parts, the west–east water track going from the camps on the plateau to the Omo River being the border.

The geological, stratigraphical, and tectonical mapping went on (the first map of the whole Plio-Pleistocene Shungura deposit was produced by the second field season (Arambourg et al., 1969), and the fossil collection (vertebrates including hominid, invertebrates, pollens, and woods) grew quickly. The first year observations on the stratigraphic succession (and not repetition) of the strata was confirmed everywhere by everyone from the two parties, and the succession of the hominids, of the faunas and of the floras, became particularly interesting, especially when tephra and paleomagnetic dating and cross-checking the biostratigraphical scale gave us the first figures of 1, 2, or 3 m.y.

This was the matter of the first biozonation (Arambourg et al., 1972), enlarged to the whole continent (Coppens, 1972; 1974), and later formally named (Coppens, 1978a). This was the matter of the first ecological and climatological tentative interpretation of this sequence, pointing out a progressive climatic change, more humid before, less humid after, and this was, of course, the reason for the first statement on the evidence of the link between the drought and the emergence of the genera *Homo* and *Zinjanthropus,* this emergence having possibly been a consequence of the drought (Coppens, 1975a,b, 1976, 1978b,c, 1979, 1980, 1983a,b,c,d, 1984, 1986, 1989, 1994).

I later called this important event the (H)Omo Event to remind one of the fundamental role played by the Omo sequence in the discovery of the possible influence of the climate in the appearance of our own genus.

How Do These Conclusions Emerge?

Since the 1960s, with the development of the use of the absolute dating, paleoanthropologists were guessing that the filiation *Australopithecus-Homo* must have happened between 2 and 3 m.y.

Before 1980, the unique site documenting this time span was the Shungura formation in the Omo area; the site of Olduvai in Tanzania was too late; the sites of Laetoli in Tanzania and Hadar in Ethiopia were too early, and the rich and extended site of Koobi Fora in Kenya had a gap in its sedimentation for this period. We had, then, the privilege to study during 10 years the particularly important period of emergence of humans in the region where this event could have happened.

In the field, it was easy to see immediately that between about 3 m.y. and 1 m.y., the fauna and the flora were not stable, that they changed, and that their speed of changing was not constant. It was equally easy to see that the change was a trend from more humid to less humid—meaning warm, humid, and dry meaning cool—thus this change was a cooling. It was easy as well to see that the hominids were also changing and that they were changing when the fauna and the flora changed. As a logical consequence, it appeared obvious that there has been a strong correlation between the climatic change from more humid to less humid and the emergence of the two new hominids, *Zinjanthropus* and *Homo.*

The association in the Omo of the "gracile" australopithecines with some humidity in the environment (what we called *Australopithecus aff. africanus* in 1975) and the association of the robust ones as well as *Homo* with much less humidity (what we called *Australopithecus aff. robustus* and *Homo aff. habilis* in 1975) was convincing enough to let us organize a first scientific meeting on that topic. We invited the different specialists who helped us in determining and studying the fossils and, for the first time with them, South African colleagues. The meeting took place in Paris in 1983 and was published in 1985 (Coppens, 1985). The statements of 1975 were confirmed.

Since these early days, many authors from different disciplines have considerably amplified the debate and developed arguments in favor or against the first claim, and several important scientific meetings have dealt partially (Delson, 1985; Grine, 1988) or completely (Vrba et al., 1989, see also Bromage and Shrenck, this volume) with the topic. This volume and Vrba et al.'s (1989) are, as far as this question of environmental influences on evolutionary process are concerned, particularly dense and rich.

With this environmental reading of the history of the Hominids, let us tell the way our family could have spent the last 10 million years (Berggren & Van Couvering, 1974; Bromage & Shrenk, 1995; Tobias, 1985). We will see successively seven episodes of a tentative scenario, the first four of them being in the so-called early hominid frame of this volume:

1-The East side story	8 m.y.
2-The first expansion	4 m.y.
3-The (H)Omo event	Between 3 and 2 m.y.
4-The second expansion	2 m.y.
5-The two exceptions	Btwn 2 and 0.05 m.y.
6-The third expansion	0.05 m.y.
7-The next expansion	To come

The East-Side Story, Approximately 8 Million Years Ago

According to their anatomical, physiological, ethological, cytogenetical, and molecular degrees of proximity, it is clear that the living organisms nearest to humans are the African apes. This means that we share ancestors with the African apes. The first question is then, who are these ancestors, and where and when did they live?

There is no definite answer to the first part of the question. But from about 8 m.y. to now, there exists in eastern Africa without any interruption (from 8 to 3.5 m.y. in eastern Africa only) remains of Hominine (we consider the half maxilla of Suguta valley in Kenya to be Homininae; Hill & Ward, 1988), successively "prehumans" and "humans." It seems then that the time of separation between the strict African apes line or lines and the Hominine line or lines occurred in this area of the world about 8 m.y. We must then point out why these common ancestors were splitting into two populations, to become true exclusive ancestors of the African apes or true exclusive ancestors of the Homininae.

Around 10 m.y. (before this important date of 8 million years), the common ancestors of African apes and Homininae were living in equatorial Africa, in a low country, from the Atlantic Ocean to the Indian Ocean, covered by rainforests, gallery forests, and wooded savanna, the area being warm and wet (Andrews and Van Couvering, 1975; Jacobs and Kabuye, 1987; Andrews, 1992a; Andrews and Humphrey, this volume). There was already at that time a general increase of grassland and a general trend of cooling worldwide because of the beginning of an ice cap over the Antarctic (Denton, 1985, 1995, this volume; Cerling, 1992a), but locally some important event came to magnify the global progressive change: it was a tectonic event, composed of a rifting process and of an uplifting of one of the western shoulders of the rift. As a consequence of the emergence of this topographic rainshadow, the western side, between the Atlantic Ocean and the wall, remained covered, whereas the eastern side, between the wall and the Indian Ocean, became more and more xeric, a province of patchy vegetation (Axelrod & Raven, 1978; Pickford, 1990; Ruddiman & Kutzbach, 1991; Raymo & Ruddiman, 1992; Partridge et al., 1995a,b). The eastern limit of the rainforest in eastern Zaire is, for instance, precisely the beginning of the uplifting of the rift.

These ecological changes obviously had consequences in the evolution of the ecosystems; some genera becoming extinct, others moving into or out of the area, and still others evolving in situ, giving birth, on the east side, to an endemic ecosystem, which is called among paleovertebrists, the Ethiopian fauna. The australopithecines are part of the endemic Ethiopian fauna (Bromage, 1990; Opdyke, 1995).

Ecological changes are probably the reason for the natural division of the unique population of common ancestors of African apes and Men, and the origin of their new destiny—the African apes on the western side, where they still are, and the Homininae on the eastern side, the site of all their most ancient remains without exception (dated from between 8 and 3.5 m.y.) (Kortlandt, 1972; Coppens, 1983c; Partridge et al., 1995a,b). This is a typical application of Vrba's (1995b,c, this volume) habitat theory.

Eastern Africa, isolated for a while by the complex rifting–uplifting process 8 m.y., became a true island for several million years. Its populations thus had to support typical peripatric genetic drifts; the very first australopithecines appeared as part of this new fauna,[1] with some at least of them initiating derived features such as erect posture, bipedalism, small canines, and large cheek teeth with more cusps and thick enamel, and, very probably, a new structure of the brain (Holloway & Kimbel, 1986), which means new behavior.

Given the late miocene reactivation of the rifting, the East African endemism and the earliest Homininae being of the very same age, I suggested naming this event the *East side story* (Coppens, 1994).

The First Expansion, About 4 Million Years Ago

It seems that some new change in the climatic evolution of the whole continent (Denton, this volume; Denys, this volume) obliged the rainforest to reduce its territory to its remaining core along the Gulf of Guinea, allowing the savanna to extend around the forest, as a sort of concentric aureole. When the savannah expanded, the organisms inhabiting it naturally expanded; when a habitat moved, its inhabitants moved with it (Vrba 1995b,c, this volume).

The prehumans, the australopithecines, inhabiting the East African savannah (their probable birthplace), developed two branches inside eastern Africa (Kimbel, 1995; Collard & Wood, this volume): one inhabiting more covered savannah (Bonnefille et al., 1987) and practicing bipedalism and arborealism (*Australopithecus afarensis;* Johanson et al., 1978), and the other inhabiting more open savannah and practicing bipedalism exclusively (*Australopithecus anamensis;*

M. G. Leakey et al., 1995). These australopithecines started their first expansion, probably in every direction except the east because of the sea, 3000 km toward the south (the most ancient hominid remains from South Africa are 3–3.5 m.y. old, found at Sterkfontein, level 2) and 3000 km toward the west (where remains of *Australopithecus,* 3–3.5 million years old, have just been found in Chad) (Brunet et al., 1995, 1996).

The (H)Omo Event, or the Turnover Pulse Hypothesis, from 3 to 2 Million Years Ago

Because of its historical importance, I opened this introduction with this prestigious (H)Omo event (Coppens, 1975; Bonnefille, 1976, 1983, 1984, 1995; Boaz, 1977; Bonnefille & Vincens, 1985; F. H. Brown et al., 1985b; Vrba 1985a,c,e; Wesselman, 1985a,b, 1995; McKee, 1991, 1995, 1996b, this volume; Rayner et al., 1993; Turner and Wood, 1993b; F. H. Brown, 1995; deMenocal 1995; deMenocal and Bloemendal, 1995; Geraads and Coppens, 1995; A. Turner, 1995a; Ramirez Rozzi, this volume), which was responsible for the emergence of the robust australopithecines, *Zinjanthropus* and *Paranthropus,* independently in East and South Africa, and for the emergence of *Homo* probably in East Africa (except if *Homo rudolfensis* emerged independently in the Zambezian province) (Schrenk et al., 1993, 1995; Bromage et al., 1995a)

We must remember how important this statement is, as it deals with our own origin and deals with some of the reasons of our emergence. Even if among the Hominids some evolutive trends have begun before this event and were only going on at that time, it is between 3 and 2 m.y. that the volumetric development of the brain, the emergence of culture, the emergence of consciousness, the development of social behavior, and the emergence of speech occurred.

The Second Expansion, 2 Million Years Ago

For several reasons, not only environmental, the genus *Homo,* as soon as it became *Homo,* moved. The autocatalytism could probably be used here, if not elsewhere (McKee, this volume)

Homo moved because, with his new omnivorous diet, which means *pro parte* carnivorous, *Homo* must have had, by necessity, more mobility than its ancestors (Peters & O'Brien, 1984; O'Brien & Peters, this volume). He had to run after the game to have some chance to catch it and to eat it. With a bigger brain, *Homo* must have had better thought processes, at least more associative and deductive and increasing curiosity. With stone and bone tools, *Homo* was in possession of an equipment that gave him more freedom and more boldness in the face of new environments, new ecological niches, and the possibility of living in them. *Homo* moved because, apparently having succeeded in adaptating to the new, very dry conditions, *Homo* must have had developed his demography (Foley, 1984, 1994, this volume).

If we add to these factors a probable extension of the open savanna beyond the previous intertropical areas, it becomes logical that *Homo* should have instantaneously (geologically speaking) moved toward the whole African continent first, and through the Middle East, to Europe on one side and Asia on the other, as soon as 2 m.y. And the possible datings obtained recently, but still discussed, in Java (1.8 m.y.) (Swisher et al., 1994), in China (1.9 m.y.), and in Georgia (1.8 m.y.), are bringing some credit to this prediction.

The Two Exceptions, between 2 Million Years Ago and 0.05 Million Years Ago

We are now at the point at which humanity covered the whole ancient world, Africa, Asia, Europe. As far as evolution of *Homo* is concerned (Wood, 1991, 1992), paleoanthropologists have given special names to different successive grades: *Homo rudolfensis, Homo habilis, Homo ergaster, Homo erectus, Homo sapiens.* It is, in a way, a source of confusion, this evolution of the genus *Homo* being a clear continuity without borders.

This evolution means that humans became *sapiens* all over the ancient world at the same time (around 0.5 m.y.) (Li and Etler, 1992), except in two peninsulas, becoming from time to time insula, where *Homo sapiens* is no more than 0.05 m.y. old: the far west of the ancient world, Europe, and the far southeast of the ancient world, Indonesia.

During the last 2 million years, there was a cyclic succession of glaciations and interglaciations (Denton, 1985, 1995, this volume). During the glaciations, Europe was isolated; ice covered the north of the peninsula and the Alps, leaving only narrow corridors along the Mediterranean coast and through the center of the continent. At the same time, Indonesia was linked to the Asian continent because the sea level was low, the water being retained in the ice. In contrast, during the interglaciations, Indonesia was isolated and Java an island; the sea level was high. Europe was then linked to its Eurasian body, the ice having decreased.

The populations of *Homo* in Europe as well as in

Indonesia, since at least 2 m.y. were thus trapped: as any species of animal or plant evolving in islands, they have had to support a genetic drift. This phenomenon, called neanderthalization in Europe (Demars & Hublin, 1989; Hublin, 1990), could be called pithecanthropization in Indonesia. Humanity was thus modern everywhere in the ancient world, in Africa, in Asia, except in Europe, where was living the famous Neanderthal man (who cannot be called *sapiens*), and in Indonesia, where was living the famous Java man.

The Third Expansion, 0.05 Million Years Ago

Modern man, Neanderthal man, and Java man were evolving, side by side, in Africa and Asia, Europe, and Indonesia, respectively. But *Homo sapiens* populations had experienced, around 50,000 yr ago, an environmental influence all over their territory, and this pressure pushed the species to expand again wherever it was possible.

It was not possible from Africa, all the borders being coasts, but it was possible from Asia: toward the northeast, humans reached America on foot and seemed to people it from north to south; toward the south from China, humans reached Australia by boat and peopled New Guinea, Australia, and Tasmania linked; toward the west, humans reached Europe, already peopled, leading to confrontation and coexistence with the Neanderthals would last perhaps 10,000 years. Modern man there is called Cro-Magnon, and the last Neanderthals were dated from less than 30,000 yr ago (Mercier et al., 1991; Hublin et al., 1995, 1996). Toward the southeast, humans reached Indonesia, already peopled, leading to confrontation and coexistence with Pithecantropes. Modern man there is called Wadjak and the last the Pithecanthropes were dated from 27,000 to 50,000 yr ago (Swisher, 1996).

The Next Expansion

The next expansion to come will concern the settling of humanity in the solar system. We are probably waiting for the right pressure to really start that new experience.

Paleoanthropologists have a lot to learn from the paleontologists, paleovertebrists (Vrba, 1982, 1988, 1995b; Wesselman, 1985 a,e, 1995; Bernor et al., 1987a; Denys et al., 1987, this volume; Owen-Smith, 1988a, 1992, this volume; Bishop, 1993, this volume; Benefit & McCrossin, 1995; Turner, 1995a, this volume; Benefit, this volume; Bernor & Armour-Chelu, this volume), paleoinvertebrists, and paleobotanists (Bonnefille, 1976, 1983, 1984, 1995; Bonnefille & Vincens, 1985; Williamson, 1985; Bonnefille et al., 1987; O'Brien & Peters, 1991; Livingstone, 1993; Scott, 1995; Maley, 1996). We have to listen to what some close but separate disciplines do in their fields, such as biology, biogeography, ecology, physiology (Grubb, 1973, 1978, 1983; Rosenzweig, 1987, 1991, this volume; Rosenzweig et al., 1987; Ruff, 1991, 1993, 1994). We have to also work in much closer collaboration with the scientists dealing with understanding the complex process of evolution (Schaffer & Rosenzweig, 1978; Foley 1984, 1994, this volume; Vrba, 1985a,c, 1988, 1993b,c, 1994a, 1995a, 1996a; Foley & Lee, 1989; Szalay and Bock, 1991; Henneberg, 1992; Szalay, 1993, this volume; McKee, this volume). We have to integrate in our information the magic isotopic reading of the evolution of the climate (Cerling et al., 1977; Ericson et al., 1981; Keigwin, 1987; Shackleton, 1987, 1995; Prentice & Denton, 1988; Hodell & Venz, 1992; Kingston et al., 1994; Sikes, 1994, 1995b, this volume; Kennett, 1995).

But what a big change in a little less than a quarter of a century, between the tentative claim that climatic change may have had a role in early Hominid evolution and these multidisciplinary developments bringing into the picture such an amount of definitive data and such a quality of debate. It is obvious that, by the way the hominids emerged, 8 m.y., by the way they complexified their phylogenetic tree and expanded, 4 m.y., and by the way the genus *Homo* emerged 3 m.y. that Hominids were clearly influenced by the evolution of the environment, just as the elephants, the suids, or the equids were during the same periods. We are partly the fruit of an astronomic event, helped by a tectonic one, which produced a dramatic drought in periequatorial eastern Africa. But we have to remind ourselves that what is obvious today was not known, nor even thought of nor easily understood or accepted in the 1970s.

It is as well quite consistent to see the typical derived way the European and Indonesian peripatric populations evolved in 2 m.y., side by side with the Afroasian main patric population. It is amazing to realize that Africa and Asia were modern a long time before the rest of the world. The genetic drifts were clearly influenced by the circumstances and by the environmental situation, just as the lemurs in Malagasi or the marsupials in Australia.

Acknowledgments I am happy and honored to have been invited to open with the first chapter of this book and to express my warm appreciation to the editors, Timothy Bromage and Friedemann Schrenk.

Note

1. The australopithecines can be much more diversified than expected. *Ardipithecus,* for instance (T. D. White et al., 1994, 1995; WoldeGabriel et al., 1994) had simple teeth with thin enamel and probably an erect posture and, like *A. afarensis,* "bilocomotion," climbing and biped.

2

Habitat Theory in Relation to the Evolution in African Neogene Biota and Hominids

Elisabeth S. Vrba

Hominids speciated repeatedly and also evolved markedly in morphology and behavior during the late Pliocene and earliest Pleistocene. Major novelties that first appeared then and thereafter evolved progressively include the "hypermasticatory" morphology of the "robust" australopithecines (Wood, 1995), advances toward obligatory bipedalism (Brown et al., 1985), and increased encephalization in at least one lineage. A momentous advance in behavioral evolution is reflected by the first appearance of stone tools. It is not yet clear how closely this event near 2.5 m.y. was linked to the commencement of encephalization (in the crude sense of relative brain enlargement), as there is a gap in the record of hominid crania between 2.6 and 1.9 m.y. But the onset of tool manufacture was presumably associated with some kind of evolutionary advances in brains and toward more open behavioral programs.[1] What forces initiated these events and drove their progressive subsequent evolution?

The elaboration of mental complexity that led to the beginnings of consciousness was a singularly dramatic change in the history of life. Consider the similarity between mental process and natural selection (see Lamarck, 1809, for an early discussion; Simon, 1962; Popper, 1978, for more recent ones). Natural selection is trial and error with eventual survival of combinations that are at least temporarily stable in the face of environmental challenges. The "trials" are variant bodies. The "errors" are those that die without reproducing. The mental processes of learning and hypothesis testing (such as implied by making and using tools) also involve trial and error until some ideas survive as internally consistent mental constructs—or as stable constructs, at least until additional tests force their disassembly and revision. Poincare (1913:387) expressed this in his analysis of mathematical invention: "Ideas rose in clouds. I felt them collide until pairs interlocked, so to speak, making a stable combination." This is also selection, but this time conducted among hypotheses in the mind. Seen simply from a biological perspective, the evolution of human consciousness was a radical step because it replaced the energetically more wasteful, selective matching of bodies against the environment by selective matching of hypotheses within the mind—by purely mental trials and errors that pit past memories and new experiences against each other to arrive at a course of action in the mind. The fauna of the Burgess Shales represents the celebrated Cambrian evolutionary explosion of organic bauplans and shapes. The evolution of consciousness also led to an explosion of designs and shapes, but this time expressed in the cultural products of evolved minds rather than in bodies. This evolutionary step was no less radical than the one evidenced in Cambrian strata.

One might expect such a singular evolutionary trend, and the marked cladogenesis within which its origin is embedded, to have had an unusual set of causes. Specifically, what were the initiating causes of speciation events that led to the origins of robust australopithecines, to the first stone tools, and to the onset of hominine encephalization? All causes of the evolution of new characters and new species result from the interplay of three elements:

1. *The inherited properties of lineages:* Inherited morphogenetic rules in lineages define how ontogenies can respond to gene mutations in a given environment (e.g., Arnold et al., 1989), and inherited aspects of fertilization systems, population structure, and ecology place limits on the ways in which speciation can occur.
2. *The physical context:* The physical context has a strong influence on which organisms can live in a given place and on how evolution can proceed there.
3. *The biotic context:* Competitive interactions constitute major selective forces. I here mean competition *sensu lato:* competition in the broad Darwinian sense is implied by all kinds of biotic interactions, whether these involve direct competition for food, space, mates, and so on, predation (including herbivory and parasitism), and being preyed upon, or mutualist behaviors.

The interactions of these three factors not only cause the evolution of new characters and new species, but also directly determine the nature of the habitats of organisms and species—namely, the ranges of physical and biotic resources that they need for life. Thus the theory of the dynamics that bind organisms and species to their habitats (or habitat theory) provides an exceptionally promising basis for macroevolutionary research as a whole. I previously explored and tested several habitat-related hypotheses and predictions (Vrba, 1985d, 1987b, 1992, 1994, 1995a,b,c, 1998). Here I give only a brief resume of the background concepts of habitat theory and of several of its hypotheses and predictions. Then I summarize some evidence and implications for evolution of the African Neogene biota including the Hominidae.

The Background of Habitat Theory

All biological systems, from genes through cells and organisms to species function in particular environments, or "habitats" at different scales, and are vulnerable to changes in their environments. Not only do organisms and species function differently in different environments, but so do genes and cells. For instance, cancer cells will not metastasize unless they are in the preferred microhabitat of a "preferred organ" (Rusciano & Burger, 1992). All of these living entities at different levels can evolve. But all are also capable of net stability in their essential organizational structures through long time if their habitats allow it. All have abilities to repair themselves and to return to stability in other ways, in the face of perturbation. At the genomic level, it is known that batteries of enzymes can move along the DNA duplexes, repairing defects and acting to reduce the frequency of mutation (e.g., Metzler, 1977). Gene–gene and cell–cell interactions can buffer morphogenetic processes such that different mutations can yield similar phenotypic results (e.g., Alberch, 1980; Zuckerkandl, 1983; Fabian, 1985; McDonald, 1990; Rusciano & Burger, 1992; Michaelson, 1993). Similarly, at the population and species levels there are processes that can in principle maintain continuity of species' fertilization systems and habitat specificities for millions of years (Lande, 1980, 1985, 1986; Paterson, 1985), and there is abundant evidence that habitat adaptations commonly have been constant features of entire clades for millions and even billions of years. Rigorous evidence of this kind comes from mapping the limits of species' habitat specificities with respect to temperature, food, etc., onto cladograms with radiometrically dated branching nodes (Vrba, 1992, 1995b,c). The facts that habitats either change or stay constant, while the living entities that define them either evolve or stay in dynamic equilibrium, lead directly to a central proposal of habitat theory: that strong habitat changes are causally associated with significant evolutionary changes, and more nearly constant habitats are associated with stable evolutionary lineages.

The application of this proposal to species has already received much attention and support from diverse biological subdisciplines. Such theory and observations relate to genetic, morphogenetic, behavioral, and ecological processes that are held to promote net equilibrium of species as long as their habitats persist (although the habitats may move over geography). Disruption of habitats and species by geographical fragmentation and by qualitative changes within habitats are needed for speciation or extinction. There is general agreement that equilibrium in constant habitats and speciation in changing habitats are perfectly in accord with established genetic and ecological theory (Paterson, 1978, 1982, 1985; Lande, 1980, 1985, 1986; Wright, 1982, 1988; Charlesworth et al. 1982; Wake et al., 1983; Carson & Templeton, 1984; Newman et al., 1985; J. R. G. Turner, 1986; Maynard Smith, 1989). But these theories vary in how they combine three factors: (1) The resilient genetic and/or epigenetic bases of many characters normally confer stasis, yet may evolve rapidly given the environmental stress, strong directional selection, and population structures resulting from habitat change. (For examples involving genome–phenotype interactions, see McDonald, 1990; for instance, thermal, chemical, and other stresses can activate increased rates of transcription and transposition in retroviral-like transposable elements and induce aberrant phenotypes that remain suppressed in the unstressed genomic habitat.) (2) Stabilizing selection acts in constant habitats, and directional selection acts in changing habitats. (3) The effects of genetic drift differ between large, demographically stable populations and those that are recurrently undergoing vicariance and shrinking in response to environmental changes (following S. Wright, 1932). My own hypothesis is that all three classes of factors play a part in evolution, particularly during speciation. The initial proposal of punctuated equilibria (Eldredge & Gould, 1972) differed from what I explore here in at least two ways. First, it did not explicitly relate equilibrium to stable habitats and punctuation to changing habitats. Second, it did not consider that some kinds of changes might occur gradually (e.g., neutral mutations, or simple body size change with allometric shape change within the same growth allometry), while others might evolve by punctuation (e.g., mutations in regulatory genes, and evolution of new dissociated growth allometries).

Habitat theory predicts that vicariance into allopatric or separated populations, induced by physical change, was causally associated with most speciation events. There is widespread consensus among evolutionists today that allopatric speciation has indeed strongly predominated in the history of life. Allopatry can result from dispersal over preexisting barriers, but most often it results by in situ fragmentation of populations by physical changes (Vrba, 1995a). Theoretical treatments of sympatric speciation (e.g., Maynard Smith, 1981) and claims by Bush (1975) and M. J. D. White (1978) raised the possibility that populations might commonly diverge to speciation while in contact. But most later authors agreed that the various nonallopatric models are neither supported by empirical evidence nor theoretically likely. The stress on predominant allopatric speciation originated with Mayr (1942, 1963), and was later supported by Paterson (1978, 1982), Carson (1982), Futuyma and Mayer (1980), and Templeton (1981). Cases that suggest sympatric speciation continue to appear occasionally (e.g., the monophyletic cichlid species in Lake Barombi Mbo in Cameroon; Schliewen et al., 1994). Nevertheless, it is fair to say that the notion of strongly predominant allopatric speciation is consistent with the evidence as a whole and continues to enjoy widespread consensus.

Differences between evolutionary views arise precisely from different emphases on the three factors whose interactions underlie both the nature of species' habitats and evolution: the inherited properties of lineages, the physical context, and the biotic context. Darwinian theory of the origin of novel features and species has strongly emphasized the biotic context, particularly selection from biotic interactions. Darwin was the first to realize fully that the physical, behavioral, and mental evolution of humans, like that of other species, must be understood within the phylogenetic context implied by the inherited properties of lineages. In notebooks dated 1838 he wrote: "Origin of man now proved. . . . He who understand baboon would do more towards metaphysics than Locke," and "the problem of the mind cannot be solved by attacking the citadel itself . . . the mind is function of body" (quoted in Wilson, 1983: 545). Yet he emphasized biotic interactions as the overriding causal forces of speciation and extinction: "Each new species is produced by having some advantage over those with which it comes into competition; and the consequent extinction of the less-favoured forms almost inevitably follows" (Darwin, 1859:320). In contrast, a theory focused more directly on the dynamics that bind organisms and species to their habitats has to emphasize all three of the above elements because all three determine habitat specificity. That is, habitat theory emphasizes inherited properties of lineages and physical context more strongly than Darwin did and than traditional neo-Darwinism still does. Under habitat theory, physical change is hypothesized to be the "kick" that initiates speciation, while the nature of the change depends on interactions among the inherited properties of lineages and the physical and biotic contexts.

The Darwinian view that selection dominates among evolutionary causes tends to see each evolutionary advance as an independent piece of history. Under habitat theory, common rules give coherence to the evolutionary responses in different lineages. Such common rules are expected for several reasons. First, common climatic changes, which act across all areas inhabited by members of the same clade, can contribute to coherence of phylogenetic patterns in that clade, as further discussed below. Second, in the words of Arnold et al. (1989:408): "The expression of genetic mutations at the phenotypic level is constrained. . . . The fact that the same morphology [e.g., the teratology of two heads known throughout vertebrates] appears recurrently in distantly related lineages is simply a reflection of the generative properties of a developmental process shared by all vertebrates." That is, morphogenetic rules are inherited properties of lineages that determine limits on how ontogenies can respond to gene mutations in a given environment. Although the mutations themselves, and the repeated morphogenetic tendencies expressed as parallel phenotypic novelties in a given clade, are random in the sense that they are independent of what natural selection might prefer, they can nevertheless impart a coherent pattern to what can and does evolve in that clade (Alberch, 1980; Oster & Alberch, 1982). One set of hypotheses below posits that particular kinds of heterochrony can appear repeatedly in parallel in particular clades and that such parallelism reflects shared, inherited morphogenetic responses to common environmental causes. Third, there are general rules that govern the ways in which speciation can occur, and more limiting rules in the case of any particular clade that shares aspects of population structure and ecology and that evolved under the same basic conditions of physical change. Finally, natural selection can act in similar ways across species faced by similar climatic changes, especially when those species share inherited aspects of the genome, morphogenesis, and population structure.

The Background of Physical Changes during the Late Neogene

The physical changes that are relevant to turnover of species include both the local to widespread effects of tectonism such as Neogene uplift and rifting in eastern Africa (Partridge et al., 1995a,b) and the globally distributed effects of astronomic climatic cycles (my references to "climatic change" include both sources). The record of global climatic change (e.g., Denton, this volume; Vrba et al., 1995) shows that a major trend of net cooling affected at least parts of each ocean and landmass during the past few million years. Within this overall trend, there were major temperature increases and decreases (roughly 0.9–0.7 m.y. between minima) in the mean of the astronomic Milankovitch climatic cycles with shorter periods, about 19–23, 41, and 100 k.y. cycles. One such major net decrease in temperature is evident after about 3.0 m.y. toward 2.5 m.y. ago, after which subsequent cooling trends reached progressively lower minima (e.g., deMenocal & Bloemendal, 1995; Shackleton, 1995). This cooling trend after 3.0 m.y. was remarkable not so much for the rate at which its mean changed, but for three other reasons: First, the net downward trend in temperature was severe relative to the previous pattern. Second, there was a rare shift about 2.8 m.y. ago toward climatic dominance of the 41 k.y. cycle, from previous dominant influence at 23–19 k.y. variance (see deMenocal & Bloemendal, 1995, who also document a second shift toward increases in 100 k.y. variance after 0.9 m.y.). Such rare changes in the global climate system are expected to have large effects on the durations of phases during which species' geographic distributions remain continuously fragmented, or allopatric, and therefore on the incidence of speciation and extinction events (Vrba, 1995b). Third, 2.5 m.y. marks the first onset of extensive Arctic glaciation (Shackleton et al., 1984), during which land biomes, including in the tropics, changed greatly. This time is often referred to as the start of the modern Ice Age.

Some Hypotheses and Predictions of Habitat Theory

We already know that changes in geographic distributions by vicariance and/or latitudinal and altitudinal shifting have been much more prevalent responses of species to climatic changes than have speciation and extinction. Many studies have traced fossil morphologies through parts of the Milankovitch cycles, particularly for the Late Pleistocene. These show that cadres of different kinds of species in concert underwent shifting (by up to thousands of kilometers in many cases) and fragmentation, alternating with reconstitution, of their geographic distributions, while maintaining habitat fidelity (for plants, see Huntley & Webb, 1989; Hooghiemstra, 1995; Dupont & Leroy, 1995; for beetles, see Coope, 1979; for marine plankton, see Howard, 1985; for mammals, see Sutcliffe, 1985). General episodes of transcontinental migrations, induced by Neogene physical changes, have been documented in horses (Bernor & Lipscomb, 1995) and in bovids (Thomas, 1984; Vrba, 1995c). Turner (1984), Turner & Wood (1993b), Schrenk et al. (1993), and Bromage et al. (1995a) have all analyzed major changes in distributions of early hominids and documented that such shifts of hominid distributions occurred together with those of other mammals in consistent relation to physical changes. The question is how these physically induced changes in geographic distribution, particularly the episodes of vicariance that affected many lineages together at particular times, relate to speciation. The following six hypotheses and predictions address this question. They, together with additional predictions from habitat theory, have been previously discussed in more detail, and some of them have been tested (Vrba, 1985d, 1987b, 1992, 1994, 1995a,b,c, 1998).

The "Eco-shuffle" Prediction

The "eco-shuffle" prediction is illustrated in figure 2-1. To some extent the ecological associations of taxa are

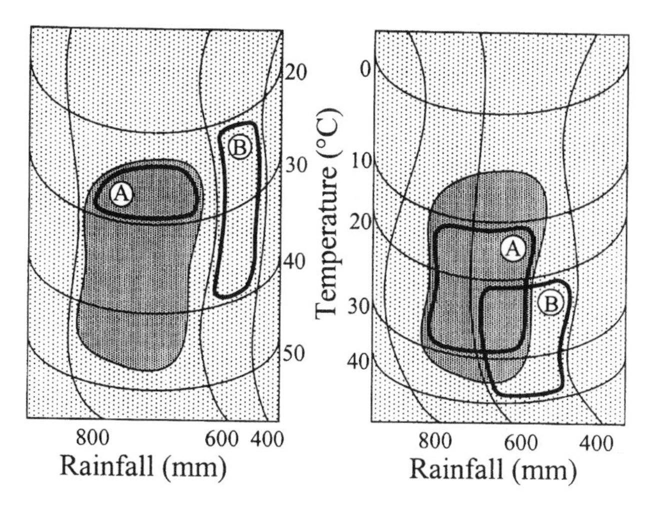

Habitat Variables	Tolerance Range	
	Species A	Species B
Temperature	10° - 30°C	20° - 40°C
Rainfall	400-800 mm	400-600 mm
Substrate	■ only	■ and □

Figure 2-1. Components of the habitat specificities of related species commonly have overlapping ranges: hypothetical species A and B, due to overlapping specificities for three habitat variables, have disjunct distributions today, yet had sympatric ones in the past. (Adapted from Vrba, 1995a: fig. 1.4.)

predicted to be different at different extremes of the climatic cycles. That is, if species are moved about as climatic change moves their habitat components, then their communities and ecosystems are expected to be shuffled in terms of species content.

One reason for this shuffling is that each species has a particular tolerance range for each habitat component that may or may not be identical to that of other species and a combination of such ranges for all its requirements that is unique. Due to the complex interactions of factors such as topography with climatic regimes, the associations of different environmental variables underwent recombination during past times, and some recombination among taxa in communities necessarily followed suit (fig. 2-1). In agreement with this prediction, there is some evidence that extant species did occur in past taxonomic associations different from their modern ones (review in Vrba, 1992). For this reason, biome islands that have resulted from vicariance are unlikely to be true community refugia in the strict sense of providing refuge for an association of taxa that is identical to that of the more widespread parent community. A biome may persist (such as the tropical forest biome) even though the associated ecosystems undergo recurrent changes into different ecosystems in terms of both their precise species content and geographic locality. Thus physical changes do not only promote vicariant population structure, but also another element that is relevant to speciation: the recurrent shuffling of ecosystems with attendant new competitive interactions among organisms.

Long-term, Continuous Vicariance, or Allopatry

Long-term, continuous vacariance is needed to initiate speciation. The new regimes of selection and drift characteristic of isolated and shrinking populations (Wright, 1932) are more likely to bring about the divergence that results in speciation if the periods of isolation are longer.

Tectonism is a major cause of long-lasting vicariance. The Neogene history of eastern Africa was marked by strong tectonism, uplift, and rifting (Partridge et al., 1995a). This hypothesis predicts increased rates of species diversification, particularly in eastern Africa during the Neogene. A second cause of vicariance is habitat change and fragmentation by widespread climatic changes. It is clear that most species during the late Neogene were frequently "attacked" and fragmented into isolated populations by global climatic changes. Yet, many such species survived numerous climatic cycles and bouts of vicariance without net change and without noticeably giving rise to new species. In other words, although allopatry is usually necessary for speciation, it is apparently not sufficient. Habitats must not only change, they must change enough to bring about disjunction of previously continuous populations for a sufficiently long interval if the irreversible changes implied by speciation are to occur. Thus, we need models of speciation that invoke particular kinds of allopatry promoted by particular kinds of climatic change. The previous and the next two proposals are of this kind.

Major Shifts in the Periodicity Pattern of the Astronomic Cycles

Given that prolonged vicariance is important for turnover, major shifts in the periodicity of astronomical cycles are predicted to promote bursts of speciation and extinction (Vrba, 1992). Changes in dominance from a cyclic pattern of shorter period to one of longer period, together with progressive severity of the cyclic extremes, are especially significant for prolongation of vicariance across many lineages and therefore for increased speciation and extinction rates.

As mentioned earlier, deMenocal and Bloemendal (1995) documented examples of such major global shifts in deep sea cores off West and East Africa: a shift occurred prior to 2.8 m.y. from dominant climatic influence at 23–19 k.y. periodicity to dominance at 41 k.y. variance thereafter, with a further shift to dominance at 100 k.y. variance after 0.9 m.y. The authors interpreted these records as reflecting changes in Africa as well as elsewhere on earth.

The Relay Model of Turnover

The relay model of turnover in response to climatic trends (figs. 2-2, 2-3) assumes that prolonged vicariance is important for turnover. Larger trends in the mean of the climatic cycles are predicted to elicit strong turnover in a particular sequential and frequency pattern. Long-term warming elicits events roughly in the following sequence: (1) extinction of cooler-adapted species; (2) speciation of cooler-adapted species; (3) extinction of warmer-adapted species, and (4) speciation of warmer-adapted species. Overall, a higher proportion of warmer-adapted species should appear, and more cooler-adapted species should be last recorded. Long-term cooling elicits the converse turnover pattern in which warmer-adapted species turn over before cooler-adapted species and, overall, a higher proportion of cooler-adapted species appear, and warmer-adapted species become extinct. Models of this kind that directly address features of the global climatic record have not been considered in the past. I therefore discuss this model in more detail (after Vrba, 1995b) than I do the others. The prediction is that lineages in contrasting habitat categories differ in timing of vicariance and therefore have new species starting up (by speciation) and others ending (by extinction) at displaced times along a climatic trend, rather like runners in a relay race. I term this the "relay model" of turnover.

It is well documented that the geographic distributions of some species are largest and most continuous around glacial extremes and smallest and most prone to vicariance during interglacials (e.g., muskoxen and collared lemmings; Sutcliffe, 1985). The situation is converse for other kinds of species (e.g., *Cercopithecus* monkeys in the African rainforest; Grubb, 1978). A similar contrast between cooler- and warmer-adapted species can be explored in cases that differ less extremely than do Arctic and rainforest forms. Thus, in Africa, hartebeests, springbuck, and other antelopes in the contingent adapted to more arid, open and seasonally cooler grasslands and open woodlands had more extensive geographic distributions during glacials, whereas taxa like lechwes and duikers within the more warmth-, moisture- and wood-cover–loving category were more widespread during interglacials into areas where they do not occur today (e.g., Klein, 1984). One can express such information on habitat continuity versus fragmentation for species directly in terms of the climatic signals in long records that reflect the astronomic cycles (e.g., $\delta^{18}O$ values in deep sea cores; loess and soil changes as in China, Kukla, 1987; or changes in pollen frequencies, e.g., in western Africa; Dupont & Leroy, 1995). That is, one can specify the predicted upper and lower limits along a given axis of climatic variation over which each species' habitat and geographic distribution is maximally continuous and beyond which increased vicariance occurs. I will refer to these upper and lower limits for each species as the "vicariance threshold" and to the range between them as the "optimal range." In relation to any actual or hypothetical climatic curve (such as those in figs. 2-2 and 2-3), the specified vicariance thresholds for species directly predict durations of vicariance for those species.

Figure 2-2 illustrates the model only with respect to the timing of speciation events. Figure 2-3 offers a more general overview of many lineages that differ in

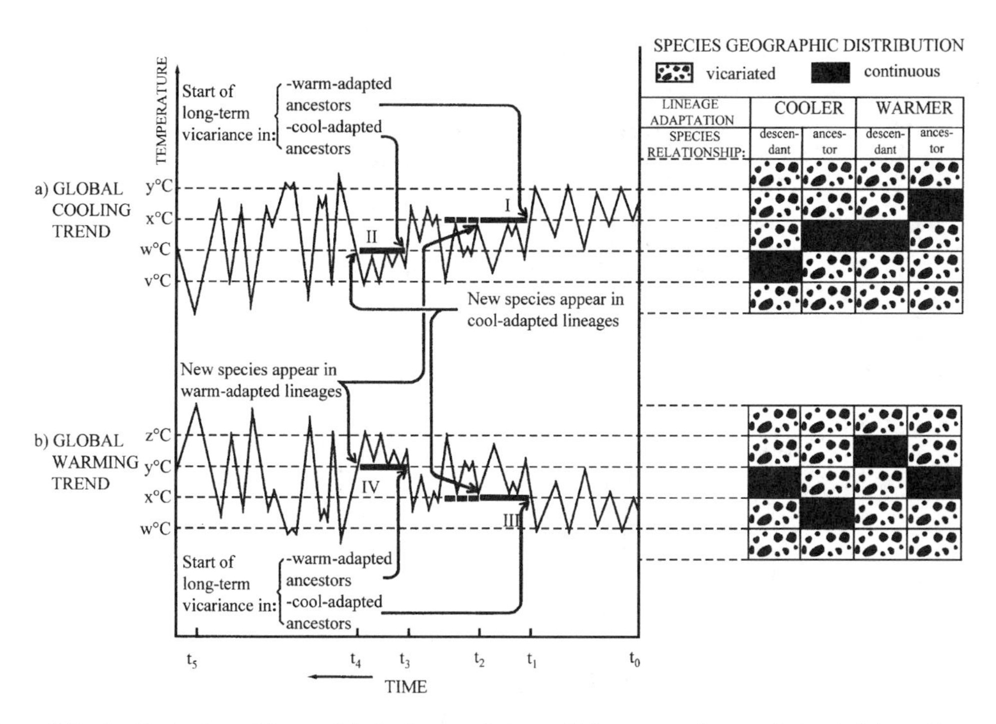

Figure 2-2. An illustration of the model of relay speciation, which assumes that speciation is initiated by continuous, long-term vicariance. The cooling curve in (a) is adapted from the 3.1–2.4 m.y. data of Shackleton (1995). The warming curve in (b) is hypothetical, being obtained by rotation of the curve in (a). We can start by looking at the earliest segment, $t_0 - t_1$, on the right side, over which interval the vicariance events are too brief to culminate in the genetic and phenotypic divergence necessary for speciation. I chose two categories of species with temperature tolerance ranges at the start ($t_0 - t_1$) as follows: continuous geographic distributions in warmer-adapted species occur only over x°C − y°C and in colder-adapted species over w°C − x°C (compare the key for geographic distributions, at the right, which is linked by dashed lines to the temperature scale). The minimum time of continuous vicariance needed for speciation in both categories of species is here modeled as the length of each of horizontal bars I–IV (ca. 80–100 k.y. in this example; the bold dashed lines produced from bars I and III indicate the total continuous vicariance phase relative to the ancestral species). Only at the end of such intervals of long-term continuous vicariance do new species appear, with new tolerance ranges evolved toward the next temperature division, cooler in the case of the cooling trend. Note that during long-term warming (b), speciation of cooler-adapted species precedes that of warmer-adapted species, whereas during long-term cooling speciation of warmer-adapted species precedes that of cooler-adapted species. See text for further discussion. (Adapted from Vrba, 1995b: fig. 2.)

speciation and extinction rhythms across successive warming and cooling trends. The model assumes that a minimum interval of continuous vicariance is necessary for most speciations, and precedes most extinctions. That is, within that interval speciation cannot occur, although extinction can occur with a given probability, and at the end of that interval new species with evolved (changed) vicariance thresholds will originate with a given probability. One can run the model using various combinations of species and climatic data and specifications of minimum intervals of continuous vicariance, etc., and predict the differences among groups of species in timing of vicariance episodes and in timing of turnover events. The model can be used to explore the internal consistency of combinations of particular postulated conditions and to compare the predictions with real patterns.

The predictions are exemplified by the warmer-

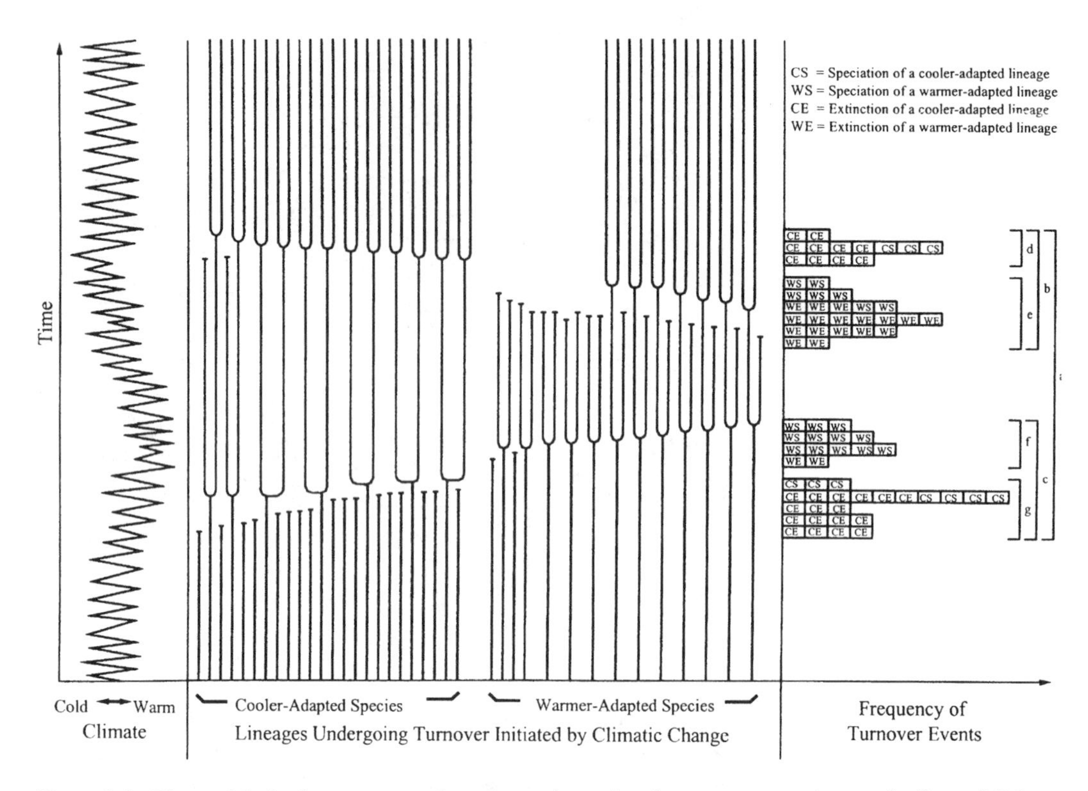

Figure 2-3. The model of relay turnover: a broad overview using the same assumptions as for figure 2-2 (see also text). The diagram shows the expected sequence of extinctions and speciations of two categories of lineages, warmer- and cooler-adapted species, during a major warming followed by a major cooling shift of the astronomical climatic cycles. Note the smaller turnover pulses nested inside larger ones. (Adapted from Vrba, 1995b: fig. 3.)

adapted and cooler-adapted lineages of species relative to the cooling curve in figure 2-2a. During such a major cooling trend, the warmer-adapted species are the first to enter long-term vicariance (vicariant phase I, time $t_1 - t_2$). During this interval, the allopatric populations in each warmer-adapted species shrink or disappear with cooling and expand with relative warming, yet never coalesce again because the maxima remain below x°C, the threshold temperature for distributional continuity of the warmer-adapted ancestral species. Long-term vicariance of many small populations is hypothesized to promote two kinds of processes important to speciation, in accordance with parts of the models proposed by Wright (1932), Mayr (1942), Carson (1987), and others: (1) Size bottlenecks associated with the decreasing populations result in random genetic sampling and increased genetic diversity among the vicariant populations. (2) A shrinking and shifting biome patch is expected to change in community structure ("eco-shuffle" prediction and fig. 2-1), and the unusual population constitution resulting from chance genetic sampling is further expected to alter the selective environment. Around its periphery, the new environmental conditions intrude in the form of new selection pressures. Thus, the environmental changes exert strong selection pressure for a shift in habitat specificity toward the new conditions within and surrounding each vicariant population. The new selection regime has an unusually high genetic and phenotypic variation to act upon, much of it apportioned among the numerous vicariant populations. As a result, a few such previously warmer-adapted populations (adapted to spread during the x°C–y°C climatic phases) are likely by the end of the vicariant phase $t_1 - t_2$ to have adapted sufficiently to be able to achieve widespread and continuous distributions during the cooler times with temperatures between w°C and x°C. If such adaptation involved divergence of the fertilization system in a population (Paterson, 1985), speciation will have resulted.

During the same cooling trend (fig. 2-2a), the cooler-adapted species enter long-term vicariance only later, vicariance phase II starting at time t_3, once the climatic maxima remain for longer below w°C, the lower threshold for distributional continuity of the cooler-adapted ancestors. Previous to time t_3 those species still resumed geographic continuity regularly over short time intervals. Similar processes to those outlined above are then hypothesized to affect the persistently vicariated populations of the cooler-adapted ancestors, culminating in new, cooler-adapted species by the end of vicariant phase II.

Vicariance precedes not only speciation but also extinction. Most extinction should precede most speciation in any one turnover episode. Thus, the first extinctions during a cooling trend are those of warmer-adapted species (in vicariance phase I in fig. 2-2a, and before new warmer-adapted species appear at the end of that phase). Similarly, the least cold tolerant among the cooler-adapted species should become extinct during phase II, preceding the speciations in other cooler-adapted lineages near the end of that phase. Figure 2-3 shows the predictions over longer time. Overall, many more extinctions and fewer speciations in warmer-adapted than cooler-adapted lineages are expected during a major cooling trend.

During a longer warming trend (fig. 2-2b), similar but inverse arguments imply that the cooler-adapted species should enter long-term vicariance (at time t_1) before the warmer-adapted species do (t_3), and the cooler-adapted species should also speciate (at t_2) before the warmer-adapted species do (t_4). Also, during the same warming trend, extinctions within each habitat category should precede speciations, and the cooler-adapted species should not only enter vicariance earlier, but also suffer earlier extinction events than the warmer-adapted species. Overall, many more extinctions and fewer speciations in cooler-adapted than warmer-adapted lineages are expected during a major warming trend (fig. 2-3).

Note three points. First, even if a cooling or warming trend has a gradual mean, the model that long-term vicariance is needed for speciation predicts a punctuated response in the responding biotic variables via threshold effects. Second, the larger-scale turnover relays with lower frequencies among successive major cooling and warming episodes have the smaller-scale relays at higher frequencies embedded within them (fig. 2-3). Third, over the Cainozoic and particularly the Late Neogene, there has been a strong emphasis on episodes of global cooling (e.g., Kennett, 1995). Thus the relay succession has been very unequal, heavily favoring diversification in increasingly cooler-adapted lineages. The first appearance data for African antelopes after 2.9 m.y. supports this prediction for a major cooling trend (Vrba, 1995c).

The Turnover Pulse Hypothesis

If physical changes initiate most speciations, extinctions, and migrations (which together comprise turnover), then turnover involving different kinds of organisms should be concentrated nonrandomly in time and in predictable association with the climatic record for that area.

This hypothesis is more complex than suggested by this statement and includes the following elements (Vrba, 1985d, 1995b): global climatic and tectonic changes are causal agents of turnover; pulses occur at different scales of time, geography, and numbers of lineages involved; biotic factors crucially influence the nature of turnover, although the initiation is reserved for physical change; and lineages respond differently depending on intrinsic properties.

Similar phenomena at different scales are relevant to this hypothesis. Scientists are accustomed to recognizing hierarchical nesting of similar phenomena, or similarity among phenomena at different scales, especially recently with the growing exploration of fractal structures and processes—from those involved in the sutures of ammonites, through tree growth, to coastlines (Mandelbrot et al., 1979; Gleick, 1987). There is a hierarchical geographical layering relevant to turnover; "habitat plates" shifting rapidly over the continental plates that drift more sedately beneath them. There is also hierarchical nesting of climatic change, which contains several tiers of higher-frequency cycles that are nested within lower-frequency cycles. The variation among taxa in breadth of habitat tolerance dictates that the more ecologically specialized should respond with turnover to smaller climatic changes and more often than the more generalized taxa (see the resource-use hypothesis, Vrba, 1987b). This, and the differences among taxa in placement of their habitat specificities along the scales of environmental variation (such as between cooler- and warmer-adapted categories) dictates that turnover pulses have particular structures, such that larger turnover pulses have nested within them smaller subparts. That is, both the multiscaled pattern of the climate signal itself and the variation in habitat tolerance among taxa suggest that biotic responses to climate should present a hierarchically nested pattern at different scales as well (fig. 2-3).

Let us consider further this last notion that is simply exemplified in figure 2-3. Organismal phenotypes are not distributed evenly throughout phenotype space, but are markedly clustered in limited parts of that space (e.g., Raup, 1966). We should expect this to be also true of those phenotypes that determine habitat thresholds. Thus, one cannot expect a simple one-to-one correspondence between the magnitude of climatic variation and turnover response: Parts of the climatic scale are likely to be very thinly populated by turnover events, or by turnover events of a particular kind (e.g., change to very extreme climate may elicit mainly extinctions and few speciations). At the same time, the fact that habitat specificities of taxa differ in being situated at different parts of the temperature scale (or of another habitat-determining scale) predicts that the turnover response has a particular structure in relation to decreasing and increasing climatic variables. This is exemplified by earlier speciation of warm- than cooler-adapted taxa during a cooling trend (figs. 2-2, 2-3). And as noted above, in spite of the lack of a simple one-to-one correspondence, turnover pulses, like climatic events, are predicted to be hierarchically nested phenomena. Thus, viewing the turnover frequency pattern in figure 2-3 from a broad perspective, one can refer to the bimodal aggregation marked **a** as a turnover pulse. With increasing resolution, the component pulses **b** and **c** (with their different contents in terms of kinds of lineages) within **a** become evident, and the subpulses **d** and **e** within **b,** and **f** and **g** within **c,** etc. Thus, in discussions of whether an observed pattern in the fossil record is a turnover pulse, the scale of resolution in the fossil data relative to that in the climatic data needs to be considered.

In the current context it is of particular interest whether the Late Neogene mammalian record supports this hypothesis. The extent to which the Plio-Pleistocene African record does so will be reviewed below. Barry et al. (1995) analyzed mammalian turnover in the Neogene Siwaliks of Pakistan. They concluded that their data do support the Turnover Pulse Hypothesis: "faunal change in the Siwaliks was episodic, occurring during short intervals with high turnover followed by longer periods with considerably less change" (Barry et al., 1995:209).

A General Model on Climate-Associated Heterochrony

The model on climate-associated heterochrony assumes that clades commonly share inherited morphogenetic rules that determine similar ecophenotypic ranges, or norms of reaction,[2] in response to variation along a given environmental axis. Reaction norms commonly evolve by heterochrony that shifts, narrows, or extends the ancestral reaction norm while retaining the same vector of correlation with environmental variables. The prediction is that particular kinds of heterochrony will appear repeatedly in parallel in related lineages in response to particular widespread climatic changes.

For instance, mammals appear to share the basic vector of a norm of reaction for body size relative to temperature. Within single species, ecophenotypes that grow longer to reach larger body sizes are generally expressed in climates that are at least seasonally cooler (Bergmann's rule; Bergmann, 1846) and smaller ecophenotypes in warmer environments. This shared basic vector of a reaction norm predicts that particular kinds of heterochrony will occur repeatedly in parallel in mammals in response to widespread climatic changes. For instance, global cooling is predicted to be associated with the evolution of longer-growing, larger-bodied new species in many mammalian lineages. Another shared norm of reaction in mammals is that bodily extremities (horns, limbs, ears, muzzles) show relative decrease as temperature declines (Allen's rule; Allen, 1877). Allen phenotypes probably evolve by the kind of paedomorphosis that occurs by reduction in net rate of shape growth relative to size growth. It results in descendants that appear juvenilized relative to the ancestor because in the ancestor the juveniles had relatively shorter extremities than the adults (Vrba, 1994, 1998). Bergmann's and Allen's rules are known both as ecophenotypes within species and as genetically-based (or evolved) differences between sister taxa living in differing environments. Although exceptions to these rules have been noted, in general their predictions are upheld (see review in Vrba, 1994). Both rules are upheld in modern humans (Roberts, 1978).

I have elsewhere presented a model (Vrba, 1994; 1998) that explains how heterochrony can result in a descendant that combines Bergmann's and Allen's rules with hyperpaedomorphosis[3] and with hyperadult characters evolved by classical hypermorphosis. The reason this can happen is as follows. Not only does growth prolongation, termed "time hypermorphosis" occur, but a special form thereof occurs, in which the ancestral ontogeny is multiphasic, such that the character develops through at least two phases of distinct exponential growth rate. Growth prolongation in the descendant then proportionally extends the growth time, while retaining the ancestral growth rate, of each of these ontogenetic phases. This evolutionary process was termed proportional growth prolongation in Vrba (1994) and sequential time hypermorphosis in McKinney & McNamara (1991). In sum, the same evolutionary event of growth prolongation in response to climatic cooling as it acts on characters with different ancestral growth profiles in the same body plan, can result in a major reorganization or shuffling of body proportions such that some characters become larger and others smaller, some hyperadult and others more juvenilized. I have suggested that this hypothesis applies to major features of hominid evolution (Vrba, 1994, 1998). The results of quantitative tests applied to brain growth in humans and chimpanzees are consistent with the hypothesis that hominine encephalization (fig. 2-5) evolved by proportional growth prolongation (Vrba, 1998).

In fact, many mammalian lineages did change as predicted here during the Pleistocene glacials (Kurten, 1957). Also, many of the new species that appeared in Africa and globally near the start of the modern Ice Age, 2.9–2.5 m.y., show similar suites of integrated character complexes, including larger bodies, relative reduction of some body parts, together with disproportionate enlargement of other features including brains. I cite some of these cases below (see Vrba, 1994, 1998, for more extended discussion).

Hominids in the Context of Fauna and Climate: Are There General Patterns?

Let us start with the evolution of Hominidae. Hominid cladograms from different analysts usually differ in exact branching topology and usually feature polytomy, as does the example in figure 2-4a. Most such cladistic analyses either explicitly identify *Australopithecus afarensis* as the potential direct ancestor of robust autralopithecines and *Homo,* or they leave open that possibility (e.g., Wood, 1992b, on whose cladistic analysis fig. 2-4a is largely based). If the inference is correct that *A. afarensis,* now recorded to 2.9 m.y. (Kimbel et al., 1994), is the direct ancestor, then rapid lineage splitting occurred in this lineage delimited between 2.9 and 2.5 m.y. It is, of course, possible that *A. afarensis* became terminally extinct, leaving no descendants, and that lineages of robust australopithecines and of taxa currently included in *Homo* arose earlier in the Pliocene with their immediate ancestors as yet unrepresented in the fossil record. But a more parsimonious inference, given the data we have, is that an increased rate of speciation between 2.9 m.y. and the earliest Pleistocene did occur in the hominid clade.

Hominids also underwent marked changes in morphology and behavior during the late Pliocene and earliest Pleistocene. The hominine trend of progressive encephalization appears to have started then (fig. 2-5). Disagreements remain on how to estimate hominid body weights (McHenry, 1991, 1992), the encephalization quotient, EQ (Martin, 1983; McHenry, 1988), fossil taxonomic and sexual identities, and taxon branching sequence and chronology (Howell, 1978; Wood, 1992a). Also, there is a gap in the record of hominid crania about 2.6–1.9 m.y. As a result, the precise Late Pliocene timing of onset of this encephalization trend remains uncertain. Nevertheless, all agree that this change is fundamental to human evolution, that the magnitude of encephalization increase even to the earliest-known crania of *Homo* was considerable (fig. 2-5), and that the setting was the African late Pliocene. Figure 2-5 is not a genealogical statement. It leaves open whether early *Homo* is most closely related to *Australopithecus afarensis, A. africanus,* or to another perhaps as yet undiscovered taxon (Wood, 1992b). But it does suggest strongly that the crucial time period for onset of this trend was 2.3–3.0 m.y.—precisely the time during which major global climatic changes occurred (fig. 2-4e).

Note that *Australopithecus afarensis,* the potential direct ancestor of *Homo* and other later hominids, and last recorded near 2.9 m.y., was hardly more encephalized than the chimpanzee (fig. 2-5). Neither were the robust australopithecines much more encephalized. In contrast, the monophyletic *Homo* lineage (including taxa *sapiens, neanderthalensis, erectus, ergaster, rudolfensis,* and *habilis* according to various cladistic analyses) embarked on a sustained net encephalization trend. I have argued that the start of this trend occurred when one hominoid lineage, *Homo,* switched to a particular evolutionary mode, the kind of heterochrony that occurs by growth prolongation to larger body size (Vrba, 1994, 1998). In contrast, there is evidence that trends in size increase in lineages of *Paranthropus* and other hominoids occurred by growth rate increase (rate hypermorphosis). Vrba (1994) reviewed the evidence for this and discussed why time hypermorphosis in *Homo* could result in a progressive EQ trend, while rate hypermorphosis in *Paranthropus* and other hominoids kept those lineages closer to ancestral EQ values (fig. 2-5). We can realistically expect that a fundamental ecological resetting in the *Homo* lineage accompanied the evolution of encephalization and new open behavioral programs (see endnote 1). The start of this trend in *Homo* was a major departure from ancestral modes of development, evolution and ecology—a good example of the evolution of a new key adaptation and a new ecological "fitness-generating function" or "G-function" (Rosenzweig, this volume). I suggest that this trend in *Homo* commenced in response to the start of the modern Ice Age and was fueled by the progressive intensification of global cooling minima since then.

Does the hominid pattern show any predicted consistency with those for other groups and with paleoclimatic records? Three questions arise, as discussed below.

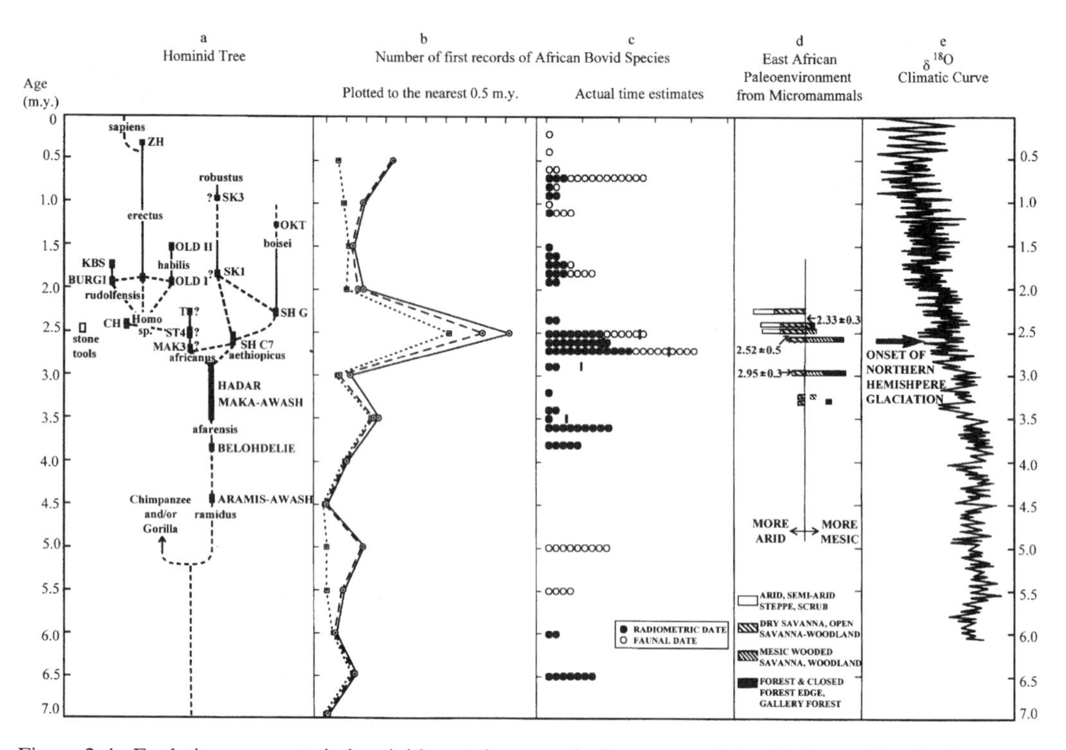

Figure 2-4. Evolutionary events in hominids, antelopes, and micromammals in relation to climatic change. (Adapted from Vrba, 1995c: fig. 11.) (a) Hominid genealogical tree (adapted from Wood, 1992, with modifications). Solid rectangles = dating estimates for first appearance data (FADs) and last appearance data (LADs) of hominid taxa; dating for East Africa after Brown and Feibel (1991; also T. White, personal communication for middle Awash); for South Africa, Vrba (1982, 1995c). Solid rectangles with question mark indicate biochronological estimates. Dashed lines and question marks on lineages denote uncertain branch connections. In South Africa: SK1-3 = Swartkrans Members (M) 1-3; ST4 = Sterkfontein M4; MAK3 = Makapansgat M3; T = Taung; in Ethiopia: HADAR = Hadar Formation; BELOHDELIE, MAKA-AWASH, ARAMIS-AWASH = Maka and Aramis units in middle Awash; in Kenya: BURGI = Burgi Member at Koobi Fora, SUB-KBS = stratum below KBS tuff at Koobi Fora; CH = Chemeron; OLDI, II = Olduvai Beds I, and II Level BK; ZH = Zhoukoutien in China. (b) Earliest African records of antelope species (Bovidae) from Vrba (1995c: appendix 2). (1) Solid line: based on all records as interpreted in Vrba (1995c: appendix 2); (2) dashed line: same data set as (1), except that all the alternative interpretations by other authors as cited in Vrba (1995c: appendix 2) are used; (3) dotted line: same data set as (1), except that all biochronological records have been removed. (c) Earliest African records of antelope species (Bovidae). Solid bars show magnitude of histogram peaks when all the alternative interpretations are used as in panel b (2). (d) Relative micromammal frequencies in different habitat categories (rectangles) indicate paleoenvironmental change in the Shungura Formation, Ethiopia (after Wesselman, 1995). (e) Deep-sea oxygen isotope data from a North Atlantic site, adapted from Shackleton (1995); ice volume increase to left.

Is There Late Neogene Evidence of Major, Widespread Climatic Changes in Africa Including in Hominid Habitats?

The start of the modern Ice Age (fig. 2-4e) appears to have initiated the spreading of the Sahara 2.8–2.7 m.y. (based on pollen cores off western Africa; Dupont and Leroy, 1995), and of more open and desert landscapes in the Horn of Africa (based on deep sea records in the Gulf of Aden showing increased dust influxes; deMenocal & Bloemendal, 1995). The changes over this period in hominid habitats included strong vegetation change in Ethiopia (see Bonnefille, 1995;

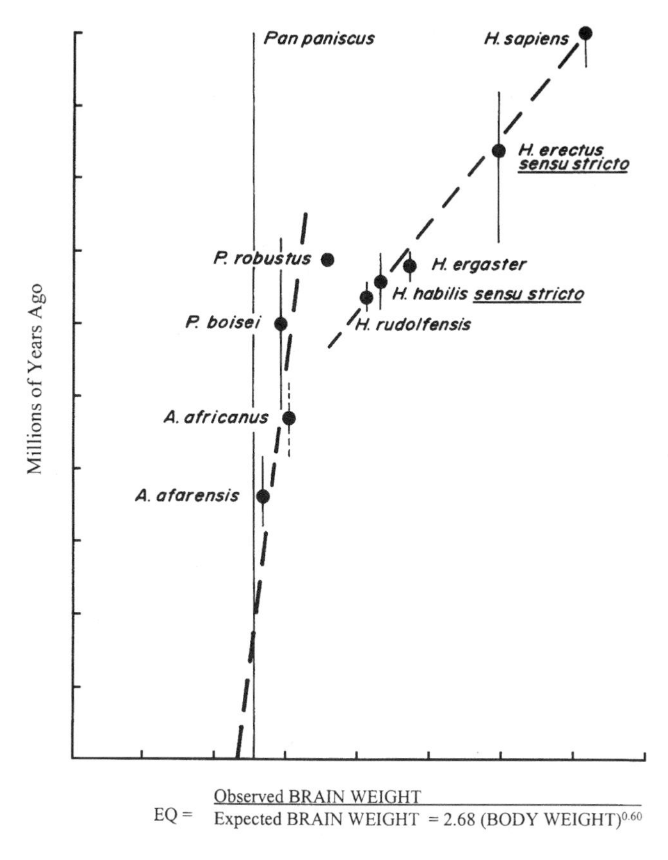

Figure 2-5. Evolutionary trend in hominoid encephalization quotients (EQs) using Martin's (1983) formula for Cercopithecoidea. The graph does not represent genealogy. Estimates of hominid body weights from McHenry (1991 1992), of cranial capacities from Holloway (1970, 1972, 1978), Tobias (1971), Johanson and Edey (1981), McHenry (1988), and of values for *Pan paniscus* from Martin (1983). Wood (1992a) and others he cites view the traditional *Homo habilis* as comprising at least two taxa, *H. rudolfensis* and *H. habilis,* the EQs for which are based on individual specimens as classified in Wood (1992a) and McHenry (1992). A similar procedure was used to subdivide *H. erectus* by separation of *H. ergaster* (Wood, 1992a). Relative to the EQ axis, solid dots are at mean EQs. Relative to the time axis, solid dots for *A. africanus* and *P. robustus* are placed at age estimates after Vrba (1982, 1985c), and for other taxa at mean ages for crania within age ranges (solid lines) as reviewed in Wood (1992a). (Adapted from Vrba, 1995c: fig. 12.)

Wesselman, 1995, who inferred a transition from rainforest to grassland–semidesert between 2.9 m.y. and 2.5 m.y., represented in fig. 2-4d). Not all studies agree on this. One hominid-associated record in Kenya suggests little local effects from global changes over the entire past 15 m.y. (Kingston et al., 1994). My own results on Bovidae, one of the most abundant, speciose, and habitat-sensitive groups among large mammals, strongly suggest that many African areas did respond to global climatic trends. These results are further discussed below.

Is There African Evidence of Turnover Episodes

The available studies have arrived at various conclusions to the question of whether there is evidence of turnover in different faunal groups in response to major climatic trends and of similar morphological novelties across these groups that are consistent with major cooling or warming. One "no" for the southern African hominid record came from McKee (1993b). He considered the pattern in South African assemblages to be consistent with random simulations of turnover, although also with simulated turnover pulses. A major problem with this study is the total lack of radiometric dates for South African sites. Concerning the start of the modern Ice Age near 2.5 m.y., Bishop (1993) showed that 13 African species of pigs do not show a pulse of first appearance data (FADs) at that time, although they may do near 1.9 m.y. Two "yes's" come from studies of micromammals and bovids. Graphs from these are compared in figure 2-4 with Shackleton's (1995) climatic curve and with the hominid record.

Wesselman's (1985b, 1995) micromammalian record (rodents, bats, etc.) from the Shungura Formation, Ethiopia, shows a remarkable change in the kinds of taxa and in their frequencies during the late Pliocene (see also Denys, this volume). It occurred after late Member B (after ca. 2.9 m.y.) and was in place probably already by Member E near 2.5–2.4 m.y. and confirmed by Member F and early Member G dated 2.4–2.3 m.y. The turnover event includes terminal extinctions, immigrants from Eurasia such as a species of *Lepus,* and global first appearances of species, such as a new species of *Heterocephalus,* the genus of naked molerats, and a new species of *Xerus,* the genus of ground squirrels (Wesselman, 1995). This time also marks the first African and global debuts in the record of several species of bipedal, saltatory desert-adapted rodents, such as the genus *Jaculus* of desert jerboas in Shungura Member F and of the genus of living springhares *Pedetes* in Taung, South Africa (Broom, 1937).

The highly specialized rodent body plan associated with bipedal locomotion is today strongly associated with open, arid, mostly desert habitats, and appears numerous times independently in rodents (in 24 genera in 8 families according to Hafner & Hafner, 1988). The fossil record of first appearances of bipedal rodent taxa (Lavocat, 1978; Savage & Russell, 1983) suggests that most may have appeared during times of global cooling and land aridification (which is consistent with their modern habitat-specificities) either during the late Miocene, or at the late Pliocene start of the modern Ice Age (like *Jaculus, Pedetes, Gerbillus* in Africa, and *Alactaga* in Asia). Hafner and Hafner (1988) point out that these bipedal forms share suites of characters including elongated hindfeet and enlarged heads, brains, eyes, and auditory bullae, and many of them also have larger body sizes by prolonged growth. I suggest that the enlarged hindfeet, heads, and brains in these rodents evolved analogously to features in hominines, notably our enlarged brains—all evolved by hyperpaedomorphosis (Vrba, 1994, 1998). Thus, there are suites of apparently integrated character complexes that suggest similar heterochronies in response to common climatic influences in mammalian groups as different as rodents and hominids.

The African record for 147 bovid species shows a strong pulse of first appearances between 2.8 and 2.5 m.y., and smaller ones 1.9–1.8 m.y. and 0.9–0.6 m.y. (see fig. 2-4b, 2-4c; Vrba, 1995c). Bovidae evolved and speciated more rapidly during the Late Neogene than did other large mammals. We have used several different approaches to document and date this radiation. Cladograms based on mtDNA sequences (e.g., Gatesy et al., 1992) were compared with cladograms based on skull characters (e.g., Vrba & Gatesy, 1995; Vrba, 1996) to check which cladistic nodes are well supported. The skull characters were used to incorporate the numerous fossil taxa together with the living ones and to list all hypothetical direct ancestors of splitting (speciation) events. Extinct taxa that in cladistic results lack autapomorphies are potential direct ancestors of their sister taxa. Thus, one can use cladograms of extinct and extant taxa, together with radiometric dates for the last appearances of the potential ancestors and the earliest appearances of their hypothesized descendants, to estimate the time intervals during which speciation occurred. For instance, using such methods I estimated that at least one-half of the 44 bovid species that appeared in Africa 2.5–2.9 m.y. (fig. 2-4c) represent speciation events during that interval, and about one-sixth are Eurasian immigrants. Only about one-quarter of these 44 first appearances are estimated to be species that were present previously but unseen due to gaps in the record. Also, the bovid FAD pulse 2.8–2.5 m.y. shows a strange, statistically significant pattern: By 2.7 m.y. much higher proportion of warmer- and closed-adapted taxa appear (80%), while the 2.6–2.5 m.y. FADs include far more cooler- and open-adapted forms (73%). This supports the sequence expected within a turnover pulse initiated by a net cooling trend. I know of no hypotheses on purely biotic initiation of speciation, or on taphonomic causes, that predict such a pattern. In sum, based on the results of several statistical and comparative tests, I have concluded that the FAD pulse 2.8–2.5 m.y. in antelopes closely follows the various predictions for a turnover pulse initiated by a strong trend of global cooling (Vrba, 1995c).

Most of the antelope species that appeared 2.8–2.5 m.y. are larger than their earlier relatives (or outgroups in cladistic terms). For instance, this time marks the advent of *Pelorovis,* the lineage of giant buffaloes, *Megalotragus,* the giant hartebeests, the giant extinct hirola *Beatragus whitei, H. equinus* (the roan antelope), and the extinct *Hippotragus gigas,* which is the largest known Hippotragini (the clade also includes oryxes and sable antelopes), and the lineage of *Kobus ellipsiprymnus,* the waterbuck which is the largest member of Reduncini (the group also includes lechwes, kobs, pukus, and reedbuck). The first five of these are among the taxa for which potential direct ancestors are known that have FADs since 2.9 m.y. (Vrba, 1995a: fig.10). Also, several of the new antelope species appear to have become encephalized over this time. For instance, the Alcelaphini account for more new species over this interval than any other mammalian clade of comparable taxonomic rank, and species of this clade as a whole are exceptionally encephalized among Bovidae (Oboussier, 1972). The enlarged bodies with disproportionately enlarged brains in the bovids that appeared near the start of the modern Ice Age parallel the similar features in the *Homo* lineage and in rodent taxa. Thus, there is support from three different groups for the hypothesis that particular kinds of heterochrony can appear repeatedly in parallel in different groups reflecting shared, inherited morphogenetic responses to common environmental causes.

In addition, the cladograms for several different groups of antelopes show a typical pattern of an ancestor ending at or after 2.9 m.y., with new descendants branching off between 2.8 and 2.5 m.y. (Vrba, 1995a: fig. 10), and in each of these cases, the interval 2.9–2.5 m.y. is one of unusually high speciation rate. These concordant genealogical patterns among different groups of antelopes strongly resemble that in the hominid tree (fig. 2-4a), which also shows an inferred direct ancestor last seen near 2.9 m.y. with rapid lineage splitting delimited between that and 2.5 m.y. This concordance in cladistic and chronological patterns among bovids and hominids, together with the turnover observed in rodents (Wesselman, 1995), strongly suggests the common causal influence of the start of the modern Ice Age (fig. 2-1d). In sum, there is some support for the notion that common causal rules connect the climate system with evolution of different biotic groups. Some of these rules govern how organisms grow—and therefore their evolutionary responses by heterochrony—across the different biotic groups. I expect that other general rules concern the responses of species' ecologies and population structures to environmental change and therefore general modes of speciation.

Mechanisms of Speciation

A major question before us is what proportion of speciation events in the history of life have been initiated by biotic interactions, in the absence of physical change. Some suggest that proportion was major (Van Valen, 1973; Foley, 1994). Others have claimed that there are no possible mechanisms by which climate might initiate speciation (e.g., McKee, 1993b, personal communication). A different perspective is that much of our theory and data, such as the extensive theoretical and empirical support for the notion of allopatric speciation (cited earlier), strongly suggest physical initiation of speciation. This body of theory opens the way for testable models that link climate with speciation (such as the relay model I suggested). From this side comes the converse challenge: Articulate compelling models of precisely how biotic interactions should alone initiate speciation in the absence of physical change, and show how we can test them.

Summary

The divergent answers to the three questions reflect some of the issues and the low level of resolution and coherence that has been achieved to date with respect to these issues. I suggest that the notion of turnover pulses in Africa remains viable for reasons additional to those implied by the positive responses on climate and on turnover pulses near 2.7–2.5 m.y. (bovids, rodents, hominids), 1.9–1.8 m.y. (bovids, pigs, hominids), and 0.9–0.6 m.y. (bovids):

1. Is it a coincidence that the three intervals with suggested FAD pulses are situated close to the minima of three successive larger global cooling trends, and that two of these—2.9–2.5 m.y. and 0.9–0.7 m.y.—also mark times of changes in dominance from a cyclic pattern of shorter period to one of longer period? At least for the 2.9–2.5 m.y. interval, some of the needed tests have been performed on the African bovid record, and the results suggest that the bovid FAD pulse cannot be accounted for by taphonomic factors alone.
2. The statistically significant pattern within the bovid FAD pulse 2.8–2.5 m.y. is exceedingly strange unless the present models have some validity: By 2.7 m.y. a much higher proportion of warmer- and closed-adapted taxa appear, while the 2.6–2.5 m.y. FADs include far more cooler- and open-adapted forms. This supports the sequence expected during a net cooling trend. I cannot think of any hypotheses on purely biotic causes of speciation or on taphonomic causes that predict such a pattern.
3. It is possible that the differences between areas and faunal groups (as reviewed above) tell us little about models of speciation. The fossil records of groups may differ in the signals they contain because they vary in richness of species and specimens, in the methods and detail of analyses, or in taphonomic biases. Another possibility is that these differences do reflect evolution and ecology and can help us to refine our hypotheses. For instance, is it possible that the relay model (fig. 2-3) could explain why the generally warmer-adapted pigs, monkeys, and giraffids tend to have more first appearances before and up to about 3.0 m.y., close to major warming (reviewed in Vrba, 1995a), whereas in open-adapted forms, like most bovids and rodents, waves of first appearances occur during the cooling trend toward 2.5 m.y.? The relay model predicts that these two categories of taxa should have highest speciation rates that are climatically out of phase with each other.
4. Hominids and other mammals share features of phylogenetic branching topology; for example, in the 3–2 m.y. interval, the hominid tree (fig. 2-4a) and several bovid trees (Vrba, 1995a: fig. 10) show a similar pattern: a potential direct ancestor last recorded about 2.9 m.y., with descendants first recorded between 2.8 and 2.5 m.y. After the first branching, several such cases show additional cladogenesis up to 2.7–1.8 m.y. The common phylogenetic patterns suggest common initiating causes.

5. Many of the morphologies that appear in the FAD pulses seem to be consistent with the nature of the hypothesized climatic causes. Examples include changes in masticatory morphology that appear during Late Neogene times of global cooling in horses (Bernor, this volume), in pigs (Bishop, this volume), in rodents (Denys, this volume), and in hominids and other large mammals (Wood, 1995, this volume; Turner, this volume). Additional examples are the parallel appearances in diverse lineages during climatic cooling and aridification of larger bodies (Bergmann's rule) by prolonged growth, together with reduced extremities (Allen's rule) and hyperpedomorphosed features: the same evolutionary event of growth prolongation, as it acts on characters with different ancestral growth profiles in the same body plan, resulted in a major reorganization, or shuffling, of body proportions such that some characters became larger and others smaller, some hyperadult and others more juvenilized, precisely as predicted for cooling (Vrba, 1994b). I suggest that this hypothesis applies to major features of hominid evolution. Recent analyses (Vrba, 1998) support its applicability to the remarkable trend towards increased encephalization in *Homo* since the Late Pliocene (fig. 2-5).

Testing of the models I have presented has just begun, and rigorous quantitative comparisons are still lacking for most of these proposals. Nevertheless, I suggest that these hypotheses remain viable and worthy of testing. Already, the available data strongly hint at general rules, acting across organismal groups, by which small alterations in parameters (such as lengthening of the growth period) can result in major novelty (such as encephalization). These parameter changes and novelties have consistent relations to regularities of physical change that influence evolution through the medium of changing habitats of species. And I suggest that our own evolution was a part of it: The beginning of the evolutionary trend that led to consciousness was as radical an evolutionary change as any in the history of life. Perhaps this trend was "in the cards," so to speak, once our ancestors, with their primate inheritance of a particular kind of brain ontogeny, became caught in the start of the modern Ice Age.

Acknowledgments I thank the organizers, Friedemann Schrenk and Tim Bromage, for inviting me to such an imaginative, productive, and thoroughly enjoyable conference, and the Wenner-Gren Foundation for making the conference possible. Ray Bernor, Alan Turner, and Bernard Wood gave useful comments on this manuscript. I am grateful to Bob Naviaux for pointing out to me, based on his cancer research at the Salk Institute, San Diego, California, that habitat theory is as likely to apply to cells as to organisms and species. Susan Hochgraf prepared the illustrations.

Notes

1. In closed behavioral programs, all behavioral steps are determined by instinctive responses. An open behavioral program is one that does not prescribe all the steps in the behavior of an animal, but leaves open certain choices (Mayr, 1976). Species with more open behavioral programs are expected to be the ecological generalists in nature. Such species are more varied and complex in behavior because they can flexibly adjust their behavior based on sensory cues. The advent of stone tools indicates evolutionary advances in such open behavioral programs.
2. A norm of reaction is the range of ecophenotypes that a given genotype can produce in different environments. Different ecophenotypes, expressed in differing environments by organisms sharing the same norm of reaction, are not based on genetic differences; but different norms of reaction do reflect genetic differentiation, and norms of reaction can evolve (e.g., Stearns, 1982). To say that the reaction norms of sister-taxa are 'similar' means that they show the same vector of correlation with environmental variables (e.g., larger bodies in colder and smaller bodies in warmer climates) although they may vary in position along that vector and in breadth.
3. Hyperpaedomorphosis involves the evolution, by prolongation of ancestral positively-allometric early growth phases, of proportions that are relatively larger ('hyper') in the descendant adult, while at the same time being juvenilized (paedomorphic). Examples are the enlarged hindfeet of saltatory rodents (Hafner and Hafner, 1988) and the expanded brain in humans.

3

Paleontology and Macroevolution

On the Theoretical Conflict between an Expanded Synthesis and Hierarchic Punctuationism

Frederick S. Szalay

It is difficult to define where macroevolution begins and microevolution ends (Levinton, 1988), and therefore who is or is not a proper macroevolutionist is equally contentious. Since a zenith of macroevolutionary understanding provided by the Modern Synthesis, and well represented by Simpson's two tomes (1944, 1953) and subsequent writings (also by Rensch's 1947 and 1959 independent contributions), enormous amount of new data has surfaced, along with new subdisciplines, all of which continue to demand theoretical treatment in the framework of taxic and other macroevolutionary perspectives. In Fitch and Ayala (1995), based on a 1994 symposium paying homage to G. G. Simpson's (1944) "*Tempo and Mode,*" widely ranging contributions examined the change in the past 50 years. Opinions varied a great deal on the relevance of the Synthesis, although it is probable that advances in genetics, molecular, and developmental biology went beyond even of what Simpson might have fantasized. Basic theoretical issues put forward in "*Tempo and Mode*" about the evolutionary dynamic, however, have not been shown to be wrong or even "outdated".

Subsequent to what was then merely called punctuated equilibrium, S. J. Gould declared in 1980 that the Synthesis was dead (see Mayr & Provine, 1980). The simple question then and now remains: Beyond a far more detailed knowledge of patterns in the fossil record, have we learned enough about causal explanations and evolutionary process through time to justify Gould's declarative statement? In particular, is the claim justified that macroevolution can and should be causally decoupled (beyond the important and unique description of various processes and interactions of organisms and environment) from the basic mechanisms that affect population level change (Gould, 1982a,b)? I briefly reexamine here some theoretical positions that surround the issue of decoupling versus continuity and mutual dependency of micro- and macroevolutionary analysis. In spite of its earlier rhetoric, punctuationism claims to retain most if not all the tenets of the Synthesis, while paradoxically it claims to replace its foundations with a new ontology and advocates the practice of methods based on the latter (Eldredge, 1995). Interesting quasi-Darwinian claims about autocatalytic evolution (Kauffman, 1993, 1995; Weber & Depew, 1996) add other issues to future theoretical examinations of macroevolution.

Any view advocating causal decoupling of levels above the organism obviously invites careful scrutiny, and this process has begun by a variety of evolutionists. To strive for a more complete understanding of specific patterns does not necessarily mean, of course, an obligatory commitment to the decoupling of causalities affecting organisms from the various explanation of broader taxonomic patterns. Macroevolutionary analysis has always been difficult because the patterns are neither self-explanatory, nor is their explanation restricted to the levels at which they are observed (W. J. Bock, 1991; Szalay & Bock, 1991). But such analyses that encompass the uniqueness of history coupled with an understanding of mechanisms are far more than a mere general application of microevolutionary mechanisms. The recognition of guiding, but also *facilitating and evolving,* constraints present in organisms through largely, but not exclusively, natural selection is not a new perspective. Lineage-specific conditions do set the context for the causes driving evolutionary change from those points onward. Simpson (1944, 1953), like Darwin before him, was clearly cognizant of this. To understand patterns with the aid of mechanisms and in light of process within the context of contingencies requires factoring in both selectional and other causalities.

So what is the problem, and why is there widespread conflict and confusion over concepts in the evolutionary literature? As much as possible, under the length limitations of this chapter, I hope to examine some conceptual controversies that followed the taxic ideas and axioms of hierarchic punctuationism—the ones that led to the claimed necessity for the causal decoupling of micro- and macroevolutionary explanations. Punctuationism, a supraorganismic selection-oriented hierarchical approach, not only challenged an expanding Synthesis, but also led some of its practitioners from its inception, through the use of monotonic logic and deductive reasoning from the premise of punctuationism, to challenge such core notions of Darwinism as the efficacy of natural selection for traits and competition theory (e.g., Gould, 1985, 1995c; Vrba, 1982, 1985d, 1992, 1993b,c, 1994; Benton, 1987; Hulley et al., 1988; G. H. Walter, 1988; Masters & Rayner, 1993; Rayner & Masters, 1995). While these attacks (some subtle, others bluntly honest) and proposed macroevolutionary models such as the habitat theory (most clearly formulated in Vrba, 1992), climate-driven turnover pulses (Vrba, 1993a,c, 1996a), and variance-shifted punctuated trends (Vrba, 1980a; Gould, 1988) are closely relevant to the issues raised here, they cannot be discussed here due to space limitations.

Perhaps the most difficult aspect of this chapter is the need to concentrate on a limited number of tasks that one can adequately cover. While some of the issues addressed deal with the nature of macroevolutionary hypotheses in general and specific ones as well, any examination of these topics without a scrutiny of the sundry relevant assumptions suffers in an inverse proportion to effort devoted to basic issues of evolutionary theory. It will be obvious that this is not merely a polite customary disclaimer. It is, however, a call for the realization that most theories beyond the highly tested foundational neoDarwinian ones do not usually stand on their own. But is punctuationism a paradigm about to displace an expanding Synthesis as a model of the evolutionary dynamic?

Punctuationism

Origins and Consolidation

Briefly, punctuational core theory (Eldredge & Gould, 1972, Gould & Eldredge, 1977), in contradistinction to what it claimed to be purportedly "ultra Darwinian" or "gradualist" views (Eldredge, 1995), conceives of most evolutionary change as the consequence of speciational evolution followed by lineage stagnation. Anagenesis without punctuated speciation is claimed to be of minor importance for evolutionary change or for the formation of sundry patterns in the fossil record. Punctuationism is based on the notion that the species taxa of the fossil record, "individuals," are an accurate reflection of the mechanism that produced them. Simply put, punctuational theory is a speciational evolutionary theory that attributes the varying durations and apparent stasis of fossil species taxa to the prevalence of "geologically instantaneous" (punctuated) peripatric speciations, which are based on "genetic revolutions" (see discussions in Nevo, 1989; Provine, 1989).

The literature on punctuationism after 1972 is enormous, many faceted, and repetitive with modifications and clarifications. Basic issues, however, remain contentious, particularly those related to the testability of ontologies proposed, those that are in irreducible conflict with foundational and tested Darwinian concepts. These issues have come to include (1) the units and levels of selection debates, (2) the attempted characterization of anagenesis in unsplit phyla as distinct from the processes that occur in cladogenesis, (3) the conceptual issue of species taxa versus lineages, and (4) the nature of evolutionary causality in general. While punctuationism has undergone its own evolution (Hoffman, 1989), the avalanche of criticism from paleontologists, evolutionary theorists, and neontologists of differing backgrounds has been so profuse that most current arguments and fine points repeated in the literature are almost guaranteed to be unoriginal. Nevertheless, there are a number of outstanding reviews and books that deal in a scholarly fashion with various phases of punctuationism (or speciational theory in general) and other theories based on it (or contained within it). Within these works, of course, the appropriate literature critiqued is fully cited. The papers by W. J. Bock (1979), Boucot (1978, 1990), Charlesworth et al. (1982), Grant (1982), Hecht and Hoffman (1986), Nevo, (1991), Schopf (1981a,b, 1982a,b, 1984), and the books by Hoffman (1989) and Levinton (1988) are a few of the critiques of various aspects of punctuationism.

While the justifiably significant influence of punctuationism on evolutionary dialogue was certainly established, it is not clear if anything has been significantly modified by punctuational theory concerning the major outlines of the Synthesis. What kind of paradigm shift, if any, is happening? It is obvious that the tremendous post-Synthesis strides in field studies, genetics, molecular biology, and developmental biology, particularly combinations of all of these (e.g., Nevo, 1989, 1991, 1996), are altering and refining evolutionary biology, but largely without any input from punctuationist theoretical constructs. What starkly stands out in the most recent reaffirmations of punctuationism, however, is that no amount of theoretical or empirically based criticism has created much conceptual change in the propositions of that school or was seriously considered in spite of endless literature debates. *Facts hardly entered into this conceptual upheaval.*

What did emerge was an ever clearer hardening of the core of punctuationism. This was, I believe, partly the result of the debate-based clarifications of its previous assertions, but also a result of the logical extension of its basic premises and, equally, perhaps, because of the complex attitude of the punctuationists expressed as hyperbole and public commitment. In Gould's (1985) view, a series of Second Tier causalities, overshadowing the Darwinian First Tier, was considered to be responsible for most evolutionary change (the sundry patterns). This is a not-so-subtle rejection of the significance of organism–environment-based driving mechanisms of the Synthesis. All of this can be substantiated by numerous quotations from the prolific and forceful writings of Eldredge, Gould, and Vrba.

Another conceptual theoretical shift relevant to the evolutionary debate began shortly before the germination of punctuationism through the discovery of Hennig in the English-speaking world, followed by an increasingly axiomatized school of cladistics, a movement in taxonomy that is difficult to define because phylogenetic analysis is not synonymous with cladistics. Through Eldredge a strong connection was established between cladistics and punctuationism (Eldredge & Cracraft, 1980), although Gould has remained luke-warm about cladistics (Szalay, 1991). Beyond Eldredge's historical role, however, an important relationship exists between punctuationism and cladistic conceptualization of the species-level taxic reality and in the testing of both punctuational theories and cladistic hypotheses via taxic evidence, issues I examine below.

Beginnings of Discord

As punctuationism hardened (and became clearer in its core tenets), there was and is an interesting and increasing disagreement emerging among its pioneer proselytizers over the very issues that make up the seemingly cohesive framework. While obvious latent disagreements between Eldredge and Gould over the importance of natural selection remain undiscussed, others have surfaced. For example, the concept of ecological hierarchies has been one of Eldredge's major and individuated contributions to the "hierarchy"-steeped punctuationist literature. This is so, curiously, in spite of the general silence of Eldredge (and Gould) concerning the historical–ecological context of punctuation as a process (if not mechanism) as such, beyond Mayr's model of peripatric speciation (after which, so the model goes, some epistatically shackled, prevalent genotype is tied up in stasis for reasons not quite explained; see Mayr, 1989, on his professed but qualified soft punctuationism). But a recent remark by Gould is significant. In one deft and brief paragraph (Gould, 1994:6769) he rejected Eldredge's ecological hierarchies:

> "How shall the items and units of selection be identified and defined? Two major contributors to this debate on hierarchical selection—Eldredge . . . and Williams . . . —have tried to establish parallel hierarchies of equal causal import: genealogical and ecological for Eldredge, material and codical for Williams. I believe that these efforts are ill advised and that only the genealogical and material sequences should be viewed as causal units participating in Darwinian selection."

While he continued discussing Williams's (1992) view on the codical domain, he did not mention again his disagreement with Eldredge on the ontological reality of ecological hierarchy, although he obviously implied that epistemic corroboration is lacking for it. This criticism of Eldredge is not related to the fact that Eldredge is more comfortable with natural selection than Gould is and that Eldredge may generally favor an adaptationist view of species origins (e.g., Eldredge, 1989).

The important role of punctuationism, more than just a midwife (but rather a conceptual foundation) for subsequently developed macroevolutionary hypotheses, has also come back to haunt these proposals. Any way one slices the dynamics of evolution in light of explicitly stated punctuationism (not a straw man, like Darwin's rate-miscast gradualism), the minimal role of Darwinian mechanisms outside of the duration of the "release" period of speciation (so the ontology goes) creates serious problems that the major proponents of punctuationism try to handle differently.

While I am loath to attribute simple motivations to the rich and complex writings of Gould, I have long realized, since my shared days of graduate school with Gould (Szalay & Gould, 1966), and also noted by others (e.g., Ruse, 1989; Dennett, 1995), that there existed (prior to punctuationism) a long established, if not overt, discomfort with many Darwinian explanations, namely the overwhelmingly adaptive nature of evolution (hence any and all forms of cladogenesis). The latter, I believe, is a significant (and troubling) component in Gould's conceptualization of the evolutionary dynamic. This is so in spite of his often expressed and eloquent homage to Darwin and Simpson (but also to the fairly discredited evolutionary works of Goldschmidt, Schindewolf, and deVries, among others) and in spite of his recent rousing restatements of the verity of Darwinian natural selection (Gould, 1995a). In all fairness, Gould's mercurial, complex, rhetorical and contradictory, emotion-tugging and liberal, and unabashedly agenda-laden discussions of various espoused causes, other evolutionists and their theories or their Zeitgeist-related social predilections, is difficult to interpret accurately by most of his readers.

Gould's evolutionism, more so than the views espoused by Eldredge or Vrba, gravitates to vaguely conceived internalist mechanisms summed up often as some all-overriding structural constraint. It manifests itself in Gould's increasing reluctance to attempt reduction of paleontological, "species"-related problems to Darwinian natural selection, even when such issues are obvious to others (see also Horgan, 1995b). This is the reason, I believe, that in his various treatments of constraints Gould sounds as if the Synthesis (or even Darwin) were not fully aware of the guiding hand of heritage and a host of mechanisms that comes with the genotype and the epigenetics of development. It is astonishing to see attempts to "reinvent Darwin" in the guise of some recent restatements and axiomatic distortions while claiming vast advances over the theoretical sweep offered by Darwin and the Synthesis (e.g., Eldredge, 1995). It is difficult to see how Darwin could have contemplated the scientific study of natural groups, of descent, or of homologous attributes used differently in taxa if his view was only about "optimum adaptations" (Sober, 1985). Gould, in particular, among other punctuationists has been queried about this by others, who have pointed to the apparent contradictions in his prolific popular and theoretical work. Although the conflict with Darwinian mechanisms is always rather brilliantly "handled" by Gould, others do not manage this particularly well.

The Role of Conceptual Terminology in Evolutionary Theory

Redefinitions of Darwinian Selection, Dilution of Basic Concepts, Misuse of Emergence, and the Creation of Metaphors for Taxic Assumptions

Adding new concepts or reformulating old ones if evidence warrants is one of the more important aspects of evolutionary biology. For example, the reformulation of the concept of gene was a natural consequence of empirical advances. Similarly, there is a clear difference between the meanings of the concept of natural groups before and after 1859. There are, however, categories of notable exceptions regarding redefinitions of well-tested concepts. In general such exceptions are attempts either to explore new ideas (and only unintentionally confuse), or to studiedly marginalize the original or tested meanings of redefined words through the apparent conceptual "inclusiveness" of the newly described meanings. In search of greater generality (but not necessarily deeper meaning), philosophers and some evolutionists did redefine or "clarify" concepts such as individual, adaptation, fitness, and selection. (Some other terms, such as interactors, replicators, more-makers, species adaptation, and species fitness are not specifically relevant here.) The aim was either to attempt an expansion of the resolving power of the tested mechanism of natural selection or to create other nomothetic theories. The results, in my view, however, were not only often of equivocal value, resulting in a growing avalanche of revisionist use of previously clear terminology, but they also ushered in a fundamental confusion about well-tested aspects of evolutionary dynamics. These efforts also obscured the meaning of the retrospective patterns in time in terms of "causality." The changes labeled as "clarifications" (e.g., in Vrba, 1989) do not work beyond a linear, monotonic, and semantic accounting of the evolutionary process. Many biologists who are not part of the debate remain justifiably confused concerning differences in meaning between evolutionary change and taxic change, as the recasting of a few basic concepts has become the engine toward a taxic reformulation of evolutionary theory.

More important than other concepts, those related to the reformulations of "selection," "fitness" (as used in the Synthesis), and "individual" had some far-reaching consequences in evolutionary theory. In a thoughtful and thoroughly revisionist synthesis related to hierarchical selection, Vrba (1989:113) has given the following redefinition of selection: "Selection is the interaction, between heritable, varying, emergent characters of individuals, and the common environmental elements experienced by the variant individuals, that causes differences in birth and/or death rates among them." But any redefinition of fitness that includes levels above the organism ("individuals") and its contained levels has to grapple with the contextual notion of differential reproduction that does not apply to either demes, species-lineages, and clades. Selectional theory has always referred to natural selection, and any redefinition that finesses the concepts of "individual," its "reproduction," "fitness," "offspring," and "birth" and "death" is a dilution of tested theory, rather than an achievement of valid greater generality. Wright's shifting balance theory of demes has individual organisms as units effected by selection, in spite of the handle of "interdemic selection" (see Provine, 1983, 1986; Wade, 1992). Species do not reproduce, are not born, nor do they die; organisms do. The origin of multidimensional species taxa in space and time is due to anagenetic change with or without a speciational event, so the concept of a taxonomic species as such has no ontological reality beyond the lineage. While terminal species (of lineages) go extinct if all their constituents die, the concept of reproduction, birth, and death for supraindividual entities is a less than useful clarification. It is particularly harmful because the claimed need "to refer collectively to similar processes at several levels," (Vrba, 1989:113) is the Trojan horse of Darwinian legitimacy that, as a result of "clarifications," renders Darwinian theory sloppy and diffuse and not truly "general" for all levels. Such are not rigorous theoretical clarifications but only semantic extrapolation that serves theories that depend on such vague or untestable concepts. (For a reminder on basic concepts, see last section of this chapter.)

The concept of emergence is often evoked when dealing with levels. Only mythical reductionists have problems with the concept of emergent. Atoms and molecules are emergent beyond their constituent elements, materials beyond matter, and structure is emergent beyond these, so buildings (or a liver) and other constructs are emergent beyond their components, and so on. Individual organisms are emergent in their phenotypic attributes. Populations and species that (usually) consist of several populations also have emergent attributes, but in a very different sense than the phenotypes do. The general conceptual core of emergent properties is that they are functionally different from those of their constituents. The materials that go into fiberglass or bricks are different from these composites that can be used in a wide variety of constructions. The various constructions (structures) made from the same materials may have fundamentally different properties depending on structural design. But at this point the linear and semantic analogy borrowed from the physical world breaks down, I believe. How could this emergence of structures be applied metaphorically to the concept of species? Individual chimps and humans are constructed differently both from other individuals of the same and of the other species. But the differences between particular evolutionary units (species) depend only on the summed particulars of the individuals that make up these units. Unlike bricks that can be used to build a variety of designs, the "same" organisms cannot be a different species than what they are the constituents of, in spite of such specious arguments (e.g., Lieberman & Vrba, 1995:397) that any compositional difference in a species is a "different entity" (true, but not quite enough). So the concept of species, while tested reality, is based on the reproductive behavior of its constituents, *isolated* from other units. It is at least debatable if the attributes of such a grouping are really emergent or can have different causality acting on them than those that affect the constituent individual organisms (for a contrasting, philosophically detailed account, see Lloyd, 1988). At best, the postulated causality on the species level compared to those affecting the constituent organisms and kin groups in realtime is simultaneous and a truism (*fide* Charlesworth, 1995). But where are the emergent traits? Adaptive variants, fitness, and selectional outcome *within a population* have no proper equivalents on the interspecies and interclade levels except as semantic summary expressions of evidence of Darwinian selection.

Populations and species evolve in a number of ways that may be beyond the explanatory power of natural selection acting on the constituent organisms, and sorting is a way of describing the temporal and spatial distribution of essentially taxic patterns (populations and species). The causally inert term "*sorting*" is increasingly used for describing evolutionary patterns in such a neutral way. Ironically, however, its usage was probably an outgrowth of species selectionist aspirations. It appears that it was carefully embraced and correctly promoted out of a failed initial enthusiasm for the concept of "species selection" (e.g., Vrba & Eldredge, 1984; Vrba & Gould, 1986). But sorting is rendered little meaning while front-loaded metaphorically as a "description of birth and death" (Vrba, 1989:118). Consequent "fitness" of supraorganismal units, resulting from natural selection of *organisms,* from random events (causes in a real sense), or from differential persistence of lineages, like any process of sorting, is at best vaguely defined. Such semantic maneuvering, however, can obfuscate tested or testable meanings. For example, Vrba's (1989:120) macroevolutionary definition of "fitness" as "an expected outcome of differential representation as a result of sorting" is vague and circular. Does it mean that evolutionary or taxic patterns are the result of "sorting?" If that is the only meaning of the extended concept of "fitness," then it is a trivialized one.

The neoDarwinian paradigm maintains that evolutionary patterns usually represent the adaptive consequences of differential fitness of organisms with their unique combination of heritage and habitus modification but with ample room for other causes. The resulting supraindividual entities are the sums of additive individual fitness differences in specific spatial and temporal circumstances and are not the outcome of the "fitness" of the antecedent lineage segments. In general, I believe it is widely understood that "sorting" may be a comfortable descriptive shorthand when discussing differential representation of species or lineages of organisms without the invocation of specific causal mechanisms. But so are the plain terms "persisting" or "evolving." Understanding what particular attributes may have contributed to differential survival/reproduction (natural selection) is an exceptionally difficult enterprise (see Williams, 1985; Cain, 1988). But because such adaptive interactions are often not readily apparent (particularly for fossils), it does not mean that such causality was absent. Whether natural selection, Dresden effects (random extinction), global extinctions, or other chance changes in population structures are the motors of change, sorting is another heuristic concept that covers all possibilities. It stands for evolutionary change and persistence.

As another example of reformulated meaning, it is of some interest to note the essentially rhetorical parallel drawn between a politically correct term change and that of a profound conceptual change promoted by Gould (1990:24; see also 1995c) in his popular restatement of the "individual" issue in relation to his disagreement with Darwin:

> I think we have to construe the nature of the word individual far more broadly than we have. We have, after all, refined our language before. We have learned to say human instead of man for example so that we can include all, not half, of us. And I think we have to do a similar thing to the word "individual." We have to expand the concept of individual, beyond its focus on bodies.

A similar thing? Hardly. The term "man," which while sexist and outdated as a stand-in for humankind, was not unclear about being an alternate vernacular description of *Homo sapiens.*

Realtime and Retrotime in Evolutionary Discussions

I use here two terms that, at least conceptually, are not in any way novel in evolutionary theory (or in everyday experience). Yet I believe that there is need for such specifications about time as they relate to discussions of contextual causality in the origins of biological and paleontological patterns in time and space. The present and the past are conceptually obvious, and we also know that anytime we attribute life to some fossil entity at a particular place and time where it occurred, or an "instant" in earth history, we mean a highly specific "moment." There is probably a strong consensus that the nature of evolutionary causality, always specific beyond a few law-like regularities that govern the material interactions, should be tied to space and time.

Natural selection (the full process, causal, processual, and consequential) also has a definite time component (inasmuch as its processual reality depends on surviving and consequentially reproducing organisms), although it is properly emphasized that it is a nomological-deductive explanation. When organisms are alive, it is their *realtime* in the continuum, and it is during instants of this time at specific locales that causalities of differential survival and reproduction are played out. All relevant causalities that occurred and that modified the antecedents of the effected organisms in *retrotime,* before specific realtime events, are "summed" in the genotype of individual organisms of a population (constraint; that is what remains of the influence of these past contingencies, both phylogenetic and developmental). Retrotime is the time during which the total, or most, of genetic change found in the genotype was accumulated. This retrotime period may be designated (arbitrarily) to be of a specified duration, as for taxa. As the succession of the relative frequencies of genotypes and phenotypes is managed by ecology or effected by other factors, so is the succession of ephemeral realtime of populations and species of a lineage.

There is, of course, a realtime and spatial correlate for any fossil entity that we examine, so one can speak just as well of the realtime of fossil species as of living ones and of their subjection to certain causal dynamics. A preceding *lineage,* however, "exists" in retrotime only. It is the sum total of evolutionary change that results in a *species* in realtime; it is history. Multidimensional paleospecies, modeled on living analogues, clearly include thousands of successive generations. It is often confusing to hear theoretical discourse, particularly about causes, about a species and lineage in the same breath, as time contexts vary.

For all practical purposes the realtime of an individual organism is the length of its life. In neontological practice, however, realtime also stands for a short stretch of population and species duration. So realtime is the frame of duration for either (1) the living organisms that are subjected to the causes that shape an evolving entity such as a population or species to which the organisms belong, or (2) the species-populations, the "instantaneous" segments that are doing the evolving compared to antecedents stages. The concept of realtime for specific populations and species, of course, can rapidly become both a theoretical and empirical greased pig. Temporal (not "hierarchic") downward causation is not reality, at least not until interfering time machines are invented.

On Causality in Evolutionary Dynamics

The genotype of individuals and gene pools of populations, while they may be considered Aristotelian material "causes" (i.e., they constrain the causal process acting on the individual organisms), are not the "initiating cause" for the cascade of events that results in selection of organisms that change the populations and species. Obviously both the genotype (the packaged "ultimate causes") and physical environment play an important role for both organisms and their populations. The genotype and the total environment should be considered, however, as *initial and boundary conditions*. Based on these latter considerations, the various laws of the material universe, biological causal mechanisms, and contingencies constrain (or facilitate) phenotypic development and fate and subsequently evolutionary change in an ever unique historical context. "Ultimate" causes before organismic and populational realtime are often considered as a nearly interminable chain of ghost causes of the past—these are the general and rarely specified causes (phylogenetic constraint, true of all lineages) alluded to by sundry authors (e.g., "effect macroevolution"; Vrba, 1983). But viewed as part of the initial and boundary conditions, they become part of the historical and environmental context to causality and evolutionary change.

Evolutionary processes are driven by causes and mechanisms, but the outcomes are also heritage and contingency dependent, and these processes are constrained (as well as facilitated) within the context of initial and boundary conditions. Highly specific causalities are mediated through nomothetic mechanisms such as natural selection (with its "random" component of variation and the "directed" differential survival/reproduction), drift, and various historically contingent genetic and behavioral mechanisms. Similarly, if one does not specify the particulars of very extended, nearly endless, world of "physical causality" as part of the initial and boundary conditions (like the "causal heritage" of organisms), then understanding the mechanisms of the evolutionary dynamic becomes diluted. That dynamic has to do with the genetics, development, and interactions of the organisms with their *total* environment in realtime. The various conditions of the physical environment obviously do causally effect the fate of individuals, but they are not *the* "initiating causes." This perspective is, I believe, in contradistinction to Vrba's recent stance on causality (1993a,b,c).

In Sober's seminal view of causality (e.g., 1984), evolutionary theory, specifically natural selection, must be conceived as containing both source and consequence laws. Although the consequences of natural selection can be simplistically summarized here as the various avenues through which selection for traits results in selection of organisms and differences in fitness, the source laws that generate the "forces" in natural selection are very many—staggeringly so. There are no simple ways to universalize (i.e., make a valid general theoretical statement) what specifically causes evolutionary change, either through speciational evolution, or the "kicks" of the physical environment. The genetic, selectional, competitive, physical, biotic, demographic, behavioral, and physiological circumstances that can provide the causes for change are legion. Who can contemplate, at least now, the number of avenues through which the physical and biotic environment in concert with physiology, behavior, and morphology can change the trajectory of the selection process? So source laws in evolutionary theory abound, in contrast to the "austere beauty of a desert landscape" (Sober, 1984:50–51) displayed by the source laws of the physical universe (gravitation, beta decay, electromagnetism, and nuclear energy).

A view that contrasts the significance of the physical environment *alone* to the biotic one has long been shown to be wrong—individual behaviors and interactions have enormous additional inputs into the evolution of phenotypes and can also confer great resilience to morphology (usually only hard parts in fossil species). All sorts of critical species-specific behavioral and physiological strategies can short circuit morphological change. The reverse is also true: individual behavior will initiate evolutionary change (Lamarck's enduring insight). Butting heads by rams for female access (differential reproduction), the outcome of a real arms race initiated by behavior, will select for increasingly thicker bone in the appropriate places, regardless of the mean annual temperature or whether the physical environment is pulsing or not. Causalities such as sexual selection do not in any way negate the importance of the physical environment as having causal (selectional) importance. The close matching required (e.g., for turnover pulses), however, between a general hypothesis of physical environmental causality and fossil evidence necessitates unacceptable simplifications with dubious deductive explanatory power for a generally uncooperative fossil record. The complex and many faceted causality of evolutionary change, at least for mammals, may be rarely reducible to an initial causal kick of the physical environment. Yet if it is shown to be a paramount cause in instances, it is certainly a fully Darwinian conception.

But it is significant that causality stemming from competition, as such (a certain corollary to any climatically induced change), in contrast to the physical environment, appears to be controversial for some (Hulley et al., 1988; Walter, 1988; Masters & Rayner, 1993; Rayner & Masters, 1995). Benton (1987), in his review dealing with competition theory on the macroevolutionary landscape, for example, specifically eschews the notion of interspecific competition. He seems to be unaware of studies that show animals as widely distinct as ants and rodents (J. H. Brown et al., 1979) or insects and hummingbirds (J. H. Brown et al., 1981) to be competitors. Yet this is an issue rife with concepts related to individual and intraspecific competition, niche overlap, and so on (see W. J. Bock, 1972). It needs careful examining case by case rather than redefinition-based rejection. How does one explain the massive evidence for all sorts of coevolutionary interactions between lineages?

Hierarchic Connections to Punctuationism

Complexity and Hierarchy Theory

The number of papers and books on hierarchy theories or works dealing with complexity are increasing. These works differ in their conceptualization and approach to the diffuse topic of "hierarchy" (e.g., Allen & Starr, 1982; Salthe, 1985; Dyke, 1988; Grande & Rieppel, 1994; see also Edel, 1988). Although the theoretical issues surrounding hierarchy and complexity concepts are usually unclear, as the reading of the various major books indicates (see also Horgan, 1995a), these theories are often major catalysts for a panoply of either pan-selectionist and/or punctuationist or structuralist views of evolutionary change or their rejection. This highly stimulating ontological landscape is as yet theoretical with little epistemic dimension. In the most general terms, levels of organization roughly correspond to what is meant by "hierarchy" by most biologists.

Dyke (1988) in particular, a philosopher concerned with biosocial complexity, dealt with "hierarchy" in a most illuminating manner. He ends by stating that "But talk of hierarchy seems extremely dangerous" (p. 70), a fitting caveat and position that explains why he entitled that particular chapter "Layers, loops, and levels." Dyke (1988:60) offers the following:

> "I suggest that we think of a 'nest' of structures, each of which imposes constraint on the behavior of the other constituents of the nest as a level-interactive-modular-array (LIMA). In general, no straightforward supersession can be established in a LIMA and the relations between its substructures can be extremely tangled and complex. It is necessary to use coordinated research strategies to gain multiple access to pry out the determinate structures involved and to investigate their interpenetration. Hierarchies would then be sorts of LIMAs. Interestingly if the [Roman Church] genuinely is a paradigm [for the concept of hierarchy], then hierarchies are organizations of constituents at a single level ordered with respect to higher level structures. That is, the church hierarchy is a hierarchy of *men* given precedence by the rules, traditions, and practices of an institution, the church. The church is not *part* of the hierarchy." (emphasis in original)

The general view of hierarchy is currently nebulous, if not hopelessly vague, and perusing Dyke's (1988) discussion of the topic confirms one's worst premonitions about hierarchy theory. Numerous opaque and semantic conceptualizations have failed to yield clear generalizations. Hierarchy theory, a part of the family of "level-concepts," is not really definable, although the multilevel complex system that organisms form with their environment (organism-ecological LIMA, as discussed by W. J. Bock, 1991) is a fundamental issue in evolutionary biology. Attempts to include inclusive (nested) hierarchies under the same "generality" as exclusive or external (non-nested) hierarchies (e.g., Allen & Starr, 1982) have not been successful, and the conflation of levels of organization with "hierarchy" have been and remain confusing. In spite of such problems, "hierarchy theory" has been cited often as essential in two prominent paradigms, one in taxonomy and the other in macroevolutionary theory, cladism, and punctuationism. The former was hailed (at least for a while) as a universal method because of its use of a simplified view of linear (nonoverlapping) hierarchy, a logic that axiomatized the application of inclusive hierarchies for cladogeny cum classification (taxograms; see relevant papers and references in Grande & Rieppel, 1994) and allowed easy computer manipulation of data. The nonoverlapping linear model is of course a small portion of the complex overlapping and feedback-looped hierarchy models of a more complex ontology. "Hierarchic" aspects of punctuationism, on the other hand, were and are viewed by some as the cornerstone not only of evolutionary understanding (beyond Gould's Darwinian "First Tier") but also as some master key in viewing history and/or the social dynamic in a global "hierarchic" context. In particular, the essential connection that punctuationists like Eldredge, Gould, Stanley, and Vrba have made in their unified (but by no means monolithic) front about punctuated macroevolution resides in a connection with "hierarchic" and "complexity" related research, namely, hierarchical and emergent complexity.

The current trend of often referencing everything to diverse notions of "hierarchy theory," an obligatory designation of one's quasi-metaphysics (or group affiliation), or complexity, is obviously tied to the philosophically legitimized expansion of a postpunctuationist macroevolutionary ontology by some. References to complexity and to hierarchic approaches are pervasive and have become routine and obligatory in papers on punctuational macroevolution. It is often implied (unfortunately sometimes smugly, and wrongly) that before or outside punctuational thinking, biologists simply did not operate in terms of a proper "hierarchy." What is usually meant by such rhetoric, I believe, is that few before seriously proposed hypotheses about consequential causality for evolutionary dynamics emanating from above the organism level, or thought little or none about "downward" causation from species or clades. There is, of course, a whole history of group selection theorizing and "species advantage" thinking in the works of some prominent biologists (see Wynne-Edwards's 1962 book on group selection, papers and books by K. Lorenz in ethology). All of these, however, disregarded the pivotal concept of natural selection for traits and of individuals and consequently adhered to a mostly nonDarwinian view of selection. I am omitting any mention of Hamiltonian kin selection that is a real causal mechanism (even if it does not always select for real kin). The serious issues tied to group selection, particularly in insects, are clearly far more complex than this brief aside can reflect (see Williams, 1966).

Distaste for Natural Selection and Darwinian Explanations

I believe that the way specific aspects of the new "hierarchic" tapestry has been introduced into evolutionary biology and employed by some of its weavers in the past two decades was tied in various ways not only to attempts to push and expand the frontiers of evolutionary biology. It was also linked to what were either conscious or unconscious efforts to marginalize the relevance of natural selection (core Darwinism) to humans in order to "protect" society from its biopolitical misuse (e.g., Salthe, 1989), a pernicious practice that such nobly motivated behavior, ironically, perpetuates.

One may ask the rhetorical question: Is Darwin really responsible for the universal human predilection to be less than completely altruistic? Clearly not. Recently Dennett (1995) has made a contribution to the role of adaptational analysis in his widely ranging defense of the Darwinian algorithm (one that drives evolutionary change and constructs "cranes," i.e., adaptations). He exposed some of the possible roots of a distrust by some professionals of "Darwin's dangerous idea" whose preference is for strategically designated "skyhooks" instead of "cranes" in order to avoid Darwin. To make my point, consider Salthe's (1985) often penetrating, various theoretical efforts that culminated in a comprehensive book on hierarchic theory and evolution. It was also Salthe (1989:176–177), the pure structuralist, who in one of the most overtly nonscientific (for a well-known evolutionary biologist) and embarrassingly metaphysical (for my personal taste) "critiques" of the Darwinian, selectionist, perspective betrayed his deepest motivations, revealing deep seated antipathy and out of context judgment about natural selection, one that was subsequently sublimated in the form of a rigorous theoretical expression in Salthe (1993).

Noteworthy in a different manner is Stanley's (1979:192) consideration of natural selection, who noted that "I tend to agree with those who have viewed natural selection as a tautology rather than a true theory . . . it is essentially a description of what has happened." This is a revelation, I believe, about how an early and prominent champion of fully rigged punctuationism thought about this most tested of evolutionary mechanisms. For someone who understood the core of the Synthesis that way, alternative mechanisms for evolutionary change must have loomed as a necessity.

Gould, whose knowledge of the literature of Darwinian and neoDarwinian evolutionism is justifiably legendary, in a paper (1993) entitled "The inexorable logic of the punctuational paradigm: Hugo de Vries on species selection," in essence claimed to be a Darwinian in order to show how selectionist logic should be united with punctuation and stasis, and extended to yield what he considered the real or full explanation of evolutionary patterns. Yet I see this, again (see Gould, 1988, on "punctuated trends") as a not particularly disguised attempt both to acknowledge but also to marginalize Darwinian selection in the very extension of the concept. Punctuationism and its recurrent and attendant message about the nature of change and causality in evolution restricts and supposedly contains in the sealed black box of "punctuated species origins" (and the preponderance of evolutionary change associated with it) the paramount tested mechanism of change (i.e., natural selection for traits of organisms)—when and if such credit is given to it by Gould in the shadow of serendipity and "internal causes" for the origin of new lineages.

Gould writes in that paper (1993:12), inexplicably (to me), that

> "[de Vries] has slain natural selection in its own domain, yet he understands selectionist logic better than any contemporary (except Weismann), and he is fiercely committed, by bonds of personal loyalty, to further the work of his guru and intellectual hero. What is he to do? De Vries found a brilliant exit to his dilemma, one open to him alone (for no one else had both the mix of concepts and the commitment). In short, he jacked up the selectionist argument by a level, applying all its principles to the explanation of macroevolutionary trends. In other words, he developed a convergent version of the punctuational paradigm."

To be sure, Gould (1993:15) disclaims:

> "I should hardly need to add that neither Eldredge nor I accept de Vries's version; we have never been saltationists, and the punctuated paradigm requires no such argument since ordinary allopatric speciation, *properly scaled into geological time, yields adequate speed* for the origin of species. The paradigm requires punctuation, *and is agnostic about the explanation offered for this evident empirical phenomenon.*" (emphasis added).

So what does this mean? Is this astutely suggestive rhetoric that leaves specifics up to the reader's imagination? There is no explanation for the recurrent "adequate speed" (one that is almost always masked by the fossil record, not to mention the assumptions of the theory about paleospecies), or a rate that is always "properly scaled into geological time." This is so mainly because the actual origin (at the "adequate speed") of a phenon (a taxonomic species) is really not documentable because of the ambiguity of the "proper scale." One person's "proper" samples are not those of another worker. Gould's statement, "agnostic about the explanation offered for this evident empirical phenomenon," is another cautious way of saying that no mechanisms are known for punctuation, while also presumably implying that a *taxonomic pattern,* while its motor is unknown, is really an *evolutionary one* for whole species (lineages). The model that accounts for the "evident empirical phenomenon" is punctuation-stasis. There are, of course, sundry explanations for "stasis" of both characters and taxa beyond the hardened confines of punctuationism. Ecologically contextualized quantum evolution early in radiations has been recognized by Simpson (1944) and pursued contextually by Boucot (1978), and it is a different mode from the punctuational model.

But in reference to species selection, reading de Vries (1906:6) made it obvious to me that his alleged understanding of Darwinian natural selection, quite overblown by Gould, was severely limited to its "pruning" aspect ("[natural selection] is the sifting out of all organisms of minor worth through the struggle for life. It is only a sieve, and not a force of nature, not a direct cause of improvement."). Furthermore, his concept of total geological time (2–3 m.y.) in combination with his narrow and distorted conception of natural selection and the general lack of knowledge at that time about the nature of "mutation" probably drove him to his saltationist and species selectionist views (see also Falconer, 1992). I guess it makes sense for Gould to so endorse de Vries's view of natural selection because it appears that he is increasingly committed to the creative role of punctuated trends through shifting variance (versus natural selection) to account for the tapestry of the evolutionary landscape (Gould, 1988; but for a mercurial boost for the "First Tier" to reclaim Darwinian credentials, possibly as a reaction to Dennett's 1995 criticism, see Gould, 1995a).

The history of misuse of Darwinian selection to bolster a variety of political positions continues to force some to attempt to relegate selectional theory to a subservient position in the great evolutionary dynamic. Others (certainly not Gould) simply may not fully understand the complexity of selectional theory. At any rate, punctuation and some hierarchic theories (or dialectic biology) are obviously not free of a diversity of deep political and social shackles that some of their proselytizers bring with them to evolutionary theorizing. Salthe is certainly not alone in his irrational opposition to Darwinian theory among those high-profile evolutionary biologists and taxonomists who aim to exorcise what they consider the Darwinian menace to evolutionary theory or, really, the social contract. One of the leaders of the cast iron version of cladistic taxonomy, G. Nelson (1992:146), for example, has succinctly expressed similar views in opposition to Darwinian phylogenetics (the latter so called because it is based on functional-adaptively focused character analysis; Szalay, 1994) when he stated that: "Indeed, one of . . . [cladistics'] greatest merits in my eyes is the fact that it occupies a position of complete and irreconcilable antagonism to that vigorous and consistent enemy of the highest intellectual, moral, and social life of mankind—Darwinism."

Hierarchy Theories and Macroevolution

It is not possible to present a detailed evaluation of the copious literature on this subject. My perspective here is influenced by the suspicion that while simultaneous multilevel "selection" is certainly a reality, a truism (Charlesworth, 1995:507), the notion of species or clade selection is probably entirely dependent on semantic definition of fitness, or on catchy encapsulations of such vague concepts like "interactors," "replicators," "more makers," or a reformulated notion of "selection," to cite a few of these (see above). While such conceptualizations can lend vigor and new fuel to genuinely interesting philosophical questions and discourse, the credentials for claiming discoveries of causal evolutionary mechanisms based on definitional perspectives without empirical tests are suspect. A tally of species sorting (i.e., what there is) is a descriptive device in chronicling the details of comparative lineage evolution in time. The literature dealing with the analysis of taxic evidence, the relevance of extinction to evolutionary opportunities of the survivors, adaptive or nonadaptive reasons for success of lineages, and meanings of trends are a few of the old, venerable, and continually exciting issues of macroevolution. But how can the "extension" of causal selection to species or clades be accomplished without extending it to whole phyla or the entire clade of Life?

Is the "hierarchically" boosted notion of species "fitness" in any way analogous to fitness as an organism's property (in a given context) beyond a semantic approach? Is it merely a metaphor with no causally meaningful foundations, or is it a concept with causal significance beyond the summed values of fitnesses of individual organisms at any moment in time in different lineages? While extending the concept of fitness to "demes," Wright repeatedly maintained that the shifting balance theory is causally dependent on individual selection and not population level "fitness" (Provine, 1983, 1986). The analogy is completely metaphorical and semantic and not material in terms of dynamics and details of the process of natural selection. It may be a serious mistake to believe that various "hierarchic" theories, relevant as they certainly are to mulling over all issues, catapult us conceptually into discovering evolutionary causality beyond or before realtime or at increasingly higher levels beyond organisms. Level-based thinking helps to narrow causal explanations because of the *contextual* constraints they offer for consideration. Furthermore, on whatever levels sorting (a mere consequence) or selection (causality) occur in a strict sense (i.e., reducible to the point where explanation at that level is sufficient to account for a phenomenon), such consequences and events, respectively, are bound up with the existence of living organisms. Organisms are the *inseparable* constituents of higher level groupings in realtime, but these levels have different causal significance for evolution. As all the individuals of a species or population die without replacement progeny, so goes that particular unit, and the ecosystem simply changes.

An ecosystem is made up of organisms that are connected in various ways well below the species or deme level, but also above the species level with other individual organisms. The shorthand that "species make up an ecosystem" is not a careful delineation, but merely a useful phrase until it, in a given context, begins to violate the theoretical issues that in some ways transcend language. Perhaps more soberly than anyone else, it was Damuth (1985) who pointed out that communities, natural functional units, and not species, that may be candidates for "unit" status in Darwinian selection theory. While Damuth (1985:1143) states that "There is no way to pursue selection of species in clades as a higher level version of . . . theories of selection. . . . Clades are not populations," he also notes that population or avatar selection is congruent with that "at the lower level" (p. 1132). Damuth's use of "congruence" expresses an equivalence to individual selection, and it appears to me that his model is based on the shifting balance theory. As repeatedly noted, Wright's ultimate emphasis in that model is on individual selection, so both the selection of Boucot's ecological–evolutionary units or Damuth's natural functional units (a kind of complex adaptive system) can be properly couched in terms of avatar dynamics that are in my view inseparable from causalities acting on organisms. Some populations (i.e., the individual organisms, the constituents in these) will do better against competitors or predators than others in communities of different make up, and the more vigorously stressed and altered populations will invigorate or outlive other populations of the same species. Thus, while synchronous, causal, multiple-level selection is still a truism, the patterns so well exposed by Boucot (1975, 1978, 1982, 1983, 1990) and theoretically supported by Damuth (1985) reveal important connections between natural selection of individuals and community and taxon longevity patterns. These are some of the most important aspects of macroevolutionary analysis.

Even if one was to grant that "fitness" can be extended as a property of various levels (and it should not be), and consequently on whatever level in a hierarchy such a commensurate, extended concept of selection should operate (from genes to clades, to cite the two extremes), it all has to occur in realtime—at the "same" time. Upward causality is always constrained by the "rules" of genetic, developmental, and mechanical properties, *temporal downward causality* is science fiction, and irreducible "downward" causality from well above the level of individual organism (not kin groups) such as species selection will have to be demonstrated to be considered a meaningful evolutionary force. Although populations and species are equally real, most, if not all, supraindividual entities can be broken down into their constituent parts (i.e., organisms, the recipients of evolutionary causality). All evolutionary causality, selection for attributes (and selection of linked attributes and individuals), or differential, or random extinction—in other words, all material and relational processes that occurred and occur—have and do so in the framework of the sequential passage of time even if the analytical perspective is macroevolutionary. Causes on levels are "simultaneous," in a realtime sense, realtime specified by the level as noted above. Darwinian selection is certainly "hierarchic" in the sense that selection for traits is meaningful only in the *context* of all other aspects of the organism in its total environment.

Taxic Approaches in Macroevolutionary Analysis

Ontological Foundations and "Species Individuals"

Misunderstanding and misrepresentation continue regarding such fundamental concepts as species, speciation, lineage, species category, and the species taxon in spite of copious and detailed explications and discussions in the literature (e.g., Mayr, 1942, 1963, 1976, 1989, 1992; see also W. J. Bock, 1986, 1995, and references therein). Aspect of these issues need to be commented on again, but there is no substitute here for the voluminous literature on the subject.

Although macroevolution usually deals with patterns in deep time, and historical patterns are of specific importance, the ontological reality of fundamental significance in evolutionary theory is the realtime tip of a lineage, the species or populations, and not the much argued-over unit, the multidimensional taxonomic species. All organism, whether clones or species (both of these can be delineated as *species taxa*), have unbroken connections that merge back in time to the beginning of life. This simple consequential fact of evolution should have fundamental theoretical relevance for any deductively significant causal analysis of patterns, including taxic ones. It equally applies to the processual interpretation of patterns, in addition to the understanding of mechanisms responsible for these.

It is widely accepted that implicitly or explicitly espoused models of evolutionary change often drive the taxonomic analyses of samples. These are often expressed as an adherence to one or another species concept and are intimately related to evolutionary paradigms. Given the need to consider evolutionary tempo, mode, and the fossil record, however, explanations should deal not only with the taxonomic units encountered in neontology and paleontology, but with the concept of evolutionary continuity, or lineages. The taxonomic endeavor itself, however, is fundamentally constrained by the assumptions accepted by practicing systematists whose pragmatism formally dominates (as it should) their evolutionary perceptions in the form of published species taxa. Evolutionary perceptions, however, even when actively sought by taxonomists, give way in practice in order to delineate "species-level divergences" of the organisms studied. This is difficult enough an enterprise for the living fauna and flora with their extensive populations of either superspecies with their ecologically similar allospecies (see Grubb, 1978, this volume; Nevo, 1991) or of zoogeographical (or "synspecies") that represent an ecological unit (a more inclusive concept), but it is even more challenging for the fossil record, although often for different reasons. For example, fossil mammals offer more "gaps" in space and time which facilitate delineation, but unfortunately they have fewer attributes than extant forms, thus making it more difficult to test such species taxa. This formal taxonomy is by necessity phenon centered and is based on limited numbers of traits. Unfortunately, purely cladistic approaches to species level taxonomy for both extant and fossil populations result in small and highly fragmented (split) taxa that often cannot be tested by sympatry criteria. The issue of zoogeographical species (see especially Bock and Farrand, 1980) has particular bearing for the analysis of species clusters and distribution in the fossil record, a macroevolutionary problem barely glanced at by most analyses of the fossil record.

Simpson often spoke of (and demonstrated) the greater value of using genera (often of superspecies equivalence for mammalian fossil taxa) for macroevolutionary analysis of all types, as he knew well that "real" species are difficult to sort out among similar morphs and their patchy spatial and temporal distribution in the fossil record (e.g., Simpson, 1960). This may seem a loose methodology to those who adhere to an ontology tied only to the present, particularly in light of the decidedly more rigorous work possible sometimes in neontology. Yet this approach is solidly in concert with the epistemic limits imposed on the real world of fossil data and the complex process of *estimation* of phylogenetic relationships of supraspecific taxa. "Rigorously" oversplit (zoogeographic or biostratigraphic) taxa render no general service when real tests (sympatry) for how speciose a group may have been are not available. Although some taxonomists, and punctuational theory, consider all phena as "species," the common processes of regional differentiation, hybridization (intercalation), and range contractions and expansions of populations with local extinctions are likely to be parts of most lineage histories (see also Grubb, this volume). As these phenomena are undoubtedly reflected in the fossil record within the additional context of an enormous missing spatial and temporal record, the necessarily phenon-based taxonomy makes "lineage" versus "species" diversity estimates extremely vulnerable. Such a record should also demand the strictest standards for the documentation of stratigraphically precise contexts of sympatry–synchrony for the proposals of more than a single lineage of basically similar organisms.

The philosophical notion of the *species individual* (based on pre-Darwinian Greek metaphysics) was born from a taxonomy-based concern, presumably to facilitate dialogue about taxa. It was bolstered in evolutionary theory by the curious metaphysical and semantic claim that "only individuals evolve." I find it reassuring that the philosopher Dyke (1988:4) bluntly notes that his decision to bypass many arcane issues in his book "was made easier by my feeling that much of the traditional philosophical agenda is a waste of time—idle talk inherited from groping intellectual ancestors and by now obsolete." The "individual-class" issue is certainly one of these. The role of multidimensional populations and species in evolution has been much debated in the literature, and I simply state here that calling species (and other) taxa individuals helps some philosophers concerned with the ancient variety of metaphysics of essentialistic classes versus individuals but solves no real evolutionary problems. The issue is obviously not about delineating synchronous and sympatric species—such taxa are clearly distinct lineages, proper units distinct from other such sympatric units. Even the practice of vertically "individuating" species taxa (giving names) of well-sampled lineages inadvertently helps a variety of perspectives when it comes to recasting what the evolutionary dynamic of taxa through geological time is supposed to be about. But the pivotal evolutionary concept of lineage (any or all previous and subsequent history of a realtime species) is irreconcilable in any theoretical (as opposed to taxonomic) sense with vertical species individuation. In a record-keeping sense, one does individuate phena that may be either new lineages or geologically instantaneous anagenetic bursts. In the case of perfect knowledge of splitting histories of lineages we may even individuate individual lineages. But these latter goals are not easily obtainable. *Hypotheses of speciation, tested merely against the evidence of minor shifts in morphology in the fossil record without accounting for stratigraphy or sympatry, are clearly invalid in light of living polytypic species or the reality of anagenesis.*

Additions to the literature on the ontological aspects of species has not stopped and doubtless will continue. But considering species to be a "complex system" in her discussions (Vrba, 1994b) is not particularly helpful, as cells, organs, and organisms are certainly also complex. As I noted above, species are not complex adaptive systems, although organisms are. I do not believe that complexity has relevance to the ontology of species, whereas the *consequential* aspect of the system in a very special sense of "emergence" discussed above (pseudo-emergence) does. Vrba (1995a:37) relentlessly redefined and misconstrued the biological species concept (BSC), then attacked it, declaring that: "Perhaps the time has come to lay to rest the 'ancestral' Isolation Concept of species and to acknowledge fairly that it has evolved into the Recognition Concept, via replacement by new theoretical 'apomorphies' of some old misconceptions." Her most significant criticism of the BSC in a preceding section of her paper was that the "Isolation Concept" has some "connotations," failing to mention that the testing of the BSC requires the establishment of reproductive discontinuity (isolation) for the validity of a species under circumstances where this should not be a problem if all individuals belonged to the same species. "Isolation" is clearly not all that the BSC is about, but it is the cornerstone of its testability (Bock, 1986). Vrba staunchly continued to defend a tradition by Paterson (1978, 1985) and argued for the superiority of the "Recognition Concept" without any attempt to answer objections to that redundant and burdened notion. Paterson tied habitat and punctuation to his specific mate-recognition system (SMRS), but such associations still do not shield the recognition concept from the fact that SMRS are clinal through space and certainly gradual (rate neutrally speaking) through time, even if a "punctuated" origin and "habitat specificity" are loaded on the concept. Its theoretical testability dissolves in clinal continuity and ring-species realities where the SMRS are clearly different, notwithstanding the definitional strictures that are invariant, and basically taxonomic rather than evolutionary. (For some bizarre and amazing connections of the SMRS to the fossil record and to sexual selection, see A. Turner, 1994, and Eldredge, 1994, respectively.)

Species Taxon, Often Mistaken for Species Concepts

Like that of many other paleontologists, my own work has been primarily oriented toward analyzing and establishing the origin, adaptive evolution, and distribution of (mammalian) taxa. The theoretical and empirical analyses of phylogenetic patterns coupled with sundry macroevolutionary and taxic issues are of central concern to such efforts. One communicates through recognized, named, and delineated taxonomic patterns—morphological patterns (at least for fossils) constructed into species taxa guided by both theory and the contingency of samples (Szalay, 1993, and references therein). Also, taxonomists estimate monophyletic groups (supraspecific taxa and lineages) the best they can as far as information, understanding of biology (and stratigraphy in case of paleontology), and methods allow under a broad umbrella. Such supportable and probabilistic estimates, classificatory and phylogenetic analyses, of species become the taxic and phylogenetic patterns. Species taxa are in fact (if defined with any real temporal dimension) lineages of very short or longer duration.

The database of taxonomy, however, becomes burdened by the limitations of the taxonomic enterprise. The latter, while not an invention at least on the species level, is a formalized version of all that we record, *hypothesize, and test* about Life. Everyone agrees that this record of species taxa is limited, even for most living species and populations. Therefore any, even implied, claim that such data can fully test a proposed, nearly universal model of speciation and species taxon ontology (a theory) is an extreme position that some taxic approaches take (see below). While the nature of taxonomic ambiguity of allopatric populations today is well known and hybrid zones are eagerly studied, this problem is multiplied exponentially in the fossil record. Species are real, and they evolve (to state the trivially obvious), but the finite taxa (the units) in the literature are packaged based on theoretical, group specific, and always pragmatic assumptions that vary among practitioners. This is obvious, and it cannot be discussed here. *Species taxa are formalized and at least temporally artificial chunks of lineages—samples of the consequences of the evolutionary process.* Fossils give a measure of morphological (hence taxonomic) diversity, but not necessarily of lineage multiplicity. Theories, therefore, such as those based on assumptions of punctuationism, need to be tested against evidence other than the few suggestive taxonomic patterns from which they are inductively derived.

For example, Vrba's frequent references to the fossil record, testability, species taxa, and its association with SMRS through horn cores of antelopes is seriously suspect. Horns and antlers are epigamic features more for displays in agonistic interactions in artiodactyls than for recognition by or being attracted to mates. The males heavily rely on scent for tracking both proceptivity and receptivity. In the polytypic African buffalo *Syncerus caffer,* a widely acknowledged single species, a great range of horn shape and size, body size, and the near extremes of a range of African habitats are characteristic, as in the larger and blacker grassland specific *S. c. caffer* and the smaller and reddish forest variety *S. c. nanus* of the Congo Basin. The biological species concept comfortably and harmoniously incorporates testability of species taxa (distinction of sympatric species through their reproductive or genetic isolation), their reproductive continuity (not any more invariant throughout a species than other attributes), and ecological distinction (the core of the cohesion concept). The fossil record only offers samples that must be clustered phenetically at first, then interpreted through models of real species of today.

Lineages

Although the biological species concept represents one of the cornerstones of the Synthesis in a moment in time (particularly its relational aspect that allows the testing of the concept; W. J. Bock, 1986; Coyne,

1992, 1996; Szalay, 1993, & references therein), the foundational Darwinian idea of gradual (i.e., bridgeable but not rate specific) continuity of species finds its expression in the concept of lineage (i.e., all or partial past history of a species in successive realtime). Simpson (1951), in fact, has defined the species as a lineage segment and was followed by Wiley (1981) with a virtually identical definition, and this view is also reflected in the "fuzzy set" notion of Van Valen (1988), a taxonomically delimitable unit, a species taxon. The concept of species in realtime was not served by these notions. Simpson's evolutionary definition, however, reflected his general concern. He was keenly aware of the prevalent tendency of neontologists and some paleontologists to see species primarily as closed entities not only in regard to other species in realtime, but also as discrete in deep time (Simpson, personal communication, 1974). He emphasized the conceptual adaptive constraint on the notion of a species *taxon* in his evolutionary species concept as an entity that makes empirical sense through time. This notion was not appreciably different from the multidimensional species taxon of Mayr (W. J. Bock, 1986; Szalay & Bock, 1991; Szalay, 1993).

The concepts of lineage and anagenesis seem to have been loaded with a variety of meanings that are difficult to appreciate unless one realizes how significant the different, highly refined, ontological definitions have become for various systematic and taxic-evolutionary theorizing. Furthermore, it is almost impossible to discuss the various concepts of lineage unless one strays into species concepts as well. The most recent process-oriented ontological discussion of Nixon and Wheeler (1992) centered on their concepts of extinction (several meanings) and lineages (several meanings) and dealt with character cum individual, as well as population/species level problems of ontology along with some issues of testability. Far from attempting a thorough (and warranted) critique of their contribution, I merely use it to illustrate the effects that redefinitions can have on stability of communication about ontological entities. These authors supplied many redefinitions and expansions or contractions of existing definitions of various concepts. All of this depended on their understanding how these might reconcile evolutionary concepts with cladistics. Nixon and Wheeler attempted to legitimize, like many before, the multidimensional species taxon as a species concept, but they called it a "phylogenetic concept" to distinguish it from the "Hennigian" one. In this attempt they managed (p. 136) to lay down some amazing definitional taboos in order to equate taxonomy with evolutionary theory, such as: "Finally, from the standpoint of phylogenetic species concept, character transformations always produce new species, so anagenesis cannot occur within species. . . . Thus within-species anagenesis is clearly envisioned by Hennig, ambiguous with Mayrian species concept, and interpreted as between-species character transformation with a phylogenetic species concept" (Nixon and Wheeler, 1992:136).

The graphic representation of Nixon and Wheeler (1992:138, fig. 4.7) of speciation from a particular phylogenetic perspective summarizes the view that a multidimensional species taxon is the equivalent of their species notion. They show three options for their species notion: "(a) 'Anagenetic' speciation, with simultaneous extinction of ancestral species. (b) and (c) 'Cladogenetic' speciation, resulting in one (b) or two new species (c), with the extinction of ancestral species." Their "temporal reproductive boundary" that equals the local or global extinction of plesiomorphs is simply a practical boundary that can be drawn between samples. In this schema species are "fuzzy sets," simply empirical units based on the presence or absence of either a plesiomorphic or apomorphic character state. As cladists, the certitude of these authors about character polarities assumes a knowledge of taxon phylogeny.

Lineage Stasis and the Multidimensional Species Taxon

It is no accident that the concept of lineage is given a pariah-like status both in cladistics and punctuationism, particularly in the ontological models of cladograms or in punctuational iconography. These are flawed visual metaphors, although in fairness, most, if not all, graphic representations of the evolutionary dynamic are likely to be so, but in significantly different ways.

Punctuationism, based on the notion of near instantaneous appearance and persistence of "species" and "species-individual" building blocks, is unclear about a lineage. The idea of lineage at first became vaguely equated by punctuationists with either the Hennigian (= phylogenetic) species concept (contra the sense of Nixon and Wheeler, 1992), or with the persistence of the mother species that was the source for another punctuated entity. The notion that segments of lineages can be axiomatized as one of the two varieties of "monophyletic species" (para- or holophyletic species) became part of the repertoire of assumptions, often implicit, in many taxonomic and evolutionary papers. As a consequence, a major restriction was placed on evolutionary theory through the focus on "species" and its stasis. Although degrees of stasis (a continuum, like most other evolutionary phenomena) are reality for many aspects of morphology in the fossil record (how else were we to recognize homologies, or even more specifically, synapomorphies?), at least in the genome the theoretical notion of full species stasis is difficult to accept. In reference to stasis, Dieckmann et al. (1995:98) noted in their modeled theoretical analysis of coevolutionary cycling, concerned with population dynamics and the Red Queen, that: "Clearly, there is no general rule in nature to say that phenotypic evolution would lead to an equilibrium point in the absence of external changes in the environment."

In a recent paper Lieberman, Brett, and Eldredge (1995) have put a new twist on the notion of stasis as negative evidence for punctuation. They have provided both data on middle Devonian brachiopods and new definitions that make it obvious that "stasis" is undergoing a metamorphosis. They seem to want to retain the concept but they also want it, in an oxymoronic sense, both ways: Stasis is change. Punctuationists like Gould and Eldredge (1993) seem to want to cover all bets for any explanation of stasis when they maintain an aura that stasis is beyond selection-based explanations (in the broadest Darwinian sense). They state that "because species often maintain stability through such intense climatic change as glacial cycling, stasis must be viewed as an active phenomenon, not a passive response to unaltered environments" (Gould and Eldredge, 1993:223–224). Breadth and flexibility of adaptive strategies, particularly those mediated by unfossilized behavior, override data available in the fossil record, well beyond their "preference for viewing stasis in the context of habitat tracking or developmental constraint." (Gould and Eldredge, 1993:224). Their view is then that if species remain "stable" through extreme fluctuations of climate and habitat (i.e., not morphologically tracking the habitat, then the explanation should be, by default, a developmental or phylogenetic constraint-based one. It appears that they have not come close to exhausting other Darwinian explanations for explaining *morphological stability* of *some* traits. It is never made clear how, when genetic-developmental constraints are invoked for "species," these constraints are overcome, and then how they become tyrannical again against change. By sidestepping such problems and accepting speciational evolution as the prevalent evolutionary mode, the reality of anagenesis without speciation is subtly exorcised. This appears to be a necessity based on the somewhat dogmatic and rigid theoretical structure of both punctuationism and cladistics (see views of Nixon & Wheeler, 1992, above). Habitat tracking and constraint, while legitimate processes of the evolutionary dynamic of specific lineages, are not believable general recipes for the supposed geographic and temporal "integrity" of a fossil morphology-based idea of species.

Differences in Taxic Approaches

There is more than one taxic approach in evolutionary analysis. Taxic approaches use taxic data to uncover the consequences of mechanisms and describe process. One speaks of a taxic approach as soon as the notion is accepted that certain (not necessarily all) phena in the fossil record mirror evolutionary change, given that these represent well-tested taxonomic species. Analytical taxic approaches differ when the notion of the species taxon is said to stand for specific ontological entities. In one approach the species taxon is unencumbered by the strictures of punctuationist theory and by notions that paleospecies (phena) must all represent distinct lineages. In the punctuationist approach, however, the notion of species is front-loaded—namely, that species taxa or static lineages represent the punctuation–stasis–extinction (either pseudo- or real extinction) model of evolutionary change, and furthermore that evolution is largely speciational.

Species taxa of the fossil record *are* mere samples of their lineages as opposed to being discrete "individual" entities (the ontological rhetoric notwithstanding; e.g., Nixon & Wheeler, 1992; Rieppel, 1994). This is both a phyletic (i.e., transformational) and a taxic perspective on evolutionary change, one that can be legitimately used together with all sorts of taxic investigations. Such is an evolutionary taxic approach of the first variety in which: a) ghost (i.e., unwarranted) assumptions are not made about the mode of origin of taxa, or b) species concepts are irrelevant for species *delineation,* except for the use of tested models of living species. A combination of such taxic and phylogenetic (transformational and cladistic) approaches are obviously complementary to one another. The issue of contention, then, is about the extent to which taxic practices should intrude into the understanding of the causal mechanisms of evolutionary change. Species-level taxonomic or phylogenetic patterns yield no mechanisms, they are to be explained by them. Such a view does not make macroevolutionary patterns either "subservient" (*contra* Gould, 1983) or "dominant," or any less interesting or less revealing.

What Eldredge (1979) dubbed *the* taxic approach (the second, or punctuational variety listed above) was a genuinely new emphasis. Eldredge, much less ambivalent, at least philosophically, about natural selection and adaptations of species than Gould, has subsequently reinforced this well coded and loaded philosophical stand on *punctuational taxic analysis* (Eldredge & Cracraft, 1980). But it is a hopeless, if not impossible, balancing act to try to be a pan-punctuationist and simultaneously adhere to foundational Darwinian tenets of the Synthesis. The distortions that have crept in are obvious. The primacy of taxic and phylogenetic patterns for understanding causal mechanisms and process in evolution is a message that recurs in not only Eldredge's but all of Gould's, Stanley's, and Vrba's papers.

In order to make justifiable causal connections between the practice of taxonomy and punctuated evolution, a number of significant and difficult-to-accept taxic assumptions need to be defended: (1) species taxon "individuality;" (2) lack of significant evolution outside of the realm of speciation (whenever speciation actually "begins" and "terminates"), (3) species (cum taxon) "selection", and (4) punctuated trends (mere shifting variance of phena; explicitly decoupled from the indivisibly time- and context-bound natural selection of organisms). These concepts are all dependent on the punctuationist taxic view. But rapid evolutionary change, quantum evolution, with or without speciation, and subsequent stagnation, *as one in the spectrum of rates and modes of phyletic change or speciations* is perfectly well within the domain of Darwinism and the Synthesis, as repeatedly stated before. It does not require a restricted taxic theory for its support on a case-by-case basis and with the numerous caveats that such empirics always entail. Rapid peripatric speciation, as noted above, certainly does not need to entail "genetic revolutions" (Nevo, 1989, 1991). Problems are not with the reality of bursts and stasis, but with punctuationism as a majority paradigm for the evolutionary dynamic based on a narrowly defined model of speciational evolution.

Anagenesis, Cladogenesis, and Attendant Conceptual Pitfalls

Taxonomic theory can be axiomatically divorced and practiced, sealed from any input into its methodology from various aspects of tested evolutionary theory. Witness how evolutionary theory as such is considered in phenetic or cladistic taxonomy. In these disciplines the chosen general (and obvious) aspects of evolutionary theory are axiomatized to stand for the whole process (divergence in the case of phenetics and sistergroup relationships, not Darwinian descent, in the case of cladistics). Although it is only marginally relevant here, it should be added that consequent to such tenets the analyses of morphological attributes of living organisms and those supplied by the fossil record are often subjected to taxonomic analysis without regard to causal analysis of character transformation (see W. J. Bock, 1981, for an account of what character analysis should entail). Such issues as levels of precision claimed for numbers of species taxa, lineages, or the nature of tests for proposed evolutionary relationships, and so on, all influence macroevolutionary accounts through specific taxonomic practices.

A taxonomic paradigm itself, because of the acceptance of its axioms, will push a particular view of the evolutionary dynamic. And such a view may be embraced in order to accommodate the axioms of that taxonomic philosophy and bolster its evolutionary validity, and vice versa. Speciational evolution, as Mayr (1989) describes evolutionary change associated with the speciation process (but by no means *the* evolutionary change, as he points out), and his endorsement of what he dubbed soft punctuation, has come to define a particular evolutionary dynamic, properly recognized for its importance. Focus on the generation of taxonomic diversity, however, shifted the theoretical base for the analysis of evolutionary patterns almost exclusively to speciation and its seemingly proper connection to *species taxa*. This is so largely, I suspect, because the horizontal (synchronous) comparisons of various entities, from populations to closely related and/or similar species of today, afford opportunities for scrutiny that are far more readily available than that for paleontological units. This is due to the fact that even a vertically "perfect" fossil record suffers from distributional problems in space and quite often lacks critical data on ecological dynamics. Although neontology may lack direct access to lineage splitting (real speciation, i.e., full cladogenesis), the seemingly inexhaustible supply of the kinds of species, superspecies, hybrid zones, and a cornucopia of species-related genetic, molecular, and morphological phenomena offer a laboratory for modeling and for deducing generalities that the fossil record cannot match in kind (e.g., Nevo, 1991).

I feel compelled to stress, however, that to consider speciational evolution, as Mayr (often) and the punctuationists (axiomatically) do, as a mechanism or process different in realtime from anagenesis is the single most unwarranted theoretical stance, in my view, of what we may glean of evolutionary change in general. It is certain that not all full cladogenesis involves morphologically detectable phenotypic change (e.g., sibling species). It is also certain that not all detectable local change between the connected or clinal and synchronous populations of the same, fully tested geographically extensive biological species is due to "speciational evolution" but is simply the result of lo-

cal anagenesis (the core of all levels of cladogenesis). It is, therefore, certain that rate-varying anagenesis occurs with or without incipient or populational cladogenesis or full or complete cladogenesis, i.e., speciation. Only in purely semantic and taxic ontological constructs can the reality of anagenesis be removed from the process leading to differentiated demes, populations, or full species.

Basic Concepts—a Reminder of Assumptions

Tested Darwinian theory has consequences that cannot simply be brushed aside by words if new proposals are to avoid being quasi-metaphysical. A lack of rigorous conceptual discipline in theorizing or in the formulation of relatively new concepts of a school of thought is understandable at its inception. In fact, I believe that it is to such an early phase of the Modern Synthesis that Gould referred to when he spoke of its early "pluralism." Conversely, Gould's widely quoted critical notion of "hardening" of the Synthesis really meant that the paradigm was finally "getting it together" about issues it addressed, at least in terms of clarifying the very basis of evolutionary dynamics. These were strongly tested against empirical evidence of all sorts. As noted, a conceptual narrowing, a hardening, in punctuationism today is also evident (related to ontological clarification, if not testing). The stage is set, then, for a theoretical debate of real issues, and it is therefore timely to have certain words stand for unambiguous concepts, without redundancies and studied ambiguities. I reject the philosophers Weber and Depew's (1996:39) cavalier stand on conceptual terminology when they essentially espouse questionable definitional stands by stating that "Such is the price that must sometimes be paid for a more general theory." The hypothetico-deductive consequences of imprecise concepts and a "general" theory should be obvious.

The theoretical literature is huge in this area, and much of it consists of restatements (with variation) and reaffirmations as well as debate about several concepts that are found under the same word. This problem will never go away, but it can be ameliorated. There is need for the exegeses and proper connections of the foundational concepts of descent, natural selection (both its Darwinian and the modified Synthesis version), adaptive and geography-bound evolution and other aspects of ecological and competition theory, and geographic (rate independent) speciation to proposed theories (see W. J. Bock, 1993). I list here a number of definitions in alphabetical order that should make clear my usage of these terms in this chapter. They should also reflect and clarify the underlying approach to my theoretical perspective on macroevolution.

Anagenesis: evolutionary change in populations or species, or phyletic evolution together (but without the restrictions placed on its reality because of its concurrence) with the whole range of cladogenesis (namely, that it occurs as part of cladogenesis).

Cause: historical event(s) & energy transfer (force, etc.) that when acting on an object (or a complex adaptive system such as an organism, but not on merely groups or classes as entities) will result in the change of the state of that object in time (see discussion in Bock, 1993). These states, in relation to evolutionary dynamics, can be differences in longevity and fitness of organisms. Causality that affects objects making up entities such as populations or species, does, of course, result in processual change on the level of these units (non-individual entities) with different outcomes at different points in time during the process. The often stated "ultimate causes" that reside in the history of the organism are, I believe, reducible to the Aristotelian material "cause," i.e., the genotype (and zygote) of the organism. The genotype represents part of the initial and boundary conditions, and it is not a cause. What remains theoretically significant for evolutionary biology out of the Aristotelian material, efficient, formal, and final causes is really only the efficient cause: Formal and final causes are not Darwinian causes, in spite of past parallels drawn between final causes and biological roles.

Causal mechanism: series of ordered causes through a particular period of time (e.g., natural selection as an N-D E; see below). Such mechanisms are often referred to as "ultimate causal mechanisms." The specific causes in such chains of events are nevertheless specific events in realtime with specific initial and boundary conditions, rendering the nomological causes often inseparable from historical ones.

Chronospecies: result of a taxonomic subdivision of the record of a lineage in paleontology into successive species taxa based on a variety of eclectic and heuristically workable criteria. It is emphatically not "The successive species replacing each other in a phyletic lineage which are given ancestor-descendant status," as Vrba (1994b:38) defines it. Vrba's definition equates species with the artificial (vertical) delineation of multidimensional species taxa and implies ontological reality for vertical taxic delineation.

Her definition of "species" is operational, which conflates tests with theory and is therefore unacceptable.

Cladogenesis: 1. (incipient, demic, or populational) when basic evolutionary units are isolated and are undergoing evolutionary (anagenetic) change; 2. (full or complete) speciation resulting from reproductively (behaviorally, or genetically, or ecologically) contained anagenesis.

Competition: a process during which individuals of the same or different species exploit some resource that is limiting the number of these individuals. Competition for resources nonagonistically has been called "scramble," "consumptive," or "passive" competition. When individuals directly fight or attempt to influence one another in some way, the terms "contest" or "active" competition have been employed. Competition is neither predation nor parasitism, although scenarios may be conjured up when these come close to definitions of competition.

Diversity: restricted here to any measure of either morphological or genetic differentiation of samples, populations, or species *between* any given section of time and space. (This term is commonly used, however, in a sense that diversity is the number of existing species at a given moment in time. This is problematical for the fossil record in general as asynchronous morphological diversity is not necessarily lineage diversity, and synchrony becomes contestable with increasing geographical separation in paleontology; see definition of multiplicity below).

Evolutionary progress: 1. mechanical and adaptational improvement of an attribute related to a particular biological role that spreads through a lineage (predator escape, ability to cope with some food resource, etc.), without any implication, however, that through time this improves the competitive ability of the organisms of a lineage compared to those of other contemporary lineages (it can signify the maintenance of status quo compared to other lineages while improvement occurs, as in Red Queen); 2. may result in decisive competitive advantage of some lineages over others through the adaptive superiority of individuals in their competition for necessary common resources (e.g., pterodactyl "niches" may have been overtaken by various bird lineages; nesting site- and food-based competition among small mammal species of similar "niches" may direct progressive change, etc.); 3. the mere change in a lineage (a sense in which it is rarely used).

Evolutionary rate: measure of relative or absolute change in an attribute in any aspect of the genome or in its products (phenotype) from proteins to "traits" in a lineage, in various segments of a lineage, or in designated related lineages or taxa.

Habitat: The physical environment in which individuals or populations of species occur naturally (i.e., not due to human interference). Defining a *fundamental habitat* as Vrba (1994b:40) does assumes some knowledge about the potential organism–environment interaction or the adaptive plasticity (or lack of) that is simply not known for a species. The definition of the *realized habitat* by Vrba exists only as an aspect of an organism's existence, like the definition of the niche, below. Neither habitat nor ecological niche have ontological realities, and they are definable only in relation to organisms.

Hierarchy: no meaningfully general definition of hierarchy exists in the literature (a most comprehensive discussion dealing with the "layers, loops, and levels" of hierarchic thinking is presented by Dyke, 1988: chapter 5). Beyond the relatively unambiguous inclusive hierarchy of classificatory systems and the direct administrative command structures of institutions (causal hierarchy), the concept of levels of organization (in the sense of level-interactive-modular-array or LIMA of Dyke, 1988, without any clear supersession within a LIMA) continues to serve as the most meaningful representation of the increasing and variable complexities in biological organization (see also discussion in Bock, 1991, and his examples for the use of LIMAs).

Historical constraint: (virtually the same as genomic constraint) the sum total accumulation of a genetically encoded recipe for the development of an organism (a result of phylogeny) and the subsequent limitations that the unfolding entails. Historical or genomic constraint is not a cause (although often the history that packaged it is called "ultimate causation") but represents the initial and boundary conditions together with the total environment; these set the stage for genetic (mutational), developmental, populational (drift, founder effect, etc.), and contingent selectional causes that shape the differential survival and reproduction of individual organisms.

Lineage: a particular designated section (duration in retrotime) of the historical path of a species, or, in practical taxonomic terms, the combined record of two or more successive (whole or partial), ancestral-descendent multidimensional taxonomic species taxa.

Mode in evolution: causal means through which

species evolve and/or multiply into two or more lineages. Simpson (1944) considered anagenesis, cladogenesis, and quantum evolution as the predominant ones—these are defined in this section.

Multiplicity: 1. the total number of lineages (i.e., species) in existence at any one time; 2. the proliferation of descendent lineages from a particular ancestry; 3. relative and absolute measures of lineages that make up one or more clades.

Natural selection: a causal mechanism and a process with critical temporal components (sequential in realtime), fully nomothetic while operating on properties of historically (and therefore uniquely) defined organisms; both mechanism and process consist of causal survival and reproductive phases that result, as a consequence, in fitness differential (in a broad sense including inclusive fitness) between heritably variable individual organisms. In a life-cycle context, survival-related attributes are usually causal to reproductive consequences in that the latter are fully dependent on the former (a fuller definition of selection depends on either its causal, processual, or outcome aspects, and it is not given here but only its causal and temporal aspects are emphasized; for detailed discussions, see Sober, 1984; Bock, 1993). Fitness is a consequential and not causal aspect of natural and sexual selection.

Niche: the sum total of interactions of an organism (individual organism or a whole species) with its environment (Bock and von Whalert, 1965). The concept of "adaptive zone" is a similar summation of the interaction of more than one species with the environment.

Outcome: the condition of an entity at the end of a process driven by one or more causes (see discussion in Bock, 1993).

Paleospecies: a taxonomic entity in retrotime, one that we cannot be certain to represent either a species, or only part of the morphological diversity of a real species in *its* realtime, or a mistakenly delineated taxon that contains more than one species in realtime.

Process: the course of change of an entity from one condition to another driven by a cause (or causes) acting on the objects that are or make up an entity and the initial and boundary conditions that prevail (see discussion in Bock, 1993).

Quantum evolution: the rapid, paleontologically largely undetectable change of a population between distinct adaptive peaks via an adaptive phase that is unstable (Simpson, 1944; quantum evolution does not axiomatically imply taxic stasis following it. Contrast this to punctuated equilibrium where adaptation as an aspect of species evolution is usually omitted as a factor, or eschewed).

Species: 1. a reproductively, or genetically, or ecologically bounded (isolated) unit in realtime (one or a group of populations of organisms) distinct from other sympatric species that has evolved and is capable of evolutionary change (as is a population); the concept is, emphatically, neither tied to speciation as such because that "event" is largely untestable in a human historical frame or in the fossil record, nor is it exactly the same as the empirically delineated multidimensional species taxon with its temporal duration. Clinal variation in geographically extensive species or ring species provides numerous examples that "common fertilization systems" are not necessarily shared by all populations of a species, nor are these punctuated, unless we define a species in a taxonomy-driven manner, based on invariant attributes.

Speciation: see under (full or complete) cladogenesis.

Species category: (usually) the lowest rank in the Linnean hierarchic classificatory system.

Species selection: differential persistence of lineages due to some causally efficacious supraorganismic properties of that lineage that cannot be accounted for on the organism level.

Species taxon: a specifically delineated part of a lineage; geologically speaking, a "chunk" of a lineage; an empirical unit of taxonomic communication; a multidimensional (in space and time) taxon. A fossil species taxon is a formalized and vertically artificially delimited unit, and for living taxa at least a lower boundary is designated.

Stasis: 1. relative constancy of one or more *characters or traits* in one or more lineages (a rate of change that is zero); 2. in reference to *seemingly unchanging species taxa* of variable duration (hence their classification in one species) viewed in retrotime. Simpson (1960:167) referred to the phenomenon as adaptive stabilization (see also Schopf, 1981a,b, 1984). Various homology hypotheses are, in fact, expressions of different levels and degrees of the stasis–change continuum. Strictly speaking, synapomorphies are taxonomic properties that are identical traits shared uniquely because they remained in stasis in two or more lineages from their last common ancestor. Transformational homologies, on the other hand, are properties that were not in stasis.

Tempo in evolution: rate of evolutionary change in specific traits or species taxa through geological time, although rates of change for characters have been successfully studied within historical time. Simpson (1944) provided widely used terminology for the spectrum of rates from slow to fast (bradytely, horotely, and tachytely).

Summary

Various causal mechanisms acting on entities above organisms have been proposed for explaining taxonomic and other evolutionary patterns. In particular, punctuationism claims that a "hierarchic" macroevolutionary science can be causally decoupled from microevolutionary mechanisms and processes. The propositions are largely based on redefinitions of such concepts as individual, fitness, selection, birth, and death. The macroevolutionary accounts of Simpson in 1944 and 1953 reflect an emended but fully Darwinian perspective of the Synthesis for the explanation of evolutionary patterns. Punctuationism challenged that paradigm and offered another. There are irreconcilable differences between the two types of explanations for patterns. To focus on these differences, the following were topics briefly examined: (1) key concepts, (2) approaches (ontological and epistemological) to the explanation of the morphological patterns of living organisms and those represented in the fossil record; (3) macroevolutionary theory-building based particularly on the redefinition of fitness in regard to species, the ontological status of the latter as "individuals," expansion of the semantically and philosophically reformulated notion of selection, and the attendant problems in testing such concepts and the theories based on them; (4) the relevance of levels of organization ("hierarchy") for macroevolution; and (5) those aspects of evolutionary theory that often have been influenced by taxonomic practices in the past, but which have gained axiomatized, paradigm-scale prominence in a punctuational taxic approach to evolutionary analysis.

In punctuationism, a particular taxic framework, based on a morphological (taxonomic) species notion, coupled with punctuated stasis, is advocated for macroevolutionary analysis of sundry patterns. In contrast, Synthesis-based taxic analysis is centered on phylogenetic (phyletic, i.e., lineage related, and cladistic) assumptions about evolutionary change of traits and taxa, on the relevance of genetic, epigenetic, or adaptive evolution to ascertain the causality of these patterns. The two approaches make critically conflicting assumptions about the nature of morphology and evolutionary change and about the meaning of lineages and species taxa based on morphology in the fossil record.

The ontological kernels of punctuational theory and their alleged tests are extremely difficult to separate. It is likely, however, that the fossil record and studies of the living will easily accommodate causality from anagenesis to full cladogenesis for a whole range of evolutionary patterns from the static, the most gradual, and the quantum (punctuated) tempos of evolutionary modes. But the notion of speciation as *the* predominant or paramount mode of change cannot be tested in the fossil record. Essentially stable communities through time give strong testimony of anagenesis in several lineages through the existence of successive species taxa in such contexts. There is even evidence for anagenetic origin of higher taxa without cladogenesis as well.

The rise of punctuational macroevolution and its purported more complete accounting for evolutionary phenomena, coupled with the incorporation of selected aspects of philosophical analyses of evolutionary theory, was based on independent but mutually reinforcing general perspectives. The first was the proper, curiosity-driven extension of new definitional concepts and their use as axioms in taxic theorizing and testing—sundry efforts to see how far ontological propositions can be sustained in order to advance a science. The other perspectives were more the outgrowths of the personal tastes and the pragmatic and ideological motivations of the founders of this paradigm. Eldredge accommodated the rise of exclusively cladistic taxonomy, whereas Gould was deeply influenced by a perceived relevance of humanistic social philosophy to evolutionary science. This coincidence of talents, motivations, and ideas has had the salutary effect of reminding the scientific community of the importance of evolutionary science. It also highlighted the dangers that conceptual simplification and taxonomy-based testing of improbable axioms entail. It should not be overlooked that taxonomic patterns alone do not yield tested evolutionary theory and that unconscious sacrifice of tested concepts for ideology and other values is not uncommon.

The rhetorical question remains: How do we establish the reality of new ontologies beyond polemics? Furthermore, can a fundamentally taxic, decoupled view of hierarchic punctuated macroevolution justify new explanations of patterns or drastic redefinitions of the tested concepts that continue to be the firm foundations of a vigorous, expanded Synthesis? Gould's tiers and the Kuhnian dismissal of the Synthesis notwithstanding, punctuationism is plagued by major theoretical problems.

Acknowledgments Discussions with the conference participants, particularly with Peter Grubb, Jeff McKee,

Norman Owen-Smith, and Mike Rosenzweig were especially stimulating and are much appreciated. Critical reading of the manuscript by Walter J. Bock, Arthur J. Boucot, Peter Grubb, Eviatar Nevo, and Eric Sargis is gratefully acknowledged. A PSC-CUNY doctoral faculty grant 666443 has facilitated the completion of this chapter. This chapter is an abbreviated version of the paper prepared for the volume.

4

The Autocatalytic Nature of Hominid Evolution in African Plio-Pleistocene Environments

Jeffrey K. McKee

Causal explanations of hominid evolution may be derived from a synthesis of paleontological data and evolutionary theory. Paleoecological perspectives gleaned from the fossil record may contribute the critical element of time depth for testing hypotheses derived from interdisciplinary research on contemporary evolutionary and ecological phenomena. The challenges for paleoecologists are to correctly interpret fossil data in light of the inherently incomplete nature of the fossil record and to reconcile long-term perspectives gained from interpretations of the past with what is known from the present. The theoretical perspective of autocatalytic evolution, viewing evolutionary change as a result of intraspecific changes independent of external events, may provide an appropriate yet counterintuitive solution to these challenges.

In this chapter I present a fundamental approach to mammalian paleoecology based on the notion of autocatalytic evolution. Such an approach appears to be consistent with theories derived from population genetics, the developmental basis of morphology, and mammalian ecology. The theoretical approach of autocatalysis is also consistent with emerging data and analyses of African fossil fauna, including the early hominids. A hypothetical model of hominid evolution can be developed from this basis, with implications for the evolutionary processes governing the origins of mammalian species in general. To build a theoretical model of the ecological causes and conditions of mammalian evolution, a number of basic premises must be borne in mind.

As an initial and most important premise, it is clear that the genetic basis of morphological and physiological novelties arise at random with respect to events in the external environment. In this sense, "random" does not imply that all possible mutations within a genome are equally likely but that environmental influences do not induce novel directional changes in that genome. Within an environmental context, natural selection then may act on those mutations in a directional fashion if the phenotypic consequences of the new allele somehow affect the relative fitness of an animal. Alternatively, alleles with selectively neutral effects may be subject to other forces of evolution such as genetic drift.

Novel phenotypes, when arising from genetic mutations or new combinations of alleles, will only be of selective value if they are consistent with the developmental and physiological processes of an organism. Thus there are inherent structural constraints limiting the range of possibilities that may have selective value. For example, a giraffe's neck cannot become longer without the proper anatomical and physiological mechanisms to control blood flow to the head (Mitchell & Skinner, 1993). Likewise, evolving features, such as the human head and face, must conform to the limits imposed by birth processes and antenatal conditions. Alternatively, changes in the geometry and physiology of developmental processes may lead to evolutionary opportunities (Thompson, 1942).

Central to the theory of autocatalytic evolution is a

further premise, perhaps less clear, that the most important part of a mammal's environment, with respect to natural selection, is its own species. At birth mammals must receive nurturing. This usually comes from the offspring's mother (to varying degrees), but often directly or indirectly from other conspecifics as well, and requires appropriate feeding morphology and behavior that may be unrelated to adult adaptations. Sometimes a newborn must compete with siblings for that nurturing, thus initiating a further mode of natural selection. Among social mammals, a developing individual must somehow fit into norms of behavior to benefit from existing mechanisms of group survival and to successfully find a mate. The relative importance of these factors may vary depending on the behavioral patterns of the mammal, but they are inevitably important to some degree in determining Darwinian fitness (which in mammals depends primarily on mating success).

Ecological factors of a mammal's selective environment are of less consequence than the intrinsic selective constraints, as represented by the heuristic diagram in figure 4-1. Each concentric circle represents the ever-widening sphere of existence for an animal subjected to natural selection. Taken from the point of view of the animal whose task it is to survive and reproduce, the progression of subsequent spheres is fairly straightforward. Each species would take a characteristic pathway from the center, out through the various levels of selective constraints and adaptive opportunities. Some of the phenotypic consequences of the pathway may then be reflected in the morphology of the fossils we find. Each ring is successively more peripheral to the genetic basis of a phenotype and places less stringent constraints on the microevolution of the population.

Conversely, what is not so obvious is that the pathway from the outside–in, the effect of the environment on the genetic resources of organism, is not so easily determined. Starting from the broad sphere of a regional environmental context, potential ecological determinants of how an animal's morphology relates to that environment must accurately penetrate each successive layer of selective constraints before changes in the gene frequencies take place. The selective values of an animal's adaptations within a particular environment are largely buffered by the more immediate constraints within the more central rings. This is due not only to the intrinsic nature of genetically controlled development arising at random, but also to the greater importance of epigenetic effects on an animal's relation with the environment (e.g., adaptable behavior and physiology throughout varied temperature and rainfall regimes). Thus the relative effect of each ring will depend on the peculiarities of an animal's adaptations and adaptability within an environmental context.

Once these fundamentals of evolutionary theory are explored, the difficulties encountered by a paleoecologist become clear, as well as the difficulties with theories that posit environmental change as a major causal force in the evolutionary origin of species. Questions arise as to how much of a species' physiology and morphology, including that detected in fossils, is relevant in the context of broader environmental features such as climate, vegetation, competitors, and so on. If that environment then changes, and potentially adaptive features are arising at random, what

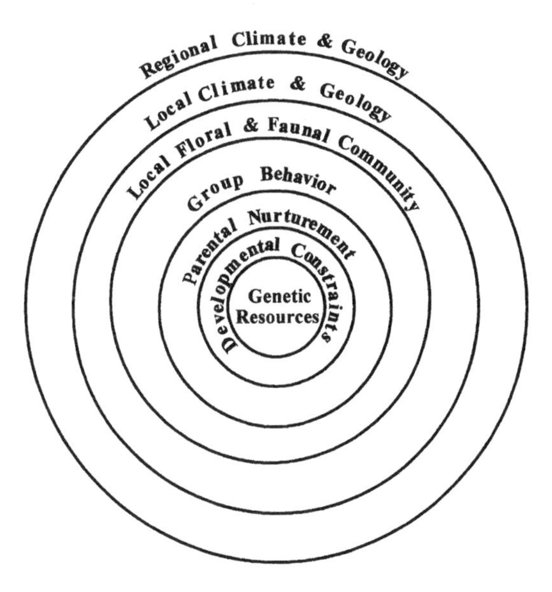

Figure 4-1. Schematic diagram of selective constraints on the genetic resources of a mammalian species population.

determines whether or not species have the genetic resources to adapt? Do some species have more genetic potential, making them more evolvable than others? Do adaptable species necessarily need to adapt morphologically to survive, or should they be treated differently in paleoecological studies from specially adapted animals? These questions may be answerable by an approach centering on the theory of autocatalytic evolution.

Autocatalytic Evolution

In the theory of autocatalytic evolution, most evolutionary change among mammals is catalyzed by the inherent nature of intraspecific variability (genes, morphology, and physiology), and not by changes in the external environment. Autocatalysis is responsible for origins of morphological novelties recognized by paleontologists as "species" characteristics as well as the origin of true "species"; in other words, it applies throughout microevolutionary and macroevolutionary levels. Thus evolution and perceived speciation would proceed with or without changes in climate or in the plant and animal community with which a species interacts.

Notions of autocatalysis depend on chance relations of evolving features with the successive rings of the selective milieu. "In this sense chance means that a variation having appeared, *chanced* to find a suitable environment" (T. H. Morgan, 1910:203; emphasis in original). Furthermore, genetically based phenotypic novelties do not necessarily become fixed in a population by natural selection alone, for the contribution of genetic drift could have the same effect on changing the character of a population. Due to these elements of chance, the timing of an evolutionary progression need not correspond to, or be catalyzed by, a change in the external environment of the evolving lineage. Change in the broader selective environment may shape an evolutionary pattern and cause evolution to proceed in one direction rather than another, but is not necessary to initiate or sustain the process.

Cogent arguments for evolutionary autocatalysis among hominids have previously focused on the autocatalytic acceleration of increasing encephalization (Mayr, 1963; Bielicki, 1969; Holloway, 1972a; Godfrey & Jacobs, 1981; Tobias, 1981, 1994; Henneberg, 1987, 1992). Such a model is based in the mutual reinforcement of selective advantages between increased brain size and increased behavioral plasticity. The model of autocatalysis presented here, however, does not necessarily lead to an accelerating feedback loop. The loop is limited to that allowed by genes and morphology and thus could temporarily stall or completely halt at any time, depending on the availability of appropriate genetic resources on which the forces of evolution may act.

In models of autocatalysis, an evolving species derives evolutionary opportunity from a greater amount of variability within a successful lineage. This differs from models of environmental causation, in which the conservative forces of natural selection act during times of environmental change. When a species is under stress from a changing environment, natural selection works within the bounds of existing variation. Variation is thus limited by the selective process, whether the stress comes from climatic changes, as in the turnover-pulse hypothesis (Vrba, 1985a,e; 1988, 1993c), or from changes in the biotic community, as in the Red Queen hypothesis (Van Valen, 1973; Foley, 1984). If that variation is not sufficient to allow the species to adapt, then extinction is more likely than an evolutionary progression (Felenstein, 1971; Foley, 1994).

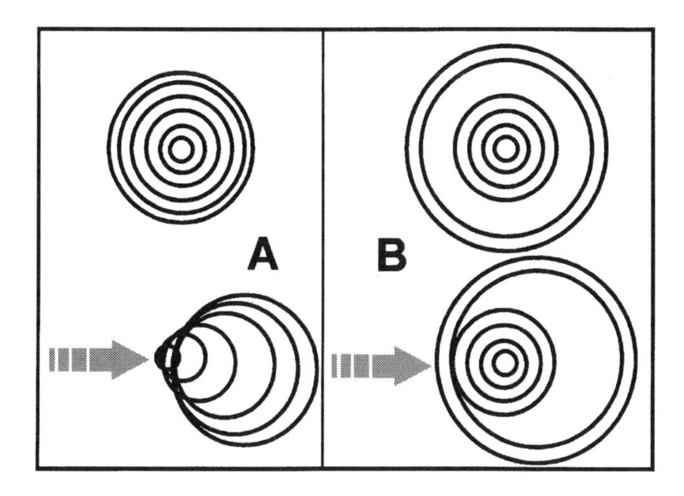

Figure 4-2. Environmental change effects (arrows) on highly adapted (A) versus adaptable (B) populations. Concentric circles represent selective constraints on the genetic resources of a population.

Difficulties with models of catalysis by environmental change can be illustrated by extrapolating the concentric circle analogy of morphological adaptation. In figure 4-2, two different types of mammalian reactions to the environment are represented. The first set (A) represents an animal that is ecologically specific with rings that are tightly bound. This would be a highly adapted animal with specific ecological requirements. Should the environment change, the outer circle impinges on various levels of morphological adaptation. The animal will only evolve if there is sufficient genetic variability already in place to allow it to shift with the environment; certainly this can happen, albeit rarely. However, should the environmental alterations severely stress the species, as in vicariance models (see Vrba, this volume), smaller selected populations with reduced variability (limiting evolvability) and possibly eventual extinction is the more likely outcome.

Effects of environmental change on ecologically adaptable animals (such as those with considerable behavioral and dietary plasticity) are represented by the second set of concentric circles (fig. 4-2B). Such mammals' relationships to the external environment are much less tightly constrained, so that a shift in the environment may not impinge upon their genetic variability at all. Environmental change would neither lead to morphological adaptation nor to the danger of extinction. I suggest that early hominids, indeed most mammals, fall into this category.

Natural selection in autocatalytic evolution, however, is initiated from the genetic and developmental source. It works from within, largely on the basis of selective advantages of niche expansion and population growth rather than selective pressure reducing and fragmenting populations. Any shift of the inner circle is possible, as long as it is consistent with the successive rings around it. Other options may be eliminated by natural selection. The selective advantages of any shift are more likely to be those dealing with changes in the inner circles, especially if reproductive success is accelerated. Sexual selection would be a good example, in which a species would evolve, perhaps dramatically (Fisher, 1958), regardless of fluctuations in the external environment. Any shift that gives a reproductive advantage to an animal carrying a novel variant would be favored by natural selection. Likewise, many variants could take hold due to genetic drift, as long as they are consistent with the developmental and ethological patterns of the species.

Continuous and potentially accelerating evolution is possible through autocatalysis. As features arise that allow niche expansion, and/or greater success at reproduction and survival, a population may grow. With a larger population, genetic variants are more likely to arise and, if they provide a reproductive advantage, to subsequently proliferate throughout the population. Initially a single populous, variable species would evolve and disperse. Subsequently, in some cases, divergent species could evolve through the process of centrifugal speciation.

Even extinction can be autocatalytic. McCann & Yodzis (1994) produced an ecosystem simulation in which top predators went suddenly and inexplicably extinct without any perturbation to the ecological system. Likewise, Lande (1994) has demonstrated that extinctions of small populations can ensue with the chance fixation of a mildly deleterious mutation, also in the absence of an environmental stimulus.

Autocatalytic evolution may be particularly relevant to the origins and spread of hominids. Here I present a hypothetical autocatalytic scenario and then support the model with paleoecological and morphological data, primarily from southern Africa. Although many aspects of the model are in need of further testing, the approach raises new questions about hominid evolution and the role of natural selection in an environmental context.

The Hominid Scenario

Narratives of autocatalytic evolution among hominids may begin with a hypothetical arboreal primate as the common ancestor of the lineages that ultimately led to *Australopithecus* and an ancestral form of *Pan*. The evolutionary process was initiated by a genetically controlled change in morphology which arose by chance (in the sense defined by Morgan, 1910). This morphological change allowed the arboreal primate to augment its locomotor and dietary behavior such that the early primates could exploit the resources on the ground as well as in the trees. There would have been a selective advantage in widening the formerly limited arboreal niche, especially if resource availability in the trees was limited either by competitors or saturation of the niche by expansion of the evolving population.

Derived features of the modified terrestrial locomotor and dietary morphology would have been preadapted to the exploitation of a niche outside of the forest. When further genetically induced modifications allowed locomotor and dietary behavior beyond that which was adapted to the former confines of morphology and environment, the primate could then exploit the savanna (where available). In at least one population the accumulating advantages of niche exploitation led to selection for bipedality and orthograde posture. Other populations, which by chance hit upon a different mode of morphology, led to the knuckle walking and brachiation still employed by *Pan*.

Consequences of bipedality included freed hands, which has often been considered to provide potential selective advantages (Darwin, 1871). Bipedality also may have freed the cerebrum for further expansion. Orthograde posture set the initial conditions for cerebral expansion, as it occurred among hominids. The developmental requisites of orthograde posture included a considerable degree of basicranial flexion (Gould, 1977b; Ross & Ravosa, 1993). Developmental flexion at the base of brain resulted in greater developmental arching of the cerebral peripheries, consequently expanding the cerebral neocortex. A slightly expanded brain was thus an indirect morphological consequence of bipedal locomotion. This brain allowed even greater adaptability through behavioral plasticity, and the autocatalysis of hominid brain evolution began accelerating. (This notion elicits the intriguing possibility that the primary selective advantage of orthograde posture was in cerebral expansion rather than in the bipedal mode of locomotion.) The development of omnivory, eventually including the consumption of meat, allowed the reduction of the face and further expansion of the brain (Ross & Henneberg, 1995); the same opportunity was not afforded to the robust lineage(s).

Hominid morphological evolution, with one feature leading to another, was catalyzed by the advantages and increased variability allowed by niche expansion and population growth rather than by environmental change. It is true that had the environment not changed, and savanna had not become available, the process may have taken a different direction, but such processes would not have been halted. In other words, the direction of evolution may be guided by the environment, but the impetus for change is autocatalyzed independently of environmental events.

Fossil Evidence Regarding the Ecological Correlates of Hominid Evolution

Autocatalytic models of hominid evolution yield predictions distinct from those of environmental forcing and/or environmentally induced vicariance. Given the random nature of the process, evolutionary events (as recognized paleontologically by changes in skeletal morphology) should be random with respect to the timing of climatic and environmental change. I have argued elsewhere that, at least from the southern African perspective, species origins among all mammals may have occurred randomly with respect to time (McKee, 1995). Analyses by Bishop (1993, this volume) Reed (1995, 1996), and Behrensmeyer et al. (1997) also show no evidence of an increased pace of evolution or turnover at times of climatic change.

Origins of bipedality and associated hominid features would not necessarily be related to climatic shifts in the autocatalytic model. This prediction is difficult to assess, given the lack of confirmed fossil data bearing on the origins of bipedality. But if *Ardipithecus (Australopithecus) ramidus* (T. D. White et al., 1994, 1995) does turn out to be a biped (or even an incipient biped), and the environmental reconstructions (Wolde Gabriel et al., 1994) are correct, then the origins of bipedality would have appeared in a wooded environment and would not be associated with major ecological shifts in climate or terrestrial biota.

Southern African data appear to support a further prediction of the autocatalytic model that early hominids should have exploited a variety of environments. By the time *Australopithecus africanus* emerges in the fossil record of southern Africa, sometime around 3 m.y. (and perhaps earlier), the hominid lineage had achieved complete habitual bipedality, albeit with vestiges of arboreal adaptations. Evidence suggests that the environments surrounding the caves in which these hominids were deposited were quite diverse, ranging from subtropical forest at Makapansgat to a more xeric environment at Taung.

Makapansgat

Makapansgat, the oldest of the established *A. africanus* cave sites (McKee, 1996a), has been characterized by Rayner et al. (1993) as being surrounded by subtropical forest at the time of deposition, largely on the basis of pollen analysis (Cadman & Rayner, 1989) and geomorphology. Although the palynological analysis may be dubious due to the inclusion of modern exotics in the sample sediments, the fauna of Members 3 and 4 (McKee et al., 1995; Reed, 1995) are certainly consistent with a forested environment (table 4-1). Makapansgat has the greatest number of bovids, which were probably primary browsers, and the highest biodiversity of any Transvaal Pliocene cave site. The presence in the sample of numerous grazers, as well as *Acinonyx jubatus* (cheetah, currently known to prefer open habitats), however, suggest that a mosaic environment may have characterized the general area. Nevertheless, Makapansgat appears to be the wettest and most densely vegetated area at the time of deposition as compared to any other southern African Plio-Pleistocene site.

Table 4-1. Mammals identified from Makapansgat Members 3 and 4 (MAK3, MAK4), Taung Hrdlička and Dart deposits (TAUH, TAUD), Sterkfontein Member 4 (STS4), and current extant species in southern Africa (EXTA).

	MAK3	MAK4	TAUH	TAUD	STS4	EXTA
Artiodactyla						
Bovidae						
Antidorcas bondi					•	
Antidorcas recki					•	
Cephalophus parvus				•		
Gazella gracilior	•					
Gazella vanhoepeni	•					
Hippotragini gen et sp nov	•					
Hippotragus equinus					•	•
Hippotragus cookei	•				•	
Makapania broomi	•	•			•	
Oreotragus oreotragus	•	•	•	•	•	•
Parmularius sp nov	•					
Parmularius braini	•					
Redunca darti	•	•			•	
Syncerus acoelotus			•		•	
Tragelaphus angasii	•		•		•	•
Tragelaphus pricei	•					
Wellsiana torticornuta	•					
Suidae						
Notochoerus scotti	•	•	•			
Potamochoeroides shawi	•	•			•	
Hippopotamidae						
Hippopotamus amphibius	•					•
Proboscidea						
Elephantidae						
Anancus kenyensis		•				
Elephas recki					•	
Perissodactyla						
Rhinocerotidae						
Ceratotherium simum	•					•
Diceros bicornis	•					•
Equidae						
Hipparion libycum	•					
Chalicotheriidae						
Ancylotherium hennigi	•					
Primates						
Cercopithecidae						
Cercopithecoides williamsi	•	•	•		•	
Papio izodi			•		•	
Parapapio broomi	•	•		•	•	
Parapapio whitei	•	•			•	
Parapapio jonesi	•	•			•	
Parapapio antiquus			•			
Theropithecus darti	•	•				
Hominidae						
Australopithecus africanus	•	•		•	•	
Carnivora						
Canidae						
Canis mesomelas	•	•	•		•	•
Vulpes chama	•	•				•

(continued)

Table 4-1. (*continued*)

	MAK3	MAK4	TAUH	TAUD	STS4	EXTA
Carnivora (*continued*)						
Felidae						
Acinonyx jubatus	•				•	
Dinofelis barlowi	•			•		
Felis issiodorensis	•					
Felis libyca	•					
Felis serval	•					•
Homotherium crenatidens					•	
Homotherium nestianus					•	
Megantereon cultridens					•	
Panthera leo					•	•
Panthera pardus	•		•		•	•
Sivatherium maurusium	•					
Hyaenidae						
Chasmaporthetes silberbergi					•	
Chasmaporthetes nitidula					•	
Crocuta crocuta		•			•	•
Hyaena brunnea					•	•
Hyaena hyaena	•	•				
Pachycrocuta bellax	•				•	
Viverridae						
Cynictis penicillata	•	•				•
Hyracoidea						
Procaviidae						
Gigantohyrax maguirei	•	•				
Procavia antiqua	•	•	•	•	•	
Procavia transvaalensis			•			
Tubulidentata						
Orycteropodidae						
Orycteropus afer	•					•
Insectivora						
Soricidae						
Crocidura taungensis			•			
Crocidura bicolor			•			•
Diplomesodon fossorius		•				
Myosorex robinsoni	•				•	
Myosorex varius		•				•
Suncus varilla	•	•	•		•	•
Suncus infinitesimus		•			•	•
Vespertilionidae						
Eptesicus bottae		•				
Eptesicus hottentotus		•				•
Chrysochloridae						
Amblysomus hamiltoni		•				
Calcochloris hamiltoni		•				
Chlorotalpa spelea					•	
Rodentia						
Batherigidae						
Cryptomys hottentotus		•				•
Cryptomys robertsi	•		•		•	
Gypsorhychus darti			•	•		
Gypsorhychus makapania	•					
Gypsorhychus minor			•			

(*continued*)

Table 4-1. (*continued*)

	MAK3	MAK4	TAUH	TAUD	STS4	EXTA
Rodentia (*continued*)						
Cricetidae						
Mystromys antiquus	•		•		•	
Mystromys hausleitneri		•				
Proodontomys cookei	•	•	•		•	
Stenodontomys darti		•				
Desmodillus auriculatus			•		•	
Gliridae						
Graphiurus monardi					•	
Hystricidae						
Hystrix africaeaustralis	•		•		•	•
Hystrix makapanensis	•					
Xenohystrix crassidens	•					
Macroscelididae						
Elephantulus antiquus	•	•	•		•	
Elephantulus brachyrhynchus		•				•
Macroscelides proboscideus	•		•			•
Muridae						
Acomys cahirinus	•		•			•
Aethomys chrysophilus		•				•
Aethomys namaquensis		•			•	•
Dasymys sp. nov.			•			
Dendromus mesomelas					•	•
Malacothrix typica			•			•
Mastomys natalensis	•		•		•	•
Otomys gracilis	•	•	•		•	
Otomys sloggetti		•				•
Prototomys campbelli			•			
Tatera brantsii			•		•	•
Thallomys debruyni			•			
Pedetidae						
Pedetes gracilis				•		
Petromyidae						
Petromus minor			•			
Chiroptera						
Rhinolophidae						
Rhinolophus clivosus		•				•
Rhinolophus darlingi		•	•		•	•

References list can be found in McKee et al. (1995), updated by Reed (1995) and Watson & Plug (1995).

Sterkfontein

Analyses of Sterkfontein Member 4 suggest a Pliocene environment that was at least partially wooded, as compared to later sites in the Blaaubank river valley. This is evident from analysis of the Bovidae (Vrba, 1975; 1976; 1980b), which reveals a large number of buck adapted to dense vegetation (particularly *Makapania broomi* and cf. *Hippotragus cookei*) as compared to greater proportions of alcelaphini and antilopini at later Pleistocene sites. Notwithstanding the possible differences in taphonomic agents that may have influenced these proportions (McKee, 1991; McKee et al., 1995), the faunal assemblage of Sterkfontein Member 4 is distinctive enough to reveal that at least local environmental changes to increased savanna in the Transvaal postdated the deposition of Member 4 and thus postdated a bipedal and slightly encephalized *Australopithecus africanus*. Yet the faunal diversity and composition appears to indicate an environment less rich than that of Pliocene Makapansgat.

Taung

Depictions of the Pliocene environment at Taung are a bit more difficult to interpret due to the destruction of most of the *A. africanus* type site by quarrying. After seven years of excavation and research at the site (McKee & Tobias, 1994), some conclusions can be drawn with a measure of caution. Most of the fauna known from Taung comes from deposits of the Hrdlička pinnacle and are unlikely to be associated directly with the hominid fossil, perhaps being separated by as much as 200,000 yr. The hominid was probably derived from a deposit temporally and spatially associated with the remaining Dart deposits (McKee, 1993a,b; McKee & Tobias, 1994). There are clear taphonomic differences between the Dart and Hrdlička deposits to compound the possible effects of the temporal separation, but knowledge gained from comparisons of the deposits at least suggest a paleoecological similarity.

Excavations of Hrdlička deposits have yielded a fauna dominated by cercopithecids, particularly *Parapapio antiquus* and *Papio izodi,* neither of which reveals a particular habitat affinity. The few bovid fossils recovered consist of mainly medium-sized alcelaphines that are unidentifiable to species level, but which tend to indicate the presence of extensive savanna grassland. The presence of the genus *Syncerus,* a probable primary grazer, would support a grassland reconstruction of Taung at the time of deposition.

Hrdlička deposit excavations have yielded two browsers, *Oreotragus oreotragus* (sensu Watson & Plug, 1995), a klipspringer, and *Tragelaphus angasii,* the nyala. Habitat requirements of both of these modern species would be consistent with Taung being surrounded by considerable savanna, as long as there was sufficient plant growth (Skinner & Smithers, 1990). The klipspringer prefers rocky outcrops, which certainly would have been provided by the tufa accretions at the edge of the Ghaap escarpment, not unlike Taung today. The nyala is known to be a mixed feeder that prefers thickets in dry, savanna woodland. The springs at the edge of the escarpment which apparently fed the tufa accretions would have been sufficient to promote localized thickets, as they do in certain areas along the Ghaap escarpment today, literally at the margin of the Kalahari desert.

Observations of the rodent fossils from southern African by Denys (1992a) also show the Taung fauna to be distinct from Transvaal sites representing the Plio-Pleistocene. The rodent evidence supports the notion of a relatively more arid environment at Taung.

Mammalian fossil evidence thus points to the probability of a savanna or perhaps savanna woodland, with small, localized oases supporting somewhat more dense brush, characterizing Taung at the time of deposition in the caves of the Hrdlička deposits. The Dart deposits and hominid-associated breccias reported by Broom (1934) do not have sufficient faunal remains to make a comparable reconstruction (table 4-1). It is tempting to extrapolate the environmental reconstruction of the Hrdlička deposits to the Dart deposits, albeit with considerable circumspection. For example, caution may come from the presence in the Dart deposits of an extinct member of *Cephalophus,* for the modern species of the genus is limited to forest habitats; other closely related members of the subfamily cephalophinae, however, such as *Sylvicapra grimmia,* occur in savanna woodland or grassland (Skinner & Smithers, 1990). (Note that *Sylvicapra* has been identified from the loose breccia over the Dart deposits, and is probably associated with them, but due to the disturbed context it has been omitted from the faunal list.)

Investigations of geological evidence from the limestone tufa in which the caves formed (McKee, 1993b), however, provide more compelling evidence for associating the environmental reconstruction of the Hrdlička deposits with those of the Dart deposits. Both are within the Thabaseek tufa; most of this tufa is consistently thinly laminated, having been formed largely by algal and moss concentrations of lime, much like the tufa forming at Taung today. There is no significant erosional valley either upstream or downstream from the Thabaseek tufa that would suggest a former river flow, so it appears as though the tufa formed from spring water emerging near the edge of the Ghaap escarpment. The nearby Oxland tufa accretion, however, which formed after the Thabaseek tufa, is quite distinct in its composition, geological association, and fossil inclusions, leading me to postulate that the tufa itself is indicative of the general environmental setting. The Oxland tufa has consistent inclusion of fossil branches and leaves, much like the tufas forming today in the wetter, subtropical montane forest environments of the eastern Transvaal. The Oxland tufa is associated with deep erosional valleys (including one that cut through the Thabaseek tufa; Peabody, 1954) and clearly formed from a substantial river. The fossils in the caves of the Oxland tufa (McKee, 1994), best represented by Equus Cave (Klein et al., 1991), yield fauna that certainly indicate a considerably wetter environment with more bush cover than exists today. The consistency of associations between tufa type and environment, past and present, imply that the nature of tufa can be used as part of an environmental reconstruction. Had the environment been substantially different between the times of deposition of the

Hrdlička and Dart deposits, then the difference should be revealed by distinct tufa types. But there is no difference, implying that relatively dry conditions prevailed during the formation of the Dart and Hrdlička deposits.

Southern African Environments and the Evolution of Hominids and Other Mammals

Now the best evidence from the Pliocene gives the appearance of a gradient of environments from Taung to Sterkfontein to Makapansgat. This is comparable to the gradient that exists today, but all of the sites would have been shifted toward somewhat warmer and wetter environments than those that currently characterize these areas. More than three decades of research have confirmed what Dart (1964:52) noted:

> "The distribution of South African sites from Taung to Makapansgat and the climatic variation to which these sites were subjected geologically during their occupation by australopithecines show that the South African types had adapted themselves to almost, if not quite, as wide a range of climatic, soil and vegetational variation as modern mankind faces outside the arctic circle."

It would thus appear that *A. africanus* used and perhaps regularly inhabited a range of habitats. Our ancestor was probably therefore a generalist rather than a specialist, adaptable rather than adapted. The morphological variability of what is currently considered to be *A. africanus* is consistent with this notion. At Makapansgat and Sterkfontein there is enough variation in craniofacial morphology to elicit proposals of the presence of more than one species (Aguirre, 1970; Clarke, 1985); alternatively, the robust elements noted in some *A. africanus* specimens and the tremendous variability in the growing sample from Sterkfontein may attest to the variability of an evolving lineage without highly specialized facial adaptations. Likewise, whereas all *Australopithecus* fossils are undeniably orthograde and bipedal to a large degree, there are substantial vestiges of arboreal adaptations, particularly in the upper limb (Berger, 1994) and pelvis (Broom & Robinson, 1950).

Diverse habitat ranges are not limited to hominids. Removing the anthropocentric bias of this analysis, the same generalized level of adaptation should be true for *Oreotragus oreotragus, Tragelaphus angasii, Canis mesomelas, Cercopithecoides williamsi, Parapapio broomi, Panthera pardus, Procavia antiqua,* as well as six species of rodent, a bat, and a shrew, all of which appear at Taung, Sterkfontein, and Makapansgat (table 4-1). Half of those species still exist in southern Africa today, suggesting that past environmental contingencies of evolutionary change seem to be irrelevant to this adaptable group.

Eastern African paleoecological data from *A. afarensis* sites demonstrate a broad habitat range for the hominid species (T. D. White et al., 1993). Together with the southern African data, the fossil record now appears to support the notion of a generalized adaptability and a wide niche for even the earliest known hominids. Thus one aspect of the scenario presented earlier, that of niche expansion, garners some support. In an autocatalytic model, this would propel further novelty as long as the genetic and phenotypic variability was available, and indeed the hominid fossil remains indicate such variability.

Virtual coincidence of environmental changes with the emergence of the genus *Homo* in southern Africa requires careful consideration. First, the morphology of *Homo* seems to be limited to a few changes in the face and a slight increase in cranial capacity, none of which is inconsistent with a gradual change from *A. africanus* given the high degree of morphological continuity, at least in craniofacial morphology (McKee, 1989). Second, taphonomic considerations and the geographical isolation of Transvaal sites obscure the reconstructions of subcontinental environmental change in southern Africa, and may make them appear more profound than they really were (McKee, 1991). Finally, when considering all of the fauna, the first and last appearances of species around that time cannot be differentiated from that expected by chance under a model of stochastically regular turnover (McKee 1995, 1996b). Only after about 1.8 m.y., following Swartkrans Member 1 and following the establishment of *Homo* in southern Africa, is there any unusual faunal change, and even this is not distinguishable from chance.

Origins of the genus of *Homo* in an apparently changing environment cannot be clearly distinguished as being anything more than mere coincidence. *Homo habilis* (sensu lato) at this stage appears in both eastern and southern Africa, seeming to indicate a proliferation of a successful lineage rather than a forced adaptation. In East Africa, the delineation of an event, as opposed to a continuous autocatalytic process, is dubious given the apparent gradual changes of ecological parameters (Cerling, 1992a; Bishop, 1993; Kingston et al., 1994; Denton, this volume). Only the East African bovids seem to show an evolutionary pulse (Vrba, 1985c), and this appears much later in southern Africa, following the introduction of *Homo;* the bovid scenario could thus be characterized as an autocatalytic adaptive radiation rather than an environmentally induced event of vicariant speciations.

Discussion

Legitimate tests of autocatalytic evolution are difficult to formulate, as with models of climatic determination of the timing and pace of evolutionary origin of species, due to the limits of the fossil record. To support one model or the other, one must distinguish the effects of chance from true cause and effect, given a multiplicity of evolutionary forces. Although the chance nature of autocatalysis appears to have the edge in current analyses of fossil data, the apparent trends do not prove autocatalytic evolution; the African fossil record is as yet too incomplete. Thus the fossil record must be complemented with an appeal to known mechanisms of evolution.

Underlying models of either autocatalysis or climatic causation are the same genetic prerequisites, and these necessarily rely on chance origins of appropriate adaptive features. In summary, the selective forces in the two models differ in that climatic forcing depends on selective pressures decreasing variability, whereas under autocatalysis selective advantages are favored in highly variable populations. The selective advantage in the autocatalytic model comes with the advantages of niche expansion rather than reliance on what could have been untimely forcing. Environmental change, for specifically adapted mammals, is often more likely to lead to extinction than evolution, largely due to unlikelihood of the appropriate genetic basis for new phenotypes existing when the population of a species is reduced by environmental stress. For ecologically generalized mammals (and the evidence suggest that early hominids were just that), environmental change would have a negligible effect in initiating or sustaining evolutionary change.

The mechanisms and concepts of autocatalytic evolution are rooted in microevolutionary theory, but paleontologists usually try to observe them at a macroevolutionary level. Perceived macroevolution must be reconcilable with known microevolutionary mechanisms. Part of the problem with this reconciliation is the focus on species for perceptions of macroevolution. Paleontological "species" are notoriously difficult to define and recognize, for they are transient entities at best at any moment of geological time. The mosaic nature of hominid morphological evolution (McHenry, 1994a) makes species definitions particularly difficult. But perhaps species questions should be the last, not the first, part of fossil analysis. To understand the ecological relevance of evolving lineages such as hominids, we must ask the right questions about the fossil data.

In the context of autocatalytic evolution, hominids may provide an example of an evolving lineage that poses legitimate research questions. A primary question in autocatalytic evolution would concern correlations of morphological features, such as orthograde posture, basicranial flexion, and encephalization. The tempo and mode of the evolution of features, not species, can then be assessed.

One must first find a means of identifying evolving features that may have affected the fitness of the evolving animals before addressing paleoecological questions. This is more difficult than it may appear, for directional trends do not always imply that natural selection has directly influenced traits. The confounding effects of other evolutionary forces (e.g., McKee, 1984) and what Vrba (1983) has termed the "effect hypothesis" limit the adaptive nature of many traits. The objective is not to fit traits to an environment, as is often done, but to discern the potential effects of traits within an environmental milieu. Once potentially selective significant features are recognized, then each of the levels of selection outlined in figure 4-1 must be considered. Thus only a few evolving features will be ecologically relevant, although certain features, such as encephalization, may act upon fitness at more than one level.

New paleoecological questions arise from the analysis of ecologically relevant morphologies. The key question is whether a trait or trait set is specifically adapted to an environment, and this is testable by circumspectly comparing paleoenvironmental clues from sites in which the animals possessing those traits are found. By then looking at lineages of related animals, one can approach questions of adaptiveness versus adaptability, evolved versus evolvable traits, and autocatalyzed versus environmentally induced evolution. We may even find unexpected general trends across species, such as one of increased adaptability in the evolution of mammals. Principles of the causes and consequences of evolution can then be gleaned within an environmental framework.

Acknowledgments I am most grateful to Tim Bromage and Friedemann Schrenk for the invitation to the Malawi conference and their superb organization of the discussions, as well as to Wenner-Gren for generously sponsoring the event. Thanks are also due to Jean and Nathan McKee for timely autocatalytic inspiration.

5

In Search of Paleohominid Community Ecology

Michael L. Rosenzweig

The Hominidae of several million years ago appear to have constituted two separate species in many places. Did they interact ecologically? If so, what can community ecology suggest about their competitive relationships? How might these relationships have influenced their geographical distributions and evolutionary histories?

The answers to those questions may well come from viewing ancestral hominids as members of an ecological community, as species that had to carry on their ecological business much as other species do today. Hypotheses generated by that view may help. Let us therefore search among the patterns we see in such communities. Perhaps some will match what little we do know about early hominids. Perhaps knowing these patterns and their associated mechanisms will help us study and understand the prehuman condition. Perhaps they will even help us discover how we have come to be.

In this chapter I outline and apply some of the features of modern community ecology. I do so with all the biases of any human being: the issues I cover will be those that both capture my attention and seem potentially useful for the study of human evolutionary ecology. I hope that at least some of them are new to paleoanthropology.

Competitive Guilds

What Are Competitive Guilds?

Community ecology has long recognized that species constitute more than a hodgepodge assembly of elements. Many seem to be members of functional groups that tackle the world in quite similar ways, yet differ in some consequential detail. Different species of canids, for example, all run after their prey rather than ambush them. But different species of canids have different body sizes (e.g., wolves and red foxes) or hunt in different habitats (e.g., arctic foxes and red foxes) (Rosenzweig, 1966). Thus, they rely on different species of prey (which also have different body sizes and occupy different habitats). So the various member species of a community manage to coexist.

Although I once suggested we call such functional groups "hunting sets" (Rosenzweig, 1966), ecologists wisely rejected this term and settled on a much more elegant one: "guilds" (Root, 1967). Like so many other terms in ecology, this one has resisted attempts to clearly define it. Ecologists out in the field however, could not help but use it. "Guilds" made too much order out of chaos in too many cases. "Guilds" simplified too many descriptions of maddeningly complex community processes and helped organize our thoughts and our investigations time and again. But, in fact, a formal definition of "guilds" may exist.

Lurking in the theoretical literature of evolutionary ecology is a concept called the "fitness-generating function," or "G-function" (J. S. Brown and Vincent, 1987). Coming from the ideas of continuous game theory, the G-function is austere, abstract, and mathematical. Every G-function that exists was invented by a theoretician for purposes of computer simulation. No G-function has ever been measured in a real organism. And yet I believe that G-functions have a biological face. I believe that, in particular, they allow us to pin down and thereby fathom the notion of the ecological guild. But what exactly is a G-function?

The fitness of a phenotype depends on its environment. That environment includes both biotic and abiotic elements. Put a penguin in the everglades, and you will surely depress its fitness. Put one alone among a group of sea lions, and you will likely do the same. If we had an equation that accurately predicted the fitness of a particular phenotype in whatever environment it found itself, that would be its G-function. The variables of the function would be a list of abiotic environmental variables plus the densities of all other phenotypes—those exactly like itself, others of its species, and still others of other species. Thus G-functions extend Levins's concept (1962) of the adaptive function—adaptive functions also predict fitnesses but use only the proportions of habitat types to do it.

To work, G-functions must embody the trade-off rules that constrain an organism's fitness. For example, attain an adult size of 50 kg and profit by feeding on large species such as deer, but lose the ability to profit much from mice. Grow tubercles on your paws and scamper freely among the rocks, but lose the ability to run as surely or swiftly over a sand dune. Evolutionary ecology offers many such examples.

Because G-functions include the densities of the organisms, a single G-function can predict the simultaneous existence and success of many phenotypes. When the 50-kg predator strategy becomes common, the fitness associated with it declines. If mice abound, a 5-kg predator strategy will bring higher rewards. Thus, the community may evolve to include many strategies all subject to the same trade-off rules. If so, we say that all the species adopting these various strategies fit the same G-function.

The fact that G-functions have never been directly observed hardly makes them an unusual scientific phenomenon. No physicist ever saw an electron. No demographer ever met a family with 2.3 children. Scientists persist in such concepts, however, and they detect them by looking for their traces, their effects. The track of the electron in the cloud chamber. The growth rate of the population with an average of 2.3 children. G-functions are the same.

Hyrda of various sizes constitute a good example of a set of species sharing a single G-function. Hydra capture their prey with tentacles, but some supplement their nutrition by harboring colonies of algae. The algae fix energy from the sun and add to the sustenance of the hydras that support them in their bodies. But the algae can grow out of control and overwhelm the hydra's cells, killing it. Hydras have thus evolved a mechanism to control algal growth. The equation that governs this control depends for success on the size of the hydra. Little hydras can successfully control algal growth; big hydras, subject to the same equation, cannot. The cost of algae outweighs their benefit. That is why there is a cutoff point among hydras. Those smaller than this threshold size have algae and are green. Those larger than this threshold size lack algae and are brown (Slobodkin et al., 1987).

Some guilds of birds provide more quantitative and continuous examples. A fine case compares various species of nectarivorous/insectivorous honeyeaters and chats in Australia (Wooller, 1984). Species that eat a diet of both nectar and insects come in many sizes but maintain a fixed ratio of bill length to body size. That is what we expect of members of a single guild subject to the same G-function. We do not know the exact nature of the trade-off in this case. Presumably, longer bills help a phenotype to probe deeper corollas and extend the range of flowers from which it can feed. Longer bills must also be more delicate and cumbersome for smaller individuals. If a species' strategy is pure nectarivory however, a proportionately longer bill would bring greater rewards. Thus, pure nectarivores form a different guild. For their body sizes, they have longer bills than the mixed feeders (fig. 5-1). Tyrant flycatchers also exhibit a set of guilds based on how they hunt (Fitzpatrick, 1981). It is possible to imagine many such cases, some based on morphology, some on behavior, some on habitat. Let us therefore examine the various ways species of a guild can subdivide a G-function.

How Are Competitive Guilds Organized?

Thirty years ago MacArthur (1964) suggested that all competitive coexistence among natural species has three possible roots: resource differences; temporal differences; spatial differences.

Resource Differences If competing species eat different food species, they may coexist. They may

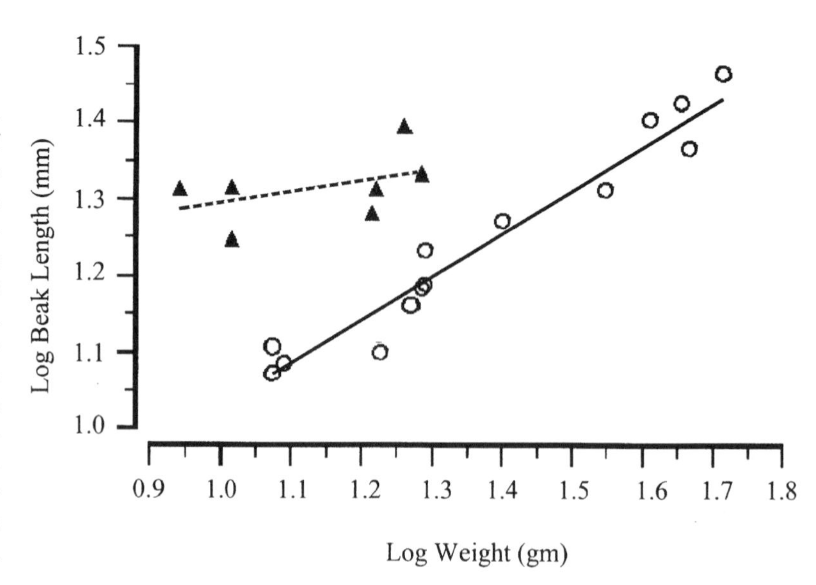

Figure 5-1. Honeyeaters of Australia belong to two separate guilds. Each conforms to its own G-function and thus arrays its species along its own allometric curve. The triangles show species that feed primarily on nectar. The circles represent insectivores. Data and concept from Wooller (1984; fig. 3); used with permission.

also coexist if they are the food species being differentially chosen (Holt, 1984; Jeffries and Lawton, 1984). Ecologists have found many examples of resource differences. For example, as we have seen, some guilds have species that differ in body size. But resource differences need not rest upon body size differences. Examples of other differences include two species of kangaroo rat in the United States. Although they weigh the same, these two species differ in the structure of their incisors. *Dipodomys ordii* has pointed incisors and eats seeds. *D. microps* has chisel-shaped incisors and eats halophytic leaves (its specially shaped teeth strip away the poisonously salty outer cell layer of such leaves) (Kenagy, 1972).

Many plants differ in the toxins they contain. Insects require special metabolic machinery to detoxify a plant poison. But the required machinery differs depending on the toxin (Fox, 1981). So the plants themselves can coexist and can also support coexistence among the insects that eat them.

Some pairs of marine or terrestrial molluscs differ in the strength of their armor. Thin-shelled species, which are easily eaten, can coexist with those that have thick, formidable shells (Schmitt, 1982).

Temporal Differences Environments differ temporally. Some sets of species coexist because the special environment of each one occurs at different times (Chesson and Huntly, 1988, 1993; J. S. Brown, 1989a). If such environments occur predictably throughout the year, we call the species "seasonal." Many other patterns of temporal variation are known. The environments may recur daily, leading to diurnal, nocturnal, and crepuscular specializations. Or they may occur quite infrequently and unpredictably (often true in arid and semiarid regions). Even such rare phenomena as abundant desert rainfall may support both ephemeral flowering plant species and the critical reproduction that leads to success in some animal species. Some lacustrine species (e.g., some lung fish and brine shrimp) retreat for decades while waiting for their dry lake to fill with water.

Spatial Differences Spatial habitat variation offers endless opportunities for specialization. Forests require different adaptations from savannas, salt water requires different adaptations from fresh. The list is infinite. Many species coexist by using different habitats.

Discussing habitat differences requires much more than a catalogue of examples. What may be of interest is that habitat differences occur in categories that generate contrasting patterns of specialization and geographical distribution. I believe the most important of these categories flow from the differences between distinct and shared preference organization of guilds (Rosenzweig, 1991).

Distinct preference organization. Most ecology texts assume that, in a guild whose species coexist by habitat specializations, each successful species has a habitat in which it outperforms all others. Because natural selection will make the individuals of such species prefer their best habitat, this method of organization is called "distinct preference organization."

We do know of a few cases of distinct preference habitat allocation. For example, two species of mussel in California subdivide their environment distinctly (Harger, 1968, 1970a,b, 1972). One species, *Mytilus californianus,* has a thick shell and thick, strong byssal threads. The byssal threads hold the mussel fast to rocky surfaces and help it to bind its aggregates of individuals ("clumps") in the face of pounding surf. In the wave-exposed habitats that support it, individuals of *M. californianus* on the outside of a clump are more exposed and more likely to get washed away. Meanwhile, the thick shell helps the species to avoid being crushed by strong waves.

The second species, *M. edulis,* is much disadvantaged in habitats exposed to heavy surf. Compared to *M. californianus,* it has a thinner, more brittle shell and more gracile byssal threads. It also tends to crawl to the outside of a clump or up to the surface when buried in silt or gravel. Its finer threads and more exposed position in a clump tend to make it susceptible to strong wave action, the waves tearing it free and leaving the mussel adrift and doomed to die. But its fine byssal threads and strategy of crawling to exposed positions do serve it well in the quiet waters of bays. There, deposits accumulate rapidly and tend to suffocate mussels living in the cores of clumps. *M. edulis* can keep from suffocating in bays because its threads grow fast and it seeks the surface and the edges; it can, as it were, keep its "head" above the mud. In bay habitats, *M. californianus* suffocates while *M. edulis* thrives.

Shared preference or tolerance organization. Ecologists rarely discover habitat allocation that depends on distinct preferences. More often they see an entirely different sort called "shared preference" or "tolerance organization." Shared preference organization depends on the existence of one species that tolerates a wide variety of habitats and another that does well in only a restricted subset of those same habitats. However, in tolerance organization, all species (in the absence of competitors) do best in the same habitat type. That is what gives this mode of organization the name "shared preference." Instead of specializing on different habitat types, species specialize on being able to use a larger or smaller variance of habitat types. The species that require the best habitat are termed "intolerant." Those that can profit almost as well from secondary habitats are "tolerants."

Often both the shared preference and the tolerance difference come about because the environment varies quantitatively—some places are relatively rich, others poor—but they each contain the same sorts of resources (Rosenzweig, 1987a). That is why both species do best in the same habitats. As an alternative to richness differences, environments may support the tolerance mode if some habitats are relatively stressful compared to others.

If the richest, least stressful habitats are best for everybody, how can two species ever coexist? Shouldn't the more tolerant always defeat the intolerant? The answer is trade-off. Being tolerant must cost something of the ability to use the better habitats. Tolerant species may be chased away from better habitats by intolerants that are larger or more aggressive. Or travel among patches of habitat may cost tolerants too much energy or expose them to too much predation risk (J. S. Brown, 1989b). That leaves the intolerant to tour the habitats, selecting only those that are best.

We know hundreds of examples of tolerance organization. Only a few, however, have been as well studied as the pair of crayfish species in the rivers that drain into the upper Mississippi (Bovbjerg, 1970). *Orconectes virilis* and *O. immunis* have the same body size and share a preference for rocky substrate. Such a substrate means rich, well-oxygenated water. But *O. virilis* is much more aggressive than its congener. In the presence of *O. virilis,* 80% of the *O. immunis* use muddy, oxygen-poor waters. What keeps the *O. virilis* from this secondary habitat? It suffocates in the oxygen-poor water due to the same high metabolic rate that sustains its high level of aggressiveness and allows it to dominate the well-oxygenated riffles.

The nature of tolerance organization leads to some interesting contrasts between tolerant and intolerant species (Rosenzweig, 1989; 1991). Intolerants tend to be larger or more aggressive (i.e., fiercer) than tolerants. They are more likely to exclude tolerants from the best habitats or even to kill them. Intolerants also tend to restrict their own foraging more, both in space and time. Yet, because they need to do more habitat searching, intolerants are often more mobile as individuals. When foraging becomes less productive (because of the time of year), intolerants are more likely than tolerants to retreat metabolically, waiting for a better season (or even a better year). One might say that intolerants tend to be picky bullies. But one must also add that they can afford to be picky because when the best habitats are available, their bad manners or efficient mobility (or both) secure them most of the opportunities in those best places.

Interrelationships of Causes Life is never as neat as science. Environmental differences often combine to make classification of community interactions blurry. For instance, different species of bumblebees have proboscises of markedly different lengths. Each length is suitable for probing flower nectaries of a certain length (Pyke, 1982). Thus, the bees can coexist by using the nectar of different flower species—a resource difference. But in a very real sense, the nectaries themselves are the habitats of the bees, i.e., the places from which they take their resources. So bees also coexist by habitat differences.

Fire succession constitutes both a predictable spatial difference and an unpredictable temporal difference. We can arrange a list of plant species according to whether they require a recently burned, recovering, or mature association. And, although we cannot predict exactly when a fire will clear a particular patch of land, we can predict that at any time many patches will be found at each stage of recovery. A plant (as a seed) or an animal able to move about can thus search for its temporal specialty in space. Australian aboriginals in the more arid parts of Australia apparently experienced their best hunting in patches that had burned several years previously. (Being human, they then went a step farther than other species can by setting the fires themselves and returning at the appropriate interval.)

The Correlation between Abundance and Distribution

Students of human evolution often ask about the geographical distribution of this or that extinct hominid. No doubt having such information would help us understand the dynamics of our development as a species. Where did we evolve? When did our ancestors spread out over the Old World? Did this occur only after they could be said to be *Homo sapiens* (as current genetic research into questions of ancestral Eve and Adam presupposes)? Or did we spread out previously? And if we did, what transformed us into living mercury and set us out on our collective career to coat the earth with our presence? Does this transformation have ecological roots or correlates?

In the past 15 years a pervasive pattern of community ecology has emerged with considerable potential for helping us interpret paleohominid distributions. It turns out that population densities and geographical breadth are connected somehow (Hanski, 1982). Moreover, their connection may not be independent of guild structure (Rosenzweig, 1989). If we can link guild structure to the pattern of density and geographical breadth, we can also use the position of a species in that pattern to interpret its relative position in its guild. Confronted with such opportunity, it seems worth at least describing the pattern and its possible connection to guild structure.

Some species have dense populations of individuals. Some species occupy large regions of the earth's surface. At first one might not guess that there would be a correlation between those properties. One might even guess that, because extinction rate climbs both as population falls and as geographical range diminishes, any correlation would have to be negative; a narrowly distributed and sparse species would doubly suffer and not appear in the real world. Yet, in the past 15 years ecologists have been noting quite the opposite (C. E. Bock & Ricklefs, 1983; J. H. Brown, 1984; C. E. Bock, 1987; Hanski et al., 1993; Gaston, 1996). Widely distributed species tend to be dense. Narrowly distributed species tend to be sparse. What could account for such a contrary pattern? Quite possibly, tolerance organization does (Rosenzweig, 1989).

Because intolerant species depend on the best habitats, we expect them to be both narrowly distributed and sparse. Locally, the best habitats generally form a small, scarce subset of those available. Regionally, those habitats may be entirely absent. A paradox results. The species whose individuals get the richest places is rare and confined to the few corners of a region where that habitat exists.

We have discovered a typical example of the result of tolerance organization among hummingbirds (Pimm et al., 1985). In southeastern Arizona, the blue-throated hummingbird is the largest and fiercest. It rapidly evicts smaller members of the family, such as the black-chinned hummingbird, from its territories in the 50-m–wide strip of riparian vegetation that provides the best habitat for all hummers. Even so, it is less common than the black-chinned which it restricts to the common, drier slopes of the mountains. The blue-throated hummingbird also has a tiny geographical range compared to the black-chinned.

Red Queens and Evolutionarily Stable Strategies: Punctuated Equilibria

Since George Gaylord Simpson, evolutionists have appreciated the variety of evolutionary rates. And, since Eldredge and Gould, they have focused on the possibility that much evolutionary change is fitful, proceeding rapidly, then not at all for a rather long interval. Often some have guessed that the short bursts of change happen simultaneously with speciation. But, if pressed, they will admit that speciation is not a part of the paradigm's core. Fitful evolution—more commonly known as punctuated equilibrium—is defined by its stop-and-start nature, not its relationship to speciation (J. R. G. Turner, 1984).

Why should punctuated equilibrium concern us in this chapter? Because the temporal patterns of hominid evolution strongly suggest punctuational events at two scales. Most apparent is the considerable variation in evolutionary rate. The experts at this conference seem to agree that sometimes our ancestral populations were racing at warp speed along their evolutionary trajectory, and sometimes they hardly changed for hundreds of thousands of years. But there is another, less apparent scale of punctuated equilibrium relevant to our evolution. It is the scale at which whole guilds exist. Evidence presented at the Malawi conference leads me to believe that the foundation of our twig of the anthropoid clade was in fact the foundation of a new guild, albeit a guild with only a single species. I turn now to the fossil data presented during the conference.

It seems evident that the ancient controversy in human evolution has long been settled: first we became fully bipedal, then we got smart. One notices that advance in intelligence by focusing, not on brain size per se, but on the allometry of brain size. To be considered advanced, brains have to be relatively large for the body they belong to.

Wood, this volume, points out that using this criterion, one can most reasonably sort hominids into their two genera. Most species have brains of ordinary relative size, but, starting with *Homo rudolfensis,* our brains evolved to be more and more outsized for our bodies.

Vrba, this volume, using this criterion too, has placed us on a provisional graph of relative brain size plotted against time. The graph shows that among our predecessors and close relatives, most species maintain and have maintained a relative brain size quite typical for mammals in general. It also shows that with the establishment of *Homo rudolfensis,* our direct ancestors and their closest relatives set out on a path of relative brain expansion culminating in our current state about 100,000 years ago.

Vrba's graph resonates with that of Wooller for honeyeaters and chats (fig. 5-1). True, Vrba's graph includes time and Wooller's does not. But that is inconsequential. What matters is the bifurcation in allometry that she is tracking. In Vrba's graph, the G-function rule that determines the net fitness value of relative brain size appears to be the same for species after species. And then *Homo rudolfensis* comes along. It breaks the allometric rule, setting out on a novel path that eventually establishes a new allometry with much higher relative brain size. Such a departure suggests that humans and their immediate ancestors are subject to a different G-function and that *Homo rudolfensis* experienced the breakthrough that founded it. If we are going to appreciate the phenomenon of human evolution at more than a trivial level, we need to put it the context of such breakthroughs. What keeps things evolutionarily static for so long? And what finally changes them, setting life racing for new vistas? In other words, what could account for punctuated equilibria? Here is one reasonable hypothesis. I believe it is the best we now have to offer.

What Produces Stasis?

The Red Queen Interacting species adapt to each other. But the improvement of one species raises the fitness value of counteradaptations held by other species. Thus, any species temporarily left behind by the improvement of another is pressed to evolve faster and catch up. No species should ever get ahead in its interactions (Rosenzweig, 1973; Van Valen, 1973; Schaffer and Rosenzweig, 1978). This is the rat race or arms race or Red Queen principle of coevolution.

Trade-off Organisms perform many different functions for themselves. Often these conflict with each other, making organisms a bundle of compromises. For instance, a larger organism may have a better chance of overcoming a smaller one, but its size alone requires it to find more food to keep from starving. This is the trade-off principle already discussed in connection with the organization of ecological guilds.

ESS The compromises resulting from trade-offs bring evolution to a standstill for long periods. The species so halted are said to have achieved an evolutionarily stable strategy, ESS (Maynard Smith and Price, 1973). This causes the stasis phase of punctuated equilibria (Rosenzweig, 1973; Rosenzweig et al., 1987). In a perverted, but significant way, such species also satisfy the Red Queen because they are all evolving at the same rate (i.e., zero).

What Causes Stasis to Breakdown?

Compromise Changes The compromises causing an ESS can change in two ways. First, the trade-off rules embodied in the G-function may change. Second, the mix of environments may change, altering not the G-function, but its prediction of which strategy constitutes the ESS. In either case, we expect an evolutionary response away from old ESS values.

Key Adaptations There is nothing immutable about trade-off rules. The penalty or reward for having a certain ability may change greatly. Perhaps, for example, a marine snail evolves the ability to build varices on its shell. That would allow it to defend itself against nibbling predators, such as crabs, without paying the high cost of thickening its entire shell. Or perhaps a species evolves a new enzyme that increases its metabolic efficiency and so reduces the cost of large size. In neither case (and both are realistic) does the species rid itself entirely of the constraints of trade-off. It merely relaxes those constraints (Rosenzweig and McCord, 1991). But relaxation will mean a new value for the ESS. And a new value for the ESS means evolution to it.

Species that evolve to loosen their internal conflicts are said to improve their versatility (Vermeij, 1973a, b, 1987). They can now improve simultaneously in more than one way. Adaptations that improve versatility (key adaptations) can provide the impetus for the bursts of evolutionary speed that punctuate long periods of stability under an ESS.

Environmental Change. Despite their importance, key adaptations cannot be the only source of evolutionary speed bursts. Optimal compromises also depend on the abiotic environmental circumstances built into G-functions. Thus, a climate change may also cause ESS shifts. For example, a new climate that produces generally richer habitats would reduce the cost (to fitness) of needing more food. So, organisms in such richer habitats should experience an ESS shift and may evolve a larger size.

An increase in environmental variability can shift an ESS toward less specialization. Suppose, for instance, that a species' best habitat in an environment is *H*. Suppose also that there is trade-off: improved ability to deal with *H* diminishes the ability of an individual of that species to deal with other habitats. Such a trade-off can be ignored by natural selection as long as individuals can always easily find *H*. The species can become a complete *H*-specialist because it pays no cost in being able to deal poorly with the other habitat types; it never uses them and so never suffers any loss in fitness because of its inability to use them (Rosenzweig, 1987b). On the other hand, if variance is high, making *H* more expensive to locate, then an individual may be better off accepting a less ideal habitat. Such individuals, forced by natural selection to use less than their optimal habitat, will also be forced by it to retain more of their ability to use lesser habitats—to be more generalized.

Now suppose there are two competing species and that they coexist by tolerance organization. Now imagine that their environment becomes more variable in space. The intolerant, locked into specialization on the rich but scarce habitats, will find its special opportunities even scarcer. Nevertheless, keep in mind that its old ESS stripped it of the flexibility that once would have both allowed and forced it to adapt to a wider variety of habitats. Now it is so specialized that it must pay too high a cost to use the poorer habitats. Its phenotype has locked it into its specialized strategy. Paradoxically, natural selection may thus force the specialist to suffer reduced population and reduced geographical range.

Meanwhile its tolerant competitor can and should undergo an ESS shift. The tolerant species, everywhere somewhat successful (at least as a species), may grow even more successful. As a consequence, it may be able to occupy places formerly under the control of its competitor. And it may also be forced to occupy places and take resources formerly unavailable to either species.

A Speculative Scenario

Here I offer a wholly speculative retelling of human evolution. It is based on the principles of community ecology collected in this sketch. Rather than worrying about whether it is correct, I designed it to provoke refinement and critical data gathering.

Prehuman Hominidae at one point in space and time often consisted of two species: one tolerant, the other intolerant. I suspect that the robust species was the intolerant one; the gracile species, the tolerant. The intolerant depended on rich, reliable habitats; the tolerant could make do with lesser places. Somehow, the intolerant had to have been able to dominate those richer places. If it was larger or fiercer, it may have discouraged or prevented the tolerant's use of the best habitats. (It seems unlikely that its advantage was based on superior mobility.)

Both species may have had relatively small geographical ranges, but that of the intolerant would have been especially circumscribed. Rich, highly predictable habitat patches tend to be limited, even in the tropics. At some point, the range restriction of the tolerant relaxed. Perhaps the change had to do with increased seasonality and the presence of seasonally abundant concentrations of ungulates, as Owen Smith (this volume) suggests. Such a change could have been brought about by a climate change some 2.5 m.y. at just about the time *Homo rudolfensis* appears (Bromage et al., 1995b; Vrba, this volume). As we have seen, the same change could well have further restricted the range of the intolerant.

Whatever the cause of the change, it seems to have been associated with a modest increase in relative brain size. That in turn could easily have given *Homo rudolfensis* a new G-function and ignited the kind of self-fueled evolutionary race envisioned by McKee (this volume). How is this possible?

Any geographical expansion only increases the value of collecting and remembering and transmitting environmental information because the species as a whole finds itself in a wider variety of circumstances. Enlarged environmental variation reduces the value of inflexible, predetermined responses to environmental conditions. Moreover, recall that the fitness value of a G-function depends on the nature of the competition. Once hominids crossed the threshold into a new G-function, the value of their phenotype depended on the strategies played by others sharing that G-function. There is no reason to suspect that they instantly evolved a new ESS. Further improvements could have made the first steps along the new G-function—its founding strategies—obsolete. If having more informational talents was good, having a lot more could have been even better. As a consequence, the evolutionary process would go into a state of temporary positive feedback, promoting a rapid and sustained increase both in our brain functions and our range.

Soon, the very adaptations that were allowing us to become a more tolerant species must have reversed the dominance roles of the two prehistoric hominid species. Lacking any other ecological advantage even in its now greatly diminished special habitat, the intolerant species disappeared.

What exactly was the key adaptation that gave us a new G-function and started the evolutionary conflagration? We may never know. There are so many possibilities that it is hard to identify the oldest. Perhaps the tolerant discovered tool making. Or perhaps our forebears experienced a wholly fortuitous and serendipitous breakthrough in the hominid capacity to symbolize or transmit information. Or maybe there was some subtle but monumentally important change in the way our brains process or store information. Social groups allow for collective memory and could have been all or part of the key adaptation (or they could have been merely the seedbed required to nourish it). In fact, any of these could have been the key adaptation. Any could have been the match that set our evolutionary rates ablaze.

Summary

Basic patterns of community ecology suggest new ways to look at the distributions and ecological relationships of paleohominids. At times and in regions where two such species existed, one may well have been much more tolerant of suboptimal conditions than the other. That would have allowed it to use a wider range of habitats and perhaps given it larger populations and a larger geographical range, typical of similar relationships among modern species. To maintain its existence, the less tolerant species would have had to prevail in the highest quality habitats. Such tolerance–intolerance guild relationships are common among animal species and often form the basis of their coexistence. I hypothesize that the more tolerant paleohominid species was the more gracile one.

Tolerant species should be able to adapt to deteriorating environments more readily than their less tolerant congeners. In fact, natural selection may prevent intolerants from adjusting to such deterioration and so may lock them into diminished ranges and abundances.

About the same time as a significant climatic shift occurred (ca. 2.5 m.y.), the more gracile species appears to have experienced a key adaptation. This would have effected a change in its fitness-generating equation, establishing a new guild of hominid. The change appears to have loosened the constraints on its relative brain size and subjected it to strong, perhaps even autocatalytic, evolutionary expansion of its brain. Eventually the less tolerant lost its advantage even in the best habitats and disappeared. We do not yet know the identity of the key adaptation.

Acknowledgments Ron Wooller supplied the data from his elegant study and allowed its use. Thanks to Tim Bromage and Friedemann Schrenk for inviting me and to Elizabeth Vrba for suggesting that I might be useful to the workshop. I took great pleasure in learning the latest aspects of hominid paleobiology and related paleontological matters. Thanks to the Wenner-Gren Foundation for affording us all the opportunity to visit such a significant evolutionary site as Lake Malawi. It was a thrill to view the monumental species flock of haplochromine cichlids in it and to hold in my own hands, collected from its erstwhile shore, the oldest known fossil specimen of *Homo rudolfensis*.

6

Evolution in the African Plio-Pleistocene Mammalian Fauna

Correlation and Causation

Alan Turner

How should we conduct the study of human evolution? There can be no disputing the need to examine the hominid fossils and their archaeological associations, but what of the larger context within which such evidence occurs? In their recent discussion of new hominid material from the Malawi Rift, Bromage et al. (1995b) argued for a paradigm shift in paleoanthropology away from narrower, character-based taxonomic analyses toward a wider, ecologically oriented understanding of hominid evolution. Viewed from that perspective, they suggested, habitat and habitat change in particular may be "as relevant to the characterisation of a paleospecies, its evolution, and its speciation, as are details of the simple morphological characters themselves" (Bromage et al., 1995b:105). The Malawi meeting was a logical outcome of efforts to set hominid evolution into that larger, paleoecological picture.

As a framework for the meeting, the organizers identified a short list of seven questions and asked participants to structure their contributions around them. Although several days of debate made it clear that such questions provided only a starting point for discussion that became at times both heated and wide ranging, it seems to me that they continue to offer a useful framework for efforts to combine the theoretical and the empirical, the combination with "the ultimate power to convince us of scientific reality" (Rosenzweig, 1995:5). They also offer a useful framework for attempting to integrate some of the seemingly divergent views that emerged at the meeting. By structuring this paper around those seven questions, I also hope to cover aspects of the topics that I was asked to discuss (pan-African Plio-Pleistocene faunal endemism and vicariance, habitat specificities, extinction and speciation rates, distribution drift and climatic change) before dealing in the eighth section with some of these matters within the particular context of the larger members of the Carnivora.

The Influence of Paradigm Shifts in Evolutionary Theory on Taxonomy

Any efforts to provoke a paradigm shift must be based on a clear understanding of what is being shifted from and to, and the circumstances in which the new framework may offer a true (or at least a better) picture. When this shift involves a change or a potential change in taxonomic procedure we must therefore start with a definition of the basic taxonomic unit, the species.

Taxonomy continues to be a stumbling block in paleontology, as witnessed by the differences in basic viewpoint expressed in the recent volume edited by Kimbel and Martin (1993). The essential problem may be distilled down to four main questions: what are species genetically, how does speciation occur, how do we deal with fossil taxa, and what do we do about the time element? My preferred definition of what species are is that of the Recognition Concept: a species is that most inclusive population of sexually reproducing organisms that share a common fertilization system—a definition with clear paleontological and paleoecological implications as discussed in detail elsewhere (Paterson, 1985, 1986; Turner & Chamberlain, 1989; Turner & Paterson, 1991; A. Turner, 1993, 1995b; G. H. Walter & Paterson, 1994). The advantage we gain from this particular concept is a definition (I would argue *the* definition) of the species as a genetic entity, as the effect of delimitation of the field for genetic recombination. Moreover, the concept readily leads to an understanding of speciation as an effect of disruption to, and changes in, the Specific-Mate Recognition System (SMRS) of the fertilization system. Such disruption is only likely to be achieved in a small, allopatric subset of the population, and an external, abiotic agent of change, such as climate, is likely to be the only significant means of inducing such vicariance events in most cases, as discussed in some detail by Paterson (1985, 1986). Grubb (this volume) provides a useful summary of the patterns of endemism, vicariance, and cosmopolitanism in extant eastern and southern African mammals, the end product of such earlier events. It is also worth noting that arguments in favor of sympatric speciation, whether categorized as polyploidy or competitive speciation (Rosenzweig, 1995), generally fall short of providing a significant theoretical or empirical challenge to allopatric speciation, as Paterson (1981, 1988) has pointed out.

When we turn to fossils the same definition of species must apply if we are to retain any link with the present, so that taxonomic allocation in any strict sense must be based on a single concept of the species. There is simply no point in being side-tracked into alternative concepts or definitions of species based on paleontological utility: the argument that we may benefit from such a move is seductive but will, in my view (see also Vrba, 1995a), ultimately lead us astray (for an alternative view see Foley, this volume). However, it may be extremely difficult to deal with species paleontologically, and it may be perfectly reasonable to argue that such an allocation simply cannot be achieved for certain groups where we cannot even begin to use the criteria by which we identify living species. Indeed, as has been pointed out (Turner & Chamberlain, 1989; Turner & Paterson, 1991; A. Turner, 1995b), one of the advantages of the Recognition Concept for the paleontologist is that it offers a clear basis for understanding why such problems arise and why certain fossil groups may always pose problems for the taxonomist.

At that point we may choose to side step the issue and attempt to deal with fossil data in terms of, say, ecological, morphological, or body-size groupings. We *can* do that, but only if we retain the sense of what is being grouped and only if we firmly avoid the temptation to invoke other definitions of what are then called species but which may have no point of congruence with such a genetic entity. In short, then, we may very well choose to deal with "ecocentric" groupings, but while they may be considered some kind of "taxonomic segment," they cannot be called species in any real sense. If that caveat is accepted, then I would agree with Foley (this volume) that we can use such segments in pursuing a variety of questions. Vrba (1995c) offers one such example in analyzing what she refers to as FADs (first appearance data) of *distinct bovid morphologies recognized in the African fossil record* (my emphasis). However, as she also stresses (Vrba personal communication) all such investigations are likely to be greatly enriched by well-researched phylogenetic contexts, a point well made in her own investigations of the relationships between male body weight, degree of sexual dimorphism, and extent of vegetational cover in members of the Bovidae. Only when genealogical information is included do the patterns make sense (Vrba 1994: fig. 18–9).

A case in point where one can deal with morphological groupings, and one where a close phylogenetic relationship is not in question, would be the various segments of the lineage of *Elephas recki* in the Omo sequence and Koobi Fora deposits, placed by Beden (1983, 1985) in five time-successive subspecies through extension of an earlier taxonomy proposed by Maglio (1973). This arrangement is supported by a similar pattern of morphological change with time identified by Kalb and Mebrate (1993) among Elephantinae from Awash Group deposits in Ethiopia. That lineage exhibits a clear trend toward higher tooth crowns with larger numbers of plates and thinner and more folded enamel, and was interpreted by Beden as evidence of a dentition increasingly well suited to coping with tougher vegetation in more arid conditions based on comparisons with European mammoths. Such an interpretation, and the morphological evidence for it, stands and can be addressed in relation to an overall assessment of climatically induced change within the fauna without necessarily considering the taxonomy of the elephants and deciding whether they should be identified as time-successive species or subspecies (Turner & Wood, 1993a; A. Turner, 1995a).

We can therefore look at the ecological implications of changing constellations of characters over time, provided that we do not fall into the trap of assuming that we are necessarily dealing with species when we do so. One answer to the first question would therefore be that a paradigm shift in evolutionary theory should have no effect on the practice of taxonomy from a genetic perspective, but that taxonomy (and particularly species-level taxonomy) is not everything in the fossil record, as Foley clearly underlined at the meeting.

One other aspect of taxonomy may be worth mentioning. There has been much attention in the literature to the question of species longevity in the fossil record (see, for example, Kurtén, 1959a,b). More recently still, we have seen studies of the probability of extinction with increasing age of a taxon, and the search for generalized explanations of turnover based on the concept of evolutionary rates (Van Valen, 1973). The implication, of course, is that some intrinsic property is being observed and that it would be appropriate to judge taxonomic issues within a chronological framework. This aspect of hominid evolutionary studies in particular has been examined most recently by Foley (1991) in an effort to assess the likely accuracy of various species number estimates by an entirely reasonable assessment of comparative data across diverse lineages. We can now see that any such efforts must be undertaken with great care because if speciation and extinction are externally driven by events that occur episodically (see below), it is hard to see how there can be a general expectation of any particular evolutionary rate across taxa and clades. This is so because, although there may be similarities in the tempo of speciation (and indeed of turnover, as discussed in the next section) in lineages of stenobiomic (resource and habitat specialists) as compared with eurybiomic (resource and habitat generalist) taxa, there is unlikely to be any meaningful average of either speciation or extinction rates in any unit of time across lineages and clades as a whole.

When we turn to the time element in paleontology, we face an aspect of the problem of species that still awaits a satisfactory solution. Clearly, any neontological, genetic concept of species is based on the situation in a single time slice—the present—a difficulty for paleontology recognized by Simpson (1951, 1961) with his Evolutionary Species Concept and by many such as Wiley (1978), who put forward his own amended version of it. We see something of a derivative of this approach in the Phylogenetic Species Concept, as recently summarized by Kimbel and Rak (1993) and effectively criticized by Vrba (1995a). In the end, however, these attempts to deal with time represent simply an answer to the paleontological problem of definition, as expressed, for example, by Aguilar and Michaux (1987) with their suggestion for stages in a lineage to be accorded specific rank, without providing any coherent statement of what species are genetically and therefore without any means of addressing the issue of speciation. A better approach to the question of time might be the informal recognition of chronocline segments along the lines suggested by Krishtalka and Stucky (1985) and recently supported by Martin (1993) as a "chronomorph." Such a system would have obvious application, for example, to the case of the *Elephas recki* sequence mentioned above.

Climatic Change, Turnover Pulse, and Evolutionary Theory

The causal role of climatic change in the evolution of the African mammalian fauna has been suggested several times in the past quarter century, often within the particular context of studies of hominid evolution (Vrba, 1974, 1975, 1976, 1985a,d, 1988, 1992; Coppens, 1975b, Brain, 1981a; Cooke, 1984, Prentice & Denton, 1988; Vrba et al., 1989; Turner & Wood, 1993a; Bromage & Schrenk, 1995; Bromage et al., 1995a,b; A. Turner, 1995a). One obvious way to evaluate a probable motor of evolutionary change is to examine the geographic and taxonomic makeup of the fauna over time and analyze the extent to which events clump together in the fossil record in relation to shifts in climate or major tectonic events. This approach was summarized during a review of progress in the meeting by Rosenzweig, who asked (1) Are there climatic excursions? (2) Are there pulses in the biota? (3) Do the excursions and pulses correlate? (4) What drives the correlations and why?

The idea of a "turnover pulse" was proposed by Vrba (1985d, 1988), based on her own investigations of African Bovidae. This idea suggests that lineage turnover in the form of speciations, extinctions, and dispersions should correlate across clades and regions in response to an external motor of change such as a climatic shift. More recent studies of the African Plio-Pleistocene large-mammal fauna as a whole have tended to confirm this idea (Turner & Wood, 1993a; A. Turner, 1995a), particularly in relation to a major mid-to-Late Pliocene global climatic event that occurred between 3.0 and 2.5 m.y. (Shackleton et al., 1984; Zagwijn & Suc, 1984; Bonnefille & Vincens, 1985; De Jong, 1988; Bonnefille, 1995; deMenocal, 1995; Shackleton, 1995; Denton, this volume), which is also reflected in changes to the terrestrial faunas of Eurasia (Azzaroli et al., 1988; Flynn et al., 1991; Azzaroli, 1995). Denton (this volume) has described the shift, often referred to in the literature as the 2.5 m.y. event, as "the most important climatic and environment change of the late Neogene."

It is clear from the discussions cited above, as well as from the data presented in some of these papers, that change within lineages, especially stepped change, may also be a part of the overall response to a particular climatic shift whether or not speciation is formally involved (Turner & Wood, 1993a; A. Turner, 1995a). The answer to the second of this particular trio of questions is therefore that we can define a "turnover pulse" as a correlated burst of evolutionary change across diverse clades, where the breadth of the observation enables us to rule out effects of taxonomic error or taphonomic bias. Detailed modeling of the more precise responses across lineages is clearly possible and should be sought in the contributions by Vrba (1995b, this volume), but a broad definition will suffice for our purpose here. With the turnover defined in this way, it should be possible to overcome some of the objections to the idea of a forcing role for climatic change voiced by T. D. White (1988) in particular (see also below). However, it becomes important to ask just what kind of time scale is involved in such definitions, and I shall return to this point below.

The questions of what constitutes climatic change in this context and how evolutionary theory might help us to understand its effects are interlinked. A seemingly obvious test for the causal role of climate would be to examine the pattern of evolutionary change against the independently known framework of Plio-Pleistocene glacial–interglacial stages. T. D. White (1988) has suggested that the frequency of environmental episodes alone would necessitate some level of noncausal correlation with events in hominid evolution, thus making it difficult to establish cause and effect, whereas Foley (1994) has recently proposed a test employing time bands of 100,000 and 500,000 yr, arguing that the results show no evidence of a continual pattern of climate-related change in the Hominidae. Instead, he argues, the motor of evolution must be sought in interactions between the members of the terrestrial ecosystem. However, the conclusion that climate is not causal, whereas aspects of some kind of (unspecified) community interaction somehow are, is open to question because it is unclear whether the test proposed is able to address this issue. We should not expect evolution in terms of speciations and extinctions to track relatively short-term oscillations of climate. Rather, we should expect correlations between significant steps in environmental variables and evolutionary changes in stenobiomic versus eurybiomic lineages, as discussed in detail by Vrba (1987b, 1992). Climatic change in this context therefore means such a step, and the best step that we can currently identify, particularly in the African context, is the 2.5 m.y. event.

Nevertheless, we clearly cannot expect a biota made up of different lineages and species to exhibit a simple, single response to any climatic event. As Walter and Paterson (1994) and G. H. Walter (1995) have stressed, communities (whether faunal or floristic) are made up of individual species each with its own relationship to the environment, calling into question the entire notion of a community ecology motor for evolutionary change. [Put most simply, biotic interactions are likely to be local and cannot be shown to extrapolate "across whole species or the causes of net evolution or extinction of species" (Vrba, 1993c:427).] Rather, what we should see is a series of effects logically related to the causal event. What was shown by the analysis of Turner and Wood (1993a) and A. Turner (1995a) was the widespread nature of the changes in the African terrestrial large mammal fauna following a cooling event that culminated at 2.5 m.y. but that may have begun around 3.15 m.y. (Shackleton et al., 1984, 1995; Ruddiman & Raymo, 1988, by deMenocal, 1995; Denton, this volume). Many of those changes, such as the increased hypsodonty of the *Elephas recki* lineage, are best seen between 2.4 and 2.0 m.y., but the point is that all are reasonably interpreted as responses to cooling during Denton's "most important climatic and environment change of the late Neogene" (this volume). They are not well explained by what Vrba (1985a) usefully categorized as the "weak environmentalism" of concepts such as interspecific competition, dispersal and radiation into available niches. The whole issue of community-level ecology implied by such concepts has been reviewed by G. H. Walter (1995), who argues that biotic communities in this sense have no objective reality because their delineation is arbitrary. As I have emphasized elsewhere (A. Turner, 1990, 1995a), it would be reasonable to expect a series of events following a major climatic step, as the vegetation responded and in turn provoked a reaction among the ungulates dependent on it and thus, eventually, among the larger carnivores. Only if the effect at each stage was sufficiently marked would it be felt further down the chain, a point to which I shall return later.

However, it is fair to point out that some of the evidence for vegetational change appears to be in conflict. In particular, there seem to be some differences between the vegetational patterns inferred from palynological analyses (see, e.g., Bonnefille & Vincens, 1985; Bonnefille, 1995) and those derived from analyses of palaeosol stable carbon isotopes that are used to reconstruct proportions of C_3 and C_4 plants and thus the extent of closed versus open landscapes since the Miocene (Cerling, 1992a; Kingston, et al., 1994, Sikes, this volume). Cerling and Kingston et al., in particular, use their results to argue against major steps in the transformation of the vegetation and therefore against the possibility of correlation with independently derived patterns of environmental change,

Sikes points out that events recorded in paleosols do not appear to mirror precisely those documented in the marine record, although she stresses that there does seem to be an interesting agreement between soil isotope results and those derived from African terrestrial dust plume records in marine cores at about 1.7 m.y. as shown by deMenocal (1995).

Such conflicts may be less real than is apparent once we recognize that differences in interpretation can be affected by scale and by the focus of study (see the useful discussion of this topic by Sikes, this volume). Cerling (1992a), for example, appears to cast the argument in terms of the extent of a relatively pure C_4 grassland and takes the absence of evidence for such vegetational structure until the later Early Pleistocene as significant evidence for there being no Late Pliocene event. Kingston et al. (1994) similarly argue for a mosaic of C_3 and C_4 plants since mid-Miocene times based on their data from the Tugen Hills of Kenya. However, it is not clear that any argument for a Late Pliocene climatic event as an evolutionary motor has seen the change in vegetational structure in such strict terms as woodland to grassland, nor denied that vegetational mosaics have continued to be a feature of the African biota. Vrba (1987b), for example, has stressed that any changes inferred from her studies of bovid tribal composition through time refer to relative shifts in the vegetational physiognomy and not to all-or-nothing appearances of grasslands. It is also clear from the discussion presented by Owen-Smith (this volume) that habitat fragmentation may be (and therefore is likely to have been) relatively greater in tropical as compared with temperate latitudes. In the tropics, habitat types are more geographically restricted in their possible migrational responses to climatic shifts compared with those occurring in temperate regions, which are able to shift both equatorward and/or poleward as necessary.

What is notable about the Tugen Hills study cited here is that it fails to pick up not only the 2.5 m.y. event but also any apparent perturbations associated with the Messinian crisis, a point stressed by Kingston et al. (1994), although without elaboration. What is perhaps significant, however, is the authors' passing reference in the same sentence to the "*putative* cooling at about 2.4 m.y." (Kingston et al., 1994:957; emphasis added), a stance on the issue hard to equate with the evidence for the global nature of the event cited by Denton (this volume) based on a plethora of independently derived data exemplified most recently by those of deMenocal (1995) from marine records of African terrestrial dust plumes. It may also be stressed that the localities discussed by Cerling (1992a) and cited as depleted in C_4 indicators until after 2.5 m.y. are almost entirely water margin localities, areas where C_4 grasses would surely have been relatively less frequent. That such localities do record a more open pattern of vegetation by around 1.7 m.y. in accordance with dust plume records is itself a strong indication of a wider pattern of change, the inception of which must have pre-dated events at the chosen localities.

Sikes (1994, this volume) has pointed out that paleosol stable carbon isotopes are in reality a useful method for deducing site-specific floristic microhabitat diversity, especially when other evidence is lacking (see below). Significantly, she mentions plant macrofossils within the category of "other" evidence. Plant macrofossils are considered useful precisely because they may refer to local conditions in a way that the pollen spectrum, a regional indicator, may not, but we would hesitate to infer a regional picture from plant macrofossils without a suitably dense pattern of sampling across the landscape. Ambrose and Sikes (1991) made this point in some detail in the context of studying isochronous horizons across modern and Holocene landscapes, and the potential problems posed by scale are reiterated by Sikes (1994) and discussed in some detail in her chapter in this volume. I suspect that we are at present in danger of using paleosol isotope data to make just such inappropriate larger-scale inferences and that we are getting a truer picture of regional and continental events from less tightly focused data sets such as pollen sequences or from marine records of dust plumes with their implications of changes in the continentwide pattern of aridity (deMenocal, 1995). The studies reported by Cerling (1992a) and Kingston et al. (1994) made some effort to include lateral sampling, and in time it may be possible to combine the data from several basins to provide a more regional picture, but at present the record is very spotty (Sikes, this volume). For this reason it may be a mistake for Andrews and Humphrey (this volume) to argue for an unchanging pattern of ecological opportunities from mid-Miocene to Pliocene times based on the isotope analyses of Kingston et al. (1994).

Whatever the arguments about the precise time scale of evolutionary motors, we can see why such steps are likely to be the causal agents when we turn again to evolutionary theory, a point underlined by G. H. Walter (1995; Walter & Paterson, 1994). Under the Recognition Concept, a species originates in what may then be considered its preferred habitat. In the short term, and that would include oscillations in climate on a time scale of tens of thousands of years, a sufficiently motile species is then likely to track any subsequent geographic shift in the distribution of that habitat, and speciation or extinction is then unlikely unless a subsection simply becomes isolated in conditions where enough selective pressures operate to alter the original SMRS. The subsequent return to the original conditions will bring the species back to its original area of distribution.

Of course, a species may also have enough tolerance to endure the habitat shift and may be able to adjust to features of the new environment by some other change such as increased hypsodonty without speciation necessarily occurring. But if the climatic change is a real step, a permanent, nonoscillatory shift of the kind we now see reflected at various points in the deep-sea oxygen isotope record (Shackleton, 1995), then some more fundamental change may occur because the species may simply be unable either to track the geographic shift in the habitat or to endure the change. Speciation or extinction may then follow, but the point for the paleontologist is that there will be no clear and inevitable pattern of major evolutionary change with every oscillatory climatic alteration. Walter and Paterson (1994) and Vrba (1995a) cite several published examples of such tracking responses of organisms in the fossil record, and these are also a clear feature of the European mammalian faunal record of the Pleistocene (von Koenigswald, 1992).

In other words, the pattern predicted in the fossil record is one of punctuation, with stasis of complex adaptations between speciation events (Paterson, 1985, 1986; Turner & Paterson, 1991; G. H. Walter, 1995). However, use of the term punctuation in this context need not provoke the reaction of the antipunctuationists who object to it so vociferously (Szalay, this volume). As argued elsewhere (Turner & Paterson, 1991; A. Turner, 1995b), because the original formulation of the notion of punctuated equilibrium by Eldredge and Gould (1972) did not go back to first principles and deal with the question of species, it is clear that the logical underpinnings to the notion of punctuation in their original scheme are quite different from those that stem from the Recognition Concept. For that reason, I would agree entirely with Szalay's argument in the pre-conference manuscript that "theories such as those packaged in punctuationism need to be tested against more than the few suggestive taxonomic patterns from which they are inductively derived." By approaching the problem from first principles, the Recognition Concept avoids such circularity and ensures that punctuation and turnover pulses need not therefore be seen as heretical threats to the notion of Darwinian evolution as Szalay appears to believe. That said, whether we need to be quite so slavishly fundamentalist in our evolutionary doctrinal adherence to Darwin, as Szalay has implied, is open to question.

Geographic Dispersion of Fossil Taxa

I suspect that we can achieve a broad but useful picture of distributions and their changing patterns, although the details will vary from place to place and from period to period. The fossil record is best suited to the study of patterns and the inference of processes that took place over longer time periods and across wider geographical areas: the dispersion of *Equus* is perhaps a prime example in the present context in view of its origin in the New World and subsequent immigration into Eurasia and Africa (Lindsay et al., 1980; Eisenmann, 1983; Azzaroli et al., 1988). Multiple localities with faunally diverse assemblages, where available, should ensure that local problems of accuracy in identification and dating, of the taphonomy of certain groups and any other ambiguities in the evidence can be minimized. I would thus argue that, in broad terms, we do have a sufficiently good idea of the pattern of larger mammalian faunal turnover in the Plio-Pleistocene of eastern Africa to ask sensible questions that use first and last appearances, and that this is broadly true of South Africa from the time span of Makapansgat Member 3 onwards.

I think we also have some idea of the broad regional pattern of dispersion in some elements of that fauna (Turner and Wood, 1993b; Bromage et al., 1995a). This pattern shows evidence of movement between eastern and southern Africa, although there is also a high degree of regional difference, especially in the distributions of some of the bovid taxa (Turner & Wood, 1993b). Endemism and vicariance are therefore both implicated as factors in the production of the overall pattern. What we probably do not have, however, are data suited to inferring changes in African distribution patterns on a cyclical basis. Such information is much better preserved in the European Pleistocene record, for instance, where the glacial–interglacial sequence of mammalian faunas and vegetational associations is (relatively) clear from Middle Pleistocene times onward (Huntley & Webb, 1989; Zagwijn, 1989, 1992; von Koenigswald, 1991, 1992). Nor does it seem likely that the South African data can stand on its own because the problems of chronology and the patchiness of the record are hard to overcome. Thus, although it is possible to integrate the faunal record of the southern part of the continent and to ask whether similar longer-term patterns occur there and in eastern Africa (as undertaken by Vrba in her various studies), it seems less reasonable to take the southern record and use it alone to test notions of correlation between faunal and climatic change as attempted by McKee (1995; this volume).

One aspect of dispersion that we can infer from the Recognition Concept is that it simply must have occurred at some level, since species have a localized origin, and the appearance of a new species must be followed by dispersion into what is later recognized as its range, a point emphasized by Tchernov (1992). As Grubb (this volume) stresses, any contact between sister taxa can then only be secondary, and he quotes a figure of only 18 out of 101 living African large-mammal genera having sympatric species. This implies that any new species of hominid must also have originated in a single area and that vicariance has therefore been a fundamental aspect of our distributional history. Such an expectation of a localized origin seems to be clearly met in earliest members of the genus *Homo* and is equally apparent in the distributions of species of *Australopithecus* and *Paranthropus,* which seem to have remained confined to regions of eastern and southern Africa (Turner & Wood, 1993b), a conclusion also reached by Bromage and Schrenk (1995) and supported by the work on rodent assemblages summarized by Denys (this volume). This picture is not altered by the more recent announcements of two more hominid taxa (T. D. White et al., 1994, M. G. Leakey et al., 1995), whether or not the first of the two is considered to belong to a new genus (T. D. White et al., 1995). Although it is not an issue in the context of the present discussion, it is also evident that *Homo sapiens* cannot have had a multiregional origin if it is considered to be a new and distinct species.

As a final point, it is worth stressing that vicariance biogeography based on cladograms is a major tool in any analyses of geographic dispersal patterns, and that in such cases a phylogenetic rather than a simple morphological context for the taxa employed is therefore extremely important (Vrba personal communcation).

Heritability of Habitat Specificities

It would follow from adoption of the Recognition Concept that species are habitat specific because the SMRS has originated within a certain habitat. Vrba (1995a) gives an extensive and compelling summary of the arguments for this point, and her ideas are supported by the recent review of species concepts in relation to ecological diversity provided by G. H. Walter (1995). Vrba (this volume) points out that all biological systems can be seen to function in particular habitats and to respond to changes in their environment. While the SMRS remains the same, the species will therefore be occupying the same habitat, with natural selection operating as a conservative agent. It would therefore also follow that general features adequately discriminating between living taxa occupying distinct habitats are likely to be inherited.

So far as tracing such features back into the fossil record is concerned, our attention must obviously focus on skeletal details. Where this has been attempted for African antelopes the results suggest heritability, as the vegetational habitat specificities of living reduncines and alcelaphines, for example, cluster in nested sets that match the taxonomic allocations (Vrba, 1992, 1995a). Provided we realize that we are probably looking at relatively broad aspects of habitat specificity, I see no reason that such analyses should not be conducted for fossil taxa.

Various efforts toward the determination of habitat from skeletal characters have been undertaken, perhaps most notably for equids by Eisenmann (1984), for cervids by Geist (1987) and Scott (1987), for bovids by Vrba (1984, 1985c,e 1988, 1992), Scott (1985), Kappleman (1988), and Plummer and Bishop (1994) and, more generally, for ungulates by Janis (1982) and for ruminants by Köhler (1993) and Solounias and Dawson-Saunders (1988). So far as the Carnivora are concerned, Werdelin and Solounias (1996) have distinguished six adaptive types, or ecomorphs, among the Hyaenidae, categories adopted by Werdelin and Turner (1996) in their revision of African hyaenas, to which I shall return later. All of these studies suggest that taxon-free analyses can have some advantages, perhaps allowing fragmentary or incomplete material to be used in circumstances where taxonomic allocation is difficult or even impossible, and one should not forget that such circumstances may be extremely common when dealing with fossil material.

Many of the vast range of African fossil bovid species, to take but one example, are identified cranially or dentally (Gentry, 1985a,b; Vrba, 1987a; Harris, 1991d), but allocation of isolated skeletal elements, even if complete, is much more difficult. Even incomplete material may still be identifiable as bovid, however, and might be assigned to a morphological group or to a size class. Bovid size classes have been used extensively in studies of African Plio-Pleistocene paleoecology (Brain, 1981b; Vrba, 1982, 1985e), and may even tell us something about the predatory agent of accumulation so that not only bovid paleoethology is being inferred. Moreover, even fragmentation patterns of unidentifiable bone may give some clues to the taphonomic history of the assemblage.

However, it is possible to demonstrate conflicts between various lines of evidence for the ecology of a fossil taxon and for the environmental implications that follow. Bishop (1994; this volume) has pointed to the apparent inconsistency between the evidence of open habitat exploitation indicated by the hypsodont teeth of some suid lineages in contrast with the limb-bone evidence for more closed habitats, and Plummer and Bishop (1994) have pointed to differences in interpretation that arise when Olduvai bovid limb-bones are examined morphologically in place of tribal groupings of the material. One possible explanation is that the conflicting evidence falls into two categories.

To take the suid material first, one could argue that the limb-bone morphology and proportions may, for instance, bear more strictly upon the evolutionary history of the taxon and may be constrained more by phylogenetic inertia while indicating a broadly similar habitat over that history. The form of the dentition itself is clearly genetically determined, and a steady state in dental parameters may also indicate both history and a broad similarity in habitat over time. However, the dentition, as part of the digestive system, may also be forced to respond more rapidly to changing habitats, and the suid dentitions, like those of *Elephas recki,* show changes in the degree of hypsodonty that appear to correlate with climatic shifts. In addition, as Bishop (this volume) points out, at least part of the conflict in evidence may also reflect the continuing mosaic nature of the vegetation because it is apparent that no argument for an environmentally induced evolutionary trigger is advocating the simple dominance of grasslands after the 2.5 m.y. event, as pointed out above. In the case of the Bovidae, we should bear in mind that the evidence for habitat changes is also based on changes in the dentition, including an increased emphasis on the molar sector at the expense of the premolars in the Alcelaphini (Vrba, 1984), so that the same arguments apply.

So far as hominid remains are concerned, and in particular those of the African Pliocene taxa, it seems clear that questions about the precise form of bipedalism will continue to dominate skeletally based inferences of habitat relationships and the extent of facultative or obligate arboreal and terrestrial locomotion, as seen in the recent papers by Berge (1994) and Arbitol (1995). Any inferences about locomotion, especially those implying foraging behavior in more open, savanna environments, will necessarily involve quantified considerations of thermoregulatory mechanisms of the kind undertaken by Wheeler (1994). Such calculations, although they may not permit us to identify the actual habitats that were occupied, may delimit the kind of habitats that could have been occupied with any constancy.

Ecosensitivity of Early Hominids

We have the same basic paleontological evidence from which to draw inferences about hominid ecology as we have for other groups: distributions in time and space and patterns of skeletal morphology. Thermoregulatory aspects of living in certain environments can be investigated, as mentioned above. We can investigate the extent to which hominid distributional changes match those of other taxa and draw broad conclusions as already mentioned, although we should beware of efforts to point to habitat preferences on the basis of observed distributions, as Vrba (1988) and T. D. White (1988) have pointed out (see below). Sikes (this volume) is careful to refer to hypotheses of habitat preference. We can also examine the hominid remains themselves and look at dental wear patterns and bone chemistry as indicators of diet. Any consistent match between such indicators and taxonomy or distribution patterns might then point to some aspect of habitat specialization (see, e.g., Grine, 1981; Kay & Grine, 1988; Grine & Martin, 1988b; Lee-Thorp et al., 1994).

So far as the sites of African hominid activity are concerned, the evidence derived from more traditional biotic and sedimentary studies is now augmented by studies of palaeosol stable carbon isotopes that can be used to reconstruct site-specific vegetational structure, as mentioned above and discussed in detail by Sikes (1994, this volume). As Sikes points out, this line of evidence is precise and, by virtue of its application to activity sites rather than to sites of isolated hominid fossils, capable of shedding light on longer term paleoecological patterning. But it is localized. It shows that Plio-Pleistocene hominids may not have been predominantly associated with open C_4 grassland at least until after 1.7 m.y., but while it is possible to show that localities are or tend to be located in particular vegetational settings, it is apparent that inferring constraints on their activities will require a wider investigation to establish real absence from certain habitats, work that is currently underway (Sikes, 1994).

Mention of activity sites underlines the point that hominids do have one unique attribute as fossils: a record of material culture, most notably (and most durably) in the form of stone tools. Once we incorporate these items into what we take as the fossil record we have an extra dimension, particularly for recording distributions, although the relationship between any one hominid species and any particular category or categories of tool is unclear (see below). Here we should bear in mind that hominid fossils are a relatively rare commodity, and the vast bulk of the evidence for hominid presence in Middle Pleistocene Europe, for example, is based on archaeological traces. I think there can be no doubt that the development of a material culture would have affected ecosensitivity. To take the most extreme case, it is impossible to envisage occupation of fully temperate and Arctic regions without the availability of a material culture to provide clothing, fire and food-processing capability. It is at least noteworthy that a wider range for individual hominid species, including movement both within and out of Africa, seems to have occurred not only with the appearance of the genus *Homo* but also in parallel with the development of stone tools (Turner & Wood, 1993b).

However, the extent to which the various species, not only of *Homo* but perhaps also of *Paranthropus* (Susman, 1988, 1993) may have had a lithic technology is less clear. The fact that material assigned to *Homo rudolfensis* (Wood, 1991, 1992) shows some interesting parallels in degree of hypermasticatory development with *Paranthropus* may suggest that the former taxon at least had a less sophisticated technological repertoire, or even no technology at all, and was following a completely different evolutionary path from other members of the genus *Homo* (Turner & Wood, 1993b).

Effect of Climate Change on Origins, Encephalization, and Tool Making

Climatic change has been explicitly implicated as a causal factor in hominid evolution by several authors (see, e.g., Brain, 1981a; Denton, 1985, Prentice & Denton, 1988; Vrba, 1985c, 1988). Turner (1995a) and Turner and Wood (1993a) summarized the evidence for correlated change in the African large-mammal fauna with the climatic event at 2.5 m.y., and included a number of the evolutionary events in the Hominidae referred to earlier. Vrba (1994a, this volume) has stressed the likely correlation between cooling, prolonged growth, and an increase in brain size in early *Homo,* a development with parallels in other lineages and one that she argues may prove to be a major feature of the fossil record. This point is strongly made by her illustration of hominid encephalization quotients over time (Vrba, 1994, this volume), where an event in the period between 2.3 m.y. and 3.0 m.y. marks an encephalization trend within the genus *Homo* that is profoundly different from that of the australopithecines and appears to qualify as what Rosenzweig (this volume) terms a new ecological "fitness generating function," or "G-function."

Against such a background I can see no obvious reason to reject the suggestions that the appearance of hominids and the divergences manifest by the *Homo* and *Paranthropus* clades in particular were likely to have been precipitated by climatic shifts and consequent habitat changes. It is difficult to see how such a similarity in the timing and patterning of phylogenetic events in diverse clades can have had other than a common initiating cause (a point also made strongly by Vrba, this volume) and what kind of cause other than an external, environmental shift is likely to have had the capability to operate across so much of the biota?

Once more, theoretical considerations based on the Recognition Concept of species strongly support this view. We have already seen that species are likely to have a localized origin and that vicariance into allopatric subsets of the original population is therefore the necessary precursor of speciation. It is difficult to see how such vicariance, at least for larger mammals, will have been provoked in the absence of major changes of the kind most likely to have included climatic changes involving steps, although Denys (this volume) has discussed the likely importance of tectonic events in restricting small-mammal populations.

Geographic Variations versus True Species Differences

In the case of the genera *Australopithecus* (and perhaps *Ardipithecus*) and *Paranthropus,* the available evidence suggests a geographic split, with different species in the two regions. In the case of *Homo* the evidence may imply the appearance of at least one of the various taxa now proposed (Wood, 1991, 1992) in both regions, although the pattern is far from clear. Bromage and Schrenk (1995) have proposed an interesting scenario to explain this patterning.

The case for specific similarities or differences between eastern and southern African hominids, as for all fossils, must ultimately rest on taxonomic principles that apply to such material as we have, however scarce that may be. Any scheme of taxa and their relationships is only as good as the underlying analysis and the potential tractability of the fossil samples, although all such schemes can be criticized if we show any danger of elevating them beyond hypotheses of relationships. But it is surely legitimate to examine the wider context in which hominids occur. The evidence for distributions of African Plio-Pleistocene larger mammals has been summarized elsewhere (Turner & Wood, 1993b) and suggests that there would be nothing unusual in a hominid biogeographic patterning based on true species differences despite the evidence for contacts and dispersions between the two regions throughout the Plio-Pleistocene. This view is clearly supported by studies of small mammals (Denys, this volume).

Evidence from the Large Carnivore Guild

The broad outline of the Plio-Pleistocene history of the African large carnivore guild has been reviewed elsewhere (A. Turner, 1990), as has the pattern of their distribution drift (Turner & Wood, 1993b), but a consideration of the Hyaenidae and Felidae in particular may be relevant to the present discussion. In particular, changes in some aspects of the structure of the guild may be informative.

The Carnivora as a whole have the greatest similarity in taxonomic composition between eastern and southern Africa of all the larger mammals during the Plio-Pleistocene, and this is particularly true for the Felidae. However, the distribution of the Hyaenidae between the two regions also now appears to have been somewhat more similar than Turner and Wood (1993b) thought, as more recent evaluation shows the presence of *Parahyaena* (the brown hyaena) and perhaps *Pachycrocuta* in eastern as well as southern Africa (Werdelin & Turner, 1996).

The African Felidae have a relatively good Plio-Pleistocene record but are poorly represented in Miocene deposits. Although specimens that may be referred to the living *Felis caracal* and *Felis serval* are present in earliest Pliocene deposits at Langebaanweg, the extant large cats first appear between 4.0 and 3.0 m.y. By around 1.5 m.y. the living species are the sole representatives of both families (Turner, 1990), but the intervening period (effectively the Pliocene) had a diverse range of machairodont (*Megantereon* and *Homotherium*), false machairodont (*Dinofelis* spp.) and conical-toothed extant feline taxa (*Panthera* spp. and *Acinonyx*). Over much of the same period, the living hyaenas were augmented by the presence of the gigantic *Pachycrocuta brevirostris* and the so-called hunting hyaenas of the genus *Chasmaporthetes.* As many of the machairodont cats were probably best adapted to ambush-type hunting in more closed terrain (Turner & Antón, 1997, in press), the demise of such taxa by 1.5 m.y., together with the disappearance of at least two hyaenid taxa, may well be telling us something about changes in the structure of the vegetation as well as the taxonomic and structural character of the prey fauna by that time. Evidence of changes in the Bovidae provided by Vrba (1995c) shows more open-country groups entering the record after 2.0 m.y., most notably Caprinae of the tribes Ovibovini and Caprini. In this context the fact that palynological data (Bonnefille, 1995) as well as both paleosol stable isotope research and ocean records of aeolian dust indicate local, regional and global changes pointing to increased aridity and more open vegetation around 1.7 m.y., as summarized by Sikes (this volume) attains considerable significance.

This significance may be enhanced, and rescued from dismissal as simply another artifact of correlation, when we realize that a similar sequence of structural alteration occurs in the large carnivore guild of the European Middle Pleistocene in conjunction with climatic and ungulate faunal changes. It takes place at a different time, but it has interesting parallels with the African pattern. Among the European Cervidae and Bovidae, for instance, there is a marked increase in the incidence of larger taxa in the period after 1.0 m.y. within a much larger-scale faunal turnover often referred to as the "end-Villafranchian" (Azzaroli et al., 1988; A. Turner, 1995c). This occurs around a time, 0.9 m.y., marked by a further major climatic event with the shift to dominance of the 100 kyr component of the astronomical cycle and the transition to the full glacial–interglacial sequence of the last several hundred thousand years (Shackleton, 1995; Denton, this volume). The machairodont cats of the genera *Megantereon* and *Homotherium,* together with the giant European cheetah, *Acinonyx pardinensis,* the European jaguar, *Panthera gombaszoegensis,* and two large hyaenas, *Pachycrocuta brevirostris* and *Pliocrocuta perrieri,* all disappear from the European fauna by 0.5 m.y., leaving a large carnivore guild quite similar in taxa, and markedly similar in its structure, to that of Africa today (A. Turner, 1992a, 1992b, 1995c). These changes, like those in Africa, are not reversed, and the European ones occur separately from the background sequence of glacial–interglacial faunas that becomes a marked feature of the later Middle and Late Pleistocene there. What we appear to be seeing is similar kinds of events in different parts of the world, sometimes involving the same widely distributed taxa but essentially involving structural changes in predators which are responding to changes in the structure of the prey fauna triggered by vegetational changes and, ultimately, by climatic shifts. These two occur at different times, although there are other important changes in the European biota at 2.5 m.y. and between 2.0 and 1.5 m.y. (Azzaroli et al., 1988; A. Turner, 1992b, 1995c; Azzaroli, 1995), but they have been highlighted here because together they throw light on an important change in the structure of the large carnivore guild in both continents.

I want to end this section by taking a somewhat wider perspective provided by the African Hyaenidae. The family has a long but patchy African record, with some 17 taxa now identified (Werdelin & Turner, 1996). Like the cats, they are poorly represented in Miocene deposits, where they are first recorded around 14 m.y. at Fort Ternan, but they appear in greater numbers in Pliocene localities (Werdelin & Turner, 1996). The patchy nature of the record is obviously a problem at some levels of resolution, particularly if one is concerned with detailed taxonomic issues or with distributional information. Many of the known taxa are present only at a few widely dispersed sites, and at this stage we cannot even be clear about the direction of original dispersions of the Hyaenidae. But what is available offers a good illustration of the potential of the fossil record to deal with some larger questions in taxon-free analyses. Six adaptive types have been distinguished in the family as a whole by Werdelin and Solounias (1991, 1996), which may be summarized as: (1) civetlike insectivore/omnivore; (2) mongooselike insectivore/omnivore; (3) jackal- and wolflike meat and bone eaters; (4) cursorial meat and bone eaters; (5) transitional bone-crackers, and (6) fully developed bone crackers.

The point of interest for those involved in evolutionary studies is that the majority of Miocene hyaenas throughout the Old World belong in types 1–4, whereas modern hyaenas nearly all belong in type 6. In Eurasia, the transition occurred at the Miocene–Pliocene boundary when nearly all type 1–4 hyaenas (and some type 5 taxa) became extinct. This pattern also holds true for Africa, where the earliest hyaenas belong to types 1 (*Protictitherium*) and 2 (*Proteles* lineage), whereas Late Miocene faunas are dominated by type 3 (*Ictitherium* and *Hyaenictitherium*) and 4 (*Lycyaena*) forms. However, Africa does differ somewhat from Eurasia in the continued presence of type 3 (*Hyaenictitherium*) and primitive type 4 (*Hyaenictis*) hyaenas in the earliest Pliocene at Langebaanweg (Werdelin et al., 1994). Nevertheless, these forms soon disappeared, and the link between Langebaanweg hyaenas and those of younger sites such as Laetoli lies in the derived type 4 (*Chasmaporthetes*) and perhaps type 5 (*Ikelohyaena*) taxa.

One notable feature of the African Hyaenidae section of the guild pointed out by Werdelin and Turner (1996) is the absence of a large, bone-cracking (type 6) hyaena from sub-Saharan localities of the earliest Pliocene. The ubiquitous Eurasian species *Adcrocuta eximia,* the likely sister taxon of the spotted hyaena, *Crocuta* (Werdelin & Solounias, 1991), and a robustly built, bone-cracking species in its own right, is known only from Sahabi. The living *Crocuta* only appears after 4.0 m.y., some time before the known appearance of *Pachycrocuta, Parahyaena,* and *Hyaena,* the striped hyaena. This absence of a Miocene–Lower Pliocene bone-cracking form among sub-Saharan hyaenas is particularly evident in the composition of the Langebaanweg hyaenid guild, with its type 4 and type 5 species but the complete absence of any type 6 taxon.

This absence casts considerable doubt on traditional, community co-evolution interpretations of hyaenid evolution, which have stressed the role of the sabre-toothed cats in creating an "empty niche" for scavengers able to exploit the nutrient-rich carcases likely to have been left by such specialist species (Ewer, 1967). It seems clear that the presence alone of sabre-toothed and false sabre-toothed cats in the Miocene of Africa, felids which are well-represented at Langebaanweg, was not sufficient to induce the local evolution of a major bone-cracking hyaenid as Ewer had suggested. The "empty niche" explanation for evolutionary change and speciation seems scarcely compatible with the direct evidence of the fossil record, which implies that the evolution of the Hyaenidae in both Africa and Eurasia was part of a larger-scale response to changes at the Miocene–Pliocene boundary.

Summary

It is clear from recent developments in evolutionary theory that a strictly taxonomic treatment of paleontological specimens must logically attempt to deal with real species founded on a genetically based concept. The only real candidate is the Recognition Concept, which offers a definition of species based on the delimitation of the field for genetic recombination. However, it may prove difficult to apply any such neontologically derived definition to fossil material—the concept is not a panacea, although it does help to clarify the reasons the species in particular clades or lineages may be more or less easily recognizable to the external observer. Nevertheless, any attempt to redefine the species at that point on the grounds of paleontological utility will be fruitless because there will then be no possibility of comparing like with like in seeking to understand "one history and one unified set of processes" (Vrba, 1995a:29).

Efforts to side step the problem of fossil taxonomy and use taxon-free groupings can successfully employ ecological categorizations, but only if it is clearly understood that any such categories are not species. If this caveat is borne in mind we may then look at larger patterns of evolution (i.e., extinction, within-species change, and perhaps speciation) in relation to environmental and habitat shifts because speciation and extinction in particular are only likely to be induced by such shifts via vicariance events.

Speciation under the Recognition Concept will occur in allopatry, and such vicariance is only likely to be induced by significant habitat changes resulting from environmental shifts of a steplike nature. While relatively localized geological processes may induce vicariance in some given instances, any larger events that operate across a diversity of lineages over a wide area are most likely to reflect wider scale steps in climate and concomitant effects on habitat distribution. In Africa in particular, with its unique distribution of land across the equator, the existing pattern of habitat mosaicism identified as a feature of tropical latitudes by Owen-Smith (this volume) is likely to have been altered only by such a major step. Shorter term cyclical swings in climate are more likely to have induced similarly cyclical changes in the distribution of habitats and of the component members of ecological communities (Walter & Paterson, 1994), without necessarily inducing speciation or extinction, and efforts to "test" the causal effect of climatic changes on evolution by reference to such shorter term, cyclical events are therefore inappropriate. It is in this context that stepped global changes such as the 2.5 m.y. event are highly significant—Denton's "most important climatic and environment change of the late Neogene." The fact that evolutionary events across a wide spectrum of the African biota, including large and small (Denys, this volume) mammals correlate in a logical manner with this shift cannot be dismissed as insignificant, particularly in view of the fact that similar patterns of change may be seen in contemporaneous Old World terrestrial faunas of Eurasia (Azzaroli et al., 1988; Flynn et al., 1991; A. Turner, 1995a). To return to Rosenzweig's point quoted at the beginning of this chapter, here we are able to combine the theoretical and the empirical with testable predictions in a way that advocates of biotic interaction (Van Valen, 1973; Foley, 1994, this volume) or autocatalysis (McKee, 1995, this volume) as mechanisms for speciation have so far been unable to do.

Habitat specificity is an inherent character of all species and appears to be reflected in certain skeletal elements on the basis of studies of living and fossil antelopes, equids, and cervids. Taxon-free analyses based on morphometric characters of families or guilds may therefore produce useful ecological insights, even down to the level of fragmentation patterns in bone unidentifiable to lower taxonomic levels, provided that we can tease apart the effects of phylogenetic history and more direct adaptations to contemporaneous conditions. However, this habitat specificity is acquired by the species as a result of the allopatric nature of the speciation process and stems from the fact that species are an effect of local adaptive evolution to an environment different from that of the parent population. Speciation and habitat specificity do not occur because the species has only arisen through the occupation of an empty niche (contra Mayr, 1963), a concept of dubious utility closely tied to notions of competitive exclusion and the alleged need for any new species to be protected from competition with other species (G. H. Walter, 1995).

So far as early hominid evolution in Africa is concerned, it seems that a paradigm shift toward a more ecologically and paleoecologically oriented perspective (if not understanding) has indeed occurred and has been in the process of doing so for the past 20 years or so since Vrba (1974) and Coppens (1975b) expressed their independently derived views. Of course, it is true that other components of the biota have always been included in multiauthored overviews and conference proceedings, although they have generally taken their place largely as providers of a dating or paleoenvironmental framework. They have thus remained as much set apart from the study of the hominids as, say, the radiometric datings or the sedimentological analyses (see, e.g., the contributions in the volume edited by Bishop and Clarke (1967) with the significant title of *Background to Evolution in Africa*). Even since Coppens and Vrba wrote, it is only as potential prey or predators that the African Plio-Pleistocene mammals have, with notable exceptions (e.g., Vrba, 1985c,e, 1988, 1994a; Foley, 1987, 1991, 1994), been admitted to discussion alongside the usually less abundant but allegedly more interesting hominids. I think that has now changed. The publications of the Malawi Rift Project (Bromage & Schrenk, 1995; Bromage et al., 1995a,b) set the entire discussion of the new hominid material firmly within a paleobiogeographic framework. The scope of that framework may be judged by the fact that Wood and I (Turner & Wood, 1993b) listed only seven eastern African hominid species in a total large-mammal fauna of 144 taxa for the period between 4.0 and 1.0 m.y., whereas Vrba (this volume) lists 147 bovid species alone over the past 7.0 m.y. in her study of taxon durations. Hominid taxon numbers may have been augmented slightly within the past year or so, but they are still a small fraction of the known fauna. It is now abundantly clear that we can only begin to evaluate and understand the pattern of hominid evolution if we make correct use of that larger data set, a point amply confirmed by the Malawi meeting.

Acknowledgments I am grateful to Tim Bromage and Friedemann Schrenk for their invitation to attend the Malawi meeting and to the Wenner-Gren Foundation for financial support and superb organization. I thank Laura Bishop, Christiane Denys, Nancy Sikes, and Elisabeth Vrba for helpful discussions and for useful comments on and corrections to earlier versions of the manuscript; the mistakes that remain are my own. Much of the research that underlies this contribution was undertaken with grants from The Leverhulme Trust.

II

GEOLOGY, ECOLOGY, AND BIOGEOGRAPHY

7

Introduction

Jonathan Kingdon

By its very nature, anthropology is anthropocentric. So are the derogatory implications of anthropocentrism in abeyance, excusable, or absent for every dimension of hominid prehistory?

In the chapters that follow the answer is an emphatic no. Plea bargaining will not deliver a special status for hominids within the complex ecosystems in which they evolved. Thus anthropologists cannot view exotic landscapes, unfamiliar ecosystems, and climatic changes as the mere stage setting for an ongoing soap opera, the story of human origins. Instead, they must engage in the unfamiliar task of finding what patches, within the vast, disintegrating ecological tapestries of Africa, might have been habitable for hominids and what climates might have triggered their rapid evolution over the last several million years. There is the humbling challenge of reconstructing archaeo-niches for a handful of interacting hominid species and other organisms. Fortunately, the task may well be narrowed by the supposed narrowness of early hominid niches, but there is still a substantial mismatch between our preoccupation with extricating a few rather insignificant species from their parent matrices, which, for the most part, continue to be all but ignored.

The authors of the chapters that follow can but hint at the enormity of the tasks ahead. They are specialists in the dynamic of climate, phytogeography, and the distribution and ecology of living mammals. All are involved in the larger quest of finding more accurate and less biased interpretations of physical and biotic patterns, both global and African. For anthropologists the value of these chapters may lie as much in the invitation to look through specialists' eyes and adopt their less anthropocentric viewpoint as in any specific viewpoints they offer. As the authors are at pains to explain, their papers are tentative and exploratory, but they are also conscious of being pioneers of a new orientation in the study of hominids. The emergent facts of global climatic history and the surviving relics of Africa's biota deserve as much to be treated as sources of hard evidence as do surviving fossils and artifacts. Both are fragmentary and incomplete, but study of the former has lagged far behind, in spite of being the greater part of the phenomena that interest us all so much—the processes that generated hominids and, ultimately, ourselves.

The historical reasons for our love affair with fossils and our neglect of the physical world in which they evolved are obvious and need no elaboration here except to remark that intense public excitement over new discoveries, especially skulls, has led to strong shifts of interest, even within the anthropological community. Sometimes the focus has been geographical—Europe to China or Java, South Africa, Olduvai or Hadar. Sometimes the emphasis is on erects or Neanderthals, then the pendulum swings to australopithecines or moderns. These will-of-the-wisp enthusiasms seem to have made it especially difficult for our science to develop programs that address perennial problems. A preoccupation with proximate phenomena and immediate causes aggravates our still poor grasp of evolutionary and ecological processes.

A skull or a skeleton is one comprehensible, material entity that is open to a limited number of interpretations, using some well-tried methodologies, which include sophisticated methods of description and well-developed procedures for relating form to function. A much more primitive state of affairs pertains when it comes to studying such a skeleton's ecological pedigree. As the product of numerous interlocking environmental and biological processes, this presents a more difficult challenge, and furthermore it requires skills still in their infancy. For example, who could have predicted that the evolution of *Homo* in Africa might have been less influenced by geological activity in the Rift Valley than by tectonic uplift on the floor of the North Atlantic? The possibility that changed Atlantic water circulation may have been the trigger for global cooling (and hence human adaptation) between 2.95 and 2.52 m.y. is but one of the startling new ideas presented by Denton. As the ultimate motor for environmental and adaptive change, the dynamics of climate deserve and have received serious treatment in this volume.

The chapters that follow are part of a growing effort to find new, more truthful paradigms to explore hominid origins. Some are also syntactical and conceptual experiments. Doubtless, many of these conceptual experiments will disappear, just as Esperanto has, but along the way we will have learned which paths lead nowhere and which advance our understanding. More important for the subject at hand is the emphasis given to physical and ecological context. These are essentially precursor essays that tell us that sooner, rather than later, a lot more attention is going to have to be directed at details of geophysiology and to an ambitious program of ecological analysis.

That changing climates and environments might elicit corresponding adaptive change in organisms is a pre-Darwinian observation. Subsequently, its development as a thread of evolutionary thinking has progressed by fits and starts, and its systematic application to the slim record of hominid evolution is quite recent. The essentially new recognition that every major climatic pulse of the Plio-Pleistocene was likely to have precipitated equivalent pulses in the evolution of many organisms is now a concept so pregnant with possibilities for biology that it is a major preoccupation in this volume. Denton shows that the reasons for climatic changes are still matters of intense controversy, but there is much less ambiguity about the dating techniques which invite the application of new dates to a variety of older biological landmarks. For example, Denton notes how the relatively mild climates of the mid-Miocene ended with progressive glaciation in the Northern Hemisphere between 7 and 6 m.y., a period of some importance for the emergence of new radiations and close to the putative date for the *Pan/Homo* divergence. It is going to be important that these seductive new techniques become applied to, and do not distract from, the detailed reconstruction of credible African paleo-environments that would have generated new species, including hominids, rather than provide an ornamental backdrop for them.

Because place, habitat, climate, and diet are inseparable parts of an animal's day to day existence, it is refreshing to find dietary potentials (expressed in the relative ratios of edible-fruit-providing plants in different African vegetation formations) put in a biogeographic context. This is but one innovation in O'Brien and Peters's discussion of African landscapes. They do not pursue the discussion of diet in any detail, but they flag the importance of feeding opportunities for our understanding of early hominids. They ask about the importance of kinds of foods, the ecology of the places in which they were found, changes imposed by seasons, specific or class preferences, and, perhaps most important, the techniques by which they might have been collected.

It is seldom appreciated that the broad spectrum of dietary opportunities that existed in the Late Miocene can still be retrieved from extant environments because, although plant communities have changed, they have retained many more Late Miocene elements than animal communities have. Plants provide evidence every bit as hard as the hardest fossil. It is, of course, important to remember, as Owen-Smith reminds us, that fire or more/less rainfall have, almost certainly, become more pervasive than they were in the past. Notwithstanding such qualifications and fluctuating boundaries, the spectrum of African ecotypes that is strung out, even today, between the extremes of desert and rainforest is of very long standing.

O'Brien and Peters are pioneers in presenting biomes as the generators, not the mere settings, of the early hominids, and this attitude leads to special insights. For example, they identify the Zambezian regional center of endemism as having conditions particularly favorable to early hominid biodiversity. I am personally convinced that this has to be true, although the particular forms of hominid that it might have nurtured is a topic that will exercise students' interest for many years to come.

For the very earliest stages of hominid evolution, there is no alternative to the reconstruction of paleolandscapes and their inhabitants. There is a hiatus, for fossils, of more than 3 m.y. between the putative *Pan*–

Homo divergence (if the molecular clocks are right) and the earliest hominid fossils. It is conceivable that no fossil from this period will ever turn up. Furthermore, it is urgent that we develop much better analytical skills for interpreting ecosystems than those we practice at present because many ecotypes will soon be as extinct as the hominids that once evolved in equivalent "chrono-landscapes."

Nowhere is the urgency more apparent than with the mammal groups that are most diagnostic for Grubb's analyses: colobus and guenon monkeys, drills, gorillas and chimpanzees, duikers, buffaloes, and elephants. All are now under threat from a burgeoning bush-meat trade; there will be little point in arguing for dispersal versus stasis if the mammal populations under discussion (and the only ultimate arbiters of the argument) have all ended up in the sewers of big cities and towns. The survival of Grubb's model, which he calls "periodic," will be as precarious as the rare and endangered species on which it is based unless scientists are willing to express their concern.

Grubb's discussion of mammalian distributions provides both a rationale and some evidence for major dispersions in their past history. At the broader level (and of relevance for the proliferation of hominids), ebbs and flows have been part of an entire continental biota's response to the climatic vicissitudes of the past. We need reminders that there are many more major and significant events experienced in the real history of any population than our reconstructions can ever hope to encompass. Every population has suffered irregularities or fluctuations, whether these derive from cycles in the environment, in climate, in predation, in competition, or in diseases. Grubb calls for a more holistic, perhaps more traditional approach to these questions. I hope too, that while we move toward more effective, eloquent, and more multifaceted explanation, there will remain a healthy awareness of the gap that must always remain between models and the realities they seek to mirror and a skepticism, too, about the capacity of language to measure up to the phenomena it attempts to grapple with. We owe it to our descendants to preserve as much of the Pleistocene landscapes of Africa as is possible. They will not forgive us for the clumsiness of our first steps at understanding our prehistory if we also stood by and let Africa's greatest assets, their inheritance, be consumed.

8

Cenozoic Climate Change

George H. Denton

The hypothesis that climatic changes initiated major biotic evolutionary events carries the prediction that patterns of macroevolution should display temporally concentrated pulses correlated with climatic and environmental change (Vrba, this volume). For example, Vrba (1985b,c,d) argued that turnover pulses of many lineages of mammals were triggered by major Cenozoic environmental changes. One such major burst of evolution was roughly coincident with a marked mid-Pliocene paleoclimatic change that culminated about 2.5 m.y. when Africa became cooler and drier (Vrba, 1975). One possible conclusion from Wood's (1992a) results is that two hominid lineages evolved from *Australopithecus afarensis* about 3.0 m.y. near the beginning of the mid-Pliocene climatic deterioration. One of these lineages led to *Homo,* the other to the robust australopithecines. In this regard Bromage et al. (1995b) suggested that *Homo rudolfensis* and *Paranthropus aethiopicus* evolved from *Australopithecus afarensis* prior to or about 2.5 m.y. If climate changes were indeed coeval with evolutionary pulses, then this opens the possibility that the basic mechanisms driving paleoclimate and evolution may be linked. As a background for testing this hypothesis, I outline here the major Cenozoic climatic changes that culminated in a succession of ice ages beginning in mid-Pliocene time. I divide the discussion into two parts. One part involves long-term Cenozoic climatic cooling; the other part concerns high-frequency oscillations superimposed on this long-term cooling trend.

Cenozoic Climatic Cooling

Before middle Eocene time the planet was much wetter and warmer than at present, climate was more evenly distributed with latitude, and evergreen and deciduous forests were widespread (Ruddiman & Kutzbach, 1991). Trees grew on Arctic land masses, and temperate rainforests even existed in Antarctica (Askin, 1992). Glaciers and sea ice were limited in extent. Deserts and grasslands were rare because of a general lack of aridity and strong seasonality (Ruddiman and Kutzbach, 1991). The equator-to-pole temperature gradient in the atmosphere was much smaller than that of today. Deep-water temperatures in the world's oceans were about 10°C warmer than those of today (Savin et al., 1975; Shackleton & Kennett, 1975a,b; Shackleton & Boersma, 1981), and extensive bottom water may have been produced by the sinking of highly saline waters formed in low-latitude marginal seas (Brass et al., 1982).

As pointed out by Ruddiman and Kutzbach (1991), such a warm, wet, equitable climate now persists only in limited geographic areas in the tropics, southeast Asia, and the U.S. Gulf Coast. The planet has become cooler since middle Eocene time. Deserts are now common. Many regions show annual extremes in rainfall. Sea ice is widespread in polar regions, and immense ice sheets cover Greenland and Antarctica. In the oceans deep-water temperatures are cold, and most bottom water originates at high latitudes. The planet

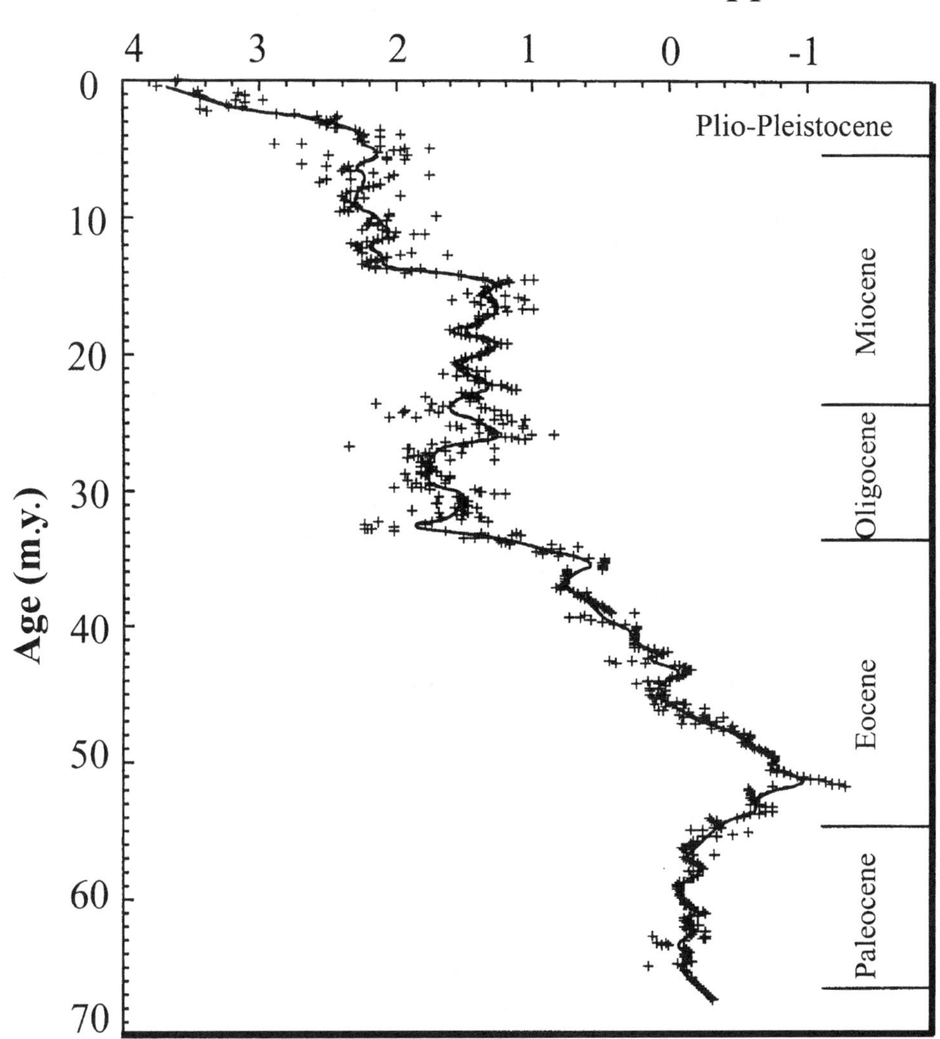

Figure 8-1. The benthic $\delta^{18}O$ record for the past 70 m.y. compiled from Deep Sea Drilling Program sites in the Atlantic Ocean. The long-term increase in $\delta^{18}O$ is due primarily to cooling of the deep ocean and to the growth of continental ice sheets. Adapted from Miller et al. (1987), with time scale from Berggren et al. (1995).

now rests in an intermission within a long train of ice ages that began with an important global paleoclimatic change at 2.95–2.52 m.y. (Shackleton, 1995). During each pronounced glacial maximum of the past 0.95 m.y., the planet was drier and on average about 5°C colder than at present; large ice sheets developed in North America and Eurasia (Denton & Hughes, 1981); Africa was drier than now; and tundra extended over much of northern Europe.

Thus Cenozoic environments have undergone major changes that culminated in a succession of ice ages coeval with the emergence of humans in Africa and their subsequent spread to Eurasia and the Americas. The long interval between the warm Cretaceous Period and the ice ages of the Late Pliocene and Pleistocene epochs was dominated by a cooling trend that began about 51 m.y. and was punctuated by three major, steplike events. This Tertiary cooling trend is depicted by several paleoclimatic indicators, the most common of which is shown in figure 8-1, adapted from Miller et al. (1987). Figure 8-1 shows a composite of the $\delta^{18}O$ record of benthic foraminifers from deep sea drilling sites distributed over the Atlantic Ocean. This record can be interpreted as a decrease in deep-water temperatures, combined at times with the buildup of continental ice sheets (Miller et al., 1987). Much of the decline in deep-water temperatures occurred in three major steps, each involving a cooling of 2°–3°C (Broecker, 1995a); one was at about 33 m.y., another at 15.6–12.5 m.y., and the third at 2.95–2.52 m.y. I now discuss several aspects of this isotopic record to illustrate the major events between Cretaceous and Pleistocene times.

The Cretaceous climate, particularly the warmest interval between 120 and 90 m.y., is of special interest because the earth at this time is widely considered to have been ice free (Crowley & North, 1991). Dinosaurs lived north of the Arctic Circle (Colbert, 1973), and coral reefs and floral provinces extended 15° latitude farther poleward than now (Barnard, 1973; Habicht, 1979). Coal and bauxite deposits also occurred farther poleward of where they now exist (Crowley & North, 1991). Sea level was about 300 m higher than it is at present and, as a result, epicontinental seas were widespread (Kominz, 1984; Haq et al., 1987).

Climate was still equitable at the time of the Eocene thermal maximum, which was the warmest time of the Cenozoic Era (fig. 8-1). Tropical conditions then extended to about 45° N paleolatitude, on the basis of rich floras in western North America and in the London Clay (Reid & Chandler, 1933; Wolfe, 1980). Laterite developed to 45° paleolatitude in both polar hemispheres (Frakes, 1979). High-latitude warmth is indicated by the finds of fossil alligators and flying lemurs in Ellesmere Island at a paleolatitude of 78° N (Estes & Hutchinson, 1980). Vegetation featuring *Nothofagus* and podocarpaceous conifer as prominent components existed in far-southern Antarctic latitudes (Truswell, 1991; Askin, 1992). The remainder of Eocene time after this thermal maximum was marked by a slow but significant cooling trend in which the temperature of near-surface waters close to Antarctica dropped to 6°–10°C by 40 m.y. and to 5°–8°C by 36 m.y. (Mackensen & Ehrmann, 1992). Ice caps developed in East Antarctica and in places reached the sea by 45.5 m.y. on the Berggren et al. (1985) time scale (Ehrmann & Mackensen, 1992; Mackensen & Ehrmann, 1992).

The earliest Oligocene isotopic event, so well marked in figure 8-1, is the most significant of the Cenozoic. From the percentage of plant species with entire-margin leaves, Wolfe (1978) estimated a drop in mean annual atmosphere temperature of 12°C in North America in earliest Oligocene time, by far the largest change in the Tertiary. Accompanying this marked climate deterioration was one of the greatest faunal turnovers of the Cenozoic (Corliss et al., 1984; Raup & Sepkoski, 1986; Stanley, 1988). A continental-scale ice sheet emerged for the first time in East Antarctica at about 35.9 m.y. on the Berggren et al. (1985) time scale or 33.6 on the newer Cande and Kent (1992) and Berggren et al. (1995) time scales (Hambrey et al., 1991; Ehrmann et al., 1992; Ehrmann & Mackensen, 1992; Hambrey & Barrett, 1993), accompanied by the development of colder and more vigorous deep-water circulation (Kennett & Shackleton, 1976). This striking event is registered by an abrupt increase in marine benthic $\delta^{18}O$ values (fig. 8-1) (Shackleton & Kennett, 1975a,b; Zachos et al., 1996) that is correlated with a sharp pulse in the deposition of ice-rafted sand and gravel on the Kerguelen Plateau near Antarctica (Ehrmann & Mackensen, 1992; Mackensen & Ehrmann, 1992; Zachos et al., 1992). This ice-rafting pulse was accompanied by a change from smectite-dominated mineral assemblages to illite- and chlorite-dominated assemblages, indicating a switch from temperate to cooler climates (Ehrmann & Mackensen, 1992). A similar timing for the onset of continental glaciation comes from the age of waterlain till in cores from the Antarctic continental shelf (Barrett, 1989; Ehrmann et al., 1992). The volume of the continental ice sheet that developed in East Antarctica in earliest Oligocene time was at least 50% of the volume of the present-day East Antarctic Ice Sheet (Zachos et al., 1996). That this early ice sheet was temperate rather than polar in character until at least the late Oligocene is indicated by the persistence to that time of *Nothofagus* in Antarctica (Barrett, 1989; Askin, 1992) and by the relatively warm surface waters of the surrounding Southern Ocean (Ehrmann & Mackensen, 1992).

An important question is whether the volume of the Antarctic ice sheet diminished substantially between early Oligocene time and the middle Miocene thermal maximum at 17–15 m.y. Ehrmann et al. (1992) did not find evidence for major fluctuations or disappearance of the ice sheet during this interval. From drilling records in Prydz Bay on the Antarctic continental shelf and in the southern Indian Ocean, they postulated that the East Antarctic ice sheet remained as a permanent feature after it initially formed in earliest Oligocene time. However, a synthesis of the Oligocene benthic foraminiferal $\delta^{18}O$ record strongly suggests changes in the global ice budget (Miller et al., 1991). Evidence from Miocene sequence stratigraphy on the U.S. middle Atlantic coastal plains (Miller & Sugarman, 1995; Miller et al., 1996), along with coeval changes in the marine $\delta^{18}O$ records (Miller et al., 1991, 1996), indicates eustatic sea-level changes.

Likewise, high-frequency sea-level changes are recorded in the Miocene sequence stratigraphy of shelf depositional systems of the northwestern Gulf of Mexico (Ye et al., 1993, 1995). It is highly probable that the Antarctic ice sheet was involved in these sea-level oscillations.

During the Early Miocene climax of Neogene warmth from about 17–15 m.y. (Kennett, 1995), marine $\delta^{18}O$ values were relatively low (Miller et al., 1987). This Neogene climatic optimum is widely recognized from molluscan and foraminiferal assemblages in southern latitudes (Hornibrook, 1992), from planktonic foraminiferal fauna (Jenkins, 1973), and from South Pacific and Atlantic calcareous macrofossil assemblages (Edwards, 1968; Haq, 1980).

Figure 8-1 shows an important $\delta^{18}O$ event in the deep ocean in the early phase of middle Miocene time between 15.6 and 12.5 m.y. (Miller et al., 1991). The total benthic $\delta^{18}O$ increase associated with this event took place in two discrete steps, one at 14.5–14.0 m.y. and another at 13.5–12.45 m.y (Flower & Kennett, 1993a). Previously, this two-step event was thought to be linked with the establishment of a permanent polar ice sheet in East Antarctica (Shackleton & Kennett, 1975a,b; Kennett, 1977). But such a linkage could not have been the case if one interpretation of the evolution of the Antarctic ice sheet is correct (Denton et al., 1993). By this interpretation early temperate conditions in East Antarctica (including a continentwide temperate ice sheet) had been replaced by polar environmental conditions (including a large polar ice sheet) before the end of Early Miocene time. The change from a temperate to a polar ice sheet left small sectors of the Transantarctic Mountains free of ice under hyperarid polar conditions. The $^{39}Ar/^{40}Ar$ chronology of in situ (or near in situ) volcanic deposits shows that ice-free sectors of southern Victoria Land have not been covered by the Antarctic ice sheet since the end of Early Miocene time (Denton et al., 1993; Marchant et al., 1993a,b,c). These areas are in key locations deep in the Transantarctic Mountains between the huge East Antarctic and the much smaller West Antarctic ice sheets. The East Antarctic ice sheet could not have exceeded its current thickness without affecting these areas, nor could grounded ice have extended into the Ross Embayment any more than it did at the height of late Quaternary ice ages. Not only were these ice-free areas not covered by glaciers since Early Miocene time, but there are numerous indications that polar desert conditions extended back into Early Miocene time (Denton et al., 1993). Overall, these new Antarctic results imply that the important early middle Miocene marine $\delta^{18}O$ events between 15.6 and 12.5 m.y. could not have featured the buildup of a large polar ice sheet, as is commonly assumed. Rather, the prior existence of polar climate implies that a polar ice sheet, not far different in dimensions from that of today, already existed in Antarctica before the key middle Miocene isotopic events. The apparent reaction of this ice sheet at the time of the middle Miocene $\delta^{18}O$ events was to expand into the Ross Embayment (and presumably the Weddell Embayment) in response to sea-level lowering (Anderson & Bartek, 1992). This grounded ice backed up Transantarctic outlet glaciers which then existed in Wright and Taylor valleys, resulting in the deposition of the Asgard and Quartermain tills in the western Dry valleys between 15.2 and 13.6 m.y. (Denton et al., 1993; Marchant et al., 1993a,b,c). The limitations on ice-sheet dimensions afforded by the $^{39}Ar/^{40}Ar$ chronology of volcanic deposits in southern Victoria Land suggest that middle Miocene expansion of Antarctic ice grounded in the Ross and Weddell embayments could have caused relatively modest sea-level lowering.

The implication of these Antarctica data is that a large fraction of the middle Miocene marine isotopic events represents a major change in deep-water temperature or circulation rather than a buildup of a polar ice sheet in East Antarctica. Kennett (1995) argued that these isotopic events marked a critical threshold in the evolution of Cenozoic environments, namely, the onset of the modern mode of ocean circulation of late Neogene time, dominated by high-latitude deep-water sources and strong meridional and vertical thermal gradients. As a result, the characteristic late Neogene deep-sea benthic foraminiferal assemblage became established (Woodruff et al., 1981). Cooling of the Southern Ocean is indicated by the fact that calcareous macrofossil and planktonic foraminiferal assemblages became monospecific (Kennett, 1995). Another change involved a significant increase in the planetary temperature gradient (Loutit et al., 1983). On land, aridity increased and open grasslands developed in Australia (Stein & Robert, 1985; Locker & Martini, 1985). Likewise, grasslands developed in South America (Pascual & Juareguizar, 1990) and in Africa (Retallack, 1992).

The Late Miocene has commonly been highlighted as a time of important climatic deterioration and sea-level lowering, both thought necessary to cause the Messinian salinity crisis in the Mediterranean basin. The commonly quoted value for sea-level drop of 50 m or more at this time (e.g., Kastens, 1992) due to build-up of the Antarctic ice sheet in excess of its current dimensions is highly unlikely, as discussed above and by Denton et al. (1993). Any glacio-eustatic sea-level drop of more than 10–12 m must have come from Northern Hemisphere glaciers. In this regard, it is now known that full-scale glaciation of southeastern Greenland had occurred by 7.0 m.y. in Late Miocene time, following the relatively mild middle Miocene climate (Larsen et al., 1994). Likewise, significant mountain glaciation, with some glaciers reaching sea level, had occurred in Scandinavia as early as 5.5 m.y. (Jansen & Sjoholm, 1991). Finally, tidewater glaciation in the northern Pacific region began 6–5 m.y. in Late Miocene time, again following a regionally warm middle Miocene climate (Lagoe et al., 1993).

The most recent explanation of the Mediterranean Messinian salinity crisis does not invoke substantial climatic or sea-level change (Clauzon et al., 1996). Two phases of evaporite deposition are identified in the Mediterranean basin. The first phase involved deposition in marginal areas and is dated on the astronomical time scale to 5.75–5.70 m.y. Therefore, it correlates with glacial isotope stages TG22 and TG20 in figure 8-2 and in Shackleton et al. (1995a). The magnitude of sea-level fall during these stages is unknown, but it could have been as little as 10 m (Aharon et al., 1993). In any case, Atlantic water still entered the Mediterranean basin during deposition of these marginal evaporites. The second evaporite phase involved widespread desiccation that accompanied a huge sea-level drop within the Mediterranean basin between 5.60 and 5.32 m.y. The isolation of the Mediterranean basin necessary to cause this dramatic event was attributed to the dominance of continuous tectonic activity over eustatic sea-level changes in the sill area separating the Mediterranean Sea from the Atlantic Ocean. The marine $\delta^{18}O$ peaks of glacial stages TG14, TG12, TG10, TG8, and TG6 in figure 8-2 do not reach the values of TG22 and TG20 and hence cannot alone explain the desiccation of the Mediterranean basin. Although the reflooding of the Mediterranean is coincident with a sea-level rise, it cannot be caused by such a rise during interglacial stage TG5 at 5.32 m.y. in figure 8-2. This is because reflooding did not occur during the earlier but equivalent isotope stage TG 9 at 5.48 m.y. Overall, it now appears that the Messinian salinity crisis is due more to local tectonic activity than to eustatic sea-level changes (Clauzon et al., 1996). Hence, the Messinian salinity crisis does not necessarily point to a fundamental shift in the size of the Antarctic ice sheet.

A decrease in mean $\delta^{18}O$ values occurred between 5.7 and 5.5 m.y. (fig. 8-2). Isotopic values then fluctuated about this new lower mean until 3.5 m.y., when they rose slightly to a new mean that persisted until isotope stage G17 just after 3.0 m.y. (fig. 8-2; and Shackleton et al., 1995a). Between 5.5 and 3.0 m.y., the climate system occupied a mode in which interglacial isotope values were consistently about 0.4‰ lighter than late Quaternary interglacial values; in fact, during this interval even glacial values (with the exception of isotope stages M2 and MG2) were close to Quaternary interglacial values (Shackleton et al., 1995a).

From such isotopic records, it has been suggested that a warm interval with fluctuating climate extended through the portion of Pliocene time prior to 3.0 m.y. Much evidence supports this inference. Particularly noteworthy is the well-documented Pliocene warmth in the Arctic Basin and North Atlantic region. For example, Pliocene sea-surface temperatures in the North Atlantic Ocean near Iceland were as much as 4.0°–6.2°C warmer than now (Dowsett & Poore, 1991; Dowsett et al., 1992). At least at times in the Pliocene, forests existed in Iceland and trees grew in Greenland and Canada alongside the Arctic basin at 82° N latitude (Funder et al., 1985; Matthews, 1990). Molluscan data from northern Iceland show nearly continuous warmer-than-present surface ocean waters between 5 and 3 m.y. (Einarsson & Albertsson, 1988), with cold water first appearing between 3.0 and 2.4 m.y. (Cronin, 1991). Fossil Pliocene marine ostracods from the North Sea region indicate Pliocene seawater temperatures 4°–5°C higher than those of today. Warmth-loving plants in Europe grew as far north as The Netherlands (Suc & Zagwijn, 1983). Arctic sea ice was greatly reduced and boreal Pacific molluscs were able to migrate from the northern Pacific through the Arctic basin to Iceland and, in turn, Arctic/Atlantic molluscs and ostracodes migrated to the eastern North Pacific (Gladenkov et al., 1991). In the Southern Hemisphere, subantarctic sea-surface temperatures were no more than about 3°C above today's values during the warmest interval of the Pliocene (Kennett and Hodell, 1993). The Antarctic polar front at 3.1–3.0 m.y. was about 6° latitude south of the present-day position in the southeastern Atlantic and Indian oceans, but at approximately the present-day position in the southwestern Atlantic and Drake Passage (Barron, 1996). There is some evidence that parts of the Antarctic polar front migrated northward and sea ice expanded in the Southern Ocean between 2.68 and 2.47 m.y. (Hodell and Venz, 1992).

There are two contrasting explanations for the generally low but fluctuating marine $\delta^{18}O$ values of the Pliocene before 3.0 m.y. (Shackleton et al., 1995a). If the temperature of the deep Pacific Ocean remained constant, then the Antarctic ice sheet must have fluctuated so extensively that its volume was greatly reduced during Pliocene interglacial marine isotope stages. Conversely, if Antarctic ice volume remained essentially unchanged, then the temperature of Pacific deep water must have varied by about 2°C.

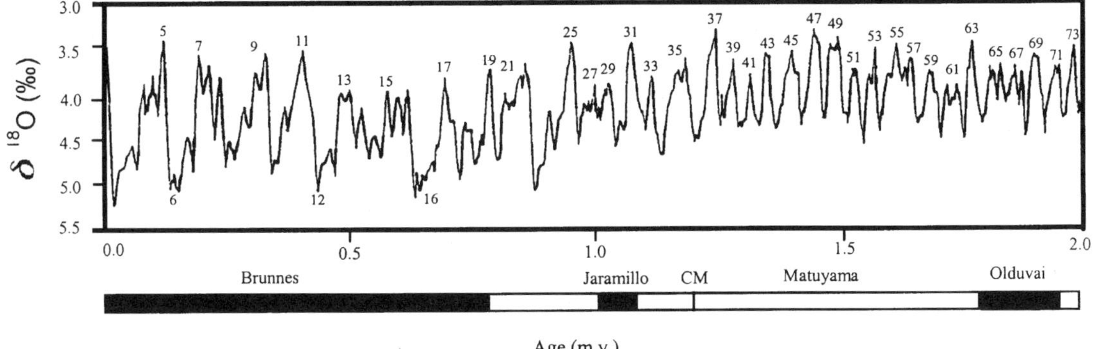

Age (m.y.)
Magnetic polarity time scale (SBP90)

Oxygen isotope record for the SPECMAP stock for the interval 0-0.63 m.y. and ODP Site 677 for the interval 0.62-2.0 m.y. The SPECMAP record is scaled to have the same mean and variance as data from the same time interval in Site 677.

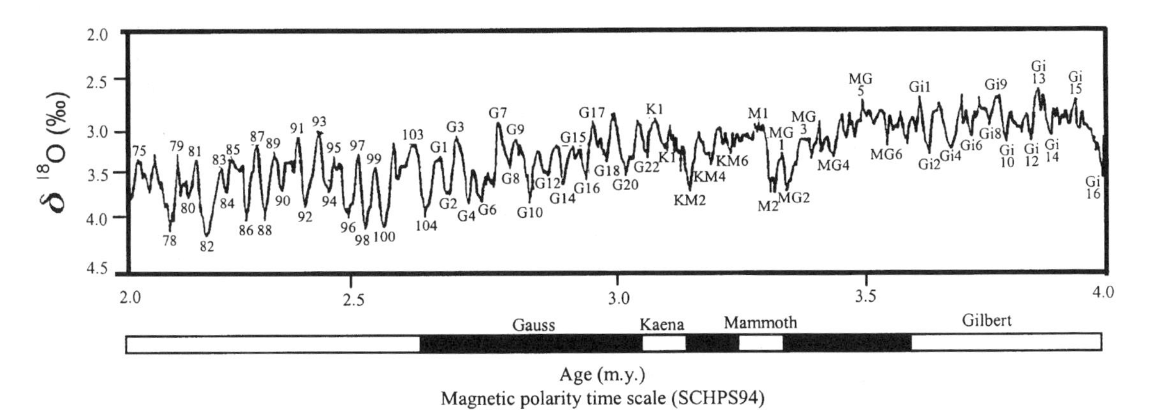

Age (m.y.)
Magnetic polarity time scale (SCHPS94)

Oxygen isotope record for benthic foraminifera from ODP Site 846 for the interval 2.0-4.0 m.y.

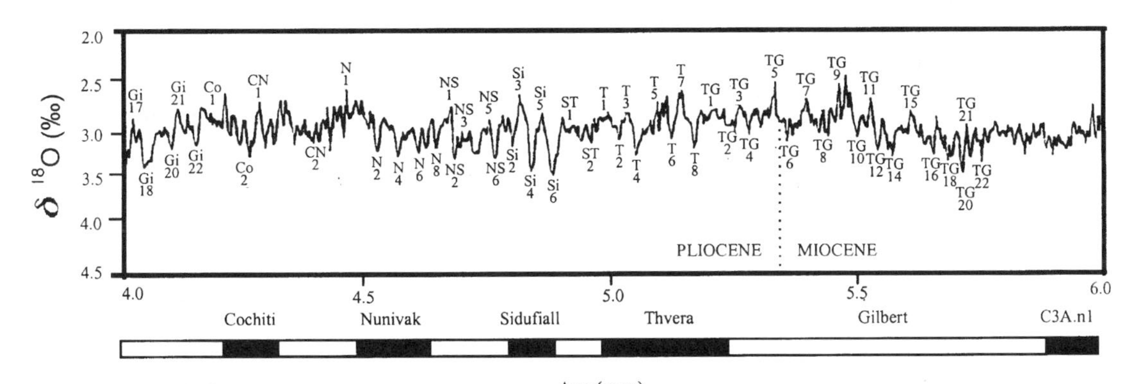

Age (m.y.)
Magnetic polarity time scale (SCHPS94)

Oxygen isotope record for benthic foraminifera from ODP Site 846 for the interval 4.0-6.0 m.y.

Figure 8-2. The benthic $\delta^{18}O$ record for the past 6 m.y. from the SPECMAP stack, ODP Site 677 and ODP Site 846 (from Shackleton, 1995a). The time scale in the upper panel is from Shackleton et al. (1990); in the middle and lower panels from Shackleton et al. (1995b). Isotope stages are denoted. (Used with permission of N.J. Shackleton and Yale University Press.)

The first of these possibilities is in accord with the conventional view that raised Pliocene marine deposits in northern Alaska and the southeastern U.S. coastal plain represent extensive melting of the East Antarctic ice sheet under warm climatic conditions. However, the overall Antarctic climate and consequent behavior of the East Antarctic ice sheet prior to about 2.5 m.y. in Pliocene time is a matter of considerable controversy, as reviewed by Denton (1995) and outlined briefly in a later section of this chapter. By one hypothesis, temperate Pliocene climatic conditions are postulated for Antarctica, such that the entire ice sheet was free to oscillate under mean annual temperatures as much as 20°C or more above present values. The inferred persistence of *Nothofagus* in the Transantarctic Mountains implies that polar conditions did not set in until Late Pliocene time. A contrasting hypothesis is that polar conditions with a stable ice sheet nucleus in East Antarctica persisted throughout the Pliocene (Denton et al., 1993). Mean annual temperatures in East Antarctica are thought to have warmed to only about 3°C above today's value in Pliocene time. Under such polar conditions, the dynamic sectors of the ice sheet that could have caused a rise in Pliocene sea level were confined to the marine-based appendages in West Antarctica and coastal East Antarctica. For now I do not discriminate between these two hypotheses. I only note that an accurate estimate of the glacio-eustatic component of Pliocene sea-level records from passive continental margins requires adjustment for tectonic uplift and isostatic rebound from denudation. No such correction has been made for the Alaskan north coast, and only a preliminary correction has been made for the southeastern U.S. coastal plain. Hence, it is my opinion that the Pliocene glacio-eustatic component of these far-field records is sufficiently unconstrained so as to be compatible with either temperate or polar conditions in Antarctica, as I now go on to illustrate.

Glaciological modeling results by J. Fastook (personal communication, 1995) indicate that Pliocene climatic warming in Antarctica of 21°C, which would have produced temperate climatic conditions compatible with the growth of *Nothofagus*, is required to melt an ice volume equivalent to 30 m of sea-level rise. The total rise would have increased to 36.5 m if the Greenland ice sheet also melted. Because it must have had extensive surface ablation zones, a temperate ice sheet in Antarctica could have responded sensitively to climatic changes and hence could have produced benthic $\delta^{18}O$ oscillations in the far-field marine record. But a polar ice sheet could also have produced a Pliocene glacio-eustatic sea-level rise. This is because the marine-based appendage in West Antarctica could have collapsed from internal ice-sheet mechanisms triggered by sea-level change. There also could have been slight marginal recession of East Antarctic ice. Thus shrinkage of polar Antarctic ice could have contributed up to 9 m of sea-level equivalent, for a total of 15.5 m if Greenland melted at the same time. The difference in sea-level equivalent in excess of today's values in these two scenarios is only about 20 m. Pliocene sea-level records from far-field passive continental margins may never be accurate enough to distinguish between these two possibilities given the corrections for crustal movements necessary to extract a glacio-eustatic component from these records. The point to be made here is that Pliocene records of high relative sea levels cannot be used to infer temperate Antarctic paleoclimate conditions. If Pliocene climatic conditions in Antarctica remained polar, as I argue below, then the most realistic explanation for the Pliocene marine $\delta^{18}O$ oscillations before to 3.0 m.y. is that they represent a combination of ice-volume fluctuations and deep-water temperature changes but are not indicative of major deglaciation in Antarctica. This conclusion is in accord with the interpretation of Hodell and Venz (1992) of a high-resolution stable isotope record from the Southern Ocean. Also, Shackleton (1995) pointed out that it is difficult to envision massive deglaciation on Antarctica without some warming of the Southern Ocean, but it is easier to envision a stable East Antarctic ice sheet with warmer deep-water temperatures resulting from increased heat input from the North Atlantic Ocean or the Mediterranean Sea.

The most important climatic and environmental change of late Neogene time began at the end of isotope stage G17 about 2.95 m.y. and culminated with isotope stage 100 at 2.52 m.y. (fig. 8-2; Shackleton, 1996). This transition is characterized by a slow shift in the isotopic mean , with superimposed 41 k.y. oscillations. Shackleton et al. (1995a) pointed out that during this mid-Pliocene transition the declining glacial values showed a near-linear trend that culminated in three large isotope events (glacial stages 100, 98, and 96); the $\delta^{18}O$ signal then reverted to low-amplitude variations until the recurrence of high-amplitude changes at isotope stage 82. Ice sheets adjacent to the North Atlantic were large enough to shed ice-rafted debris as far south as 40° N latitide by about 2.65 m.y. (Raymo et al., 1989). In contrast, the interglacial $\delta^{18}O$ values during the mid-Pliocene transition show a marked change at interglacial isotope stage G17; after this time the interglacial values were close to those of the present day, whereas before this time they were generally about 0.4‰ lighter than the modern value (Shackleton et al., 1995a).

In accord with the isotopic record, major mid-Pliocene environmental change is recorded in many localities on the planet. In the Transvaal of southern Africa, changes in fossil bovids suggest a fundamental shift from relatively mesic and bush-covered environments to more arid and open environments between 2.5 and 2.0 m.y. (Vrba, 1975, 1985c). In East Africa, micromammals (Wesselman, 1985), pollen assemblages (Bonnefille, 1976, 1983), and bovids (Vrba, 1985b) all show expansion of arid and open grasslands close to 2.5 m.y. The Transvaal and East African changes probably represented the same marked spread of open grassland near 2.5 m.y. across sub-Saharan Africa in response to global cooling and associated rainfall changes (Vrba, 1985d; Van Zinderen Bakker & Mercer, 1986). Janecek and Ruddiman (1987) showed from sediment composition changes in the equatorial Atlantic Ocean that the African continent became increasingly arid about 2.5 m.y., a result in accord with terrestrial evidence. Less significant environmental changes preceded the mid-Pliocene shift (Williamson, 1985; Bonnefille et al., 1987).

More than 30 yr ago, Zagwijn (1960) recognized pronounced mid-Pliocene cooling in north-central Europe at the base of the cold Praetiglian stage, when tundralike, open vegetation first replaced the transitional elements of the Reuverian stage and warm-temperature elements of the older Brunssumian stage. During the Praetiglian, arctic conditions were reached for the first time in north-central Europe, and the ice-age cycles of the Quaternary began. In fact, in north-central Europe the base of the Praetiglian stage has long been considered to mark the beginning of the Quaternary Period. The Reuverian/Praetiglian boundary is close to 2.4–2.5 m.y. (van der Hammen et al., 1971; Boellstorff, 1978; Zagwijn & Doppert, 1978; Grube et al., 1986). This runs counter to international stratigraphic convention, which places the base of the Quaternary in the Vrica section of Calabria, southern Italy, at 1.6 m.y. (Berggren et al., 1985).

In central China, widespread and thick loess rests stratigraphically on the Red Clays, a formation of Late Pliocene age. The boundary between the Red Clays and overlying loess is transitional. The initiation of loess deposition, which marks a fundamental change from warm and humid environments to harsh continental steppes, is dated between 2.3 and 2.5 m.y. (Kukla, 1987). Finally, a 357 m-long core recovered from the high plain of Bogotá, Colombia, contains evidence for a significant lowering of the forest line and high plain temperature at 2.5 m.y. from the relatively inflated levels of the preceding 0.5 m.y. (Hooghiemstra, 1984).

Proposed Explanations of Cenozoic Cooling

Distribution and Size of Continents

All proposed explanations of the Cenozoic cooling trend shown in figure 8-1 originate in some aspect of global tectonics. One common explanation involves the paleodistribution and size of continental land masses. From model simulations of Cretaceous climate, it is recognized that there is a large climate sensitivity to paleogeography (Barron et al., 1980). But Barron (1985) argued from the results of climate model simulations that large-scale variations in paleogeography could not have been the primary cause of the long-term Cenozoic cooling trend or of the onset of Antarctic glaciation. Paleogeographical changes by themselves do not explain the globally averaged decrease in surface temperature of 6°–12°C between the Cretaceous and the present day (Barron, 1985). Likewise, the effect of topography is not a viable explanation for the overall Cenozoic cooling trend. For example, there is almost no change in zonally averaged surface temperatures in a sensitivity experiment in which the Tibetan and Rocky Mountain topography is removed (Barron, 1985). Similar sensitivity experiments by Ruddiman and Kutzbach (1989) showed that plateau uplift in Tibet and the American West has only a modest effect on direct global temperature changes and cannot by itself explain the magnitude of the Cenozoic cooling trend. What the plateau uplift does accomplish, however, is to accentuate regional differences of climate and thus move the climate system away from the earlier equable, moist climates. Sensitivity experiments for uplifted plateaus in Tibet and the American West also showed important regional effects from increased orographic diversion of Northern Hemisphere westerly winds, from differing surface atmospheric flow as a result of seasonal heating and cooling of uplifted plateaus, and from the intensification of vertical circulation cells in the atmosphere. Therefore these climate-sensitivity experiments suggest that factors other than large-scale changes in geography caused the long-term Cenozoic climatic deterioration.

Ocean Gateways

Another possible mechanism for Cenozoic climatic change involves the opening and closing of gateways for the flow of ocean currents. This mechanism can change the redistribution of heat on the planet, isolate polar regions thermally (causing the growth of ice sheets and sea ice, thus resetting the equator–pole temperature gradient), and cut off or increase the supply of warm, saline bottom waters produced in mid- to low-latitude marginal seas.

Ocean modeling experiments suggest it is unlikely that poleward heat transport by the ocean could have maintained the warm equable Cretaceous climate shown by paleoclimatic data (Barron, 1983; Schneider et al., 1985). However, it is possible that changes in poleward heat transport from variations in ocean gateways, together with the resulting feedbacks, influenced Cenozoic cooling, and this may help explain some of the sharp isotopic steps in figure 8-1. Caution is necessary in interpreting changes in ocean heat transport in terms of global climate, however, because an opposite sense of heat transport in the atmosphere may compensate for oceanic changes (Stone, 1978; Covey & Barron, 1988). One modeling study showed that as oceanic heat transport increased, the tropics cooled, while surface temperatures increased at high latitudes (Covey & Thompson, 1989); globally averaged temperatures decreased because of the dominance of the tropical cooling and the transport of heat to areas with less atmospheric water vapor that can trap thermal infrared radiation.

A particularly important issue concerns the effect of the development of a deep-seated Antarctic circumpolar current on the climate of Antarctica. To address this issue, I return to the subject of the evolution of Antarctic climate during the Tertiary. From the beginning of the Tertiary into the middle Eocene, a land bridge connected South America with Antarctica (which was then in about the same geographic position as it is today) and Australia. Hence a nearly continuous land barrier to ocean circulation extended south from near the equator across the South Pole and then well northward into the Pacific. This land distribution may have been responsible for a Southern Hemisphere ocean circulation quite different from that of today (Kvasov & Verbitsky, 1981). Longitudinal circulation may have brought western boundary surface currents from the tropics far south to Antarctica. As a result, surface water temperatures could have been about 28°C in the tropics and 12° C near Antarctica (Kvasov & Verbitsky, 1981).

Macrofossils in the Antarctic Peninsula and pollen in erratics in the Transantarctic Mountains afford reasonable evidence that temperate forests featuring *Nothofagus* and podocarpaceous conifers as prominent components existed at far-southern Antarctic latitudes at least through the Eocene (Truswell, 1991; Askin, 1992). As mentioned previously, a continent-wide ice sheet emerged in East Antarctica in earliest Oligocene time close to 36 m.y. on the Berggren et al. (1995) time scale (Hambrey et al., 1991; Ehrmann & Mackensen, 1992; Zachos et al., 1992; Hambrey & Barrett, 1993). Based on the Tertiary paleovegetation record of Antarctica, this early ice sheet probably developed in a temperate climate. This inference comes from the persistence of vegetation consistent with cool temperate rainforests into the late Oligocene in the Antarctic Peninsula (Truswell, 1991; Askin, 1992). Numerous recycled pollen grains from near the base of glaciogenic sediments in DSDP core 270 in the Ross Sea (Kemp & Barrett, 1975), along with a *Nothofagus* leaf and concentrated pollen recovered from late Oligocene mudstone in the CIROS-1 core from western McMurdo Sound (Barrett, 1987, 1989; Mildenhall, 1989), likewise suggest persistence of rainforest with *Nothofagus* as a prominent component. It is known that *Nothofagus* has stringent environmental tolerances, and therefore its presence in Antarctica until late Oligocene time gives important information for the climatic environment of the early ice sheet in East Antarctica. Present-day species of *Nothofagus* require abundant rainfall and temperate climatic conditions. For example, the modern *Nothofagus* treeline in southernmost South America corresponds with a mean annual temperature of about +3°C (V. Markgraf, personal communication), and modern *Nothofagus* species cannot survive temperatures below -19°C for more than a few hours (Sakai et al., 1981). Thus the persistence of *Nothofagus* in Antarctica until late Oligocene time suggests a temperate environment for the early Antarctic ice sheet.

When this early temperate ice sheet reorganized into a polar ice sheet, under severe conditions similar to the arid polar climate of today, is a matter of controversy (Sugden et al., 1995; van der Wateren & Hindmarsh, 1995). A critical question is how the two contrasting scenarios of this reorganization tie in with the growth of a deep-seated Antarctic circumpolar current. If polar conditions in Antarctica developed concurrently with a circumpolar current, then the implication is that the development of the deep gateways necessary for this current was important in the climatic evolution of the Southern Hemisphere because of the resulting thermal isolation of Antarctica and the Southern Ocean. But if temperate climatic conditions and the growth of *Nothofagus* continued in Antarctica until mid-Pliocene time, then the implication is that the formation of a circumpolar current had little effect on Cenozoic climatic evolution in Antarctica and the Southern Hemisphere.

The present-day thermal isolation of Antarctica reflects the difficulty for water bodies to mix (and thus carry heat) across the barrier of the Antarctic circumpolar current except in deep boundary currents, large eddies, and wind-induced northward movement of surface water (Gordon, 1988). In this regard, the Drake Passage is essential for the thermal isolation of Antarctica because by connecting the Atlantic and Pacific oceans it affords a continuous zonal band of water and thus has a great effect on the thermally driven overturning in the Southern Hemisphere. Ocean modeling experiments show that with the Drake Passage closed, the warm upper surface branch of thermal overturning extends far southward, introducing a large advective heat flux into Antarctica (Gill & Bryan, 1971; Toggweiler & Samuels, 1995). An open Drake Passage results in a circumantarctic latitudinal ocean band without continental barriers to support the east–west pressure gradients necessary for meridional flow of surface water (Warren, 1990; Toggweiler & Samuels, 1992, 1995). Under these circumstances thermally driven surface currents cannot pass south of the latitude of Drake Passage (Toggweiler & Samuels, 1995). Rather, the surface winds in the circumpolar westerly belt cause a northward Ekman drift in the surface water. That portion of the northward drift that occurs at and south of the latitude of Drake Passage is compensated by southward flow of deep water (Imbrie et al., 1992). This southward deep flow occurs below topographic ridges that afford the necessary east–west pressure gradients to support meridional geostrophic flow. In this way North Atlantic deep water is drawn into the Southern Ocean and upwells in zones of Ekman divergence (Toggweiler & Samuels, 1995).

According to one hypothesis, temperate rainforests featuring *Nothofagus* (with elements of a Magellanic Moorland flora) are thought to have persisted in the Transantarctic Mountains, even close to the South Pole, until mid-Pliocene time about 2.5 m.y. (Webb & Harwood, 1993). Likewise, a temperate ice sheet was envisioned to have persisted until mid- to Late Pliocene time in East Antarctica, collapsing repeatedly under temperate climatic conditions to cause eustatic sea-level rises. In this reconstruction, the continent could not have been isolated thermally by development of the Antarctic circumpolar current, because a polar ice sheet did not emerge until about 2.5 m.y. (Webb & Harwood, 1993). The single focal point of this hypothesis is the assumption that reworked marine Pliocene diatoms in the glaciogenic Sirius Group were glacially transported into the Transantarctic Mountains from marine basins now beneath the East Antarctic ice sheet. This hypothesis is weakened severely if the assumed origin of the diatoms and therefore the glacial transport mechanism proves to be erroneous. For example, the chronological constraints on the Sirius group (which contains *Nothofagus* wood fossils at Beardmore Glacier) would be removed if the key Pliocene marine diatoms turn out to have been airborne from the Southern Ocean rather than glacially transported from interior East Antarctica. Nor would there be any basis for postulating Pliocene warmth and ice-sheet collapse. In this regard, Kellogg and Kellogg (1996) discovered numerous marine and nonmarine diatoms in ice cores from central Antarctica, demonstrating widespread eolian transport of diatoms to interior Antarctica. Burckle and Potter (1996) found Pliocene–Pleistocene diatoms near the surface of Paleozoic and Mesozoic sedimentary and igneous bedrock, concluding that they must have come to their present sites through the air. Stroeven et al. (1996) showed that diatoms were strongly concentrated near the upper surface of the Sirius outcrop at Mt. Fleming. If this is the case, then marine diatoms cannot constrain the age of the Sirius group because they were introduced after its deposition. These new results are consistent with the demonstration by landscape analysis that the Pliocene marine diatoms in the Sirius outcrop in Mt. Feather (an original site used to postulate Pliocene deglaciation by Webb et al., 1984) could not have been emplaced by glacial transport, but rather must have been introduced by atmospheric processes (Denton et al., 1993).

A contrasting chronology for the initiation of polar ice is derived from landscape analysis of Pliocene surfaces in Antarctica. The conclusion is that a polar-desert climatic environment (even at sea level) extended back at least to Early Miocene time on the basis of (1) more than 50 $^{39}Ar/^{40}Ar$ analyses of Miocene and Pliocene volcanic ashes that rest within thermal contraction cracks (sand wedges that form only in cold, dry continental climates; Péwé, 1959) and on desert pavements; (2) the stability of in situ Pliocene and Miocene-age drift sheets and relict colluvium on steep rectilinear (28° to 36°) slopes; (3) the texture of Pliocene and Late Miocene alpine drifts that fringe modern glaciers; and (4) the lack of features produced by running water on old land surfaces above 800–1000 m elevation in the western dry valleys (Denton et al., 1993). These results suggest early thermal isolation of Antarctica by the Antarctic circumpolar current.

Existing data on the development of the East Antarctic ice sheet could be reassessed if the chronologic constraints on Sirius group glaciogenic deposits were removed and if instead polar conditions existed back at least until Early Miocene time. The wood-bearing Sirius deposits in the Beardmore Glacier valley could then date back to Oligocene time when it is known that *Nothofagus* was widespread in Antarctica. The glacial history of Antarctica could then have featured an early temperate continental ice sheet that formed about 33 m.y. on the Berggren et al. (1995) time scale (Kvasov & Verbitsky, 1981; Hambrey et al., 1991; Bartek et al., 1992; Ehrmann & Mackensen, 1992) and that persisted until late Oligocene time. A polar ice sheet could then have come into existence after 24 m.y. (the youngest known *Nothofagus* fossil) and before the end of the Early Miocene (the oldest known polar-desert geomorphic features). The most likely time for this to have occurred would have been when the Drake Passage opened and deepened sufficiently to allow the existence of an Antarctic circumpolar current, probably about 23–20 m.y. (Barker & Burrell, 1977, 1982).

This sequence of paleoclimatic events is consistent with the notion that development of an ice sheet on Antarctica was concurrent with the spreading of ocean basins and a redistribution of southern continents. In this model Paleocene and Eocene temperate climates at far-southern latitudes would have been reinforced by meridional surface ocean circulation (due to continental configuration) that allowed tropical and subtropical currents to reach far south (Kvasov & Verbitsky, 1981). A major perturbation of this circulation system would have occurred in earliest Oligocene time when Australia sufficiently fragmented from Antarctica and when thermal subsidence was sufficient to form a deep submarine channel south of the South Tasman Rise (Exon et al., 1995). By allowing partial circumantarctic flow and decoupling some southward-flowing warm currents, this early Oligocene event could have set the scene for the development of a temperate ice sheet in East Antarctica (Kvasov & Verbitsky, 1981; Bartek et al., 1992). The subsequent reorganization from a temperate to a polar ice sheet could then have occurred when the Drake Passage opened and deepened enough to allow circumpolar flow and thermal isolation of Antarctica about 23–20 m.y. (Barker & Burrell, 1977, 1982).

By this postulated sequence for development of the Antarctic cryosphere, the prominent early middle Miocene marine $\delta^{18}O$ increase shown in figure 8-1, as well as the mid-Pliocene climatic deterioration between 2.95 and 2.52 m.y., are both too young to have resulted from the development of a permanent polar ice sheet in Antarctica. It is important to note that thermal maxima occurred just before both of these prominent increases in $\delta^{18}O$ depicted in figure 8-1. The thermal maximum at 17–15 m.y. was the greatest since the Eocene. During the Pliocene thermal maximum at 4–3 m.y., the Arctic basin was so warm that forests grew to its shores, sea ice was greatly reduced relative to today, and boreal conditions linked the Pacific and Atlantic through the Arctic Basin. Strengthened North Atlantic drift is the probable cause of the elevated Pliocene sea-surface temperatures in the North Atlantic Ocean (Dowsett & Poore, 1991), the warm faunas in the Pliocene sediments of the North Atlantic Ocean (Cronin, 1991), and the Pliocene warmth of the Arctic Basin. Therefore, any explanation of the positive $\delta^{18}O$ events must also account for the preceding thermal maxima. Moreover, any explanation must involve a mechanism for sharply curtailing the inflow after 2.95 m.y. of warm and saline North Atlantic water into the Greenland and Norwegian seas and the Arctic Ocean. Otherwise, it is difficult to understand how large ice sheets could have developed adjacent to these oceans in Late Pliocene and Pleistocene times. To address this issue, Kvasov and Verbitsky (1981) attributed the onset of large-scale Northern Hemisphere glaciation to Late Pliocene growth of the Iceland-Faeroe submarine plateau. More recently, both the middle Miocene and mid-Pliocene $\delta^{18}O$ events superimposed on the long-term Cenozoic cooling (fig. 8-1) have been attributed to changes in the depth of the Greenland-Scotland Ridge in the North Atlantic Basin (Wright & Miller, 1996).

The potential effect of the Greenland-Scotland Ridge on the production of North Atlantic deep water was recognized more than a decade ago (Vogt, 1971, 1972; Schnitker, 1980), along with the possible existence of an Iceland mantle plume (Vogt, 1983). Wright and Miller (1996) have correlated changes in the depth of the Greenland-Scotland Ridge with the varying contribution of northern component water to the deep ocean in Neogene time. The correlation depends on a reconstruction of Neogene deep-water circulation patterns (from benthic foraminiferal $\delta^{13}C$ records) compared with a reconstruction of mantle plume activity under Iceland (based on the ages of escarpments and ridges superimposed on the Reykjanes Ridge). The topographic swell associated with high mantle-plume activity has an influence of 1000–2000 km and hence affected the Greenland-Scotland Ridge. The results show that both the Miocene and the Pliocene thermal maxima in figures 8-1 and 8-2 were times of low mantle plume activity and high flux of northern component water (and presumably high northward flux of warm, saline surface water), whereas both of the positive $\delta^{18}O$ excursions corresponded with times of high mantle-plume activity and low flux of northern component water (and presumably low northward flow of warm, saline surface water).

Shoaling and closure of the Panama isthmus in the late Cenozoic could also have strengthened the North Atlantic drift current, thus increasing heat import to the Arctic (Maier-Reimer et al., 1990). There is some indication that significant shoaling was underway by 6 m.y. and that the isthmus was uplifted above sea level before to 3 m.y. (Kaneps, 1979; Keigwin, 1982; Collins et al., 1996). Thus any increase in the North Atlantic drift current resulting from the closure of the isthmus could have been coeval with the lower mean $\delta^{18}O$ values that set in about 5.5 m.y. after the Messinian Stage (fig. 8-2).

Atmospheric Carbon Dioxide and Water Vapor

Changes in the concentration of atmospheric trace gases can affect global climate and are a strong candidate to explain the overall Cenozoic trend in figure 8-1. Chief among these trace gases are carbon dioxide and water vapor. The two can be closely linked because atmospheric water vapor is the dominant feedback in model simulations of climatic warming from increased atmospheric carbon dioxide. A main source for atmospheric carbon dioxide is from thermal degassing reactions, which take place in association with mantle plumes and both divergent and convergent plate boundaries. A major sink is from chemical weathering that occurs when atmospheric carbon dioxide, dissolved in rainwater and groundwater, attacks Ca- and Mg-bearing silicate materials to produce bicarbonate. The bicarbonate is then carried to the sea in the dissolved load of rivers, incorporated into shells of marine biota, and buried in marine sediments on the sea floor. The cycle is complete when seafloor spreading carries these marine sediments into a subduction zone and the carbon is released to the atmosphere through mantle degassing processes (Berner, 1990). The long-term amount of carbon dioxide in the atmosphere depends on the balance between sources and sinks not only in this carbonate-silicate cycle but in the organic carbon cycle (Berner et al., 1983; Berner, 1990). The organic carbon cycle involves the production and burial of organic matter (carbon dioxide sink) and the oxidation of organic matter associated with the exhumation of sediments (carbon dioxide source). A link between the carbonate-silicate and organic carbon cycles involves the enhanced production of organic matter brought on by nutrients delivered to the ocean through chemical weathering of silicate materials.

By relating carbon dioxide input linearly to the rate of seafloor spreading (involving both production of seafloor at ridges and destruction at trenches), Berner et al. (1983) suggested that Cretaceous warmth and the Eocene thermal maximum both were caused by elevated atmospheric carbon dioxide. In both cases, the model results indicate that high atmospheric carbon dioxide concentrations are source-driven by seafloor spreading rates inferred to be more rapid than at other times during the Cenozoic. The higher sea levels (up to 300 m above present) of Cretaceous and Paleocene/Eocene time were related to these faster seafloor spreading rates because they increased the volume of mid-ocean ridges. Chamberlin (1906) and Brass et al. (1982) suggested that the resulting epicontinental seas in low and middle latitudes were the source of the warm, saline bottom waters of Cretaceous and early Cenozoic time. By flooding extensive inland areas, these seas also reduced the area of continents susceptible to chemical weathering and hence reduced the output term for atmospheric carbon dioxide.

Recently, Beck et al. (1995) suggested that oxygen and carbon isotopic excursions during the Eocene thermal maximum can be attributed to the uplift and weathering of organic-rich sediments during the initial collision between India and Eurasia. Drawdown of atmospheric carbon dioxide has long been postulated as a mechanism for the long-term cooling that followed the Eocene thermal maximum (Chamberlin, 1899; Berner et al., 1983; Raymo, 1991; Raymo & Ruddiman, 1992). A contentious issue is whether the postulated drawdown is source driven (reduced degassing) or sink driven (forced by increasing mountain building and associated exhumation, during the later Cenozoic). Orogenic activity not only exposes silicate materials to the atmosphere because of exhumation but the associated surface uplift induces orographic and even monsoonal precipitation, which increases weathering rates and hence drawdown of atmospheric carbon dioxide. Pertinent processes include rapid exposure of silicate basement by extensional detachments, mechanical disintegration of rock in areas of high relief, and transport of erosional products to lower elevations with high temperature and moisture. Although rainfall and temperature have long been considered to be first-order controls on the rate of chemical weathering (Dunne, 1978; Berner et al., 1983; White & Blum, 1995), Edmond et al. (1996) contended that relief and lithology are the most relevant factors.

The prime candidates for forcing Cenozoic drawdown of atmospheric carbon dioxide by silicate weathering are the Himalayas and the Tibetan Plateau, along with the Andean and Eurasian alpine belts. In support of the importance of topographic relief in forcing the carbon cycle is the fact that a large fraction of the solutes now reaching the ocean comes from rivers draining the Himalayas, the Tibetan Plateau, the northern Andes, and Indonesia. The Himalayan collision began in earliest Cenozoic time and therefore, in conjunction with partial development of the Antarctic circumpolar current, may have contributed to the growth of an ice sheet in Antarctica by about 33 m.y.

The Miocene cooling event (Flower & Kennett, 1993b) has been linked to anomalous deposition of organic matter in the Monterey Formation by Vincent and Berger (1985); subsequently, Raymo (1994) implicated a spurt of exhumation in the Himalayas between 20 and 16 m.y. as the cause of this anomalous organic carbon burial (release of nutrients by chemical weathering).

The effect of varying rates of mountain uplift and associated changes in chemical weathering on atmospheric carbon dioxide is also implicated as the cause of the major late Neogene climatic pulses (Raymo et al., 1988; Raymo & Ruddiman, 1992). In this regard, the overall Pliocene warmth before 3.0 m.y. could have involved higher-than-present atmospheric carbon dioxide related to a tectonic event. For example, climate modeling experiments suggest that a doubling of carbon dioxide could explain this Pliocene warmth (Crowley, 1991). The implication of this argument is that the subsequent mid-Pliocene cooling between 2.95 and 2.52 m.y. reflects drawdown of atmospheric carbon dioxide from enhanced chemical weathering related to an unidentified pulse of mountain uplift and denudation. However, new proxy records show that atmospheric carbon dioxide during the warm Pliocene interval probably was only 35–50% higher than preindustrial levels (Kurschrer et al., 1996; Raymo et al., 1996). This modest increase is much smaller than the doubling of atmospheric carbon dioxide proposed to explain Pliocene warmth (Crowley, 1991). An alternative explanation is that increased meridional ocean heat transport, probably from strengthening of thermohaline circulation (Raymo et al., 1996), produced Pliocene warming, not only at high latitudes but also globally because of the consequent reduction of sea ice (Rind & Chandler, 1991). A possible mechanism involving variations of the depth of the Greenland-Scotland Ridge in controlling North Atlantic deep water production (Wright & Miller, 1996) was outlined in the previous section. The preferred explanation of Crowley (1996), Dowsett et al. (1996), and Raymo et al. (1996) is that a combination of higher atmospheric carbon dioxide and enhanced thermohaline circulation produced the Pliocene warmth inferred from geologic records. The positive feedback effects among atmospheric carbon dioxide, thermohaline circulation, and sea-ice extent suggested by this interpretation of the geologic record is in contrast with the results of atmospheric modeling experiments, which predict that weakened North Atlantic thermohaline circulation (from increased precipitation) will accompany rising atmospheric carbon dioxide levels (Manabe & Stouffer, 1993; Raymo et al., 1996). However, the two situations may not be comparable because of Pliocene changes in the depth of the Greenland-Scotland Ridge (Wright & Miller, 1996).

The independent role of water vapor has generally been ignored in discussions of long-term Cenozoic cooling, but water vapor is potentially the most important greenhouse gas. In this regard, Wright and Miller (1996) suggested that changes in atmospheric water vapor may have been the dominant cause of long-term Cenozoic cooling. In support of this argument, they pointed to the long-term decrease in the area of the tropical and subtropical oceans, the major source for water vapor, due to the progressive closure of the Tethys Sea from the Eocene to the middle Miocene.

Orbital Pacing of Cenozoic Paleoclimatic Oscillations

From figure 8-2 it is evident that oscillations in the $\delta^{18}O$ content of benthic foraminifera in the marine record are superimposed on overall Cenozoic climatic trends through at least the past 6 m.y. These oscillations reflect some combination of changing ice-sheet volume and bottom-water temperature. The weak oscillations between 5 and 6 m.y. do not seem to have a concentration of power at orbital frequencies (Shackleton et al., 1995a), but the oscillations through the last 5.0 m.y. appear to be linked to variations in the eccentricity of Earth's orbit, as well as in the tilt and orientation of its spin axis. The unequivocal demonstration of such an imprint of orbital geometry on the deep-sea sediment record (Hays et al., 1976) was made possible by a reliable time scale based on the simultaneous measurements of magnetostratigraphy and $\delta^{18}O$ stratigraphy in core V 28–238 from the western equatorial Pacific Ocean (Shackleton & Opdyke, 1973). In figure 8-2 the benthic $\delta^{18}O$ oscillations between 5.0 and 2.52 m.y. are concentrated in the orbital obliquity bandwidth (Shackleton et al., 1995a). The variability between 2.52 m.y. and isotope stage 25 at about 0.95 m.y. is also concentrated in the obliquity bandwidth (Shackleton et al., 1995a); within this interval at site 607 in the North Atlantic Ocean the 41 k.y. period is dominant, but some precession effects appear to be present in the record between 1.65 and 1.80 m.y. and also between 1.25 and 2.15 m.y. (Raymo et al., 1989).

The overall amplitude of these 41 k.y. variations is considerably less in the interval between 2.52 and 5.0 m.y. than it is in the interval between 0.95 and 2.52 m.y. (fig. 8-2). The small variations of amplitude of the oscillations within each of these latter two intervals mirrors the changing strength of the obliquity component of seasonality (Tiedemann et al., 1994). The important mid-Pliocene transition between these two intervals begins about 2.95 m.y. and culminates about 2.52 m.y. (fig. 8-2). During this transition, the cold extremes show a steady increase from isotope stage G 22 to isotope stage 100; the warm extremes show a shift at isotope stage G 17 from lower mid-Pliocene values to higher Late Pliocene and Pleistocene values similar to those of today (fig. 8-2; Shackleton et al., 1995a). After isotope stage 25 about 0.95 m.y., the amplitude of the $\delta^{18}O$ cycles increases significantly and the dominant frequency changes from about 41 k.y. to about 100 k.y. (fig. 8-2; Shackleton & Opdyke, 1976). However, it should be noted that the amplitude of the 41 kyr component of the overall signal continues nearly unchanged across this boundary and throughout Late Pleistocene time (Pisias & Moore, 1981). Also, it is important to note that a significant 23/19-k.y. precession component appears in the Late Pleistocene record (Imbrie, 1985). The important change in the frequency and amplitude of the benthic $\delta^{18}O$ signal at about 0.95 m.y. was first noted in the studies of cores V 28–238 and V 28–239 from the western equatorial Pacific Ocean (Shackleton & Opdyke, 1973, 1976) and subsequently emphasized by Pisias and Moore (1981). The characteristic shape of the 100-k.y. glacial cycles of Late Pleistocene time after 0.95 m.y. features long buildups to a maximum, followed by sharp terminations. Superimposed on the buildups are oscillations that reflect obliquity and tilt effects. During at least the last cycle, Dansgaard-Oeschger events, Bond cycles, and Heinrich events were also superimposed on the long buildup phase (Broecker, 1995a,b).

Subtropical African arid–humid cycles revealed in the dust content of deep-sea cores also evolved during Pliocene-Pleistocene time (Tiedemann et al., 1994; deMenocal, 1995; deMenocal & Bloemendal, 1995). Strong aridity with pronounced cycles extended back to at least 5 m.y. (Tiedemann et al., 1994). There were marked changes in the variability of the dust record at 2.8 m.y. and 0.9 m.y. Before to 2.8 m.y., eolian dust had a dominant 23 k.y. variability that showed strong amplification during times of high eccentricity (Tiedemann et al., 1994). Intensification of glacial arid cycles and an increase in the long-term dust flux rates occurred between 2.8 and 2.0 m.y. (Tiedemann et al., 1994), followed by a decrease in dust flux between 2.0 and 1.5 m.y. This shift was also marked by a change in dust variability to a dominant 41 k.y. frequency, although the precessional dust cycles still remained reasonably strong until 1.5 m.y. There was a subsequent increase to a dominant 100 k.y. variance after 0.9 m.y., with this interval marked by the highest level of dust flux rates (Tiedemann et al., 1994). However, it should be noted that rhythmic $CaCO_3$ bedding and layers of sapropel in sections and cores in the Mediterranean basin show the persistence of 23 k.y. arid–humid cycles in parts of subtropical Africa (the Nile River controls these sedimentary sequences and it drains subtropical highlands in East Africa susceptible to monsoonal variations) even after 2.8 m.y. (Hilgen, 1991a,b), despite the obliquity-forced variations of dust input into nearby deep-sea cores.

The development of the Northern Hemisphere ice sheets can be tied to the benthic $\delta^{18}O$ record in figure 8-2. The variations in benthic $\delta^{18}O$ before about 2.95 m.y. are small, with several exceptions such as isotope stages M 2 and MG 2. Moreover, the dust oscillations in Africa before 2.95 m.y. responded strictly to precession forcing. The implication is that the Northern Hemisphere ice sheet/North Atlantic system was too weak to affect climate in northern Africa. But the situation changed markedly after about 2.95 m.y. The increased amplitude of the benthic $\delta^{18}O$ signal beginning at that time heralds the growth of intermediate-sized ice sheets that subsequently fluctuated with a regular 41 k.y. pulsebeat, as did the North Atlantic Ocean thermohaline circulation (Dwyer et al., 1995). A critical point in this evolution was reached when ice sheets began to shed widespread ice-rafted detritus in the North Atlantic Ocean at least by 2.65 m.y. (Raymo et al., 1989). That the high-latitude ice sheet/North Atlantic system had become strong enough to dominate North African climate is shown by the evolution of a 41 k.y. cycle of the dust flux to deep-sea cores (deMenocal, 1995). Finally, the change in the amplitude and frequency of the benthic $\delta^{18}O$ oscillations, beginning in isotope stage 25 and culminating in isotope stage 22 (fig. 8-2), signals the first appearance of the great Northern Hemisphere ice sheets of Late Pleistocene time.

The dominance in Late Pleistocene time of the asymmetric 100 k.y. glacial cycle is a major problem in paleoclimatic research. In this regard it should first be noted that throughout Late Pleistocene time the amplitude history of the tilt-induced changes in seasonality matches that part of the variability in the benthic $\delta^{18}O$ record associated with the tilt frequency. Likewise, the amplitude history of the precession-induced changes in seasonality matches that part of the variability in the benthic $\delta^{18}O$ record associated with the precession frequency (Imbrie et al., 1984). In this way the response of the climate system of Late Pleistocene time to orbital forcing is similar to its earlier response. The major difference, however, is that the Late Pleistocene climate signal is dominated by an asymmetrical 100 k.y. signal featuring long glacial buildups and abrupt terminations. This 100 k.y. cycle is not an expected consequence of orbital forcing, which has the greatest power at the 19/23 k.y. and 41 k.y. frequencies. Therefore, it seems to be a nonlinear response to orbital forcing. The spacing of the terminations of 100 k.y. cycles appears to be linked in a loose fashion with the 100 k.y. component of eccentricity (Imbrie et al., 1984). The 100 k.y. cycle is dominant despite the fact that the amplitude of the direct eccentricity radiation forcing is too small and too late in phase to drive it directly (Imbrie et al., 1993). Also, it is striking that the Late Pleistocene $\delta^{18}O$ record does not exhibit the 400 k.y. eccentricity signal, which is present in older climate records. For example, the last termination (isotope stage 2/1 transition in fig. 8-2) and that about 420 k.y. ago (isotope stage 12/11 transition in fig 8-2), both at times when the 400 k.y. eccentricity signal (and therefore the amplitude of precession) was low, had the same magnitude as the intervening terminations when this component of eccentricity was high. Thus during terminations the ocean–atmosphere system appears to jump between stable modes of operation (Broecker and Denton, 1989). Although the jumps may be triggered by orbital forcing, as discussed below, their amplitude is relatively constant and independent of the magnitude of orbital forcing. Such a behavior of the climate system may explain why the 400 k.y. eccentricity signal is absent in Late Pleistocene climate records.

It is also important to note that there are key differences between the Late Pleistocene 100 k.y. oscillations and those earlier in Pliocene-Pleistocene time. Before about 0.95 m.y., the climate system appears to have responded to orbital variations in a linear fashion; where present the 100 k.y. component was related to the amplitude of precessional forcing, which in turn depended on the value of eccentricity. A common result is that sedimentary records show bundles of prominent layers spaced at about 100 k.y. intervals, and they also show the 400 k.y. component of eccentricity. One such example of rhythmic sedimentation tied to orbital forcing comes from terrestrial sections and marine sediment cores in the Mediterranean Basin, where sapropel-bearing sequences of Late Pliocene and Pleistocene age can be related to a precession-related cooling (Hilgen, 1991a,b). The sapropels, which occur in bundles, can each be tied to a correlative peak of a precession index. Therefore, sapropels are taken to reflect orbital forcing at the precession frequency. The bundles of sapropels generally occur at about 100 k.y. intervals, which are related to high values of eccentricity that increase the amplitude of the precessional signal. A further tie to eccentricity forcing is indicated by the fact that sapropel bundles die out each 400 kyr, in concert with diminished eccentricity. Another example of orbital forcing involving dust fluxes from northern Africa to the eastern equatorial Atlantic was given previously (Tiedemann et al., 1994). Between 2.95 and at least 5.0 m.y., the dust flux to ODP site 659 near Africa shows variations linked to precession, with bundles of high-amplitude peaks related to eccentricity-driven, high-amplitude precession variations spaced at intervals of about 100 k.y.

It is not yet clear from the paleoclimatic record whether variations of such indicators as benthic ($\delta^{18}O$ and terrestrial dust represent regional or global climatic events prior to Late Pleistocene time. However, there are two important points to make in this regard concerning the Late Pleistocene climatic changes shown in figure 8-3.

First, the stratigraphic record indicates that Late Pleistocene climatic variations were global in nature. For example, a comparison of marine $\delta^{18}O$ records with terrestrial paleoclimatic sequences at widely differing latitudes shows a striking correspondence of paleoclimatic changes in both polar hemispheres through at least the last four 100 k.y. glacial cycles (fig. 8-3). The key records are (1) the planktonic SPECMAP isotope stack (largely Northern Hemisphere ice volume, which reflects mass balance and hence atmospheric processes between 37° and 80° N latitude) (Imbrie et al., 1984), (2) the equatorial records of benthic $\delta^{18}O$ from ODP site 677 (Shackleton & Hall, 1989; Shackleton et al., 1990) in the eastern Pacific Ocean and of arboreal pollen from the high plain of Bogotá in the eastern cordillera of Colombia at 4° N latitude (correlation of these records, using independent chronologies, implies synchrony at least back to 1.2 m.y.) (Hooghiemstra, 1984, 1995), (3) the combined benthic $\delta^{18}O$ and arboreal pollen records in DSDP site 594 from the southwestern Pacific Ocean at 45° S latitude east of the South Island of New Zealand (Nelson et al., 1985; Heusser & van der Geer, 1994); and (4) the Vostok ice-core isotopic records from 78° S latitude placed on the SPECMAP time scale and compared with V19–30 benthic oxygen-isotope record (Bender et al., 1994).

These records span the planet from nearly 80° N to 80° S latitude. They are linked through a common marine benthic $\delta^{18}O$ stratigraphy. They all reflect atmospheric changes, with mean annual temperature probably being the most important variable. They all show the same patterns and therefore almost surely represent a global signal. This signal is at least near-synchronous between the polar hemispheres. Such interhemispheric symmetry is reinforced by a new radiocarbon chronology for the last glaciation that shows a synchrony of climatic changes between the

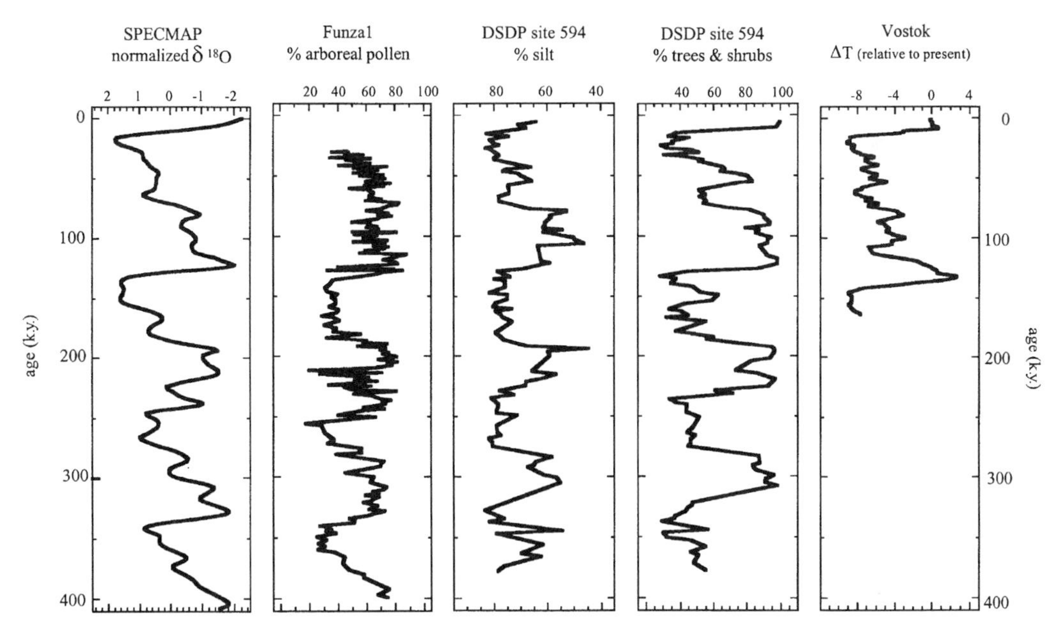

Figure 8-3. Paleoclimatic records for Late Pleistocene time. The time scales of all five records have been adjusted to the SPECMAP time scale by matching to marine benthic $\delta^{18}O$ records. The SPECMAP $\delta^{18}O$ isotope stack largely represents Northern Hemisphere ice volume between 37° and 80° N latitude and is from Imbrie et al. (1984). The Funza I pollen core is from the high plain of Bogotá in the eastern cordillera of Colombia at 4° N latitude and 2550 m elevation (Hooghiemstra, 1995). DSDP site 594 is from the southwestern Pacific Ocean off the eastern coast of the South Island of New Zealand at about 45° S latitude (Nelson et al., 1985; Heusser and van der Geer, 1994). In this case the percentage of silt reflects alpine glaciation in the southern alps, and the percentage of trees and shrubs tracks vegetation on the South Island. The Vostok ice core is from 79° S latitude in East Antarctica; the temperature record is inferred from the isotopic signal (Bender et al., 1994).

polar hemispheres during the last glacial–interglacial hemicycle (Lowell et al., 1995).

The second important point is that these Late Pleistocene global changes seem to be of roughly uniform magnitude over the entire planet except in the interior of large continents. For example, the mountain snowline and vegetation records in a north–south transect along the western cordillera of North and South America show approximately equal climatic cooling of 5–8°C below present-day values at the last glacial maximum in both polar hemispheres (including tropical regions) (Broecker & Denton, 1989, 1990). Likewise, it has been recognized that tropical temperatures a low elevations may have varied by as much as 5°C during Late Pleistocene glacial/interglacial transitions (Stute et al., 1995) and that monsoonal wetness in East Africa was tied in closely with global temperature change during the last glacial/interglacial transition (Roberts et al., 1993).

A climate linkage to orbital frequencies is also common in the older geologic record, not just that of Pliocene-Pleistocene time. An example is the discovery of orbital forcing of mid-Cretaceous black shale rhythms in the Tethyan region of central Italy (Herbert & Fischer, 1986). Most of the sedimentary variability of this sequence can be tied to orbital precessional and eccentricity forcing. In northern Italy, lower Cretaceous rhythmic stratification shows an obliquity control. Moreover, the dominant frequency at these sites changes from 41 k.y. in a lower Cretaceous sequence to 20 k.y., 100 k.y., and 400 k.y. in mid-Cretaceous rocks. Hence, the low-latitude Tethys Sea of an ice-free planet shows stratigraphic sequences that illustrate many of the orbital frequencies. An illustration that this is not a unique situation comes from an analysis of sedimentary cycles of the Mesozoic-age Newark Supergroup, which was deposited at 10°–23° N paleolatitude in eastern North America. The sedimentary rhythms, here produced by rises and falls of large lakes in low-latitude closed basins, also show dominant frequencies corresponding to precession, obliquity, and several components of eccentricity (Olsen, 1986).

Hence, it is becoming increasingly evident that paleoclimatic records through a long segment of geologic time show climatic variations linked to orbital frequencies. In fact, the orbital signal is so pervasive that it serves as a chronologic tool within cyclic stratigraphic sequences (Imbrie et al., 1984; Ruddiman et al., 1986a, 1989; Raymo et al., 1989; Shackleton et al., 1990, 1995a,b; Hilgen, 1991a,b; Tiedemann et al., 1994) and has been used to tune the time scale in figure 8-2. The strong implication is that orbitally paced climatic oscillations must have been superimposed on the entire long-term Cenozoic climatic deterioration shown in figure 8-1. The major change that stands out is that the older oscillations appear to be linear responses to orbital forcing but that Late Pleistocene time features a dominant asymmetrical 100 k.y. cycle which is not a linear response to orbital forcing.

Despite the enormous recent advances in demonstrating the presence of orbital frequencies in geologic records, the physical mechanisms that link paleoclimate and orbital changes remain elusive. A leading possibility is that global thermohaline circulation lies at the heart of this linkage (Broecker & Denton, 1989, 1990). This possibility is explored exclusively in the following discussion. However, this focus on thermohaline circulation may well be incorrect. A more complete review of the overall problem appears in Imbrie et al. (1992, 1993).

Waters sufficiently dense to sink to the abyss and thus ventilate the deep sea form in the northern North Atlantic Ocean and in the Southern Ocean near Antarctica. Thus marginal seas in polar regions are important regulators of thermohaline circulation because here water becomes dense by cooling and by salt enrichment. But the potential for cooling to near the freezing point exists in all polar regions. Hence salinity is the critical factor in dictating the sites of deepwater formation. Today warm, salty waters move northward near the surface in the North Atlantic Ocean to the vicinity of Iceland, where in winter they cool, sink, and spill over the Greenland-Scotland Ridge as Denmark Strait overflow water and Norwegian sea overflow water (which entrains Labrador sea water and Mediterranean overflow water). Together these overflow waters from the Nordic Seas form the major component of North Atlantic deep water. Because it entrains warmer waters as it flows southward, North Atlantic deep water is a major contributor of heat to the Southern Ocean and hence an important regulator of the extent of Antarctic sea ice. At the other major site of deep-water formation near Antarctica, brines formed as the by-product of the growth of winter sea ice cause dense water to flow into the abyss from marginal shelves. Cooling in the area of polynyas also causes deep convection. This Southern Ocean water flows northward along the seafloor into the North Atlantic region. In addition, intermediate waters form in subpolar seas and spread equatorward at shallower depths.

Because salinity is such an important influence on the density of sea water, changes in water vapor transport or freshwater flux into sensitive parts of the ocean can lead to changes not only in the rate but also in the pattern of ocean circulation. Thus the atmosphere and the ocean are tied together in a combined system susceptible to change. The North Atlantic region is the most important source of deep water that drives thermohaline circulation, and the Atlantic conveyor appears to be the most vulnerable part of the system. Therefore, salinity changes in this part of the sea are most likely to create the instabilities that cause the ocean–atmosphere system to reorganize (Broecker & Denton, 1989, 1990). This is compatible with the results of computer simulations showing that the thermohaline circulation in the North Atlantic Ocean can operate in different modes, depending on freshwater fluxes (Stocker & Wright, 1991; Maier-Reimer et al., 1993; Rahmstorf, 1994). Recent studies show that the export of Arctic sea ice by the East Greenland current can critically influence these freshwater fluxes into critical areas of the Nordic Seas where thermohaline sinking occurs (Aagaard, 1988; Schlosser et al., 1991). During glaciations, discharge of icebergs from large ice sheets peripheral to the northern North Atlantic Ocean also was vital to the surface freshwater budget and hence the strength of thermohaline circulation.

To explain global climate changes registered in terrestrial records, it must be postulated that ocean–atmosphere reorganizations in some fashion cause simultaneous changes in the greenhouse gas content of the atmosphere (Broecker & Denton, 1989; Broecker, 1994, 1995a,b). A likely candidate is that the reorganizations in some fashion affect the strength of the tropical tradewinds, thus changing upwelling, sea-surface temperatures, and hence the production and distribution of tropical atmospheric water vapor (the most effective greenhouse gas).

Within the context of thermohaline forcing of climate, the linkage between orbital pacing and global climate would thus focus in large part on the North Atlantic region (Broecker & Denton, 1989). Indeed, Imbrie et al. (1992) identified Arctic sea-ice production and freshening of the surface waters of the Nordic Seas as the initial response to orbital forcing during each glacial cycle. The consequences of this initial response are then rapidly transmitted as a climate signal to the rest of the planet through the resulting ocean–atmosphere reorganizations that feature changes in thermohaline circulation and tropical production of atmospheric water vapor.

In figure 8-2 the obliquity frequency of 41 k.y. is dominant until about 0.95 m.y. and is still important after that time. The implication is that Arctic sea ice and sea-surface temperatures fluctuated with that frequency, thus regulating thermohaline circulation. The difference in the amplitude of the oscillations before and after 2.52 m.y. would reflect the effect of the changing depth of the Greenland-Scotland Ridge (Wright & Miller, 1996) on the production of North Atlantic deep water (and the associated northward flow near the surface of warm, saline water). The Pliocene thermal maximum before 3.0 m.y. would reflect extensive northward flow of warm, saline water associated with low mantle-plume activity near Iceland and consequently a deepened Greenland-Scotland Ridge. The result would be the warm Arctic sea-surface temperatures recorded by paleoclimatic indicators discussed previously. As a consequence the 41 k.y. oscillation of sea-ice, sea-surface temperatures, and thermohaline circulation would have had a low amplitude. The recorded mid-Pliocene decrease in the production of North Atlantic deep water (Raymo et al., 1992) would reflect increased mantle-plume activity near Iceland with the resultant rise of the Greenland-Scotland Ridge and decrease in northward-flowing warm surface water. This change in the production rate of North Atlantic deep water would have triggered the fundamental climate shift between 2.95 m.y. and 2.52 m.y. seen in figure 8-2. The increased sea-ice cover over a colder Arctic Ocean induced by the overall decline in the production (of North Atlantic deep water) would then have permitted the higher-amplitude 41 k.y. oscillations seen in figure 8-2 after 2.52 m.y. The widespread atmospheric climatic changes induced by the switches in thermohaline circulation and associated sea-ice extent and sea-surface temperatures would have caused the buildup and decay of Northern Hemisphere ice sheets about half as large as those of Late Pleistocene time (Raymo et al., 1989).

Any explanation of the development of asymmetrical 100 k.y. glacial cycles of Late Pleistocene time must address the question of why large Northern Hemisphere ice sheets appeared in the first place. In other words, why, at 0.95 m.y. did intermediate-sized ice sheets with symmetrical 41 k.y. oscillations give way to large ice sheets with asymmetrical 100 k.y. oscillations? The previous large changes in the Cenozoic climate system (earliest Oligocene, middle Miocene, mid-Pliocene) involved shifts of the $\delta^{18}O$ mean, as well as of the values reached during the peaks and troughs of the oscillations on both sides of the major change. Moreover, the frequency of short-term oscillations does not seem to have changed across these major shifts in the mean of the climate system. But the situation at 0.95 m.y. is quite different because the amplitude and the frequency of the climatic oscillations both change at this time. Also, the impression is that the subsequent interglacial values of the climate system remained about where they were before 0.95 m.y. In contrast, the glacial values become more extreme, and these colder glacials were accompanied by the growth of Northern Hemisphere ice sheets about twice larger than those before 0.95 m.y. (Raymo et al., 1989). The Late Pleistocene glacial cycles were also accompanied by major differences in the flux and source of North Atlantic deep water as deduced from measurements of carbon isotopes in benthic foraminifera. During the interglacial mode, a strong flux of North Atlantic deep water occurs as today. During the glacial mode, only a weak flux of North Atlantic deep water forms in the open northern North Atlantic Ocean south of Iceland but north of the glacial-age polar front. Active formation of an intermediate water mass occurs south of the polar front.

As one possible explanation of the emergence of the asymmetrical 100 k.y. glacial cycles, Maasch (1988) argued that the mid-Pleistocene change in the climate system was global and abrupt, both in the mean and in the variance of the changes in the benthic $\delta^{18}O$ record, thus suggestive of a bifurcation in the climate system in the absence of any known tectonic forcing. This concept of a rapid transition is also favored by Prell (1985), Shackleton and Opdyke (1976), and Pisias and Moore (1981). According to the model of Saltzman and Maasch (1990), which is a generalization of previous models (Saltzman and Maasch, 1988; Maasch and Saltzman, 1990), this abrupt bifurcation occurred when the long-term Cenozoic cooling, perhaps the result of drawdown of atmospheric carbon dioxide from weathering of uplifted mountains, was sufficient to cause the climate system to become unstable and thereafter to undergo asymmetrical 100 k.y. oscillations. By this model the 100 k.y. cycles are thought to represent an auto-oscillation of the global climate system, paced by orbital forcing, in which trace gases have a prominent role. The onset of mid-Pleistocene instability of the climate system caused by this inferred bifurcation is coincident with the first appearance of large Northern Hemisphere ice sheets in the Cenozoic.

But it is still possible that the important mid-Pleistocene climatic transition was caused by a tectonic event. This is likely if the mid-Pleistocene change turns out to be complicated and transitional, as suggested by the evolution of the 100 k.y. rhythm in Late Pleistocene time (Imbrie, 1985; Ruddiman et al., 1986b, 1989). During this transition, the production rate of North Atlantic deep water during glacial maxima again declined (Raymo et al., 1990), and significant precession power emerged in the benthic $\delta^{18}O$ record (Imbrie et al., 1984, Imbrie, 1985). Both of these developments are consistent with sea-surface temperature decline in the Arctic and a southward growth of Arctic sea ice into regions strongly affected by the distance component of seasonality. A likely but undocumented tectonic cause would be further uplift of the Greenland-Scotland Ridge, similar to that which occurred earlier during the mid-Pliocene climatic transition. Wright and Miller (1996) related uplifts of this ridge to reduction in North Atlantic deep water because sill depths are now generally less than 500 m, and the deepest is less than 1000 m. Thus the volume of waters that spill over the ridge to feed North Atlantic deep water are potentially sensitive even to small uplifts of the ridge crest.

The impression from the paleoclimatic record is that the changes in the northern North Atlantic Ocean (cooler sea-surface temperatures, spread of sea-ice field) during the mid-Pleistocene climatic transition had a fundamental effect on the response of thermohaline overturning to orbital pacing such that the production of the important component of North Atlantic deep water produced by spillover from the Nordic Seas became increasingly easy to turn off and difficult to turn back on. For example, at the end of the penultimate interglaciation, the deep waters of the ocean underwent a 2°C cooling, and the air temperature over Antarctica cooled significantly (Broecker & Denton, 1989). Both events were probably linked to a reduction or cessation of North Atlantic spillover waters triggered by an orbitally induced drop in summer radiation at the isotope substage 5c/5d boundary. The deep-water cooling came from an expansion of Antarctic deep water at the expense of North Atlantic deep water, and the Antarctic air temperature came from a spread of Antarctic sea ice due to less rising North Atlantic deep water in the Southern Ocean. The North Atlantic thermohaline circulation did not subsequently succumb to radiational forcing in the opposite sense. In fact, it never recovered to an interglacial mode until the Oldest Dryas/Bölling transition nearly 100 k.y. later. The result was that the buildup of Northern Hemisphere ice sheets could begin during substage 5d. A correction of the $\delta^{18}O$ benthic marine record for the temperature effect at the substage 5c/5d boundary results in a nearly triangular shape for the last 100 k.y. ice-volume cycle following the initial buildup (Shackleton, 1987; Broecker & Denton, 1989). By as early as isotope stage 4 this growing ice-sheet complex peripheral to the North Atlantic Ocean was larger than the maximum ice sheets produced during the earlier 41 k.y. cycles. By isotope stage 2, near the culmination of the last 100 k.y. cycle, the Northern Hemisphere ice sheets were nearly double the size of their Late Pleistocene predecessors.

Thus it is postulated that the necessary condition for the dominance of 100 k.y. cycles in Late Pleistocene time is the insertion into the climate system of a northern North Atlantic thermohaline circulation that even in an interglacial mode is close to the threshold of switching toward a glacial state. Once this initial mode switch is triggered by orbital forcing, the resulting thermohaline circulation does not switch back to the interglacial mode from subsequent orbital forcing in the opposite sense as it did before 0.95 m.y. when the system cycled regularly with a 41 k.y. pulsebeat. Such a change in behavior of the thermohaline circulation served to decouple the Late Pleistocene climate system from the dominance of orbital forcing and hence allowed the glacial mode to become increasingly more robust for nearly 100 k.y. The weakening and strengthening of thermohaline circulation from orbital pacing is seen only as an overprint on this buildup to maximum glacial conditions. Under this scenario a flip of the thermohaline circulation system is required to initiate ice-sheet growth in the Northern Hemisphere. Broecker and Denton (1989) postulated that the reduction in ocean heat transport and the reduction in greenhouse gas capacity that are the consequence of the change in the mode of ocean–atmosphere operation may well be prerequisites for ice-sheet growth. Once they reach a critical size, the ice sheets can help lock in the glacial mode of thermohaline overturning in the North Atlantic from iceberg calving and downwind cooling effects. But what brings the 100 k.y. cycles to an abrupt end? To answer this question, I turn to a discussion of Heinrich events in the North Atlantic Ocean.

Massive, short-lived discharges of icebergs into the North Atlantic Ocean occurred each 7000–10,000 years during the gradual buildup phase of the last 100 k.y. cycle (Heinrich, 1988). These dramatic outbursts of icebergs left prominent layers of ice-rafted debris that were deposited rapidly on parts of the North Atlantic sea floor (Bond et al., 1992, 1993; Broecker et al., 1992; Bond & Lotti, 1995; Manighetti & McCave, 1995; Manighetti et al., 1995). Such outbursts have become known as Heinrich events after the discoverer of the layers of ice-rafted debris (Heinrich, 1988). The region of maximum sedimentation, which features carbonate detritus from Canada, is along the southern margin of the glacial-age North Atlantic Ocean. Here grains from the penultimate Heinrich layer deposited about 20,500 ^{14}C yr BP show a lead isotopic composition consistent with derivation of ice-rafted debris from eastern Canada (Gwiazda et al., 1996). In the eastern part of the main iceberg track, as well as to the north of the main track, Revel et al. (1996) showed that Scandinavian, British, and Icelandic ice caps contributed to the iceberg flux and thus concluded that all the ice caps surrounding the North Atlantic, not just the Laurentide ice sheet, experienced major discharge of icebergs during Heinrich events. From two important cores at the eastern end of the maximum iceberg track, Bond and Lotti (1995) found evidence of synchronous discharges of icebergs from several sources, not only during Heinrich events but also during millennial-scale climatic oscillations known as Dansgaard-Oeschger events in the Greenland ice-core records. A comparison of the Greenland and North Atlantic paleoclimatic records reveals another important characteristic of Heinrich events: they occurred at or near the culmination of short cooling hemicycles superimposed on the overall buildup phase of the last 100 k.y. cycle (Bond et al., 1993). Following each Heinrich event, the sea-surface temperatures warmed as the North Atlantic Ocean made a short but failed excursion toward the interglacial mode of circulation. The situation is even more complex because millennial-scale Dansgaard-Oeschger oscillations are superimposed on each cooling hemicycle (Bond et al., 1993).

The cause of the huge discharges of icebergs during Heinrich events is not clear. Hence it is questionable whether these discharges are a trigger for climatic change or whether they are simply a response to climatic forcing (Broecker, 1994, 1995b). There are at least three possible sources for these massive iceberg discharges into the North Atlantic Ocean. One is from surges of the Laurentide ice sheet (Broecker et al., 1992; MacAyeal, 1993). This mechanism has the advantage of explaining the rapid deposition of Heinrich layers. One disadvantage is that clear geologic evidence of down-trough flow of an ice stream draining the heart of the Laurentide ice sheet has not been found in Hudson Strait (Miller et al., 1993). Another disadvantage is that all ice sheets, not just Laurentide ice, contributed icebergs during Heinrich discharge events (Bond & Lotti, 1995; Revel et al., 1996). A second source for the icebergs is simply increased calving from expanded margins of ice sheets on North Atlantic continental shelves during cold pulses. The main disadvantage of this mechanism is that it cannot easily account for catastrophic outbursts of icebergs or for the sharp lower boundaries of the Heinrich layers that buried organic horizons on the ocean floor within a few hundred years (Manighetti & McCave, 1995; Manighetti et al., 1995). The third source for the icebergs is from collapse of unstable marine-based marginal ice from the North Atlantic continental shelves as occurred, for example, on the Barents shelf during the last Heinrich event (Bischof, 1994). Each cooling hemicycle superimposed on the long buildup phase of the 100 k.y. cycle would then involve expansion of marine-based ice onto the continental shelves, and each would end with collapse of destabilized ice from these shelves during a Heinrich event. This mechanism can explain the dramatic outbursts of icebergs from all ice caps during Heinrich events. But it is important to note that this mechanism requires a pulse of climatic warming to trigger each collapse.

I prefer this third option and shall assume it to be correct in the following discussion. The reason for my preference is that our southern South American paleoclimatic record suggests the existence of such a global climatic trigger for the youngest Heinrich event (Denton et al., 1999). This South American record shows climatic cooling that culminated at 14,600–14,900 ^{14}C yr BP and was followed immediately by climatic warming (Denton et al., 1999) that is coincident with the increased discharge of icebergs into the North Atlantic during the initial phase (14,800 ^{14}C yr BP) and peak (14,100 ^{14}C yr BP) of the youngest Heinrich collapse. This warming in southern South America is the most extensive since sometime prior to isotope stage 4 (Denton et al., 1999). It is mirrored by similar warming in New Zealand (Newnham et al., 1989). Climatic warming at 14,500–14,700 ^{14}C yr BP in the Northern Hemisphere is indicated by the recession of the Barents Sea ice sheet (Bischof, 1994; Svendsen et al., 1996), the Scandinavian ice sheet (Lehman et al., 1991; Haflidason et al., 1995), the Swiss alpine glaciers (Schlüchter, 1988), and the initial reduction of Northern Hemisphere ice volume as deduced from the benthic $\delta^{18}O$ record in an eastern Pacific equatorial core (Shackleton et al., 1988). By this scenario, a pulse of global warming at about 14,500–14,700 ^{14}C yr BP triggered the massive collapse of glacier ice from North Atlantic continental shelves during the youngest Heinrich event.

The consequence of selecting this third option is that the magnitude of each Heinrich collapse would be linked to a combination of the overall size of the ice sheets peripheral to the North Atlantic Ocean and to the magnitude of the warming that triggered each collapse. The final Heinrich event that peaked at 14,100 ^{14}C yr BP would involve the largest collapse of peripheral grounded ice because by this time the ice sheets had reached their overall maximum of the 100 k.y. cycle, as indicated by the detailed $\delta^{18}O$ record derived from benthic foraminifers in the eastern tropical Pacific Ocean (Shackleton et al., 1988). It is postulated that this last collapse caused such a large influx of icebergs into the North Atlantic Ocean that the Atlantic salinity conveyor belt was crippled (Sarnthein et al., 1994). It is further postulated that when the conveyor circulation began to recover after the massive Heinrich iceberg flux cleared the North Atlantic Ocean, it reorganized into its interglacial mode of operation, thus initiating the crucial massive ocean–atmosphere reorganization of the last termination. The probable reasons that the North Atlantic was nudged into an interglacial mode of operation is that orbital forcing had by this time produced an interglacial radiation field and that the massive Heinrich event had flushed out marine-based ice peripheral to the North Atlantic, thus reducing surface salinity and hence sea-ice formation. By this scenario, an asymmetrical 100 k.y. cycle of the Late Pleistocene begins when orbital forcing produces an ice sheet/North Atlantic thermohaline system so robust that it cannot be destroyed by subsequent orbital forcing in the opposite sense, but rather begins to dominate climate. The ensuing long and gradual growth phase ends abruptly when the ice sheets become sufficiently large to produce a Heinrich event massive enough to cripple the glacial mode of North Atlantic conveyor-belt circulation. The role of orbital forcing is then to nudge the system into its interglacial mode as it undergoes reorganization during recovery from this final and massive Heinrich outburst of icebergs.

But what caused the climatic warming that triggered a Heinrich event massive enough to cripple the glacial mode of thermohaline circulation in the North Atlantic Ocean? First, it is argued that this warming began at 14,700 ^{14}C yr BP at the beginning of the Oldest Dryas. The widespread, synchronous nature of this warming implicates a trace gas, most probably atmospheric water vapor produced in the tropics. In this regard, McIntyre and Molfino (1996) showed that abundance cycles of the marine alga *Florisphaera profunda* in high-resolution deep-sea core records from the equatorial Atlantic Ocean correlate with each of the last five Heinrich lithic peaks in the North Atlantic Ocean. The *F. profunda* variations result from moderation of the zonal intensity of tradewinds and hence equatorial upwelling. The timing of *F. profunda* peaks matches that of Heinrich lithic peaks in the North Atlantic Ocean. Furthermore, the *F. profunda* maxima increase in amplitude along with those of the Heinrich lithic spikes, until a dominant peak is reached during the youngest Heinrich event. The likely reason for the timing of the *F. profunda* peaks is that they appear to be locked into the half-precession forcing characteristic of the tropics. The reason for the steady increase in the successive *F. profunda* maxima is that the amplitude of precessional cycles is modulated by eccentricity, which increased toward a maximum about 11,000 calendar years ago. Thus it is tempting to speculate that the Heinrich collapses superimposed on the last buildup hemicycle were each a response to a warming pulse created by the increased production of atmospheric water vapor due to changes in the zonal intensity of the trades, tropical upwelling, and hence sea-surface temperatures. The tendency of terminations to occur during times of higher eccentricity has long been noted (Imbrie et al., 1984). Perhaps the correspondence is a result of precession-forced atmospheric warming driven from the tropics during high eccentricity triggering a massive collapse of glacier ice into the North Atlantic Ocean, which eventually leads to the reorganization of the thermohaline circulation that is the key event of the last termination.

9

Landforms, Climate, Ecogeographic Mosaics, and the Potential for Hominid Diversity in Pliocene Africa

Eileen M. O'Brien
Charles R. Peters

We assume that hominid biodiversity in Pliocene Africa was somehow related to the continent's mosaic of evolving landscapes, vegetation, and changing climate. But how? As yet, we do not know the answer. Our knowledge of the continent's physical geography and biotic environments during the Pliocene and earliest Pleistocene is limited. The development of evolutionary theory, as it concerns both biodiversity and adaptation, is still weak. What we can be glad of is a fairly well-developed descriptive understanding of landforms, climate, and vegetation as they exist today. Theoretical considerations and alternative interpretations of the paleorecord suggest that the environments of the Pliocene were richer than usually assumed, certainly richer than today. We explore the potential of Pliocene Africa to support an array of hominids within a broad theme of environmental dynamics and changing scales of geographic heterogeneity across the continent. In sum, our chapter attempts to sketch the possibilities of hominid biogeographic potential, from deathlands to potential heartlands of relatively high species richness, interrelated through mosaics of contrasting adjacent environments at the regional and subregional scales. First we introduce landforms, climate, and African phytogeography. Then we explore some ecogeographic possibilities for continentwide biodiversity.

Physical Environment: Setting the Stage

Landforms

Africa includes 20% of the world's land surface (30.3 million km^2), stretching 8000 km from Cape Blanc (37°51′ N) in the Northern Hemisphere to Cape Agulhas (34°51′ S) in the Southern Hemisphere. It is 7200 km at its greatest width, from Ras Hafun (51°50′ E) to Cap Vert (17°32′ W). Unlike all other landmasses, it virtually straddles the equator, extending almost equidistant into both the Southern and Northern Hemispheres. Its present-day geographic location and configuration relative to bordering landmasses and oceans (including currents) has been almost the same since the Early Pliocene—northward movement of the African plate having shifted the equator only slightly southward since then (see discussion in F. H. Brown, 1995).

The main factors/processes involved in the physiographic evolution of the continent have been tectonic activity (rifting, epeirogenesis, etc.), volcanism, erosion, and fluvial dynamics. The resulting terrain forms and land systems are referred to as the physiographic divisions of Africa. Lobeck (1946) provides a broad-scale analysis of the continent down to the sub-

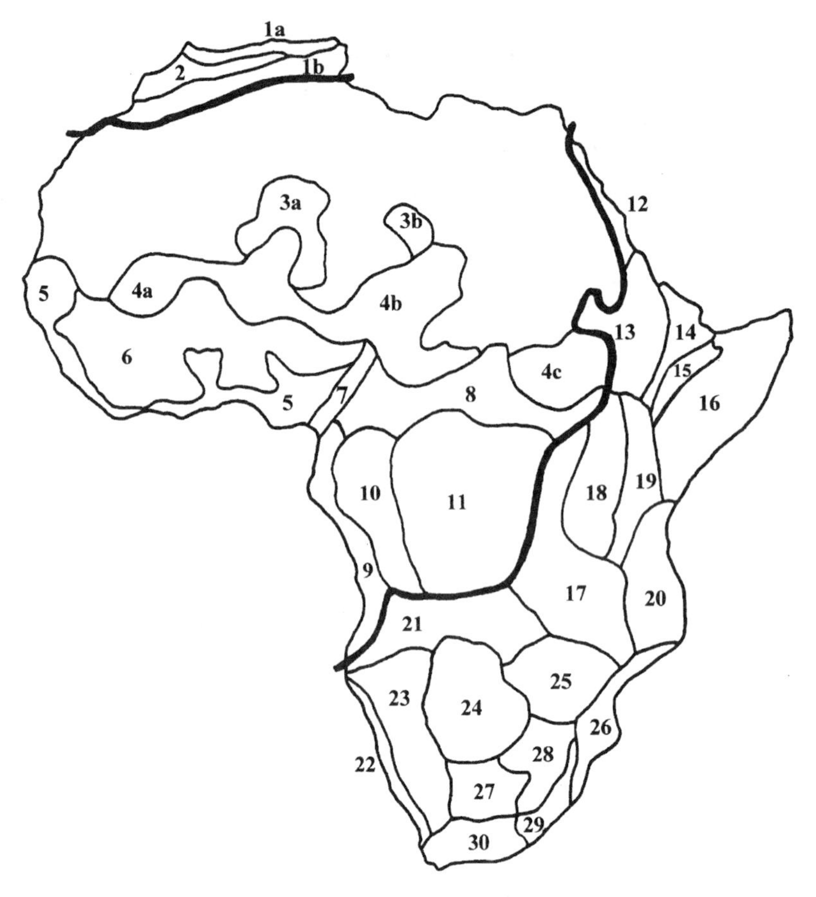

Figure 9-1. Physiographic regions of Africa (cf. Lobeck, 1946). Bold lines discriminate the Alpine System, Low Africa, and High Africa. The regional sections are as follows. Alpine System: **1a,** Mediterranean atlas; **1b,** Southern atlas; **2,** high plateau. Saharan Section: **3,** central Saharan Domes (**3a,** Adrar Des Foras, Ahaggar, Air; and **3b,** Tibesti). Sudanian Section: **4a,** Niger Basin; **4b,** Chad Basin; **4c,** Sudd. Mid-African Section: **5,** Guinea Coast; **6,** West Guinea Highlands; **7,** Cameroon Mountains; **8,** Ubangi-Shari Upland; **9,** West-central Coastal Lowlands; **10,** South Guinea Highlands; **11,** Congo Basin. East African Highland Section: **12,** Etbai Range; **13,** Ethiopian Massive; **14,** Abyssinian Graben; **15,** Harar Massive; **16,** Somali Plateau; **17,** Western Rift Belt; **18,** High Eastern Interior Plateau; **19,** Eastern Rift Belt; **20,** East-central Coastal Belt. South African Platform Section: **21,** Lunda Swell; **22,** Namib Desert; **23,** Damara-Nama Upland; **24,** Kalahari Region; **25,** Matabele Upland; **26,** Mozambique Plain; **27,** High Karroo; **28,** The Veldt (high, middle and low); **29,** Natal Giant-Terrace Belt; **30,** Cape Mountains.

regional level. The following general points are based on his analysis (see fig. 9-1 and appendix 1).

Africa consists of two major physiographic divisions: the Alpine System, nominally part of Eurasia, and the African Massif, Africa proper, so-to-speak. The former is represented by the Atlas Mountains. The rest of Africa, south of the Atlas Mountains, is part of what was once Gondwanaland. This is the African Massif, divided into Low Africa (<900 m) and High Africa (>900 m) (see fig. 1). Low Africa consists of three major subcontinental sections: the Saharan (including 3a,b), the Sudanian (4a-c) and the Mid-African (5–11, below the Sudanian). High Africa consists of two major subcontinental sections: the East African Highlands (12–20) and the Southern Platform (21–30). As can be seen, each of these subcontinental sections is in turn made up of one or more smaller land systems. Lobeck (1946) recognized 43 of these subdivisions. We have simplified his account (see fig. 9-1 and appendix 1), modifying it somewhat, and empha-

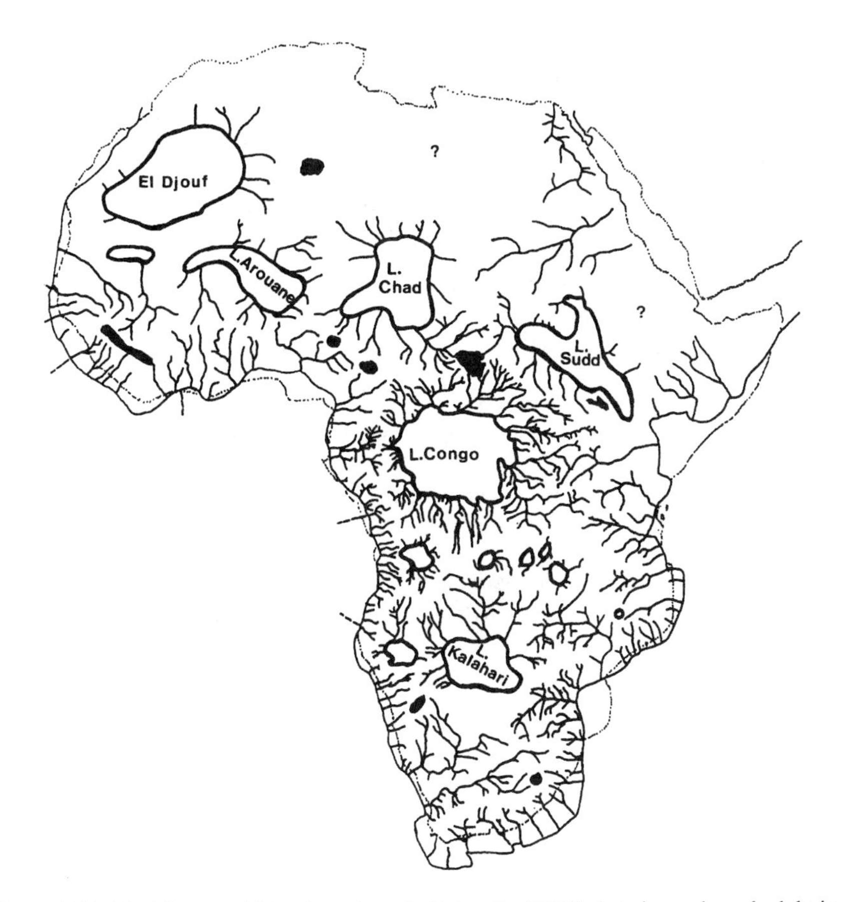

Figure 9-2. Pre-Rift Mio-Pliocene Africa; based on de Heinzelin (1963). Interior, unbreached drainage basins are conjectured to have contained seasonal, if not perennial water; these waterbodies are boldly outlined and larger basins named (cf. Pritchard, 1979). Blackened areas designate some ancient massifs. Dotted outline of continent denotes present-day coastline of Africa. Question marks indicate that data were unavailable or unobtainable (buried, e.g., Ethiopia). Currently submerged west coast river canyons are indicated by dashed lines.

sized the physiographic land systems south of the Sahara. Numbers appearing in parentheses in the text refer to these subdivisions.

Numerous aspects of the modern-day physiographic character of Africa were apparently in place by the end of the Miocene (King, 1978; Pritchard, 1979; Pettars, 1991). Other aspects were developed in the Pliocene (see below) and Pleistocene. On the basis of current knowledge, we do not know the exact timetable, magnitude, and relatedness of these later changes (see F. H. Brown, 1995; Partridge et al., 1995a,b; Feibel, this volume). Therefore any model is an interim model.

The most significant difference between the present-day and the Pliocene physiographic character of Low Africa (elevation <900 m) (1–11) is that during most of the Pliocene, all of its interior basins may still have lacked outlets to the sea. The interior basins and associated drainage systems hypothesized for pre-rift Africa (Miocene-Pliocene) by de Heinzelin (1963) are depicted in figure 9-2. Today, these basins are etched by old terraces, strandlines, deltas, and alluvial deposits, supporting the possibility of at least seasonally expansive internal lakes in the Arouane (Niger) (4a), Chad (4b), and Sudd (4c) basins during the Pliocene (and Early Pleistocene). Most important, Paleolake Congo (covering most of region 11) was likely to have been a perennial water body up to at least the Late Pliocene (as recently as the Early Pleistocene). Its eastern catchment extended to the High Interior Plateau (18) and volcanic highlands of the Eastern Rift Belt (19) (Pickford et al., 1993). The demise of

Paleolake Congo began when the basin's western rampart, the Southern Guinea Highlands (10), was breached by the headward erosion of a coastal river (9) or by a fault (Cahen, 1954; Beadle, 1981; Pritchard, 1979).

In High Africa (11–30), the ongoing, tectonically driven African Swell that began in the Oligocene was active again in the Late Pliocene (and Pleistocene) (Pettars, 1991). It encompasses both eastern and southern Africa but with differential effects.

The East African Highlands Section (12–20) experienced increasing subregional and local environmental fragmentation during the Late Pliocene. Up to about 3 m.y. (in some cases as late as the Early Pleistocene), the rift grabens associated with doming were probably shallow and at somewhat higher elevations than now (F. H. Brown, 1995; Partridge et al., 1995a; Feibel, this volume), making the terrain surrounding these domed areas less fragmented during the Early as opposed to the Late Pliocene. The High Eastern Interior Pleateau (18) may have been higher than now, while the Ethiopian Massif (Afar Dome; 13–15) was probably lower than now (see Feibel et al., 1992, this volume; F. H. Brown, 1995). Beginning around 3–2 m.y., there was major subsidence of rift grabens and concommittant uplift of rift shoulders by 1000–1500 m—first in the Afar (13–15) and Eastern Rift Belt (Kenya Dome) areas (19); somewhat later in the Western Rift Belt (17) where some of the uplifted flanks rose to 4000+ m above sea level (e.g., Ruwenzori Mountains; Pickford et al., 1993; Partridge et al., 1995b). From this point onward, the Western Rift Belt escarpment mountains (in 17) became the eastern drainage perimeter for the Congo Basin and diverted some of the westward drainage from the Interior Plateau (18) and Eastern Rift Belt (19) into the Sudd (4c). Throughout the Pliocene (and Pleistocene), the East African Highlands Section (12–20), and especially the Rift Valley Domain (17–19) within it, were perforated by active volcanoes, some of which (e.g., Mt. Kenya) were high enough for glaciation to have occurred by at least 2 m.y. (Mahaney, 1990).

The Southern Platform (21–30), in contrast, has been characterized by broad-scale regional uplift or subsidence events that have resulted in little environmental fragmentation, except in areas immediately adjacent to uplifted/subsided margins (e.g., upland margins surrounding the Transvaal depression). At the end of the Miocene and during much of the Pliocene, the Southern Platform was apparently lower in elevation than today—by as much as 1000 m in the extreme southeast (28, 29), and by as little as 200 m in the southern escarpment mountains (30) (Partridge et al., 1995a,b). The Transvaal (in 28) was somewhat higher than the present, perhaps an additional 400 m.

Of particular interest to hominid ecogeography in the Southern Platform is the fluvial dynamics and paleolakes (e.g., Makarikari) of the Okavango/Kalahari Basin (24), before the capture or tectonic diversion of its major tributary rivers (e. g. Cunene) in the Late Pliocene (and Early Pleistocene) (Cooke, 1980; Dingle et al., 1983). In the East African Highlands, the grabens of the Eastern and Western Rift belts (17 and 19; 14?) contained intermittent lakes and rivers (e.g., Feibel, this volume). However, Lake Malawi approached its present state only at the end of the Pliocene; Lake Tanganyika approached its present state not until the Pleistocene (Pritchard, 1979; Beadle, 1981).

Climate

Almost the entire continent of Africa is within the global annual heat surplus zone, making water the primary limiting factor for biochemical processes (reviewed in O'Brien & Peters, 1999). The prevailing wind patterns, the migration of the Intertropical Convergence Zone (ITCZ) over Africa, and a variety of physiographic factors determine the distribution, amount, and seasonal variability in Africa's rainfall (Griffiths, 1976; Tyson, 1986; for maps and climate diagrams, see Jackson, 1961; Walter et al., 1960–67). The Southern Hemisphere oceans are virtually the sole source of rainfall for Africa. The prevailing equatorial easterlies, southeasterly trades, and equatorial westerlies originate over the southern Atlantic and Indian oceans and advect warm, wet air, and thus monsoonal rainfall, to Africa. The prevailing northeasterlies out of Europe and Asia, which dominate the northern subtropics and influence the Horn of Africa, are essentially dry winds (and have been so since the demise of the epicontinental seas of Eurasia in the Miocene). The subtropical high pressure cells in the northern and southern Atlantic Ocean generate westerly airflow into subtropical Africa that is also dry, except in winter, when it contributes to winter rainfall in the northern and southernmost fringes of the continent. The resulting distribution of annual precipitation is depicted in figure 9-3.

Africa's present-day climate exhibits four main seasonality-of-precipitation regimes (fig. 9-4). Outside the humid-equatorial diurnal climate, precipitation rarely meets the annual environmental demand for water (measured as potential evapotranspiration, PET). The equatorial bimodal-rain climate covers a significant portion of eastern Africa. Here the short rains often fail, resulting in a brief single rainy season. The summer-rain climate covers the largest portion of Africa. Subhumid conditions occur in a significant

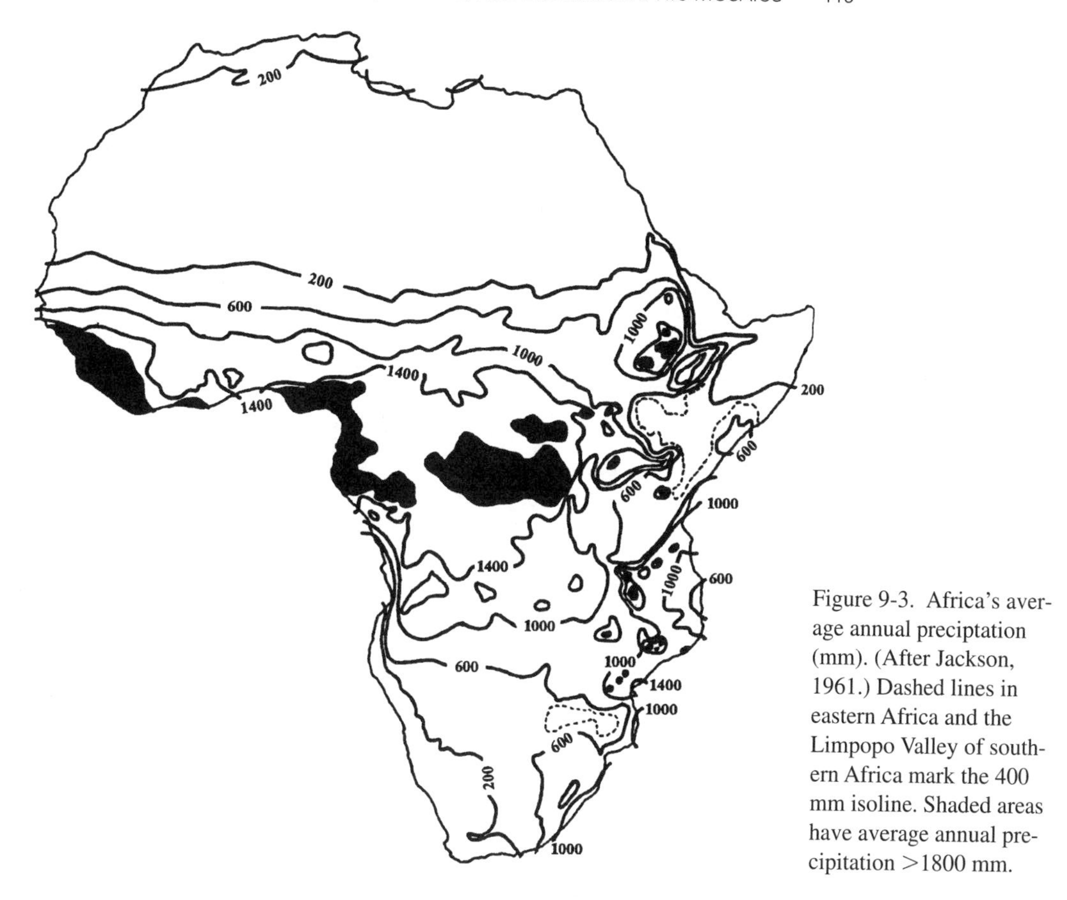

Figure 9-3. Africa's average annual precipitation (mm). (After Jackson, 1961.) Dashed lines in eastern Africa and the Limpopo Valley of southern Africa mark the 400 mm isoline. Shaded areas have average annual precipitation >1800 mm.

portion of summer-rain Africa south of the equator. Here rainfall can meet or exceed the environmental demand for water during the growing season. Finally, note the complex mosaic crossing the Rift Valley Domain (fig. 9-1, 17–19) in eastern Africa, centered on the High Eastern Interior Plateau and the Western Rift Belt west of Lake Victoria.

The same factors and processes contributing to variations in climate over space today cause or contribute to variations in climate over time (Crowley & North, 1991). They include changes in the configuration and physiography of continents and oceans, plus changes in the atmospheric-oceanic circulation/composition (e.g., Ruddiman & Kutzbach, 1989, 1991; Molnar & England, 1990; Raymo & Ruddiman, 1992; Bradley et al., 1993; Burckle, 1995; Kukla & Cilek, 1996). In addition, there are the cyclical changes in earth–sun relationships, especially the Milankovitch cycles (for details see Hays et al., 1976; Crowley & North, 1991; House, 1995; Denton, this volume). These cycles describe changes in the amount/duration of insolation intercepted by the earth's outer atmosphere. They arise due to the gravitational affects of the sun, moon, and other planets on the earth's axial tilt, rotation, and orbit (see Schwarzacher, 1993; House, 1995, for details) and are now documented as far back as the Triassic (Olsen & Kent, 1996). Because these earth–sun relationships are ongoing throughout geological time, it is not clear why the effects of a particular cycle dominate at one time period but not another, nor why ice ages occur at some geologic time periods but not others. Some think it may be a function of processes or cycles that are internal, rather than external, to the planet (atmospheric composition, orogeny, etc.; see various reports in Eddy & Oeschger, 1993). One thing is certain: the potential impact of earth–sun relationships on tropical (ca. 0°–15° N/S) and subtropical (ca. 15°–30° N/S) climates and vegetation is significantly less than their impact on the temperate-to-polar regions.

As suggested by Milankovitch and supported by findings for tropical Pangaea (Olsen & Kent, 1996), the obliquity cycle (41 k.y.) does not have a significant impact on tropical climates (Imbrie et al., 1993). The

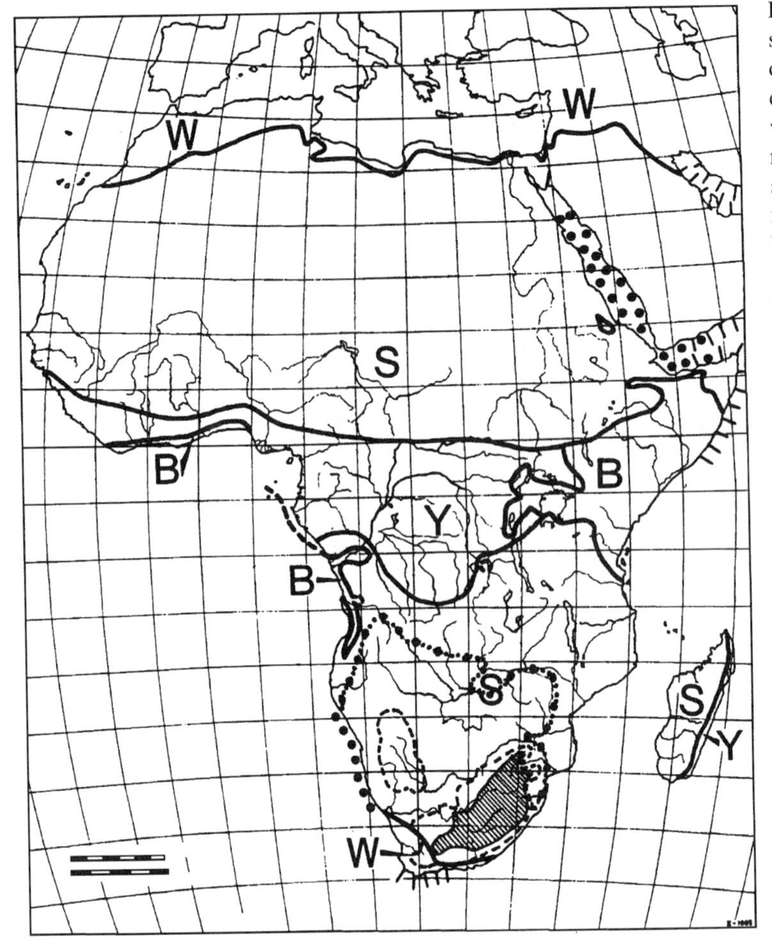

Figure 9-4. Africa's seasonality-of-precipitation climate regions: **Y,** humid-equatorial diurnal climate with rain more-or-less year-round (only a short dry season invariably occurs in most years); **B,** equatorial bimodal-rain climate with two marked dry seasons, particularly pronounced in semi-arid East Africa; **S,** tropical summer-rain climate (subhumid to desert); **W,** winter-rain climate of the Cape and circum-Mediterranean. Coastal black dots denote littoral climates with seasonal mists (fog) chiefly in winter; coastal spikes denote littoral climates with seasonal mists chiefly in summer. From Peters (1990). Frost isolines for southern Africa have been added, based on Werger & Coetzee (1978): the hatched area designates 3 or more months of frost annually; the dashed outlines indicate the areas that have frost annually; the dotted outline indicates the frost-susceptible subregion, where frost is known to occur, but not annually.

greater the tilt, the greater the seasonal difference in insolation, but in tropical regions, the distance to the sun remains essentially the same year-round, regardless of axial tilt. In subtropical regions, some seasonal changes in daylength/insolation occur, especially beyond the Tropics of Cancer and Capricorn. At higher latitudes, typical departures from present-day differences can be ±5% (Imbrie, 1985).

The precession cycle (ca. 19–23 k.y.) effects seasonal changes in the distance between the sun and earth due to changes in the earth's rotation. The resulting wobble in the earth's axial rotation and changes in the orbit produce the precession of the equinoxes, whereby the direction of the earth's axial tilt relative to the sun changes over time. According to Short & Mengel (1986), over oceans, the effects of the precession cycle on temperature can be relatively minor. Over land, it can be significant, especially in the mid- to high latitudes of the globe. But for Africa, it is estimated that only a 0°–1°C difference (2°C in northeastern Africa) occurs between the warmest and coolest maximum monthly (summer) temperatures across a precession cycle (Short & Mengel, 1986).

The eccentricity cycle (106 kya) involves a shift in the shape of the earth's orbit from almost spherical to an ellipse, causing seasonal differences in the distance of the planet from the sun. At present, the earth's orbit is close to spherical, and the upper atmosphere receives about 3% more insolation at perihelion than at aphelian. During an extreme phase of the elliptical orbit, this intra-annual difference can be as much as 30%, and absolute annual insolation can decrease (5°–10°C) (Imbrie et al., 1993; House, 1995). The cycle most likely to alter Africa's vegetation is the eccentricity cycle.

The effect of Milankovitch cycles on Africa's climate/vegetation depends on how they alter the realized heat–water balance during the growing season. In Africa, the environmental demand for water is generally high to extremely high during the growing season (except in some winter rainfall areas), making water the limiting factor. However, because a reduction in insolation would tend to reduce the environmental demand for water during the growing season, the eccentricity cycle is more likely to positively rather than negatively affect vegetation everywhere in Africa except at high elevations (especially >2000 m) and at Africa's highest latitudes where seasonal cold can be a limiting factor (note frost areas in fig. 9-4). Elsewhere, decreases in the environmental demand for water during the growing season should increase productivity, especially if the associated precipitation regime remained the same or increased. Furthermore, species richness, productivity, and/or vegetation type could remain the same even if precipitation decreased, as long as the decrease was on par with the decrease in the environmental demand for water (see O'Brien, 1998). This is dealt with further below.

We know very little about the actual climate of Africa during the Pliocene. With the exception of a few documented or inferred changes during the Pliocene, the oceanic and atmospheric circulation patterns relevant to Africa are virtually the same today as during the Late Miocene and Pliocene (Crowley & North, 1991). In southwestern Africa, the stabilization and strengthening of the south Atlantic Subtropical High Pressure Cell (ca. 3.2–2.5 m.y.) resulted in the prevailing wet warm westerlies becoming the arid cool westerlies of the present by the mid- to Late Pliocene (Crowley & North, 1991; Deacon et al., 1992; also supported by palaeobotanic evidence for Namibia and the Cape; Axelrod & Raven, 1978; Cowling, 1992).

For northern Africa, Malay (1980) has reviewed findings for the Sahara that suggest deserts first appear in the northern Sahara during the Early Pliocene, while humid conditions prevail (at least seasonally) in the southern Sahara (Sahel), with Paleolake Chad being an extant waterbody during the Pliocene. Malay notes the appearance of aeolian dust in the lacustrine deposits of the northern portions of Lake Chad beginning around 2.2–1.8 m.y.

What other evidence exists suggests northern and equatorial Africa's climate was generally wetter during the Pliocene than during the Middle and Late Pleistocene (e.g., Williamson, 1985; Yemane et al., 1985, 1987; Bonnefille & Hamilton, 1986; Leroy & Dupont, 1994; Dupont & Leroy, 1995). Less mesic, and in some cases arid, conditions appear after 2 m.y. (e.g., Vincens, 1987; Leroy & Dupont, 1994). At present, almost all prehistoric changes in local/regional vegetation tend to be attributed to changes in global climate. For most of Africa, this appears unlikely. There is little, if any, evidence of vegetation change during the Pliocene (including the Late Pliocene) that cannot be more simply and parsimoniously attributed to increased tectonic and/or volcanic activity and/or changes in interior drainage basin dynamics, rather than to global climate change.

An apparent exception is suggested by the marine record of aeolian dust deposition off northeastern and northwestern Africa. During the middle to Late Pliocene (3.2 and 2.4 m.y.), spikes of increased dust deposits are correlated with global ice volume fluxes (e.g., Crowley & North, 1991; deMenocal & Rind, 1993; Tiedemann et al., 1994). This is consistent with a steepened temperature gradient between the poles and the equator causing an increase in the velocity of the prevailing dry northeasterlies and with the hypothesized evolution of the Indian monsoonal rainfall patterns associated with the Himalayan uplift (ca. 2.8 m.y.) (see Prell & Kutzbach, 1992). Because the potential for airborne dust (and pollen) load increases with wind velocity, an increase in subsequent marine aeolian deposition is to be expected. Because the dust can be from the same source areas, however, this does not demonstrate that there was a concommittant increase in aridity elseswhere in Africa.

Alternative Hypotheses about Landform–Climate Interactions

The presence of major interior paleolakes may be the most significant factor affecting Africa's general-to-regional climates during the Pliocene. These lakes would have been internal buffers, dampening or eliminating the potential effects of anomolous or cyclical variability in the amount and distribution of ITCZ-driven rainfall, as well as offsetting or minimizing the effects of the arid northeasterlies. Being sources of surface water for evaporation and thus the hydrologic cycle, these lakes could provide atmospheric humidity, generate rainfall/clouds, contribute to water table and soil moisture stability, and thereby reduce seasonal stress on vegetation, especially during periods of reduced monsoonal rainfall.

In effect, Africa was probably wetter, and in some places warmer (e.g., the Southern Platform), during the Pliocene than today. Given the presence of

Paleolake Congo and the perpetual presence of the ITCZ in equatorial regions (reviewed in O'Brien & Peters, 1999), high year-round rainfall (and cloud cover) is expected to have persisted in this region throughout the Pliocene, with an eastward extension of wet equatorial westerlies as far as the interior plateau and volcanic highlands of eastern Africa (fig. 9-1; 18, 19). Given the presence of large interior waterbodies in the Sudanian basins and Southern Platform, the subhumid summer rainfall region (>1000 mm) is expected to have extended beyond its present-day perimeters (north and south of the equator) even during the Late Pliocene. The less mesic to semiarid summer rainfall region of today was probably wetter than present and located further northward/southward of its present-day limits.

However, some regions and subregions of Africa were probably more sensitive than others to even the low magnitude global climate flux of the Pliocene. Keeping in mind that Africa is in an area of the world that suffers a surplus of heat, arid lands would have been supersensitive areas. In particular, arid regions resulting from dry northeasterlies (northern Sahara) are likely to have been supersensitive to changes in the moisture content of these winds. Arid lands resulting from dry westerlies related to upwelling (Namib) or the temperature of west coast cold currents are also likely to have been supersensitive to global climate change. Highest elevations throughout Africa (especially >2000 m), but particularly in Africa's higher latitudes, are likely to have been supersensitive to temperature changes.

Outside of these marginal areas, moderately small semiarid regional transition zones are likely to have occurred where precipitation was solely ITCZ driven, seasonal, normally insufficient to meet the annual environmental demand for water, influenced by the strength and location of the continental high pressure cell over southern Africa, and/or subject to marked interannual fluctuations in timing or amount of rainfall (semiarid summer-rain Africa and the drier bimodal rainfall regions). These semisensitive areas would have been primarily the northern fringes of both the Sudanian lakes and Saharan dome regions of northern Africa, the southern Kalahari Basin in southern Africa, and the Horn of eastern Africa.

But most of Africa would have remained insensitive to mid- to high-latitude changes in climate and accompanying atmosphere/oceanic circulation. These would have been primarily the equatorial humid region and the heart of subhumid summer-rainfall Africa; regions associated with large interior lakes; coastal margins associated with warm currents; and the mid-elevation Afromontane areas (<2000 m?), especially on the windward side of the tallest mountains. Energy (heat/light) is not a limiting factor in these areas, and they are likely to have had high predictable rainfall (at least 1000 mm) and annual water surpluses, or at least surpluses during the growing/rainy season. Thus, even if ITCZ-driven precipitation were to decrease somewhat, they are likely to have received sufficient amounts to have maintained their extant vegetation.

Biogeography: Neontological Patterns in the Distribution of Vegetation

Climate's relationship to hominid biodiversity and paleoecology is likely to be a function of its relationship to plants, which were the primary source of food and shelter. First, we describe the present-day spatial variation in the ecophysiognomy, chorology, and diversity of Africa's vegetation and plant species. Then this variability is discussed in terms of the wild edible plants exploited by humans today in Africa. This is followed by a hypothetical reconstruction of Africa's Pliocene vegetation patterns.

Five general points can be made about climate's relation to the vegetation formations of Africa at the continental to regional scale of analysis. First, the ecophysiognomic character of vegetation types and biomes, their distribution, and the plant life forms associated with them (tree, grass, succulent, etc.) have long been known to be related to climate (e.g., Koppen, 1931; Thornthwaite, 1933, 1948; Raunkaier, 1934; Walter, 1971). Second, much of the vegetation is well adapted to seasonal changes in precipitation and to drought. Third, climate exerts a primary control over the quality and duration of the growing season—i.e., the amount and duration of photosynthesis and subsequent biological activity (e.g., Woodward, 1987), as well as the distribution of species and subsequent spatial variations in species richness (e.g., O'Brien, 1993). However, the relationship is not linear. Both productivity and richness, as well as rainfall, display a parabolic curve when graphed against increasing potential evapotranspiration (heat/light; environmental demand for water), emphasizing that temperature is both positively and negatively related to vegetation, and that too much is as limiting as too little (O'Brien, 1998). Fourth, changes in climate that result in decreases in the effective moisture regime (some function of heat–water balance, or actual evapotranspiration) are more likely to result in decreases in the quality/length of the growing season before they effect changes in constituent plant species and, thus, vegetation, especially in subhumid and semiarid regions where most species have broad climatic tolerances. Fifth, changes in climate that occur during the dormant season are of little or no relevance to vegetation, unless they result in frost or alter its seasonal duration. This is a limiting factor only on the interior plateau of southern Africa (see fig. 9-4) and the highest elevations of mountainous Africa.

In addition to climate, physiography determines the diversity of habitats/communities within a region or subregion. In other words, given the same regional climate, flat terrain generates a more homogeneous vegetation cover (e.g., Somali-Masai Steppe) than physiographically diverse terrain due to the latter's increased micro-climate and habitat diversity (e.g., Mt. Kilimanjaro). High Africa has a more diversified physiography and climate, reflected in a wider range of subordinate vegetation types forming more complicated mosaics (F. White, 1983). After climate and terrain, edaphic factors are the most important influence on the subregional to local physiognomic character of Africa's vegetation. The most well-known example is perhaps the carbonatite-rich, volcanic-ash grasslands of the eastern Serengeti. Less well known, but more important because of their broad distribution, are the wetland vegetation formations of Africa. The example of dambo grasslands is presented below.

Africa's Major Ecophysiognomic Vegetation Types

Africa's broad-scale ecophysiognomic vegetation types represent formations of regional extent (White, 1983) that are broadly related to climate (Walter, 1971) (cf. figs. 9-3, 9-4, 9-5). This is the first step in phytoecological analysis for African biogeography, but it is highly simplified. The original 100 mapping units presented at 1:5 million by F. White (1981) have been reduced to 14 mapping units in figure 9-5 (cf. White, 1984). As White (1984) points out, even at the original scale of 1:5 million, the mapping units rarely correspond to single ecophysiognomic vegetation types. So the mapping units are more complex than shown here. The more important of these complexities are summarized below.

Figure 9-5 depicts the major vegetation formations under dry late Holocene (<3000 yr BP) climate conditions. It is partially a reconstruction (see F. White, 1983) because much of the current vegetation is a degraded product of anthropogenic influences, as witnessed by numerous historical observations, iron age archaeology, ecological studies of secondary succession, and palynological studies.

Tropical Lowland Rainforest: this is moist tropical lowland rainforest and drier peripheral (partly deciduous) transition-zone rainforest. Most of the drier transition forest has been destroyed. The mapping unit also includes extensive areas of swamp forest.

Broadleaved Woodland: most of this mapping unit is probably climax deciduous woodland; some is secondary where semievergreen dry forest has been destroyed; much of the landscape is now wooded farmland.

Thorn Bushland, Wooded Grassland, and Semidesert Vegetation: (1) Sahel—northern portion (<250 mm average annual precipitation) is semidesert grassland (on deep sand), southern portion (250–500 mm average annual precipitation) is *Acacia* wooded grassland (on deep sandy soils), while *Acacia–Commiphora* bushland is much more restricted (mostly confined to rocky outcrops); (2) Southern Kalahari—*Acacia* wooded grassland on the thick mantle of Kalahari sand and *Acacia* bushland on the stony soils; (3) East Africa—most of this area is covered with *Acacia–Commiphora* bushland and thicket, which grades into and is replaced by an ecotone of evergreen bushland on the lower slopes of the mountains above the deciduous bushland, but below the montane forest. In East Africa, there are also relatively small areas of semidesert shrublands (on stony soils) and semidesert grasslands (on deep sand) in the driest parts of Kenya and Somalia. In the wetter parts of this mapping unit (ca. 500 mm average annual precipitaion) *Acacia* woodland can occur, but in East Africa large bushes usually remain dominant with only scattered trees present (e.g., *A. tortilis* trees 6–8 m tall at a density of about one per hectare).

Deserts: some notes on the Sahara and the Namib can be found at the end of this section and in appendix I.

Mediterranean Sclerophyllous Forest and Cape Sclerophyllous Shrubland: most of the former has been destroyed in North Africa, and the Cape Shrublands (Fynbos) have not fared much better.

Karoo-Namib Semi-Desert Vegetation: this area has a number of unusual shrub and grass lifeforms that give the landscape a somewhat less than barren appearance.

Highveld Grassland (Highveld): extensive grasslands where woody plants are rarely seen, except in sheltered ravines, on rocky hills, along streams, and on the escarpment.

Eastern Coastal Forest: this has been largely destroyed, degraded to secondary wooded grassland and grassland.

Afromontane Vegetation: this includes moist forest, dry forest, bamboo, ericaceous bushland, afroalpine shrubland, and various types of secondary grassland. Unfortunately, much of the

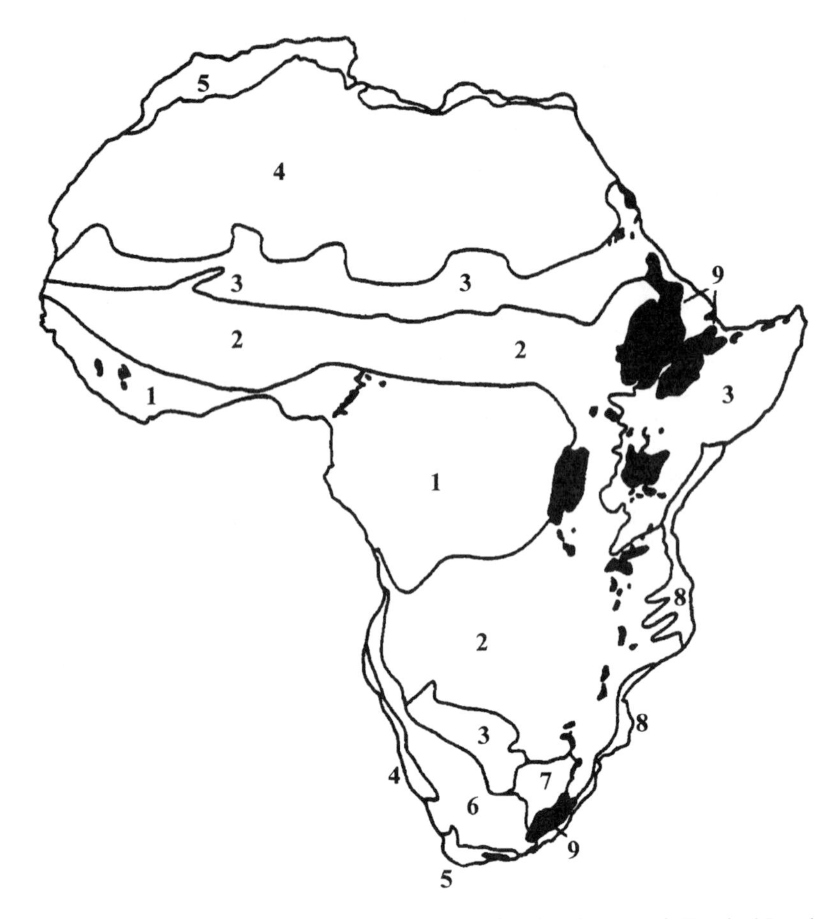

Figure 9-5. African ecophysiognomic vegetation formations of regional extent: **1,** Tropical Lowland Rainforest; **2,** Broadleaved Woodland; **3,** Thorn Bushland, Wooded Grassland and Semidesert Vegetation; **4,** Desert; **5,** Mediterranean Sclerophyllous Forest and Cape Sclerophyllous Shrubland; **6,** Karoo-Namib Semidesert Vegetation; **7,** Highveld Grassland; **8,** Eastern Coastal Forest; **9,** Afromontane (including Afroalpine) Vegetation. Based on F. White (1984), modified somewhat by reference to Central Intelligence Agency map 800630(547147)6–86.

forest, and especially the transition vegetation between the montane forest of the lower slopes and the lowland vegetation, has nearly everywhere been destroyed by over-harvesting of wood, by fire, and by cultivation. The mapping unit also includes submontane evergreen bushland in eastern Africa (including Ethiopia) and submontane forest in eastern Zaire.

Some of these vegetation formations (or ecophysiognomic types) are of greater importance than others in our modeling of hominid diversity. Perhaps the most misunderstood vegetation component is that of the grasslands.

Although grasses play a significant role in much of the vegetation that covers the Sahel, Sudano-Zambezian, and Kalahari portions of the continent, they are probably not the natural dominants in these areas. F. White (1983) recalls H. Walter's (1971) conclusion that wooded grassland is the zonal vegetation on sandy soils covering flat terrain receiving 250–500 mm annual rainfall in the summer-rain regions. White (1983) notes these requirements are found extensively only in the wetter half of the Kalahari portion of the Kalahari-Highveld transition zone, and the wetter half of the Sahel transition zone (these transition zones are defined in the section on phytochoria, below). These zones stand in contrast to East Africa. For East Africa, in the bimodal-rain region, where the annual rainfall is 250–500 mm, bushland is the dominant vegetation (perhaps in large part because of the near absence of extensive flat areas mantled with deep sands). Grasses are present, but they are physiognomically subordinate. Generally, where the annual rainfall exceeds 500 mm, woodland replaces wooded grassland and bushland as the natural vegetation. Where the annual rainfall is between 250 mm and 100 mm, pure grassland occurs on sandy soils. Most extensively, this would be the Saharo-Sahelian grasslands (Le Hourou, 1993), the Namib desert grasslands (Tainton & Walker, 1993), and portions of the Somalia coastal plain grasslands (Herlocker et al., 1993). Elsewhere in Africa, the extensive grasslands we commonly see, lacking any significant growth of woody plants, are the consequences of local edaphic and/or pyric conditions (Herlocker et al., 1993). The Highveld Grassland subregion of South Africa is apparently a product of frost, fire, and soil conditions (Tainton & Walker, 1993).

What then of the potential role of grasslands in hominid evolution? The point that we would make here is that the widely distributed edaphic grasslands of Africa (flood plains, and especially dambos, see below) were probably of much more importance to hominids throughout their evolution (and to the grazing large herbivores as well) than the limited zonal grasslands that are under more direct climatic control. (See Walter & Breckle, 1985, for a discussion of grasses and woody plants as competitors, the competitive balance between them, and the ecophysiological factors that favor grasses as the dominant zonal vegetation in arid summer-rain climates.)

Dambos are treeless, seasonally waterlogged, shallow, channelless drainage lines (a few tens to hundreds of meters in width) at the upper end of plateau drainage systems (Acres et al., 1985; Thomas & Goudie, 1985). Typically, immediately adjacent woodlands include an edge community comprising a number of important edible fruit trees (O'Brien & Peters, personal observations in Zambia and Zimbabwe). In the south-central portion of the southern dambo heartland (fig. 9-6), two species of proteinaceous oil seeds of special dietary significance are also associated with the dambos: *Schinziophyton* (syn. *Ricinodendron*) *rautanenii* at the sandy heads of dambos and *Parinari curatellifolia* on the poorly drained soils marginal to the seasonally waterlogged drainage lines (Peters, 1987a,b, 1993). This type of critical food resource complements that provided by the other fruit trees. In addition, the herbaceous community occupying the dambo wetland includes species with edible rootstocks (Peters, 1990, 1994, 1996, in preparation). Moreover, there exists the possibility of hominids scavenging from large carnivore kills in the relative safety of this fine-scale mosaic of woodland and grass (see fig. 9-7), not to mention the possibilities of hunting. This woodland component of the vegetation mosaic is lacking in the vleis of the South African Highveld. In all cases, the herbaceous vegetation (primarily hydromorphic grasses) on the poorly drained moist soils of the bottom lands provides important dry season pasture for grazers. For a variety of reasons, then, dambo environments were probably important habitats throughout hominid evolution. These natural glades of grass permeate the entire Sudanian and Zambezian woodland regions, especially the Zambezian (fig. 9-6). Therefore, increased aridity to produce grasslands in what are now woodlands is not necessary to accommodate the movement of grazers from northern/eastern Africa to southern Africa.

Africa's Phytochorial Regions (Endemism) and Species Richness Patterns

Phytochoria (for our purposes) are most usefully based on species richness and the endemism of the flora at the species level. In F. White's (1983) treatment of Africa, the following (18) phytochoria are recognized: 7 regional centers of endemism; 1 archipelagolike center of endemism (afromontane); 1 archipelagolike center of extreme floristic impoverishment (afroalpine); 6 regional transition zones; and 3 regional mosaics. A regional center of endemism (CE) is a phytochorion that has both more than 50% of its species confined to it and a total of more than 1000 endemic species (White, 1983). These phytochoria are separated by regional transition zones (TZ) and regional mosaics (M), which are named and given equal rank if they are of comparable areal size to the regions they separate.

Figure 9-8 shows the main phytochoria of Africa. There is a general concordance between chorology and ecophysiognomy (F. White, 1983), as can be seen in a visual comparison of their patterns of variation in figures 9-5 and 9-8. But because they have been independently analyzed, continua, transitions, and mosaics also can be discerned (White, 1983).

Some notes on the flora of each of these mapping units follows, based on F. White (1983), to give the reader a feel for the nature of these floristic divisions. The regional centers of endemism are Guineo-Congolian having about 8000 species, with more than 80% endemic; Zambezian, at least 8500 species, with about 54% endemic; Sudanian, possibly no more than 2750 species, with about 33% endemic (a traditionally recognized regional center of endemism whose status is still uncertain); Somalia-Masai, about 2500 species, with possibly 50% endemic; Cape, about 7000 species, more than 50% endemic; Karoo-Namib, about 3500

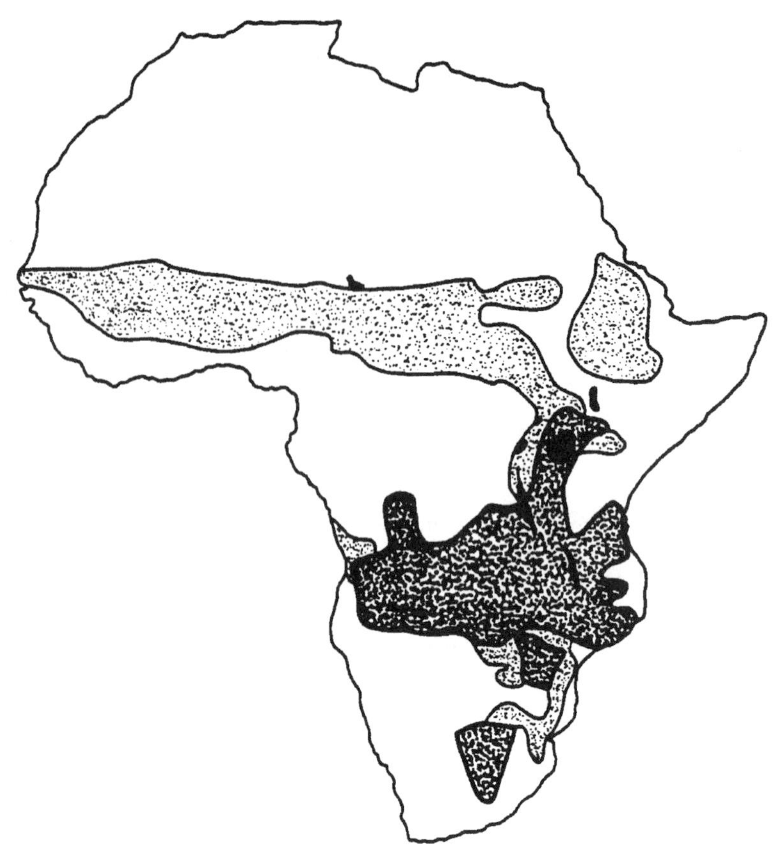

Figure 9-6. Distribution of dambos in Africa. Dark stippling identifies main areas of occurrence; light stippling, areas of sporadic occurrence. After Acres et al. (1985).

species, with more than 50% endemic; North African Mediterranean, about 4000 species, with about 72.5% endemic to the Mediterranean, but only about 20% confined to North Africa; the Afromontane archipelagolike center of endemism: at least 4000 species, with about 75% endemic; in the tropics they are found mostly above 2000 m, whereas in the Cape region exclaves of Afromontane forest are found only a few hundred meters above sea level; the West African mountains west of Cameroun and the highlands of Angola are not included because of the presence of many lowland species in those highlands. Regional Transition Zones include Guinea-Congolia/Zambezia, probably no more than 2000 species, very few endemic; Guinea-Congolia/Sudania, probably fewer than 2000 species, very few endemics; Sahel, about 1200 species, probably less than 3% endemic; Sahara, about 1620 species, with about 11.6% endemic; Mediterranean/Sahara, possibly no more than 2500 species, few endemics; and Kalahari Highveld, possibly about 3000 species, very few endemic. Regional Mosaics include: Lake Victoria, no more than 3000 species, very few endemic; Zanzibar-Inhambane, about 3000 species, at least several hundred endemic; and Tongaland-Pondoland, about 3000 species, with about 40% of the larger woody plants endemic (probably less for herbs and smaller woody plants).

Distribution of Wild Food Plants: Broad Climatic, Phytochorial, and Phytoecological Patterns

Wild food plants generally have broad geographic distributions—most are regional climate and phytochorial transgressors. The genera of edible plant species tend to have broad African, sometimes Afro-Eurasian, distributions (Peters & O'Brien, 1981; cf. Willis, 1973, Mabberley, 1987; Peters et al., 1992). The species also tend to have broad distributions within Africa (e.g., *Adansonia digitata*, baobab; *Sclerocarya birrea,* marula), or to be geographic species of widely distributed genera (e.g., *Cordia, Grewia, Lannea, Rhus*). Some species have marked edaphic preferences but still cover a broad geographic area (e.g.,

Figure 9-7. Dambo grasslands associated with the upper Kafue River system north of the Lukanga marshlands (Zambia). After Michelmore (1939).

Schinziophyton (Ricinodendron) rautanenii, mongongo nut) (Peters, 1987a).

Analyses for southern Africa show that the distribution and species richness of woody plants are strongly related to climate (O'Brien, 1993). Moreover, the distribution of woody plant species-richness mimics the pattern of variation in vegetation ecophysiognomy and chorology in southern Africa (O'Brien, 1993), as well as the pattern of variation in species richness for all vascular plants across South Africa, Namibia, and Botswana (compare with Gibbs Russell, 1987). The same applies to woody plants that provide edible fruit (O'Brien, 1989; O'Brien & Peters, 1998). It is noteworthy that although only about 5% of Africa's plant species provide edible parts, the proportion of woody plants that provide edible fruit is usually much greater. That percentage ranges in southern Africa from <10% in desert and arid-adapted vegetation (Karoo-Namib) to 10–20% in sclerophyllous vegetation associated with winter rainfall (Cape), 10–30% in evergreen/semievergreen subtropical coastal forest associated with high and sometimes year-round rainfall (e.g., Tongaland-Pondoland), to 30–40% in forest and woodland associated with subhumid summer rainfall (e.g., Afromontane, mesic Zambezian, and Zanzibar-Inhambane), to 40–50% in bushland and woodland associated with semiarid summer rainfall (e.g., semiarid Zambezian and Kalahari-Highveld) (O'Brien & Peters, 1998).

Similar patterns are seen when climate is used to predict the macroscale pattern of variation in woody plant species richness for Africa as whole (of both all species and only fruit-providing edible species) (O'Brien, 1998; O'Brien & Peters, in preparation). Again, species richness increases in accord with shifts in vegetation from desert to semiarid vegetation, to shrubland, bushland, woodland, and forest, reaching its greatest number in equatorial rainforest (Guineo-Congolian). When predicted values are used to generate relative proportions of fruit-providing edible

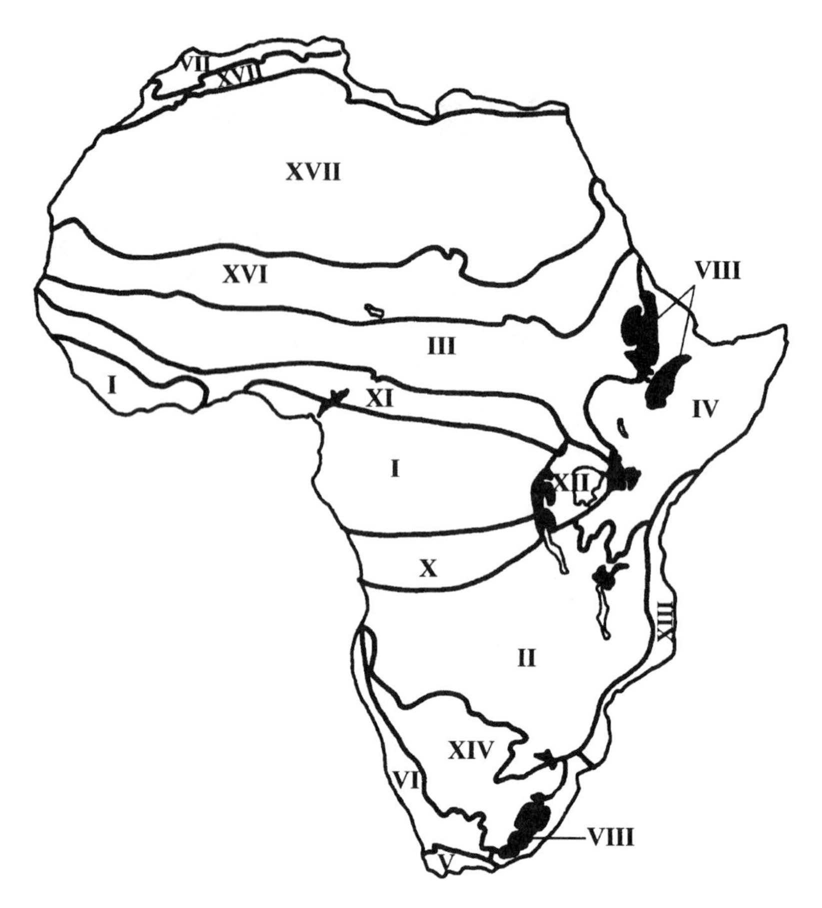

Figure 9-8. The major phytochoria of Africa—regional centers of endemism (CE), regional transition zones (TZ), and regional mosaics (M): **I,** Guineo-Congolian (CE); **II,** Zambezian (CE); **III,** Sudanian (CE); **IV,** Somalia-Masai (CE); **V,** Cape (CE); **VI,** Karoo-Namib (CE); **VII,** North African Mediterranean (CE); the archipelagolike; **VIII,** Afromontane center of endemism is solid black (includes IX, Afroalpine); **X,** Guinea-Congolia/Zambezia (TZ), **XI,** Guinea-Congolia/Sudania (TZ); **XII,** Lake Victoria (M); **XIII,** Zanzibar-Inhambane (M); **XIV,** Kalahari-Highveld (TZ); **XV,** Tongaland-Pondoland (M), **XVI,** Sahel (TZ), **XVII,** Sahara (TZ), and **XVIII,** Mediterranean/Sahara (TZ). After F. White (1981, 1983), conserving his numbering system.

species, the patterns are similar to those obtained for southern Africa. For example, in the equatorial rainforest zone the proportion of fruit-providing species is 20–30% and similar to the wettest portions of the Tongoland-Pondoland phytochorion. The Sudanian phytochorion is similar to the Zambezian; the Mediterranean phytochorion, to the Cape; the Sahelian and Somali-Masai phytochoria, to the Kalahari-Highveld transition zone. Moreover, the spatial discreteness of predicted values (within a 75 km radius of a climate station) emphasizes the mosaic of vegetation types that can occur at the subregional to local scale of analysis within broadscale phytochorial and ecophysiognomic vegetation categories (O'Brien, 1998; O'Brien & Peters, in preparation). The foregoing emphasizes the potential importance of climate as a predictive factor with regard to hominid dietary ecology during the course of human evolution and the role that climate change (regardless of cause) has probably played in influencing biodiversity across the African continent.

Toward an Ecogeographic Model of Pliocene Africa

According to Axelrod and Raven (1978), the modern flora of Africa was virtually extant by the Late Miocene/Early Pliocene. Figure 9-9 is their map of

Africa's vegetation for that time period. It is modified here to reflect a more conservative estimate of the areal extent of high elevation afromontane and afroalpine vegetation in eastern Africa (>2000 m). Only in the Late Pliocene-Early Pleistocene is this vegetation type expected to have approached its present-day distribution (or exceeded it, perhaps, as originally depicted by Axelrod & Raven (1978).

Apparently, nearly all of Africa was vegetated in the Pliocene. The extensive deserts we know today apparently did not exist. Deserts were present only in parts of northern Africa and the Namib, and perhaps the Horn. Coastal forest is not discriminated in figure 9-9, but it probably extended from Somalia to South Africa.

In the equatorial region, rainforest associated with the humid equatorial–diurnal climate and the Guineo-Congolian phytochorion is thought to have extended uninterrupted from West Africa (no Dahomey Gap) to what is now the Eastern Rift Belt (fig. 9-1, 19), plus farther north and south of its present-day position. This type of forest probably also extended southeastward into parts of southern Tanzania and northern Mozambique, where it graded into coastal forest.

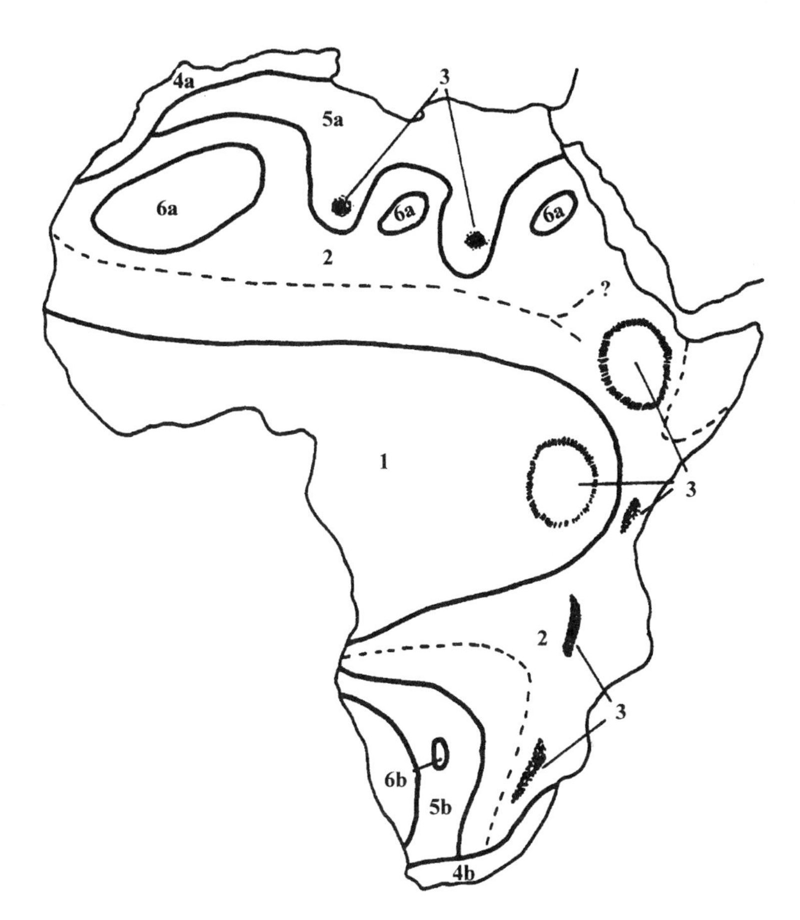

Figure 9-9. Inferred vegetation ecophysiognomy for Africa during the Late Miocene–Early Pliocene. Based on Axelrod and Raven (1978), modified somewhat in terms of montane rainforest. Numbers identify ecophysiognomic vegetation types of regional extent: **1,** Tropical Lowland (and mid-elevation montane) rainforest; **2,** broadleaved woodland; **3,** (high elevation) montane rainforest to Afroalpine (?); **4,** Subtropical laurel forest—a, Canarian, b, Cape; **5,** sclerophyll vegetation—a, Tethyan, b, Cape; **6,** Thornscrub-succulent woodland—a, Sahelian, b, Kalaharian. Note numbering does not match that of Axelrod and Raven (1978). The hatched circles outline areas subjected to tectonic doming and rifting prior to the Pliocene that were not uplifted to their present-day elevations until the Late Pliocene–Early Pleistocene. We have added dashed lines to suggest a hypothetical division of broadleaved woodland by the Late Pliocene into subhumid woodland/forest versus more seasonal or semiarid summer and bimodal rainfall associated woodland/bushland.

Today, the flora of this coastal region (fig. 9-1, 20) still shares strong affinities with the Guineo-Congolian phytochorion. Exclaves probably occurred elsewhere, especially in the Ethiopian cul de sac, as well as in mountains east of the volcanic highlands (e.g., Usambara Mountains).

Broadleaved woodlands and drier transition forests, associated with subhumid summer rainfall and the Sudanian and Zambezian phytochoria, surrounded this humid equatorial core, extending farther north, east, and south of their present-day positions, well into the Southern Platform and to the eastern coast. Again, relatively small, isolated exclaves outside of this broad area probably occurred in mountainous regions within both the humid equatorial–diurnal climate region (rainshadow effects, volcanic mock-aridity) and semiarid portions of the summer and bimodal rainfall areas (e.g., the Saharan Domes). In the Ethiopian highland cul de sac, broadleaved woodland and drier transition forest probably formed an apronlike belt separating montane rainforest from the surrounding more semiarid lowland vegetation. In north and southwest Africa, vegetation associated with semiarid to arid summer rainfall prevailed, grading from deciduous bushland to sclerophyllous vegetation in accord with gradients of decreasing rainfall.

Sclerophyllous vegetation is apparently Africa's most floristically derived and ecologically as well as geographically altered flora. Associated originally with summer rainfall in the Kalahari and Sahara during the Pliocene, it is found today only in the winter rainfall regions of the Cape and Mediterranean areas (with some relicts in the Saharan Domes) (Axelrod & Raven, 1978). This change was associated in part with the Subtropical High Pressure Cells (in the north and south Atlantic) becoming geographically fixed about 3.2–2.5 m.y. in positions that blocked summer rainfall (Deacon et al., 1992). Only those species able to persist in low productivity winter rainfall environments survived (Axelrod & Raven, 1978). Moreover, in southern Africa this vegetation has been reduced ecophysiognomically from woodland to shrubland (*i.e., Fynbos*).

Considering the foregoing in light of our earlier environmental sections, some key points with regard to Pliocene Africa can be suggested:

- Throughout most of the Pliocene, Paleolake Congo and the waterbodies associated with the Sudanian and Okavango-Kalahari basins were sources of water for evaporation, cloud cover, and precipitation in the interior of the continent, independent of ITCZ-driven rainfall.
- The Pliocene climate was wetter than present, with less seasonal variability in the amount and duration of rainfall (and cloud cover). This would have promoted relatively greater productivity, with more reliable growing seasons, in the humid equatorial–diurnal, summer, and bimodal rainfall regions. It also may have contributed to reduced ambient temperatures (environmental energy demand for water) during the dry season.
- The lower elevations of both the Southern Platform section (especially 28) and portions of the East African Highlands section meant these areas were warmer than present. In the Southern Platform, it also meant the area susceptible to frost was smaller (increasing the length of the growing season, etc.).
- Until the Late Pliocene, there were no major mountain ranges (no rainshadows) in what is now the Western Rift Belt (fig. 9-1, 17). Thus, wet prevailing westerly windflow would have been uninterrupted in the equatorial region up to the volcanic highlands of the Eastern Rift Belt (fig. 9-1, 19).
- Although incipient lakes might have been present in the shallow grabens of the early rift system, only Lake Malawi reached its present extent by the end of the Pliocene.
- In regions with volcanic activity, subregional vegetation mosaics were promoted by the local dominance of mock aridity, synergistic with rainshadow effects.
- The eastern coastal forests extended farther inland than present. Relict forests from this period are still found in the Uluguru and Usambara Mountains and in the environs of Iringa in Tanzania.
- The Somali-Masai phytochorion probably occupied a smaller area in basically the same location—eastern Kenya and Somalia (fig. 9-1, 16).
- Afromontane vegetation probably was more restricted geographically before Late Pliocene uplift events than after them.
- Seasonal wetland associations and, in particular, dambos were the dominant grasslands in sub-Saharan Africa.
- Areas dominated by xerophytic plant types were probably of minor significance to the dietary ecology and biogeography of hominids.

Early Hominid Eco-Hyperspace

Basic Ecological Requirements

One way in which changes in landforms, climate, and vegetation are likely to have affected Pliocene hominid ecology is via the distribution and diversity of critical natural resources. Dietary resources for a water-dependent small, chimp-sized ape to large, human-sized ape include potable water, and animal-based food resources and plant-based food resources, with the plant diet providing more food than the animal diet. Nondietary resources would include safe sleeping sites and raw materials for tools.

In terms of potable water, the highest quality and most reliable sources of clean water would be mountain springs/streams and the major regional rivers. Prime breeding sites for hominids would be associated with reliable clean drinking water, given the vulnerability of the young to waterborne infectious diseases (see Peters & Blumenschine, 1996).

In terms of animal-based food resources, most invertebrates and virtually all vertebrates are edible by humans. And, because their exploitation is not restricted to particular species, they can be lumped into ecosystem- or community-based groups or guilds (e.g., grazers associated with edaphic grasslands). This is not the case with plant-based food resources.

Most plants are not edible, and most parts of edible plants are not edible. Because we cannot automatically lump types of plants or their plant parts together as we do with animal foods, we must know a good deal about Africa's wild plant food resources today before we can even think about reconstructing what was available in the past. Fortunately, our analyses and syntheses of information on Africa's edible wild plants are now relatively well-advanced, especially for eastern and southern Africa. Overall, it appears that perhaps 5% of Africa's flora may be considered edible: about 1300 species (Peters et al., 1992) out of about 40,000 identified taxa (Gibbs Russell, 1985).

With regard to the edible plant parts, the range of potential dietary items includes nutlike oil seeds, fruit pulp (fleshy and dry), other types of seeds, seed arils, flowers, flower nectar, leaves, petioles, shoots, stems, bark, pith, cambium, exudates (gum, sap, resin), and rootstocks (tubers, corms, bulbs, etc.), as well as mushrooms (Peters & O'Brien, 1981, 1984, 1994; Peters, 1990, 1994; O'Brien & Peters, 1991).

Some general attributes pertaining to the nutritional value of the different types of dietary items have been identified (Peters & O'Brien, 1984, revised in O'Brien & Peters, 1991). The proteinaceous nut-like oil seeds are the most important source of both plant protein and (especially) lipids. The trees that provide these remarkable seeds are keystone species in a great variety of African environments (Peters, 1987a,b, 1988, 1993). The Sudano-Zambezian species package these seeds in an edible fruit, unlike the nuts of the temperate zones of Eurasia and North America. A variety of vegetative reproductive plant parts (including rootstocks) provide relatively small amounts of protein, but more typically they provide carbohydrates and vitamins. Virtually all edible plant parts contribute minerals to the diet.

Most wild food-providing species are woody plants (i.e., trees, bushes, and shrubs). This is especially so for those species providing edible fruit. The second most common food plant-type is perennial herbs, including wetland forms and dryland geophytes that provide edible rootstocks. Being long lived, both of these plant types ensure a relatively high level of predictableness over time in the type, supply, and distribution of food resources.

In terms of how the diversity of food items varies across the landscapes of Africa, the number of different food-item categories tends to remain about the same from one region or subregion to the next. However, the number of species per category, their plant types, and their relative abundance (described earlier) varies dramatically.

The Ideal Environment

In light of the foregoing discussion, an ideal environment, would be a topographically complex landscape mosaic, an area of landform diversity in the mesic or subhumid summer rainfall region, and an area with a mosaic of vegetation types within the Zambezian phytochorion. Figure 7 depicts one example of this type of environment. The ecophysiognomic vegetation type is primarily broadleaved woodland, but a great variety of wetlands are also represented, including the dambos introduced earlier. Because of the large wetlands component in the Pliocene landscapes of tropical/subtropical Africa, we can expect at least partial parallel wetlands adaptations to have occurred in the African great apes and hominids for feeding on the monocotyledons (herbaceous food plants). [With regard to probable hominid food plants in wetlands, see Peters (1990, 1994, 1996). Also see Peters & Blumenschine (1995, 1996) for theoretical applications to one Late Pliocene basin.]

An environment such as that depicted in figure 9-7 would have provided the greatest diversity of dietary resources available to hominids in Africa. The plant-based diet would have been synergistic—a power and brain food diet because of the lipid, protein, carbohydrate, and micronutrient diversity; but also a diet resulting in decreased deleterious chemical side effects (from secondary compounds) through diversification of intake, substitutions, decreased consumption of any particular food item, and ameliorating affects of medicinal foods. We can speculate that this would have been true across the ontogenetic spectrum, from weaning foods to the adult diet. As one moved out of this optimal type of environment into the more monotonous landscapes of the semiarid regions, plant-food diversity would decrease overall, most dramatically in the dry season. We have partially documented this for food-providing trees and shrubs in extant landscapes (Peters & O'Brien, 1994), and any notable decrease in wetlands would obviously result in dramatic decreases in the plant foods associated with those habitats.

In addition to this great diversity of plant foods, the type of environment depicted in figure 9-7 would also provide an optimal mosaic of animal foods. High diversity and relative abundance of the microfauna would range from invertebrates to small mammals, but the wetlands would also offer spawning fish and migratory bird life in great profusion. Moreover, the drylands in this type of dambo mosaic would offer nonconfrontational scavenging opportunities for large-mammal carcasses along the dambo edges and in the closed woodlands. Finally, we note that the value of this type of environment might be augmented even more by increasing local topographic relief, providing for locally increased gradients in habitat diversity.

Demographic Sources and Sinks

The Environmental Mosaic The demographic dynamics of a species depends on the mosaic of landscapes and habitats that it occupies across the full range of its distribution. The interaction of the local populations across this mosaic determines demographic dynamics at the level of the species. This is true because within this mosaic some landscapes/habitats are demographic sources, while others are demographic sinks (Pulliam, 1988). Geographic areas that are sources are rich enough to support reproductive rates greater than that needed to sustain their local populations. These areas provide the migrating (surplus) individuals that recolonize adjacent areas where reproductive rates are below replacement (i.e., not self-sustaining). A large (perhaps the largest) fraction of the individuals that make up a species may regularly (or episodically) occur in sink landscapes/habitats, where local reproduction is insufficient to balance local mortality. Here breeding sites may be abundant, but they are of poor quality. Pulliam (1988) has argued that active dispersal from source areas can maintain large sink populations. If the reproductive surplus of the source areas is large and the reproductive deficit of the sink areas is small, the great majority of the species population may occur in the sink. This might correspond on a transgenerational time scale to periods of environmental amelioration when species reach their maximum geographic extent.

We should also note that, although demographic sources/sinks are usually thought of as different (adjacent) geographic areas, they can be temporally successive characterizations of the same area as it experiences marked flux in the quality of the breeding sites.

Sources, Sinks and Deathlands Sources would be characterized by mosaics of extensive woodlands and tall bushland, with local groundwater forest; riparian or drier montane forest and medium to small-sized wetlands; footslopes of well-watered mountains in the subhumid tropics and subtropics; numerous small streams with high water quality and fewer crocs; high productivity and species richness of edible wild plants; and safe sleeping sites.

Sinks would be characterized by drier semievergreen peripheral transition-zone lowland rainforest; extensive semiarid bushland; hot, broad valleys, coastal forest, and woodlands; limited potable water; and increased competition for and/or lower productivity and species richness of edible wild plants.

Deathlands would be characterized by desert; arid grassland; arid, wooded grassland; shrubland; extensive wetlands; extensive moist tropical lowland rainforest; and afroalpine and ericaceous mountain uplands.

Regional Mosaics as Settings for Early Hominid Biodiversity

The southern Saharan, Sudanian, and northernmost edge of the Mid-African sections of Low Africa form an interactive ecogeographic system (fig. 9-1, 3–8). In the east this ecogeographic system links up to High Africa through narrow zones or portals connecting it with the Abyssinian cul de sac (12–15) and the northern portions of the proto-East African Rift Belts and High Belts Interior Plateau Belts (17–19). The latter (including 20) form another ecogeographic system that interacts broadly with the northern portion of the Southern Platform (21, 25, 26), which in turn interacts with a number of landsystems of theoretical interest for hominid evolution farther to the south.

Taking into account what we know or have been able to infer about the topography, vegetation, hydrology, and climate of Pliocene Africa, we can offer some working hypotheses toward a biogeographical model of regional hominid biodiversity.

Hypothetical Region(s) of Lowest Hominid Diversity (One-Species Regions) Deathlands would be regions dominated by desertic conditions, arid grassland, arid wooded grassland, and arid shrubland, plus the interiors of extensive wetlands (marshes) and extensive tropical lowland rainforest (Guineo-Congolian).

Regions for potential incipient speciation include:

Central Saharan Domes (3a,b): lie within a seasonal, relatively low rainfall area likely to be subject to climatic anomalies (drought, etc.). Can become isolated from demographic source areas to the south.

Damara-Nama Uplands (23): lie within a highly seasonal, relatively low rainfall area subject to climatic anomalies and perturbation. Can become isolated from demographic source areas to the north in the western arm of the Lunda Swell (21).

Mountains of Northern Somalia Plateau (16): lie within a highly seasonal, relatively low rainfall area subject to climatic anomalies and episodic environmental perturbation. Can become isolated from demographic source areas to the west in the Ethiopian Highlands.

Hypothetical Region(s) of Intermediate Diversity [two species regions] Regional sinks include: (1) the northern edges of the Sudd (4c), which are subject to climatic perturbation. Demographic source areas are to the east and west–southwest; (2) the southern edges of the Kalahari Basin (24), which are subject to climatic perturbation. Hominid occupation would be restricted to woodland habitats. Demographic source areas are to the north and east.

Moderately fluctuating broad-scale environmental mosaics include Ethiopian Massive Apron (13–15): species not indigenous to the region may come in from the east or the west, but are less likely from the south.

Sudanian Section (4a–c): narrow zones of connectedness within the subsections of this region; bordered by tropical rainforest to the south and semiarid or arid vegetation to the north (excepting Central Saharan Domes). Species not indigenous to the region may come in from the Central Saharan Domes (3a,b), the Ethiopian Apron (13), and/or the northwestern East African domain (17–19).

The Veldt (28): bordered by semiarid and/or sclerophyllous vegetation to the west, southwest, and northwest. Species not indigenous to the region can come in from the north.

Hypothetical Region(s) of Highest Diversity (three-plus-species regions) There are only two areas that appear to qualify as regions of highest diversity. One is the East African hypermosaic (Rift Valley Domain) where the Somali-Masai, Sudano-Zambezian, and Guineo-Congolian phytochoria intersect on the High Interior Plateau (18), between what will later be the Western (17) and Eastern (19) Rift Belts. This is an area of potentially fluctuating environments, especially if we include the volcanic uplands of the Eastern Rift Belt as part of the north–south corridor system connecting northern and southern Africa. This East African hypermosaic can be hypothesized to have been largely a sink for all hominid species.

The second, undoubtably more stable area, is the South-Central African centripetal heartland, a broad geographic region centered on southeastern Zambia, and encompassing the following regions: the southern half of the Western Rift Belt (17); the southern East-Central Coastal Belt (20); the Lunda Swell (21); the northern perimeters of both the Damara-Nama Upland (23) and Kalahari Regions (24); and the Matabele Upland (25). Throughout the Pliocene, this was probably the richest environment for hominids, an area of diverse landscapes and complex vegetation mosaics that included the dambo grasslands. This area would be dominated by the Zambezian phytochorion, which probably has the highest diversity (number and relative abundance) of wild food plants in Africa. It can be hypothesized to have been largely a demographic source area for eastern as well as for southern Africa. It could have supported the greatest diversity of hominid species in Africa throughout the Pliocene, well into the Pleistocene. It was probably isolated by distance, more or less completely, from the western Sudanian hominid populations throughout the Pliocene.

Summary and Conclusions

During the Pliocene, Africa's physiographic evolution made it possible for the continent to support a number of hominid species. Regional and subregional environmental heterogeneity contributed to speciation and the development of multispecies heartlands in some regions of the continent. Truly great interior lakes, especially Paleolake Congo, buffered most of Africa from external (global) climate changes during the Pliocene. We hypothesize that only a limited area of Africa was influenced by the moderate global climate changes of the Pliocene and Early Pleistocene. In Africa, climate was wetter and vegetation was lusher and its productivity greater and longer than expected otherwise (then and now). There were no great deserts in the Pliocene (or Early Pleistocene?). In the Pliocene, rainforest or broadleaved woodland (associated with year-round or subhumid summer/bimodal rainfall) dominated the continent, extending uninterrupted from West Africa to the eastern coastal forests, from the Sudanian lakes and Ethiopian cul-de-sac south to the southeastern coast of southern Africa. Semiarid summer rainfall and associated vegetation surrounded this continental core, shifting from bushland/wooded grassland to sclerophyllous woodland to patches of semiarid succulent vegetation and then to subtropical laurel forest at the continent's highest latitudes. In the mid-Pliocene, these high-latitude laurel coastal forests began to be replaced by winter rainfall-adapted sclerophyllous woodlands/shrublands. Portions of western southern Africa (Namib), northern Sahara, and the Horn developed increasingly more semiarid climates in the Late Pliocene that were sensitive to changes in global climate. However, landform changes were still more significant influences on regional climate changes over most of the continent than global climate change. Late Pliocene/Early Pleistocene intensification of tectonic and volcanic activity promoted increased regional to local environmental fragmentation of High Africa (especially eastern Africa). When coupled with changes in interior drainage due to river capture and the breaching of interior basins, these landform changes probably account for most of the changes in vegetation (and regional to local climate) evidenced for this time period.

The ideal habitat for African hominids was probably the woodland/forest and dambo/grassland mosaic, a bioregion associated with the Zambezian (Sudanian) phytochorion. Centered more or less over south-central Africa, this bioregion was probably the most stable ecogeographic heartland for Pliocene (and Early Pleistocene) hominids—a major demographic source area, where reproduction exceeded replacement and where diversity was greatest. Portions of northern and southern Africa also probably were important ecogeographic regions for biodiversification. The Eastern Interior Plateau and the Eastern and Western Rift Belt regions constituted a major subcontinental demographic sink, also expected, because of its geographic location, to have had a high diversity of hominid species.

Appendix 1: The Physiographic Divisons of Africa

This appendix accompanies figure 9-3 and provides brief descriptions for the physiographic regions, **3–30**, of the African Massive (Africa excluding the Alpine System, regions **1** and **2**), following Lobeck (1946). Additional notes from a variety of sources are also included (e.g., Pritchard, 1979; F. White, 1983, for deserts; Beadle, 1981, for inland waters; Pettars, 1991). Physiographically, the African Massive is divided into Low and High Africa, each of which are divided into subcontinental sections or domains, and then into regions. The boldface numbers associated with regional descriptions correspond to the regional numbers in figure 9-3. Most of the elevations were originally given in feet.

THE AFRICAN MASSIVE

I. LOW AFRICA

Excluding the narrow coastal plain and a few high plateaus and mountain peaks, the average elevation is between 200 m and 900 m.

A. The Sahara Section

Descriptions are limited to the Saharan Uplands. The other landforms found here include basins; level Cretaceous limestone plains and sandstone pediplains with outcropping inselbergs; great sand seas (Algeria, Mauritania, and Libya); and flat stretches of desert pavement and bare, windswept hammadas.

3 Central Saharan Domes. Maturely dissected massifs.

3a, Adrar Des Foras, Ahaggar, and Air (eastern domes). Exposed Precambrian crystalline rocks (600 m) with volcanic necks and lava plateaus being the dominant features, ringed by broad lowlands with inward-facing sandstone scarps and shale valleys (with oases).

3b, Tibesti (western dome). Entirely of volcanic origin; deeply weathered lava flows 180–210 m thick, forming towers, turrets and pinnacles or extensive lava-capped plateaus. Elevation approximately 900–1800 m, Emi Koussi being the highest peak at 3300+ m. All domes are associated with permanent and semipermanent water (springs, oases, lakes, seasonal streams). All were watersheds for quaternary fluvial systems.

B. The Sudanian Section

4 Three interior basins along the southern border of Sahara.

4a, Niger Basin. Fed by headwaters of upper Niger in the Guinea highlands. Niger becomes a braided river when it reaches the basin; marshes form (Timbuktu). A paleolake (Arouane) basin; inland deltas and heavy alluvial deposits. Outlet to sea developed when basin was breached by lower Niger in the Late Pleistocene/early Holocene. Today marshland.

4b, Chad Basin. Enclosed by the Air Dome (**3a**; northwest), the Tibesti Dome (**3b**; northeast), and the highlands of mid-Africa (**8**; south). Present-day Lake Chad is freshwater; interior delta for the Chari, Logone, and Komadougou rivers; shallow, 250 m above sea level. A paleolake basin. Old lake bed now huge, reed-grass–covered basin; flat, a marsh after rains. No outlet to the sea.

4c, Sudd Basin. Extremely flat. A paleolake basin breached by lower Nile at Khartoum in middle Pleistocene (Pettars, 1991). Today Nile is channelless within basin; extensive marshes; mass papyrus/reeds, shallow lagoons; dry season grassland pasture.

C. The Mid-African Section

5 Guinea Coast. Low-lying marshy plains and abrupt promontories where interior highlands reach sea; deltaic plains, lagoons; Niger delta, mangrove forests. Extensive rainforest up to highland plateau.

6 West Guinea Highlands. Drained by upper Senegal, Gambia, Upper Niger, and Volta river systems. Remarkably flat, peneplaned Precambrian complex of Africa; average elevation <610 m, with monadnocks rising another 610–900 m; numerous granite tors and ant hills; grass or scrub. Rainforests cover southern, western, eastern flanks.

7 Cameroon Mountains. Double chain of volcanic peaks separated by a graben, Cameroon Peak, approximately 4070 m above sea level, estuaries at foot; basalt plateaus drained by Benue River.

8 Ubangi-Shari Upland or Usande/Banda Swell. Upwarped great African Precambrian platform, forms divide between the Sudan and Congo drainage systems. In the west (Cameroon, Ngaoundere, and Yade plateaus), elevation reaches 900 m; headwaters of the Benue River. Northern slopes gradual, drain into Chad Basin; open grassland, dense riparian forest. Dafur upland is a gentle upswell, 3050 m; many inselbergs; separates Chad and Sudd basins.

9 Coastal Lowlands. Narrow coastline, with a sharp rise to interior uplands. Fringed with bars, tidal lagoons, and mangrove swamps; estuaries. The Congo River has no delta, but a deep channel canyon that extends 160 km out to sea.

10 South Guinea Highlands. Western rampart of Congo Basin, breached only by Congo River, via a narrow 490-m deep gorge; part of the great African Precambrian platform, plateau-like, so flat that the headwaters of rivers are marshlands; relatively recent rejuvenation—all streams typified by cascades, rapids, and waterfalls. Generally, the elevation is >900 m, with peaks rising another 610+ m. The entire region is rainforest, except south of the Congo River, where conditions can be somewhat dryer.

11 Congo Basin. Drainage basin for Congo River and all its tributaries. Former Pliocene lake (remnants include Lakes Tumba and Maindombe); now a mosaic of vast marshlands and wet alluvial plains into which rivers are only slightly entrenched. Unbroken mantle of rainforest. Flanked to the south by a semi-circular platform, 450–610 m above sea level, made up of sedimentary beds, still covered in places by lake deposits, into which the rivers are deeply entrenched. South of this, the land rises gradually over deeply weathered crystalline rocks to the rimming uplands, forming divides between the main tributary river systems. Farther east and south are the highlands of High Africa, approximately 1830–2135 m above sea level. The west and north rims of the basin have average elevations of 900 m. Rainforests in the basin give way to degraded forests/woodlands and scattered-tree grasslands on plateau uplands.

II. HIGH AFRICA

A. The East African Highlands Section

1. The Abyssinian Domain

12 Etbai Range. The western scarp of the Red Sea granite graben; descends approximately 1525–2135 m. Backslope to the Nile River is more gradual, overlapped by sandstone formations. Red Sea coast: narrow alluvial terrace and upraised coral reefs.

13 Ethiopian Massive. Great highland northwest of Abyssinian Graben. Extensive flood basalt-covered plateau. Streams flowing northwestward to the Nile River have deeply dissected this upland into a multitude of steep-walled gorges and flat-topped plateaus. Elevation 3050+ m. Uplands once completely forested.

14 Abyssinian Graben. Continuation of the East African Rift system beyond Lake Turkana (see **19**), north of which the graben splits into two branches, one west toward the middle Nile region, the other east, forming the Abyssinian graben (extending more than 800 km). The latter in turn widens and bifurcates, becoming the Red Sea and Gulf of Aden grabens, also broken up into small graben estuaries (sometimes below sea level) and salt basins. Isolated volcanic peaks. Oases and springs border margins between the uplands and graben floor. The upper portion is a narrow, steep-walled trench, partly filled with recent lava flows; many active and extinct volcanos, fumaroles, hot springs; with several large lakes (some fresh, some salt) in the depressions.

15 Harar Massive. Deep layer of flat flood basalts capping underlying Somali Plateau beds, 3050–3350 m above sea level; deeply dissected and divided into flat-topped blocks by the gorges of rivers flowing to the Indian Ocean.

16 Somali Plateau. Slopes southeastward from the base of the Harar Massive to the Indian Ocean; crystalline rocks overlain by Jurassic and Cretaceous limestones. Low coastal plain is fringed by high dunes (or raised barrier islands), deflecting the passage of rivers to the sea and forming marshes and alluvium-filled lagoons. Escarpment mountains rise approximately 1500 m above sea level along the northern edge of plateau bordering the Gulf of Aden.

2. The Rift Valley Domain

17 Western Rift Belt. A north–south belt, called the Central African Swell, extending approximately 2400 km. Associated grabens and scarps, plus lakes, run in three directions: from Lake Tanganyika northward to Lakes Edward and Albert and to the Nile River; westward to the Congo and Atlantic; southward to Lake Nyasa/Malawi, the Zambesi River and the Indian Ocean. The western graben wall (up to 2440+ m) is almost unbroken; the eastern wall is much breached. The grabens have many offshoots and bifurcations.

Katanga region is a rectangular area bordered to the northwest by the Lualaba River (headwaters of the Congo River), to the northeast by Lake Tanganyika, to the southeast by the Luangwa River Valley, and to the southwest by the Kafue drainage system. Flat-topped mountains of Paleozoic rock (largely sandstones) descend in 1525 m scarps to deep, down-faulted depressions where hot springs and marshes form. Gallery forests on the rivers, open woodlands at stream heads; the higher ground is bushland or grassland.

North of Lake Tanganyika, the rich tablelands of Ruanda-Burundi, between Lake Edward and Lake Turkana. Ruwenzori Mountains (4000+ m above sea level): glaciated; not of volcanic origin.

18 High Interior Plateau. Relatively intact portion of the African Precambrian platform; approximately 1220 m above sea level, composed of downwarped basins: Uganda, Ruanda, Unjamwesi, Serengeti Plain, and the Masai steppe; as well as Lake Victoria. Inselbergs/kopjes rise above these plains and basins; large marshlands (Malagarasi) or undrained depressions common. Woodlands to the south, thorn bushland to the north.

19 Eastern Rift Belt. Extending approximately 1600 km from Lake Nyasa/Malawi to Lake Turkana (Rudolf); divergent rift valleys, scarps, and evaporite lake basins. Mt. Kenya and Mt. Kilimanjaro (approximately 6000 m above sea level) glaciated. Great physiographic diversity. Arid floors of the main trench (the Naivasha graben) contain Lake Turkana, Lake Nakuru, Lake Naivasha (freshwater), and Lake Natron; contrasts with wooded scarps and uplands.

20 Coastal Belt. Along the Indian Ocean: east of Lake Nyasa/Malawi, between Zanzibar and Mozambique. Crystalline rock basement, approximately 900 m above sea level, and proximate to the sea. Abundant rains from the onshore tradewinds. Typical features of submerged coastline: alternate stretches of rock headlands and drowned estuaries; river deltas, lagoons, coral reefs.

B. The South African Platform Section

21 Lunda Swell or Benguela Ridge. Approximately 1220+ m above sea level Woodland-covered plateau: undulating relief, rounded outcroppings of resistant rock. Northern slope steep, entrenched headwaters of the Congo River tributaries, thickly wooded valleys, rapids, and waterfalls; southern slopes more gradual.

22 Namib Desert. Narrow coastal strip (150 km wide), from the Cunene River (Angola) to the Orange River (South Africa) and farther south. True desert of massive sand dunes; no running streams for more than 1500 km. Offshore is the cold Benguela Current, over which mists form (fog) as warm air is cooled. No rainfall. Going inland across the dunes and crystalline hills, desert is replaced by arid grasslands, then bushy grasslands, then scattered shrubs and trees.

23 Damara-Nama Upland. Highland belt from

the Cunene River of Angola to the Cape, forming the western rim of the Kalahari Basin. Includes the Kaoko, Damar, and Grand Nama plateaus. Little Nama Plateau is a continuation of this plateau system south of the Orange River. Sedimentary rock overlying granite basement. Descent to the west is abrupt (natural terraces) to a narrow coastal plain. Elevation approximately 1800–2100 m. Semiarid in the north, arid in the south.

24 Kalahari Region. Interior basin, about 1300 km in diameter; average elevation approximately 900 m.

24a Northern Kalahari. Rivers and marshes, high groundwater table. Tributary rivers, the Cuando and Okovango rivers, descend from the Lunda Swell (**21**) into the Northern Kalahari. The Cuando feeds into the Zambezi River via the Chobe Swamp (marshland), while the Okovango River ends in the Okovango delta (marshland), with a partial (ephemeral) distributary joining the Chobe (Selinda spillway). Southern distributaries from the Okovango delta periodically feed into Lake Ngami and via the Botletle River into the Makarikari salt pans and Lake Dow.

24b Central Kalahari. Arid with unpredictable summer rains, numerous ephemeral (or fossil) stream beds and depressions in limestone (vleis); (salt)pans and salt plains; shallow basins with seasonal water.

24c Southern Kalahari. Desert (red sands), fixed, long dune ridges, crossed by two tributaries of the Orange River that flow out of the eastern highland boundaries of the basin.

25 Matabele Upland. Between the Zambesi and Limpopo rivers. Upland of Archean rocks, elevation 1220–1520 m; rivers drain the central watershed north, south, and east–southeast (the Save River). Rolling upland dominated by granite hills rising from valleys of crystalline schist. Rainfall greater in the north and especially in the mountainous east; generally frost free. Southern part drier; winters with regular frost.

26 Mozambique Plain. Low, marshy coastal plain <300 m elevation, up to 320 km wide, composed mostly of alluvium from the Limpopo, Save, and Zambesi rivers; lakes, marshes, and lagoons.

27 High Karroo. Southern continuation of Highveld (see below). An extensive plateau of sedimentary rock, dipping northward; southward a steep escarpment descends to narrow coastal plain. Dry tableland, 1830–2440 m above sea level through which the Orange River is deeply entrenched, forming a canyon; a network of canyons in Little Namaland.

28 The Veldt. Between the Limpopo and Orange Rivers; Highveld in the south, Low Veldt in northeast, separated by the Vaal River. Transvaal or bush veldt in the north-central and northwest. Middle Veldt occupies the area between the Low and Highvelds and the Kalahari Basin. Elevation is lowest (approximately 900 m) in the Low Veldt near the Limpopo River. It increases southward, reaching 1800 m in the Transvaal (>1800 m in the surrounding Soutpanbergs Mountains). Elevation continues to rise across the high sedimentary plateaus of the Highveld, ultimately reaching approximately 3050 m on the the lava-capped Basutoland uplands.

29 Natal Giant-Terrace Belt. Region approximately 300 km wide between the Drakensberg scarp (Great Escarpment extension) and the narrow, often rocky, coastal plain. Typified by giant staircase of sedimentary strata with two pronounced terraces. Top of the Great Escarpment 2440–3050 m above sea level; streams descend through steep gorges, waterfalls. Dense woodland and coastal forest from the coast to the first terrace, approximately 24 km inland: sandstone, elevation 610 m. Second terrace, farther inland (elevation 1220–1520 m); an undulating, mostly treeless, plain. North of Natal and Swaziland, the terraces are modified by basaltic flexure of the Lebombo Mountains along the border of South Africa (Kruger National Park) and Mozambique.

30 Cape Mountains. Trellis drainage pattern typical of folded mountains; trending east–west valleys alternate with narrow gorges. In the north, at the foot of the Great Escarpment, there is a broad, plateaulike valley known as the Great Karroo. A second longitudinal valley is called the Little Karroo, dry and similar to Great Karroo.

Acknowledgment The first author is grateful to the Wenner-Gren Foundation for their support to participate in the symposium "African Biogeography, Climate Change and Early Hominid Evolution."

10

Ecological Links between African Savanna Environments, Climate Change, and Early Hominid Evolution

Norman Owen-Smith

In this chapter I consider the ecological factors that potentially influenced the temporal association between periods of major climatic change during the Late Miocene and Plio-PleistocLene, turnover pulses among large mammalian herbivores, and evolutionary trends in hominid clades. I propose that certain features peculiar to the African continent promoted the adaptations leading to early hominids and ultimately the first humans, including the following:

1. Most of eastern and southern Africa is high in elevation.
2. Precipitation is relatively low, and seasonally restricted, over much of the tropics and subtropics.
3. Savanna conditions predominate over most of this region.
4. Continental uplift activated erosion and exposed various geological substrates, leading to much spatial heterogeneity in soils and hence in vegetation.
5. Widespread vulcanism enriched soils, while low rainfall restricted leaching, thereby promoting nutritious grasses over extensive areas.
6. The favorable forage quality supports a high abundance and diversity of large mammalian herbivores, especially medium-sized grazers.

The critical episodes of global climatic cooling occurred during the Miocene between 15.6 and 12.5 m.y. and around the Pliocene–Pleistocene transition between 3.0 and 2.5 m.y. (deMenocal, 1995; Denton, this volume). Significant turnover pulses among large mammalian herbivores took place between 10 and 5 m.y. and around 2.5 m.y. (Vrba, this volume). The earlier period saw the origination of most of the modern bovid genera, as well as the appearance of the first autralopithecines. The later period was associated with speciations, extinctions, and dispersal events among bovids and equids (Vrba, this volume; Bernor, this volume), and the divergence of the hominid lineage from that leading to the robust autralopithecines (Collard & Wood, this volume). Lesser turnover pulses occurred among bovids around 1.8 m.y. and 0.9 m.y. (Vrba, this volume).

Vrba (this volume) writes of "warmer-adapted" ungulate species giving way to "cooler-adapted" species during these turnover pulses. These adaptations would have been largely a response to changes in vegetation, and hence in habitat conditions and food resources, brought about primarily by changes in precipitation. Accordingly, my approach is first to outline how climate, geology, and soils influence vegetation formations at geographic scales, with emphasis on the savanna biome. Next I describe how vegetation patterns affect habitats and resources for large mammalian herbivores. Finally, I consider how increasing seasonality in rainfall might have led to the evolutionary changes that occurred in hominid clades.

Savanna Biome

Definition

As a biome type, the term "savanna" refers to geographic regions typified by the co-occurrence of trees and grasses. The grass layer must be sufficiently continuous to support periodic fires. Forms of woody vegetation include open canopy woodlands, tall tree parklands, and low tree and shrub bushland (Pratt & Gwynne, 1977; Cole, 1982, 1986). Within regional landscapes there may be areas of forest (e.g., along riverlines), and open grassland (e.g., vleis or dambos). To be mapped as savanna, more than 80% of the vegetation cover should be concordant with the above definition (Scholes & Walker, 1993). Fairly narrow transitional zones separate the savanna biome from the adjoining forest, grassland, and arid shrubland biomes.

The savanna biome today covers about 40% of Africa's surface, a greater extent than on any other continent. It encompasses most of eastern and south-central Africa and extends in a broad band through West Africa (fig. 10-1). In Australian and South American savannas, the predominant trees are commonly evergreen (e.g., eucalypts), whereas in African savannas they are invariably deciduous (although certain species may retain their leaves late into the dry season).

South and Central America and India contain extensive areas of seasonally dry forest or closed canopy woodland. In Africa such vegetation is limited in extent, perhaps because rainfall is generally lower, or fire more pervasive, than in the other continents. This woodland type may have been more widespread in Africa in the past.

The grassland biome is currently restricted to the east-central region of South Africa, including the Lesotho highland. Within this region forest patches occur in moister pockets, and shrubs are localized on rocky hillsides. Areas of treeless grassland elsewhere in Africa are limited in extent and mostly depend on local soil conditions (e.g., the Serengeti Plain of Tanzania).

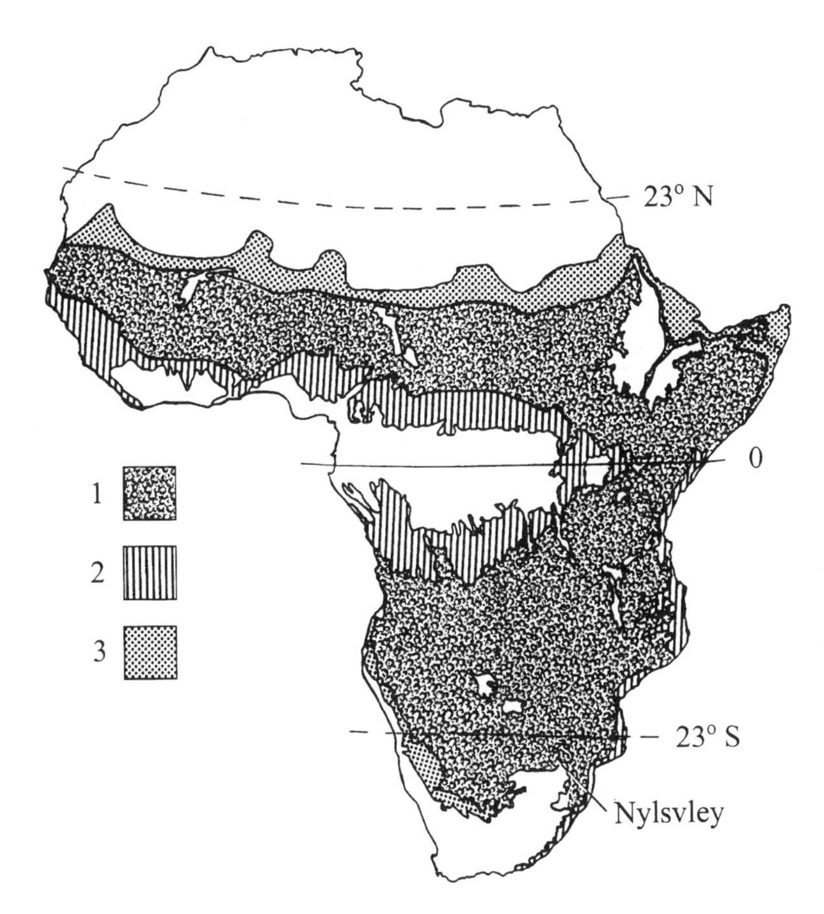

Figure 10-1. Distribution of the savanna biome in Africa. The core savanna zone (1) has >80% of the vegetation as savanna. Narrow transition zones between savanna and forest (2) and between savanna and arid shrubland (3) are indicated. (From Scholes & Walker, 1993.)

Climate

Within the savanna biome, annual rainfall total varies from a maximum of about 1500 mm down to about 250 mm. Temperatures tend to be hot, so that potential evapotranspiration exceeds precipitation. In southern Africa, about 80% of the rainfall typically falls during the 6 months of summer. In equatorial Africa, two rainy seasons follow the passage of the intertropical convergence zone, separated by dry seasons lasting about 3 months. In western parts of South Africa where rainfall is low but relatively evenly distributed through the year, savanna grades into low shrubland, with a variable grass component, or evergreen shrub thicket.

Grassland prevails in cooler, slightly moister regions than savanna. In South Africa, winter temperature minima tend to be 5°C cooler in the grassland biome than in savanna (Rutherford & Westfall, 1986). Precipitation covers a slightly higher range in grassland (400–1800 mm per year) than in savanna (250–1200 mm per year), but likewise occurs mostly during the summer season. Savanna grasses generally show the C_4 photosynthetic pathway, adapted for rapid CO_2 uptake in hot environments where soil moisture is available during restricted periods. In the grassland biome, both C_3 and C_4 grasses occur.

The seasonally dry conditions promote regular fires in both savanna and grassland. Fire promotes grassland at the expense of woody vegetation. Lightning strikes are a natural source of ignition in some areas, but the frequency of fire is greatly increased by deliberate or accidental human ignitions. The latter may transform forest into savanna in the region bordering these two biomes. Frequent hot fires may also give savanna regions the aspect of open grassland (e.g., Masai Mara region of southern Kenya).

In South America savanna occurs in regions with somewhat higher total rainfall than typical of African savannas, up to 2500 mm per year (Medina & Silva, 1991). In India areas having the appearance of savanna have been derived recently from woodlands through deforestation, cultivation, and burning (Yadava, 1991).

Soil Texture and Fertility

A distinction between moist/dystrophic (or infertile), and arid/eutrophic (or fertile) savanna types (Huntley, 1982) derives from the influence of soil texture, fertility, and moisture retention on ecosystem pattern and processes (figs. 10-2 and 10-3; Frost et al., 1986). The particle size distribution of soils determines the extractability and persistence of soil moisture, as well as the retention of soil nutrients. The geological substrate, in interaction with rainfall, influences the texture and potential nutrient status of soils. Fine-grained volcanic rocks (e.g., basalt) or sediments (e.g., shales) tend to yield fertile clay soils. Coarse-grained igneous rocks (e.g., granite) or sediments (e.g., sandstone) usually form infertile, sandy soils. Rainfall modifies inherent soil fertility through nutrient leaching. Hence wetter regions have less fertile soils than drier regions with the same soil substrates. Clay soils are vulnerable to waterlogging in the rainy season but dry out intensely during the dry season.

The link between geology and fertility is clearest where the annual total rainfall is between 400 mm and 1000 mm, as it is within much of the African savanna biome. The distinction between eutrophic and dystrophic savanna has not been recognized on other continents. Almost all areas of savanna in South America appear highly dystrophic (Medina & Silva, 1991), while in Australia fertile soils are limited in extent (McKeon et al., 1991).

Infertile savannas in Africa typically have broadleaf savanna woodland characterized by tall trees of the legume subfamily Caesalpinioideae (e.g., *Brachystegia* spp.), or low trees of the family Combretaceae (*Terminalia* or *Combretum* spp.). Fertile savannas are characteristically dominated by fine-leaved, commonly spinescent trees of the legume subfamily Mimosoideae (e.g., *Acacia* spp.). The overriding influence of soil nutrient status was highlighted in the study site of the South African Savanna Biome Study, where the two savanna types coexisted under the same total rainfall (623 mm per year). *Acacia* savanna replaced broadleaf *Burkea* savanna in the nutrient-enriched sites of ancient human settlements, without any marked change in soil texture (Scholes & Walker, 1993). In western Africa, both Guinean (moist) and Sudanian (dry) savanna zones are underlaid by generally infertile soils, and mimosaceous shrubland predominates only in the arid Sahel zone.

Geomorphology

The continental uplift that occurred during the Miocene led to the high elevation of the interior plateau of much of eastern and southern Africa. Consequent erosion caused bedrock to have a strong influence on soils and vegetation (Cole, 1986). Over much of south-central Africa, basement igneous rocks are exposed, yielding sandy soils of low fertility except in bottomlands. Erosion of the continental interior produced the infertile Kalahari sands that cover much of the western part of south-central Africa. Rifting, probably as-

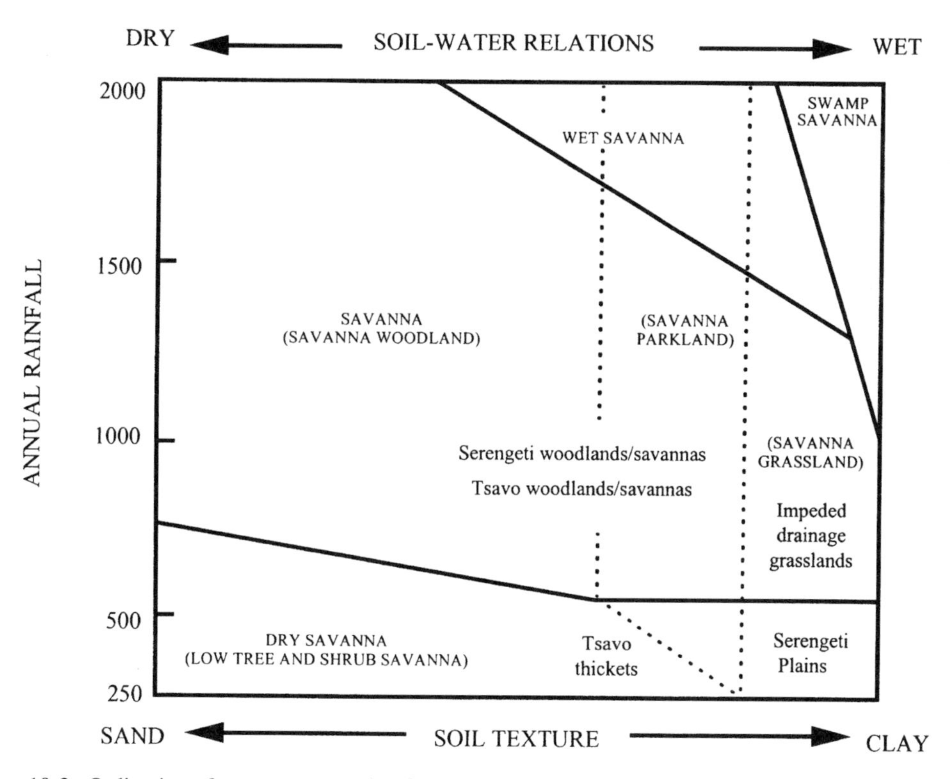

Figure 10-2. Ordination of savanna vegetation formations in relation to annual total rainfall, soil texture, and soil–water relations (Johnson/Tothill model, modified by Belsky, 1990).

sociated with the uplift, led to extensive vulcanism over much of eastern Africa, with resultant deposits of alkaline ash greatly enhancing soil fertility. At a much earlier time, following the breakup of Gondwanaland, massive flows of basaltic lava covered most of southern Africa from the Cape to the Zambezi River. Although the basalt surface has mostly been eroded away, feeder dykes and sills of dolerite intrude extensively through the underlying Karoo sediments that cover much of interior South Africa. Hence volcanic enrichment of soils occurs widely through eastern and southern Africa, while western and west-central Africa lacks this influence.

Local transitions between savanna and grassland are influenced by soil texture, structure, and water retention within landscapes. Extensive grassland occurs on the floodplain of the Kafue River in Zambia. The grassland of the eastern Serengeti Plain is maintained by an impermeable clay layer at shallow depth in the volcanic ash-derived soil. A clay accumulation layer in soils may be a factor in the prevalence of grassland on the oldest land surfaces in interior South Africa (Tinley, 1982).

Herbivory

The grasses growing on nutrient-deficient soils tend to be high in fiber, and low in nitrogen and mineral contents. Likewise, woody plants prevalent in infertile savanna commonly have unpalatable leaves, containing high levels of phenolics or other plant secondary metabolites (Owen-Smith & Cooper, 1987; Owen-Smith, 1993a). Accordingly, the biomass of large mammalian herbivores, relative to rainfall, is two to three times higher in savanna regions underlaid by nutrient-rich substrates than the biomass of these mammals in infertile regions (fig. 10-4; Bell, 1982; Fritz & Duncan, 1993). In dystrophic savannas, herbivores are concentrated in localized areas where nutrients accumulate, such as in floodplains. Plateau regions support a relatively low herbivore biomass, predominantly made up by elephants.

In fertile savannas, a substantial fraction of the herbaceous layer is consumed by large herbivores, except where surface water is limiting. The consequent reduction in fire incidence can lead to an increase in the woody component at the expense of the grasses.

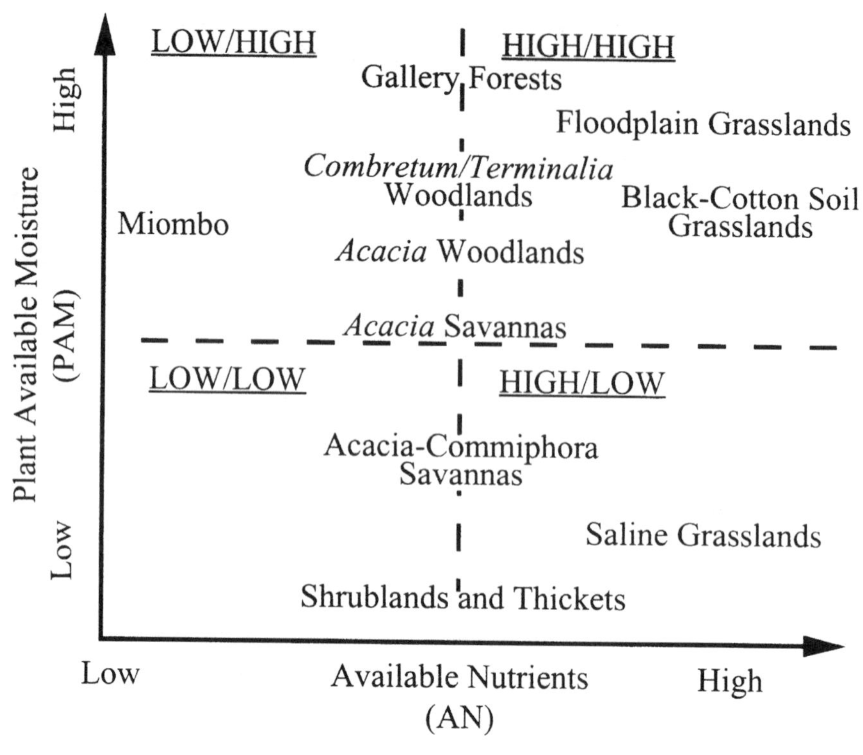

Figure 10-3. Functional classification of savanna forms in relation to plant available moisture and available nutrients (from Frost et al., 1986; modified by Belsky, 1990).

Elephants in sufficient numbers can suppress this tendency, through their destructive impact on woody plants (Dublin, 1995). In infertile savannas with sandy soils, woody plants retain a large fraction of their biomass underground, making them resistant both to fire and to elephant damage. High elephant abundance promotes a shrub coppice woodland (Bell, 1981).

In South America and Australia, African species of grass have been introduced because indigenous grasses seem unable to withstand the grazing impacts of cattle. This may be largely because the native grasses are adapted to tolerate nutrient-deficient conditions, where levels of herbivory are generally low.

Herbivore–Habitat Relations

Species Richness and Abundance

Large mammalian herbivores and associated carnivores are more abundant and species rich in Africa than elsewhere in the world today. This was not always the case. The megafaunal extinctions that occurred at the end of the Pleistocene, between 15,000 and 10,000 yr ago, eliminated 75% of the genera of large herbivores in both North and South America and 45% of European and Australian genera, while Africa lost only 13% of its genera (Owen-Smith, 1987). The unique feature of the African fauna is the diversity of large ruminants, most notably among grazers. Among those bovids resident in savanna, 17 species representing 9 genera are strict grazers. Another 8 species from 5 genera are mixed grazer–browsers, and 12 species from 9 genera are browsers (Owen-Smith, 1982).

The biomass levels attained by African wild herbivores in nutrient-rich savannas vastly exceed those exhibited by indigenous herbivores elsewhere in the world today. For intact communities retaining abundant megaherbivores (e.g., elephants, rhinoceroses, and hippopotamuses) this biomass rivals that achieved by managed livestock under comparable rainfall (Owen-Smith & Cumming, 1993). The population abundance attained by grazing ungulates is an order of magnitude greater than that reached by browsing ungulates of similar size, because of the greater year-

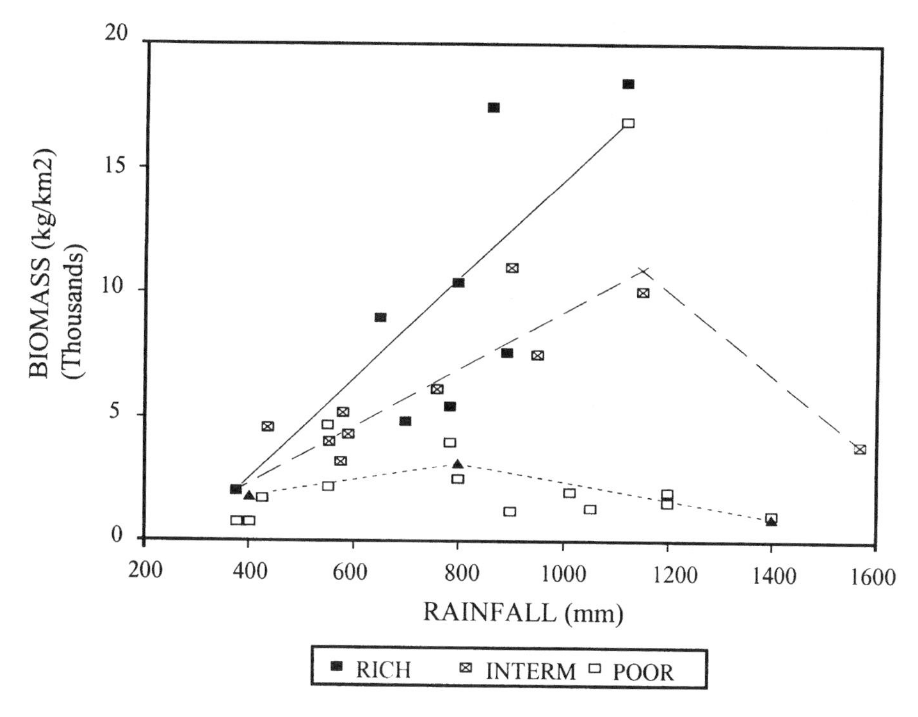

Figure 10-4. Relationship between total biomass of large mammalian herbivores and rainfall for African savanna areas, showing the influence of the nutrient status of the regional geological substrate (modified and updated from Bell, 1982). Filled squares (solid line), nutrient-rich; squares with crosses (dashed line), intermediate nutrient status; open squares (dotted line), nutrient poor.

round accessibility of grass relative to browse (Owen-Smith, 1992).

Habitats and Niches

Greenacre and Vrba (1984) found cover and height of woody vegetation and soil nutrient status to be the main factors distinguishing the habitats occupied by African bovids. Some species are restricted to nutrient-rich savanna, notably wildebeest (*Connochaetes taurinus*). Other species, notably roan (*Hippotragus equinus*) and sable (*H. niger*) antelope and Lichtenstein's hartebeest (*Alcelaphus lichtensteini*) are distributed predominantly in nutrient-poor regions (Owen-Smith, 1982; East, 1984). Woody cover influences vulnerability to predation, depending on the specific adaptations of prey species to escape capture. Grazers such as topi (*Damaliscus lunatus*) are built for rapid locomotion in open country. Browsers such as the tragelaphine antelope and mixed-feeders such as impala (*Aepyceros melampus*) are built for jumping and so will be more successful at evading predators in wooded habitats. The restriction of grazers such as roan antelope, which are not fast runners, to infertile savannas may be largely a consequence of their vulnerability to predation. Because of generally low prey abundance, lions tend to be uncommon in these regions.

As nonruminants able to handle somewhat poor and fibrous grass, zebra (*Equus* spp.) are widely distributed through both fertile and infertile savanna regions. Megaherbivores such as the elephant, rhinoceros, and hippopotamus are likewise abundant in both savanna forms but tend to be most prominent in biomass in nutrient-poor regions (fig. 10-5; Bell, 1982; Owen-Smith, 1988a).

The trophic separation between grazing and browsing ruminants is associated with differences in body size, dentition, and digestive physiology (Owen-Smith, 1985, 1989a) and supported by differences in gut anatomy, liver mass, and salivary gland size (Hofmann & Stewart, 1972; Owen-Smith, 1982; Hofmann, 1989). These features seem designed to handle differences in fiber content, fermentation rate, and al-

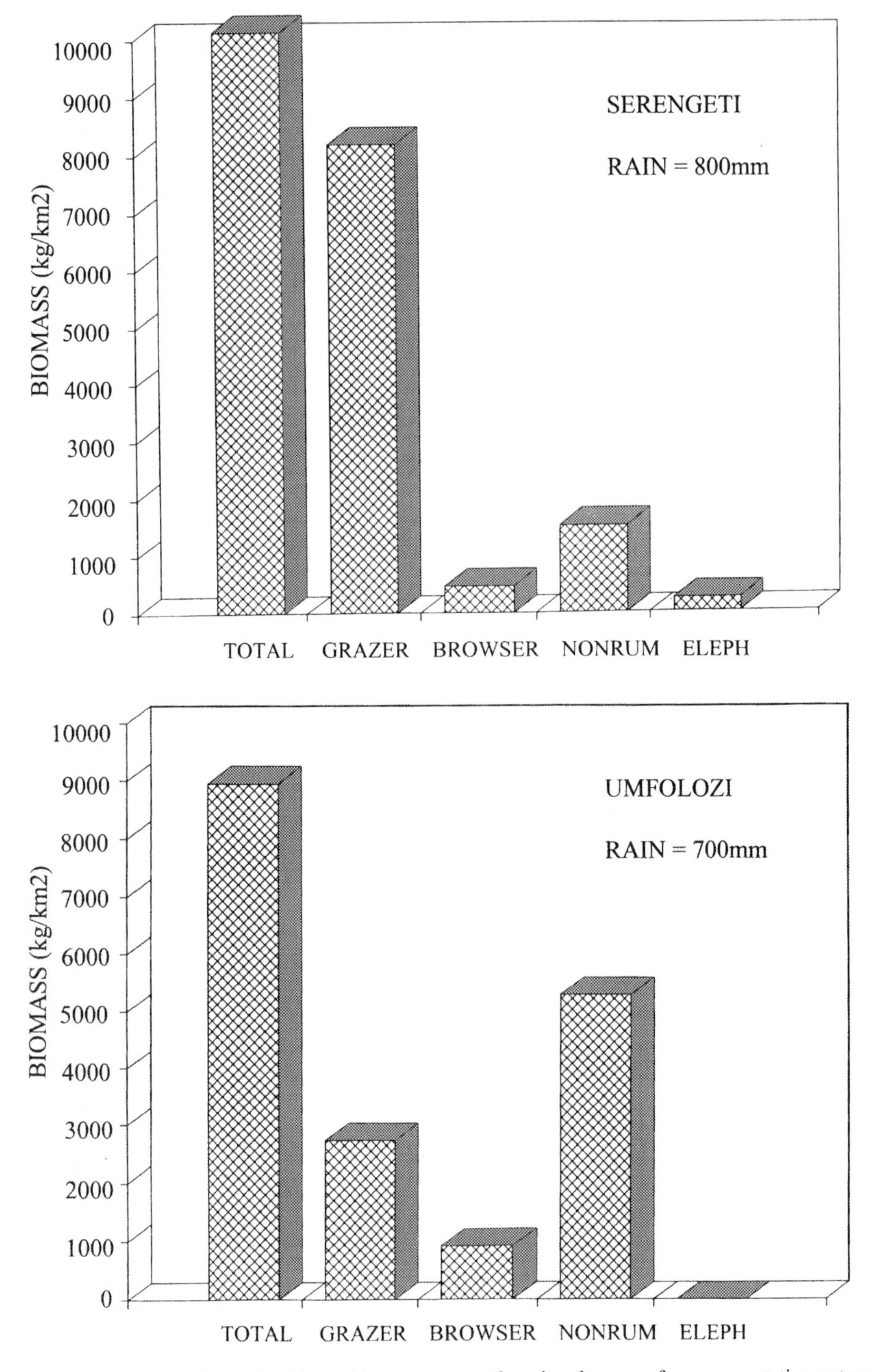

Figure 10-5. Distribution of large herbivore biomass among functional groups for representative eutrophic savannas: (a) Serengeti, Tanzania, and (b) Umfolozi, South Africa, and dystrophic savannas: (c) Chobe/Hwange, Botswana/Zimbabwe, and (d) Kasungu, Malawi. Grazer = grazing ruminants, browser = browsing ruminants, nonruminants = rhinoceroses, zebras, hippopotamuses, and warthog, elephant = African elephant only.

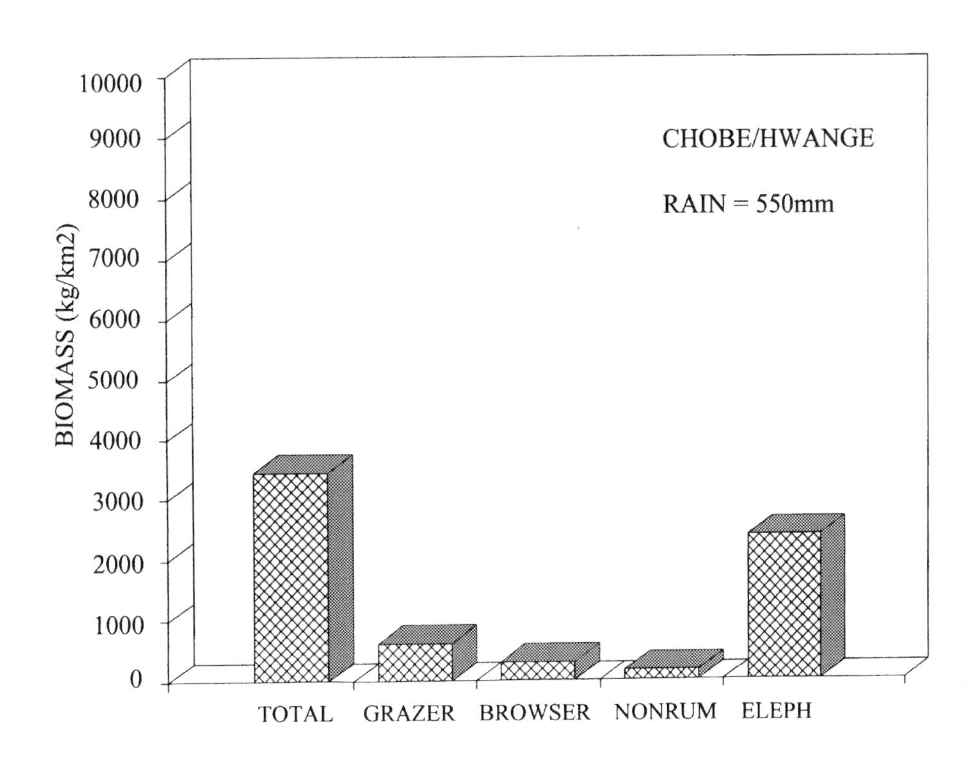
CHOBE/HWANGE
RAIN = 550mm
BIOMASS (kg/km2)
10000
9000
8000
7000
6000
5000
4000
3000
2000
1000
0
TOTAL
GRAZER
BROWSER
NONRUM
ELEPH

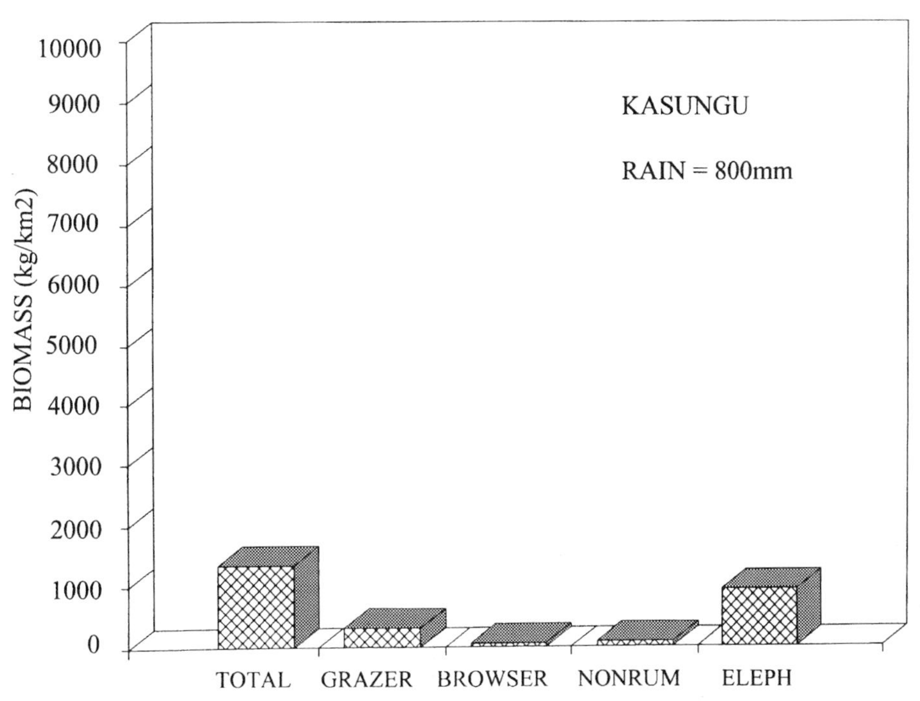
KASUNGU
RAIN = 800mm
BIOMASS (kg/km2)
10000
9000
8000
7000
6000
5000
4000
3000
2000
1000
0
TOTAL
GRAZER
BROWSER
NONRUM
ELEPH

lelochemical levels between graminoids and dicotyledonous plants. The dichotomy between grazers and browsers is more distinct among African bovids than among ungulates inhabiting other continents (Owen-Smith, 1992), perhaps because C_4 savanna grasses are characterized by especially high fiber contents when mature. During the Pleistocene, the grazers in other faunas were mostly large nonruminants (e.g., proboscideans, edentates and equids), able to tolerate poor-quality forage (Owen-Smith, 1988a).

Among grazing bovids, differences in muzzle width influence ability to feed on grasses of different heights (Bell, 1969, documented in Owen-Smith, 1982, 1989a; Gordon & Illius, 1988; Murray & Brown, 1993). Species with relatively wide mouths and procumbent incisors can feed successfully on short, highly nutritious grass. Species with narrow muzzles are better able to select green leaves from within relatively tall grass swards. Among antelope tribes, the Reduncini (e.g., waterbuck, lechwe, and other *Kobus* spp.) are generally narrow-muzzled and favor flood-plain grasslands or other areas with tall grass. Among Hippotragini, both roan and sable antelope are fairly narrow-muzzled. Grazers in the Alcelaphini are wide-muzzled to varying degrees, with wildebeest being the most extreme. Notably, white rhinoceroses (*Ceratotherium simum*) and hippopotamuses (*Hippopotamus amphibius*) preferentially graze short grass, using their wide lips for feeding. Among browsers, niche differences appear to be underlaid primarily by differences in body size and in habitat selection (Owen-Smith, 1989a).

Grazing ungulates commonly aggregate in mixed-species herds during the wet season, favoring regions of the landscape where nutrients are most concentrated (McNaughton, 1988, 1990). They become more distinct in habitat occupation during the dry season, when food quality and quantity are both restricted. A grazing succession, with large species dependent on abundant tall grass, such as buffalo (*Syncerus caffer*) being replaced by species better able to handle progressively shorter grass, may be apparent in some localities (Vesey-Fitzgerald, 1960; Bell, 1971). Large species are dependent on areas where sufficient grass biomass remains, albeit of relatively poor quality. Small species seek out patches where high quality food persists, although low in amount. Mixed feeders like impala switch from a predominantly grass diet during the wet season to feeding largely on woody browse during the dry season. Warthogs (*Phacochoerus aethiopicus*) are able to dig for underground grass rhizomes. Fruits such as *Acacia* pods and the large "monkey oranges" of *Strychnos* spp. are especially valuable in helping browsers such as kudus bridge the nutritional bottleneck of the dry season (Owen-Smith & Cooper, 1989).

The availability of surface water for drinking limits the distribution of most grazing ungulates. Populations concentrate near rivers or other perennial water sources during the dry season and disperse over a much wider area during the wet season (Western, 1975). Browsers and arid-adapted grazers such as oryx (*Oryx* spp.) may be able to persist without drinking through the dry season, provided sufficient moisture is obtained in the diet (e.g., from plants with evergreen or succulent leaves or from underground tubers). More extensive migrations, such as those in the Serengeti region of eastern Africa, connect fertile, short-grass grasslands, highly productive during the wet season, with regions of higher rainfall where tall grass reserves persist into the dry season (Fryxell & Sinclair, 1988). Comparable migrations may have occurred in the western part of the grassland biome of interior South Africa, involving the endemic black wildebeest (*Connochaetes gnou*), blesbok (*Damaliscus dorcas*), and quagga (*Equus quagga*). Migrations with a somewhat different cause occur in extensive floodplains. Grazers using the floodplain grassland during the dry season are forced to concentrate in restricted areas of high ground during periods of inundation during the wet season (e.g., white-eared kob [*Kobus kob leucotis*] in southern Sudan; Fryxell & Sinclair, 1988).

Climatic Influences

The abundance of large herbivore species is affected by changes in rainfall, influencing food availability. In eastern South Africa, population changes by kudu (*Tragelaphus strepsiceros*), African buffalo, and waterbuck (*Kobus ellipsiprymnus*) were positively correlated with annual total rainfall, but for wildebeest and zebra the relationship between population trend and rainfall was neutral or even negative (Owen-Smith, 1990; Mills et al., 1995). Reduced rainfall favors the short, nutritious grass preferred by wildebeest and zebra, except under extreme drought conditions. A shift in the rainfall distribution, with relatively more rain being received during the dry-season months, was a factor promoting the increase of the wildebeest population in the Serengeti during the 1970s (Sinclair, 1979).

Browsing ungulates, including kudu, nyala (*T. angasi*), impala, and giraffe (*Giraffa camelopardalis*), are vulnerable to mortality should cold, wet weather occur at the end of the dry season, when their body-fat reserves are depleted (Owen-Smith, unpublished). This cold sensitivity seems to be related to the tolerance of these species for the high ambient temperatures that commonly prevail just before the rainy season, when extended foraging time must compensate for low food abundance (Owen-Smith. 1998). For grazing ungulates, in contrast, forage quality rather than quantity is generally most limiting, requiring prolonged digestion of food. In open grassland or shrubland habitats, high radiant heat loads may restrict daytime foraging by grazing ungulates. This may cause foraging activity to be shifted toward the night, with correspondingly increased predation risk.

Overall, the direct influences of temperature change on geographic distributions seem likely to be minor. Most ungulate species occur over quite wide latitudinal and altitudinal ranges.

Species Interactions

Competitive interactions among herbivore species are widely assumed to exist, with little supporting evidence (Owen-Smith, 1989a). Grazing ungulates with similar food requirements commonly aggregate in mixed-species herds, benefiting through mutual security against predation (Sinclair, 1985). Nevertheless, there may be situations where the presence of one species may prevent another from using certain habitats or resources during critical bottleneck periods. Niche separation among extant species may have originated from competition in past times, causing character divergence or species replacements.

Competitive interactions could arise through habitat transformations. Short-grass grazers can keep grass height too low to be used effectively by species requiring taller grass. The expansion of white rhinoceros in the Hluhluwe-Umfolozi Park in Natal, which resulted in an extensive decline in tall grass cover, was associated with marked decreases in the abundance of both reedbuck (*Redunca arundinum*) and waterbuck (Owen-Smith, 1988a). However, wildebeest and zebra, which favor short grass, seemed to be unaffected or even favored.

In the northern section of the Kruger Park, the roan antelope population declined from about 450 in 1986 to about 40 in 1994, accompanied by large increases in the abundances of zebra and wildebeest. However, direct competition for food did not seem to be responsible. Increased predation by lions, as a result of a buildup in prey populations around artificial water sources, was implicated. A contributory factor was a decline in the prevalence of tall grass, as a result of recurrent droughts (Harrington, et al., in press). This change in habitat conditions favored zebra and wildebeest but probably had a detrimental effect on roan, sable, and tsessebe populations.

Elephant concentrations can radically alter the tree–shrub–grass balance in a region, affecting habitat suitability for other large herbivore species. Densely wooded savanna may be transformed into open grassy savanna, especially on clay soils where shallow-rooted trees and shrubs predominate (Parker, 1983). On sandy soils and in wooded savanna on clay soils where mopane (*Colophospermum mopane*) woodlands occur, the impacts of elephants promote shrubland at the expense of woodland. Conversely, after the elimination of elephants thicket may replace grassland, especially in regions subject to heavy grazing pressure with consequent suppression of fire. This appears to be a contributory factor in declines of several species of grazing ungulate in the Hluhluwe Park in Natal (Owen-Smith, 1989b).

The Late Pleistocene exterminations of proboscideans and other megaherbivores throughout the Americas and Eurasia, largely through the agency of human hunters, may have contributed to the major extinction wave among other large mammals around this time (Owen-Smith, 1987, 1988a). Habitats formerly kept open, spatially heterogeneous, and productive by the disturbing effects of these very large animals reverted to the more uniform forests and grasslands that prevail today. Climatic warming at the end of the last glacial could have contributed to the extinctions by isolating populations of medium-sized herbivores in patches of receding habitat.

Grazing and browsing megaherbivores coexisted throughout the Pleistocene. In Africa, the primarily grazing *Elephas recki* lived alongside ancestors of the modern African elephant (*Loxodonta africana*), which is a mixed feeder. White rhinoceroses were abundant and widely distributed throughout eastern and southern Africa until recent times. The combined impacts of such megagrazers and browsers on vegetation would have been far more severe than observed in African environments today. As a result, fire would have been restricted in effects on vegetation, except where surface water availability limited herbivore presence.

Linking Climate Change and Hominid Evolution

Habitat Transformation

Models of macroevolution entail fragmentation and isolation of habitats and divergence of conditions in the fragments from those formerly prevalent (Vrba, 1992, this volume). Populations stranded in unfavorable habitats either go extinct, or evolve into distinct taxa in the process of becoming adapted to the new conditions. Climatic change may be the initiator of concurrent turnovers among associated taxa through its effects on habitats.

Global temperature minima were associated with reduced rainfall in tropical Africa (deMenocal, 1995). Global warming led to the replacement of open, grassy vegetation by moist forest on the slopes of the Rwenzori Mountains in central Africa around 12,700 yr ago (Livingstone, 1967). The forest plant species must have been present nearby to have increased in abundance so rapidly. Hence the montane vegetation pattern during hypothermal times was probably a grassland–forest mosaic, with forest patches localized in wet valleys, such as occurs on the slopes of the Drakensberg escarpment in South Africa today. In plateau regions, cool and seasonally dry conditions would have promoted open grassland, such as prevails in the South African highveld, perhaps even in tropical latitudes. However, in the valley regions where sediments and fossils accumulate, vegetation patterns are likely to have been locally heterogeneous and dependent on features of topography and soils. This obviously applies to sections of the African Rift, where local vegetation may have varied from low-tree savanna to moist forest or flood-plain grassland over a short distance.

Miocene-Pliocene Transition

A period of major global cooling occurred during the middle Miocene, between 15.6 and 12.5 m.y. (Denton, this volume). By the beginning of the Pliocene around 5 m.y., C_4 grasses had become widespread, and numerous mammalian herbivores had evolved high-crowned teeth to cope with this abrasive food source. Most of the modern bovid genera originated during this interval (Vrba, 1987b). In Africa, faunal assemblages typical of deciduous forest gave way to species associations similar to those found today in savanna woodlands, except for a greater predominance of browsers (Andrews & Humphrey, this volume). Among primates, baboons (*Parapapio* spp.), geladas (*Theropithecus* spp.) and large semiterrestrial colobines made their appearance (Benefit, this volume). The earliest hominids appear in the form of *Australopithecus (Ardipithecus) ramidus* from Ethiopia and later *A. afarensis* from both Ethiopia and Tanzania (Andrews & Humphrey, this volume).

The lack of fossils dating to between 10 m.y. and 5 m.y. in Africa is associated with the major continental uplift that affected both eastern and southern Africa. The rise of the interior plateau blocked moisture from the Atlantic Ocean in the west, leading to a rain shadow effect over much of eastern Africa (deMenocal, 1995). Seasonality in precipitation was accordingly accentuated.

The morphological changes distinguishing the earliest hominids from the hominoid apes include limb modifications facilitating terrestrial locomotion and increases in surface area of the molar dentition (Collard & Wood, this volume). These appear to be adaptations for exploiting tougher fruits and seeds in more open, and also more seasonal, habitats than inhabited by their ape predecessors. Brain capacity increased relatively little. The early hominids probably needed to forage over extended ranges, including a mosaic of open and closed woodland or savanna patches, in order to encounter an adequate food supply (Foley & Lee, 1989). This demanded more efficient terrestrial locomotion.

Pliocene–Pleistocene Change

Between 3.0 and 2.5 m.y. there was a sharp decline in global temperatures, leading to extensive glaciation in the Northern Hemisphere (Denton, this volume). After 0.9 m.y., regular climatic oscillations with a period of about 0.1 m.y. become predominant, with global temperatures decreasing to even more extreme minima during glacial phases. In Africa the appearance of large grazers among bovid taxa around 2.5 m.y. suggests that open savanna or grassland conditions had become more prevalent (Vrba, 1985a, 1992; this volume). *Equus* species entered and spread rapidly through Africa at about this time (Bernor & Armour-Chelu, this volume). Among geladas, a specialist grazer replaced a leaf-browsing form (Benefit, this volume). The australopithecine clade split into two divergent lineages, both showing increased bipedality and both associated with bone or stone tools (Brain, 1981b). The lineage leading to *Paranthropus robustus* and *P. boisei* exhibited greatly increased tooth area, but only a minor increase in relative cranial capacity. This form became extinct around 1 m.y. The other lineage, leading to *Homo* spp., retained moderate tooth area with increased body size and relative brain capacity, with the latter change accelerating after 0.9 m.y. (Collard & Wood, this volume).

These two hominid forms evidently adapted to the accentuated seasonality in food supplies in contrasting ways. The robust dentition of *Paranthropus* suggests increased effectiveness for chewing hard seeds. This food type could potentially be stored to bridge the seasonal period when soft fruit supplies were low. Why did this niche cease to be viable after about 0.9 m.y.? Baboons (*Papio* spp.) persisted, exploiting underground plant storage parts during the dry season (Benefit, this volume). Squirrels gnaw the hard kernels of fruits, such as the drupes of the marula (*Sclerocarya caffra*), to extract the endosperm (Viljoen, 1977; 1983). Perhaps suitable fruiting trees become too sparse in the increasingly open landscapes to supply the food needs of a large, group-living omnivore such as the robust australopithecines.

The form that became *Homo* apparently turned toward animal flesh to bridge the dry season (Foley & Lee, 1989). Meat could have been obtained from carcasses left by carnivores, and weakened ungulates may even have been killed directly. This food source was most readily available during the bottleneck period for fruits and other plant products, as a result of the concentrations of ungulates around water sources in the late dry season. The abundance of meat was greater in fertile savanna regions of Africa than in any other part of the world.

Foraging for meat required expanded home ranges and extended daily ranges and hence more efficient locomotion (Foley & Lee, 1989). It was also essential that foraging groups be effective in defending themselves against the carnivores with which they were competing. This required multimale alliances and the employment of stone weapons, demanding increased brain capacity.

Corresponding adaptations in life-history features would also be needed. The rates of growth and development of the early hominids were probably similar to those of the modern great apes (Bromage, 1990). The latter show ages at first reproduction and birth intervals resembling those of megaherbivores like elephants, rather than those of similar-sized ungulates (Harvey & Clutton-Brock, 1985; Owen-Smith, 1988a). Elephants show an effective maximum population growth rate of about 5–6% per annum, associated with natural mortality rates of only 1–2% per year among prime-aged adult females. Kudu females, twice the weight of a female gorilla, experience an annual mortality amounting 8% per year even among prime-aged animals, due mostly to predation (Owen-Smith, 1993b). Hence it was essential that early *Homo* retained low mortality rates similar to those of megaherbivores while exploiting habitats conferring predation risks similar to those experienced by the larger ungulates. Increased body size and relative brain capacity would facilitate antipredator defense but reduce potential reproductive rates (Hofman, 1983). The slow and apparently internally generated evolution within the *Homo* clade (McKee, this volume) may have been due to the manifold adjustments in life history features needed as brain capacity expanded.

Summary

Climate change induces evolutionary change through its effects on vegetation patterns, habitat conditions, and resource supply. Cladogenesis among ungulates and hominoids following climatic cooling during the mid-Late Miocene and around the Pliocene-Pleistocene transition was fostered by special features of African environments. These include (1) high elevation of eastern and southern Africa, (2) somewhat low and seasonally restricted rainfall, (3) wide prevalence of savanna vegetation, (4) high spatial heterogeneity, (5) mineral nutrient enrichment through vulcanism, and (6) high abundance and diversity of large mammalian herbivores.

Cooler conditions promoted the spread of open savanna or grassland vegetation at the expense of forest or close-canopy woodland. The critical aspect was increased seasonality in precipitation. This resulted not only from global temperature change but also from continental uplift in the interior of eastern and southern Africa. Aided by the vulcanism associated with the development of the African Rift Valley, this led to the appearance of extensive areas of mineral-rich soils, supporting nutrient-rich and highly productive grasses. In turn this fostered high abundance and diversity among large mammalian herbivores, particularly grazing bovids.

Increasingly seasonal vegetation growth caused certain frugivorous apes to become both more bipedal and dentally adapted to utilize, and perhaps collect and store, hard, durable seeds; this trend led to *Paranthropus* spp. This lineage became extinct when oscillations in temperature and hence rainfall became more extreme during the middle Pleistocene. The separate lineage leading to *Homo* spp. became dependent on the procurement of meat to bridge the seasonal bottleneck in food resources. Predator kills and weakened animals were abundantly available near water sources during the late dry season, especially in nutrient-rich savanna regions. This required collaborative foraging groups and effective use of stone weapons to fend off carnivores, both of which favored enhanced brain capacity. Concomitant adjustments in life history features needed to be made to make this niche viable.

Acknowledgments I thank the Wenner-Gren Foundation for their support to attend the Malawi conference and Friedemann Schrenk, Tim Bromage, and Laurie Obbink for all their work in putting the meeting together. For helpful comments on the manuscript, I thank Tim Bromage, Peter Grubb, and Peter Andrews.

11

Evolutionary Processes Implicit in Distribution Patterns of Modern African Mammals

Peter Grubb

Organisms have evolved through anagenesis and cladogenesis. The latter process, particularly in its geographical context, remains a source of controversy. Studies of African mammals have hardly contributed to our knowledge of cladogenesis, so it may be constructive to look at the fauna afresh with this subject in mind. This exercise should provide parallels with the evolution of hominids.

Extant African mammals form numerous clades, most known from the continent no earlier than the Miocene (Maglio & Cooke, 1978) and now comprising 1000 biological (conventional) species (Wilson & Reeder, 1993). Of these, 900 are sub-Saharan, dispersed through rainforest (260 species), savanna (325), arid (190), and montane biomes (135). There should be many examples from which to draw inferences and develop generalizations and theories concerning Plio-Pleistocene cladogenesis, leading even to reappraisal of textbook speciation-paradigms. The larger mammals (245 species of anteaters, primates, carnivores, and ungulates) are likely to provide the closest evolutionary analogues to hominids. However, knowledge of phylogenetic relationships still relies on a traditional and provisional taxonomy. More systematic and biogeographical data are needed to test any preliminary evolutionary hypotheses.

Cladogenesis has not always been a favored term (see e.g., Simpson, 1953) but conveniently names the evolutionary process whereby taxa proliferate (e.g., M. J. D. White, 1978; Sober, 1993; Dennet, 1995). It is used here in contradistinction to '"speciation": not all cladogenesis is speciation, and the entities involved are not necessarily species.

This chapter is based on the extensive but diffuse systematic and regional literature, collated by Dorst and Dandelot (1970), Kingdon (1971–1982), Meester and Setzer (1971–1977), Skinner and Smithers (1990), and others and on my studies of museum material.

Cladistic and Vicariance Analyses and Cladogenesis

It should be possible to develop our understanding of cladogenesis in African mammals with the assistance of cladistic and vicariance analyses. Though the latter are said to be theory-free, cladogenetic hypotheses emerge implicitly from their methodological principles, which rest on the rejection of supposedly untestable propositions rather than on empirical research (Hull, 1979) and which have tended to color interpretations of faunal history. A central tenet is the treatment of dispersal: "I see no reason to assume that dispersal (migration) is a necessary or even common adjunct of the splitting (vicariance) of an ancestral species" (Nelson, 1974:555); "Dispersal . . . appears to be an unnecessary assumption in speciation studies" (Lynch, 1989:551); " . . . dispersal events are not testable under generally recognized circumstances" (Bauer, 1993:262). The rationale behind this rejection is that dispersal can be invoked to explain any distribution pattern. So, like a metaphysical proposition, it cannot be refuted—it is not science. Methodologically it must be minimized, to the extent that the ancestral distribution of a clade of allopatric taxa is assumed to be the whole area occupied by the present descendants; there was no localized center of origin from which new organisms emanated. The extensive distribution or "primitive cosmopolitanism" (Nelson, 1974) is thought to have become fragmented step by step to produce new taxa. Only later did these spread to achieve cosmopolitanism once again, some acquiring sympatry in the process (Rosen, 1978; Platnick & Nelson, 1978). Thus dispersal is invoked after all, but only in the context of this obscure cosmopolitanization event - perhaps it is a logical necessity, allowing the rejuvenation of the vicariance process, for otherwise if barriers continued to arise at random in a static biota, species would come to have smaller and smaller ranges. Phases in which cosmopolitanism prevails are said to alternate with sequences of vicariance events, leading to "narrow endemism": together they are believed to account for the development of faunas (Rosen, 1978; Nelson & Platnick, 1981; Cracraft 1986, 1988).

But there are limits to the scope of cladistic and vicariance analyses: there are cladogenetic phenomena they cannot address (Hull, 1979). Their results cannot be refuted in the Popperian sense (Cartmill, 1981), though they can be disconfirmed (Sober, 1993) or corroborated, and in these respects they do not differ from other hypothetical accounts of historical events. A more holistic approach to the reconstruction of cladogenetic phenomena—perhaps a more traditional one—is called for.

Alternative Approaches to Cladogenesis: A Periodic Model

Any investigation of cladogenesis therefore attempts to recount unique historical events. Had we been present when these events were in progress, our hypotheses would amount to observations and be subject to verification or falsification. Naturally, this is not possible, and even observations require critical appraisal where they grade insensibly into hypotheses (Rosen, 1988). But if ideas could have been tested in the distant past, then in principle they can still be tested today (Kitts, 1977; Szalay & Bock, 1991). A fundamental distinction between untestable scenarios and falsifiable hypotheses (e.g., Delson et al., 1977; Cracraft, 1982) is not required.

Cladogenetic dynamics in mammals can be inferred from patterns of phylogenetic relationships and geographical dispositions of taxa. Emergent hypotheses are subjected to critical review including appropriate cross-referencing ("reciprocal illumination") and a circumspect use of parsimony (Szalay & Bock, 1991; Sober, 1993). Numerous contributors dating back to the older literature (such as Lönnberg, 1918, 1929) have suggested that evolution was stimulated by the dynamic of Africa's past. Biogeographical hypotheses have implicated fluctuating paleoenvironmental conditions, and paleoenvironmentalists rely on biological evidence in the development of their own conjectures (Hamilton, 1976; Maley, 1987). Avoiding this possible risk of circularity, paleoenvironmental data on Quaternary changes in landforms, vegetation, and climate independently cast light on biogeographical conclusions, though only a small part of the significant record contained within Africa's lacustrine sediments has yet been interpreted (Livingstone, 1993). Through the Quaternary, fluctuations of the earth's orbital eccentricity generated climatic cycles with a 100,000 yr period (Denton, this volume), causing alternating cool, dry and warm, wet conditions. One envisages that in response to these cycles, biome blocks and their constituent vegetation formations advanced and coalesced or were forced to retreat, disengage, and fragment, at times drifting past or across major physiographic features, if they were not to be rebuffed or channelled by those features. Animal populations, bound to their preferred habitats, were carried along in the drift (Vrba, 1992), experiencing alternating dispersal and disruption. Vicariance followed from fragmentation of the biome or by a topographic feature becoming a barrier after it had first been circumvented. Astronomical and climatic changes could not cause vicariance per se: the consequent biome drift was the intervening agent. Other things being equal, biome drift is likely not only to have been periodic but also to have followed similar courses from cycle to cycle, even as longer term environmental change was in progress; landforms steadily evolved at their own pace, making their independent impact on climate (Partridge et al., 1995a,b). Vicariance events were therefore associated with periodic dynamics of biomes and biota and relatively static topography. Physiographic processes unrelated to biome shift such as changes in river courses, including joining of river meanders, also led to vicariance, though the relative significance of such catastrophic events is not established.

The amplitudes of biome-drift in this model are controversial (F. White, 1993), but it seems that drift cycles were out of phase for different biomes (see the speculative paleovegetation maps of Cooke, 1962; Carcasson, 1964; Hamilton, 1976; Crowe & Crowe, 1982; and Vrba's relay model, this volume). The net effect would be to stagger vicariance events across astronomical cycles, from which periodicity in the history of individual biomes or smaller habitat units would have become decoupled.

The theme of this chapter can be described as a periodic model of cladogenesis (Grubb, 1973, 1978, 1982, 1990; Kemp & Crowe, 1985). Its origins are neontological, and it draws inferences from the living fauna. While agreeing that speciation requires forcing by the physical environment (habitat theory) and is a dynamic process occurring in allopatry (Vrba, 1985d, 1992, 1995b; this volume), the model also invokes active evolutionary responses from taxa and reappraises such topics as dispersal, barriers, vicariance events, sister taxa, and relational or dichotomous speciation concepts. It cannot readily address effects of long-term climatic change or account for major turnover pulses, macroevolutionary phenomena which are least accessible from the neontological standpoint. Testing the model is going to take time. Though the fossil record of African mammals is a splendid summary of evolution that the extant fauna cannot hope to reveal, it hardly registers the geographical detail of cladogenetic events and for many clades provides almost no information at all. But there is a corresponding lack of sufficient, systematically comprehensive, calibrated data on molecular evolution of clades within which phylogenetic relationships have already been established by cladistic analysis (Morin et al., 1994, is an exception). In view of this deficiency, many of the dispersals and divergences of extant taxa within Africa cannot be satisfactorily dated—allusion here to the age of taxa, particularly infraspecifc taxa, will necessarily be vague.

Patterns and Appropriate Inferences in the Periodic Model

Monotypic Taxa

Whether they are conventionally treated as species or subspecies, monotypic taxa are the phylogenetic species of cladistic analysis (Eldredge & Cracraft, 1980; Cracraft, 1989b), the units that have emerged from vicariance and yet may be about to experience it. In this limbo, their contribution to cladogenetic theory may be limited, but their potential dynamics should be recognized even as abiotic and biotic factors constrain their powers of Malthusian increase and dispersal; absence from a tract of suitable habitat may be because the habit is not yet accessible or because it is preoccupied by a related taxon. Former occurrences outside the present range (evinced by fossil or subfossil material) and discontinuities in distribution indicate local retreat or extinction, perhaps following shift and fragmentation of the habitat. In any case, length of occupation of the range cannot be assumed to have been uniform in view of Quaternary environmental dynamics. Continuity of distribution and variation may imply that the taxon's dispersion has not been disrupted (e.g., *Xenogale naso;* Colyn & Van Rompaey, 1994), yet discontinuously distributed taxa do not necessarily exhibit geographic variation (e.g., *Procolobus verus;* Oates, 1981); indications of recent history can be ambiguous.

Monotypic taxa of larger African mammals comprise 65 biological species and 885 nominal subspecies. Regardless of what defining criteria are adopted, many of these latter would not survive systematic revision: recent published and unpublished studies of artiodactyls, for instance, reduce a subset of 215 taxa cited by Ansell (1972) by 26% to 160. From large carnivores such as lion (*Panthera leo*) or spotted hyaena (*Crocuta crocuta*), which do not seem to form subspecies, to such localized animals as the sun-tailed monkey (*Cercopithecus solatus;* Gautier et al., 1992), monotypic taxa inhabit a wide spectrum of areas. Those with the smallest ranges occupy montane forests and other habitats of very limited extent and should not be confused with peripheral isolates of otherwise widespread taxa—such isolates do not appear to be common, though their incidence has yet to be evaluated. Range area tends to be larger among savanna than among forest taxa and among those that have few or no allopatric representatives, therefore including bats and carnivores.

Monotypic taxa may or may not exhibit geographic variation (compare the bats *Epomophorus labiatus* and *Nycteris hispida;* Claessen & De Vree, 1991; Van Cakenberghe & De Vree, 1993). Where discernible, the variation is continuous (clinal) by definition, but the way it developed over time is not self-evident. Clinal variation may be one of the properties predicting which monotypic taxa are most likely to experience cladogenesis: taxa with small ranges and pronounced clinal variation might be more susceptible to cladogenesis than those with extensive geographical and ecological ranges and weak geographic variation. But clinal variation and its relation to range-size have been underresearched in African mammals, and a categorization of monotypic taxa by these and other pertinent attributes has not been made.

Groups of Allotaxa

More information relevant to cladogenesis should be gained from species that are polytypic and/or contribute to superspecies. Within these entities, the rank-

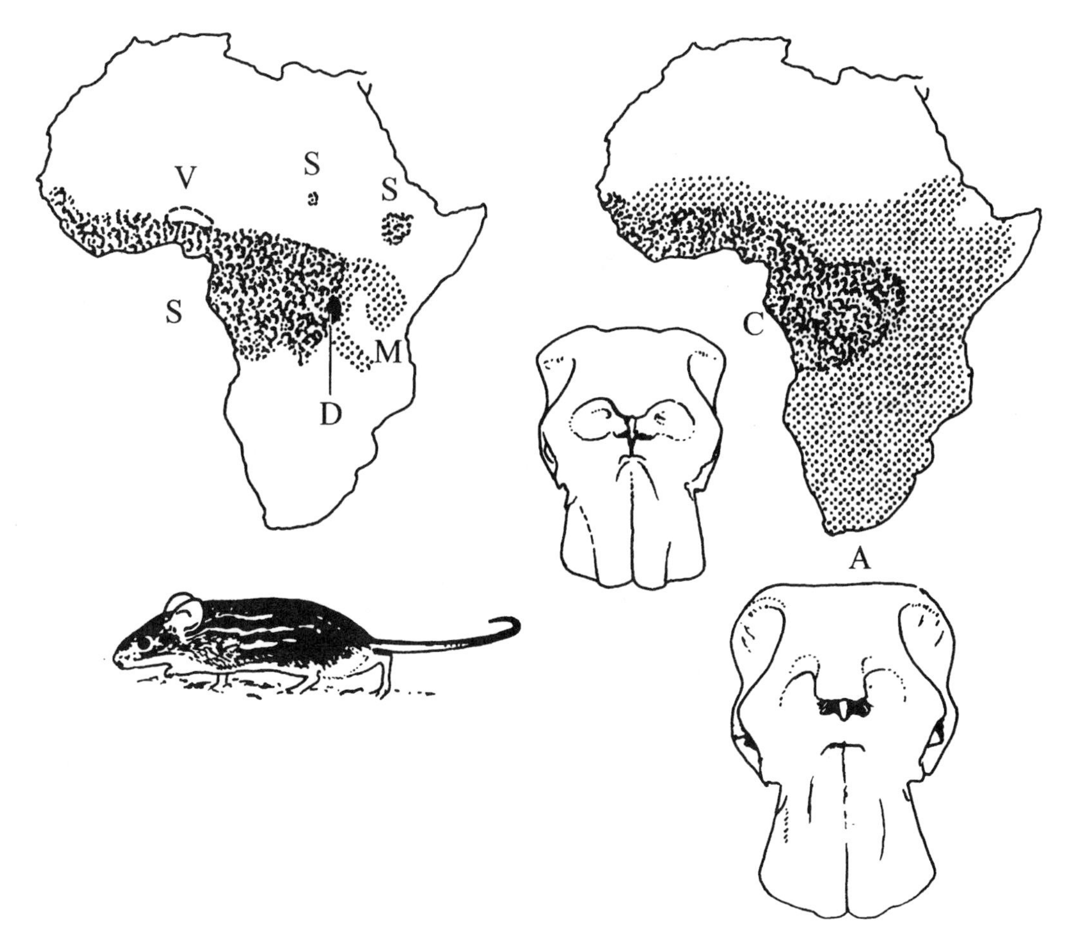

Figure 11-1. Distributions illustrating the occurrence of allotaxa in widely different African mammals. (Left) Striped mice, based on data supplied by Erik Van der Straeten (personal communication): *Lemniscomys striatus massaicus* (M), *L. s. dieterleni* (D), *L. s. venustus* (V), and discontinuously distributed *L. s. striatus* (S). (Right) Elephants (former distribution): *Loxodonta africana cyclotis* (C, forest elephant), *L. a. africana* (A, bush elephant), with skulls of adult males to scale.

ing of allopatric taxa (allotaxa) as species or subspecies continues to be controversial in general or in particular cases. Allopatric representation is certainly extensive: for larger African mammals, 62% of biological species and 97% of supposed phylogenetic species are allotaxa; between 1 and 20+ phylogenetic species are included in each of the 158 zoogeographic species, of which 79% are polytypic taxa, including 1 (41%), 2 (23%), or 3 or more (15%) biological species. The formation of polytypic species and superspecies has been a dominant feature of cladogenesis in African mammals, from mice to elephants (fig. 11-1) and among pangolins, hedgehogs, shrews, bats, primates (figs. 11-2 and 11-3), mongooses, squirrels, porcupines, or ungulates (figs. 11-4 and 11-5): it is allotaxa rather than species that have proliferated.

Allotaxa may be geographically isolated or in contact and may reveal states of interbreeding or non-interbreeding (zygostructure; Jolly, 1993) through parapatry, introgression, hybridization, or marginal overlap. The incidence of these phenomena is exemplified, if not typified, by forest monkeys (Grubb, 1990): relationships between allotaxa that are in line of sight of each other involve 47 contacts, including marginal sympatry or hybridization (rarely with extensive zones of intergradation), and 101 instances of separation by river, lake, sea, or a distribution gap. As for the taxa themselves, 5 overlap or are parapatric, 15 are in contact but relationships are not known, 42 are geographically isolated, 41 hybridize, and some may

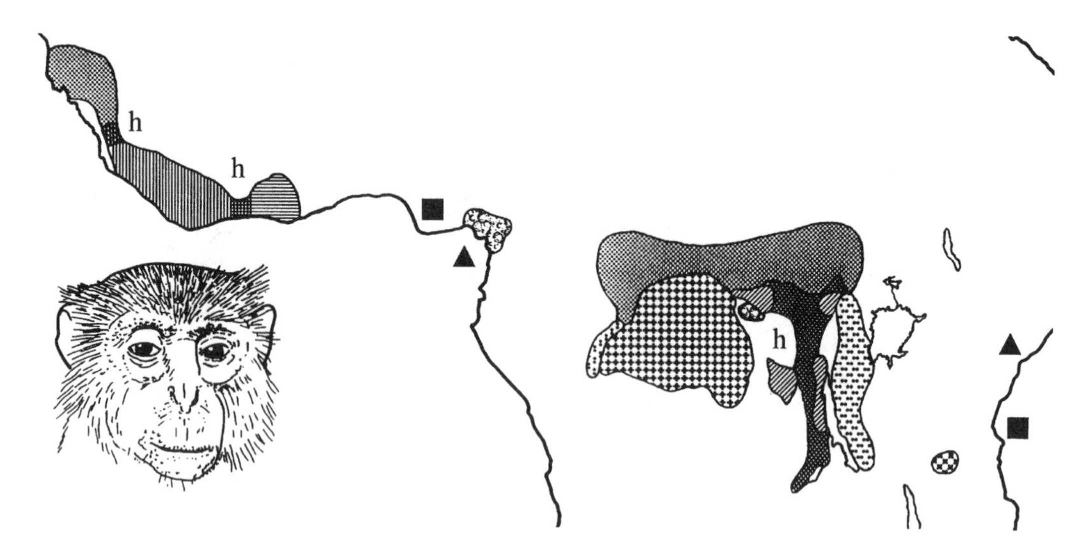

Figure 11-2. Distribution of red colobus monkeys, *Procolobus badius* superspecies, to demonstrate prolific cladogenesis resulting in 18 allotaxa (each represented by a patch of shading or a symbol), together with areas of hybridization between taxa (h).

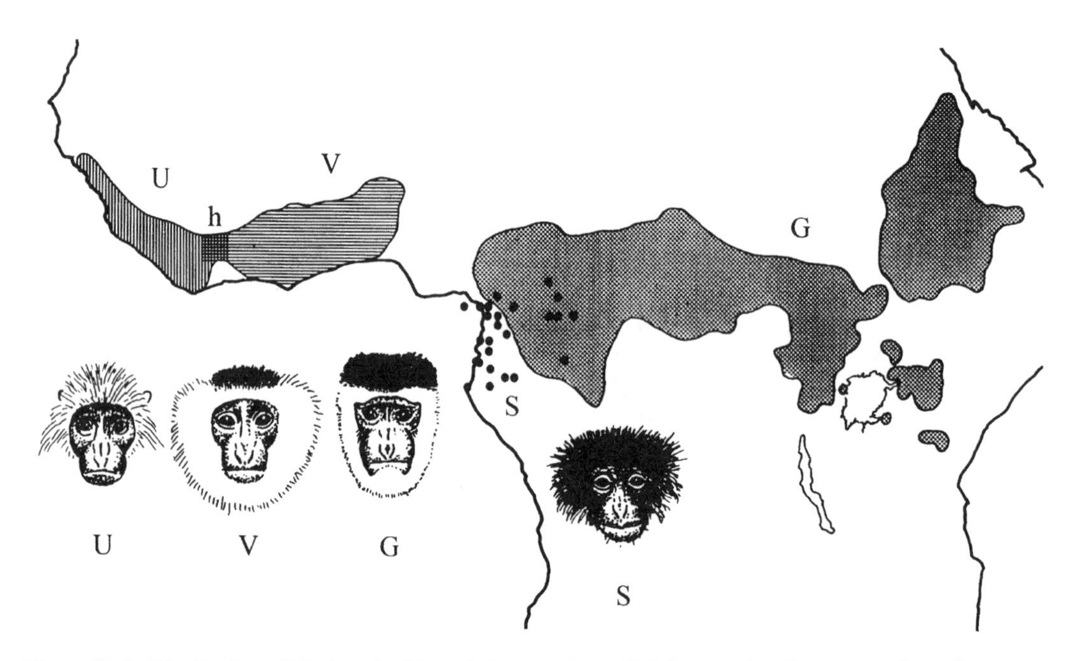

Figure 11-3. Distribution of black-and-white colobus monkeys, *Colobus* species. Taxa mapped are *C. satanas* (S), *C. ursinus* (U), *C. vellerosus* (V), and *C. guereza* (G), forming a morphocline, with secondary marginal sympatry between *C. satanas* and *C. guereza.* The morphocline concerns skull proportions, pelage color and form, and vocalizations (references in Oates and Trocco, 1983).

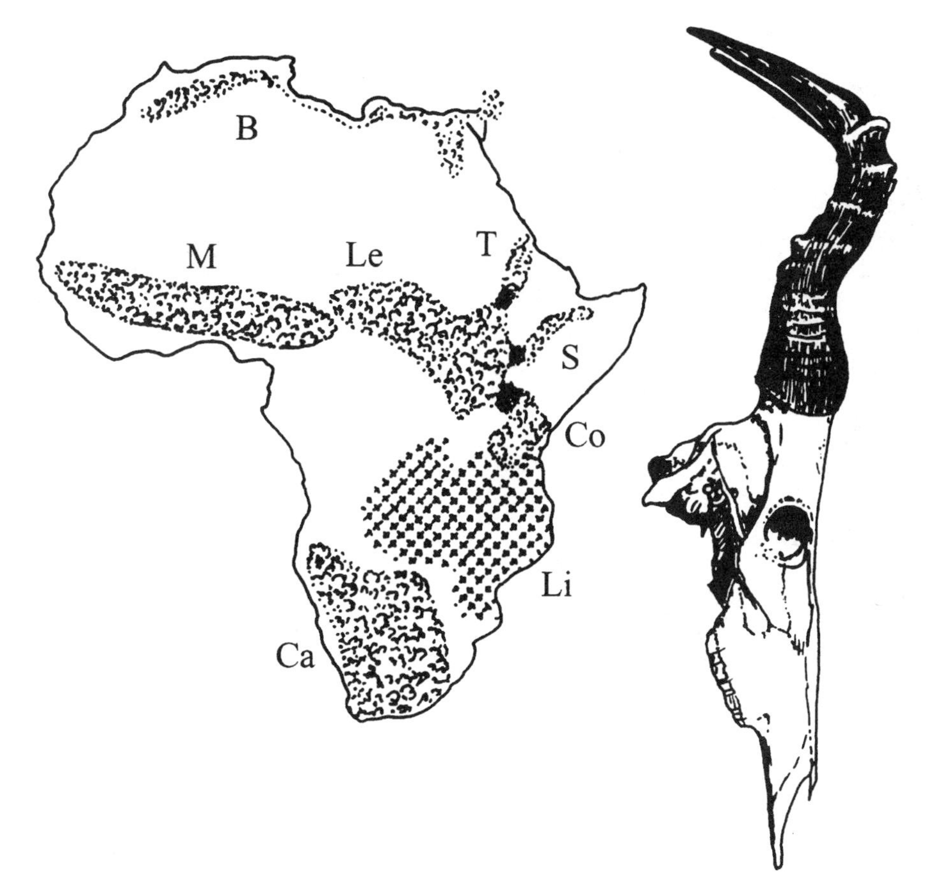

Figure 11-4. Former distributions of allotaxa of hartebeests, *Alcelaphus* species, to show vicariance, hybrid zones (black areas), and separation of synapomorphic subspecies of *A. buselaphus* by the more generalized taxon, *A. lichtensteinii* (see Vrba, this volume). Taxa are *A. b. buselaphus* (B), *A. b. major* (M), *A. b. lelwel* (Le, skull of male illustrated), *A. b. tora* (T), *A. b. swaynei* (S), *A. b. cokii* (Co), *A. b. caama* (Ca), and *A. lichtensteinii* (Li). Modified from Ruxton and Schwarz (1929).

be of hybrid origin. Hybridization could lead to reticulate evolution below the conventional biological species level, yet is not an appropriate indicator of phylogenetic proximity because the ability to interbreed is a primitive retention, not a synapomorphy (Rosen, 1979; Cracraft, 1982).

Cladistic theory deems that more than two taxa in contact cannot all be related to each other as sister taxa. If one can show that, in particular cases, adjacent taxa are not sister taxa even in the extended sense discussed below, their geographical proximity must certainly be secondary (Crawford-Cabral, 1993). These conditions will exist where synapomorphic sister taxa are geographically separated by a related plesiomorphic taxon or where adjacent taxa are distanced by two or more vicariance events on the cladogram (a ring-species situation; figs. 11-3 and 11-4). Such criteria have not always been considered in presumed instances of parapatric speciation (e.g., Endler, 1982), yet in 38 examples of hybridization among monkeys, no more than 21 (55%) are between sister taxa. At least under the conditions just cited, nonsister allotaxa cannot have acquired contact in situ. It is inferred that they have drifted relative to each other so that their geographic dispositions have become rearranged to attain their present topology. This is a challenge to the proposition that cladogenesis occurs among implicitly stationary populations (see below).

Where sister species are parapatric, it is thought that contact may not necessarily be secondary; they could have differentiated in situ (parapatric speciation, discussed by Taylor & Meester, 1993), perhaps from a geographically uniform population. Where they are not now in contact, they might have separated

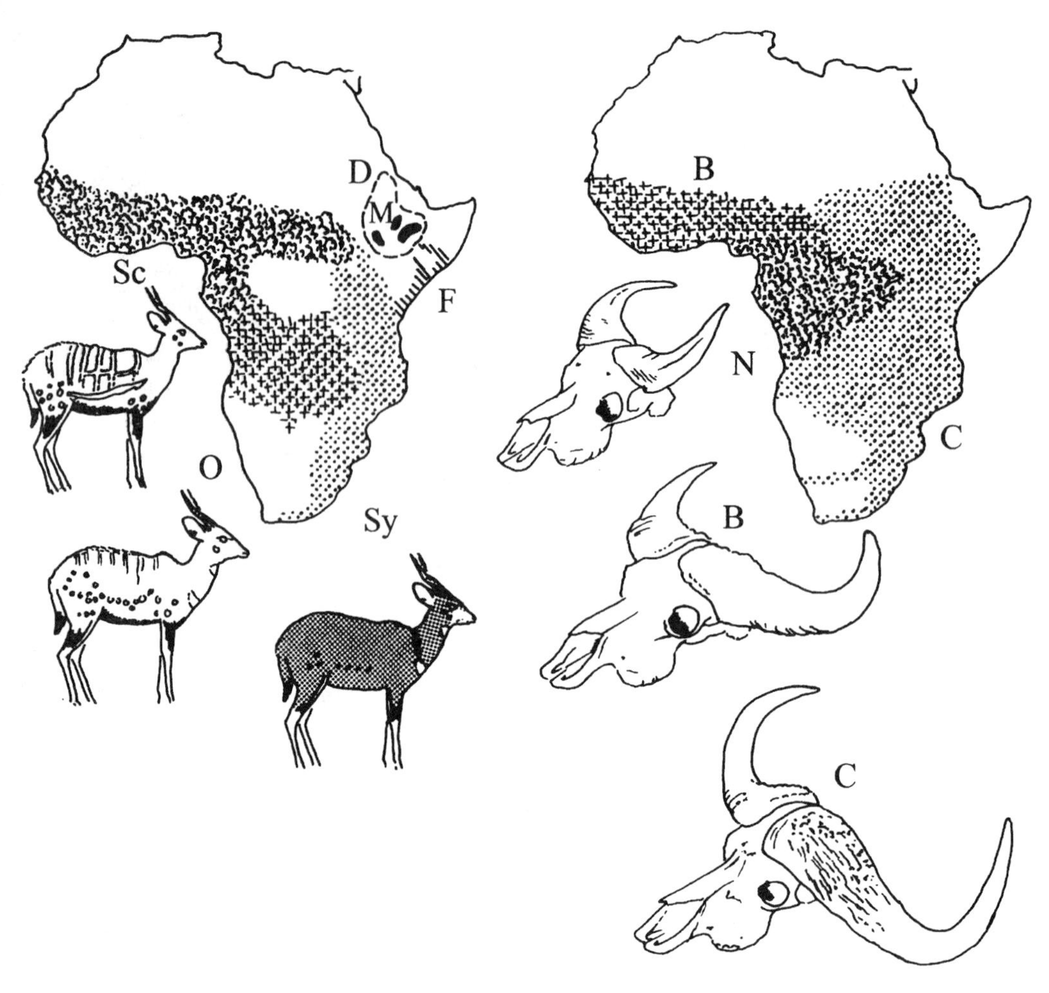

Figure 11-5. Morphoclinal patterns of distribution in two bovids. (Left) Bushbuck (*Tragelaphus scriptus*). Morphocline concerns body size, decline of spotted/striped pattern in both sexes, degree of eumelanization of pelage in males, development of a short-haired collar, and form and relative length of horns; and runs from *T. a. scriptus* (Sc) through *T. s. ornatus* (O) to *T. s. sylvaticus* (Sy). Additional subspecies are *T. s. fasciatus* (F, similar to *T. s. ornatus*) and the montane *T. s. meneliki* (M), discontinuously distributed within the range of *T. s. decula* (D). (Right) Buffalo (*Syncerus caffer*). Morphocline concerns body size, incidence and degree of eumelanization of pelage, form of skull, and form and relative dimensions of horns; and runs from forest-living *S. c. nanus* (N) through amphibiomic *S. c. brachyceros* (B) to savanna Cape buffalo *S. c. caffer* (C, including *S. c. aequinoctialis*). Skulls of males drawn to scale.

after differentiating. But no examples among Afrotropical mammals have yet lent themselves to these interpretations, and it seems that no hypothesis has been formulated showing how parapatric speciation would operate during cycles of environmental perturbation.

Vicariance and Cladogenesis

The principal mode of cladogenesis is said to concern the reverse of the parapatric schema—namely, differentiation of populations after they have become geographically separated, not before: an ancestral taxon would once have been continuously distributed but subsequently gave rise to allopatric sister-taxa through a vicariance event (Croizat et al., 1974; Nelson, 1974; Nelson & Platnick, 1981; Vrba, 1985d), the appearance of a discontinuity in the range. Any contact between the sister taxa, including hybridization, can only be secondary. Vicariance could be achieved by long-distance ("jump") dispersal of a founder population, which would thereby become a peripheral isolate (Bush, 1975); the ancestral taxon would continue to exist (Eldredge & Cracraft, 1980). This peripatry is not a model much favored by cladists because it involves dispersal as well as vicariance (Nelson & Platnick, 1981). An alternative, sometimes expressed inappropriately as the dumb-bell model, assumes (Nelson & Platnick, 1981) that barriers arose to produce vicariance within the range of an implicitly stationary population: a formerly continuously distributed taxon would be split into two taxa occupying a total area equivalent to the range of the ancestor. In this dichopatry or speciation by subdivision (Bush, 1975) (also confusingly termed vicariance), the ancestral species is said to become extinct—a methodological necessity of cladistic analysis, at times also regarded as a natural phenomenon (Wiley, 1978). Although part of received doctrine, the form in which stasis is interrupted (almost a metaphoric concept) is antithetical to the suggestions of the periodic model and surely needs reappraisal.

The peripatric and dichopatric models have nevertheless become paradigms. Peripatry is supported by numerous case histories, but not usually from continental milieus. It has little application to the African mammal fauna, although some bats have established African colonies after flying there across oceans or seas. The second model is the very simplest inference relating to sister-species. It is hardly an inference at all. To call the process "vicariance" does no more than name the resultant state. By referring to speciation as dichopatric or dichotomous, a theory concerning the nature of the ancestral taxon is implicitly built into the terminology and preempts fundamental inference. The vicariance is also distanced from the earlier attainment of cosmopolitanism, making an unsubstantiated assumption that the two are not causally related. Few would deny that splitting has occurred, but of what? Is splitting of uniform populations usually the point from which a process of cladogenesis is thought to commence? It is not obvious that this postulate has empirical support, especially as it ignores the existence of geographical diversity already established in the ancestral taxon. Anyway, if "splitting" is at least necessary for cladogenesis, it has not always been sufficient to generate discrete taxa: fragmented units of a formerly continuously distributed population have not always differentiated (Vrba, 1995b). Could splitting nevertheless be sufficient, at least usually, to cause taxonomically recognizable differentiation? After a vicariance event, populations are said to diverge, but this concept may be another unsupported idea insofar as it is doubtful that divergence would occur aided only by an interruption to gene flow (Ehrlich & Raven, 1969; Gould, 1977a; Stanley, 1979; Vrba, 1985d; Barton, 1988), so in seeking to describe how taxa differentiate it might be fruitful to consider alternative interpretations.

The full context of cladogenetic events remains obscure. Cladogenesis has been confused with speciation, for which there have been diverse theories. Perhaps for these reasons, appropriate process names for cladogenesis (metaphors analogous to natural selection) have not been coined. An alternative model to dichopatry may be worth exploring—one that takes into account patterns and processes immediately before the vicariance event.

Morphoclines and Cladogenesis

The term "sister species" contributes to an implicit but not necessarily empirically sustainable theory of cladogenesis, for sister taxa may not really be sisters, especially when they are parts of morphoclines. Morphoclinal patterns are exhibited by about 30 out of 125 (24%) superspecies of African monkeys and ungulates (figs. 4 and 5; Grubb, 1978, 1990) and there are good examples from non-African monkeys (e.g., Albrecht, 1978; Ford, 1994). Though more cases may be revealed by further research, morphoclines do not appear to be very common. However, their incidence is likely to have been reduced through the extinction of intermediate allotaxa or perhaps through reticulate evolution below the conventional biological-species level (e.g., Colyn, 1991). Seemingly divergent adjacent taxa among polytypic species or superspecies (e.g., the drill and mandrill, *Mandrillus leucophaeus* and *M. sphinx*) may be remnants of a morphocline. Out of 158 zoogeographic species of large African mammals, 33 (21%) are strictly monotypic, so without survival of allopatric sister taxa, there is no way of knowing whether they were once parts of morphoclines. Vicariance biogeography expects morphoclines to result from successive partitions of an originally cosmopolitan biota, no doubt followed by successive divergences, but would these generate morphoclinal patterns? Another reappraisal seems to be called for.

Cladogenesis and Adaptive Radiation in Habitat Choice

The occupation of different biomes has been a major consequence of cladogenesis in African mammals. Thirty-four families and subfamilies and 55 genera live in both forest and savanna, and, of these, 30 families and subfamilies and 12 genera are found also in arid zones; 13 families and subfamilies occur in savanna and arid zones though not in forest, and relatively few are confined to a single biome—8 in forest, 4 in savanna, and 6 in arid zones. Out of 44 large-mammal superspecies, 20 occur in 2 or more major biomes.

Vicarious taxa in, say, forest and savanna, could have differentiated (1) from a widespread amphibiomic taxon; or (2) once one biome was colonized from the other. If there was a pool of undifferentiated eurybiomic source taxa, one might be inclined to support hypothesis (1). But out of the 245 biological species of larger African mammals, only 3 are truly eurybiomic—occurring in forest, savanna and arid habitats—and only about 30 are amphibiomic: 10 are widespread in forest and savanna and 21 in savanna and arid habitats. They are unrepresentative of the large-mammal fauna as a whole (15 are carnivores) or are already partitioned into distinct infraspecific taxa in the two biomes. So hypothesis (1) is ruled out as a general postulate.

Lönnberg (1929) spoke of forest mammals being stranded in the savanna following the withdrawal of their natural habitat, and while no doubt small populations survived in riverine forest or forest remnants, entry into a new biome is unlikely to have been forced in this way; retreat from inimical conditions is a more likely prospect (Vrba, 1992). Geographic variation in the occupied habitat even within monotypic taxa of restricted distribution (e.g., *Colobus angolensis ruwenzorii;* Kingdon, 1971–1982), and variation in habitat between allopatric members of a clade (e.g., *Cercopithecus mona,* particularly eurytopic within its superspecies; Oates, 1988) are indications that environmental preferences evolve. A taxon characteristic of one biome may give rise to another, able to extend from this ancestral habitat into transitional environments—for instance, as the chimpanzee (*Pan troglodytes*) ranges beyond the rainforest (Kortland, 1972). From this initial colonization of a new biome, a taxon confined to the latter could arise: an ecological translation has been completed. Such events would be facilitated where appropriate structural relationships between vegetation zones encourage encounters with new ecosystems, through steep environmental gradients or fragmented and interdigitated contrasting habitats.

Presumed phylogenetic relationships and polarization of morphoclines suggest that ecogeographical directions of cladogenesis have not been random. Where new biomes are colonized, organisms have predominantly dispersed from more mesic to less equable environments (Grubb, 1978). Forest mammals have given rise to savanna representatives, and savanna taxa have evolved into arid-zone species. A theory of forest origins is 80 years old (Lönnberg, 1918), though it remains controversial. For most groups that have radiated to produce forest and savanna taxa in Africa, this hypothesis has not been overtly rejected, in some cases has been actively canvassed, is in agreement with the available data, and probably will be supported by future cladistic analysis (e.g., Griffiths, 1994). Nonforest African clades may ultimately have had a tropical forest ancestry outside the continent, while there are a few taxa that have entered forest habitats from the savanna. There must have been at least 60 crossings of the forest–savanna ecotone during the evolution of the existing fauna. Such transgressions are at odds with the concept of dichopatry, but they surely represent the kinds of adaptive trends that could over long time scales lead on to turnover pulses (Vrba, this volume).

Speciation

Vicariance events, according to cladists, produce new species. To cladists, cladogenesis and speciation are synonymous, as discrete subspecies are classed as species. Other species concepts perceive speciation as the acquisition of new recognition systems (Paterson, 1993), which may lead to reproductive isolation. The frequency of hybridization between allotaxa among African mammals suggests at the least that such conditions are not always attained in just one vicariance event—that cladogenesis and speciation, though they get confused, are by no means equivalent. Speciation in allopatry should occur when one taxon becomes reproductively isolated from another or when it acquires a different recognition system, but the problem of deciding when allopatric taxa have passed such a threshold is no better resolved by exchanging the isolation species concept for the recognition species concept (Szalay, 1993; this volume).

Achievement of sympatry, however, confirms completion of speciation. The special case where sister species coexist could result from dispersal following a vicariance event, an explanation invoked by cladists. Sympatric speciation could also account for such cases and has been preferred on the grounds that it is the most parsimonious interpretation (Lynch, 1989). This view has been challenged (Chesser & Zink, 1994), though stasipatric speciation is said to have occurred among small African mammals (Meester, 1988) and even equids (Short, 1975). But it has not yet been shown that any sympatric sibling species of African mammals are sister species, whereas sister species and chromosomal races of

Otomys (vlei rats), the target of research on chromosomal evolution, are allopatric (Contrafatto et al., 1994; Taylor et al., 1994).

So it seems likely that sympatry nearly always derives from allopatry (Vrba, 1995), as cladists suggest. Models of speciation (Bush, 1975) terminate in sister taxa attaining sympatry, and this may happen among small African mammals (Denys, this volume). But evidence from the large-mammal fauna suggests that sympatry (or, for that matter, hybridization) does not arise predominantly between sister taxa, nor is it the usual consequence of a vicariance event or even a necessary part of cladogenesis. Sympatry tends to originate between taxa separated by two or more vicariance events, as exemplified by cases within clades of otherwise mainly allopatric taxa, including monkeys (*Procolobus, Colobus* [fig. 11-3], *Cercocebus,* the *Cercopithecus mona* species group), mongooses (*Crosssarchus*), antelopes (*Madoqua, Gazella*), and squirrels (*Heliosciurus*). If gracile australopithecines gave rise independently to both *Paranthropus* and *Homo,* then sympatry between latter genera would necessarily have also concerned non-sister taxa.

Sympatric taxa of larger African mammals are often distantly related, and few congeneric taxa coexist: out of of 101 genera, sympatric species occur only in 18. How is speciation to be envisaged in the other 83? It may be necessary to reconsider our understanding of the process. Opportunities for phylogenetically proximate taxa to coexist, compete, and consequently diverge seem to have been rare, supporting the habitat theory (Vrba, 1992).

Taxon and Biome Effects on Cladogenesis

Taxon effects on cladogenesis are apparent insofar as major taxonomic groups differ in the development of vicariance, explicable in part by Vrba's (1992) resource-use hypothesis. The vagility of bats accounts for their wide distributions, leaving little room for allotaxa. For larger mammals, incidence of allotaxa is demonstrated by percentages of zoogeographic species including one, two, or three or more biological species and are, respectively 38%, 24%, and 38% for primates; 61%, 26%, and 12% for ungulates; and 77%, 18%, and 5% for carnivores. Mean numbers of taxa (nominal species and subspecies) in each zoogeographic species are 6.9 for primates, 5.7 for ungulates, and 4.7 for carnivores. The differences are maintained even within a subset, the forest fauna, where carnivores and ungulates average only 3.1 taxa per zoogeographic species, whereas primates have a greater propensity to split into allotaxa—up to 22 per zoogeographic species, with an average of 6.0 (Grubb, 1990). Extant African apes represent a small sample from which to make comparisons with other primates. Within the ranges of western gorilla (*Gorilla gorilla gorilla*) and nominal subspecies of the common chimpanzee, zoogeographic species of monkeys are differentiated into as many as five allotaxa. To this extent, African apes are geographically less diverse. On the other hand, distributions of the bonobo (*Pan paniscus*) in the Cuvette Centrale and eastern gorillas (*G. g. graueri, G. g. beringei*) in Congo are quite as limited as those of the monkey allotaxa with which they occur.

In zoogeographic species occurring throughout the savanna, there are not more than about five or six widespread taxa, though there may be additional localized representatives and further allotaxa in arid zones. Nonforest primates are less diverse than those in forest—compare savanna and arid-zone baboons (*Papio hamadryas* superspecies) with only about nine subspecies, and the much less widespread red colobus (*Procolobus badius* superspecies) of forested habitats with 18 well-marked taxa (fig. 11-2). Other nonforest mammals have distribution patterns akin to those of baboons, such as 5 taxa in the *Epomophorus gambianus* species group (fruit bats), 6 in the *Damaliscus pygargus* superspecies (topi and allies), 8 in the *Alcelaphus buselaphus* superspecies (hartebeest), and 11 in the *Tatera afra* superspecies (gerbils).

Centers of Endemism

The diversity of allotaxa reflects the degree to which biomes are partitioned into centers of endemism. The Afrotropical mammal fauna divides horizontally into zoogeographic species and vertically into the distinctive local faunas occupying these centers. There are 11 lowland centers in the main west and central forest blocks alone (Grubb, 1990), demarcated by discontinuities in the biome and by rivers, and there are other peripheral or montane centers, including several in the Eastern Forest Zone (Grubb, 1978, 1983). Savanna barriers to dispersal (major escarpments and highlands, internal drainage basins, arid river valleys) define up to 10 centers, larger in area in view of the overall extent of this biome. Arid zones each include several centers, except for the hardly divisible and least diverse Sudanese Arid Zone.

In a center, a proportion of endemics may have similar distributional boundaries, suggesting that they share limitations to population dispersal and have experienced similar recent histories, perhaps including

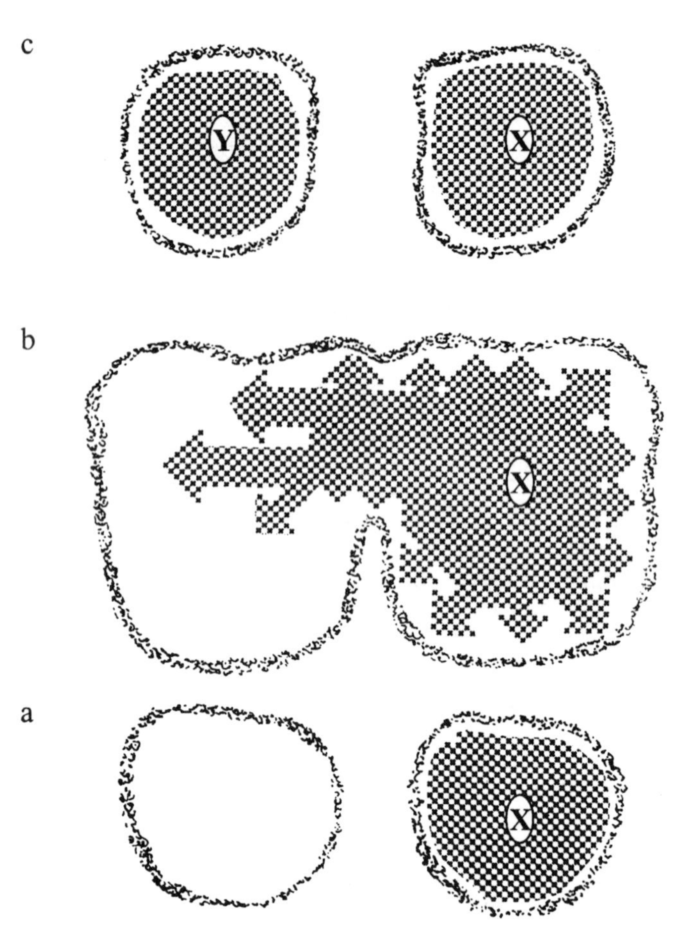

Figure 11-6. Diagram of cladogenesis over a single climatic cycle. (a) Taxon X restricted to a single biome block. (b) Ameliorating conditions allow dispersal drift and also accelerated dispersal into previously unoccupied tract of preferred biome. (c) Deteriorating conditions restore previous discontinuous distribution of biome and cause vicariance event, with derived population Y cut off from ancestral population X, through extinction of intermediate demes.

confinememt to a Pleistocene refuge. It has been hypothesized that such endemics can be the same age (Cracraft, 1983). The conclusion may seem parsimonious, but, in the context of the periodic model, centers or refuges could have formed and reformed repeatedly (fig. 11-6); taxa that share membership of a currently existing center may have differentiated at different times. Nevertheless, some clades may indeed have experienced similar histories. This is suggested when clades have similar patterns of vicariance, including similar partitioning into regional endemics and common ecogeographical polarities in distribution of allotaxa. Shared morphoclines could arise through (1) numerous contemporary vicariance events, (2) sequential vicariance events in a static biota, or (3) sequential dispersals of fauna followed by vicariance events. The first would not produce a morphoclinal pattern. The second involves cosmopolitanism, a particularly unlikely starting point for cladogenesis in the forest biome in view of riverine barriers. Even if they have meandered, the rivers were there first and could not have arisen later to partition widespread biota. The third possibility is not deemed by cladists to occur (e.g., Cracraft, 1982, 1986), as they would not expect there to be concordant dispersal in different lineages. But the objection is to coordination of what are seen as inherently uncoordinated, random jump-dispersals, not to range expansion followed by a vicariance event. Populations of different clades could have dispersed together in the same direction when topographical opportunities were available (e.g., when a suitable corridor opened or when a previously isolated mountain became accessible), and then they could all have been split by a shared vicariance event. This is what is to be expected from the periodic model and anticipates reassessment of the cladogenetic process.

Allusion to Pleistocene refuges for African mammals dates back over 35 years (Booth, 1958), and a general refuge theory for the tropics has since developed, primarily concerning the rainforest (Mayr & O'Hara, 1986). Although compatible with the periodic model, correspondence between present-day centers and past refugia is still controversial (F. White, 1993). During the cool, dry climatic period contemporaneous with the last high-latitude glacial, there were thought to be three major forest refuges in Africa located around the highlands of Liberia, Cameroon, and eastern Zaire. This hypothesis is now seen to be inconsistent with the biogeographic evidence (Colyn, 1991) on which it was supposed to rely, and at least in eastern Zaire it seems that there were more refugia than had been supposed, some confined to lowland, fluvial sites. There is substantial geographical isolation of forest taxa at the present day, as illustrated by monkeys, so the amplitude of biome shift required to disrupt contacts between allopatric taxa and to turn centers into refuges was not necessarily great. Absences of zoogeographic species from particular centers of endemism imply that they did not survive in corresponding refuges or never reached them. For example, the lowland range of the eastern gorilla is confined to an interfluvial area forming a hiatus in the distribution of lowland black-and-white colobus monkeys (Emlen & Schaller 1960; Colyn, 1991), an allopatry suggesting past occupation of different refuges.

It seems that under cool, dry climates, the forest blocks receded, yet montane vegetation and perhaps whole montane forest ecosystems advanced downslope (F. White, 1981, 1993; Maley, 1987; Livingstone, 1993). Some forest fragmentation occurred, including separation of highland from fluvial centers. But the fluvial centers continued to be partitioned wholly by rivers—every center did not become a refuge in the sense of being a discrete biome island.

Regional Effects on Cladogenesis

The structure of whole subcontinents also has a bearing on cladogenesis. Low Africa's relatively gentle relief and almost orderly latitudinal stratification of biomes contrasts with the environmental complexity of High Africa with its many mountain ranges and other topographical barriers, greater variety of habitats and intricate interdigitation of biomes (Denys, this volume). Consistent with the implications of the periodic model, these peculiarities have led to exceptional numbers of contact or hybrid zones, first noted among hartebeests (*Alcelaphus buselaphus*) from the floor of the Rift Valley just north of Nairobi (Ruxton & Schwarz, 1929). Within a circle of 1000-km radius centered on Nairobi, at least 73 contacts between allospecies or well-marked subspecies are already known (e.g., Stott, 1959; Grubb, 1972; Leuthold, 1981; Maples, 1972; Peters, 1986), including 31 hybrid zones, 27 examples of marginal overlap, and 14 cases of parapatry, quite apart from numerous instances of vicariance. In an equivalent area (West Africa, east to Nigeria), there are fewer than 20 known contacts.

A Reappraisal of Cladogenetic Processes

This narrative has already suggested that some evolutionary paradigms should be reconsidered, leading to alterations in our concept of speciation. Dispersal in particular has to be rescued from eclipse. In peripatry, dispersal is envisaged as occurring only across long distances, a rare or unlikely process for nonvolant vertebrates within a continent. In dichopatry, any prior role for dispersal is ignored, with cladogenesis tacitly dating from the vicariance event (Cracraft, 1984). This undue separation of dispersal and vicariance treats them almost as opposing philosophies instead of compatible phenomena. There appears to be no substantial evidence that dispersal is constrained except under the special circumstances designated by cladists. Cladistic analysis reveals topological relationships between taxa that can be explained only by dispersal. Often it can be inferred that mammal populations have dispersed along particular ecological or geographical pathways and so have come to occupy new areas—it is a major conclusion of paleozoogeography. Dispersal is a necessary requirement for continuing allopatric speciation, and it also seems reasonable to assume that through the activities of roaming individuals, animal populations attempt to disperse at every opportunity. Distribution drift (Vrba, 1992) may be the predominant form of population dispersal (Myers & Giller, 1988b) for nonvolant continental mammals. Populations passively track their expanding or contracting habitats; they can advance into an area or withdraw, leaving it unoccupied; demes can evolve anew or become extinct. Perhaps these conditions amount to stasis.

But controlling for such relative shifts reveals an absolute or net movement—an active, not a passive dispersal: when opportunity arises, populations newly enter drifting biome islands, previously unoccupied (Grubb, 1990), and even enter new biomes. Such incursions do not necessarily or usually involve long-distance dispersal, but on a geological time-scale they have been instantaneous (van der Made, 1992; Gentry, 1994; Vrba, 1995). The invading taxon would be an addition to the established local fauna. Its pioneering populations, released from regulatory factors, would at first increase rapidly, if not exponentially. Selection could become more directional, as suggested by Geist's (1971) dispersal theory, especially as the founder demes are likely to be descended from relatively few of the potential parental demes, which may not fully represent the parental genome. This is surely a punctuational phenomenon.

Clinal variation probably arises during dispersal, whether passive (forced) or active: what amounts to an experimental simulation is described by Yom-Tov et al. (1986). As they disperse, populations differentiate even before a vicariance event occurs. If they are then broken up by environmental perturbation, it is ancestral and derived morphotypes that then tend to get separated. Thus we are provided with a gloss on the classic vicariance event, an alternative interpretation to the traditional cladistic model, suggested nevertheless by Hennig's (1966 [1950]) deviation and progression rules—namely, that taxa have been derived serially, one from another (Schwarz, 1929; Maslin, 1952; Grubb, 1978; 1990).

Plesiomorphic and apomorphic populations tend to get separated because phenomena at first inhibiting but then facilitating dispersal, and finally leading to vicariance, tend to recur in much the same places: this is the very nature of biome-drift in the periodic model. Hence the vicariance event is more like budding than splitting: so-called sister taxa are related as ancestor is to descendant. If we cannot accept this concept, we have to adopt the null hypothesis that taxa tend to be divided otherwise—at random (suggesting there has been no order or structure in the history of change among vegetation zones and landforms); or even "against the grain," so that there was a tendency for parental and descendent populations not to get separated by vicariance events (an even more improbable situation). If a succession of dispersals and vicariance events were to unfold, then a morphocline could develop. Length of occupation of territory would be related to position in the morphocline and direction of dispersal, and center of origin could be inferred from its polarity: the morphocline can represent contemporaneous ancestors and descendants, insofar as this can be said about groups of demes. The sister relationship is not confined to the terminal dichotomy in the cladogram, but applies to any two taxa that are truly phylogenetically adjacent. And as species A could have existed for a long time before budding-off its sister species B, one should no longer expect sister species generally to have the same or even similar first-appearance dates in the fossil record.

Conclusion: The Periodic Model of Cladogenesis in African Mammals

Processes implied by biogeographic patterns can be pieced together to provide a generalized sequence of occurrences during cladogenesis (fig. 11-6). Vicariance events are repetitive, so the sequence must begin and end at equivalent points in the cycle if it is to be complete. It must cover a whole "wavelength," including not only the event and its aftermath, but also processes leading up to it, ignored in the concept of dichopatry (differences from peripatry have thereby been exaggerated). As speciation is not encompassed by a single vicariance event, at least among larger mammals, the model has to go on to consider the consequences of further events up to a point where speciation is completed.

Unless cladogenetic events were tightly bunched in time, brief stages in the process should be represented today by rare states and lengthy events by more frequent ones. It is not yet possible to define clearly what these states might be or to comment on their frequency. But a variety of phenomena are at least not inconsistent with the postulate that cladogenesis involved a brief punctuational phase (including active dispersal?) which interrupted relative stasis. Among these phenomena are the contemporaneity of primitive and derived taxa in morphoclines, the abundance of allotaxa, and the limited range of geographic variation within a cline. It may seem that over time taxa have perhaps veered toward being quanta (individuals) rather than continua, but until sufficient calibrated data become available on molecular evolution of extant allotaxa, it is hardly possible to evaluate such ideas.

Our theory also envisages the impact of environmental flux, including patterns prevailing at nadir and zenith of the local cycle and processes occurring through its upswing and downswing. Under climates adverse for the biome in question, allospecies were restricted in distribution and could be widely separated from congeners; with more favorable climates, distributions were more extensive and allotaxa more likely to be in contact.

During ameliorating regimes the biome expanded, and populations could then realize their natural propensity to disperse; new ground was colonized as newly emerging demes joined the advancing front. Populations not only dispersed with the expanding biome but also moved faster to invade virgin territories once access to these previously unoccupied sectors of suitable habitat became available. Invasion of the "promised land" led to more rapid population increase, with selection becoming more directional, generating heightened geographical variation. Differentiation could instead be thwarted through introgression following encounters with related taxa.

During deteriorating conditions, the biome contracted and populations receded, split up, and lost demes through extinction. Divergence of taxa was thereby furthered—these demes had formerly intermediated between what were now becoming remnant populations. Some of these remnants, in spite of their coherence, could ultimately have been of hybrid origin. There was no absolute requirement for any of them to be peripheral isolates.

Cladogenesis need not have occurred in every cycle. If circumstances did not provide an opportunity for dispersal, a taxon's demes would not come to lie in the path of a potential splitting event, and for the moment the taxon would not experience vicariance. But in general, as climatic cycles were repeated, opportunities to reach more and more distant unoccupied territory continued to be created, and by gradual steps new biomes were invaded, so that successive dispersals, colonizations, and breakup of populations could generate chains and branchings of morphoclinally related taxa (Grubb, 1990). Within a lineage, there was a tendency for the evolution of polytypic species and superspecies, rather than assemblages of sympatric taxa. The larger the area colonized, the greater the number of allotaxa formed, within the constraints of taxon and biome effects. The numbers of taxa within zoogeographic species seem to be significantly positively correlated with total range, at least in my preliminary studies of forest monkeys and apes. Vicarious taxa originating within a clade would inhibit each other's further dispersal and reduce opportunities for experiencing vicariance events. The superspecies would eventually come to occupy all available territory, and its representative taxa would be subjected to repeated disruption and coalescence without issue, at times leading to introgression and reticulate evolution. Only when extinction or the opportunity and ability to coexist with other clade members could provide a window for further net dispersal would material be generated for new incipient taxa to be split off and for cladogenesis to continue. Successions of dispersals in a clade eventually brought phylogenetically less closely related taxa together, leading to the formation of hybrid zones or swarms, or alternatively to sympatry and enrichment of the local fauna.

This did not always happen, as evidenced by the relative rarity of sympatric congeneric large mammals. While taxa differentiated within an evolving clade and became successively more distant phylogenetically from each other, they also became increasingly more distant in their morphology, ecology, or ethology through selection acting unpredictably, generating different adaptive changes in different environmental regimes ("adaptive shift"). Extinction of phenetically and phylogenetically intermediate entities (demes, subspecies, even nominal species) could then separate surviving taxa in both morphospace and niche space so that they would be reproductively isolated from each other and from all others in the clade. This is speciation by default, not relational speciation (where taxa that once were conspecific become, in time, specifically distinct); there would not be a particular taxon from which our organism had speciated, nor with which it could complete speciation by becoming sympatric.

So populations do not need to differentiate by independent anagenesis alone in order to diverge. Long-term, continuous allopatry (Vrba, 1995b) is not required. Extinctions would have produced the same effect on relationships between surviving taxa as any divergent evolution (e.g., Taylor & Meester, 1993). Just as evolution of new recognition systems (Paterson, 1993) may, as a secondary consequence, have provided clade members with ready-made reproductive isolation, adaptive shift may have led to ecological isolation, conducive to future coexistence.

There could also be cases where mammal taxa continued to evolve and proliferate actively without speciating at all. Each taxon in the clade would retain conspecificity with its sister taxa (in the extended sense), and reproductive isolation need never arise, yet cladogenesis could lead to a turnover of taxa through extinction of some allotaxa and emergence of others over time. No sympatry would be achieved, no completion of speciation would occur, and no new specific-mate recognition system would develop. Yet evolution in morphology and behavior would continue. Impala (*Aepyceros* spp.; Vrba, 1984) and the gracile australopithecine lineage could provide examples—the human chrono-superspecies may thereby have had an exceptionally lengthy history. Indeed, was the sympatry between *Paranthropus* and *Homo* the only completion of speciation in the australopithecine-human lineage?

The importance of speciation has been inflated at the expense of cladogenesis. Cladogenetic events may need to accumulate before it is possible to achieve speciation, the less frequent and often less tangible phenomenon. Paleontological data examined up the stratigraphic column will often fail to discriminate between inter- and intrataxon differentiation, that is whether a taxon has or has not experienced cladogenesis. Apparent anagenetic trends may have included successions of cladogenetic events.

No novel biological phenomena are required by the periodic model of cladogenesis. Clines could eventually turn into sympatric species, speciation being a fortuitous consequence of cladogenesis (Paterson, 1993). Time lapse rerun of Africa's history would reveal rapid fluctuations—biota flitting across the continent, appearing and disappearing in a kaleidoscope of cladogenesis and extinction. If long-term environmental changes were to occur above the noise of biome shift, then patterns of selection could alter in response. Incremental evolutionary changes must occur in every cladogenetic cycle, else no new taxa would be generated. Shifts in the trajectory and velocity of anagensis occurring across many cycles could culminate in radically new adaptations and new faunas. Frequencies and locations of vicariance events in these turnovers (Vrba, 1992) may be affected too, but not the algorithm of cladogenesis itself.

Summary

Cycles of disruption and coalescence among biomes, not climatic change per se, have provided the motor for highly dynamic cladogenesis of African mammals during the Quaternary. Cladogenesis has occurred through periodic dispersal and partition of mammal populations, in what can therefore be called a periodic model. Biome shift has facilitated active dispersal into previously unoccupied areas, including new biomes. Recently dispersed populations have been split off at a later stage in the cycle. Thereby taxa have multiplied by budding from ancestors rather than by splitting: so-called sister taxa can be related as ancestor is to descendant. There has been a tendency for these sister taxa to form morphoclines. Adaptive trends arising across the cycles of cladogenesis could amplify the trends seen in single morphoclines. In appropriate circumstances, these trends may herald the emergence of a new mammalian community and perhaps a turnover pulse. Vicariance events bunched in space or time have led to the formation of centers of endemism, some of which may have constituted Pleistocene refuges. This structuring of mammalian communities could also contribute to emergence of new regional faunas. Though speciation is involved in all such major evolutionary events, in certain cases cladogenesis can continue perhaps indefinitely without appearance of features characterizing speciation. Speciation remains the special case, the less frequent and more elusive phenomenon, often arising by default.

Acknowledgments This chapter was originally prepared for the Wenner-Gren Malawi conference. I am grateful to Tim Bromage, Friedemann Schrenk, John Oates, Laurie Obbink, and the Wenner-Gren Foundation for the opportunity to attend this meeting, and to Norman Owen-Smith and Alan Turner for constructive comments on the original manuscript.

III

FOSSIL FAUNAS

12

Introduction

F. Clark Howell

More than three and a half decades ago, with encouragement and facilitation of the Wenner-Gren Foundation, François Bourliére and I organized a symposium to interrelate ecological studies in Africa and the emerging development of paleoanthropological investigations on the continent. The published outcome was the volume *African Ecology and Human Evolution* (Howell & Bourlière, 1963). A subsequent massive and protracted symposium, held in 1965 under the same auspices, entitled "Systematic Investigation of the African Later Tertiary and Quaternary" (published as *Background to Evolution in Africa,* Bishop and Clark, 1967) reified and expanded the earlier proposed goals, effectively setting the direction for subsequent interdisciplinarily focused paleoanthropological and diverse associated natural science studies. *African Biogeography, Climatic Change, and Early Hominid Evolution* exemplifies this tradition and affords new approaches toward synthesis, altered perspectives, and, of course, new unforeseen problems.

Some perspectives and conventional frameworks were ill-conceived or seriously weakened and needed to be shored up or totally abandoned at the time of the first symposium more than 30 years ago. The well-known "pluvial hypothesis," as developed by Brooks and subsequently adopted and promulgated by Wayland and Leakey, had become seriously weakened, largely through the efforts of Cooke and the rigorous recent criticism of Flint and Bishop. The first results of Quaternary isotopic geochronology (conventional K-Ar) suggested a totally unexpected set of ages for some important fossiliferous and artifact-bearing formations in eastern Africa. Some pioneering paleontological studies were revealing substantial taxonomic diversity, indications of unexpected intercontinental linkages in taxa, and afforded evidence for the primacy of the Ethiopian Region for the origin and ultimate source of particular mammal higher categories and a number of lesser taxa. The impact of some intensified, multifaceted prehistoric investigations were having an effect, and the concept of paleoanthropology as a broad-scale concern with human biological and cultural (behavioral) evolution was being mooted. Simplistic conceptions and interpretations of humankind's (earlier) Paleolithic record were being challenged, and the complexity of the hominid fossil record and its potential phylogenetic implications were being reexamined in the perspective of the Evolutionary Synthesis.

In contrast with past decades we are now confronted with a surfeit of new, often in-depth data across a broad front of the natural sciences relevant to events and circumstances of the African late Cenozoic and thus, to matters paleoanthropological, broadly defined. The wider application of established methodologies and of new techniques has been pursued more vigorously as *normal* science has taken hold. Newly developed approaches and their applications have afforded expanded or even utterly new insights into previously uninvestigated or otherwise intractable situations and problem areas, such as off-shore ocean cores, lake basin cores, isotopic geochronology, isotopic geochemistry, paleomagnetics, refined geophysical studies, tephra studies, sedimentary geology, paleopedology, and taphonomy. As a consequence of these often unexpected developments and their ramifications and implications, paleoanthropology has come to be increasingly dependent on a remarkable variety of fields that were once adjunct to it but are now in many respects absolutely integral. This dependence has required development of a breadth and depth of concern with and understanding of these adjunct fields on an unprecedented scale. The central issue remains of whether and how such diverse and variously derived knowledge bears on matters properly paleoanthropological and to what extent this knowledge may result in the development of innovative testable hypotheses and in the development of new paradigms.

Investigations of protracted sedimentary successions in rift or peri-rift situations within "High" Africa (as in the Awash, Turkana, Baringo, Olorgesailie, Natron, and Olduvai basins), for which there is now temporal control through isotopic dating and paleomagnetics, have yielded rich documentation of past biotas (especially mammalian, but also invertebrate). These investigations have thus enabled establishment of biostratigraphies based on criteria developed and now widely used in northern continents. Similarly, karstic infills of fossiliferous significance (e.g., Transvaal cave breccias), the ages of which are notoriously difficult if not impossible to ascertain directly, may be both seriated biostratigraphically (among themselves) and related, through comparative analysis, to such quasi-continuous, fossiliferous successions of known geochronological age elsewhere. Such developments have encouraged and enabled inquiry into diverse issues of paleobiological and evolutionary interest that were scarcely envisioned, much less pursued, until now. These issues include taxic diversity, phyletic evolution, speciation, vicariance, extinction, endemicity, biotic provinciality, dispersal, and intercontinental exchange.

Several mammalian families (or higher categories) exhibit substantial and even remarkable taxic diversity; Cercopithecoidea (Benefit), (hipparionine) Equidae (Bernor and Armour-Chelu), Suidae (Bishop), Bovidae (Vrba), and Rodentia (Denys) are discussed in this volume. All are groups for which there is an established and often refined alpha taxonomy (at the generic level for Rodentia); both Suidae and Bovidae, which are often well-defined at the species level, and to an extent Cercopithecoidea as well, have been used extensively in both biostratigraphic analyses and in efforts to elucidate habitat specificity.

Denys's admirable, expansive, and in-depth analysis of late Cenozoic sub-Saharan rodent (generic) taxa reveals that affinities with currently recognized vegetation (and biotic) zones are usually demonstrable, that past (biotic) zones may at times have had different magnitudes and latitudinal representation, and that reflections of past zonal complexity are sometimes manifest, indicating distinctive patterns of both distribution and of taxic co-occurrence. Some zonal displacements, or at least their interfaces, were at times substantial. It is perhaps a bit premature, due to the small number of requisite comparative studies, to elaborate explication of species distributions through time and across space, but such documentation will ultimately be most revealing of past biotic zones and their temporal transformations.

Carnivora exhibit much broader distributions (over subregions) within the Ethiopian Region than the mammalian families mentioned earlier. Some important intercontinental links between Africa and Eurasia have long been recognized and have multiplied as a consequence of new discoveries, more adequate material and intensified comparative analyses. Major efforts to revise certain higher taxa has been a key factor in such advances. Links at the generic and even specific levels are most notable among Hyaenidae and Felidae, but are substantially less notable among Canidae (which is rather suprising, given the relatively recent antiquity of concerned genera). Links are now known as well among some (few) Mustelidae and Viverridae, notably within the Mio-Pliocene interval.

Africa possesses the most diverse primate fauna of any continent, now comprising some 15 genera and 46 species. Most lorisids, most colobines, many cercopithecines, and the apes are forest distributed and adapted. The few notable exceptions are among the galagines (two or three), papionines (baboons), and cercopithecine monkeys (three species) The recent taxa are, overall, either ill-represented (larger taxa) or almost unknown in the recent fossil record, such that evidence relevant to their differentiation(s) and distribution(s) is poor (both absolutely and relative to other larger species). Although some taxa (*Colobus, Cercocebus, Cercopithecus,* etc.) are demonstrably represented in the Pleistocene, or even (late) Pliocene record, the specific affinities of samples are largely unresolved. The fossil (Plio-Pleistocene) record, as well presented here by Benefit, is surprisingly one of substantial, but different diversity, of differentiation at the generic level, of some speciosity, and of seemingly distinctive patterns of distribution and vicariance. Some seven genera of Colobinae are known (or distinguishable), only two of which are Mediterranean African (Late Miocene), and encompass a dozen (if not more) species. A single taxon (*Cercopithecoides*), perhaps the same species, is cosmopolitan in distribution in sub-Saharan Africa. The bulk of the diversity is documented in eastern Africa a few degrees on either side of the equator. The fossil record is such that the origin(s) of this diversity is enigmatic, but it must largely reflect an Late Miocene radiation (the Mpesida and Lukeino formations are critical, if still obscure in this regard).

Among cercopithecines there is an unexpected diversity of papionines and of widespread representatives of the *Theropithecus* clade. *Papio* is clearly and best documented by diagnostic remains in (earlier Pleistocene) southern Africa, but is certainly present back into the Pliocene equatorward, if still undistinguished specifically. Thus, the extent to which such distribution was cosmopolitan is unknown. Some sharing of *Parapapio* species is claimed between eastern and southern Africa, the latter having the maximum taxic diversity; but, available collections are often insufficient and/or inadequately studied to resolve the issue of specific identity (or vicariance), if any. (Traces of such papionines and as well *Theropithecus,* all undetermined specifically, are known from the Chiwondo Beds, Malawi). *Theropithecus* in the form of some five "form" taxa are well delineated, abundant, and widespread within eastern Africa, and in several instances (*T. darti, T. oswaldi* ssp.) appear cosmopolitan in their distribution into southern Africa. In the (northern) Turkana Basin there is demonstrable stratigraphic overlap between two such taxa (*T. brumpti and T. oswaldi*). *T. oswaldi* had a remarkably broad latitudinal distribution and persisted into the mid-Pleistocene in high latitudes of the Maghreb and South Africa. Sympatric diversity in (mid-) Pliocene cercopithecoids is well documented at Laetoli (four species), Turkana Basin, and Afar localities. Mangabeys (*Cercocebus*) are certainly represented (e.g., Turkana Basin, Olduvai), even if undiagnosable specifically. Other small cercopithecines are known, but imperfectly and rarely, and hence the explication of presumptive monkey diversity in forest habitats remains elusive.

Hipparionine equids were Eurasian emigrants into Africa some 10 m.y. An Late Miocene diversification soon followed, but its details are obscure. Some 10 specific or form taxa have been proposed for hipparionines of the African late Cenozoic. The extent to which particular proposed taxa are valid, fall into synonomy, and/or constitute clades is a matter of current reinvestigation, as Bemor and Armour-Chelu skillfully outline. One or more such lineages of the African *Eurygnathohippus* group (radicle) persisted well into the mid-Pleistocene, unlike the situations known elsewhere in Eurasia. Several large and small forms occur in the Plio-Pleistocene, often in some abundance, but uncommonly with substantial cranial/postcranial associations, and only too rarely in the form of complete crania. The recognition and confirmation of some long-standing nomina remains a major obstacle. As a consequence, the extent of vicariance is obscure, and the composition of individual clades is often unresolved. Thus, in the Chiwondo Beds assemblage, hipparione(s), from successive stratigraphic intervals, constitute the second largest mammal group; their identity is of particular concern both for biogeographic perspectives and for further biostratigraphic elucidation.

In this respect the documentation of *Equus* in Africa is in a more advanced state. The approximate time of first continental appearance is closely delineated, particularly in the Turkana Basin (~2.35 m.y.). The nature and pattern of diversification and lineage constitution has been broadly elucidated, and major taxic issues are largely in hand. Consequently, this group and its representatives are of singular importance in respect to biostratigraphic and biogeographic issues within the continent, and still more broadly in respect to the elucidation of Eurasian intercontinental exchange and dispersals.

Two groups of suids replace the hyotherine, listriodontine, and sanitherine pigs characteristic of the African earlier to middle Miocene: representatives of the Tetraconodontinae and Suinae subfamilies. The former, presumably of Asian derivation and of earlier Late Miocene age, comprises the genera *Nyanzachoerus,* of which five species are nominally recognized, and a descendant genus, *Notochoerus,* of which three species (the type *capensis,* constituting an intermediate form) are often recognized. In fact, the former nominate genus apparently constitutes a single, only slightly diversified phyletic lineage from the Late Miocene (*devauxi-syrticus*) into the earlier Pliocene (*kanamensis-jaegeri*), and extinction (<3.0 m.y.), by phyletic evolution and cladogenesis, to *Notocherus* about mid-Pliocene times. These nyanzachoeres are generally regarded as forest or woodland mesic-adapted species. A diminutive tayassuid peccary, *Caenopithecus africanus,* is also represented (at least) in the Mio-Pliocene interval, both in southern and equatorial Africa.

Suines are of Eurasian origin and constitute dispersals into Africa around the mid-Pliocene. Some, if not all, taxa are considered to reflect open, savannalike adaptive settings. They have been a central focus of efforts at affording an independent biostratigraphy in geochronologically dated stratigraphic successions in equatorial Africa. The two principal genera, *Metridiochoerus* and *Kolpochoerus,* comprise four and five nominate species, respectively, of which four have cosmopolitan distributions in sub-Saharan

Africa. Bishop's work (this volume) has shown, based on postcranial evidence, that at least one species of each genus (and perhaps two of *Kolpochoerus*) were adapted to mesic habitats. The oldest representative(s) of *Kolpochoerus* is represented by distinctive species (*afarensis, phacochoeroides*) in eastern Africa and the Maghreb, respectively. Vicariant species of each genus occur in subsequent Pleistocene time ranges in both eastern (*M. hopwoodi, K. majus*) and southern (*K. paiceae*) reaches of the continent. An early suine, attributed to *Phacochoeroides shawi,* occurs in upper Pliocene cave fills in the Transvaal. It has been affiliated with the *Metridiochoerus* group, particularly as an early part of the *M. andrewsi* lineage. However, a related form occurs in the older, basal Pliocene of eastern Africa, suggesting an earlier than expected appearance and perhaps supporting its taxonomic individuality. The oldest recognized occurrence of the extant phacochoere lineage, *P. antiquus,* is perhaps only documented in the southern African subregion. *Sus scrofa* is a Palaearctic emigrant into northernmost Africa only well into the Pleistocene. The African suid record appears to document species origins (two) 3.0–2.8 m.y. and extinctions (two) slightly later (2.7–2.5 m.y.) a major turnover, with extinctions and originations (three each) between 2.0 and 1.6 m.y., and at least three extinctions since the base of the mid-Pleistocene.

The recent fauna of Africa is distinguished by its plethora of antelope and related bovid species. Customarily 15 tribes are distinguished, comprising some 30 genera and (minimally) 94 nominate species (table 12-1). These lineages emerge, for the most part, initially in the Late Miocene and especially in the Pliocene. There is great diversification, both overall and within particular (larger-bodied) bovid lineages, in the course of the Pliocene, as reflected in persistent Pleistocene taxa. Lesser and dwarf antelope species, especially with forest, mesic, or upland habitat adaptations, are generally much less known as fossils, and their distribution and temporal records are correspondingly sparse. This disparity affects as many as six tribes. Both immigration into and emigration from Africa played important roles in the history of antelopes. Boselaphini and Ovibovini, now confined to other continents and different latitudes, were present in the African upper Cenozoic. Overall some seven instances of Eurasian emigration into Africa are recorded: from the Late Miocene (a caprine species); basal and earlier Pliocene (two ovibovines), and early upper Pliocene (six species, including caprines, ovibovines, and an antilopine); and successively within the earlier Pleistocene (ovibovines and four caprine species). Emigrations from Africa are fewer, occurring in the earlier Pliocene (a hippotragine) and during the early upper Pliocene (two reduncines, two

Table 12-1. Taxa of African Bovidae, extant and those extinct since the Upper Miocene

	Extant Taxa		Extinct Taxa	
Tribe	Genus	Species	Genus	Species
Cephalophini	2	14	—	1 (1)[a]
Neotragini	1	3	—	—
Madoquini	1	5	—	1 (1)
Raphicerini	2	3–4	—	1 (3)
Dorcatragini	1	1	—	—
Oreotragini	1	1	—	1 (1)
Tragelaphini	3	9	—	5 (4)
Hippotragini	2	5	4	10 (3)
Reduncini	4	8	2	14 (5)
Alcelaphini	4	8	6	32 (7)
Aepycerotini	1	1	—	2–3 (1)
Antilopini	4	12	1	13 (1+)
Peleini	1	1	—	2 (1)
Bovini	1	1	4	9 (1)
Caprini	2	2	?	5–6
Ovibovini	—	—	2	3–4
Boselaphini	—	—	2	2

[a]Numbers in parentheses are of taxa attributed directly to extant species or considered as of close affinity (aff.) to extant taxa.

hippotragines, and an alcelaphine). Thus, Ubeidiya (Jordan Rift) has at least six African immigrant taxa, including hippotragine and bovine species, as well as hippo, giraffe, and a kolpochoere.

The highest levels of diversity in extinct taxa are recorded among alcelaphines (32 + 7), reduncines (14 + 5), hippotragines (10 + 3), and antilopines (13 +), and, to a lesser extent, bovines (9 + 1). This pattern of rapid evolution and extensive speciation is extreme compared with any other African mammalian higher taxonomic category. Remarkable turnover among African Bovidae has been progressively well documented, and implications of turnover have been extensively examined by Vrba (this volume). A strong pulse of first appearances has been demonstrated in the upper Pliocene (~2.7–2.5 m.y.), with almost half as a consequence of cladogenesis, and more than 15% as a consequence of Eurasian immigrant allochthones. There is a strong representation of open-habitat, cooler-adapted taxa within that interval. Lesser pulses of new appearances are manifest just before the Plio-Pleistocene boundary (1.9–1.8 m.y.) and in an interval approximating the interface of the Matuyama/Brunhes chron boundary (0.9–0.6 m.y.).

Vrba has been responsible for developing and elaborating a multifaceted theoretical framework to address patterns of faunal change, including turnover, cladogenesis, phyletic evolution, extinction, immigration, and emigration. These topics are viewed within the context of late Cenozoic paleoclimate cycles, induced by astronomical (orbital) forcing, and attendant habitat shifts and transformations as physical factors acting to constrain and modulate evolutionary change and, thus, a primary causation of biotic events. Her proposal of the Turnover Pulse Hypothesis and the correlative covering law propositions of relay turnover, eco-shuffling, resource use, distribution drift, and the off–on–off ("stop light") model of corridor dispersal constitute substantive, innovative contributions to paleobiology that have produced much discussion, and debate, and a much needed reorientation of research endeavors. Although only just over a decade old, this perspective has had a fundamental impact in phylogenetic, biostratigraphic, and paleoenvironmental studies. It stands to be tested further against the emerging paleoclimatic framework of the African late Cenozoic.

The elucidation of past environmental circumstances over the vastness of Africa has accelerated greatly in recent decades, such that much prior speculation and (often ill-founded) inference has been succeeded by an ever-expanding body of empirical documentation. The intensified investigation of sedimentary sequences has been complemented by broadening application of palynological and macrobotanical investigations, paleopedology, the study of paleosol stable carbon isotopes (Sikes, this volume) and the recovery and analysis of cores from basins of freshwater lakes and the deep-sea from pericontinental situations. The Pliocene interval has been revealed as encompassing major intervals of lowland forest and woodland expansions, both latitudinally and translongitudinally, with successive intervening peaks (3.25, 2.6–2.5, 2.2–1.7 m.y.) of aridification in both low and equatorial latitudes. Some detailed paleoenvironmental records have emerged from the Laetoli, Olduvai, Baringo, Turkana/Omo, and Awash (Hadar) basins, and further studies are continuing. The Turkana/Omo Basin continues to constitute a key reference for the equatorial Pliocene and Pleistocene due to its multiple, widespread formations, protracted fluviatile (fossiliferous) sedimentary record, (five successive lakes between ~4 and ~0.9 m.y.), multiplicity of distinctive and dateable tephra, paleomagnetic and geochronological control, and linkages to the marine (Arabian Sea) core record. The emerging, broadly comparable sequence in the middle Awash (Afar) will both complement and greatly enhance this record.

The contributions here clearly reflect how rapidly and how far paleobiological and associated biogeographic and paleoenvironmental studies have progressed in recent years. Similarly, they provide a guide to the immediate future in which multidisciplinary researches are mandatory, and cross-disciplinary communication and exchange of perspectives and goals are absolutely essential.

13

Biogeography, Dietary Specialization, and the Diversification of African Plio-Pleistocene Monkeys

Brenda R. Benefit

Fossil cercopithecoids occur at all Plio-Pleistocene hominid-bearing deposits. In many ways the geographic range, patterns of regional abundance, speciation and extinction, as well as the paleobiology of Plio-Pleistocene cercopithecoid monkeys mirror those observed for early hominids (Dunbar, 1983, 1992a; Foley, 1993). As for the hominids, several Old World monkey taxa occur exclusively in either eastern or southern Africa (fig. 13-1), and many exhibit postcranial features indicative of a predominantly terrestrial habitus (Jolly, 1970; Delson, 1975; Birchette, 1981, 1982; M. G. Leakey, 1982; Ciochon, 1986). It is likely that the larger-bodied monkeys and australopithecines shared dietary resources, fled from the same predators, and competed with each other for safe sleeping sites. It is certain that they experienced the same changes in climate and vegetation. Consequently, understanding the evolutionary history of eastern and southern African monkeys should provide insight about the circumstances that influenced the initial emergence and divergence of our bipedal ancestors.

This chapter examines the evolution of Plio-Pleistocene cercopithecoids primarily, but not exclusively, from the perspective of their dietary adaptations. Cercopithecoid communities at individual fossil sites and within geographic/temporal regions are examined in terms of diversity in species numbers, dietary preferences (inferred from their dentition), and locomotor adaptations (reconstructed from postcranial remains). Special emphasis is placed on understanding the emergence of Old World monkeys from an ancestral Miocene stock and on differences in cercopithecoid communities between eastern and southern regions of Africa during the Plio-Pleistocene.

Reconstructing Habitat and Habitus

The importance of applying principles of functional morphology to interpretation of the habitus of extinct animals associated with fossil hominids and their ancestors is not new. For example, criteria were established for reconstructing the foraging adaptations, locomotor patterns, and habitat preferences of African sabre-toothed cats (Ewer, 1954), pigs (Ewer, 1958), monkeys (Jolly, 1972), and antelopes (Gentry, 1970; Gentry & Gentry, 1978). These criteria continue to be used and refined by current scholars interested in the paleobiology of the same taxa (Lewis, 1994; McCrossin, 1983, 1987; Bishop, 1995; Birchette, 1982; Ciochon, 1986; Plummer & Bishop, 1994; Spencer, 1994).

Insight about the broader environments in which these animals lived results from such studies. For example, Gentry (1970) used the frequencies of more and less cursorial bovids at Fort Ternan to reconstruct its environment as woodland. Inferences about the paleoenvironment at sites where *Nyanzachoerus* occur, including the Pliocene sites of Sahabi, Aramis, Kanapoi, and Hadar, were altered when McCrossin's (1983; 1987) study of a complete forelimb of *Nyanzachoerus syrticus* revealed it to be highly cursorial.

Figure 13-1. Biogeographic distribution of Plio-Pleistocene cercopithecoids in sub-Saharan Africa, based on information summarized by Szalay and Delson (1979) and M. G. Leakey (1982).

Cursoriality in this suid probably indicates that it occupied open-country environments, in contrast to the forested habitats it was previously assumed to have occupied. Detailed analyses of bovid metapodials from Olduvai Gorge reveal the presence of important subhabitats not previously detected there (Plummer & Bishop, 1994).

However, an important caveat that must be taken into account when making such reconstructions is that a particular diet or locomotor behavior does not always correspond with a specific habitat. Although it is tempting to equate grazing with open country, some grazers, such as the forest hog (*Hylochoerus meinertzhageni*), are actually denizens of the forest that consume grasses in small glades or clearings. Although cursoriality is often equated with open country, the highly cursorial white-tailed deer (*Odocoileus virginianus*) of eastern North America lives in woodland environments. Moreover, the assumption that terrestrial locomotor adaptations are equivalent with open-country habitats rings false when one considers that largely terrestrial highland gorillas and semiterrestrial lowland gorillas, chimpanzees, and bonobos occupy forested areas. Dietary and locomotor adaptations must be considered together when reconstructing habitat preferences of extinct primates. Even then, such interpretations must be treated cautiously.

In spite of advances in reconstructing environment and paleobiology from the perspective of the functional morphology of fauna, some researchers continue to take the expedient path of assuming that fossil animals had the same habitus as their closest extant relatives. This is especially true for Old World monkeys. The abundance of colobine fossils at the Early

Pliocene site of Aramis has been cited as indicating that *Ardipithecus ramidus* might have preferred forest or woodland habitats (WoldeGabriel et al., 1994). However, the paleobiology of the Aramis colobines has yet to be examined from the perspective of their bones and teeth. Such a study is warranted because most Plio-Pleistocene colobines have no modern analogues. *Paracolobus, Rhinocolobus,* and *Cercopithecoides* were two to three times larger than extant African colobine monkeys and possessed an unusual array of cranial traits indicating that they were not directly ancestral to living taxa. At least one, *Cercopithecoides,* was largely terrestrial in its habitus and is commonly associated with open-country habitats (Birchette, 1981; M. G. Leakey, 1982). Diets of these animals are assumed to have been as folivorous as those of modern *Piliocolobus,* without taking into account that many other colobines, including *Colobus, Procolobus, Presbytis,* and *Nasalis,* consume as many or more seeds as leaves (Davies & Oates, 1994). Some living colobines, such as the hanuman langur, are semiterrestrial and thrive in open habitats rather than in forest.

Additional errors result when broad environmental reconstructions of hominoid-bearing deposits are used to make inferences about the adaptations of associated hominoids. For example, interpretations of australopithecines as being obligate, fully terrestrial bipeds seem to have been influenced by the fact that predominantly open habitats are represented at the southern Africa cave deposits and Olduvai, the sites at which they were first discovered. However, functional assessments of early australopithecine postcrania indicate that, while bipedal, they retained adaptations for climbing and may have stayed closer to the trees than was once predicted (Stern & Susman, 1983).

Another excellent illustration of the pitfalls of inferring a hominoid species diet or locomotor behavior from its broad environmental context is the case of *Kenyapithecus.* Studies of soil carbonate isotopes have been used to claim that *Kenyapithecus* retained primitively arboreal patterns of feeding and locomotion in a closed-canopy forest (Cerling et al., 1991, 1992). As will be discussed later, detailed analysis of the anatomy and adaptations of *Kenyapithecus* definitively refutes such speculation. New fossils of the genus clearly show it to be the first semiterrestrial hominoid, with adaptations for the consumption of hard fruits and nuts (McCrossin, 1994a,b; McCrossin & Benefit 1992a, 1994, 1996, Benefit & McCrossin, 1993a, 1995; McCrossin et al., 1998). From its own morphology and consideration of its environmental context, it can be inferred that *Kenyapithecus* probably foraged fallen fruits and nuts in seasonally dry woodlands. The failure of broad environmental reconstructions to comprehend the paleoecology of *Kenyapithecus* does not inspire confidence in the reliability of these methods to predict the adaptations of early hominid species at Aramis, Olduvai, or any other locality.

In this study inferences about the dietary preferences of fossil monkeys from the Miocene and Plio-Pleistocene of eastern and southern Africa are based mainly on the objective measurement of dental features (shear crest lengths and degree of cusp proximity) shown to be functionally correlated to diet among extant cercopithecoids (see Kay 1977, 1978, 1981, 1984; Kay & Hylander, 1978; Kay & Covert, 1984; Benefit, 1987; Benefit & McCrossin, 1990). Regression equations expressing the relationship between these dental measurements and the average proportion of fruits and leaves consumed annually by extant cercopithecoids are then used to estimate the proportions of fruits and leaves eaten by the extinct monkeys (Benefit, 1987; in press; Benefit & McCrossin, 1990). Because the method does not differentiate grass-eating from leaf-eating, patterns of wear are considered together with the measurements for differentiating browsers from grazers. The reliability of the equations for predicting the diets of extinct species is indicated by the accuracy with which they estimated those of extant species (fig. 13-2). Inferences about the locomotor adaptations of fossil monkeys are taken from the literature.

Fossil cercopithecoid monkeys were sampled from the Miocene sites of Wadi Moghara, Gebel Zelten in North Africa; Buluk, Maboko Island and Ngeringerowa in Kenya; as well as Plio-Pleistocene deposits in southern Africa (Sterkfontein Member 4, Kromdraai Members A and B, Taung, and Swartkrans Member 1) and eastern Africa (Laetoli and Olduvai, Tanzania; Kanapoi, Olorgesailie and Koobi Fora, Kenya, Areas 1–203, Kenya; Omo, Ethiopia, Usno Formation, Shungura Formation Members B–H, and Kalam Area). A complete list of specimens sampled is given in Benefit (1987).

Miocene Perspectives on Plio-Pleistocene Cercopithecid and Hominid Emergence

All known Early and Middle Miocene monkeys belong to the family Victoriapithecidae, distinguished from Cercopithecidae (colobines and cercopithecines of the Late Miocene to present) by their incompletely bilophodont upper and sometimes lower molars and deciduous premolars; the differential expression of the primitive crista obliqua (absent in cercopithecids)

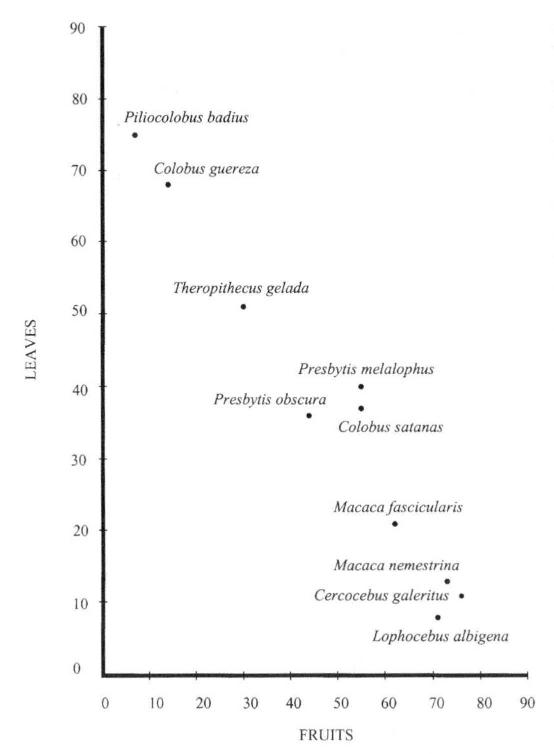

Figure 13-2. Predicted proportions of leaves and fruits included in the annual diets of extant monkeys, based on averaging estimates from regression equations correlating diet with shear-crest length, flare (a measure reflecting cusp proximity and basin size), and cusp relief.

along the upper and of hypoconulids along the lower molar row; and unusual cranial morphology with a steep and linear facial profile, an airorhynchous hafting of the neurocranium on the face, supraorbital costae and frontal trigon, tall and narrow orbits, and an exceptionally tall zygomatic and low malar root, among other features (Von Koenigswald, 1969; M. G. Leakey, 1985; Benefit, 1987, 1993, 1994, 1999; in press; Benefit & McCrossin, 1991, 1993b, 1997). These Miocene monkeys were of small to medium body size, weighing between 3 and 8 kg (Fleagle, 1988; Harrison, 1989). Only two genera, *Prohylobates* from North Africa and the Kenyan site of Buluk, and *Victoriapithecus* from Kenya and possibly Uganda are currently recognized during the early and middle Miocene, and both are attributed to the Victoriapithecidae. The oldest known victoriapithecid occurs at the radiometrically dated 19 m.y. site of Napak in Uganda (Bishop, 1968; Pickford, 1983, 1987). Victoriapithecidae are critical for reconstructing the appearance and lifeways of the earliest cercopithecoids because they represent the sister group to modern monkeys. Features shared between them and either or both the colobine or cercopithecine subfamily were probably present in the earliest true cercopithecoid.

Although the earliest Old World monkeys are usually thought to have been more folivorous than the ancestral hominoids with which they lived, such an interpretation is not supported by the fossil or neontological evidence. The broad upper central incisors, low molar cusp relief and shearing ability, formation of bowl-shaped depressions on molar cusp tips (a pattern of wear found only among extremely frugivorous monkeys), and heavily pitted enamel microwear of the victoriapithecids indicates that they were highly frugivorous relative to folivorous Miocene apes such as *Simiolus* and *Mabokopithecus,* which occur in the same deposits (Benefit, 1987, 1990, 1991, 1999; McCrossin & Benefit, 1994). *Victoriapithecus* appears to have been megadont, as indicated by visual comparison and the fact that its estimated body weight based on molar size (8 kg) is twice as large as that based on postcranial remains (3.5–5 kg) (Fleagle, 1988; Harrison, 1989). Among catarrhines, megadonty is usually associated with frugivory. Based on the regression equations used to estimate the diets of extant monkeys in fig. 13-2, the annual diet of *Prohylobates* from Buluk probably consisted of 74% fruit and 13% leaves, of the North African *Prohylobates* of 84% fruits, and of *Victoriapithecus* from Maboko 79% fruit and 7% leaves. Victoriapithecids are therefore reconstructed as being as or more frugivorous as the Tana River mangabey *Cercocebus galeritus,* the most frugivorous of extant catarrhines (Homewood, 1978). Independent assessment of nonocclusal microwear on *Victoriapithecus* molars support this conclusion (Ungar & Teaford, 1994).

As the diversity of colobine species with documented ecology increases as a result of field work, ecologists have come to realize that the dietary adaptation of colobine monkeys is as much for consuming seeds as for eating leaves (Davies & Oates, 1994). Because cercopithecine monkeys are predominantly frugivores, neontological data tells us that the ancestral cercopithecid was probably adapted for eating the seeds and/or pulp of fruit. Given that frugivory is shared by basal cercopithecids and victoriapithecids, the common ancestor of all cercopithecoids was undoubtedly a frugivore. Cranial features that distinguish all cercopithecoids from hominoids, such as the derived absence of the maxillary sinus in Old World monkeys, seem to be related to a strengthening of the upper face and of postcanine tooth roots, presumably to buttress the teeth and face against high occlusal forces produced during the initial fracturing of hard fruit or seed coatings. Such an interpretation differs significantly from previous reconstructions of the earliest monkeys as having been consistently or facultatively folivorous based on the assumption that bilophodonty evolved in Old World monkeys for the purpose of slicing leaves (Jolly, 1970; Napier, 1970; Delson, 1975). The origin of cercopithecoid bilophodonty instead seems to have been related to increasing grinding surface area and creating wedgelike cusps with the ability to produce initial fractures in hard fruits and seeds, rather than to improving shearing capacity (Kay, 1975, 1978; Maier, 1977; Benefit & Pickford, 1986; Benefit, 1987, 1999; Lucas & Teaford, 1994).

Examination of the postcranial remains of fossil and extant cercopithecoids reveals that the origin of Old World monkeys was related to a shift toward terrestriality, in addition to the consumption of hard fruits and seeds. Victoriapithecids and cercopithecids share many derived features of the postcrania that must have been present in the earliest monkeys, including a narrow distal humerus, well-defined median trochlear keel, posteriorly rather than laterally oriented medial epicondyle, well-defined ischial callosities, and restricted hip and ankle joints (Napier & Davis, 1959; Jolly, 1967; Rose, 1983; Harrison, 1989; McCrossin & Benefit, 1992b, 1994). All of these features indicate that victoriapithecids had adopted a cursorial pattern of terrestrial locomotion, most similar to that of modern vervets (von Koenigswald, 1969; Delson, 1975; Harrison, 1989; McCrossin & Benefit 1992b; McCrossin et al., 1998). Hence, the earliest monkeys were designed to exploit a broader range of resources both on the ground and in the trees.

That the semiterrestrial and seed-eating adaptations of the earliest monkeys are correlated with a preference for a particular habitat unlike that of the forest environments in which the majority of Early Miocene apes occur is indicated by their restricted distribution during the early and middle Miocene. Victoriapithecids are absent from well-documented Early Miocene rainforest localities such as Songhor, Koru, and Rusinga, where apes are diverse and abundant. The only potentially rainforest locality at which they occur is the oldest known cercopithecoid bearing deposit, Napak V (Bishop, 1968). The 17- to 16-million-year-old North African *Prohylobates* sites of Gebel Zelten and Wadi Moghara in North Africa are reconstructed as deltaic riverine forest with savanna in the hinterland (Savage & Hamilton, 1973; Pickford, 1987), but fauna from these deposits and Napak are poorly known.

The site with the greatest potential for revealing the preferred habitat of Miocene victoriapithecids is that of Maboko Island. Whereas only two dozen victoriapithecid specimens are known from all other early and middle Miocene localities combined, as of 1997 more than 3500 fossils of *Victoriapithecus macinnesi* have been collected at Maboko. All the cercopithecoid fossils come from the lower stratigraphic levels of the Maboko Formation, Beds 3 and 5, which are radiometrically dated as older than 14.7 mya (Feibel & Brown, 1991; Benefit, 1993, 1994). Monkeys apparently disappear in upper levels (Bed 12), which are dated as between 13.8 and 14.7 m.y., and are absent from Fort Ternan, which is of a similar age.

In terms of reconstructing the Maboko paleoenvironment, the impressive bird and rodent faunas and numerous artiodactyl postcrania from the post-1987 collection are especially important. Preliminary identification of the previously unknown bird fauna by McCrossin and Chandler (personal communication; Benefit & McCrossin, 1992) indicate the presence of at least 12 families, of which the majority are associated with water habitats. Flamingos may indicate the presence of a shallow alkaline lake with extensive mudflats; lily trotters may indicate the presence of swamp with floating vegetation; and darters, cormorants, and pelicans may indicate that stretches of shallow but open and probably fresh water occurred in the area. The only Maboko birds characteristic of forest or savanna habitats are hornbills, guinea fowl, passeriforms, and birds of prey. Among the small mammal fauna are dry-adapted bathyergids (Cifelli et al., 1986; Winkler, 1994), for which previously collected remains had been misidentified as the forest-adapted anomalurid *Zenkerella* (Andrews et al., 1981). Anomalurids and forest-adapted rodents are currently unknown from the Maboko deposits (Cifelli et al., 1986; Winkler, 1994). Of the four documented bovid species at Maboko, some exhibit highly cursorial adaptations of the lower limb similar to those of the modern impala and are possibly indicative of open country habitats (McCrossin, personal communication).

Before the discovery of this collection, analysis of gastropod fossils not found in direct association with mammalian fauna indicated that Maboko was similar to a "nyika" semiarid *Acacia–Commiphora* woodland with grass patches and gallery forest along streams and some nearby forest (Pickford, 1983). Analysis of the small pre-1974 mammalian sample indicated that the Maboko environment was primarily riverine woodland with thick brush cover (Evans et al., 1981). Study of the new collections indicates an even greater diversity of microenvironments in the area, with dry, open woodland adjacent to swamps and seasonally flooding streams and little evidence of forest (Benefit & McCrossin, 1992; McCrossin & Benefit, 1994). The presence of a thick vertisol in Bed 5 indicates that the sediment had been deposited during alternating wet and dry periods (C. Feibel, personal communication), such as would be expected if rainfall were seasonal.

In addition to being restricted in distribution, victoriapithecids were also limited in the type of apes with which they coexisted. Napak is the only site at which victoriapithecids occur with the presumably arboreal soft-fruit–eating apes *Proconsul* (*P. major*) and *Micropithecus* (*M. clarki*) (Pilbeam & Walker, 1968; Fleagle, 1975). North African species of *Prohylobates* did not co-exist with hominoids at all. *Prohylobates tandyi* from Wadi Moghara is associated with an archaic pliopithecid-like catarrhine known only from its distal humerus, which possesses a primitive entepicondylar foramen, but *P. simonsi* from Gebel Zelten is found with no other primate (Pickford, 1983, 1987; Simons, 1994). In Kenya, the 17 m.y. *Prohylobates* from Buluk is associated with *Afropithecus* (M. G. Leakey, 1985). Middle Miocene *Victoriapithecus* is known only from sites at which *Kenyapithecus* occurs, such as Maboko Beds 3 and 5, Kipsaramon, and Nachola/Baragoi (von Koenigswald, 1969; Benefit, 1993; Pickford, 1987; Hill, 1995; Ishida, 1984). It is unclear why victoriapithecids are not present at all *Afropithecus* and *Kenyapithecus* localities, being absent from Moruorot, Maboko Bed 12, and Fort Ternan.

Like the monkeys they co-occur with, *Afropithecus* and *Kenyapithecus* possess a suite of dental characters indicative of a hard fruit/seed-eating diet. The teeth of these apes have thick or moderately thick enamel, a feature characteristic of frugivorous primates (Kay, 1981). Discovery of a nearly complete mandible of *Kenyapithecus* at Maboko demonstrates that it possessed tall and strongly procumbent incisors like those of South American bearded sakis (*Chiropotes*) and uakaris (*Cacajao*) (McCrossin & Benefit, 1992a, 1993, 1994, 1997; Benefit & McCrossin, 1995), in contrast to the short and hominidlike incisors it was expected to have (Andrews & Walker, 1976). As for uakari and saki monkeys, it is likely that *Kenyapithecus* used its anterior dentition to break open hard coats of fruits and seeds. Both *Afropithecus* and *Kenyapithecus* possessed robust and laterally rotated canines, upper incisor heteromorphy, and an anterior position of the zygomatic root, features that also characterize the South American pitheciines and are functionally related to cracking hard seed coatings (McCrossin & Benefit, 1992a, 1993, 1994, 1997; Benefit & McCrossin, 1995). Additional indicators of their hard fruit/seed diet are the prognathic upper incisors of *Afropithecus* (degree of prognathism is unknown for *Kenyapithecus*) and simian shelf on the unusually long mandibular symphysis of *Kenyapithecus*.

Although *Afropithecus* seems to have been arboreal in habitus, *Kenyapithecus* is the first, and so far only Miocene ape, to share a semi-terrestrial pattern of locomotion with the victoriapithecids (McCrossin & Benefit, 1992b, 1994, 1996; Benefit & McCrossin, 1993a, 1995; McCrossin, 1994a,b, 1995, 1996; McCrossin et al., 1998). Discovery of approximately 45 postcranial specimens of *Kenyapithecus* at Maboko between 1992 and 1997 have been critical in determining its locomotor pattern and substrate preference. Functional analysis of the new postcrania indicates that *Kenyapithecus* shares with extant semiterrestrially adapted catarrhines a proximal extension of the greater tubercle above the humeral head, a long and retroflexed olecranon process, a transverse ridge (the metacarpal torus) on the dorsal surface of the proximal phalangeal facet of the third metacarpal, short and straight phalanges, and an adducted big toe (McCrossin, 1994a,b; Benefit & McCrossin, 1995; McCrossin & Benefit, 1997). Before the new discoveries, the only fossil that could have indicated a semiterrestrial locomotor strategy for the genus was a distal humerus from Fort Ternan. Unfortunately, the Fort Ternan distal humerus was misinterpreted as being identical to *Proconsul nyanzae* and as indicating a generalized arboreal quadrupedal pattern of locomotion (Andrews & Walker, 1976; Rose, 1993). McCrossin's analysis of the bone clearly indicates that the medial epicondyle of *Kenyapithecus* was oriented posteromedially as in terrestrially adapted cercopithecoids such as *Erythrocebus patas* and *Macaca nemestrina* (McCrossin & Benefit, 1992a, 1994, 1996; Benefit & McCrossin, 1993a, 1995; McCrossin, 1994a,b, 1995, 1997; McCrossin et al., 1998). On this basis, he suggested that *Kenyapithecus* was adaptively similar to mandrills (forest-dwelling baboon relatives) and mangabeys (references as above). This interpretation, based on detailed analysis of new fossils, substantively refutes the claim that the genus retained the primitive generalized arboreality of *Proconsul* and other Early Miocene hominoids (Andrews & Walker, 1976; Andrews, 1992a; Rose, 1993).

Functional analysis of Miocene fossil catarrhines therefore tells us that although the diverse Plio-Pleistocene baboons and bipedal hominids are often treated as if they were the first to evolve terrestrial locomotor patterns in response to climatic change leading to the emergence of open-country habitats, terrestriality first appeared among African catarrhines during the Early and Middle Miocene (McCrossin et al., 1998). The combination of hard fruit and seed eating and semiterrestriality seems to have evolved in Old World monkeys and apes for the purpose of exploiting resources in a disturbed mosaic of seasonally flooded and wooded habitats where food sources were abundant closer to the ground and were seasonally unpredictable (Benefit, 1987, 1999; McCrossin & Benefit, 1992a, 1994, 1997; McCrossin 1994a,b, 1995; McCrossin et al., 1998). It is possible that during dry seasons these catarrhines exploited dry-adapted, hard-shelled fruits such as monkey oranges, which some extant bovids rely on to survive the dry season (N. Owen Smith, personal communication).

Following the middle Miocene, catarrhines that had survived the transition from Early Miocene forest to more open woodland environments in eastern Africa largely went extinct. These included the victoriapithecids, folivorous smaller-bodied apes *Simiolus* and *Mabokopithecus,* and *Kenyapithecus.* The last occurrence of these taxa is approximately 12–14 m.y. at Fort Ternan and Maboko Bed 12. In spite of their disappearance, adaptations of the Middle Miocene monkeys and large-bodied hominoids provided the bauplan from which catarrhines of modern aspect evolved.

Unfortunately, the fossil record of African apes is too poorly known during the Late Miocene to determine if they retained semiterrestrial adaptations from *Kenyapithecus.* Molecular evidence and the absence of suspensory adaptations in any of the known Miocene apes other than the enigmatic *Oreopithecus* indicates that the last common ancestor of all recent apes, including gibbons, did not evolve until 15–10 m.y. (Sarich & Wilson, 1967; Sarich, 1971; McCrossin, 1992, 1994a; McCrossin & Benefit, 1992a, 1994; Benefit & McCrossin, 1995). Consequently, the specialized adaptations of *Kenyapithecus* may tell us only that terrestriality could have evolved more than once among African apes.

Adaptations and Habitus of Late Miocene and Plio-Pleistocene Cercopithecids

The common ancestor of all modern monkeys emerged from a victoriapithecid stock sometime between 14 and 11 m.y. Little evidence exists to support the co-occurrence of Late Miocene monkeys and apes. The oldest fossil cercopithecid is the first true colobine *Microcolobus tugenensis* from Ngerngerowa, a Kenyan site dated at approximately 11 m.y. (Benefit & Pickford, 1986). Cercopithecines do not appear until approximately 7 mya, with the occurrence of *Macaca* sp. (a form genus not necessarily indicative of its relationship to modern macaques; Delson, 1975). These Late Miocene to earliest Pliocene colobines and cercopithecines retained a small-to-medium body size and the semiterrestrial locomotor pattern found in their victoriapithecine ancestors. All of the cercopithecoid postcrania known from this time period, those of the colobine *Mesopithecus* and small macaquelike cercopithecine from Sahabi, exhibit adaptations for a semiterrestrial mode of locomotion similar to those of *Victoriapithecus* (Delson, 1973, 1975; Meikle, 1987). In addition, these early cercopithecids were predominantly frugivorous (Benefit, 1987). Shear crest lengths and other measures indicate that the earliest colobines, *Microcolobus* and *Mesopithecus,* had relatively low shear-crest lengths and consumed more fruits than leaves (Benefit & Pickford, 1986; Benefit, 1987, 1999). Their diets were more like those of *Presbytis* than of *Piliocolobus.*

In contrast to the apparently low cercopithecoid species diversity of the Miocene, numbers of cercopithecoid taxa proliferated during the Plio-Pleistocene. In Africa alone, five colobine genera and at least eight species occur during this time period. Cercopithecines are equally speciose, being represented by 7 genera and at least 18 species (fig. 13-1). Aside from the form genus *Macaca,* most of the Plio-Pleistocene genera are new. With the exception of North African *Libypithecus,* all are found in association with bipedal hominids (*Australopithecus* and *Homo*) or their ancestors (*Ardipithecus*). Whether they occurred together with Plio-Pleistocene African apes is unknown because fossils of the latter have never been found.

Plio-Pleistocene cercopithecoid faunas are dominated by the presence of larger-bodied colobines and cercopithecines. The majority of these monkeys retained terrestrial adaptations from their Miocene ancestors. Of the taxa whose postcrania are known, all cercopithecines and one (*Cercopithecoides*) of three colobine genera were terrestrial. Locomotor adaptations of rare smaller monkeys approximating modern *Cercopithecus* and *Colobus* in size are currently unknown. These cercopithecid monkeys are associated with semiterrestrial and terrestrial bipedal hominids, with the possible exception of *Ardipithecus,* whose locomotor pattern is currently unknown.

The oldest Pliocene colobine is the moderately large-bodied *Libypithecus* from Wadi Natrun in North Africa. The longer-snouted cranium of *Libypithecus,* with distinctive sagittal and nuchal crests, was once considered enigmatic, but can now be viewed as retaining basal cercopithecoid features because it strongly resembles that of *Victoriapithecus* (Benefit & McCrossin, 1997). *Libypithecus* may have given rise to the even larger-bodied long-snouted colobines *Paracolobus* and *Rhinocolobus* (Leakey M. G., 1982; Hynes & Benefit, 1995), but more complete material of the North African species is needed to prove this relationship. *Cercopithecoides* may belong to the same clade, but its shorter-snouted face and more globular crania are quite different from the other genera. Smaller-bodied colobines are currently assigned to the form genus *Colobus,* but their relationships and adaptations are little understood. Long-awaited descriptions of skeletons of smaller-bodied colobine monkeys from Hadar and recent fieldwork by Leakey should expand the known diversity of this subfamily in eastern Africa.

The regression equations used to estimate the diets of fossil cercopithecoids in this study are based on M1–M3 shear crest lengths (lengths of the occlusal margins of cusps), cusp relief (height of the crown above the median lingual notch of lower molars compared with crown height below the notch), and crown flare (degree to which mesial or distal pairs of cusp tips are closely approximated relative to crown width at the cervix). Plio-Pleistocene colobines are most derived in terms of cusp relief and fairly conservative in terms of shear-crest length. All categories of data indicate that the cosmopolitan *Cercopithecoides williamsi* (especially from Swartkrans) was a highly derived folivore, whereas *Cercopithecoides kimeui* had the lowest cusp relief of all the Plio-Pleistocene colobines and was the least derived toward folivory (table 13-1, figs. 13-3–13-5). Both terrestrial species exhibit an excessively rapid rate of occlusal wear, indicating either that they consumed ground-level browse that was laden with gritty dust or that they were grazers with fewer adaptations for resisting tooth wear than the modern gelada.

In addition to having lower shear-crest lengths, the lower cusp relief of *Colobus* from Koobi Fora and *Paracolobus mutiwa* from the Omo contribute to their averaged annual diets being less folivorous than that of the living *Piliocolobus.* On the basis of lower second molar shear quotients and flare, populations of *Paracolobus* (Laetolil and Omo), *Rhinocolobus* (Koobi Fora), and small colobines (Koobi Fora) are estimated to have included more fruits than leaves in their annual diets (table 13-1). However, when these predictions are considered together with estimates based on their high cusp relief, these colobines appear to have been slightly more folivorous than frugivorous (figs. 13-3, 13-4).

When shear crests alone are considered, fossil *Parapapio* from Kanapoi, *Cercocebus* from Koobi Fora and Olduvai, and *Papio* from Olduvai were found to be the most frugivorous of the eastern African "baboons" (a term used loosely here to represent the terrestrially adapted, large-bodied *Papio, Parapapio, Gorgopithecus,* and *Dinopithecus;* table 13-1; Benefit, 1987, 1990; Benefit & McCrossin, 1990). Shear-crest analysis revealed *Parapapio* from Laetoli and Koobi Fora to be the least frugivorous. However, when averaged estimates based on all three categories are considered, *Parapapio* from Laetolil instead appears to have been one of the more frugivorous species (fig. 13-4). *Parapapio jonesi* and *Dinopithecus ingens* were found to be the most frugivorous and *Parapapio whitei* the least frugivorous of southern African species (table 13-1, fig. 13-5). The majority of southern African baboons are reconstructed as sharing a similar diet composed of approximately 55–60% fruits and 25–30% leaves/grass (fig. 13-5). Isotopic analysis confirms that *Papio robinsoni* from Swartkrans had a C_3-based diet, which is usually associated with browsing (Lee-Thorp et al., 1989).

The dietary habits of fossil *Theropithecus* are more difficult to assess than those of baboons due to their unusual molar morphology, which includes accessory cuspules along shear crests, and due to the fact that the method used does not discriminate leaf eaters from grass eaters. The mid-Pliocene eastern African species of *Theropithecus, T. quadratirostris* (3.4 m.y.), and the population of *T. brumpti* from the Omo, as well as the southern African subspecies *T. oswaldi danieli* from Swartkrans had longer shear crests and the higher cusp relief than other Plio-Pleistocene *Theropithecus,* indicating their greater potential for shearing fibrous foods (table 13-1, figs. 13-3–13-5). Only its high degree of molar flare makes *T. oswaldi danieli* appear to be less folivorous on average (table 13-1). Isotopic analysis showed *T. oswaldi* to be more of a C_4-dependent grazer than a browser (Lee-Thorp et al., 1989). Although it has long shear crests, molars of *T. brumpti* from Koobi Fora also exhibit a high degree of flare, leading to the population's predicted diet being more frugivorous than that of conspecifics from the Omo. *T. oswaldi* from Koobi Fora, Olorgesailie, and Olduvai are predicted to have been more frugivorous than other *Theropithecus,* including the extant gelada (figs. 13-2–13-5).

Molars of *T. quadratirostrus* and *T. brumpti* are generally less massive and elaborate than those of *T. oswaldi,* with fewer accessory cuspules and fewer infoldings of enamel. That *T. brumpti* may have included leaves rather than grass in its diet is indicated by its thin tooth enamel (1.16 mm) compared with that of *T. oswaldi* (1.5 mm), with which it coexisted at Koobi Fora. Deep transverse striations are not apparent on the worn molars of *T. brumpti,* indicating that little grit adhered to its food (Benefit, 1987; Benefit & McCrossin, 1990). *Theropithecus brumpti* may have been a true browser rather than a grazer. The rapid rate of cusp deformation observed for the molars of *T.*

Table 13-1. Predicted diets for Late Miocene to Plio-Pleistocene cercopithecoids from eastern Africa based on regression equations in Benefit (1987, in press) and Benefit and McCrossin (1990).

		Fruits/leaves		
	N	Shear	Cusp Relief	Flare
Ngeringerowa (11 m.y.)				
Microcolobus tugenensis	1	54/28		
Kanapoi (4.4 m.y.)				
Parapapio jonesi	1	83/0		
Laetoli (3.5–3.8 m.y.)				
Paracolobus ado	12	39/44	12/70	55/29
Parapapio ado	25	57/29	67/17	83/0
Koobi Fora (1.5–3.3 m.y.)				
Cercopithecoides kimeui	7	47/35	30/53	
Cercopithecoides williamsi	5	23/60	3/82	34/50
Colobus sp.	7	49/34	30/53	
Paracolobus mutiwa	2	33/50	0/91	
Rhinocolobus turkanaensis	9	58/24	19/73	21/62
Cercocebus sp.	30	75/10	62/22	92/0
Parapapio sp.	11	58/24	51/33	
Theropithecus brumpti	23	52/31	28/55	
Theropithecus oswaldi	119	61/24	31/52	64/24
Omo, Shungura Formation (1.3–2.9 m.y.)				
Colobinae indet sp.	15	37/47	23/65	
Paracolobus mutiwa	38	31/54	29/53	40/45
Rhinocolobus turkanaensis	59	38/46	18/65	46/38
Theropithecus brumpti	6	35/46		13/70
Theropithecus quadratirostrus	1	28/60		
Olduvai (0.5–2.2 m.y.)				
Cercopithecoides kimeui	2	40/44	40/43	35/49
Cercocebus ado	3		74/9	
Papio sp.	5	96/0	48/35	39/45
Theropithecus oswaldi	32	64/18		100/0
Olorgesailie (0.5–1.3 m.y.)				
Theropithecus oswaldi	168	67/19	11/80	93/0
Leboi (0.5 m.y.)				
Cercopithecus sp.	5	67/17	54/27	48/36
Sterkfontein (3.2–2.5 and 1.75–1.4 m.y.)				
Parapapio broomi	12	57/25	50/33	60/25
Parapapio jonesi	17	58/33	59/24	60/25
Parapapio whitei	9	65/21	47/37	36/47
Kromdraai (2.7–1.8 m.y.)				
Gorgopithecus major	15	64/27	39/31	66/19
Papio angusticeps	11	60/24	54/29	53/31
Taung (2.3–1.0 m.y.)				
Papio angusticeps	6	46/37	63/20	
Swartkrans (1.8–1.6 m.y.)				
Cercopithecoides williamsi	1		1/82	
Dinopithecus ingens	24	69/14	63/21	66/19
Papio robinsoni	15	61/22	55/28	
Parapapio jonesi	10	67/15	53/30	95/0
Theropithecus oswaldi danieli	11	34/49	6/84	95/0

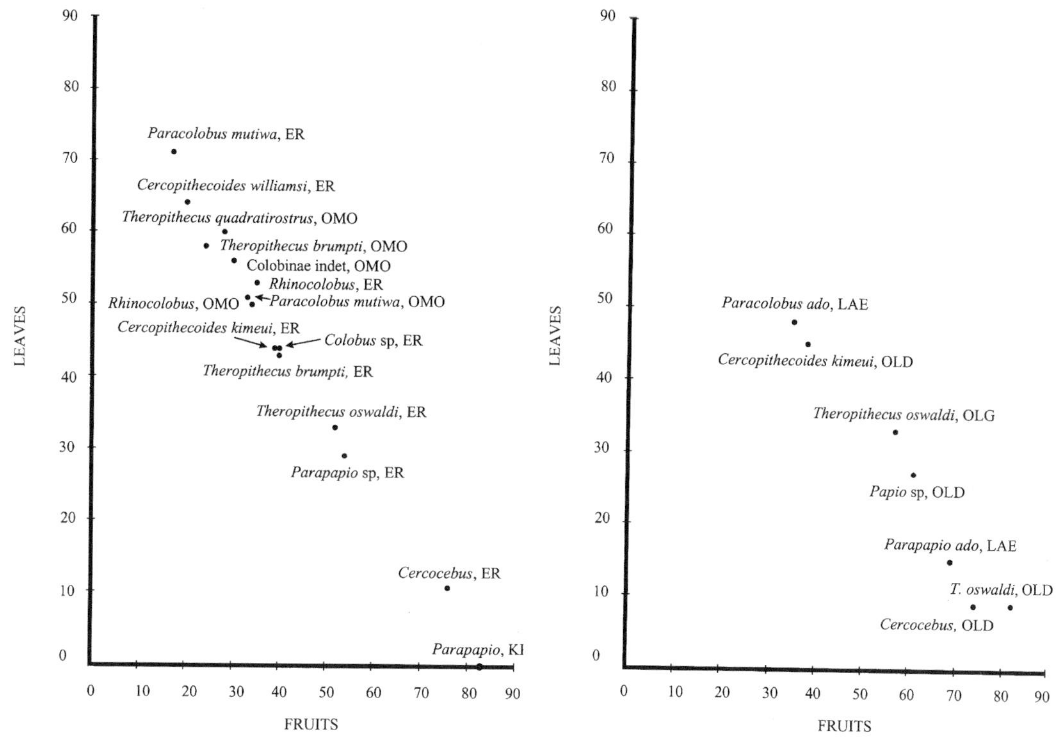

Figure 13-3. Predicted proportions of leaves and fruits included in the annual diets of extinct Plio-Pleistocene Old World monkeys from northern Kenya and southern Ethiopia, based on averaged estimates from regression equations.

Figure 13-4. Predicted proportions of leaves and fruits included in the annual diets of extinct Plio-Pleistocene Old World monkeys from southern Kenya and northern Tanzania, based on averaged estimates from regression equations.

brumpti can be attributed to the close proximity of the molar cusps or the consumption of fruits, rather than to a diet of abrasive grass (Benefit, 1987, 1990; Benefit & McCrossin, 1990). Microwear studies by Teaford (1993) indicate that it may actually have consumed more fruits than *T. oswaldi.* A similar diet is suggested for *T. quadratirostris.*

Light microscopy of wear striations revealed deep bucco-lingually oriented parallel striations and the relative absence of pits on the molars of *T. oswaldi* from Koobi Fora (Benefit, 1987; Benefit & McCrossin, 1990). These deep striations were probably caused by grit in the diet of *T. oswaldi.* They are also indicative of a heavy reliance on the transverse component of mastication, such as is associated with grass-eating in modern *T. gelada.* It is plausible that the nonfruit component of the diet of *T. oswaldi* consisted of the blades, seeds, and rhizomes of grasses, as suggested by Jolly (1972), rather than leaves. However, the diet of *T. oswaldi* is reconstructed here as having been more eclectic than that of *T. gelada,* contrary to Jolly's (1972) suggestion that the species chiefly ate grass.

Inferences about the environments in which cercopithecoids lived can be made from consideration of their dietary and locomotor diversity in eastern and southern Africa. For example, the relative proportions of fruits, leaves, and grasses in the diets of living monkeys is related to whether they occupy forest or woodland habitats versus open or treeless savanna. African grasslands are characterized by low and seasonal fruit productivity. Grasses are an important supplement to the diets of baboons living near or on the savanna during times of the year when fruits are unavailable (Dunbar, 1983). Extant baboons exploiting forest habitats tend to eat greater quantities of fruits than grasses or herbs, whereas the opposite is true of baboons living in scrub savanna (Dunbar, 1983). The diet of *Papio* baboons living in unforested (desert) areas of Namibia consists of 80% grasses (Hamilton et al., 1978). In a similar environment in Ethiopia, the

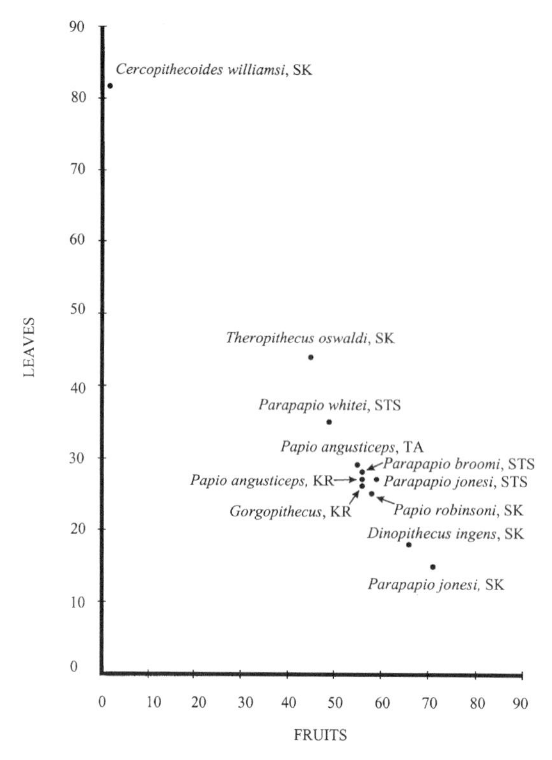

Figure 13-5. Predicted proportions of leaves and fruits included in the annual diets of extinct Plio-Pleistocene Old World monkeys from southern Africa, based on averaged estimates from regression equations.

baboon diet consisted of 40% grasses (Dunbar & Dunbar, 1974). In contrast, grass composed only 10–20% of the diet of baboons living in heavily forested areas, and 20–50% of the diet of baboons living in wooded but not heavily forested regions (Kummer, 1968; Dunbar & Dunbar, 1974; Popp, 1978; Post, 1978; Sharman, 1980). A similar relationship may have existed during the Plio-Pleistocene.

From figures 3–5 it is evident that a wider range of microenvironments from forest to savanna were represented in Omo basin deposits (Koobi Fora and Omo) than in the southern Kenya and northern Tanzania region and southern Africa. This is inferred from the greater diversity of cercopithecoid dietary adaptations in that region than in the southern region of East Africa. Of those colobine and cercopithecine species predicted to have included more than 50% leaves in their annual diet, six occur in northern East Africa, none in the southern east region, and only one in southern Africa, indicating major differences in environment between these three regions (figs. 13-3–13-5).

Of the six folivorous species in the Omo basin, two (*P. mutiwa* and *R. turkanaensis*) were arboreal (M. G. Leakey, 1982), one (*T. brumpti*) may have been less cursorial than *T. oswaldi* and may have climbed trees as often as does the extant mandrill (Ciochon, 1986; M. G. Leakey, 1993), and one (*C. williamsi*) was terrestrial (Birchette, 1981, 1982, M. G. Leakey, 1982). These folivorous monkeys may have shared a similar diet, but differences in their locomotor adaptations indicate that they may have exploited at least three different microenvironments in the region, with arboreal folivores indicative of forest or woodland, terrestrial frugivores of forest or woodland, and terrestrial grazers of open country.

Whereas habitats preferred by browsing arboreal cercopithecoids were absent in Southern Africa, the presence of terrestrial grazers and paucity of extreme frugivores (consuming more than 75% fruits annually) indicates that forest and woodland were scarce in the region. The more limited range of microhabitats in this region was able to support a large number of terrestrial baboons, six species of which shared a diet of approximately 60% fruit and 25% leaves/grass. Southern East Africa was similarly characterized by the absence of extreme folivores but supported fewer cercopithecoid species than either the Omo Basin or South Africa. Only one arboreal species (*Paracolobus*) is documented from the southern east region.

In general, inferences made from cercopithecoid data about the diversity of microhabitats in the three regions considered in this study are supported by broad environmental reconstructions. Kanapoi and the Omo group deposits are thought to represent a broad range of environmental settings from forest to open grassland (M. G. Leakey, 1976; Boaz, 1977). Within this region *Rhinocolobus* seems to occur only where there is evidence of riverine forest or woodland, and it has been suggested that the fossil baboons of eastern Africa occupied forest or woodland habitats (M. G. Leakey, 1982). *Cercopithecoides* occurs at deposits that sample predominantely open savanna habitats with little bush cover such as Olduvai, Sterkfontein, Kromdraai, and Swartkrans and is probably associated with similar habitats in the Omo Basin (Boaz, 1977; M. G. Leakey, 1982). Laetoli, where the frugivorous-folivorous *Paracolobus* and more frugivorous *Parapapio* occur, is reconstructed as being dominated by woodland to open savanna with no major rivers or lakes (Leakey & Harris, 1987). Indications that most southern African baboons were not extreme frugivores is similarly consistent with reconstructions of the southern cave deposits as representing drier, more open savanna habitats than was typical of eastern African sites such as Koobi Fora and the Omo (Boaz, 1977).

The inclusion of more fruits in the diet of fossil *Theropithecus* indicates that its habitat may have been characterized by a higher degree of tree cover than that occupied by the extant species *T. gelada. Theropithecus oswaldi,* which probably consumed grasses rather than leaves, is likely to have occupied open-country habitats adjacent to more wooded areas, such as grasslands growing along the margins of shallow lakes where seasonal flooding inhibited the growth of trees, as suggested by Jolly (1972). If the geologically more ancient of the eastern African species *T. brumpti* and *T. quadratirostris* consumed leaves rather than grasses and were less terrestrial than *T. oswaldi,* it is possible that they inhabited forested environments (see also M. G. Leakey. 1993).

Biogeography of Plio-Pleistocene Old World Monkeys

The dietary and habitat preferences of the fossil Old World monkeys almost certainly influenced the patterns of species diversity and relative abundance observed for Plio-Pleistocene cercopithecines and colobines in eastern and southern Africa. Such patterns, as well as data on species distribution and location of their earliest occurrence, provide a great deal of information from which the evolutionary history of these cercopithecoids can be reconstructed (figs. 13-1, 13-6–13-8). For example, during the earliest Pliocene of eastern Africa, colobines were more diverse and abundant than cercopithecines. *Paracolobus* occurs at the 4.3 m.y. site of Aramis, where it is 12 times as abundant as associated *Parapapio* (T. D. White et al., 1994). However, a million years after Aramis, colobine monkeys represent only 4–12% of the cercopithecoid faunas in which they occur in eastern Africa (figs. 13-6–13-8). In southern Africa this subfamily is represented only by *Cercopithecoides williamsi,* which has a consistently low abundance, representing 2–6% of the total cercopithecoid fauna in the region, although it persists there throughout the Plio-Pleistocene (figs. 13-1, 13-6–13-8). Their first appearance and greater abundance in the eastern region indicate that colobines probably evolved in eastern Africa before migrating south.

The first appearance of *Theropithecus* is in 3.5–4 m.y. eastern African deposits at Lothagam (Delson, 1993). *Theropithecus* is rare in South African deposits, composing 7% of the total cercopithecoid fauna collected before 1976 and occurring at only Makapansgat, Swartkrans, and Hopefield (Freedman, 1957, 1976; Maier, 1971). In contrast, before 1976, 84% of the monkeys collected at the Omo and 85% of those collected at Koobi Fora, both in eastern Africa, belong to the genus *Theropithecus* (figs. 13-6–13-8; Eck, 1976; M. G. Leakey, 1976). In addition, *Theropithecus* is more diverse in deposits from eastern Africa, with five species occurring there, as opposed to only two in southern Africa. It is probable that *Theropithecus* evolved in eastern Africa before migrating south.

A pattern of geographic diversity and abundance opposite to that of *Theropithecus* is observed for *Papio* and its closest relatives. Baboons first occur at the approximately 6 m.y. old South African site of Langebaanweg (Grine & Hendey, 1981) and represent 84% of the Plio-Pleistocene fossil monkeys from southern Africa (Freedman, 1976; Brain, 1981b). In contrast, baboons compose only 10% of the fossil monkeys from eastern Africa. In southern Africa at least seven and possibly nine species of baboons have been recovered, but only three species are known to have occurred in eastern Africa. The diversity and abundance of baboons in eastern Africa declines from the mid-Pliocene to the Early Pleistocene, but diminishes only slightly thoughout this period in southern Africa (figs. 13-3–13-5).

If the origin of the genus *Theropithecus* is linked to the beginnings of leaf eating in forest-dwelling baboons (Benefit, 1987; Benefit & McCrossin, 1990), the paucity of *Theropithecus* in southern Africa may be explained by the absence of forest environments in that region. This hypothesis is consistent with the absence of leaf-eating colobine monkeys other than the terrestrial *Cercopithecoides* in southern Africa. Alternatively, *Theropithecus* in southern Africa may have suffered from competition with grass-eating, savanna adapted baboons. If *Theropithecus* were endemic to eastern Africa, as seems likely, it may have arrived in southern Africa after the savanna baboons had successfully filled the grass-eating niches available to monkeys, inhibiting *Theropithecus* from "swamping" the southern grasslands with its high population numbers, as *T. oswaldi* did in eastern Africa.

The greater diversity of *Theropithecus* in the eastern region is in part attributable to reduced competition between species of the genus as a result of differing preferences for fruits, leaves, and grasses (M. G. Leakey, 1993). The overwhelming abundance of *T. oswaldi* fossils in collections from eastern Africa between 2.5 m.y. and 0.5 m.y. may be due to its having lived closer to fluvial and lacustrine depositional environments than the more forest-adapted monkeys, resulting in a higher frequency of fossilization. It is also likely that absolute population numbers of *T. oswaldi* were greater than those of the forest cercopithecines. Population densities of *T. gelada* are considerably

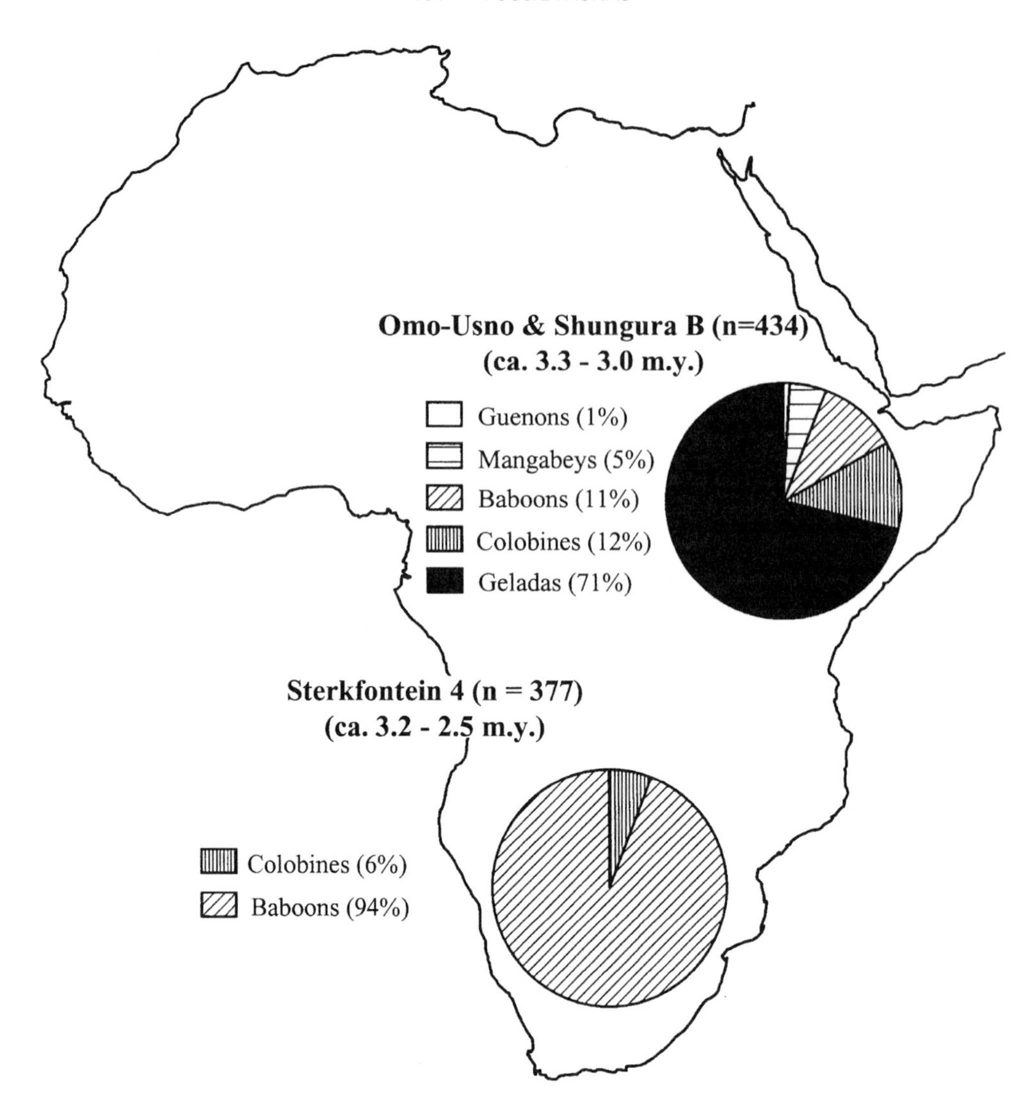

Figure 13-6. Relative abundance of cercopithecoid groups in the mid-late Pliocene, approximately 3.3–2.5 m.y. The cercopithecoid fauna of eastern Africa during this interval is exemplified by the Omo Group Usno Formation and Member B of the Shungura Formation (Eck, 1976), southern Africa is represented by Sterkfontein 4 (Brain, 1981b). Age estimates for the Omo are from Brown and Nash (1976), Brown et al. (1985), and Feibel et al. (1989), and for Sterkfontein from Vrba (1982).

higher than those of any known population of *Papio,* presumably because dense and evenly distributed grasses can support larger numbers of animals than forest resources, which are more sparsely and patchily distributed (Dunbar, 1983).

The lower species diversity and rarity of baboon fossils in eastern Africa may have resulted from their preference for forest habitats. The eastern mangabeys and baboons would have competed for forest resources with *T. brumpti* and large-bodied colobine monkeys *Paracolobus* and *Rhinocolobus,* which included almost equal portions of fruits and leaves in their diets. Competition between the forest baboons and colobines would have been more intense during the Plio-Pleistocene than it is today. As a consequence, the diversity of forest-dwelling members of both Colobinae and Cercopithecinae seems to have been affected. Because *Papio* baboons did not become the dominant savanna monkey in eastern Africa until after the demise of *T. oswaldi,* when it presumably be-

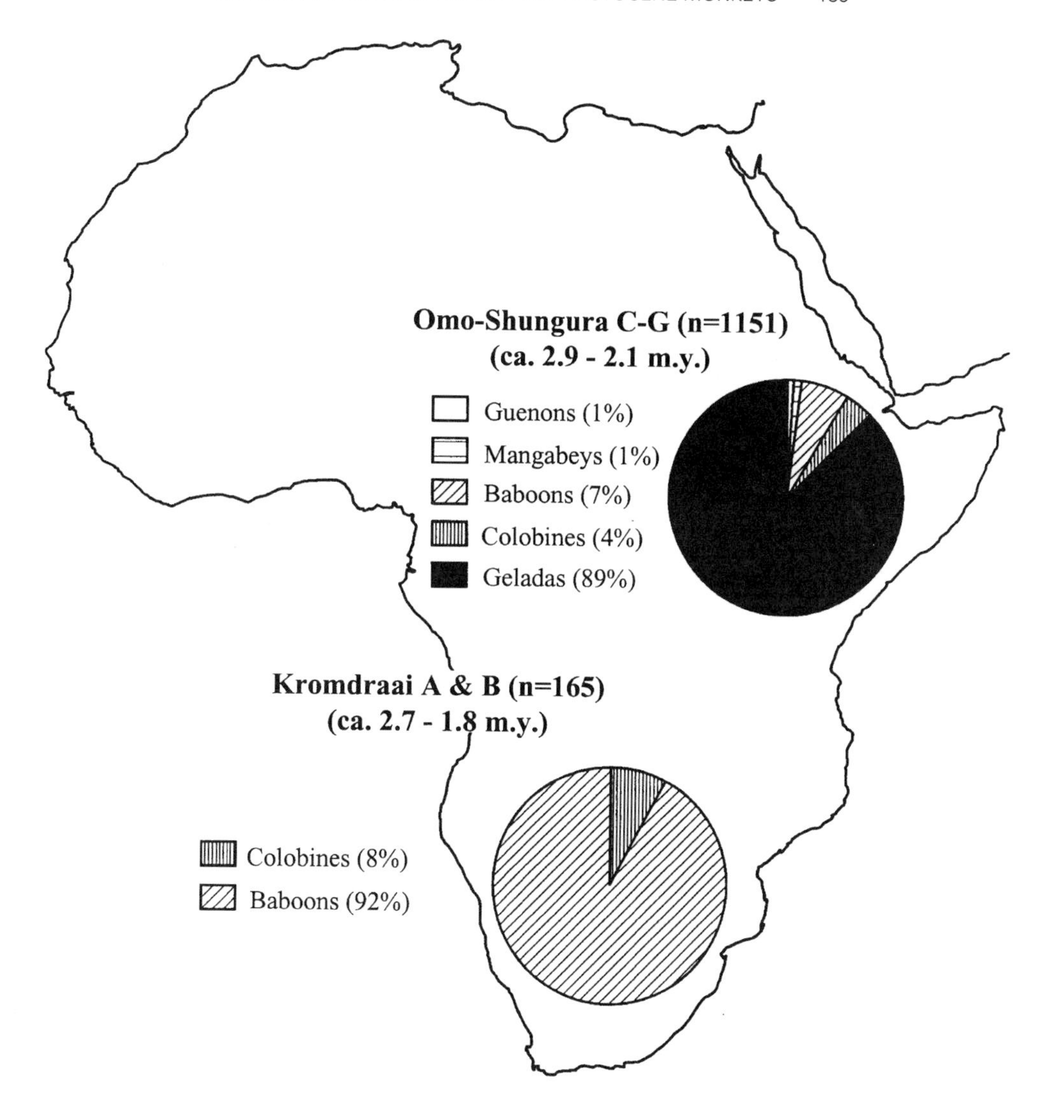

Figure 13-7. Relative abundance of cercopithecoid groups in the late Pliocene-Early Pleistocene, approximately 2.9–1.8 m.y. Eastern Africa is represented by the Omo Group Shungura Formation Members C through G (Eck, 1976) and Kromdraai A and B serves as an exemplar for southern Africa (Brain, 1981b). Age estimates from Brown and Nash (1976), Brown et al. (1985), Feibel et al. (1989), and Vrba (1982).

gan to exploit the grassland habitats for the first time, it is possible that competition with *T. oswaldi* prevented the baboons from taking advantage of grassland resources at an earlier time. Competition with other forest monkeys and lack of grass-eating savanna adaptations, combined with lower population densities in forest habitats, are probably responsible for the low numbers and diversity of baboon fossils at deposits in eastern Africa. Thus, the greater abundance and diversity of the southern baboons is attributed to the absence of and lack of competition with *T. oswaldi* and forest-dwelling colobines, to the general tendency for grasslands to support large numbers of animals, and to a diversity of dietary preferences among the baboons themselves.

Climate Change and the Evolution of Old World Monkeys

In this volume considerable attention has been placed on the possible relationship between climate change and evolutionary events such as the appearance of

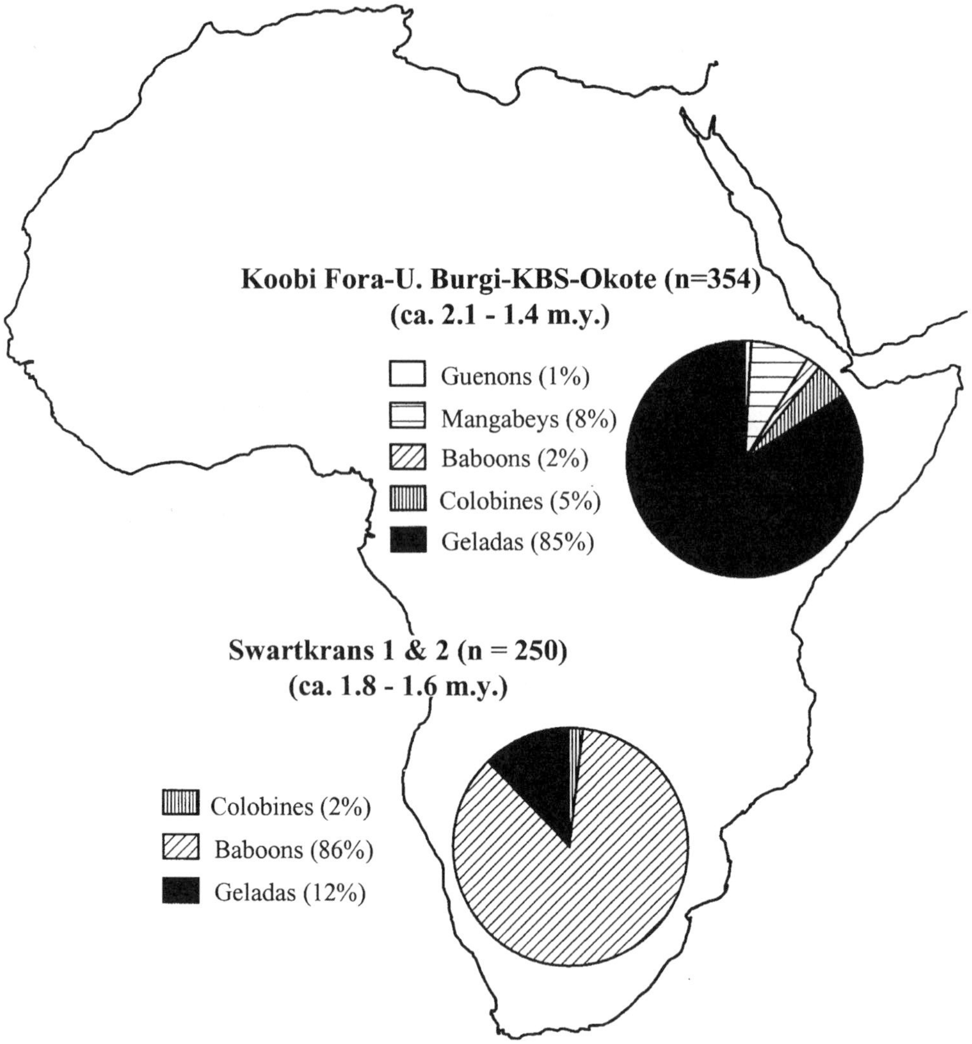

Figure 13-8. Relative abundance of cercopithecoid groups from the Early Pleistocene, approximately 2.1–1.6 m.y. The cercopithecoid fauna from eastern Africa is represented by the Upper Burgi, KBS, and Okote units at Koobi Fora (M. G. Leakey, 1976), whereas Swartkrans 1 and 2 (Brain, 1981b) demonstrates the composition of Old World monkey fauna from southern Africa. Age estimates from Brown and Nash (1976), Brown et al. (1985), Feibel et al. (1989), and Vrba (1982).

new taxa and extinction of others. Andrews and Van Couvering (1975) were among the first to discuss how the interaction of global climate change and significant regional alterations in geomorphology (such as the uplift of the eastern African peneplain and the beginning of rift formation) acted together to influence regional climate change, which in turn could influence changes in vegetation and ultimately lead to faunal turnover. Andrews and Van Couvering's main interest was in how a regional increase in the seasonality of rainfall in eastern Africa ultimately led to a replacement of Early Miocene forest by Middle Miocene woodland and eventually to the evolution of even more open habitats, such as are characteristic of modern savannas. The major faunal turnover that occurred during this time included the evolution of cercopithecoids and of semiterrestrial, seed-eating apes, the extinction of soft-fruit–eating apes, and the first appearance of bovids, giraffes, hippopotamuses, and other major mammalian groups. These origin and extinction events are intimately tied to changes in regional climate and vegetation and to adaptations related to diet and locomotor pattern.

Continuing global and regional climatic changes such as the end of the Maboko heating event and the subsequent dramatic drop in temperature may have led to the extinction of most middle Miocene catarrhines in eastern Africa (Pickford, 1987). However, the origin of colobines and cercopithecines are less clearly aligned with a particular change in climate or environment (Pickford, 1987). The interesting correspondence between late Cenozoic cooling events and hominid evolution was discussed by Brain (1981a). In particular, he noted the coincidence of hominid origins at about 5 m.y. with the Messinian Event, an episode of dessication of the Mediterranean and attendant global cooling. Moreover, the onset of Antarctic glaciation at 2.5 m.y. was seen to correspond with the divergence of the *Homo* and *Paranthropus* clades (Brain, 1981a). First appearances of baboons (*Parapapio*), geladas (*Theropithecus*), and the larger-bodied African colobine clade, beginning with *Libypithecus,* may also be related in some way to the Messinian Event.

During the Plio-Pleistocene a more reasonable case for climatic effects on cercopithecoid evolution is seen in the evolutionary history of *Theropithecus* than in other lineages. Throughout the Turkana Basin, there is evidence of a significant turnover event from faunas dominated by *T. brumpti* before 2.5 m.y. to the emergence of *T. oswaldi* as the most abundant form after this time (Leakey, 1993). This shift, from the leaf browsing of *T. brumpti* to the specialized grazing of *T. oswaldi,* may signal the establishment of widespread eastern African grasslands in association with the cooling event at 2.5 m.y. However, Pickford (1993) convincingly argued that changes in the Albertine rift were largely responsible for origin and extinction events within that genus rather than changes in climate. By plotting numbers of Plio-Pleistocene cercopithecine species against a composite oxygen isotope curve, Foley (1993) found a correlation between the degree of climatic instability and the number of *Theropithecus* species existing at any one time. However, climate alone appears to have played less of a role in the evolutionary history of other African monkeys.

Conclusions

By studying the dietary and locomotor adaptations of fossil Old World monkeys from the Miocene to Pleistocene, as well as by taking their environmental and faunal context into consideration, understanding of their evolutionary history is profoundly altered. The earliest monkeys were not the colobine-like arboreal folivores they were once thought to have been, but instead were semiterrestrial and hard-object–feeding cercopithecoids that seem to have been preadapted for entering the savanna and other open habitats that dominate much of the African continent today. The diversification of Plio-Pleistocene Old World monkeys is related to slight alterations in molar design (changes in cusp proximity and basin size, cusp relief, shear-crest length, and grinding surface area) and postcranial form, but the basic cercopithecoid bauplan (bilophodonty, lack of maxillary and frontal sinus, narrow articular end of the distal humerus with well-defined median trochler keel, and ischial callosities, among others) was inherited from their Miocene victoriapithecid ancestors.

I have tried to demonstrate the importance of understanding patterns of Plio-Pleistocene cercopithecoid species diversity and biogeography from the perspective of their locomotor and dietary adaptations. In many ways patterns of diversity and abundance of eastern African *Theropithecus* and southern African baboons can be attributed to their ancestors having been in the right place at the right time. It is obvious from the different cercopithecoid histories in southern and eastern Africa that few connections existed between these two regions during the Plio-Pleistocene. Baboons seem to have been endemic to the south and *Theropithecus* and colobines to the east. Because arboreal folivores are absent from the south, and fewer highly folivorous colobines occur in southern regions of eastern Africa, particularly during the Pleistocene, it seems that an absence of forest habitats between southern and eastern Africa, and/or within southern Africa, may have provided an effective geographic barrier between these regions. An apparently episodic exchange of terrestrial cercopithecoid taxa took place between the two regions but was limited to terrestrial species including grass in their diet such as *Cercopithecoides williamsi, Theropithecus darti, Theropithecus oswaldi,* and *Papio.*

Sometime between their last fossil occurrence and the present, extinctions of most African larger-bodied cercopithecoids took place. Virtually all of the fossil monkeys discussed in this chapter are extinct. Today *Papio* is limited to five or six predominantly allopatric species or subspecies, and *Theropithecus* to one population surviving in the highlands of Ethiopia. Whatever caused the demise of the Plio-Pleistocene monkeys may also have been responsible for the extinction of the hominid taxa with which they cooccurred. Whether the cause was related to climate, changes in vegetation, disease, or other factors remains to be proven.

Acknowledgments I am grateful to T. Bromage and F. Schrenk for inviting me to make this contribution and to the Wenner-Gren Foundation for sponsoring the conference that inspired this volume. Gratitude is expressed to the Office of the President, Republic of Kenya for permission to conduct research in Kenya, to F. C. Howell for access to the Omo collection, as well as to P. Tobias, C. K. Brain, E. Vrba, and A. Turner for assistance with the study southern African material. Special thanks are given to M. G. Leakey of the National Museums of Kenya for support with all aspects of this study, to R. F. Kay for teaching me his method of measuring shear crests, and to Monte and Fiona McCrossin for their contributions. Funding for the Plio-Pleistocene aspects of the research was provided by an I.T.T. International Fellowship to Kenya and a New York University Dean's Dissertation Fellowship; and for Miocene research at Maboko by the Wenner-Gren Foundation, Leakey Foundation, National Science Foundation (DBS 9200951and SBR 9505778), National Geographic Society, Fulbright Collaborative Research Grants, and the Boise Fund.

14

Toward an Evolutionary History of African Hipparionine Horses

Raymond L. Bernor
Miranda Armour-Chelu

Africa has long attracted the study of Neogene vertebrate fossils, particularly hominid primates. Field research programs aimed at discovery of hominids have brought the bonus of large fossil vertebrate collections and an increasing interest in the relationships between environmental change and community evolutionary response (Andrews & Van Couvering, 1975; Van Couvering & Van Couvering, 1975; Bernor, 1983, 1984; Behrensmeyer et al., 1992). Most recently, there have been attempts to test hypotheses of global climate change forcing African community evolutionary response, with various results (Vrba, 1985d, 1988; Feibel et al., 1991; Bernor & Lipscomb, 1995; Vrba et al., 1995). Given the valuable role that fossil vertebrates play in stratigraphy, correlation, taphonomic, paleoecologic, and paleoenvironmental interpretations, it is remarkable that there is a relative dearth of systematic work accompanying the growing geological and paleoenvironmental database. The problem is far more than a taxonomic one: it is a general failure to produce vertebrate phylogenetic reconstructions within a precise stratigraphic, sedimentologic, and chronologic framework. We believe that an evolutionary record so developed forms an essential basis for pursuing a full range of paleobiological interpretive studies.

Our study of the African hipparion record is nascent. However, previous studies (Bernor and Lipscomb, 1991, 1995; Bernor & Armour-Chelu, 1997; Bernor, in preparation), as well as this review, demonstrate that African hipparions had a complex history of migration, dispersion, adaptation, and evolutionary radiation, warranting study in their own right. We present our most current understanding of this record, how it compares evolutionarily, biogeographically, and ecomorphologically with Eurasian hipparions, and forward some recommendations for the general pursuit of paleobiological studies using mammalian systematic records.

Materials and Methods

We have assembled material for this study from a variety of sources. Principal among these are Bernor's database of hipparionine horses, which currently includes approximately 8000 records. Also, we have assembled measurements of the postcrania from a number of published sources, and these same data are used here (Bernor et al., 1989). When measurements are used from other sources, we cite the author and, when known, the particular specimen number(s) in the text. Our systematic analyses depend on both discrete character states and continuous variables for identifying hipparion species and assessing their evolutionary relationships. Definition and use of discrete characters that we have adopted for the skull and mandible have been developed and progressively refined by Woodburne and Bernor (1980), Bernor and Hussain (1985), Bernor (1985), Bernor et al. (1988, 1989, 1993a, 1993b), Bernor and Franzen (1997), and Bernor and

Lipscomb (1991, 1995). Measurements follow those prescribed by the American Museum of Natural History International Hipparion Conference (1981) and are figured in Eisenmann et al. (1988).

Continuously distributed data have been entered and sorted using dBase IV and statistically analyzed using Systat 5.03. Forty-nine discrete characters of the skull, mandible, and dentitions are further used to define species and reconstruct their phylogenetic relationships. Bernor and Lipscomb (1991, 1995) have recently presented a phylogenetic analysis of the *Hippotherium* Group and the "*Sivalhippus*" Complex using Hennig 86. We incorporate here the Höwenegg horse sample (Late Miocene, early Vallesian age, about 10.3 m.y.; Swisher, 1996; Woodburne et al., 1996; Bernor et al., 1997) in all of our analyses of continuous variables. The Höwenegg fauna includes a relatively large population of the hipparion, *Hippotherium primigenium* (14 partial to complete skeletons, 1 with a fetus and another being a foal), accumulated over a very short temporal interval (Woodburne et al., 1996), and as such it represents as good a "biological population" as can be expected in a fossil mammal sample (Tobien, 1986). For most of the continuously distributed variables that we have analyzed here, we have been able to calculate 95% confidence ellipses of the Höwenegg sample and graphically render the expected range of variability in a single species of hipparion. The use of the Howenegg sample further provides the basis for understanding the primitive condition of Old World hipparions and the various degrees to which other lineages evolutionarily diverged from it.

Definitions of statistical tests used here are provided in the Systat "Graphics" documentation (Wilkinson, 1990).

Evolutionary History of Eurasian Hipparions: An Ecomorphological Perspective

The first Old World hipparions occur near the base of the Late Miocene (Steininger et al., 1996). The most primitive taxon is referred to the species *Hippotherium primigenium* and would appear to be most closely related to the late middle–early Late Miocene North American taxon *Cormohipparion occidentale s.l.* The geographic extension from North America to Eurasia undoubtedly occurred across Beringia and is correlated with the terminal Serravallian regression (Bernor et al., 1988, 1989). Recently, there has been much new systematic reevaluation of Old World hipparions in general (Bernor et al., 1989, 1996c) and more specifically at a regional scale, including East Asia (Qiu et al., 1987; Bernor et al., 1990), South Asia (Bernor and Hussain, 1985), southeastern Europe/southwestern Asia (Bernor, 1985), and Central Europe (Bernor et al., 1988, 1993a,b). Accompanying reviews of the regional chronologies of hipparionine first occurrences include those for Indo-Pakistan (Pilbeam et al., 1996), Turkey (Kappelman et al., 1996), and western Eurasia and Africa (Sen, 1996; Swisher, 1996; Woodburne et al., 1996). These investigations indicate that the "*Hippotherium* Datum" (sensu Woodburne et al., 1996) is most accurately placed between 11 and 10.5 m.y. Moreover, whereas the first occurrence of Old World hipparion was erroneously viewed as having heralded the nearly instantaneous replacement of earlier Miocene forest environments with open country savanna-like environments (Berggren & Van Couvering, 1974), it is now known that the evolution of Eurasian and African ecosystems from forested to open-country "savanna mosaic woodlands" occurred diachronically and differentially in their degree through the Middle–Late Miocene interval (Bernor et al., 1979, 1996a,b; Bernor, 1983, 1984; Fortelius et al., 1996; Leakey et al., 1996).

Central Europe has one of the better records of first occurring hipparions from the standpoints of both its geologic and paleontologic contexts and associated studies of sediments, faunas, and paleobotanical data (Bernor et al., 1988, 1993a,b, in press c; Rögel et al., 1993; Kovar-Eder et al., 1996; Rögel & Daxner-Höck, 1996; Woodburne et al., 1996). These lines of evidence have revealed that first occurring Central European hipparions entered warm temperate to subtropical forested environments with highly equable conditions and appear to have contrasted strikingly with the more open-country, seasonal, savanna-like conditions of the founding North American lineage, *Cormohipparion occidentale s.l.* (Webb, 1983).

Forsten (1968) advanced the notion that *"Hipparion" primigenium* was morphologically highly variable as well as geographically and chronologically wide ranging. This view was adopted in its essential details by Berggren and Van Couvering (1974) in their proposal of an Old World "Hipparion Datum" whereby a single species of hipparion made an essentially instantaneous (within 10,000 years) prochoresis across Eurasia and Africa. Paleontologists engaged in resolving the Western Eurasian/North African Neogene time scale readily adopted the "Hipparion Datum" as a means of interprovincial correlation (Mein, 1975, 1979) and sought further subdivision of Late Miocene western Eurasian faunas using hipparion evolutionary grades (Sen et al., 1978). However, beginning with Woodburne and Bernor's (1980) and Bernor et al.'s (1980) preliminary systematic and biogeographic evaluation of Old World hipparions, and continuing through the regional revisions cited above, it was found that the morphological variability of early occurring hipparions was too great to be restricted to *Hippotherium primigenium* and that the subsequent evolutionary radiation was extensive, including several monospecific and multispecific lineages (Bernor et al., 1989; Bernor & Lipscomb, 1995). In the most recent reevaluation of western Eurasian Late Miocene–medial Pliocene hipparions, Bernor et al. (1996d) have shown that shortly after their first, earliest Vallesian (MN 9; 11–9.5 m.y.) occurrence, regional hipparion assemblages show more morphological variability than can be accepted for a single species. The Old World hipparion radiation was established by the late Vallesian (MN 10; 9.5–9.0 m.y.) and exhibits strong provinciality in species diversity across Eurasia by 9.0 m.y. Western Eurasian and North African hipparions underwent a decisive decline in species abundance during the latest Miocene (MN 13; ca. 7.1–5.3 m.y.) and a further even more precipitous reduction in their diversity at the base of the Pliocene. The rise and fall of western Eurasian hipparion diversity closely parallels that of open country savanna-mosaic herbivores in general (Bernor et al., 1996a; Fortelius et al., 1996).

Systematic revision of Old World hipparion lineages has led to greater precision in the chronology of evolutionary change, biogeographic extensions, and disjunctions. Bernor et al. (1989) and Bernor and Lipscomb (1991, 1995) summarized these and showed that Old World hipparions may be assembled into several monospecific clades and four major superspecific lineages: the *Hippotherium* Complex, a paraphyletic group of taxa which were the first to diverge from the founding Old World evolutionary stock (ca. 11–8 m.y.); the *Cremohipparion* lineage(s), a monophyletic group which first appeared in the Eastern Paratethys (former western U.S.S.R.), southeast Europe and southwest Asia and appears to have extended its range eastward into China (ca. 8–4 m.y.); the *Hipparion s.s.* group which includes taxa known to have occurred across a narrow latitudinal belt extending from France and Spain eastward through Greece, Iran, Afghanistan, and Indo-Pakistan (ca. 9.5–7.1 m.y.); the "*Sivalhippus*" Complex, a diverse group of hipparion lineages which first occurred in Asia (ca. 9.5 m.y.) and underwent distinctively Asian (8–0.7 m.y.), European and West Asian (ca. 5–2.6 m.y.), and African (8+–0.6 m.y.) evolutionary radiations. Old World hipparions exhibit a complex mosaic of evolutionary pathways which include extensive convergence between members of both the monospecific and multispecific lineages. In many cases these include anatomical complexes of ecomorphological significance. Moreover, hipparion biogeographic events often span intercontinental distances and would appear to frequently correlate with global sea-lowering events (Bernor & Lipscomb, 1995).

Morphological characters believed to be ecologically relevant include a number of facial, dental, and postcranial features. Facial characters include the relative development of facial fossae, degree of nasal notch incision, and length and width of the snout. Dental characters include incisor arcade shape, relative anteroposterior curvature and hypsodonty, cheektooth crown height, relative mediolateral curvature, and degree of occlusal enamel ornamentation. Postcranial features have largely been restricted to metapodial length and distal articular width, but certainly the morphology of podial articular surfaces will be of considerable importance in the reconstruction of locomotor behavior. There has been an attempt to correlate many of the facial and dental features with individual species' diets as measured independently by studies of microwear (Hayek et al., 1991) and to correlate postcranial morphology with locomotor behavior (Tobien, 1952; Sondaar, 1968; Hussain, 1975; Bernor et al., 1997), but these studies remain underdeveloped. We offer some ecomorphological hypotheses below to be tested by future studies of functional anatomy, dental microwear, and carbon/oxygen isotopes.

Facial Morphological Features

Primitively, Old World hipparions have shallowly incised nasal notches placed somewhat more than halfway from between I^3 and P^2. Also, there is a single well-developed preorbital fossa (POF) which is deep medially and posteriorly and forms a posteriorly excavated "pocket" deep to the lacrimal bone; the peripheral rim is prominent, giving a nearly uniform depth to the POF; the POF's posterior rim is placed 40+ mm anterior to the anteriormost orbital rim and distinctly anterior to the lacrimal bone so as to present a long preorbital "bar." The function of the preorbital fossa is still debated, but there is no doubt that it spatially occupies the area where the levator labii superioris takes its origin in hipparion (Zhegallo, 1978), as well as in extant *Equus* (R. L. Bernor, personal observation). This leads us to hypothesize that a well-developed POF supported an enlarged, expanded muscle belly of the levator labii superioris, which acted on an enlarged, highly mobile upper lip. This external component to the masticatory apparatus would have facilitated selective feeding, which could have included specific graze or a significant percentage of browse. In some hipparion lineages the POF became

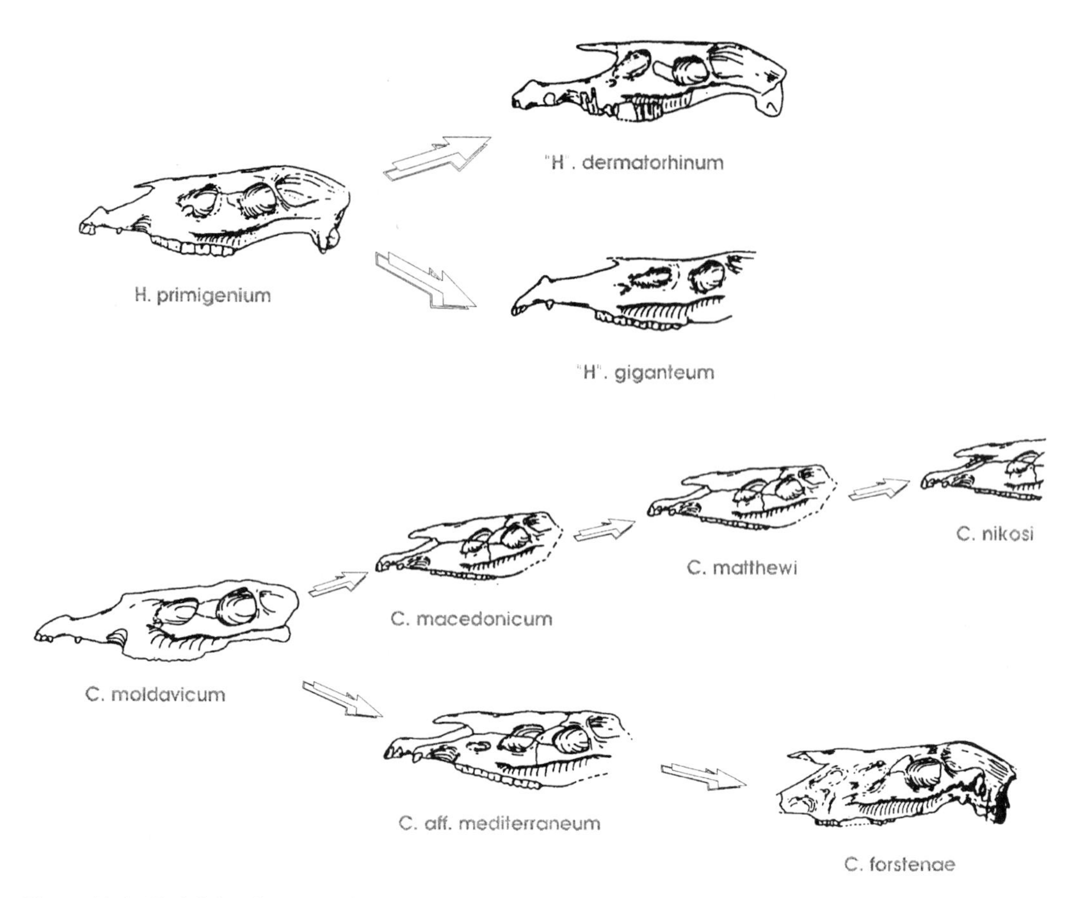

Figure 14-1. Facial development of the Eurasian hipparion taxa *Hippotherium primigenium*–*"Hippotherium" dermatorhinum/giganteum,* the small *Cremohipparion,* and the *Cremohipparion moldavicum-aff. mediterraneum-forstenae* evolutionary pathways.

hypertrophied and was augmented by further facial fossae evolution both in relative areal extent as well as in depth. In yet other lineages, the POF was lost altogether. Deepened nasal notch incision is likewise believed to be functionally correlated with support of an enlarged upper lip, or small proboscis, and selective feeding of browse. Progressive deepening of the nasal notch was commonly, but certainly not always, associated with a well-developed POF and/or the development of additional facial fossae. Retention of shallowly incised nasal notches occur in lineages that lose their facial fossae altogether. Ecomorphological groupings of several Late Miocene Eurasian and North African hipparions are provided in figures 14-1–14-4 and selected morphologic characters are cited in table 14-1.

From the primitive condition described above, the POF remained essentially the same in some lineages, while nasals retracted slightly to moderately or were progressively lost. Within the *Hippotherium* complex, "*Hippotherium*" *dermatorhinum* (East Asia; fig. 14-1; Qiu et al., 1987; Bernor et al., 1989, 1990) retained a single, well-developed POF but evolved a deeply incised (to mesostyle of P^4) nasal notch and a long, narrow V-shaped muzzle with a strongly arcuate incisor arcade (see figs. 14-5A,B). This combination of features suggests that this species was a selective feeder that may have incorporated a significant percentage of browse into its diet. *Hippotherium giganteum* (southeast Europe–southwest Asia) evolved a larger size and an elongate snout, but only slight nasal retraction; it was not far removed evolutionarily, nor probably dietarily, from *Hippotherium primigenium.*

The genus *Cremohipparion* includes four distinctive Eurasian lineages that segregate well morphologically, geographically, and temporally. Two of these are depicted in figure 14-1. The *Cremohipparion moldavicum–Cremohipparion macedonicum–*

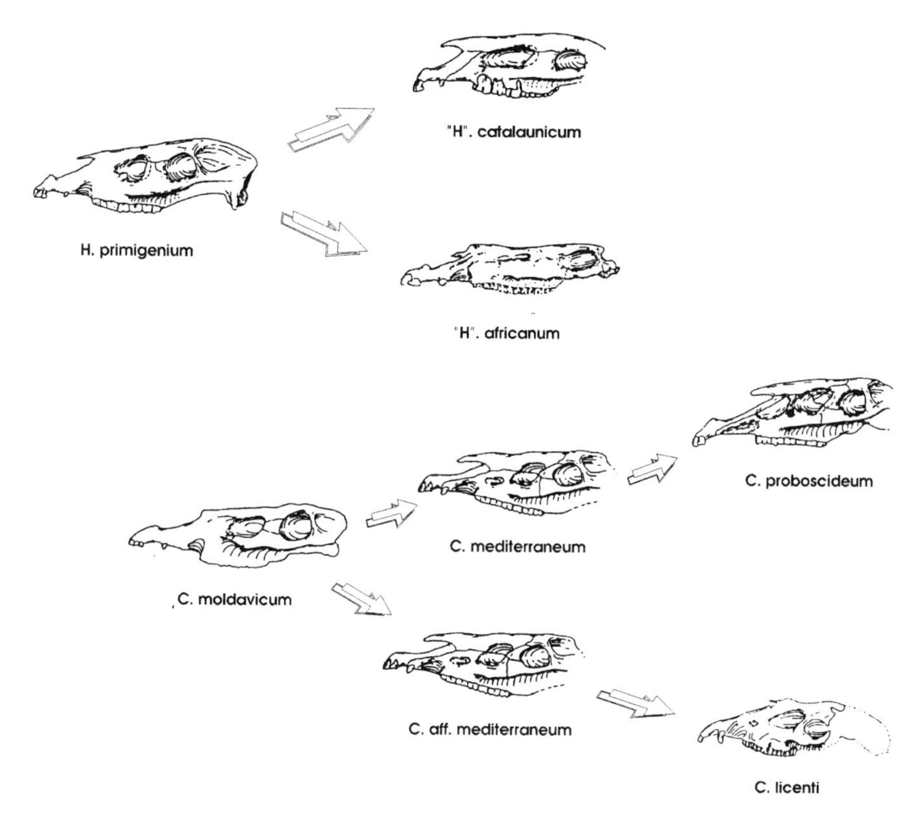

Figure 14-2. Facial development of the *Hippotherium primigenium-catalaunicum/africanum* evolutionary pathway, the *Cremohipparion moldavicum-mediterraneum-proboscideum,* and the *Cremohipparion moldavicum-aff. mediterraneum-licenti* evolutionary pathways.

Cremohipparion matthewi–Cremohipparion nikosi–Cremohipparion periafricanum (the last one not shown here because there is no known skull) lineage exhibits a progressive decrease in size and loss of the POF from the late Vallesian (ca. 9.5 m.y.) until the end of the Turolian (ca. 5.3 m.y.) in southeast Europe, southwest Asia, and Spain (Bernor et al., 1996d). This lineage of small horses maintained a conservative nasal retraction in all but *Cremohipparion nikosi* (retracted to mesostyle of P^4). As far as we know, all these taxa retained arcuate incisor arcades and snouts that were rather narrow and elongate. Based on limited microwear data, Hayek et al. (1991) and Bernor et al. (1996d) have hypothesized that this clade predominately grazed.

The second *Cremohipparion* lineage depicted in figure 14-1, *C. aff. mediterraneum–C. forstenae* is a vicariant East Asian lineage in which the facial fossae were strongly diminished in their depth after the acquisition of three distinct fossae (preorbital, caninus, and buccinator) and some nasal retraction (to mesostyle of P^2) in *C. mediterraneum* (then mesostyle of P^3 in *C. forstenae*) had evolved. The secondary loss (= evolutionary reversal) of the facial fossa complex in *Cremohipparion forstenae* is apparently correlated with a shift to a predominately grazing behavior (Hayek et al., 1991).

As depicted in figure 14-2, two members of the *Hippotherium* group, *"Hippotherium" catalaunicum* (Spain) and *"Hippotherium" africanum* (North Africa) increased the length of the POF substantially; of these two taxa, the first retained the primitive condition for nasal retraction, while the latter evolved slightly retracted nasals (to MP^2). Figure 14-2 also depicts two additional clades of the *Cremohipparion* lineage. In the first, the *Cremohipparion mediterraneum–C. proboscideum* lineage, there was an increase in size, depth, and areal extent of the POF, the development of an additional caninus fossa, and retraction of the nasals (to mesostyle of P^3 in *C. probsoscideum*). In the second, again a vicariant Late Miocene–Pliocene East Asian lineage, *Cremohipparion aff. mediterraneum–Cremohipparion licenti,* further evolved the POF, caninus, and buccinator fossae and added a distinct malar

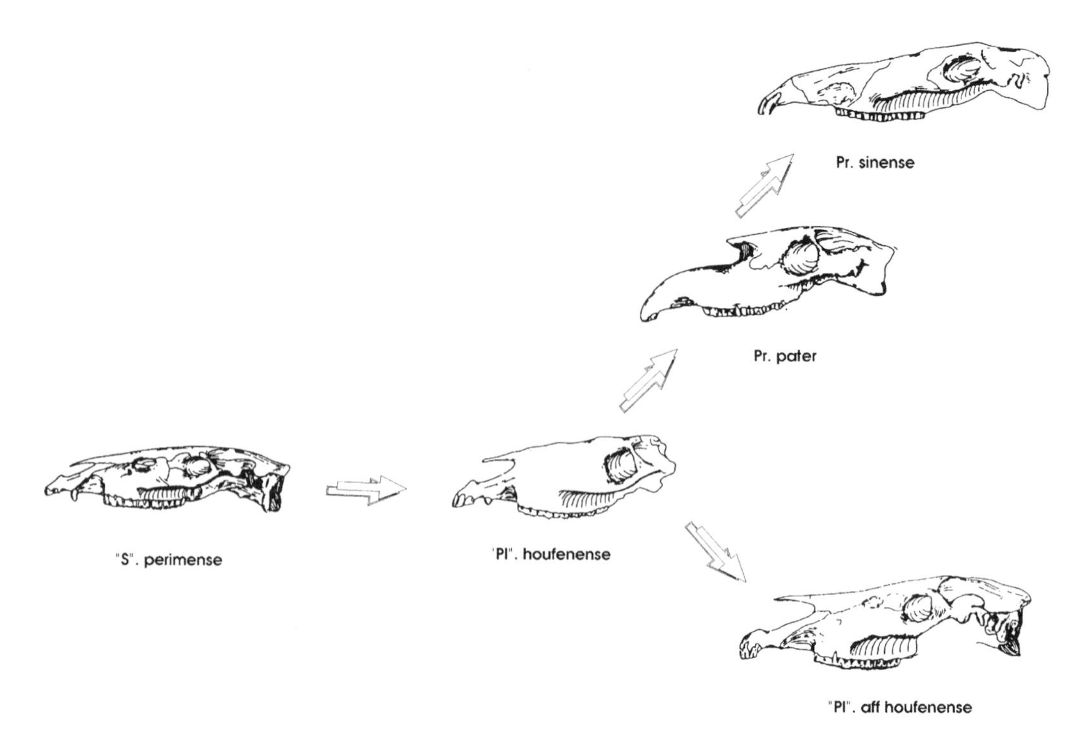

Figure 14-3. The Eurasian "*Sivalhippus*" Complex: "*Sivalhippus" perimense*–"*Plesiohipparion" houfenense*–*Proboscidipparion pater/sinense* evolutionary pathway with the contrasting "*Pleisohipparion" aff. houfenense* pathway.

fossa while retracting the nasals to the mesostyle of M^1. *Cremohipparion licenti* further developed the facial region by deepening the buccinator fossa medially and posteriorly and developing deep circular pits within the posterior aspect of the POF, presumably for supporting strong attachment of an enlarged levator labii superioris muscle belly. The development of a progressively deeper nasal notch accompanied by the evolution of multiple, deeply excavated facial fossae is believed to be related to the increased browsing component of *Cremohipparion licenti*'s diet, an interpretation supported by preliminary microwear studies (Hayek et al., 1991). Collateral to this suite of facial characters is the retention of an arcuate incisor dentition in all these *Cremohipparion* taxa.

Figure 14-3 depicts advanced East Asian Pliocene members of the "*Sivalhippus*" Complex: *Proboscidipparion pater* and *Proboscidipparion sinense.* These taxa exhibit hyperretraction of the nasal notch superior to M^1 and to a position superior to the orbits, respectively, while losing the POF entirely (Sefve, 1927; Qiu et al., 1987; Bernor et al., 1989). Bernor and Lipscomb (1991, 1995) have demonstrated that the *Proboscidipparion* lineage evolved its nasal hyperrectraction after the POF was lost and cheek-tooth crown height had increased in its ancestors. Their phylogenetic reconstruction further suggests that the evolution of a tapirlike pseudo-proboscis (after Sefve, 1927), appears to have evolved convergently between the *Cremohipparion mediterraneum–proboscideum–Cremohipparion aff. mediterraneum–licenti* lineages (fig. 14-2) and the *Proboscidipparion pater–sinense* lineage (fig. 14-3) by two distinctly different anatomical pathways. In the *Cremohipparion* species there was an unconstrained increase in facial fossa area and specifically its deep dorsoventral dimension because cheek-tooth crown height remained relatively low (50 mm or less). The *Proboscidipparion* lineage had the preexisting morphologic constraint of increased crown height (about 71 mm in "*Sivalhippus*" "*perimense*"; see fig. 14-8), which required an alternative anatomical pathway for supporting a manipulative proboscis. In *Proboscidipparion,* nasal incision was radically increased and the muscle origin was likely displaced off the side of the face to an internal position on the nasomaxillary rim. We have not yet sampled the *Proboscidipparion* lineage for microwear or carbon isotopes, but we believe it is highly probable that species of this lineage incorporated a significant proportion of browse in their diet.

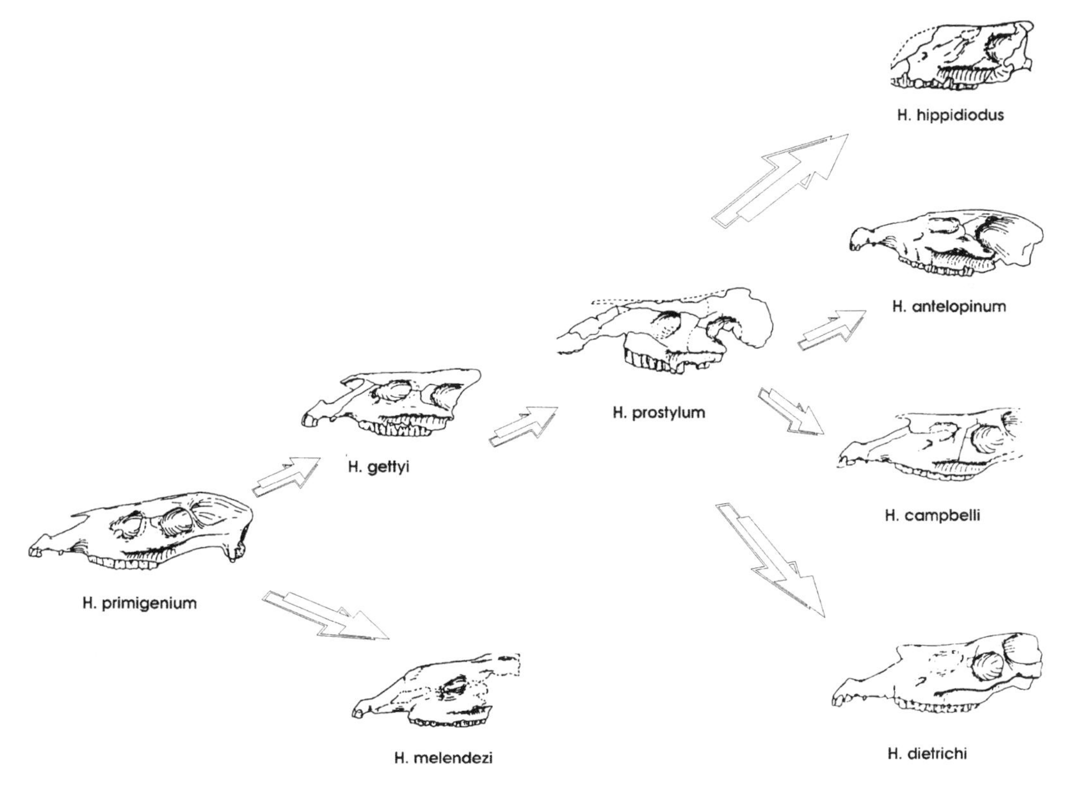

Figure 14-4. The *Hipparion s.s.* Group.

Figure 14-3 has included for contrast the *"Plesiohipparion" houfenense–"Plesiohipparrion" aff. houfenense* lineage. *"Plesiohipparion" aff. houfenense* is known to cooccur with *"Proboscidipparion sinense* and *Equus* (ca. 2 m.y.) and was *Equus*-like in a number of morphologic features: it was very large, lacked a POF, had an unretracted nasal notch, an incisor arcade that was nearly mediolaterally oriented, heavily built incisors with grooving, and high-crowned cheek teeth. Its diet remains uninvestigated, but we hypothesize that it most probably incorporated a large percentage of graze into its diet.

A number of independent lineages lost their POF, did not develop additional facial fossae, and retained shallowly incised nasal notches. The Old World *Hipparion s.s.* group (fig. 14-4) lost its single POF through a long evolutionary progression, occurring between 9.5 and 5.3 m.y., and geographically dispersed from western Europe to East Asia. Figure 14-4 depicts the known taxa: *Hipparion melendezi/ Hipparion gettyi* (with well developed POF)–*Hipparion prostylum* (reduced POF)–*Hipparion dietrichi/ Hipparion campbelli/ Hipparion antelopinum / Hipparion hippidiodus* (POF absent). Incisor arcade morphology varied from being strongly arcuate (*Hipparion campbelli*), to moderately arcuate (*Hipparion prostylum*), to broader with a mediolateral alignment of the incisors (*Hipparion dietrichi*). Hayek et al. (1991) found in their microwear analysis that *Hipparion dietrichi* apparently exhibited a purely grazing behavior. *Hipparion campbelli* is further evolved in its elongate, narrow snout (figs. 14-5A,B). These observations have led Bernor et al. (1989, 1996c) to hypothesize that the evident cladogenetic event(s) whereby *Hipparion prostylum* diverged into southeast European (*Hipparion dietrichi*), southwest Asian (*Hipparion campbelli*), Indo-Pakistan (*Hipparion antelopinum*), and East Asian (*Hipparion hippidiodus*) lineages was accompanied by divergent feeding behaviors. We believe it probable that all these taxa grazed to a large extent but may have differentially exploited various strata of graze, and/or different percentages of browse. For example, *Hipparion campbelli* may have "selected" high grass with its long narrow snout and arcuate incisor arcade, whereas *Hipparion dietrichi,* with its mediolaterally aligned incisor row, was adapted to exploit shorter grass in drought conditions (see Owen-Smith, 1985).

Table 14-1. Ecomorphological features of selected Eurasian hipparionine horses.

Species	POF	Caninus fossa	Malar fossa	Nasal notch	Incisor curve	Incisor arcade	PM curve	PM comp	PM groove
Cormohipparion									
C. occidentale	P	A	A	AP2	Y	A	MC	B	M
Hippotherium									
H. primigenium	P	A	A	AP2	Y	A	MC	C	D
H. dermatorhinum	P	A	A	AP4	Y	VA	MC	C	
H. giganteum	P	A	A	AP2	Y	A	MC	C	
H. catalaunicum	P	A	A	AP2	Y	A	MC	C	
H. africanum	P	A	A	MP2	Y	A	MC	C	D
Hipparion s.s.									
H. melendezi	P	A	A	AP2	Y	A	MC	C	
H. gettyi	P	A	A	AP2	Y	A	MC	M	M
H. prostylum	R	A	A	AP2	Y	A	MC	M	M
H. campbelli	A	A	A	MPA	Y	VA	MC	M	M
H. dietrichi	A	A	A	AP2	Y	S	MC	M	M/F
H. antelopinum	A	A	A	?	Y	?	MC	M	M
H. hippidodus	A	A	A	?	Y	?	MC	M	M
Cremohipparion									
C. moldavicum	P	A	A	AP2	Y	A	MC	M	
C. macedonicum	R	A	A	AP2	Y	A	MC	M	
C. matthewi	R	A	A	AP2	Y	S	MC	M	F
C. nikosi	A	A	A	MP4	Y	A	MC	M	F
C. periafricanum	?	?	A	?	Y	?	MC	S	
C. forstenae	R	P	A	MP3	Y	A	MC	M	
C. mediterraneum	P	P	A	MP2	Y	A	MC	C	
C. proboscideum	P	P	A	MP3	Y	A	MC	C	
C. licenti	P	P	P	MM1	Y	S	MC	S	F
Sivalhippus Complex									
S. perimense	R	A	A	AP2	Y	A	S	C	F
Plesiohipparion aff. houfenense	A	A	A	AP2	Y	S	S	M	F
Proboscidipparion pater	A	A	A	MM1	Y	A	S	VC	
Proposcidipparion sinense	A	A	A	ORB	Y	A	S	VC	

POF (preorbital fossa): P = present, R = reduced, A = absent; caninus fossa: P = present, A = absent; malar fossa: P = present, A = absent; nasal notch: A = anterior limit of tooth, M = mesostyle of the tooth as follows in table; incisor curvature: Y = yes, N = No; incisor arcade: A = arcuate, VA = very arcuate, S = straight; PM curve: MC = premolars and molars moderately curved; S = straight; PM comp: C = premolars and molars complexly plicated, M = moderately complex, S = simple, VC = very complex; PM groove: D = premolars and molars deeply grooved, M = moderately deeply grooved, F = flat.

Old World hipparions evolved facial fossae, nasal retraction, arcuate incisor arcades, and lengthened and narrowed snouts to various extents and in various combinations. We believe that despite the frequent evolutionary independence of these morphological complexes, all could and did contribute to dietary behavior. However, it is apparent that a developed POF supported the musculature required to move a highly mobile upper lip and even a short proboscis, which in turn was adapted for acquiring browse. This case contrasts with those species that had elongate narrow snouts but lacked a POF: these species were also likely selective feeders, but not having the highly mobile upper lip, these taxa may rather have been selective grazers (see Owen-Smith, this volume). These morphological nuances further exemplify the variety of anatomical pathways that evolution could, and did, follow to achieve greater selectivity in feeding behavior and further supports our conviction that a sound systematic basis is a necessary prerequisite to understanding evolutionary phenomena, especially those pursued by behavioral ecologists. The degree to which these particular horses incorporated either browse or graze into their diet requires independent testing by enamel microwear, mesowear, and isotopic analyses. Moreover, the evolution of Old World hipparionine facial and nasal morphologies is clearly subject to extensive homoplasy. We present some relevant bivari-

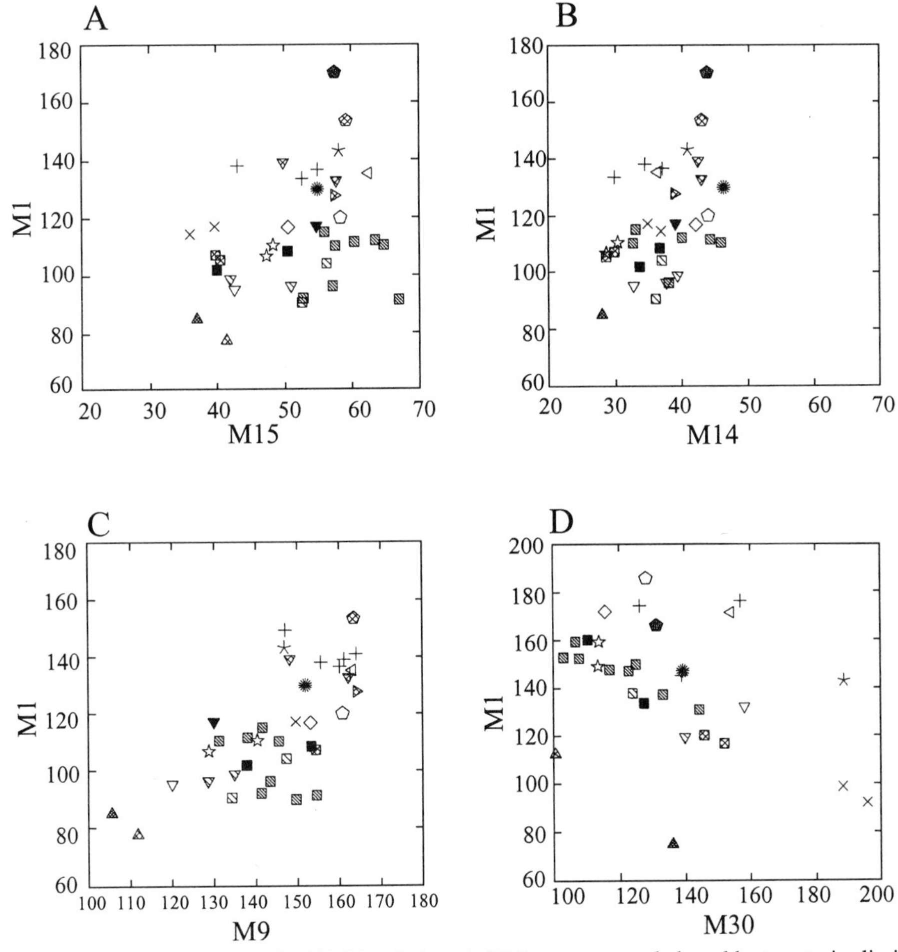

Figure 14-5. Bivariate plots on skulls. (A) Muzzle length (M1) versus muzzle breadth at posterior limit of I3s (M15); (B) muzzle length (M1) versus minimum muzzle breadth (M14); (C) muzzle length (M1) versus cheek tooth row length (M9); (D) cheek length from nasal notch to anteriormost orbit (M31) versus length from prosthion to nasal notch (M30). Note that the legend applies to figures 14-5–14-7.

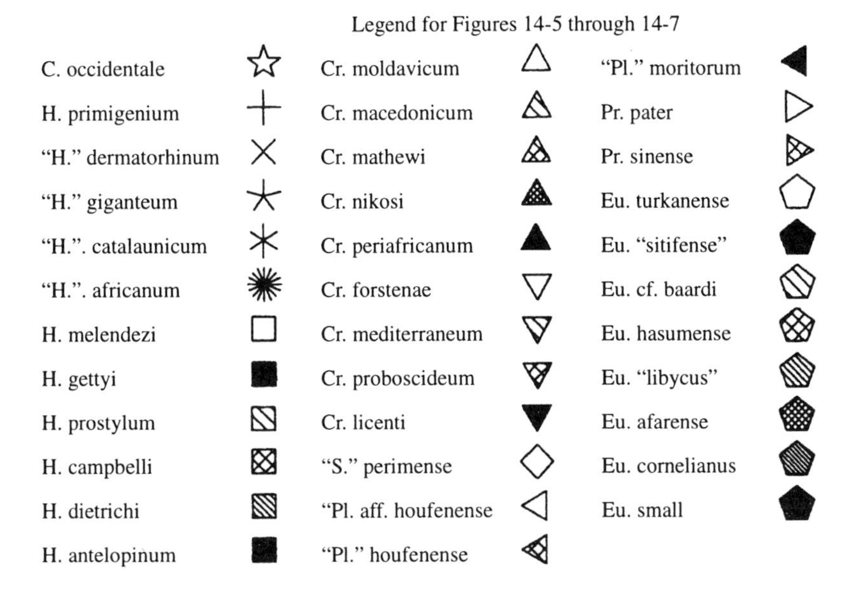

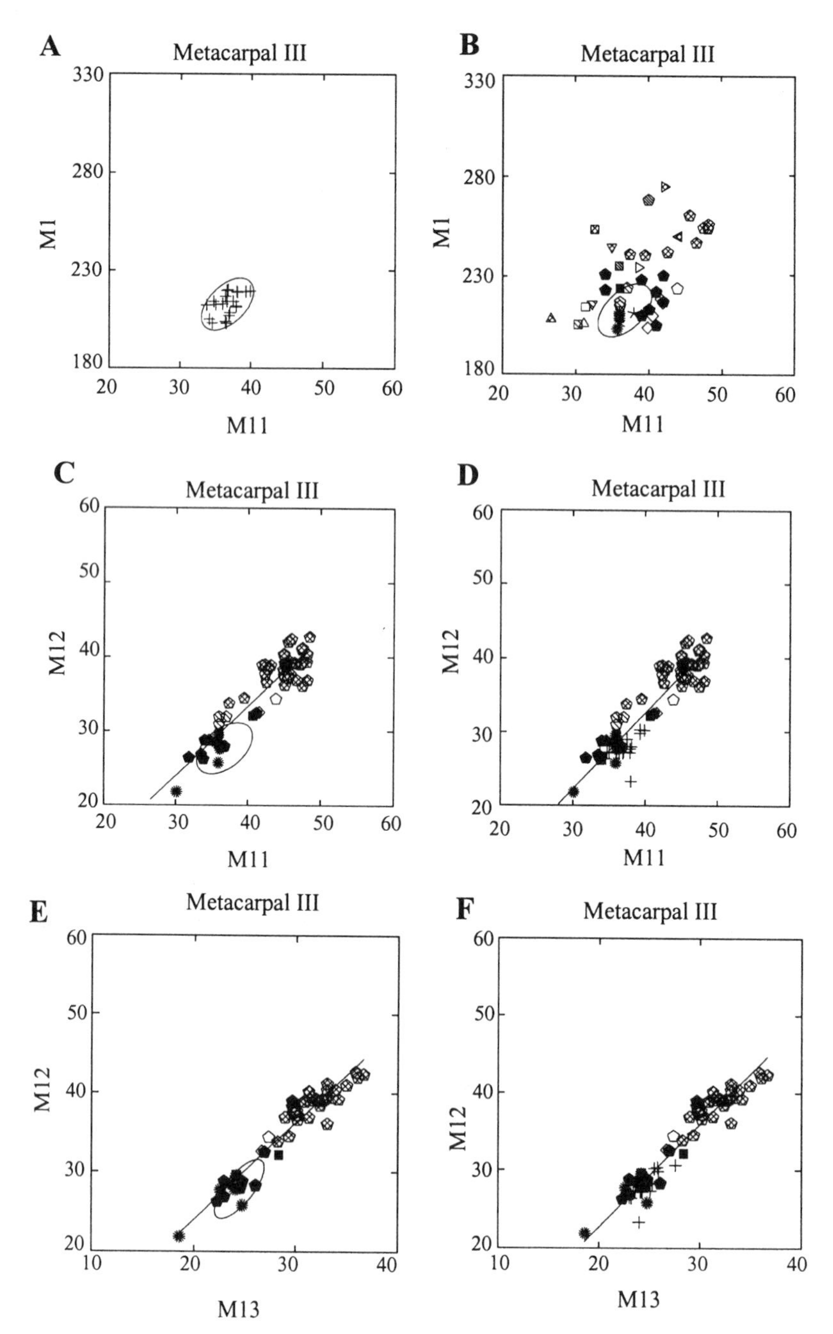

Figure 14-6. Bivariate plots on third metacarpals: (A) 95% confidence ellipse for maximum length (M1) versus distal articular width (M11) with all of the Höwenegg dimensions plotted within the ellipse; (B) M1 versus M11, again with the Höwenegg 95% confidence ellipse and all other available taxa; (C, D) distal maximal keel depth (M12) versus distal articular width (M11), with panel C excluding the Höwenegg measurements for the Pearsons's correlation coefficient and panel D including the Höwenegg sample for the correlation statistic; (E, F) M12 versus distal minimal depth of the lateral condyle (M13), with panel E excluding the Höwenegg measurements and panel F including the Höwenegg measurements for the Pearson's correlation.

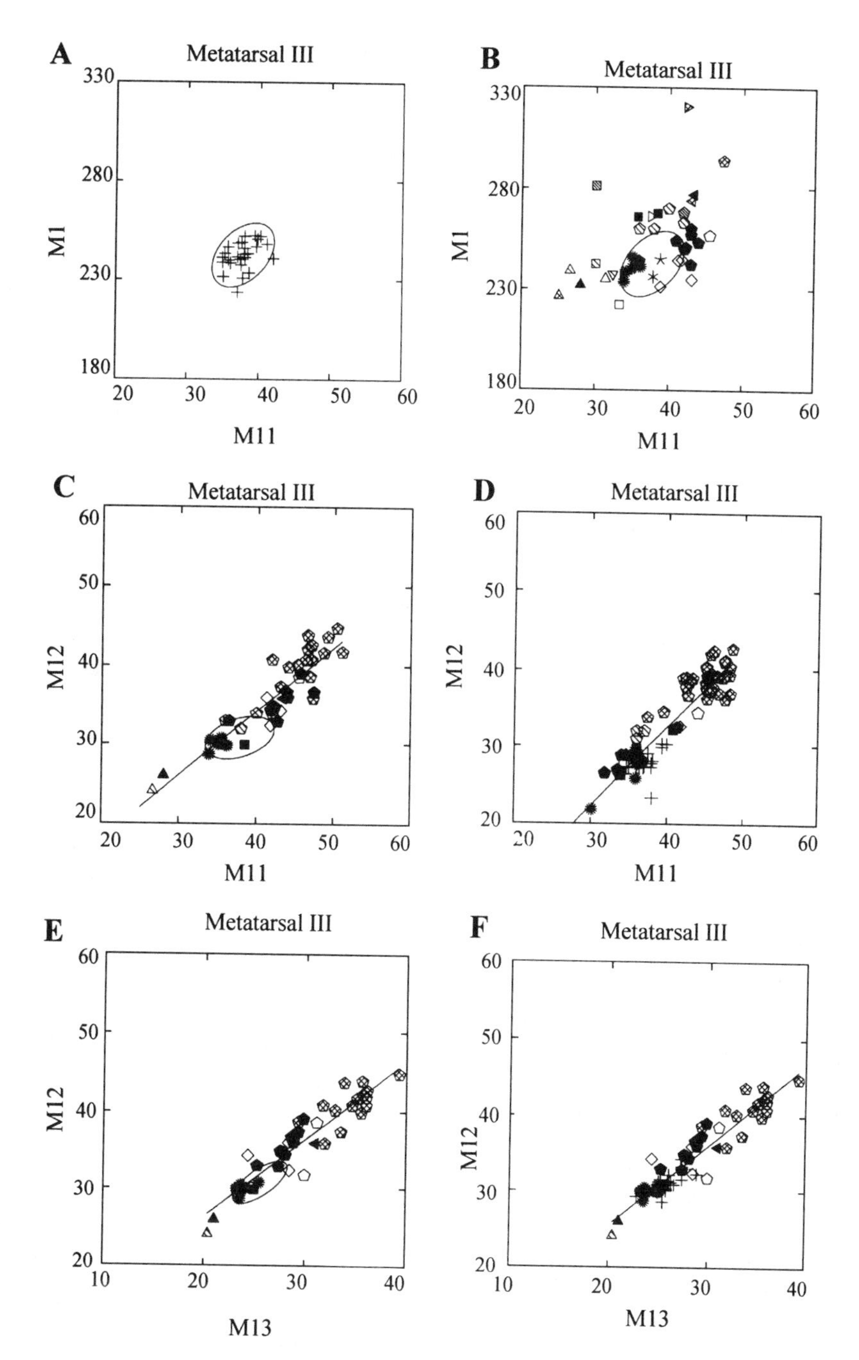

Figure 14-7. Bivariate plots on third metatarsals: (A) 95% confidence ellipse for maximum length (M1) versus distal articular width (M11) with all of the Höwenegg dimensions plotted within the ellipse; (B) M1 versus M11, again with the Höwenegg 95% confidence ellipse and all other available taxa; (C, D) distal maximal keel depth (M12) versus distal articular width (M11), with panel C excluding the Höwenegg measurements for the Pearsons's correlation calculation, and panel D including the Höwenegg sample; (E, F) M12 versus distal minimal depth of the lateral condyle (M13), again with panel E excluding the Höwenegg sample and F including it for the Pearson's correlation coefficients.

ate statistics on snout morphology later when we compare the ecomorphological evolution of Eurasian and North African hipparions with the evolution of the distinctly African clade *Eurygnathohippus*.

Dental Ornamentation and Cross Grooving

The development of a deep, mediolateral groove across the protocone-mesostyle plane of the maxillary cheek teeth has been observed to occur in most populations of Central European Vallesian-Early Turolian hipparions (ca. 10.5–8 m.y.; Bernor et al., 1997; Bernor & Franzen, 1997). Fortelius (personal communication) argues that this is indicative of a diet that incorporated a significant percentage of browse. We have thus far only been able to observe a selected number of taxa listed in table 14-1. Those with deep mediolateral grooving and presumably incorporating a great percentage of browse in their diet include *Hippotherium primigenium* and "*Hippotherium*" *africanum*. Those taxa with moderate grooving include *Cormohipparion occidentale, Hipparion gettyi,* and *Hipparion prostylum.* Those with flat occlusal surfaces include *Cremohipparion matthewi, Cremohipparion nikosi, Cremohipparion licenti, "Sivalhippus" perimense,* and *"Plesiohipparion" aff. houfenense.* Other than Central European hipparion, these observations are preliminary and in need of close examination for variability and consistency through all stages of wear.

Potential cross-correlates of mediolateral grooving are the occurrence of premolar/molar enamel complexity and mediolateral cheek tooth curvature. Gromova (1952) and Forsten (1968) both emphasized that complex enamel ornamentation of premolar and molar occlusal surfaces was indicative of a browsing component to a given hipparion species diet. Table 14-1 shows that a number of taxa retained complex plications of the premolar and molar teeth: all listed members of the *"Hippotherium"* Complex, *Hipparion melendezi, Cremohipparion mediterraneaum, Cremohipparion proboscideum,* and *"Sivalhippus" perimense. Proboscidipparion pater* and *Proboscidipparion sinense* evolved extremely complex enamel ornamentation. Taxa that evolved either moderate or simple enamel complexity include most taxa belonging to the *Hipparion s.s.* clade, the small radicle of the *Cremohipparion* lineage, *Cremohipparion forstenae, Cremohipparion licenti,* and *"Plesiohipparion" aff. houfenense.* Mediolateral curvature of the cheek teeth remained essentially moderately curved in all hipparion taxa except the Eurasian radicle of the *"Sivalhippus"* Complex, which evolved straighter-walled cheek teeth concommitant with increased crown height.

Metapodial Dimensions

A number metapodial continuous variables have been considered to be of ecomorphological significance for determining locomotor and environmental significance. Paramount among these are the relative maximum length versus distal articular width—believed to be related to increased cursoriality (long and slender metapodials) versus slower, deliberate locomotion (short and robust) in a woodland or upland setting (Tobien, 1952; Forsten, 1968). Of further purported ecomorphological significance is the development of the distal sagittal keel on the metacarpal and metatarsal (Sondaar, 1968; Hussain, 1975; Sondaar & Eisenmann, 1995). We explore the significance of these characters later when comparing Eurasian and North African hipparion metapodial evolution with that of East African members of the *Eurygnathohippus* clade.

The African Hipparion Record

Our understanding of Africa's hipparion record is based on relatively few systematic investigations. Recent studies have proceeded mostly on either a site or restricted provincial scale. The East African hipparion record has been most intensively investigated (Aguirre & Alberdi, 1974; Hooijer, 1974, 1975; Hooijer & Maglio, 1974; Eisenmann, 1976, 1979, 1980, 1983, 1985, 1994; Hooijer & Churcher, 1985; Bernor & Armour-Chelu, 1997). A few studies have been made on North African (Arambourg, 1959, 1970; Bernor et al., 1987b) and South African (Cooke, 1950; Boné & Singer, 1965; Hooijer, 1976) hipparion assemblages. There have been two general overviews of African hipparions (Boné & Singer, 1965; Churcher & Richardson, 1978), and there have been two phylogenetic reconstructions generally relating the major African lineages to the Late Miocene–Pliocene Eurasian hipparion radiation (Bernor & Lipscomb, 1991, 1995). However, the systematic methods pursued in these investigations vary so greatly that the group's phylogeny remains far from being resolved. We provide a brief review of the African *Hipparion* record below, along with the areas of controversy, as a basis for hypothesizing its phylogenetic and ecomorphological development. Table 14-2 provides a list of taxa previously applied for African hipparionine horses. Those in italics are considered to be either valid or of provisional use, and/or of taxonomic im-

Table 14-2. A list of african *Hipparion* taxa.

Taxon	Author
Hipparion libycum	Pomel, 1897
Hipparion massoesylium	Pomel, 1897
Hipparion ambiguum	Pomel, 1897
Hipparion steytleri	van Hoepen, 1930
Eurygnathohippus cornelianus	van Hoepen, 1930
Stylohipparion hipkini	van Hoepen, 1932
Stylohipparion steyleri	van Hoepen, 1932
Notohipparion namaquense	Haughton, 1932
Libyhipparion ethiopicum	Joleaud, 1933
Libyhipparion steyleri	Joleaud, 1933
Stylohipparion cf. albertense	Hopwood, 1937
Stylohipparion albertense	Arambourg, 1947 (Hopwood)
"Hippotherium" africanum	Arambourg, 1959 (species only)
Stylohipparion libycum	Aguirre and Alberdi, 1974
Hipparion albertense	Aguirre and Alberdi, 1974
Hipparion turkanense	Hooijer and Maglio, 1973
Hipparion "sitifense"	Hooijer and Maglio, 1974
Hipparion primigenium	Hooijer, 1975
Hipparion cf. baardi	Hooijer, 1976
Hipparion baardi	Hooijer, 1976
Hipparion cf. namaquense	Hooijer, 1976
Hipparion ethiopicum	Eisenmann, 1976
Hipparion afarense	Eisenmann, 1976
Hipparion hasumense	Eisenmann, 1983
Hipparion cornelianum	Eisenmann, 1983
Hipparion libycum	Churcher and Richardson, 1983
"Hipparion" macrodon	Eisenmann, 1994
Eurygnathohippus turkanense	This study
Eurygnathohippus "sitifense"	This study
Eurygnathohippus hasumense	This study
Eurygnathohippus afarense	This study
Eurygnathohippus cf. baardi	This study
Eurygnathohippus ethiopicus	This study
Eurygnathohippus libycus	This study
Eurygnathohippus cornelianus	This study
Eurygnathohippus small sp.	This study

portance. Those not in italics are mostly not discussed below because they were usually based on insufficient material and are often of uncertain stratigraphic provenance. In other cases, there is simply too little original material to warrant specific taxonomic recommendation. A formal taxonomic revision is needed but outside the scope of this chapter.

Late Miocene Hipparions

The earliest appearing African hipparions are early Late Miocene age. East Africa would appear to have documented the occurrence of a legitimate *Hippotherium* cf. *primigenium* at Chorora, Ethiopia, between 10.8 and 9.3 m.y. (Bernor et al., 1989; Woodburne et al., 1996; M. G. Leakey et al., 1996). There is a Kenyan record of this lineage identified by Bernor (personal observation) from various localities in the Baringo Basin and from Nakali (Aguirre & Alberdi, 1974: their *"Hipparion" africanum*); this material is meager compared to the Eurasian record of *Hippotherium primigenium.* By the end of the Late Miocene, genuine East African members of this lineage are sparse, with a single phalanx from Lothagam's Nawata Member representing the last plausible record we currently can verify of this lineage.

The North African record of the *"Hippotherium"* group is sparse but relatively well represented by the Bou Hanifia (Algeria) species *Hipparion* (= *"Hippotherium"*) *africanum* (Arambourg, 1959). This is a derived form believed to be endemic to North Africa (Bernor et al., 1980, 1987a, 1989; Bernor and Hussain, 1985). Bernor et al. (1987b) found some limited evidence for this lineage continuing until the latest Miocene of Libya (Sahabi), but this tenet requires a reexamination in light of recent systematic work (Bernor, in preparation).

Beyond these citations, a number of authors have referred additional assemblages to *"Hipparion primigenium."* Churcher and Richardson (1978), Hooijer and Maglio (1974), and Hooijer (1975) reported this taxon from a number of additional North African (Morocco, Algeria, and Tunisia), East African (Ethiopia, Uganda, Kenya), and South African localities. The North and South African records have yet to be reevaluated. The East African records of this lineage are doubtful beyond those mentioned above: the Mursi and Usno Formations, Ekora, Kanapoi, Kanjera, and Chemeron most likely do not include members of the *Hippotherium primigenium* lineage. Taxonomic attributions to this lineage are largely based on symplesiomorphic characters, especially the apparent lack of ectostylids in the permanent cheek-tooth dentition. It is noteworthy that nearly 20 years ago Eisenmann (1977) recognized the same problems with these referrals.

Bernor and Lipscomb (1991, 1995) and Bernor and Armour-Chelu (1997) have recognized an additional East African Late Miocene clade, *Eurygnathohippus.* The genus *Eurygnathohippus* is united by a number of skull, mandibular, and dental characters, the most pervasive being the occurrence of ectostylids (albeit variably) in adult mandibular cheek teeth. Bernor (in preparation; see also M. Leakey et al., in press) has recognized two species of *Eurygnathohippus, Eurygnathohippus turkanense* and *Eurygnathohippus "sitifense,"* from the Nawata Formation, Lothagam. Bernor (in M. G. Leakey et al., 1996) has also reported that there is some evidence of these taxa, or their direct antecedents, from the somewhat older Samburu Hills fauna (Nakaya et al., 1984, in press; Nakaya, 1993, in press). Furthermore, Bernor and Armour-Chelu (1997) recognize two similar but skeletally more poorly represented forms from the Ibole Formation, Tanzania, also believed to be Late Miocene age.

The type skull of *Eurygnathohippus turkanense,* LT-136 (Hooijer & Maglio, 1974), is virtually complete and characterized as follows: the preorbital fossa is shallow and placed well anterior to the lacrimal; the nasal notch is narrowly incised, being placed well anterior to P^2; the cheek teeth have a maximum estimated crown height of about 65 mm; the protocones are lingually flattened and labially rounded to oval in shape; the metaconids and metastylids exhibit an angular morphology with accompanying broad linguaflexids; the ectostylids are variably present, and when present often are weakly developed and have a short height; postcranial elements previously referred to this taxon are robust, with the metapodials being relatively short and very robustly built (figs. 14-6, 14-7; Hooijer & Maglio, 1974). Bernor and Hussain (1985) and later Bernor et al. (1989), Bernor and Lipscomb (1991, 1995), and Bernor and Armour-Chelu (1997) have cited the close morphologic and phylogenetic relationship shared by *Eurygnathohippus turkanense* and the Middle Siwalik species *"Sivalhippus" perimense.*

The small species of *Eurygnathohippus, Eurygnathohippus "sitifense"*, is currently known only from East Africa. Its oldest occurrence may be from the Samburu Hills, and it occurs in the younger Nawata Formation at Lothagam Hill (M. G. Leakey et al., 1996). This taxon is recognized as belonging to the *Eurygnathohippus* lineage based on the occurrence (again, variably) of ectostylids on the lower permanent cheek teeth. It is further distinguished in the Nawata Formation by an articulated partial forelimb. The metacarpal and first phalanx contrast strikingly from their homologues in *Eurygnathohippus turkanense* in being gracile and elongate in their build and suggesting a cursorial adaptation (Bernor & Armour-Chelu, 1997; Bernor, in preparation). The type *"Hipparion" sitifense* is from St. Arnaud Cemetary, Algeria (Pomel, 1897), and according to Eisenmann (personal communication) has been lost, and the existing figures are too indefinitive to validate the taxon. The Sahabi (Libya) small horse originally attributed to *"Hipparion" cf. ?sitifense* is not well represented dentally and lacks evidence of ectostylids in the lower dentition (Bernor et al., 1987b). This problem is currently under study by Bernor (in preparation).

The *Eurygnathohippus* clade is believed to be the sister taxon of the large South Asian hipparions belonging to the *"Sivalhippus"* Complex, with the middle Siwalik form *"Sivalhippus" perimense* being the sister taxon of *Eurygnathohippus.* Bernor and Lipscomb (1995) estimate the age of this group's first occurrence in East Africa to be about 8 m.y., a time when East African faunas were in transition between older Miocene forested environments and younger Plio-Pleistocene savanna-like habitats (M. G. Leakey et al., 1996).

Hooijer (1976) has reported an assemblage of hipparions from strata exposed in the phosphatic mines of Langebaanweg, South Africa. There are a series of quarries there including Baard's quarry, C quarry, E quarry, and a quarry intermediate between C and E. According to Hooijer (1976), these quarries vary considerably in age, with the E quarry being the oldest (ca. 7–4), based on biochronologic correlations with East Africa. Baard's Quarry, which includes specimens of *Equus,* is medial Pliocene age at the oldest because *Equus* does not occur in the Old World before the Gauss/Matuyama boundary (Lindsay et al., 1980), about 2.6 (Steininger et al., 1989, 1996; Bernor & Lipscomb, 1991), and does not occur in East Africa before the Omo Shungura G (or possibly Upper Member F), <2.4 m.y. (Bernor & Armour-Chelu, 1997). Hendey (1981) has further analyzed the age of these quarries based on stratigraphic correlations to global sea-lowering events and argues for quarry E's latest Miocene age and Baard quarry's medial Pliocene age.

Hooijer (1976) has referred a skeletally well-represented assemblage from E quarry to *Hipparion* cf. *baardi.* The species-type specimen of this taxon is actually based on material from the younger Baard's Quarry deposits. *Hipparion* cf. *baardi* is represented by a nearly complete skull, maxillary and mandibular cheek teeth, and a modest size, but not fragmentary, postcranial assemblage. According to Hooijer (1976), salient characters pertinent for determining evolutionary relationships include skull with a preorbital fossa, lower cheek teeth lacking ectostylids and attaining crown height of 75–80 mm. Hooijer (1976) believed that, although similar in size, this taxon is phylogenetically distinct from the Lothagam horse, *Eurygnathohippus turkanense,* exhibiting a closer relationship to *"Hipparion" primigenium.*

Although we have not yet been able to study the original Langebaanweg material, our recent work on Old World hipparions casts new light on Hooijer's phylogenetic interpretation. The skull is similar in its size, snout, and much of its dental morphology to *E. turkanense* (see figs. 14-5A,B). Comparable features include its large size, unretracted nasals, and apparently strongly reduced POF placed dorsally high on the face as in *"Sivalhippus" perimense* and distinct from *Hippotherium primigenium* (contra Hooijer, 1976; Churcher and Richardson, 1978; see Bernor et al., 1988, 1989). The maxillary incisors and cheek teeth figured by Hooijer (1976) appear to be generally similar to *Eurygnathohippus turkanense* (R. L. Bernor, personal observation). Of further importance is the apparent presence of lingual grooving on the mandibular incisors (Hooijer, 1976: specimen L20553, pl. 6, fig. 1), a synapomorphy for more advanced members of the *Eurygnathohippus* clade. However, the mandibular I_3 is not mediolaterally elongate, as found in more progressive members of the clade. The lack of ectostylids is not surprising because they are mostly diminutive when present in more primitive members of the group and are often lacking in the Lothagam Nawata hipparions. They are not even well developed in the Early Pliocene Apak or medial Pliocene Sidi Hakoma or Denen Dora hipparions. However, the maximum crown heights reported by Hooijer exceed those known from any bonafide African Late Miocene *Hipparion;* indeed, they slightly exceed the highest recorded crown heights known from the Hadar Formation, about 3.4–2.9 m.y. (see fig. 14-8; dates from Walter & Aronson, 1993, personal communication) and are equivalent to those of the Omo Shungura F Member and younger (Hooijer, 1974). Finally, metapodial proportions (figs. 14-6A, and 14-7A: maximum length versus distal articular width) are not as robustly built as in *Eurygnathohippus turkanense,* but at the same time not as elongate as early middle Pliocene hipparions from Hadar and the Manonga Valley, Tanzania (Bernor & Armour-Chelu, 1997).

The mosaic of skeletal characters exhibited in the Langebaanweg quarry E hipparion assemblage, with a Late Miocene East African horse morphology (particularly skull characters and lack of ectostylids) occurring alongside a decidedly post-Hadar later Pliocene morphology (cheek-tooth maximum crown height), would suggest that the E quarry horse might not be as temporally homogeneous as has been previously as-

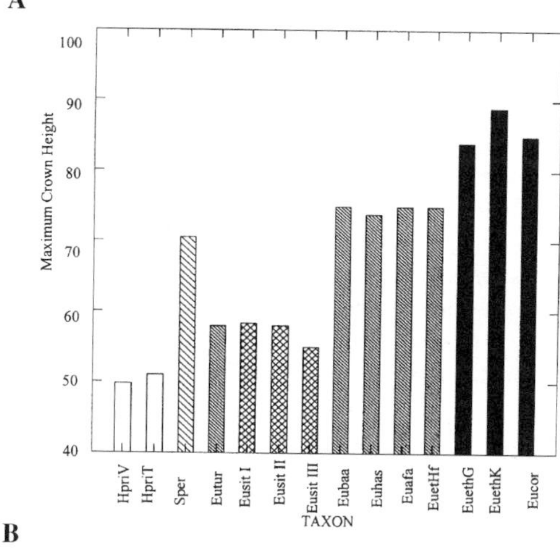

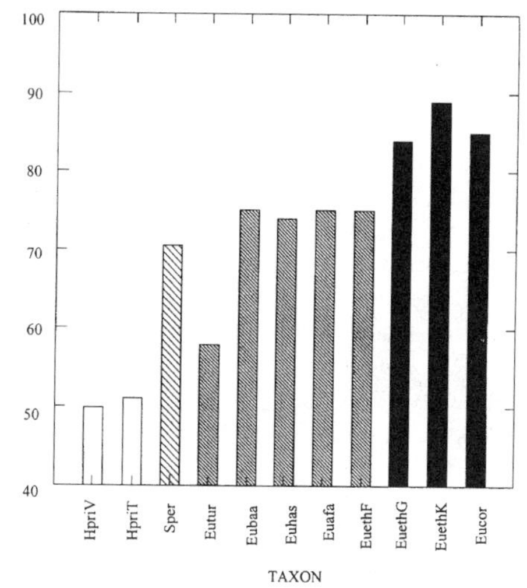

Figure 14-8. Evolution of cheek-tooth crown height in African hipparionine horses: HpriV, Central European Vallesian *Hippotherium primigenium s.s.;* HpriT, Turolian aged *Hippotherium primigenium* from Dorn Dürkheim, Germany; Sper, *"Sivalhippus" perimense* from the Potwar Plateau, Pakistan; Eutur, *Eurygnathohippus turkanense* from Lothagam, Kenya; Eusit, *Eurygnathohippus "sitifense"* from Ethiopia and Kenya; Eubaa, *Eurygnathohippus cf. baardi* from Langebaanweg, South Africa; Euhas, *Eurygnathohippus hasumense s.l.* from Hadar, Ethiopia Euafa—*Eurygnathohippus afarense* from Hadar, Ethiopia; EuethF, *Eurygnathohippus "ethiopicus"* from Omo Shungura F; EuethG, *Eurygnathohippus "ethiopicus"* from Omo Shungura G; EuethK, *Eurygnathohippus ethiopicus"* from Omo Shungura K; *Eurygnathohippus cornelianus* from Olduvai Gorge.

serted. Alternatively, this horse may represent a provincially derived clade that is most likely younger than the Nawata Formation hipparions.

The skull and associated dentition of the Langebaanweg quarry E "*Hipparion*" cf. *baardi* (Hooijer, 1976), and perhaps most of the dental material and postcrania, are plausibly derived from a Late Miocene/early Pliocene horse assemblage belonging to the *Eurygnathohippus* lineage. There is really no basis for referring this assemblage to *Hippotherium primigenium.* Although the absence of ectostylids is problematic, it should be noted that Hooijer (1976) did report permanent mandibular cheek teeth with ectostylids from a site intermediate between the C and E quarries, "*Hipparion*" cf. *namaquense.* Accompanying the ectostylids, specimen L24197 (Hooijer, 1976: pl. 7) also exhibits a more advanced, angular metaconid and metastylid morphology typical of later Pliocene African hipparions. The mix of morphologies among Langebaanweg hipparions and the occurrence of *Equus* from Baard's quarry, suggest that Langebaanweg's chronology is not well controlled. We provisionally refer the Langebaanweg E quarry skull to "*Eurygnathohippus*" cf. *baardi;* despite the apparent lack of ectostylids, there are a number of synapomorphies of the face and dentition that unite this horse with other Late Miocene-Pleistocene species of the *Eurygnathohippus* clade.

Recently, Eisenmann (1994: 290; right P^{2-4}, pl. I, fig. 1) named a new taxon of hipparionine, *Hipparion macrodon* from locality NY 75, Kakara Formation of the Kisegi-Nyabusosi region, Toro, Western Rift Valley, Uganda, correlating it with the "base of the Upper Miocene of Uganda (see also Pickford et al., 1993)". Eisenmann characterized this taxon as having extraordinarily large cheek teeth, lenticular-shaped protocones on the P^3–P^4 and rounded on the P^2. Eisenmann (1994) further referred a left P_3 or P_4 from locality KI 6, two associated lower left molars from locality NY 33, and a right P^2 from locality NY 2 to *Hipparion* cf. *macrodon.* The mandibular cheek teeth are reported to again be large, typically hipparionine in the morphology of their "double knot" (i.e., they have rounded metaconids and metastylids), lack ectostylids, and the vestibular groove is shallow in the premolar while being deep in the molars (symplesiomorphic characters for Old World hipparionines; Bernor et al., 1989). Eisenmann (1994:294) has further stated: "At present, I prefer to assume that three hipparions are present in the Vallesian of East Africa: *Hipparion aff. africanum* (small species of Samburu Hills), *Hipparion cf. primigenium* s.l. (middle sized *Hipparion* of Samburu Hills, possibly present also at Nakalia and Ngorora) and *Hipparion macrodon* from locality 75 of the Kakara Formation."

Pliocene Hipparions

The earliest Pliocene (ca. 5–4 m.y. interval) has a poorly known African *Hipparion* record. The Apak Member (Lothagam) encompasses this time interval and includes a larger member of the *Eurygnathohippus* clade. Bernor (in M. G. Leakey et al., 1996) reports that this horse exhibits morphologic advances in the mandibular dentition over the latest Miocene Nawata Formation taxa.

Beginning at 4 m.y., the East African hipparion record improves substantially in the number of localities and geographic distribution of assemblages; however, it is far from clear systematically. The Pliocene hipparion species we currently believe are potentially viable include *Eurygnathohippus "sitifense", Eurygnathohippus hasumense, Eurygnathohippus afarense,* and *Eurygnathohippus "ethiopicus". Eurygnathohippus cornelianus* may occur in the latest Pliocene and is certainly present in the Early Pleistocene of Tanzania (Olduvai Gorge, Bed II) and Cornelia (former Orange Free State, South Africa).

Hooijer (1974) reported the occurrence of a small hipparion which he referred to *"Hipparion" cf. sitifense* from Lothagam, Kanapoi, and Ekora. Kanapoi and Ekora are believed to be about 4 m.y. of age (Hooijer, 1974; M. G. Leakey et al., 1995). Hooijer (1975) further reported *"Hipparion" cf. sitifense* from Shungura Members B-G, circa 2.95–2.3 m.y. (Feibel et al., 1989, 1991; Brown and Feibel, 1991). Hooijer (1974: 26, pl. 8, figs. 3, 4) has also reported that a single specimen of this taxon from Kanapoi (KP-42) has a distinct, albeit short, ectostylid. Of greater interest still is a juvenile skull from Ekora, EK4, described and figured by Hooijer, which he refers to "*Hipparion*" *primigenium* (Hooijer, 1974: pl. 4, figs. 1–3). The skull is preserved posterior to the snout and includes a well-developed preorbital fossa set anterior to the lacrimal bone, dP^{2-4}, and an M^1 perfectly formed and unerupted in the crypt measuring 55 mm in height (fig. 14-8). Although the preorbital fossa is prominent, it is not as well developed in either its medial or posterior depth or in its peripheral rim expression as Central European *Hippotherium primigenium s.s.* (Bernor et al., 1988, 1989). The expression of this character is likely symplesiomorphic. What has not been previously reported is that this is a relatively small horse—certainly smaller than *Hippotherium primigenium.* Further evidence is needed, but it is possible that this individual is referable to *Eurygnathohippus "sitifense".*

The species *Eurygnathohippus hasumense* was erected by Eisenmann (1983) for a series of cheek teeth collected from beneath the Hasuma Tuff, and zones A, B, and C of the Kubi Algi Formation. Eisenmann (1983: pl. 5) further referred a cheek tooth assemblage from the Denen Dora Member of the Hadar Formation to this taxon. Earlier, Eisenmann (1976) referred a skull from the Denen Dora Member to *Hipparion* sp. Bernor and Armour-Chelu (1996) noted that the skull appears to be more primitive than the younger and more derived Hadar species *Eurygnathohippus afarense,* and that it and a beautifully preserved Denen Dora postcranial skeleton (AL155), as well as an abundant skeletal assemblage from the Sidi Hakoma Member, were plausibly referable to *Eurygnathohippus hasumense.*

According to Harris (1983a), the Kubi Algi Zone A faunas are correlative with horizons below the level of tuff 250/T.I., which he provisionally correlated with the Allia Tuff (204/T.II); Zone B faunas are derived from between the base of the Allia Tuff (204/T.II) and the base of the Hasuma Tuff (202/T.I., 203/T.II, 204/T.IV); Zone C faunas are from the upper part of the Kubi Algi Formation and lower part of the Lower Member of the Koobi Fora Formation, between the Hasuma Tuff and the "Suregei Tuff" (202/T.IV). According to Brown and Feibel (1991), the interval of time represented by the Zone A–C faunas would be between strata older than the Allia Tuff (ca. 3.2 m.y.) and between the Hasuma (Omo Shungura C correlative, 2.8 m.y.) and "Suregei Tuff" (no age data listed in Brown and Feibel, 1991: fig. 1.8). Bernor and Armour-Chelu's (1997) provisional attribution of *Eurygnathohippus hasumense* to the Sidi Hakoma Member would plausibly extend the range of this taxon to 3.4 m.y. based on Walter and Aronson's (1993) dating of the Sidi Hakoma Tuff.

Of the skeletal material cited above, Bernor has thus far only examined the Hadar sample, making species assignments of the relevant Kubi Algi and Hadar samples inappropriate at this time. However, if we provisionally accept the observations made above, we can begin to formulate a characterization of a broader *Eurygnathohippus hasumense* form-species. *Eurygnathohippus hasumense* is a relatively large species of medial Pliocene hipparion exhibiting a number of advanced characters: the skull is characterized by its absence of a preorbital fossa (never again to occur as more than a slight depression in members of the larger *Eurygnathohippus* lineage); the snout is long and relatively narrow, indicating that it was likely a selective feeder (Bernor & Armour-Chelu, 1997; after Janis & Ehrhardt, 1988, Solounias et al., 1988); the incisors exhibit distinct grooving on both the lingual and labial surfaces, and the third incisors are more oval shaped, evidently presaging more advanced later Pliocene and Pleistocene members of the lineage; mandibular cheek teeth have, as noted by Eisenmann (1983), "caballine" (angular to pointed) metaconids and metastylids, and ectostylids present on most mandibular cheek teeth which taper apically, not extending to the full crown height; also, the maximum crown height for the cheek teeth (P4–M1) is 74 mm (fig. 14-8); metacarpals and metatarsals are elongate and strongly built (figs. 14-6A, 14-7A).

There is a chronologic gap between the Late Miocene Nawata Formation hippparion, *Eurygnathohippus turkanense,* and *Eurygnathohippus hasumense.* As mentioned above, Bernor has observed some advances in the earliest Pliocene Apak Member hipparion at Lothagam. Hooijer (1974) has tentatively referred Early Pliocene material from Kanapoi and Ekora to *Eurygnathohippus "turkanense"*. However, there is really no basis for referring larger hipparions of the 5–3.4 m.y. level to either *Eurygnathohippus turkanense* or *Eurygnathohippus hasumense,* and the heavy-limbed morphology of the former would appear to be a postcranial specialization unique to that species and would require a major evolutionary reversal to be directly ancestral to the Denen Dora horse. Finally, it should be noted that a horse virtually identical to the Denen Dora sample of *Eurygnathohippus hasumense* has been found from the earlier Pliocene of the Manonga Valley, Tanzania (Bernor & Armour-Chelu, 1997).

The type assemblage of *Eurygnathohippus afarense* is derived from the Upper Kada Hadar Member, Hadar, circa 3.0–2.9 m.y. This species is well represented by a skull, complete posterior to the snout, a maxillary dentition and mandible with dentition (Eisenmann, 1976: AL 363–118). This horse is advanced over *Eurygnathohippus hasumense* in incisor hypertrophy and I3 outline, perhaps a broadened mandibular symphysis and ectostylids that would appear to extend higher upward on the crown's labial surface. These characters are intermediate between *Eurygnathohippus hasumense* and the Late Pliocene species, *Eurygnathohippus "ethiopicus"*. As cited above, there is a chronologic overlap between the Kubi Algi assemblage of *Eurygnathohippus hasumense* (>3.2 to <2.8 m.y.) and Hadar *Eurygnathohippus afarense* (3–2.9 m.y.). Indeed, Eisenmann (1983) has noted the presence of a more advanced member of this lineage, her *Hipparion* sp. A, from Zone C. This form has more prominent ectostylids than *Eurygnathohippus hasumense,* and moreover, Eisenmann (1983) has duly noted the difficulty in distinguishing *Eurygnathohippus hasumense* from *Eurygnathohippus afarense* due to their close morphological similarity. We believe it distinctly plausible that the transition from *Eurygnathohippus hasumense* to *Eurygnathohippus afarense* may have been relatively gradual, occurring circa 3.0 m.y. Hooijer's (1974) hypodigm for *Hipparion* sp. from the Omo Usno Formation and Shungura Members B–E (2.95–2.4 m.y.; Brown and Feibel, 1991) may include a further transition. We further point out that in our analysis of the third metacarpals and third metatarsals (figs. 14-6, 14-7) that we have arbitrarily grouped all the Hadar specimens under the nomen *Eurygnathohippus hasumense;* indeed, we expect these to represent that taxon and *Eurygnathohippus afarense,* at the very least.

"*Hipparion*" *ethiopicum* was named by Joleaud (1933; his *Libyhipparion ethiopicum*) for an assemblage of teeth derived from the Omo River Valley. Eisenmann (1983) has noted that the stratigraphic provenance of the type assemblage is not only unknown, but in all likelihood, mixed. As such, this is a nomen of dubious value. We use the nomen *Eurygnathohippus"ethiopicus*" as an informal taxon which may reflect a stage of evolution bridging *Eurygnathohippus afarense* and *Eurygnathohippus cornelianus.* Hooijer (1974) referred hipparion material from Omo Shungura Members F (2.36 m.y.) through K (1.5 m.y.; Brown & Feibel, 1991), as well as Olduvai, to this taxon. The Olduvai material ranges considerably younger in time (Bed I–Bed IV), the youngest material being perhaps as young as 620,000 years (M. D. Leakey & Hay, 1982). Hooijer (1974) applied Joleaud's (1933) nomen based on the correspondence of a single, left I_1, worn down to 35 mm height and his observation that the morphology of this tooth was similar to Olduvai specimens which he had studied. Hooijer further claimed that beginning with Shungura Member C, and continuing through Member K of that sequence, that there is no significant difference with the Olduvai collection, best represented by the Olduvai Bed II assemblage from site BK II.

Beginning with Omo Member F, there is an apparent increase in maximum crown height and length, width, and maximum ectostylid height. However, even compositely, the Omo material referred to this taxon is poor, lacking good skull and postcranial material. Moreover, the maximum crown height of this taxon varies stratigraphically from ≥75 mm in Member F (Hooijer, 1974:68) to 89 mm in Member K (Hooijer, 1974: 72). Eisenmann (1976) initially recognized the taxon *Hipparion ethiopicum* for a juvenile cranium from the *Notochoerus scotti* Zone (Eisenmann, 1983: pl. 5.3, p. 167; collecting area 105, circa KBS correlative, 1.89 m.y.), which she later referred to *Hipparion cornelianum* based on incisor morphology. There are several fundamental morphological problems with Eisenmann's attribution: (1) this is a juvenile cranium, (2) the incisors are absent, leaving only the alveoli to interpret their morphology, (3) her assertion that the third incisors would have to be atrophied to emerge from the crypt are based on the alveolar crypt of the deciduous third incisors which in all likelihood were smaller and had a different morphology than the adult third incisors. Moreover, it is only the lower third incisors that are so highly atrophied in *Eurygnathohippus cornelianus* (see Leakey, 1967: fig. 20, lower left corner figure is a premaxilla with incisors clearly showing the lesser reduction of I^3 than I_3; note other three figures here are of mandibular symphyses with their incisor dentitions). There is no secure morphological basis that we know of to relate the KBS correlative hipparion to Bed II Olduvai sample and Cornelia type specimen of *Eurygnathohippus cornelianus.*

Yet another Plio-Pleistocene age taxon is prominent in the literature: "*Hipparion libycum*" (Pomel, 1897). This taxon has been applied for most Plio-Pleistocene assemblages in Africa. Aguirre and Alberdi (1974) applied this nomen for assemblages in Kenya (Chemeron), Tanzania (Serengeti, Olduvai), Ethiopia (Omo), South Africa (Makapansgat, Bolt's Farm, and Cornelia) and North Africa (Ain Jourdel and Ain Boucherit). Together, these would range in age from about 4 to 0.6 m.y. Churcher and Richardson (1978) adopted an even broader ranging referral of "*Hipparion libycum*" listing a huge series of North, East, and South African localities of Plio-Pleistocene age, and proposing regional subspecies of no essential morphological difference. This led them to formally recognize three regional subspecies (Churcher and Ridhardson 1978: 399): "*Hipparion libycum libycum*" (North Africa), "*Hipparion libycum ethiopicum*" (East Africa), and "*Hipparion libycum steytleri*" (South Africa).

According to Eisenmann (1983: 161) the type specimen of "*Hipparion*" *libycum* (Pomel 1897: 8) is a "large lower caballine [meaning that the metaconid and metastylid have angular facing surfaces] premolar with a well developed ectostylid in the Villafranchian levels of the Carriere Brunie, Oran" (= Algeria; see Arambourg 1970:92, fig. 55). Indeed, Arambourg (1970) figured a number of specimens that he referred to *Hipparion libycum* from Garet (Lac) Ichkeul and Ain Brimba, Tunisia, and Ain Boucherit and Ain Jourdel, Algeria. According to Geraads (1987), these localities are between 3 and 2.5 m.y. of age. Arambourg (1970) figured a number of specimens from these localities from which we can identify a number of salient morphological features: (1) a right I^3 from Ain Brimba (pl. XVI: 3, 3a; 1958–14: 225) is labiolingually strongly curved, elongate, and tapers strongly distally; (2) a right P^2 (pl. XVI: 7, 7a; 1958–14: 173) from Ain Brimba is in early wear and high crowned (estimated mesostyle height = 57 mm); (3) a left P^4 (pl. XVI: 8, 8a; 1958–14: 191) from Ain Brimba is likewise in early wear and has an estimated crown height of 70 mm; (4) a right, apparently unerupted, I_2 from Ain Brimba (pl. XVII: 10, 10a; 1958–14: 227), which has a slight curvature, is not extremely high crowned but has a distinct lingual cleft which is broad at its occlusal summit and tapers strongly toward the root to give an elongate V-shape; (5) a right P_4 (pl. XViII: 4, 4a; 1958–14: 175) and right M_1 (pl. XVIII:3, 3a; 1958–14: 183) both from Ain Brimba are high crowned with strongly developed ectostylids becoming pointed at their summit and with metaconids rounded and metastylids flattened; (6) a right metacarpal III from Garet (Lac) Ichkeul (Arambourg, 1970:91 pl. XVI: 9, 9a;1948–2:19), which has a maximum length of 268 mm and distal articular width of 38 mm; (7) a right metatarsal III also from Garet (Lac) Ichkeul (pl. XVI: 10; 1950–1: 3), which has a maximum length of 268 mm and distal articular width of 42 mm. The morphology of this assemblage generally conforms in all details to East African hipparions in the 3.0–2.5 m.y. chronologic interval and does not show advances characteristic of East African hipparions younger than 2.5 m.y. Therefore, we do not advocate the taxonomy provided by previous workers (and explicitly those espoused by Aguirre & Alberdi, 1974, and Churcher & Richardson, 1978). We further believe that "*Hipparion libycum*" is best considered a nomen vanum in light of Eisenmann's (1983) observation that the type material is a single premolar, and it is likely lost. We apply it here as *Eurygnathohippus "libycus"* only as a point of reference for the Algerian and Tunisian medial Pliocene assemblages reported by Arambourg (1970) and briefly described above, and to clearly distinguish them from the distinctly different Late Miocene lineages discussed above.

Pleistocene Hipparions

Eurygnathohippus cornelianus was named by Van Hoepen (1930:23, fig. 20, 24, figs. 21, 22) for a mandibular incisor region from the early to medial Pleistocene (ca. 1 m.y.; McKee, personal communication; see McKee et al., 1995) locality of Cornelia (= Uitzoek), South Africa. This specimen is similar to other mandibular symphyses with incisors from Olduvai site BK II, middle Bed II, Olduvai (Hooijer, 1975: pl. 12, figs. 1–3). Furthermore, two skulls from Bed II (Hooijer, 1974: plates 7–10) and a good assemblage of material from the British Museum derived from Beds I–IV, and possibly the Masek Beds exhibit a number of morphological advances. These include a highly derived incisor region whereby I1 and I2 are hypertrophied, procumbent, and straight, heavily grooved with a distinct medial stylid on I_1, and with I_3 being highly atrophied, caniniform, and placed posterior to the midline of I_2 (see Hooijer, 1975: pl. 12, figs. 1–3). The skull lacks a preorbital fossa and has an elongate snout (fig. 14-5). The cheek teeth are remarkable in a number of characters: extremely high crowns [maximum crown height reported by Hooijer (1974) to be 85 mm; reestimated by Bernor to be >80 to 85 mm based on the British Museum collection]; secondarily evolved strong mediolateral and anteroposterior curvature, particularly in second and third molars; sharp serial reduction of the molars compared to the premolars, very angular facing borders of the metaconids and metastylids with accompanying broad linguaflexids; highly developed ectostylids that are high crowned, long, and wide (lenticular shape).

When one weighs the morphologic variability exhibited between the Omo Shungura F–K hipparion sequence, the Cornelia-type specimen, and Olduvai Beds I–IV hipparions, it is difficult to defend the use of a single taxon whether it be *Eurygnathohippus"ethiopicus"* (sensu Hooijer, 1974), *Eurygnathohippus libycus* (sensu Chrucher & Richardson, 1978), or *Eurygnathohippus cornelianus* (sensu Eisenmann, 1983). Churcher and Richardson's (1978) concept of *Hipparion libycum* is broader yet, making it more implausible. Moreover, there is possibly a second species of smaller hipparion from Olduvai (Eisenmann, 1983, personal communication; R. L. Bernor, personal observation) and the Pleistocene of the Middle Awash (Bernor, personal observation), further confusing the taxonomic and phylogenetic interpretations.

Ecomorphological Trends in African and Eurasian Hipparions

Table 3 provides an ecomorphological characterization of African Late Miocene-Pleistocene members of the *Eurygnathohippus* clade; the North African Late Miocene species *Hippotherium "africanum"* is included in table 14-1. The preorbital fossa is found to be highly reduced in *Eurygnathohippus turkanense* and absent in all other members of the group (when known), except for a skull from the Early Pliocene of Ekora (EK-4; Hooijer & Maglio, 1976), which we believe may be referable to *Eurygnathohippus*

Table 14-3. Ecomorphological features of selected African hipparionine horses.

Species	POF	Caninus fossa	Malar fossa	Nasal notch	Incisor curve	Incisor arcade	PM curve	PM comp	PM groove
Eurygnathohippus									
E. turkanense	R	A	A	AP2	Y	A	MC	M	F
E. sitifense	P	A	A	?	Y	A	MC	CM	F
E. cf. baardi	A	A	A	AP2	Y	A	MC	M	F
E. hasumense	A	A	A	AP2	Y	A	MC	M	F
E. afarense	A	A	A	AP2	Y	A	MC	M	F
E. ethiopicus	A	A	A	?AP2	?	?	?	?M	F
E. cornelianus	A	A	A	AP2	N	S	C	M	F

POF (preorbital fossa): P = present, R = reduced, A = absent; caninus fossa: P = present, A = absent; malar fossa: P = present, A = absent; nasal notch: A = anterior limit of tooth, M = mesostyle of the tooth as follows in the table; incisor curvature: Y = yes, N = no; incisor arcade: A = arcuate, VA = very arcuate, S = straight; PM curve: C = premolars and molars curved, MC = moderately curved; S = straight; PM comp: C = premolars and molars complexly plicated, M = moderately complex, S = simple, VC = very complex; PM groove: D = premolars and molars deeply grooved, M = moderately deeply grooved, F = flat.

"*sitifense*" (Bernor, in preparation). The caninus and malar fossae are likewise absent in the *Eurygnathohippus* clade. Where the nasal notch is known, it is always unretracted. These facial features are in common with advanced members of the *Hipparion s.s.* clade, advanced members of the "*Plesiohipparion*" clade, members of the small *Cremohipparion* clade and the Chinese form *Cremohipparion forstenae,* and suggest a shift toward increased grazing. Incisor curvature is maintained through much of the *Eurygnathohippus* radiation until the apparent occurrence of advanced incisors from the Omo Shungura F horizons, where *Eurygnathohippus "ethiopicus"* (Hooijer, 1975: pl. 13, fig. 2) exhibits some straightening and hypertrophy of the first and second incisors. Although incisor grooving is present in most Pliocene members of *Eurygnathohippus,* it is stronger in *Eurygnathohippus hasumense* and *Eurygnathohippus afarense* and is accompanied by hypertrophy of the first and second incisors and atrophy of I_3 in *Eurygnathohippus cornelianus.* Premolar and molar curvature is secondarily strong in *Eurygnathohippus cornelianus:* all cheek teeth exhibit a strong mediolateral curvature, with the second and third molars further exhibiting strong anteroposterior curvature. The strong curvature accompanied a sharp increases in crown height (see fig. 14-8), and we believe that this was adaptive in facilitating increased hypsodonty in the absence of increased height of the maxilla in this species. Complexity of the maxillary cheek teeth is predominantly moderate, and mediolateral grooving is nonexistent, suggesting an increased grazing habitus. In summary, discrete ecomorphological features suggest that African hipparions show a progressive increase in their preference for graze, particularly when compared to those Eurasian forms, which we believe incorporated various portions of browse in their diet.

Figure 14-5 gives four bivariate comparisons of Old World hipparion Late Miocene-Pleistocene skulls. The measurement numbers (M1, M14, M15, etc.) follow those of Eisenmann et al. (1988). These figures exhibit the diverse evolutionary directions that Old World hipparions pursued. The symbols provided in the legend of figure 14-5 are keyed in their shape to the major lineages discussed above. We maintain these same symbols for the sake of continuity in our analysis of metacarpal III and metatarsal III morphology.

Figure 14-5A is a bivariate plot of muzzle length (M1) versus muzzle breadth taken at the posterior limit of I3 (M15). For reference, *Cormohipparion occidentale* is rendered (by stars near the center of the field), and the evolutionary directions taken by the respective Old World hipparion lineages are shown. Species of the "*Sivalhippus*" Complex exhibit striking trends in the increase of snout length: "*Sivalhippus*" perimense exhibits slightly increased length and width measurements over *Cormohipparion occidentale,* followed in the African *Eurygnathohippus* lineage by a progressive lengthening through time, holding incisor width virtually constant: *Eurygnathohippus turkanense–Eurygnathohippus hasumense–Eurygnathohippus cornelianus.* In the last stage increased snout length is accompanied by hypertrophy of the upper and lower I1–2, atrophy of lower I_3, strong grooving and even lingual pillaring on the lower I1–2, and a sharp increase in hypsodonty (fig. 14-8). The increased snout length, accompanied by striking dental evolution, suggests that the *Eurygnathohippius* clade evolved a masticatory apparatus progressively more efficient for acquiring and processing graze. We believe it is plausible that the effective lengthening of the head by increased muzzle length and broad mandibular incisor region with procumbent, mediolaterally aligned teeth is functionally convergent with the white rhinoceros, *Ceratotherium simum,* which indicates a commensurate adaptation to low-grass feeding in drought conditions (see Owen-Smith, 1982). It should not be overlooked that while the broadening of the mandibular symphysis in *Eurygnathohippus cornelianus* is convergent with *Ceratotherium simum,* the retention and hypertrophy of incisors in this hipparion is in striking contrast to the white rhinoceros, in which the incisors have been lost altogether since the Late Miocene first occurrence of the lineage.

The Asian radicle of the "*Sivalhippus*" Complex is represented here by *Proboscidipparion sinense* and "*Plesiohipparion*" *aff. houfenense.* Both exhibit similar increases in length and width ("*Plesiohipparion*" *aff. houfenense* having the greatest M15 measurement), but neither achieved the length dimension of the advanced members of the *Eurygnathohippus* clade. We believe that *Proboscidipparion* with its extensive nasal retraction supported a pseudoproboscis for selective feeding of browse, whereas "*Plesiohipparion*" *aff. houfenense* was *Equus*-like, lacking nasal retraction and POF, and likely incorporated a high percentage of graze in its diet.

The "*Hippotherium*" group exhibits a diversity of snout morphologies. *Hippotherium primigenium* and *Hippotherium africanum* are closely comparable, exhibiting a somewhat increased muzzle width compared to *Cormohipparion occidentale,* accompanied by a greatly increased muzzle length. "*Hippotherium*" *giganteum* continued this trend evolutionarily, increasing both muzzle length and breadth. "*Hippotherium*" *dermatorhinum* is the one member of this group that strongly retracted the nasals, and commensurate with this is the reduction of both nasal length and width. When combining these statistical results with the discrete ecomorphologic characters presented in table 14-1, we believe that members of the "*Hippotherium*" group were primitively mixed feeders, incorporating both graze and browse in their diet. "*Hippotherium*" *dermatorhinum,* with its highly retracted nasals, narrow incisor region, and V-shaped snout, likely was the most browse-facultative species of this group.

The *Hipparion s.s.* lineage exhibits a different evolutionary pathway in snout morphology. *Hipparion gettyi* is an early member of this lineage and exhibits similar snout length and width proportions to *Cormohipparion occidentale,* a putative grazer (Hayek et al., 1991). The *Hipparion prostylum–Hipparion dietrichi* lineage exhibits a striking increase in muzzle width, and in some cases muzzle shortening. Some individuals referred to *Hipparion dietrichi* exhibit the straight mediolateral incisor alignment (without I3 atrophy) found in the *Eurygnathohippus "ethiopicus"–cornelianus* lineage. We believe that the observations on muzzle morphology are congruent with facial morphology (lack of nasal retraction and loss of POF) and suggest an increased adaptation to grazing; this observation is also congruent with what is known about the evolution of latest Miocene southern European and southwest Asian faunas where this advanced lineage occurs (Bernor et al., 1996a; Fortelius et al., 1996). The *Hipparion campbelli* lineage evolved in the opposite direction, maintaining muzzle length but sharply decreasing its width, while also losing the POF and not strongly retracting the nasals. This combination of morphological features suggests that *Hipparion campbelli* was perhaps a more selective, higher-grass grazer than *Hipparion dietrichi.*

The plot of muzzle length versus width at the distal incisor level exhibits yet different evolutionary directions in the *Cremohipparion* group. Among the larger group with more developed facial fossae, *Cremohipparion mediterraneum* and *Cremohipparion forstenae* exhibit reduced muzzle length and width; in *C. mediterraneum* this is accompanied by extensive preorbital fossa development and modest nasal incision, which we believe indicate the ingestion of a significant amount of browse in the diet. In *C. forstenae* there is a further retraction of the nasals, but a secondary loss in depth and definition of the facial fossae accompanied by a predominately grazing habitus (Hayek et al., 1991). *Cremohipparion licenti* and *Cremohipparion proboscideum* both exhibit increased muzzle length and width measurements accompanying strong nasal retraction and facial fossae development, which are hallmarks for increased browse in their diets. *Cremohipparion matthewi* and *Cremohipparion nikosi* are the smallest hipparions in this sample. *Cremohipparion matthewi* has a snout with the width equivalent to *Cremohipparion mediterraneum* and *Cremohipparion forstenae,* but a shorter length. *Cremohipparion nikosi* has a slightly longer but strikingly narrower snout. We believe that these small species of *Cremohipparion* with their lack of strong nasal retraction and loss of preorbital fossae evolved a preference for increased graze in their diet; the more advanced form *Cremohipparion nikosi* likely evolved greater selectivity in its grazing behavior.

Figure 14-5B is a bivariate plot of muzzle length (M1) versus muzzle width at its minimal (usually mid-premaxilla distance) point. In comparison to figure 14-5A there is less dispersion of points, but the same general species trends are found. In the "*Sivalhippus*" *perimense–Eurygnathohippus* clade we see a greater relative increase in snout muzzle width and the same increase in M1 length through the *Eurygnathohippus* sequence. The Asian lineages of the "*Sivalhippus*" Group, *Proboscidipparion sinense* and "*Plesiohipparion*" *aff. houfenense,* exhibit somewhat narrower snout width than shown in figure 14-5A. In the "*Hippotherium*" Group, *Hippotherium primigenium* and "*Hippotherium*" *giganteum* exhibit a similar increase in muzzle length and width, while "*Hippotherium*" *dermatorhinum* has moved relative to *Cormohipparion occidentale* to a position of greater muzzle width. This reflects the narrow incisor region of "*Hippotherium*" *dermatorhinum*. "*Hippotherium*" *africanum* also shifts in this figure dramatically toward a relatively broader muzzle dimension than other members of this group. The *Hipparion s.s.* and *Cremohipparion* groups exhibit a dispersion of points quite similar to that seen in figure 14-5A.

Figure 14-5C depicts the relationship between muzzle length (M1) and length of the cheek tooth row: P2-M3 (M9). We have calculated this comparison to gain further insight as to how muzzle morphology varies with this rough estimate of species size. In the "*Sivalhippus*" Complex, the "*Sivalhippus*" *perimense–Eurygnathohippus turkanense/hasumense* clade, as well as *Proboscidipparion sinense* and "*Plesiohipparion*" *aff. houfenense,* show that snout lengthening was accompanied by a substantial increase in cheek tooth length. Likewise, members of the "*Hippotherium*" Group, *Hippotherium primigenium,* "*Hippotherium*" *giganteum,* and "*Hippotherium*" *africanum,* exhibit increased size over *Cormohipparion occidentale.* Members of the *Hipparion s.s.* clade vary from having cheek-tooth row length equal to or somewhat greater than *Cormohipparion occidentale.* The *Cremohipparion* group is the most variable in size: *Cremohipparion forstenae, Cremohipparion matthewi,* and *Cremohipparion nikosi* show a progressive decrease in size away from *Cormohipparion occidentale,* while *Cremohipparion mediterraneum* has a cheek tooth range at the lower limit of *Cormohipparion occidentale* and *Cremohipparion proboscideum* has a range that exceeds the upper range of *Cormohipparion occidentale,* being close in size to *Hippotherium primigenium* and "*Hippotherium*" *giganteum.*

Figure 14-5D is a bivariate plot of the length of the cheek as measured by a straight line from the nasal notch to anteriormost orbital rim (M31) versus the length dimension from prosthion (point between the first incisors) to the nasal notch (M30). Overall, the plot exhibits a negative correlation: as M30 increases, M31 decreases, which is expected because taxa that evolve ever-increasing nasal notch incisions actually decrease cheek length as defined here. Within the "*Sivalhippus*" Complex, the "*Sivalhippus*" *perimense–Eurygnathohippus* clade exhibits elevated values in both M31 and M30, reflecting overall size increase of the species. Interestingly, *Eurygnathohippus cornelianus* exhibits an increased prosthion–nasal notch dimension with commensurate depression of the cheek length variable, reflecting overall increase in snout length. In the Asian radicle of the "*Sivalhippus*" Complex, only "*Plesiohipparion*" *aff. houfenense* could be measured and shows a strong elevation in M30, reflecting a long snout.

The "*Hippotherium*" Group exhibits a broad dispersion of points. The greatest dispersion of a species is found in *Hippotherium primigenium,* no doubt due to the mediolateral crushing of the Howenegg sample; this species exhibits elevation in both M31 and M30 over the condition seen in *Cormohipparion occidentale,* reflecting its larger size (Bernor et al., 1989). "*Hippotherium*" *gigantum* exhibits a striking increase in M30, retaining a cheek length similar to *Cormohipparion occidentale.* "*Hippotheirum*" *dermatorhinum* likewise has a large M30 dimension, but this is accompanied by a highly reduced cheek length, reflecting the stong retraction of the nasal notch. "*Hippotherium*" *africanum* falls within the lower range of *Hippotherium primigenium* for both M31 and M30 measurements.

The *Hipparion s.s.* group exhibits particularly high variability in its M30 dimension; nearly all members of this group have a reduced M31 dimension and increased M30 dimension compared to *Cormohipparion occidentale. Hipparion gettyi* and *Hipparion prostylum* are situated in the center of the *Hipparion s.s.* distribution, showing a concommitant increase in M30 and decrease in M31. *Hipparion campbelli* is the most extreme member of this clade in its length from prosthion to nasal notch and short cheek length: this supports our previous observation that *Hipparion campbelli* evolved an elongate, narrow snout for selective feeding. *Hipparion dietrichi* exhibits great variability along the M31 versus M30 regression despite the fact that all specimens are derived from Samos. This may indicate that there is more than one species included in the current *Hipparion dietrichi* hypodigm.

The *Cremohipparion* lineage is represented by *Cremohipparion matthewi, Cremohipparion nikosi, Cremohipparion forstenae,* and *Cremohipparion mediterraneum. Cremohipparion forstenae* and *Cremohipparion mediterraneum* are closely comparable with *C. matthewi* in having markedly greater M31 dimensions than *C. nikosi. Cremohipparion nikosi* has a relatively longer snout than similarly sized *Cremohipparion matthewi* (nasals retracted to MP^4) and, presumably, was a more selective feeder.

Metacarpal III exhibits a broad range of morphologies. Because of the excellent postcranial sample from Höwenegg, Germany (Bernor et al., in press d), we are able to calculate 95% ellipses for any bivariate plot. Figure 14-6A presents a 95% ellipse for metacarpal III maximum length (M1) versus distal articular width (DAW; M11) dimensions with individual Höwenegg specimens plotted in association with the ellipse. Figure 14-6B calculates the same dimensions including all the taxa under consideration. *Hippotherium primigenium* has consistently been considered to be a robust-limbed hipparion. This figure shows that "*Sivalhippus*" *perimense* is more robustly built, clearly falling within the range of *Hippotherium primigenium* in its length, but with elevated DAW. "*Eurygnathohippus*" *cf. baardi* falls in the middle of the Höwenegg sample in its M1 versus M11 dimensions. The Hadar assemblage, broadly referred to *Eurygnathohippus hasumense,* may well include more than a single species, but all third metacarpals are more elongate, and most have greater DAW dimensions than *Hippotherium primigenium.* The Lac Ichkeul sample of *Eurygnathohippus "libycus"* is elongate (Arambourg, 1970). The small species *Eurygnathohippus "sitifense"* falls at the lower portion of the range for *Hippotherium primigenium,* but is longer. More interesting yet is that this species' metacarpal III proportions are similar to the Hadar sample and in striking contrast to its co-occurring species at Lothagam, *Eurygnathohippus turkanense. "Sivalhippus" perimense, Eurygnathohippus cf. baardi,* and *Eurygnathohippus turkanense* then are more robustly built forms plausibly adapted to more wooded conditions; *Eurygnathohippus "sitifense"* and *Eurygnathohippus hasumense* arguably were more cursorial, plausibly being adapted to more open-country conditions.

The Asian radicle of the "*Sivalhippus*" complex, "*Plesiohipparion*" *houfenense* (note, a different species from "*Plesiohipparion*" *aff. houfenense;* Bernor et al., 1989) had elongate metapodials within the upper limit of the Hadar sample; this too suggests a cursorial mode of locomotion. *Proboscidipparion pater* exhibits a modest increase in length, while maintaining DAW within the range of the Höwenegg sample. The most extreme increase in metacarpal III length is reported for *Proboscidipparion sinense.* Does this mean that it was the most cursorial form, and did this species live in savanna environments? Facial morphology is incongruent with this conclusion because it suggests a highly derived facial morphology for the incorporation of browse in its diet (see table 14-1). Perhaps a more viable deduction is that the increased limb length reflected in the metapodials was convergent with the long-limbed morphology of giraffes and that this increased limb length was meant to facilitate the acquisition of higher strata browse (leaves).

The members of the "*Hippotherium*" Group, "*Hippotherium*" *africanum* and "*Hippotherium*" *giganteum,* fall directly within the Hoewenegg 95% ellipse. The *Hipparion s.s.* group either has narrower limbs with equivalent length (*Hipparion melendezi* and *Hipparion prostylum*) or elevated length measurements with equivalent DAW (*Hipparion antelopinum* and *Hipparion dietrichi*). The *Cremohipparion* group as a whole has relatively slender third metacarpals compared to *Hippotherium primigenium.* Those *Cremohipparion* species that have reduced DAW dimensions but fall within the M1 range of the Howenegg hipparion include *Cremohipparion moldavicum, Cremohipparion mediterraneum,* and *Cremohipparion matthewi. Cremohipparion proboscideum* has a substantial length dimension while falling in the lower portion of the Höwenegg sample range for DAW. In itself the metapodial morphology would suggest that this species was cursorial and adapted to more open-country conditions, but the strong nasal retraction and development of multiple and hypertrophied facial fossae again suggest that limb lengthening may have had more to do with the acquisition of browse in the forms of leaves than with running.

Figure 14-6C–F provides data on the relative expression of the distal sagittal keel on third metacarpals. Sondaar (1968), and more recently Sondaar and Eisenmann (1995), have argued that the elevation of the distal sagittal keel reflects the degree to which the distal metapodial (either metacarpal III or metatarsal III) interlocks with the proximal surface of the first phalanx III: effectively, the higher the keel, the more cursorial the adaptation of the distal limb. Figure 14-6C and D present plots of distal maximal keel depth (M12) versus DAW (M11). In both plots we have made regression and correlation analyses: in 14-6C we have calculated a Pearson's correlation of all variables excluding the Höwenegg sample; in 14-6D we have calculated the correlation including the Höwen-egg sample. We have done this to account for the differentially large size of the Höwenegg sample, and yet in both analyses the correlation of M11 and M12 is equal, with an $r = .92$: the two variables are very highly correlated.

Figure 14-6E and F similarly analyze sagittal keel height (M12) versus distal minimal depth of the lateral condyle (M13). The correlation of these variables is even higher: 14-6E, which excludes the Höwenegg sample, has an r of .95; 14-6F including Höwenegg has an r of .96. Figure 14-6C–F exhibit a strong correlation between distal sagittal keel height and either distal sagittal width or distal minimal depth of the lateral condyle. If distal sagittal keel height had the functional significance ascribed to it by Sondaar (1968) and Sondaar and Eisenmann (1975), we would expect a more dispersed array of bivariate dimensions for these variables with a resulting lower correlation of M11 versus M12 and M13. Our analysis indicates that distal sagittal keel height is size dependent: the larger the distal articular width and minimal depth of the lateral condyle, the greater the height of the sagittal keel.

Figure 14-7A–F present bivariate plots of the same dimensions, but this time for the third metatarsals. As expected, the plots closely parallel those described for the third metacarpals. We add here some additional salient points with regard to figure 14-7B. Within the "*Sivalhippus*" Complex, we have here rendered what we believe to be a metatarsal III referable to *Eurygnathohippus cornelianus,* and this is found to rest on the upper limit of the Höwenegg ellipse for M1 versus M11. In this plot we further include the Hungarian medial Pliocene species "*Plesiohipparion*" *moritorum,* which compares closely in its dimensions to "*Plesiohipparion*" *houfenense.* Again, *Proboscidipparion pater* exhibits increased M1 dimensions, still falling within the range of the Höwenegg sample for DAW, while *Proboscidipparion sinense* reveals an even more dramatically increased length dimension than the rest of the sample. *Hipparion dietrichi* and *Hipparion antelopinum,* believed on the basis of facial (and in the former case incisor alignment) evidence to have been adapted to open-country environments, exhibit sharply elevated maximum length dimensions. The small *Cremohipparion* radicle likewise stands out as being slender limbed and also likely more adapted to a cursorial mode of locomotion in an open-country context. Figures 14-7C–F produced similar correlation values to those for metacarpal III: for M11 versus M12 without Höwenegg, r = .91 (fig. 14-7C), while with Höwenegg, r = .90 (fig. 7D); for M13 versus M12 both without (fig. 14-7E) and with (fig. 14-7F) the Höwenegg sample included, r = .94.

Discussion

Whereas the basin is a fundamental unit of analysis in geology (Feibel, this volume), the species is so for paleontology. A species' morphology is fundamental for its recognition and characterization, and these are in turn the elements of taxonomy. Taxonomy may be perceived as a "stumbling block" for paleobiological study by some (Turner, this volume), but taxonomy is actually as much a building block for systematics as stratigraphy is for geology. Alpha taxonomy, the identification and naming of species, remains the starting point for reconstructing evolutionary lineages and interpreting their adaptive pathways. Testing potentially competing evolutionary hypotheses such as the "Red Queen hypothesis" (Foley, this volume) and the "Turnover Pulse Hypothesis" (Vrba, 1995, this volume) ultimately depends on establishing the relative congruence of well-resolved mammalian systematic records. Whether cladistic analyses or more traditional phyletic reconstructions are used to track lineages across time and space, the most highly resolved evolutionary records are those that reconstruct viable monophyletic lineages. Collard and Wood's (this volume) argument for classifications erected from the interpretation of evolutionary grades is misaimed in that grades are not monophyletic and only obscure our resolution of evolutionary lineages and the full complement of paleobiological interpretations derived from them (Bernor & Lipscomb, 1995). We believe it is far more accurate to first characterize and define species based on relevant morphological characters and then proceed with reconstructing changes in the function of relevant anatomical complexes across a monophyletic lineage species' transformation series. We believe that this provides the most resolved record of a lineage's adaptation and evolution. Our characterization and definition of species has not applied the Species Recognition Concept advocated by Turner (this volume) because we know of no objective means by which to operationalize this concept with a paleontological record; indeed, species recognition depends on behavioral repertoires that are usually not fossilized in the geological record. Morphological characters usually are evolutionary artifacts and do not reflect those anatomical aspects that originally maintained biological separation of sympatric species.

This volume has brought forward a wealth of biogeographic and ecological information. Africa has been tectonically active since the late Oligocene and has evolved a highly varied topography and climate as a result. This activity has created high areas in the east and portions of the north and south, sustaining a variety of habitats, while maintaining low areas in the west and central parts of Africa (Grubb, this volume). Africa's variegated topography has promoted spatial heterogeneity in vegetation, with vulcanism enriching soils that support nutritious grasses (Owen-Smith, this volume; Sikes, this volume). Although Andrews and Humphrey (this volume) argue for an essentially unchanging pattern of ecological opportunity from the

Middle Miocene to Pliocene interval, we agree with Turner (this volume) as well as Vrba (1987) that the succession of faunas over this time (Bernor & Pavlakis, 1987; Turner & Wood, 1993a,b) most certainly reflects an evolution of habitats with concommitant shifts in vegetation physiognomy. Today Africa exhibits a mosaic of environmentally distinct biogeographic provinces which may be interconnected by topographic gradients (Grubb, this volume). Provincial boundaries would shift and endemism subside readily (Grubb, this volume; Grubb et al., this volume), as it undoubtedly was in the past, by the short-interval climatic changes so exhaustively illustrated by Vrba et al. (1995), Denton (this volume), and Vrba (this volume). Neogene short-interval climatic changes would appear to have facilitated intercontinental-scale migration events (Bernor & Lipscomb, 1995), which we agree with Grubb et al. (this volume) explicitly contradicts the necessity of a "primitive cosmopolitanism" in cladistic biogeographic reconstructions (Nelson, 1974).

The Hipparion record is one of the better ones for African Neogene mammals. Yet, as this study shows, this particular evolutionary record is far from being resolved. Our current understanding of African hipparion evolution suggests that the first founding population was derived from the Eurasian *Hippotherium primigenium* Group. There would appear to be some regional differentiation of this group with the more derived form, "*Hippotherium*" *africanum,* occurring in North Africa and, at least at entry, the more primitive form, *cf. Hippotherium primigenium,* occurring in East Africa. The East African record is documented up to the latest Miocene only. Referrals of most Late Miocene and all Pliocene occurrences of this group appear to be based on symplesiomorphic characters and are, as such, suspect.

The second founding population, the "*Sivalhippus*" Complex, appears to have occurred in Africa no later than 8 m.y. from South Asia. The African radicle of this clade, *Eurygnathohippus,* acquired primitive ectostylids in the permanent mandibular dentition before to the Late Miocene cladogenesis of a larger form, *Eurygnathohippus turkanense,* and a smaller form, *Eurygnathohippus "sitifense"*. A large form and a small form persist together at least until the later Pliocene Omo Shungura G levels. A larger form, which we refer to the *Eurygnathohippus "ethiopicus"–Eurygnathohippus cornelianus* lineage, then continues up to the middle Pleistocene, about 0.6 m.y. A smaller Pleistocene form may occur in the Middle Awash and at Olduvai, but its morphology and chronologic range are poorly known (R. L. Bernor, personal observation).

The evolutionary record of African hipparionine horses would appear to be relatively simple. But is it really? Do we simply have two immigration events, the former evolving into a single North African species and a single East African species? The skeletal material is not particularly abundant, and the study and analysis of this material is currently too meager to make such an interpretation. With regard to the *Eurygnathohippus* Group, there does seem to have been a single entry during the medial Late Miocene followed closely by a cladogenetic event which yielded a small and a large lineage. At face value it would seem plausible to accept the parallel, gradual evolution of these two groups. However, the larger form may have undergone a cladogenetic event producing an East African lineage, *Eurygnathohippus turkanense,* and, plausibly, a South African lineage, "*Eurygnathohippus*" *cf. baardi.* These two lineages may prove to be distinct based principally on postcranial anatomy. Thereafter, the East African Plio-Pleistocene larger *Hipparion* lineage most plausibly was not derived from *Eurygnathohippus turkanense* because postcranial material from Hadar, Ethiopia (*Eurygnathohippus hasumense*), and the Manonga Valley, Tanzania (*Eurygnathohippus cf. hasumense*), are distinctly more gracile in their build and would necessitate a major reversal in postcranial anatomy. The evolutionary relationship between the later Plio-Pleistocene forms *Eurygnathohippus" "ethiopicus", Eurygnathohippus cornelianus,* and the *Eurygnathohippus* small species from Olduvai are too poorly understood and require further study to interpret their phylogenetic relationships.

What is evident about the relationship between African hipparion evolutionary events and purported global abiotic (paleoenvironmental) events? Bernor and Lipscomb (1995) have reported a correspondence between the two hipparion group entries and global sea-lowering events. It seems likely that sea lowering expanded the continental shelves between South Asia and Africa and likely shifted continental climates to cooler conditions supporting multihabitat corridors between Eurasia and Africa. The latest Miocene Lothagam fauna exemplifies this observation by its inclusion of several distinctly Western Eurasian taxa in its fauna (M. G. Leakey et al., 1998).

With regard to Africa's Plio-Pleistocene *Hipparion* record, Hooijer (1974) has observed a significant increase in cheek-tooth crown height and ectostylid development at the Omo Shungura F (2.36 m.y.) level (first occurrence of *Eurygnathohippus "ethiopicus"*), just when *Equus* makes its first regional appearance. Figure 14-8 presents data on the evolution of crown height in African hipparionine horses: 14-8A includes data on the small hipparion lineage; 14-8B excludes this group to better present the trend in increased cheek-tooth maximum crown height in the larger species-lineage(s). To establish the primitive condition for Old World hipparions, we present maximum crown heights for Central European *Hippotherium primigenium.* For a total of 348 cheek teeth, the maximum crown height for Central European Vallesian hipparions is 49.8 mm, whereas that for the early Turolian assemblage of *Hippotherium primigenium* from Dorn Durkheim (Germany) is only 51.0 mm. The next step, "*Sivalhippus*" *perimense,* has a maximum recorded crown height on an unworn M^1 of 70.5 mm (AMNH 19476). Following this record, the highest recorded crown height for *Eurygnathohippus turkanense* from Lothagam is 57.9 mm for a worn crown (KNM-LT26295). Figure 14-8A depicts three populations of smaller hipparion, referred here to *Eurygnathohippus "sitifense,"* with the following crown heights: Lothagam (6.5–5.5 m.y.), 58.3 mm; Sibabi (Ethiopia; 4.2 m.y.), 58.0 mm; Ekora (Kenya, 4.0 m.y.), 55.0 mm. *Eurygnathohippus cf. baardi* from Langebaanweg has a maximum reported crown height of 75 mm (Hooijer, 1976). Next, the Hadar population has no less than six unworn upper cheek teeth (five M^{2}s and one M^1) with crown heights >70 mm, the maximum being 73.9 mm (Al315–2). It should be noted that the Hadar sample does not show a net increase in crown height through the section: the second highest crown height, 73.5 mm, is of an individual from the Kada Hadar 1 Member (AL288–19), and the third highest crown, 73.3 mm, is from Sidi Hakoma 2 (AL364–3).

There is a paucity of material readily attributable to *Eurygnathohippus afarense,* but we estimate its maximum crown height to have been 75 mm. Hooijer (1975) reported increased hypsodonty in Omo Shungura Member F (2.36 m.y.; Brown & Feibel, 1991). However, the maximum actual crown height he reported from this interval is 75 mm (L92–2a and L398–1182, P^3 or P^4). In the succeeding Omo Shungura G level (2.33 m.y.) the maximum recorded crown height is 84 mm (L627–89, slightly worn M^1 or M^2). Unworn teeth are not known from the succeeding Omo H and Omo J levels, but the youngest level with hipparions, Omo Member K (1.59 m.y.), has a reported maximum crown height of 89 mm (F203–2-10, 13, 14, 15; Hooijer, 1975). Our own study of the hipparion material referable to *Eurygnathohippus cornelianus* from Olduvai in the British Museum of Natural History includes the maximum crown height on early wear teeth, for each bed, as follows: Bed I, 77.3 mm (BMNHM1435a—M^2); Bed II, 78.6 mm (BMNHM25428—M^2); Bed III, 77.7 mm (BMNHM25426); Bed IV, 76.5 (BMNHM25522—M^2). We have thus far not found any basis for distinguishing different evolutionary stages within this taxon at Olduvai, and we estimate the maximum crown height for a P^4 or M^1 to have been about 85 mm. This is in accord with Hooijer's (1975) report for other populations of hipparion, such as Makapansgat, where crown heights are reportedly >80 mm, and even approached 90 mm in the Omo Shungura Member K.

Figure 14-8 illustrates two marked increases in crown height: the first realized in the "*Sivalhippus*" Complex at the "*Sivalhippus*" *perimense* stage of evolution, about 8 m.y. (note we expect maximum crown heights in *Eurygnathohippus turkanense* to have been about 65 mm). Thereafter, crown height did not increase substantially from 6.5 m.y. until 2.3 m.y. when we see an abrupt rise in crown height in the Omo Member G population to 84 mm. This marked increase in crown height is apparently a consistent characteristic of Late Pliocene–Middle Pleistocene East and South African hipparions. Moreover, *Eurygnathohippus cornelianus* had further evolved highly derived features of the masticatory apparatus including an elongated snout with a wide symphyseal gape, hypertrophied and procumbent I1–2, and atrophied I_3, which together with its high-crowned cheek teeth reflect a shift to short-grass feeding functionally convergent with the white rhinoceros, *Ceratotherium simum* (Owen-Smith, this volume).

The striking shift in *Eurygnathohippus cornelianus'* morphology correlates well with Vrba's (1988, 1995) prediction that many ungulate lineages responded to increased seasonality about 2.5 m.y. by evolving new morphologic and behavioral repertoires adapted for more seasonal environmental conditions. However, in that this apparent shift in hipparion feeding behavior occurred concommitant with the first occurrence of *Equus* in Africa, we cannot exclude Foley's (this volume) hypothesis that the "Red Queen Effect" may have forced *Eurygnathohippus cornelianus'* evolutionary origin by making it a more dedicated, low-grass grazer than *Equus.* To test the viability of either hypothesis, we need to develop detailed systematic records of several mammalian lineages across a long chronologic interval and determine whether there is congruence in these results. According to Foley (this volume), congruent (synchronic evolutionary change) results would support the "Turnover Pulse Hypothesis", and incongruent (asynchronic evolutionary change) results would support the "Red Queen hypothesis." As has been exemplified here, one cannot depend on an assembly of faunal lists for these tests; the quality of data is just too questionable at this time.

Conclusions

The Wenner-Gren Foundation symposium was convened to consider the potential for a paradigm shift in paleoanthropology away from a focus on taxonomic analyses limited to character similarities and differences toward a paradigm that interprets characters ecologically and adaptively in the larger context of their habitat specificities. We prefer to see this "shift" rather as a broadening of the multidisciplinary paradigm that was set in place with the founding of the Omo Research Project some 30 years ago. This paradigm, adopted by virtually every late Cenozoic Old World paleoanthropology field project, seeks an empirical geologic and paleontologic record for interpreting the environmental context of ape and human evolution. This record includes an independently developed geological and geochronologic framework within which a systematically based paleontological record is developed. Moreover, the systematic record is not limited to taxonomy, but derives considerable value when evolutionary reconstructions are made for a broad range of studies including, but not limited to, biostratigraphic/ biochronologic correlation, biogeographic reconstruction, and, as exemplified here, the ecomorphologic reconstruction of chronologically long and biogeographically extensive evolutionary histories. That the field has pursued new avenues of scientific inquiry such as the relationships between regional to global scale paleoenvironmental events and vertebrate evolutionary response, and the discovery and use of ecomorphologically sensitive characters for environmental interpretation, provides further justification for pursuing a rigorous empirical scientific framework within the existing multidisciplinary paradigm.

We have attempted in the short space available here to interpret our systematic and morphologic database of Old World hipparions from an ecomorphologic perspective. In so doing, we have been able to uncover inadequacies in the fossil record and attempted to frame our ecomorphologic interpretations as hypotheses requiring rigorous testing by independent functional anatomical, paleodietary microwear and various isotopic analyses. But, most important, we want to emphasize the necessity of pursuing these studies from within a rigorously constructed systematic framework and exemplify how pursuing such reconstructions using phylogenetically ordered lineages that may extend over millions years of time and intercontinental scale distances give a far more complete understanding of the processes underlying the adaptation of behavioral repertoires and species evolution.

Acknowledgments We thank Timothy Bromage, Friedemann Schrenk, and the Wenner-Gren Foundation for providing us with the opportunity to present our preliminary observations on African *Hipparion* evolutionary biology. We acknowledge the National Science Foundation, the Joint American-Hungarian Fund, the National Geographic Society, the L. S. B. Leakey Foundation, and the Alexander Von Humboldt Stiftung for funding R. L. Bernor's research on the evolutionary history of Old World hipparions over the last 20 years. The Staatliches Museum für Naturkunde, Karlsruhe, provided laboratory facilities and support during the last phases of this research. R. L. Bernor also acknowledges Heinz Tobien's (formerly of Mainz University) interest, enthusiasm, and work on *Hipparion* evolution and the significant role he has played in formulating our methodology. We dedicate this manuscript to him in memorium.

15

Suid Paleoecology and Habitat Preferences at African Pliocene and Pleistocene Hominid Localities

Laura C. Bishop

In this chapter I investigate the ways in which analysis of the ecological preferences of nonhominid fauna can contribute to reconstructions of the paleobiology of Pliocene and Pleistocene hominids. Hypotheses that link habitat with patterns of hominid behavior and evolution involve assumptions about the environments occupied by hominids. In many scenarios, environmental change, whether driven by global or local climate causes, leads to shifts in habitat availability, particularly toward a predominance of open grasslands. A growing body of literature documents global climate change during the Neogene (e.g., Shackleton et al., 1984; deMenocal, 1995; Denton, this volume).

Hypotheses connect climate change with numerous important milestones in the evolution of Hominidae, including the development of habitual bipedal locomotion, the expansion of relative brain size, the beginnings of stone tool use and the inclusion of animal protein in the hominid diet. Although such a causal relationship between climate change and evolutionary patterns has been criticized (Hill, 1987, 1995; Foley, 1994), it still remains a prevalent model in evolutionary theory. To evaluate these hypotheses, which can be traced back to Dart if not to Darwin, we must have direct evidence of the types of habitats available to and exploited by hominids during the Pliocene and Pleistocene and how these habitats changed through time.

Numerous methods, from geological to biological, have been used to investigate Plio-Pleistocene environments (see, e.g., Sikes, this volume). Arguably, the best approach would be direct analysis of the habitat preferences of hominids themselves (Ramirez Rozzi, Bromage & Walker, this volume). Unfortunately for paleoanthropologists, the fossil record for Hominidae is relatively sparse and poorly distributed in time and space, and some museums have shown a bewildering reluctance to let bits of these precious few fossils be ground up for geochemical analysis. For other less destructive forms of analysis such as comparative morphology, hominids suffer from a lack of modern relatives for comparison.

A proxy for hominid paleoecology is required, and pigs fit the bill. Suid fossils are far more common and numerous than are hominid fossils. In addition to similarities in their evolutionary histories, Suidae and Hominidae share ecological characteristics, such as an omnivorous diet and large body size (Hatley & Kappelman, 1980). These resemblances suggest that the study of the ecology and evolutionary history of Suidae might be interesting not only in itself, but also as a heuristic device for generating and examining hypotheses about hominid ecology and evolution. Moreover, because modern suids are more taxonomically diverse than modern hominids, some analytical tools and techniques can be applied to them that might aid in ecological reconstruction. Pigs have particular relevance to problems of human evolution and paleoecology that are difficult to address directly because of the scarcity of fossil hominids.

This chapter reviews the evolution and paleobiogeography of large African omnivores. Because the distribution of past taxa is an important part of their evolutionary histories, potential taphonomic reasons for differences in paleobiogeography will be reviewed. Finally, I present results of some recent research on the paleoecology of omnivores in East Africa and discuss their implications for southern African sites.

An Outline of Large Omnivore Evolution in Africa

A Brief History of Swine

The Old World pigs, the Suidae, underwent a large radiation in Africa during the Pliocene and the Pleistocene (Cooke, 1968; Harris & White, 1979). Throughout the end of the Miocene this artiodactyl family was represented by tetraconodont species, *Nyanzachoerus devauxi* and *Nyanzachoerus syrticus* (Cooke, 1987; Hill et al., 1992a). Early Pliocene assemblages are dominated by *Nyanzachoerus* and what is generally considered to be its descendant genus, *Notochoerus.* By the middle of the Pliocene suine pigs of more modern aspect, the genera *Metridiochoerus* and *Kolpochoerus,* were beginning to appear. Throughout the Pliocene, the number of suid species was increasing (T. D. White, 1985). The level of diversity continued to rise until the Early Pleistocene, despite the extinction of the last, largest, and most derived of the tetraconodonts, *Notochoerus scotti.* By the Middle Pleistocene, suid diversity had apparently decreased significantly. In East Africa it is uncommon to find extinct suid taxa in sediments younger than the Middle Pleistocene. Figure 15-1 summarizes the phylogeny of Plio-Pleistocene East African Suidae.

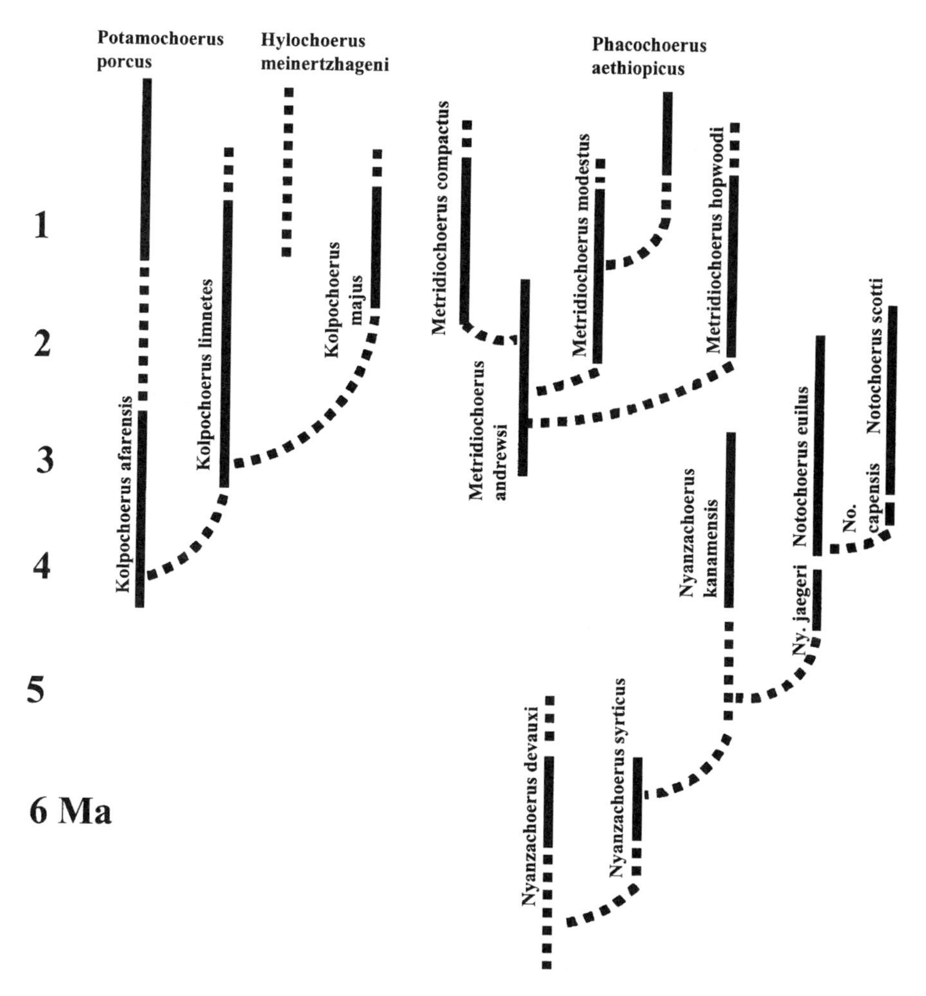

Figure 15-1. Consensus phylogeny for Suidae. Data from Harris and White (1979) and Cooke (1978, 1987).

The Hominidae

Hominidae exhibit a similar pattern to Suidae, apparently diversifying from the single identified Late Miocene-Early Pliocene species, *Ardipithecus ramidus* (T. D. White et al., 1994, 1995). Specimens from Lothagam and Tabarin in Northwestern Kenya predate the recent *A. ramidus* finds, but they have not yet been attributed to taxa (Hill, 1985; Hill et al., 1992a). Later, Early Pliocene hominids radiate into a more speciose group, with multiple species from each of the genera *Australopithecus, Paranthropus,* and *Homo* (Wood, 1992b). After more than 2 m.y. during which there are two or more species documented within Hominidae, the family seems to become monospecific in Africa by the Middle Pleistocene.

The period of maximum diversity for the Hominidae and the Suidae is roughly coincident, from 1.8–1.6 m.y. (Bishop, 1993, 1994, in preparation; T. D. White, 1995). For the pigs, however, this maximum diversity represents a coincidence of intrageneric speciation occurring over the interval from about 2.2 to 1.8 m.y. rather than the origins of new genera—variations on a theme rather than novel adaptations (see Vrba, this volume).

Species Distribution and Endemism of Omnivore Taxa

Neogene fossil sites that have yielded hominid remains are for the most part confined to two regions of Africa, the East African Rift Valley and southern Africa. Plio-Pleistocene sites in northern and southwestern Africa have yielded mammalian faunas, but hominid fossils have yet to be recovered from these regions. Two relatively recent discoveries have expanded the documented range of hominids. A hominid mandible from Malawi (Schrenk et al., 1993) has contributed to closing the geographical gap between the two main concentrations of fossil hominids; the site is still within the Rift Valley complex, although at its southernmost limit. The first Pliocene hominid fossils described from Chad extend the range of early hominids westward by approximately 2500 km (Brunet et al., 1995). Despite these recent discoveries, examination of hominid faunal communities during the Pliocene and Earlier Pleistocene is largely limited to comparisons between the two hominid-bearing regions in the eastern and southern extremes of the continent. The distribution of fossil hominid taxa is summarized in table 15-1.

The Hominidae of southern Africa are currently interpreted as being less taxonomically diverse than their East African cousins. Two Plio-Pleistocene hominids are represented in southern African cave sites, *Australopithecus africanus* and *Paranthropus robustus,* but conventional interpretations suggest that they do not occur synchronically or time successively at the same cave sites. Moreover, these taxa are known only from southern Africa and appear to be endemic there. Additional finds of early *Homo* come from Swartkrans; there is no agreement upon their taxonomic affinities, but current evidence suggests that they belong to a species known from East Africa. The East African endemic hominids are currently thought to be more speciose—one species of *Ardipithecus,* two of *Australopithecus,* two of *Paranthropus,* and two of *Homo.* Three hominid species of the genus *Homo,* including our own species *Homo sapiens,* are cosmopolitan in their African distribution (table 15-1).

A similar situation exists for Suidae. There are numerous cosmopolitan taxa, many East African endemics, and one southern African endemic, *Notochoerus capensis* (table 15-2). The endemic status of *N.*

Table 15-1. Distribution of African hominid taxa.

Cosmopolitan hominids	*Homo habilis*
	Homo erectus
	Homo sapiens[a]
East African endemic hominids	*Ardipithecus ramidus*
	Australopithecus anamensis
	Australopithecus afarensis
	Paranthropus aethiopicus
	Paranthropus boisei
	Homo rudolfensis
	Homo ergaster
Southern African endemic hominids	*Australopithecus africanus*
	Paranthropus robustus

[a]Extant taxon.

Table 15-2. Distribution of African suid taxa.

Cosmopolitan Suids	*Nyanzachoerus jaegeri*
	Nyanzachoerus kanamensis
	Notochoerus scotti[a]
	Metridiochoerus andrewsi
	Metridiochoerus modestus
	Metridiochoerus compactus
	Kolpochoerus limnetes
	Phacochoerus aethiopicus[b]
	Potamochoerus porcus[b]
East/North African endemic suids	*Nyanzachoerus syrticus*
	Nyanzachoerus devauxi
	Notochoerus euilus
	Metridiochoerus hopwoodi
	Kolpochoerus afarensis
	Kolpochoerus majus
	Hylochoerus meinertzhageni[b]
Southern African endemic suid	*Notochoerus capensis*[a]

[a]Species with uncertain distributions.
[b]Extant taxa.

capensis can be doubted, however, because specimens from East African localities now attributed to *N. scotti* and *N. euilus* have been assigned to *N. capensis* in the past (Harris & White, 1979; McKee et al., 1995). *Notochoerus capensis* is considered by workers who recognize it to be intermediate in morphology between *N. scotti* and *N. euilus,* making this issue difficult to resolve without further study and larger samples. Further difficulties in assessing levels of endemism arise from differences in the taxonomic identifications and phylogenetic interpretations offered by Harris and White (1979) and Cooke (1978).

Taphonomic Bases for Faunal Differences

There are numerous reasons—taphonomic, ecological, and geographical—that the species distributions of omnivores might be limited to particular regions or, conversely, be cosmopolitan. Of these classes of reasons, the one that has the most potential for biasing interpretation is the taphonomic differences. The ecological and geographical differences are the useful information which taphonomic biases make it difficult to ascertain. Some potential taphonomic reasons for these discrepancies may have their roots in historical, formational, and geological differences between the eastern and southern African sites (e.g., W. W. Bishop, 1980). All of the East African sites are open-air sites, with terrestrial environments represented and preserved, with the notable exception of Laetoli (Harris, 1985, 1987), by fluviatile and lacustrine sedimentary regimes. Many southern African Plio-Pleistocene sites are formed in limestone sinkholes or in other similar subterranean accumulations. These differences in geomorphology may explain some potential differences in the habitats represented because they represent different habitat types.

Differences between the collection methods used from site to site represent another possible taphonomic reason for differences between southern and eastern African faunas. Collection and excavation of the smaller, more localized southern African cave sites tended to be comprehensive, with sediment, often in blocks, removed from the site and prepared elsewhere. This leads to more complete recovery of the fauna represented. However, the southern African localities are much smaller than is typical of East African exposures, which tend to be more widely dispersed over the landscape. Fossils from East African sites have mostly been surface collected. The exception is specimens from Olduvai Gorge, where a program of archeological excavation has yielded large collections including thousands of faunal specimens (M. D. Leakey, 1971; Potts, 1988).

There are also differences in the chronological resolution of the two hominid-bearing regions. East African localities such as the Tugen Hills, the Omo, Hadar, the Turkana Basin, and Olduvai offer a relatively complete and geochronologically controlled temporal history of the Pliocene and Pleistocene epochs. Although recent dating methods offer some promise, it is difficult to evaluate the timing and duration of deposition in southern African deposits. It is therefore possible—indeed, likely—that the differences in species representation between the two regions are at least partially attributable to differences in temporal representation—periods of time in which certain faunal taxa are present might just be underrepresented in the southern African fossil record. This is another phenomenon that might be responsible for apparent differences between the faunas.

Consideration must also be given to the effects of observer bias. As mentioned above, different interpretations of the taxonomy and phylogeny of Suidae have been enunciated by workers who specialize in eastern (e.g., Harris & White, 1979) and southern Africa (Cooke, 1978, 1985). Several taxa that Cooke continues to recognize as endemic in southern Africa are collapsed into the larger range of temporal and morphological variation hypothesized by Harris and White (1979), primarily on the basis of the East African sample. Depending on which scheme is used, the interpretations of level of endemism of eastern and southern Africa changes considerably. I cannot evaluate these differences since I have, as yet, no first-hand ex-

Table 15-3. Habitat preferences for extinct suid taxa determined on the basis of partial skeletons.

Species	Habitat preference
Nyanzachoerus devauxi	*Intermediate*
Nyanzachoerus kanamensis	*Intermediate*
Notochoerus euilus	*Closed*
Kolpochoerus limnetes	*Intermediate*
Kolpochoerus majus	*Closed*
Metridiochoerus modestus	*Closed*

perience of the southern African forms. In contrast, although taxonomic schemes for the Hominidae differ between researchers, most workers would agree on the endemic status of *Australopithecus africanus* and *Paranthropus robustus* and confine arguments to issues of species diversity and specimen attribution.

Even given the numerous reasons, taphonomic and otherwise, that East African and southern African Plio-Pleistocene faunas might not be comparable, there is information that might be used to evaluate ecological and geographical differences between these ar-

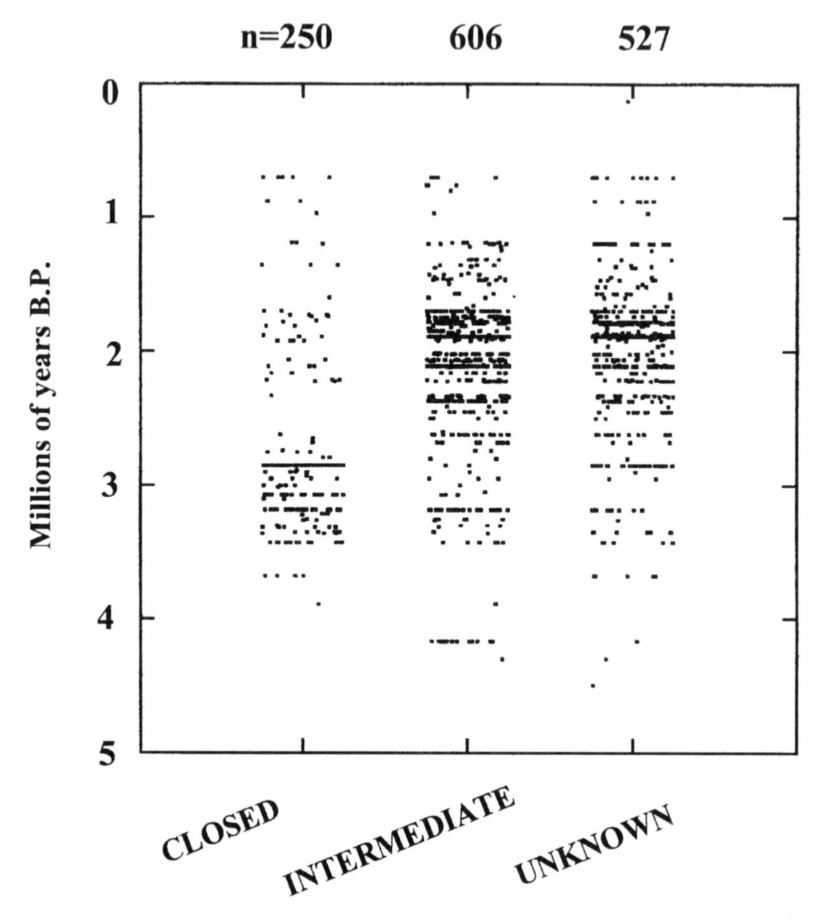

Figure 15-2. Habitat preferences of East African Suidae. Each dot represents an identified, well-dated occurrence (dental or cranial remains) of an extinct pig. The specimen information was combined with habitat preferences determined by this study. The result is a history of suid habitat preferences over the period from 5–1 m.y. Note that no partial skeletons were found to have open habitat preferences (see table 15-3). For numerous species there were no partial skeletons available; these are the occurrences in the "unknown" category. Even if all these taxa were found to have open habitat preferences, the conclusion that intermediate and closed habitats were available throughout the Pliocene and Pleistocene would remain valid. Sensitivity analysis demonstrates that, although there are significant differences in the distributions of closed and intermediate habitat taxa, fewer than 100 additional occurrences of closed-habitat animals negate any statistical significance in *t*-test results.

eas. Knowledge of the ecology of other omnivores might help us understand any differences between the ecology of eastern and southern African hominids during the Pliocene and the Pleistocene.

Ecomorphic Studies of Faunal Autecology

One of the avenues of inquiry into hominid paleoecology is the study of autecology of various other mammalian species found with fossil hominids.[1] Paleoecological studies are often based on uniformitarian principles. One particular level of actualism frequently applied is to assume that the ecological parameters of modern species provide proxies for those of their extinct relatives. This is perhaps the simplest assumption, but it masks differences between the past and the present and makes it difficult to identify unique adaptations to past environments or to investigate the ecology of animals without modern relatives and restricts adaptive scenarios to those which are extant. The applicability of this method to more remote time periods is doubtful because the relationships between ecology and phylogeny in the past are potentially different from those of the present.

An alternative approach is to use more basic levels of analogy that can be presumed to have not changed between the present and the past, those of functional anatomy and biomechanics. Several studies have already indicated the usefulness of this approach for artiodactyls and carnivores (e.g., Van Valkenburg, 1987; Kappelman, 1988, 1991; Bishop, 1994; Plummer & Bishop, 1994; Kappelman et al., 1997). The relationship of locomotor anatomy to substrate type can be explored in modern animals. The study of ecomorphology can then be extended to the fossil record to help interpret the habitat preferences of extinct taxa.[2] Although this method also relies on extension of present conditions into the past, it makes the not unreasonable assumption that the principles of mammalian biomechanics are unlikely to have changed between the past and the present. Conditions such as the distribution of vegetational types across a landscape are the variables we seek to examine, and therefore their constancy should not be assumed when seeking to reconstruct past environments. Instead, the application of ecomorphological methods allows us to test hypotheses about the relationship of taxa to habitats in the past without being constrained by the suites of adaptations in contemporary faunas.

Results of an ecomorphic study of antelope metapodials from Olduvai Gorge suggest that, even as recently as the Plio-Pleistocene, bovid taxa might have had different ecological preferences from those of their modern descendants (Plummer & Bishop, 1994). *Antidorcas recki,* a common Olduvai bovid, is a close relative of *A. marsupialis,* the springbok, which has a modern distribution confined to open grassland environments in southern Africa (Ansell, 1971; Dorst & Dandelot, 1970). *A. recki* had been reconstructed previously as preferring grassland habitats on the basis of analogy with its modern relative (Gentry & Gentry, 1978a,b). However, ecomorphic analysis of metapodial functional anatomy suggests that *A. recki* preferred a more intermediate, perhaps bushland or swamp, habitat type. Because *A. recki* is so common at Olduvai, a difference in its habitat preference significantly alters our picture of hominid paleoenvironments around the paleolake.

The deviation of *A. recki* from the modern antilopine preference for open habitats also casts doubt on the accuracy of the Alcelaphine-Antilopine criterion (AAC) for estimating relative amounts of tree cover (Vrba, 1975; Kappelman, 1984, 1986; Plummer & Bishop, 1994; Kappelman et al., 1997). This method draws parallels between the percentages of alcelaphine and antilopine antelopes in modern bovid faunas associated with particular levels of vegetation cover and extends this to paleofaunas. Although this method undoubtedly has some utility, the fact that recent work finds a different habitat preference for past members of a common antilopine group that today prefer open grassland suggests that it may prove misleading in paleohabitat reconstructions. Ecomorphic studies may also affect how we interpret and reconstruct the distribution of biotic zones in the past (see Grubb, this volume).

In addition, ecomorphic studies of postcrania decouple the ecological variables of habitat preference and diet. Because teeth are so common in the fossil record, phylogenetic reconstructions are often made on the basis of dental morphology, which is linked to diet as well as to phylogenetic relationships. Assumptions about the diets and habitats of modern animals are therefore extended into the past, such as is the case for hypsodonty in suids. Previously, interpretations of site paleoecology based on pigs have focused on dental adaptations, particularly the development of high-crowned, continuously growing (hypsodont) third molars, a common and reoccurring theme in the evolution of African Suidae (Harris & White, 1979). The sole modern hypsodont suid, the warthog (*Phacochoerus aethiopicus*), is a grassland specialist that eats grass almost exclusively (Cumming, 1975; Kingdon, 1971–1982). This observation led many researchers to equate the presence of hypsodont suids with the predominance of open grassland habitats (e.g., Harris, 1985; Cooke, 1985). The remainder of this chapter examines whether the relationship between hypsodonty and grassland habitat preference can be extended into the past.

Materials and Methods

Morphological characters that relate to habitat preference are detectable in the postcrania of modern pigs (Karp, 1987; McCrossin, 1987; Van Neer, 1989). An ecomorphic analysis was therefore performed on extant pigs and peccaries to identify characteristics that would allow their separation into distinct habitat preference groups (Bishop, 1994). Investigations on fossil suid postcrania have also been carried out to establish whether data from these omnivores can be used to reconstruct the environments available to synchronic and sympatric hominid species (Bishop, 1994; 1995, in preparation).

To perform this type of analysis, some simplification and grouping of modern habitat types is necessary. This seems desirable when examining past ecosystems because a realistic view should be taken of the loss of information between the past and the present to avoid making unjustified claims about the precision of these reconstructions. This pitfall can be avoided in analyses that rely on a broader spectrum of fauna (Andrews & Humphrey, this volume).

In this analysis of suid autecology, as in previous similar studies, the spectrum of habitat types (see Pratt et al., 1966) has been divided into three categories: open, intermediate, and closed. These correspond, respectively, to grasslands, mixed vegetation or bushland, and forest—categories based on the approximate proportion of tree cover and woody dicotyledonous plants. This categorization is a rudimentary framework to facilitate this study (see, e.g., Owen-Smith, this volume). However, the full complexity of modern environments would be useless as a framework for this type of study, which relies on statistical analysis. The gross categorization of habitats into three types, while obviously blurring some distinctions, is appropriate to a more gradistic approach to both paleontological and phylogenetic research and might be more appropriate to the scale of resolution inherent in historical data (Collard & Wood, this volume).

Ecomorphic characters were quantified using a sample of 64 extant pigs and peccaries, and this information was used to examine habitat preference in extinct swine (Bishop, 1994, in preparation). The ecomorphic characters of postcrania which are used to identify habitat preference in fossil suids are linked, in modern pigs, with biomechanical requirements that relate to joint stability/mobility in substrates of varying complexity. Thus a categorization of habitats based on vegetation density and degree of tree cover is appropriate, but it carries dangers of oversimplification.

Modern pig and peccary ecomorphic characters were quantified in a series of dimensionless ratios for each postcranial element. Ratios were regressed against body weight, and those that correlated highly with body weight ($R^2 > .30$) were eliminated from the analysis. This was necessary to ensure that classifications were due to habitat preference rather than to body weight, which is an important ecological variable (Wilson, 1975; Schmidt-Nielsen, 1984). Stepwise discriminant function analysis, using the STEPDISC procedure of SAS software (SAS Institute, 1985), was used for feature selection to reduce intercorrelation and to create a smaller variable list (James, 1986). A quadratic discriminant function analysis (DFA) was then performed for each skeletal element. When, as for this analysis, there are significant differences between covariance matrices of the groups to be compared, quadratic rather than linear DFA is appropriate (Morrison, 1976; James, 1986). Reclassification success rates were high (83–100%) for modern complete postcranial bones, testifying to the usefulness of this method.

Although DFA studies of ecomorphic characters can be considered "taxon-free" in the sense that unknowns are analyzed without reference to their species identifications, it is important to note that at many levels such analyses do depend on a species concept (for a discussion of species concepts in paleontology, see Turner, this volume; Foley, this volume). For example, the choice of which groups of modern taxa to include in each habitat preference category is based on modern ethological literature for each included species. Also, the determination of morphological characters linked to substrate use in modern fauna is based on their identification in individual modern species. At one level, then, the species concept is an implicit part of taxon-free analysis.

Fossil bones are rarely complete, however, so discriminant functions were derived in order to use fragments of limb bones that preserve some locomotor anatomy on the epiphyses. These partial DFAs were tailored to the portion of the fossil represented. The discriminant functions for complete and partial bones were used to separate extinct suids into habitat groups based on locomotor morphology. Results for a sample of 34 postcrania from 8 partial skeletons allocated to 6 different species are presented here. This is a relatively small number of specimens, but it is the largest sample available for analysis here. For these specimens, habitat preference data were examined in relation to taxonomic and temporal information to determine whether shifts in ecology were correlated with taxonomic and dietary change. In several cases, results produced by this method are in contrast with previous reconstructions of habitat preferences, which have in general been linked to dental adaptations and level of hypsodonty (L. S. B. Leakey, 1958; Cooke, 1985; Harris, 1985, 1987).

Results

The most primitive suid in this study, *Nyanzachoerus devauxi,* is thought to be the ancestor of subsequent *Nyanzachoerus* and *Notochoerus* species. Postcrania from this taxon were determined to have morphology indicative of an intermediate (bushland) habitat preference. Another species of the genus, *N. kanamensis,* has a broad temporal and spatial distribution on the African continent (Bishop, 1997). A partial skeleton of this taxon is known, and its postcrania indicate that they preferred an intermediate habitat. Some discrepancies exist in the results between skeletal elements, however. Known *N. kanamensis* metatarsals have functional anatomy indicating an adaptation to open habitats, whereas forelimb and more proximal elements indicate a preference for intermediate or closed habitats. Observations on bovid postcrania and results for the modern babirusa, *Babyrousa babyrussa,* indicate that this apparent discrepancy may be linked with locomotor requirements for a swampy environment (Scott, 1985; Bishop, 1994). The results for *Nyanzachoerus kanamensis* postcrania suggest that they might have been swamp specialists.

The known, associated postcrania of two taxa that possess high-crowned cheek teeth (*Notochoerus euilus* and *Metridiochoerus modestus*) have been found to have morphological characters associated with preference for closed habitats. Different postcranial adaptations for earlier and later *Kolpochoerus limnetes* may indicate a gradual change in habitat preference for the species, from closed toward more mixed-country habitats. A relatively late example of *K. limnetes,* KNM-ER 4576, which has an extremely elaborated third molar, has postcrania that are classified with intermediate and closed habitat forms. Despite the advanced level of hypsodonty shown by later examples of *K. limnetes,* there is no indication of postcranial adaptations for open habitat exploitation or cursoriality.

A partial skeleton of *Kolpochoerus majus* has been examined, and the limb bones possess ecomorphic characters indicative of a preference for closed, forested environments. Relative to the modern bushpig *Potamochoerus porcus, K. majus* is hypsodont, which has been linked to a more open environment by several researchers. It appears that, for Pliocene and Pleistocene suid species, hypsodont dentition cannot be unequivocally linked with a preference for open habitats. In fact, none of the partial skeletons examined by this study were classified into the open habitat preference group (table 15-3).

When the habitat preferences determined for particular taxa are combined with their known, well-dated occurrences in East Africa, a chronological picture of past habitat preference in the Suidae results (fig. 15-2). Its most important implication is the presence, throughout the period between 4.5 and 0.5 m.y., of a substantial number of suids that preferred the more heavily vegetated habitat types. No conclusions can be drawn about the presence or absence of suids that preferred open habitats because no taxonomically identified partial skeletons were reconstructed as having a preference for this habitat type.

Even if all the taxa for which no partial skeletons are known preferred open habitats, at no time between 4.5 and 0.5 m.y. would suids with open habitat preferences numerically dominate combined or individual site assemblages. This suggests that a mosaic of habitats was always present during the Plio-Pleistocene (see Grubb, this volume). Despite the increase in hypsodonty level, which is especially noticeable with the radiation of the genus *Metridiochoerus* (considered ancestral to the modern warthog), the proportion of intermediate- and closed-habitat animals through time is relatively consistent (see caption of fig. 15-2 for further details). The most primitive species, *M. andrewsi,* first appears in the fossil record at approximately 3.4 m.y., while more advanced species are first detected approximately 2.2 m.y. and dominate many fossil assemblages thereafter. The postcranial evidence suggests that at least the smallest of these hypsodont species, *Metridiochoerus modestus,* favored closed habitats. The habitat preferences of the other *Metridiochoerus* species are still open to speculation because no postcrania have been attributed to the other species of the genus.

Discussion

Current results suggest that suids that preferred intermediate and closed habitats were consistently present at Pliocene and Pleistocene hominid sites in East Africa. This in turn suggests that these types of habitats were available at or near these sites. This result is in agreement with the conclusions of a growing number of paleoecological studies that use stable carbon isotopy to investigate the vegetational structure of past environments, in particular the presence of tropical grasslands (e.g., Cerling, 1992a; Kingston et al., 1994; Sikes, 1994, this volume). The variety of suid habitat preferences found at Pliocene and Pleistocene paleontological sites supports the conclusions of these studies that habitat structure in the past was different from that of East Africa today.

Grasses, although present during the Pliocene and earlier Pleistocene, are suggested here to have been spatially distributed in a different way from that found in modern East African grassland-dominated ecosystems. One study of paleosol carbonate isotopy on samples from Olduvai Gorge suggests the presence of a significant amount of tree cover and nongrass C_4 vegetation at some Plio-Pleistocene archeological sites (Sikes, 1994). Grassland-dominated ecosystems may not have been prevalent in East Africa until relatively late, perhaps not until the Middle Pleistocene (Cerling, 1992a; Kingston et al., 1994). Tropical grasses were present, but they were most likely distributed as grassy woodlands rather than as open grasslands with little woody vegetation. A sparser distribution of tropical grasses might explain why Pliocene and Pleistocene suids with dental hypsodonty (usually considered to be an adaptation for eating grass) do not exhibit postcranial adaptations to open habitats and cursoriality.

Another possible explanation, not favored here, is that environments opened up significantly during the Plio-Pleistocene, but there is a lag between environmental change and certain morphological characters. This argument would state that the teeth, representing dietary adaptations, were more rapidly tracking environmental change than were postcrania, representing locomotory adaptations with the result that a hypsodont dentition would be visible in the fossil record of the suids before postcranial adaptations to more open environments (see Plummer & Bishop, 1994). An adaptationist paradigm requires that the organism be well suited to its current environment, however, and because there is evidence of morphological change in the postcrania studied here, there seems no reason to invoke this sort of special pleading.

It could also be argued that the association between hypsodonty and grass eating might not be so strong for pigs. It seems likely that for suids the development of hypsodonty might represent a response to pressures that are not primarily dietary. In most large ungulates, hypsodonty occurs in all postcanine teeth. However, in modern and extinct suids hypsodonty is mostly confined to the third molar. This tooth is the last to erupt and develops closer to the animal's maturity. Although most ungulates have singleton births of precocial offspring, suids have large litters of altricial offspring. These small juveniles lack the jaw depth necessary to develop large hypsodont teeth until relatively late in life. This suggests the possibility that, for ontogenetic reasons, suids might not have a clear correlation between abrasive, herbivorous diets and hypsodonty. This is a speculative hypothesis that remains untested at present, but it is clear that hypsodonty in suids is different in both ontogeny and morphology from that in other ungulates.

Results of this study indicate that all habitat types are being sampled by the fossil record of East Africa during the Pliocene and Pleistocene. Although data are sparse before 2 m.y., this study provides compelling evidence that suids with closed and intermediate habitat preferences were present in East Africa from 4.3 to 0.7 m.y. While various evidence has been used to support the hypothesis that grassland-dominated ecosystems were becoming pervasive in East Africa, the presence of animals with more closed habitat preferences does not appear to decrease markedly. The extent to which this might be a bias of the predominantly fluvial sedimentary regimes has not yet been fully explored. The presence of closed- and intermediate-habitat adapted animals remains constant, even after 2 m.y. This leads to the conclusion that a variety of habitats were consistently present through this time and were available to be exploited by the hominids, which were sympatric and synchronous with the suids.

As yet, I have been unable to study and measure many of the specimens from southern African localities, so I am unable to draw any direct conclusions about the evidence from those sites. However, if the ecological preferences are the same in East African and southern African members of the same species, some indirect observations are possible based on taxa for which partial skeletons have been examined. *Nyanzachoerus kanamensis* is reported from Langebaanweg, *Kolpochoerus limnetes* from Cornelia and Elandsfontein, and *Metridiochoerus modestus* is known from Kromdraai, Swartkrans, and Cornelia (Harris & White, 1979; Cooke & Hendey, 1992; McKee et al., 1995). These taxa are reconstructed as preferring intermediate and closed habitats, suggesting the presence of these habitat types within proximity of the hominid localities. None of the southern African suid taxa that have been analyzed are reconstructed as preferring more open habitats. Although these results are only partial and preliminary, they are in contrast with taxon-linked ecological studies of fossil bovids, which suggest the presence of extensive grasslands at or near the southern African hominid sites (Vrba, 1975, 1980b).

The results of a habitat preference study for one group of large-bodied omnivores have implications for hypotheses about foraging in the hominids. Models that advocate a theory of hominid foraging based on exploitation of any single habitat zone (such as riparian woodlands; e.g., Blumenschine, 1987) are not supported by the results of this study. Numerous lines of evidence indicate hominid involvement in the collection and processing of animal remains at Olduvai Gorge (Potts, 1988). There, the presence of pig postcrania from suids preferring a variety of habitats suggests that the full spectrum of habitat types was available for exploitation by hominids (Bishop, 1994). The same conclusion was reached in a study of bovid metapodials from Olduvai (Plummer & Bishop, 1994). This result is equally valid whether suid bones are considered as hominid food remains or simply as indicators of habitats, favorable to omnivore exploitation, which were locally available to contemporary and sympatric hominids.

It is clear that multiple species of hominid and suid coexisted in the same place at the same time and that these were associations of long duration. Ecological theory predicts that in order to survive sympatrically, these otherwise similar species must have relied on different life strategies (e.g., Hardin, 1961). The suggestion made here, for the pigs, is that they did this by exploiting a diversity of habitat types that were distributed patchily but adjacently in space, if not in time. This suggests that habitats at Plio-Pleistocene hominid sites were distributed differently than they are today, and that savanna ecosystems, used in the broadest sense (Owen-Smith, this volume) might have been more heterogeneous in the past.

Habitat preference is an important aspect of paleoecology. This analysis of the habitat preferences of extinct suids based on postcranial morphology has resulted in conclusions that were not apparent from examining dental adaptations or from phylogenetic relationships as deduced from tooth morphology. The present study demonstrates that there is a wealth of paleoecological information in the postcranial remains of nonhominid animals, analysis of which yields interesting and interpretable results bearing on past habitats. Collecting postcranial remains of nonhuman species is important, and their analysis has the potential to advance our understanding of the ecological context of hominid adaptation and evolution.

Acknowledgments I thank Tim Bromage and Friedemann Schrenk for inviting me to join this enlightening conference. I also acknowledge the support of the Wenner-Gren Foundation, which has generously funded the research upon which this contribution was based as well as my participation in the conference. Funding for this research has also been provided by the Leverhulme Trust, the National Science Foundation, the W. W. Bishop Memorial Trust, the Boise Fund, and Yale University. The staffs of numerous museums made collections available for study. Research permission was granted by the governments of Kenya and Tanzania. Tom Plummer, Bernard Wood, Ray Bernor, Alan Turner, Rob Foley, Simon Fear, Mark Collard, and Mikael Fortelius contributed comments and suggestions, for which I thank them.

Notes

1. Use of the term "autecology" here refers to the ecology of the individual species in relation to its environment.
2. "Ecomorphology" is used here as the correspondence between ecology and morphology.

16

Of Mice and Men

Evolution in East and South Africa during Plio-Pleistocene Times

Christiane Denys

Mice and men are often associated in Plio-Pleistocene sites of Africa. Mice belong to the most abundant order of mammals (Rodents) and are representatives of the most successful family of tropical rodents, the Muridae. In addition to the Muridae, tropical Africa has an impressive record of endemic rodent families and genera. Some of the representatives of Muridae have followed humans, like *Rattus* and *Mus*. It is less widely known that in Africa some other Muridae, such as *Mastomys* and *Arvicanthis,* are commensal. In western Europe and in North America important and extensive changes in the distribution of rodents during the last glaciations have been well described. In Africa, neither these events, their causes, nor their passive versus active dispersal have yet been established.

For the Plio-Pleistocene faunas of East and South Africa, Denys (1989d, 1990a) has discussed an important distribution change in *Jaculus* and some murid species. Kowalski et al. (1989) showed an important shift toward the north in the distribution of savanna rodents, birds, and reptiles at around 120,000 years BP. The research on the dispersal of rodent taxa was not exhaustive and was limited to Late Pleistocene times or to the Holocene of South Africa. The suggested causes of dispersal events have been discussed by Foley (this volume) and Vrba (1992, this volume). According to Vrba, the habitat theory suggests that climate and habitat specificity play a prominent role and minimize interspecific competition and its effects. During climatic cold events, cold-adapted species increase their distribution areas, whereas this is reversed during hot periods. Foley (this volume) developed the hypothesis that competition as well as increase in population size are the motors of dispersal. Cases of modern geographical extension of wild rodent populations are not yet described, and to better understand the nature, frequency, and origin of the dispersal events found in rodents of tropical Africa during Plio-Pleistocene times, their influence on speciation–extinction events, a general review of the modern and fossil faunas in the context of a biogeographical approach is presented in this chapter. The establishment of the biogeographical affinities of the fossil rodent faunas is of great importance to understanding speciation and extinction events and to the characterization of the general habitat specificity of some larger mammals or of extinct genera such as some species of hominids.

Presentation of the Modern and Fossil Rodent Faunas of East and South Africa

Modern Rodent Faunas of the Biotic Zones of Africa

Rodents are found in all types of habitats and vegetation in Africa. Using a transect from the North to the South of Africa, the main characteristics of the different rodent assemblages associated with the vegetation regional categories are presented. Following Menaut

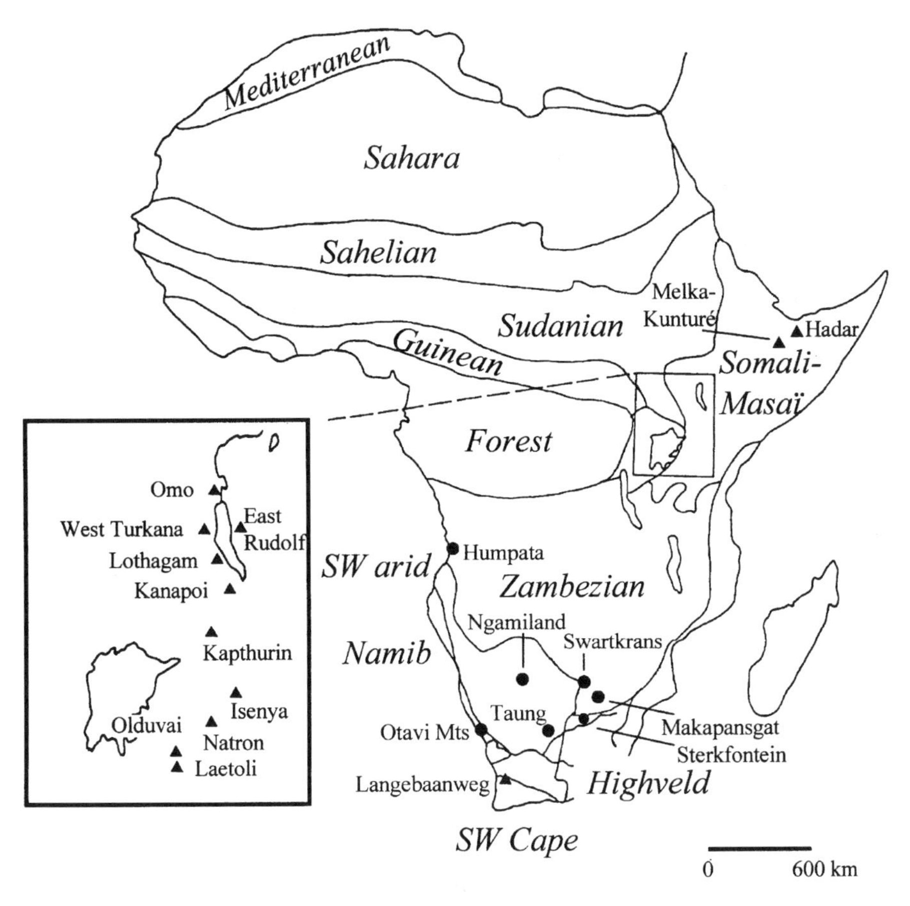

Figure 16-1. Main vegetation zones of Africa (after Menault, 1983) and geographic localization of the mentioned Plio-Pleistocene sites.

(1983) and F. White (1986), there are at least 12 main vegetation zones for Africa (fig. 16-1, table 16-1). This classification is broad and does not take into account the local environmental changes due to the presence of mosaic habitats related to soil nature or topography, as presented by Owen-Smith (this volume) and Cole (1986). This classification depends on plant endemism and reflects broadly the precipitation patterns although in some regions the evapotranspiration factor influences primary production. Grubb et al. (this volume), based on mammal (mainly large mammals) distribution patterns defined by previous author, finds at least 18 different biotic zones. Because most of the rodents are primary consumers, it is more appropriate here to use the vegetation distribution patterns. The faunal lists of small mammals have been compiled from the literature by country or by region (Perret & Aellen, 1956; Setzer, 1956; Petter, 1967, 1973; Rosevear, 1969; Petter & Genest, 1970; Kingdon, 1974; Koch, 1978; Osborn & Helmy, 1980; Gautun et al., 1985; Aulagnier & Thévenot, 1987; Happold, 1987; Ansell & Dowsett, 1988; Kowalski & Rzebik-Kowalska, 1991; Duplantier & Granjon, 1992; Yalden & Largen, 1992). When a country is crossed by different vegetation zones, especially in West Africa, the faunal lists are different.

Africa has the most extensive desert zone of the world, the Sahara, were the annual rainfall is <250 mm and in some places <50 mm. Farther south is a region of transition, the Sahel, which receives <500 mm annual rainfall with the rain season changing from one year to the other. The Sahel is composed of subdesert steppe with some scattered *Acacia* and *Commiphora* scrub. It is adjacent to the Sudanian zone, a narrow band of savannas beginning in north Senegal and ending at the bottom of the high Ethiopian plateau. It is 500–700 km wide, with a low altitude (<750 m). The dry season is well marked, with no frost, mean annual temperature between 24° and 28°C, and mean annual rainfall between 500 and 1000 mm. It is a dry

Table 16-1. Modern rodent generic composition of the zoogeographic regions of Africa (references in text).

Sahara (<250)[a]	Sahelian (250–500)	Sudanian (500–1000)	Guinean (1000–1500)	Forest (1500+)	Montane forest (1500+)	Regional mosaic of Lake Victoria (1000–1500)	Somali-Masai (<500)	Zambezian[b] (500–1500)	Highveld SSG[c] (500–1000)	Namib (<250)	Kalahari-SW arid (<500)	Cape (250–1000+)
Gerbillus	*Desmodilliscus*		*Taterillus*	*Anomalurus*	*Anomalurus*	*Anomalurus*	*Gerbillus*			*Tatera*	*Tatera*	*Tatera*
Meriones	*Gerbillus*	*Gerbillus*	*Tatera*	*Idiurus*		*Tatera*	*Tatera*	*Tatera*	*Tatera*	*Gerbillurus*	*Gerbillurus*	*Gerbillurus*
Pachyuromys	*Tatera*	*Tatera*		*Zenkerella*			*Taterillus*			*Desmodillus*	*Desmodillus*	
Psammomys	*Taterillus*	*Taterillus*	*Heliosciurus*	*Heliosciurus*	*Heliosciurus*	*Heliosciurus*	*Ammodillus*					
				Funisciurus	*Funisciurus*	*Funisiciurus*	*Lophiomys*		*Mystromys*	*Petromyscus*	*Petromyscus*	*Mystromys*
		Heliosciurus		*Protoxerus*	*Protoxerus*	*Protoxerus*	*Xerus*	*Paraxerus*			*Funisciururs*	
	Xerus	*Xerus*	*Xerus*	*Xerus*	*Paraxerus*	*Paraxerus*	*Paraxerus*	*Xerus*	*Xerus*		*Xerus*	
						Saccostomus	*Beamys*					
		Cricetomys	*Cricetomys*	*Cricetomys*	*Cricetomys*	*Cricetomys*	*Cricetomys*	*Cricetomys*		*Saccostomus*	*Saccostomus*	*Saccostomus*
		Steatomys	*Steatomys*	*Deomys*	*Deomys*	*Deomys*	*Saccostomus*	*Saccostomus*		*Steatomys*	*Steatomys*	
				Prionomys	*Dendromus*	*Steatomys*	*Dendromus*	*Steatomys*	*Steatomys*		*Dendromus*	*Dendromus*
				Dendroprionomys			*Megadendromus*	*Dendromus*	*Dendromus*		*Malacothrix*	
									Malacothrix	*Otomys*	*Otomys*	*Otomys*
					Otomys	*Otomys*	*Otomys*	*Otomys*	*Otomys*	*Parotomys*	*Parotomys*	
Mus	*Mus*	*Mus*	*Mus*	*Mus*		*Mus*	*Mus*	*Mus*		*Mus*	*Mus*	*Mus*
	Arvicanthis	*Arvicanthis*	*Arvicanthis*	*Arvicanthis*			*Muriculus*		*Mus*			
Acomys	*Acomys*	*Acomys*	*Acomys*				*Acomys*	*Acomys*				*Acomys*
				Lophuromys	*Lophuromys*	*Arvicanthis*	*Arvicanthis*	*Aethomys*	*Aethomys*	*Aethomys*	*Aethomys*	*Aethomys*
		Dasymys	*Dasymys*	*Dasymys*	*Lamottemys*	*Aethomys*	*Aethomys*	*Dasymys*				*Dasymys*
			Aethomys	*Hybomys*	*Hybomys*	*Dasymys*	*Dasymys*	*Grammomys*				*Praomys*
		Grammomys		*Hylomyscus*	*Hylomyscus*	*Pelomys*	*Grammomys*	*Pelomys*				*Myomyscus*
		Lemniscomys	*Lemniscomys*	*Malacomys*	*Malacomys*	*Lemniscomys*	*Lemniscomys*	*Lemniscomys*		*Rhabdomys*	*Rhabdomys*	*Rhabdomys*
	Mastomys	*Mastomys*	*Mastomys*	*Mastomys*		*Mastomys*	*Mastomys*	*Mastomys*	*Mastomys*		*Mastomys*	
				Oenomys	*Oenomys*	*Praomys*	*Rhabdomys*	*Rhabdomys*	*Rhabdomys*			
			Praomys	*Praomys*	*Praomys*	*Oenomys*	*Oenomys*					
		Myomys	*Myomys*	*Stochomys*		*Grammomys*	*Praomys*	*Uranomys*				
				Thamnomys	*Thamnomys*	*Thamnomys*	*Thallomys*	*Thallomys*	*Thallomys*	*Thallomys*	*Thallomys*	
			Uranomys	*Uranomys*		*Lophuromys*	*Zelotomys*	*Zelotomys*			*Zelotomys*	
				Colomys	*Colomys*	*Colomys*	*Colomys*					
Jaculus	*Jaculus*	*Graphiurus*	*Graphiurus*	*Graphiurus*	*Graphiurus*	*Mylomys*	*Stenocephalemys*	*Graphiurus*			*Graphiurus*	*Graphiurus*
Eliomys						*Uranomys*	*Desmomys*		*Graphiurus*			
Ctenodactylus			*Thryonomys*	*Thryonomys*		*Zelotomys*	*Pelomys*	*Thryonomys*	*Thryonomys*	*Pedetes*	*Pedetes*	
Massoutiera	*Felovia*	*Felovia*					*Graphiurus*	*Pedetes*	*Pedetes*		*Cryptomys*	*Cryptomys*
							Thryonomys	*Cryptomys*	*Cryptomys*		*Georychus*	*Georychus*
						Graphiurus	*Pedetes*	*Heliophobius*			*Bathyergus*	
							Cryptomys			*Petromus*	*Petromus*	
						Tachyoryctes	*Heterocephalus*					
							Tachyoryctes					
						Thryonomys	*Pectinator*					
Diversity 10	11	17	19	30	19	30	38	25	16	13	24	16
Gerbillinae 4	4	4	2	0	0	1	4	1	1	3	3	2
Murinae 2	4	8	11	17	9	16	18	12	5	4	6	7

[a]Numbers in parentheses indicate average annual rainfall, in millimeters.
[b]SSW and SSG (South Savanna Woodland and South Savanna grassland).
[c]Temperate subtropical grassland.

woodland savanna with local riverine and swamp forest as well as some steppic patches. In the valleys are some grasslands, but there are few bush or thicket areas compared with the Zambezian savanna zone. The adjacent zone is called the Guinean savanna, ranging from Senegal to Uganda. It reaches the Atlantic ocean in the Dahomey gap, which separates the forest into two blocks. The average annual rainfall is between 1000 and 1600 mm. The Sahelo-Sudanian-Guinean zones correspond to the northern savannas of Africa.

In West and Central Africa, the tropical evergreen forest is at a low to medium altitude. This region receives between 1600 and 2000 mm annual rainfall on average and is home to 25% of the rodent species of Africa (Dieterlen, 1989). The montane forest has a slightly different vegetation community. The next two regions are interesting because they correspond to the place where the different vegetation zones meet. In the west is the regional mosaic of Lake Victoria and in the east the Somali-Masai endemic and the coastal regional mosaic of Zanzibar-Inhabame. This region is peculiar, for it is situated north of the equator in the east of Africa. It is centered around the eastern Rift Valley region, which is made up of a mosaic of environments and ends in the center of Tanzania in the south. In the north, there is the subdesert steppe of the Somali coast and the Afar depression. At the bottom of the Rift Valley and on the medium-altitude plateau, the vegetation is mainly composed of grass steppe, thickets, and dry woodland savanna. On the Ethiopian Highlands, above 2000 m, there is the montane ericaceous belt and Afroalpine (above 3000 m) vegetation. The ericaceous belt is composed of moorlands of arboreal *Erica* (up to 12 m high) and the alpine belt (<4 m high) is characterized with giant *Senecio* and *Lobelia*. This vegetation occurs at the top of major volcanoes of the rift such as Mt. Elgon, Mt. Kilimanjaro, and Mt. Meru (Kingdon, 1990).

The southern savannas are constituted by the Zambezian savanna region and the coastal Tongoland-Pondoland. This region extends from Angola to the south of Tanzania in the north and to northern Botswana and the Transvaal. The mean annual rainfall is between 500 and 1500 mm. It is a woodland savanna, more or less dense with *Brachystegia* and *Julbernardia* and with some steppe patches. In South Africa, the vegetation zones are contrasted. On the plateau in the eastern part of South Africa, one finds the temperate highveld grassland region (Transvaal region, Nyika Plateau). This area has winter frost and low rainfall (between 300 and 1000 mm). The west of southern Africa, Botswana, and the north of South Africa are occupied by a subdesertic zone with steppic and dry savanna vegetation called the southwest arid region. In the west it is composed of a unique vegetation named karoo shrub and grass. In the east it is more tropical vegetation. This arid region receives <500 mm annual rainfall. A narrow band near the Atlantic Ocean in Namibia is very desertic and has an endemic vegetation. Finally, the southwest Cape Province comprises the Cape folded mountains and a narrow coastal plain. The western part of the region has a mediterranean climate with winter rainfall, while the mountain or karoo region has a more continental one. The natural vegetation is the Fynbos or Cape macchia composed of Restionaceae, Ericaceae, and Proteaceae. In some parts, there are remains of forest or scrubforest.

Fossil Faunas

The Early and Middle Pliocene (5–2.5 m.y.) is well documented in rodent faunas (table 16-2). Tanzania has yielded the Laetolil beds (3.7 m.y.) and upper Ndolanya beds at Laetoli (2.5 m.y.) (Denys, 1987a), the Ibolé fauna (around 4–5 m.y.; Winkler, 1997). In Ethiopia are the Hadar (Sabatier, 1982) and Omo Members B and C faunas (Wesselman, 1984). The Kanapoi site has yielded only *Tatera* and *Hystrix* as rodents (Behrensmeyer, 1976). In Namibia, among many karstic sites of various ages (Senut et al., 1992), two have been selected for their diversity: Jägersquelle and Nosib. In Botswana the Ngamiland site is around 3 m.y. (Pickford & Mein, 1988). In the Transvaal (South Africa) the rodent faunas come from the caves of Makapansgat (Pocock, 1987). In the Cape region the open-air deposits of the Langebaanweg are famous for their Early Pliocene faunas (between 6 and 4.5 m.y.; Pocock, 1987; Denys, 1990b).

Concerning the Late Pliocene to Early Pleistocene deposits (from 2.4 m.y. to 1 m.y.), 13 rodent faunas are known (table 16-3). In Tanzania are the Olduvai Beds I and II (Jaeger, 1976, 1979; Denys, 1990a,c) and the Natron fauna (Denys, 1987b). In Kenya, rodents are known from the Koobi Fora formation in East Turkana (Black & Krishtalka, 1986). In Ethiopia are the Omo F and G sites (Wesselman, 1984). The new site of Lusso beds (Zaire) has yielded the first occurrences of modern *Otomys, Tachyoryctes,* and *Thryonomys* at 2.3 m.y. in the north of South Africa (Boaz et al., 1992). The South African faunas come from Taung (northeast Cape region), Sterkfontein type site and Sterkfontein extension site, Kromdraai A and B (Transvaal) and have been revised and studied by Pocock (1987), Denys (1990a), and McKee (1993a). The Humpata caves faunas in Angola (Pickford et al., 1992) are dated from Late Pliocene to Early Pleistocene. Even

Table 16-2. Faunal composition of the Early-middle Pliocene faunas of East and South Africa.

Genus	IBL	UNB	LB	OMB	OMC	QSM	3AN	3AS	EXQR	MRCIS	MLWD	NGA	HAD	JAG	NOS
Xerus	0	1	1	1	1	0	0	0	0	0	0	0	1	0	0
Paraxerus	0	0	1	1	1	0	0	0	0	0	0	0	0	0	0
Gerbillinae, indet.	0	0	1	0	0	0	0	0	0	0	0	0	0	0	0
Gerbillurus	0	0	0	0	0	0	0	0	0	0	0	0	0	1	1
Tatera	0	1	1	1	0	0	0	0	0	1	0	0	1	1	0
Desmodillus	0	0	0	0	0	1	1	1	0	0	0	0	0	1	0
Taterillus	0	0	0	0	0	0	0	0	1	1	0	1	0	0	0
Saccostomus	1	1	1	0	0	0	0	0	0	0	0	0	0	1	0
Dendromus	0	0	1	0	0	1	1	1	1	1	1	1	0	1	1
Steatomys	0	0	1	0	0	0	0	0	1	1	1	0	0	1	1
Malacothrix	0	0	0	0	0	0	0	0	1	1	1	1	0	1	1
Golunda	0	0	0	1	1	0	0	0	0	0	0	0	1	0	0
Millardia	0	0	0	0	0	0	0	0	0	0	0	0	1	0	0
Lemniscomys	0	0	0	1	1	0	0	0	0	0	0	0	0	0	0
Thallomys	0	1	1	1	1	0	0	0	0	0	0	0	0	1	0
Mastomys	0	1	1	1	1	0	0	0	0	1	1	0	0	0	0
Oenomys	0	0	0	0	1	0	0	0	0	0	0	0	0	0	0
Arvicanthis	0	0	0	1	0	0	0	0	0	0	0	0	0	0	0
Aethomys	0	0	0	0	0	1	1	1	1	1	1	0	0	0	0
Acomys	0	0	0	0	0	1	1	1	1	1	1	0	0	1	0
Dasymys	0	0	0	0	0	0	0	0	1	1	1	0	0	0	1
Grammomys	0	0	0	0	0	0	0	0	0	1	1	0	0	0	0
Mus	0	0	0	1	0	0	0	0	1	1	1	1	1	1	1
Myomyscus	0	0	0	0	0	0	0	0	1	1	1	0	0	0	0
Pelomys	0	0	0	0	0	0	0	0	0	0	1	1	0	0	0
Praomys	0	0	0	0	0	0	0	0	0	0	0	0	1	0	0
Rhabdomys	0	0	0	0	0	1	1	1	1	1	1	0	0	0	1
Proceromys	1	0	0	0	0	0	0	0	0	0	0	0	0	0	0
Saidomys	1	0	0	0	0	0	0	0	0	0	0	0	1	0	0
Zelotomys	0	0	0	0	0	0	0	0	0	0	0	1	0	1	1
Mystromys	0	0	0	0	0	1	1	1	1	1	1	0	0	0	1
Stenodontomys	0	0	0	0	0	1	1	1	0	1	0	0	0	1	1
Proodontomys	0	0	0	0	0	0	0	0	1	1	1	0	0	0	0
Otomys	0	0	0	0	0	0	0	0	1	1	1	1	0	0	1
Myotomys	0	0	0	0	0	0	0	0	1	1	1	0	0	0	0
Prototomys	0	0	0	0	0	0	0	0	0	0	1	1	0	0	0
Cryptomys	0	0	0	0	0	1	1	1	1	1	1	0	0	0	1
Gypsorychus	0	0	0	0	0	0	0	0	0	0	1	0	0	0	0
Bathyergus	0	0	0	0	0	1	1	1	0	0	0	0	0	0	0
Euryotomys	0	0	0	0	0	1	1	1	0	0	0	0	0	0	0
Georychus	0	0	0	0	0	0	0	0	0	0	0	1	0	0	0
Graphiurus	0	0	0	0	0	1	0	0	0	0	0	0	0	1	1
Heterocephalus	0	1	1	0	0	0	0	0	0	0	0	0	0	0	0
Thryonomys	1	1	0	1	1	0	0	0	0	0	0	0	0	0	0
Pedetes	0	0	1	0	0	0	0	0	0	0	0	0	0	0	0
Tachyoryctes	0	0	0	0	0	0	0	0	0	0	0	0	1	0	0
Diversity	4	7	11	13	9	14	13	13	17	19	19	11	8	13	15
MNI[a]	—	53	217	85	44	843	779	565	2708	1448	291	—	165	—	—
Gerbillinae	0	1	2	1	0	1	1	1	1	1	0	2	1	3	2
Murinae	2	2	2	5	5	4	4	4	8	8	9	3	5	5	5
Endemic/extinct	3	1	1	1	1	3	3	3	4	5	4	2	4	4	3

[a]Minimum number of individuals.

Key to site abbreviations

IBL	Ibolé
UNB	Upper Ndolanya Beds
LB	Laetoli Beds
OMB	Omo B
OMC	Omo C
QSM	Langebaanweg Quartzose Sand Member
3AN	Langebaanweg Pelletal Phosphate Member North
3AS	Langebaanweg Pelletal Phosphate Member South
EXQRM	Makapansgat exit quarry red mud
MRCIS	Makapansgat rodent corner in situ
MLWD	Makapansgat Limework Dumps
NGA	Ngamiland
HAD	Hadar
JAG	Jägersquelle
NOS	Nosib

Table 16-3. Faunal composition of the Late Pliocene to Early Pleistocene faunas of East and South Africa.

Genus	Taung	STS	SE	SK	KA	KB	H2	BBI	SBI	BII	SII	NAT	ET	OMF	OMG
Aethomys	0	1	1	1	1	1	1	1	1	0	0	1	1	1	1
Arvicanthis	0	0	0	0	0	0	0	1	1	1	1	1	1	0	1
Acomys	0	1	1	1	1	0	0	0	0	0	0	0	0	1	0
Dasymys	0	1	1	1	1	1	1	0	0	0	0	0	0	0	0
Grammomys	0	0	1	1	0	0	0	1	0	0	0	0	0	0	0
Lemniscomys	0	0	1	1	1	1	0	0	0	0	0	0	0	1	0
Mus	1	1	1	1	1	1	1	1	1	1	0	0	1	0	0
Mastomys	1	0	0	0	0	0	0	1	1	1	0	0	1	1	1
Oenomys	0	0	0	0	0	0	0	1	0	1	0	0	0	0	0
Myomyscus	0	1	1	1	1	1	0	0	0	0	0	0	0	0	0
Pelomys	0	0	0	0	0	0	1	1	1	0	0	0	0	1	0
Rhabdomys	0	1	1	0	1	1	0	0	0	0	0	0	0	0	0
Thallomys	1	0	0	0	0	0	1	1	1	1	0	1	1	1	1
Zelotomys	0	1	1	1	1	1	1	1	1	1	0	0	0	0	0
Saccostomus	0	0	0	0	0	0	0	1	1	0	0	1	1	0	0
Steatomys	0	1	1	1	1	1	1	1	1	1	0	0	0	0	0
Dendromus	1	1	1	1	1	1	1	1	1	1	0	1	0	0	0
Malacothrix	1	1	1	1	1	1	0	0	0	0	0	0	0	0	0
Mystromys	1	1	1	1	1	1	0	0	0	0	0	0	0	0	0
Proodontomys	0	0	0	1	1	1	0	0	0	0	0	0	0	0	0
Tatera	1	1	1	1	1	1	0	1	1	1	1	0	1	1	1
Desmodillus	0	0	0	0	1	0	0	0	0	0	0	0	0	0	0
Gerbillurus	0	1	0	0	0	0	0	0	0	0	0	0	0	0	0
Gerbillus	0	0	0	0	0	0	0	0	1	0	0	0	0	1	1
Otomys	1	1	1	1	1	1	1	1	1	1	0	0	0	0	0
Prototomys	1	0	0	0	0	0	0	0	0	0	0	0	0	0	0
Cryptomys	1	1	1	1	1	1	1	0	0	0	0	0	0	0	0
Gypsorychus	1	0	0	0	0	0	0	0	0	0	0	0	0	0	0
Graphiurus	0	1	0	0	1	1	1	0	0	0	0	1	0	0	0
Jaculus	0	0	0	0	0	0	0	0	0	0	0	0	1	1	1
Saidomys	0	0	0	0	0	0	0	0	0	0	0	0	0	0	1
Heterocephalus	0	0	0	0	0	0	0	0	1	0	0	0	0	1	1
Xerus	0	0	0	0	0	0	0	0	1	0	0	0	0	1	1
Paraxerus	0	0	0	0	0	0	0	0	0	0	0	0	0	1	0
Pedetes	1	1	0	0	0	0	0	0	0	0	1	1	0	0	0
Thyronomys	0	0	0	0	0	0	0	0	0	0	1	0	1	1	0
Petromus	1	0	0	0	0	0	0	0	0	0	0	0	0	0	0
Diversity	13	16	16	16	17	16	11	14	15	10	4	7	9	13	10
MNI[a]	—	3551	—	—	371	—	—	111	429	—	—	14	15	69	16
Gerbillinae	1	1	1	1	2	1	1	1	2	1	1	0	1	2	1
Murinae	2	7	9	8	7	7	5	9	7	6	1	4	5	6	5
Endemics	4	3	3	3	3	3	0	1	1	1	0	0	1	2	2

[a]Minimum number of individuals.

Key to site abbreviations

TAUNG	Taung
STS	Sterkfontein type site
SE	Sterkfontein extension
SK	Swartkrans
KA	Kromdraai A
KB	Kromdraai B
H2	Humpata level 2
BBI	Olduvai Middle Bed I
SBI	Olduvai Upper Bed I
BII	Olduvai Lower Bed II
SII	Olduvai Upper Bed II
NAT	Natron Peninj site
ET	East Turkana
OMF	Omo F
OMG	Omo G

Table 16-4. Faunal composition of the middle and upper Holocene faunas of East and South Africa.

Genus	BIV	MSK	ISN	KAPTH	BTW	NGAB	KRM	MLK	WWK
Acomys	0	0	0	0	1	0	1	0	0
Aethomys	1	0	1	0	0	0	0	0	1
Arvicanthis	1	1	1	0	1	1	0	0	0
Dasymys	1	0	0	0	0	0	1	0	0
Grammomys	0	0	1	0	0	0	1	0	0
Myomyscus	0	0	0	0	0	0	1	0	0
Zelotomys	1	0	1	0	0	0	0	0	1
Lemniscomys	0	0	0	0	1	0	0	0	0
Mastomys	1	1	0	0	1	1	0	0	0
Mus	0	0	1	0	1	1	1	1	1
Oenomys	0	0	0	1	0	0	0	1	0
Thallomys	1	0	0	0	0	0	1	0	1
Rhabdomys	0	0	0	0	0	0	1	0	1
Stenocephalemys	0	0	0	0	0	0	0	1	0
Tatera	1	1	1	0	1	1	1	0	1
Desmodillus	0	0	0	0	0	0	0	0	1
Taterillus	0	0	0	0	1	0	0	0	0
Desmodilliscus	0	0	0	0	1	0	0	0	0
Meriones	0	0	0	0	1	0	0	0	0
Gerbillus	0	0	0	0	1	0	0	0	0
Gerbillurus	0	0	0	0	0	0	1	0	1
Otomys	1	0	1	1	0	1	1	1	1
Dendromus	0	0	0	0	0	0	1	0	1
Steatomys	1	0	0	0	1	0	0	0	1
Malacothrix	0	0	0	0	0	0	0	0	1
Saccostomus	1	0	1	0	0	0	0	0	1
Pedetes	0	0	1	0	0	1	0	0	0
Jaculus	0	1	0	0	0	0	0	0	0
Thryonomys	0	0	1	1	1	0	0	0	0
Cricetomys	0	0	0	1	0	0	0	0	0
Graphiurus	0	0	0	1	0	0	1	0	1
Mystromys	0	0	0	0	0	0	1	0	1
Heterocephalus	1	0	0	0	0	0	0	0	0
cf. Heliophobius/ Cryptomys	0	0	1	0	0	0	0	0	0
Cryptomys	0	0	0	0	0	0	1	0	1
Heliosciurus	0	0	0	0	1	0	0	0	0
Tachyoryctes	0	0	0	0	0	1	0	1	0
Diversity/site	11	4	11	5	15	7	15	6	16
MNI[a]			50						
Gerbillinae	1	1	1	0	4	1	2	0	3
Murinae	6	2	5	1	5	3	7	3	5
Endemics/extincts	0	1	0	0	5	1	2	3	4

[a]Minimum number of individuals.

Key to site abbreviations

BIV	Olduvai Bed IV	NGAB	Ngaloba Beds
MSK	Olduvai Masek Beds	KRM	Klasies River Mouth
ISN	Isenya	MLK	Melka Kunturé
KAPTH	Kapthurin	WWK	Wondewerk
BTW	Bir Tarfawi		

though the faunal lists are provisional and the fissures are not the same age, we included them in this study.

In East Africa, different sites have yielded rodent faunas from the Middle and Late Pleistocene to Holocene (1–0 m.y.) interval. These are the Olduvai (Bed IV, Masek beds; Denys, 1990a), the Laetoli Ngaloba beds site (about 120,000 years BP; Day et al., 1980), Isenya (Brugal & Denys, 1989), and the Kapthurin sites in Kenya (Denys, 1990a) and the Melka Kunturé (Ethiopia; Sabatier, 1978) faunas. Due to the presence of tropical African rodents in Egypt at 120,000 years, we have included the Bir Tarfawi fauna (Kowalski et al., 1989). In South Africa, one finds numerous Holocene localities, but no Late and Middle Pleistocene ones. The Klasies River mouth assemblage (southwestern Cape, 125,000–65,000 BP; Avery, 1982), which has a Fynbos bush-thicket vegetation, and the Wondewerk cave site (northern Cape region, 7000–6000 BP; Avery, 1981) in the southwest arid region have been incorporated in the present study (table 16-4).

Methods

Comparisons between the eastern and southern African fossil and modern faunas have been undertaken at the genus level because the specific level is not well known for modern rodents, and the specific determinations of the fossil faunas are still poorly known (see appendix for this volume). Species richness, as well as endemism and the number of taxa in common have been used to compare the faunas. A multivariate analyses has been undertaken through the use of Correspondence Analysis (AFC) performed on presence–absence data (Statitcf software) of taxa in each site, one of the best ways to directly compare the faunal composition of the faunas (Denys, 1985, 1990a, 1992b; Geraads & Coppens, 1995).

Characteristics of the Modern and Fossil Faunas in East and South Africa

Diversity Patterns

Modern Faunas The diversity characteristics of each biogeographical region are summarized in table 16-1. The generic diversity is generally lower in northern savannas than in southern savannas. The highest generic diversity occurs not in the forest but in the Somali-Masai region. The rodent community of the Sahara is dominated by the Gerbillinae, Dipodidae (gerboas), and Ctenodactylidae (gundis), which are well adapted to desert life. In the Sahel, one finds as many Gerbillinae as Murinae genera (table 16-1). The Sudanian zone has similar numbers and genera of Gerbillinae than the Sahel. There are seven Murinae genera in the Sudanian zone. The main characteristics of the tropical forest rodent community are the diversification of the arboreal Sciuridae (at least 13 species for 5 genera) and of the endemic Anomaluridae (3 genera and 5 species). In the tropical and montane forest, the diversity increases from east to west but decreases with altitude. There are 56 rodent species (30 genera) in the low-altitude forest and 31 species (19 genera) in mountainous zones (Dieterlen, 1989). The diversity in the Lake Victoria mosaic region is particularly high (30 genera) due to the variety of habitats (from montane grassland to moist Guinean savanna, moist low-altitude forest, dry woodland savanna, and swamp regions). This region has not been well studied taxonomical and will probably reveal a higher specific diversity.

In the south of Kenya and in northern Tanzania is the contact zone between Somali-Masai and Zambezian savannas for different species of the same genera (e.g., example *Saccostomus mearnsi* and *S. campestris; Aethomys chrysophilus* and *A. hindei*). In this zone some genera are replaced by ecological equivalents, so that *Arvicanthis* is replaced by *Pelomys* and *Heterocephalus* is replaced by *Heliophobius.* In the Somali-Masai region Sahelo-Sudano-Guinean rodent genera such as *Gerbillus, Taterillus, Arvicanthis, Lemniscomys* and Zambezian savanna elements such as *Dendromus, Pedetes, Saccostomus, Rhabdomys, Otomys, Thallomys, Uranomys, Zelotomys* occur together. Near the east coasts of Kenya, Tanzania, and Mozambique is a narrow vegetation band with a higher rainfall showing a coastal forest–savanna mosaic. This region has yielded the same fauna as in the Somali-Masai center but with a lower diversity (21 genera) and also some relict forest taxa (*Lophuromys, Praomys, Heliosciurus*).

In the Zambezian savanna region, there is a good specific diversification of the Otomyinae and Dendromurinae subfamily. The east coast of South Africa is characterized by a narrow band of more humid climate but is mainly composed of an impoverished Zambezian assemblage. The temperate highveld grassland region is where *Otomys* is most abundant in the faunas as well as another grazer, *Rhabdomys.* The Cape Province fauna is not very different from that of the Zambezian savannas. The Cape Province represents an exception of the Gerbillinae/Murinae ratio, and there are two Gerbillinae genera despite a relatively high rainfall pattern (500–1500 mm). But what differentiates this region from the surrounding desert is the abundance of Murinae genera (7).

In tropical Africa, the importance of chromosomal species is currently underestimated, and a lot of sibling species have been discovered during the last 20 years. Grubb (1978) has demonstrated the relatively large number of genera with apparently sympatric sibling species and few hybrid zones between allopatric species and cited the cases of *Funisciurus, Paraxerus, Graphiurus, Tatera, Dendromus, Malacomys, Aethomys, Lemniscomys, Arvicanthis, Praomys,* and *Mus.* Since then, the number of cases has increased and we can cite at the least *Mastomys* (Britton-Davidian et al., 1995), *Thallomys* (Taylor et al., 1995), *Otomys* (Contrafatto et al., 1992), and *Saccostomus* (Gordon, 1986).

Fossil Faunas Tables 16-2, 16-3, and 16-4 give the number of genera in the fossil faunas and, when known, the minimum number of individuals (MNI). The maximum species number is, in general, higher in South Africa than in East Africa. The species and genus number depend on the MNI. There is a good correlation between the MNI and the genera number, except for the Makapansgat Limework Dump which seems to have a high generic diversity for a rather small MNI (fig. 16-2). Between 3 and 2.5 m.y., both South and East African fossil faunas show a decrease in diversity (fig. 16-3). The South African fossil assemblages of karstic origin have yielded important concentrations of rodents, which increases their spectra of diversity. In East Africa, the rodents have been collected by screening open-lacustrine and riverine sediments, and the MNIs are lower compared to South Africa, which might reduce the diversity spectra. But the Olduvai upper Bed I, which has yielded an important MNI, has in proportion a lower generic diversity when compared to the South African faunas. Compared to modern East and South African owl pellet assemblages, whose diversity is correlated with both MNI and with the diet selection of the predator, we find that the Transvaal Plio-Pleistocene sites as well as some of the richest East African sites have, on average, a higher diversity than the modern pellet assemblages (about eight to nine species per assemblage).

Patterns of Endemism

Modern Faunas Table 16-5 reviews the modern endemic rodent genera of each vegetation zone. In table 16-1 the number of endemic genera is given.

In the Sahara, among the four Gerbillinae genera,

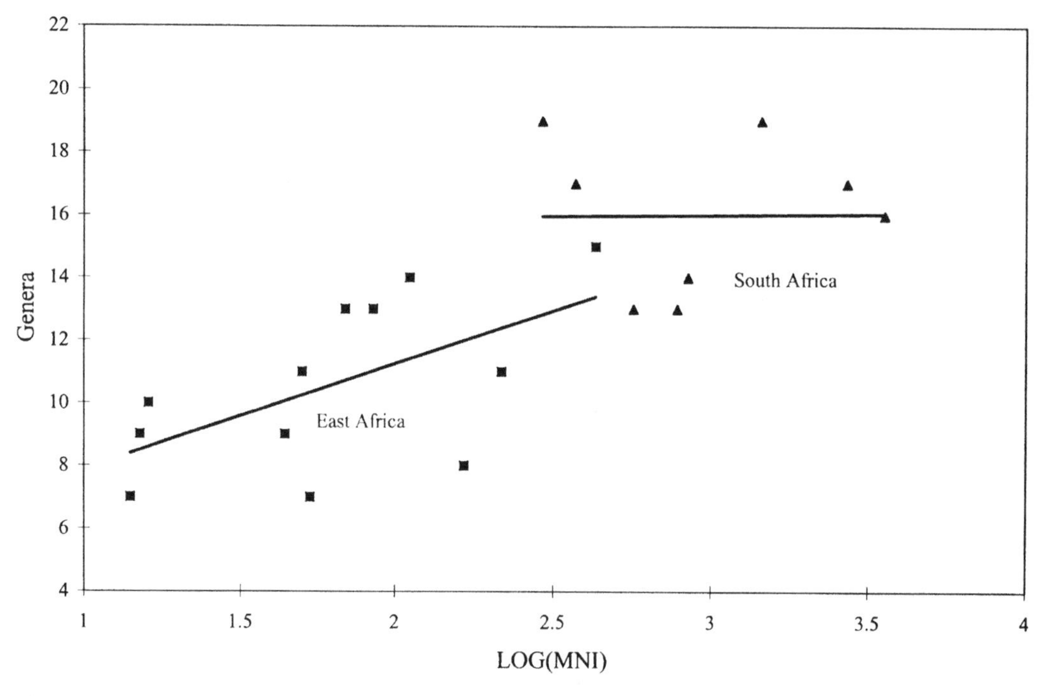

Figure 16-2. Comparison of the generic diversity between the East and South African fossil rodent faunas as a function of the minimum number of individuals (MNI; from data in tables 16-2, 16-3, and 16-4) showing the correlation between the MNI and the number of genera. For South Africa some of the faunas have a higher diversity pattern.

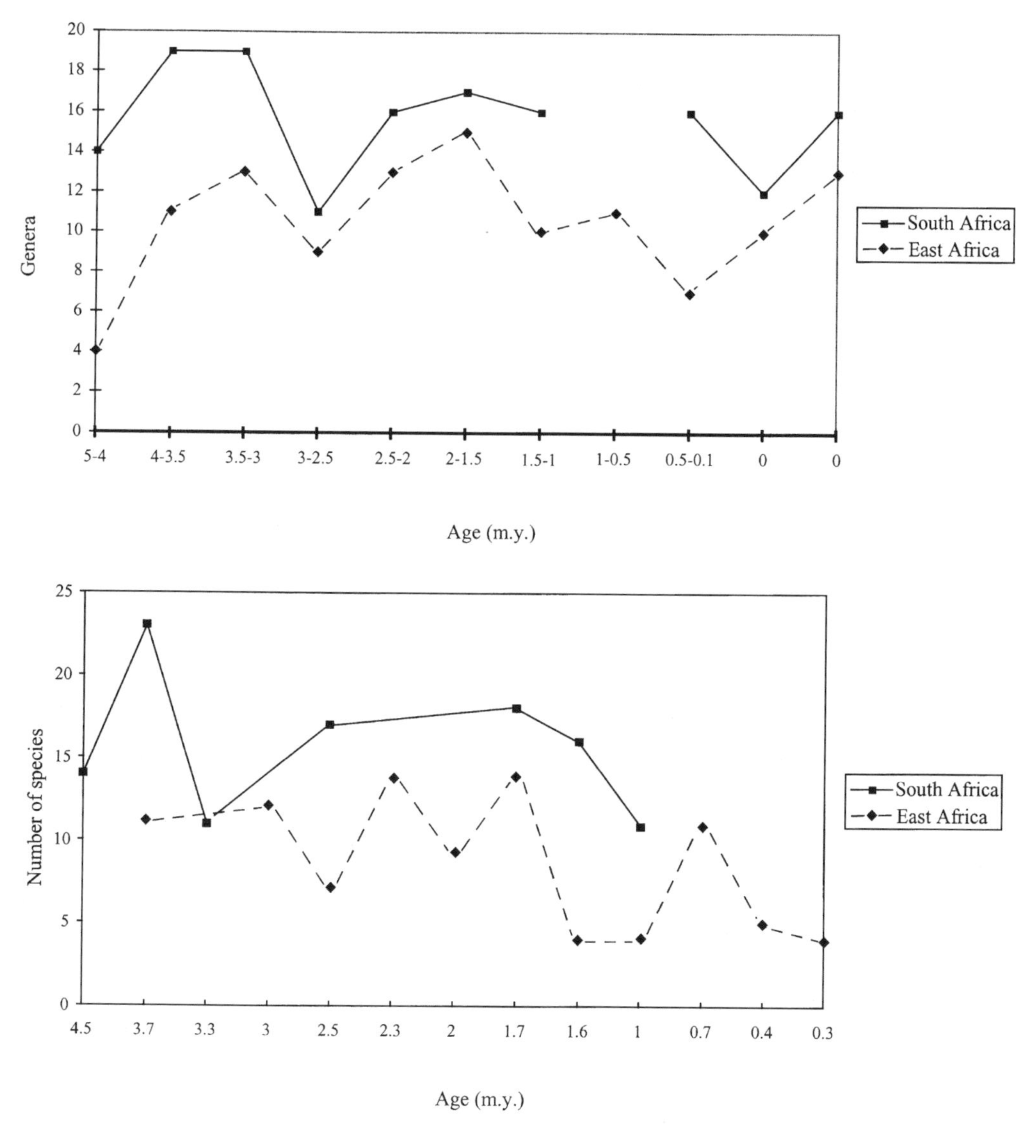

Figure 16-3. Evolution of the rodent generic (top) and specific (bottom) diversity through time in East and South Africa.

two are endemic (*Pachyuromys* and *Psammomys*). The Sahel shares with the Sahara the presence of *Jaculus jaculus*. The Sudanian savanna features the same endemics as the Sahelian zone. The same situation occurs in the Guinean region, which has no true endemic but shares with the Sudanian zone two endemics, *Taterillus* and *Myomys*. The Guineo-Congolese forest zone has 28 endemic species of rodents, compared to 15 in Ethiopia. The Congolese block has seven endemic genera (*Idiurus, Zenkerella, Dendroprionomys, Deomys, Prionomys, Myosciurus,* and *Lamottemys*). The Guinean block has no true endemic genera. The eastern Zaire forest shows some similarity to the Congolese block, with nine common endemic genera, but it has one proper endemic, *Delanymys*.

Compared to other montainous regions of East Africa, which have a low endemism and poor and recent (about 1 m.y. old) rodent diversity, the Ethiopian Highlands in the Somali-Masai region have a rich endemic fauna not only of rodents but also of other mam-

Table 16-5. Endemic rodent genera of Africa by biogeographic regions.

Region	Endemic genera
Sahara	*Pachyuromys, Psammomys, Massoutiera*
Sahelo-Sudanian	*Desmodilliscus, Felovia*
Saharo-Sahelian	*Jaculus*
Somali-Masai	*Desmomys, Megadendromus, Muriculus, Stenocephalomys, Beamys, Heterocephalus, Lophiomys, Pectinator, Ammodillus*
Northern savannas	*Arvicanthis, Oenomys, Taterillus, Tachyoryctes*
Forest	*Epixerus, Myosciurus, Idiurus, Zenkerella, Hylbomys, Lamottemys, Mylomys, Stochomys, Delanymys, Dendroprionomys, Deomys, Leimacomys, Prionomys*
Zambezian	*Heliophobius*
Southern savannas	*Rhabdomys, Georychus, Malacothrix*
Highveld	*Mystromys*
Namib + SW arid	*Petromyscus, Parotomys, Petromus, Desmodillus, Gerbillurus*
Southwest arid + Cape	*Malacothrix, Gerbillurus, Bathyergus, Desmodillus*
Cape	*Myomyscus, Mystromys*

mals (Yalden & Largen, 1992). In that region we find 4 endemic genera (*Muriculus, Megadendromus, Desmomys,* and *Stenocephalemys*) and 14 endemic species (table 16-5). The dry savannas that characterize this region also have a rather high percentage of endemics with *Lophiomys, Heterocephalus, Tachyoryctes, Beamys, Ammodillus,* and *Pectinator.*

The Zambezian savanna has one endemic genera and shares many taxa with the Somali-Masia region. The Namib fauna has many affinities with that of the South West Arid Zone, as they share the same endemic taxa: *Petromyscus, Desmodillus, Parotomys,* and *Petromus.* Two of these rodents are the monogeneric representatives of endemic African families: Petromyscinae and Petromuridae. The Cape Region shares *Mystromys* with Highveld grasslands, *Desmodillus* with Namib and southwest arid, *Malacothrix, Gerbillurus, Bathyergus,* with southwest arid and has only one true endemic: *Myomyscus.*

Fossil Faunas The number of endemic taxa is given in tables 16-2, 16-3, and 16-4. During the Early and Middle Pliocene, most of the faunas show some modern endemic taxa, especially the South African faunas. Langebaanweg, for instance, has southwest arid and Cape endemics *Desmodillus* (found also in Namib), *Mystromys* (also in Highveld), *Bathyergus* southwest. Ngamiland has southern savannas *Malacothrix* and *Georychus* southwest and *Taterillus* (Soudano-Guinean element, but I have some reservations, as the molars of *Desmodillus* and *Taterillus* can be easily confused). In the Namib, the Jägersquelle and Nosib faunas share the following endemics: *Gerbillurus* (Namib + southwest + Cape), *Malacothrix, Mystromys* (southwest). Jägersquelle is distinct from Nosib in showing *Desmodillus* (Namib + southwest arid + Cape). The Makapansgat faunas share the following endemics: *Taterillus* (if the Pocock, 1987, determination is valid), *Myomyscus* (Cape), *Malacothrix* (southwest), *Mystromys.* In East Africa, the Laetoli faunas, Hadar, and Omo C have only one endemic each: *Heterocephalus, Tachyoryctes,* and *Oenomys,* respectively.

During Late Pliocene to Middle Pleistocene times, the Taung fauna shows *Malacothrix, Mystromys* (Highveld, southwest arid, Cape) and *Petromus,* a typical genus of Namib and the South West Arid Zone. This indicates a more arid environment at Taung compared with the Transvaal sites of Sterkfontein. The latter faunas share with the older Makapansgat sites the same endemics, *Malacothrix, Mystromys,* and *Myomyscus,* and are not very different. The Sterkfontein type site has a supplementary endemic genus, *Gerbillurus* (found in Namib, southwest arid, Cape). Kromdraai A has *Desmodillus,* which is a Namib-South West Arid Zone Cape endemic today and could indicate slightly more arid conditions in that region by that time. The Natron and Humpata faunas have no endemics. In Tanzania, *Oenomys* is found at Olduvai Beds I and II and *Gerbillus* and *Heterocephalus* at the top of Bed I. In that region, the East Turkana and Omo F and G shows two endemics, *Jaculus* (Sahara + Sahel + Somali) and *Gerbillus* (Sahara, Sahel, Sudan, Somali regions).

During the Late Pleistocene, the Bir Tarfawi fauna shows two Sudano-Guinean taxa (*Taterillus* and *Heliosciurus*), a Sahelo-Sudanian element (*Desmodilliscus*), and a Saharian one (*Meriones*). The Olduvai Bed IV, Isenya and Kapthurin have no endemic taxa. The Masek Beds have *Jaculus*. The Ngaloba Beds have yielded *Tachyoryctes*, and the Melka Kunturé faunas are characterized by typical endemic genera of the Ethiopian Highlands: *Tachyoryctes*, *Stenocephalemys*, and *Oenomys*. In South Africa, the Klasies River mouth fauna has two endemics: *Gerbillurus* (Namib, South West Arid, Cape) and *Georychus* (southwest Cape, South West Arid). Wondewerk has Highveld, Namib, and South West Arid Zone elements (*Mystromys, Malacothrix, Gerbillurus* and *Desmodillus*).

FAD and LAD Occurrences

The first appearance datum (FAD) and last appearance datum (LAD) occurrences are summarized in table 16-6. We can clearly see two diachronic phases of generic differentiation in East and South Africa. The highest frequency of first occurrence of modern taxa is at 5 m.y. in South Africa and Namibia. The second peak occurs in East Africa at 3.7–3.3 m.y. (fig. 16-4). After 1.7 m.y., all of the modern genera are present. Modern rodent species are found at younger sites, despite evidence that some were already described at Olduvai Bed I (Denys, 1992a). The systematic study and comparison of the East and South African rodent species has shown that they differ and are at different evolutionary stages in contemporaneous faunas—for instance, *Dendromus, Otomys,* and *Aethomys,* which are still present today (Denys, 1989b, 1990a, 1990b). There are few extinctions both of genera and species (Denys & Jaeger, 1986). There are seven generic extinctions after 2.5–2 m.y. and only two occur in East Africa (table 6). Concerning the species extinctions, Denys (1990a) has shown that they occur between 3–1.7 m.y. and 0.8–0.4 m.y. in East Africa, whereas in South Africa is an earlier phase, between 4 and 3 m.y. In South Africa, extinctions seem to occur more recently, between 1.5 m.y. and 0.3 m.y. Looking at the dispersal direction of the taxa, at least seven South African taxa have never moved from the place they originated. Dispersal through the south is less frequent than through the north (9/52 cases). There is not an evident indication of turnover observed for the rodents faunas at around 2.7–2.5 m.y., contrary to Vrba (this volume) for bovids but accordingly to the suid fossil record (Bishop, this volume).

Common Taxa

Modern Faunas In East Africa, the Sudanian and Guinean savanna are the most similar, as they share a high number of common genera with the Victoria mosaic and the Somali-Masai region (table 16-7). In South Africa, the Zambezian and the South West Arid Zones also are quite similar in rodent composition. Of course, the montane forest and the lowland forest share much similarity with the Victoria mosaic. The comparison of the northern savannas (Sudanian, Guinean, and Sahelian) with the Zambezian ones suggests only 5–12 common genera. In contrast, the comparison with the northern savannas and the Somali-Masai indicates between 8 and 14 common genera. The comparison between the Zambezian (or southern savanna) with the Somali-Masai results in the highest value of 21 common genera. When comparing the northern savannas sensu lato (Sudanian, Guinean, and Sahelian) to the Zambezian ones, we observe *Desmodilliscus, Taterillus, Arvicanthis,* and *Myomys* as typical taxa for the north. *Otomys* is absent from northern savannas as *Saccostomus, Cryptomys,* and *Dendromus.* The Somali-Masai region is also characterized by two Saharan taxa that do not occur farther south, *Gerbillus* and *Jaculus.* The southern savannas sensu lato (Zambezian, Highveld, South West Arid, and Cape) are characterized by *Pelomys, Malacothrix, Mystromys, Parotomys,* and *Myomyscus.* The Somali-Masai region comprises all the elements of the two savanna regions plus a high endemism in the Ethiopian Highlands.

Fossil Faunas During the Early and Middle Pliocene, the number of common genera between East and South Africa is rather low, from 0 to 5 (table 16-8). The Laetoli Bed faunas have the most genera in common with the Makapansgat and Jägersquelle faunas. The Ethiopian faunas never have more than three genera in common with any of the South African ones. The correspondence analysis diagram of the first two axes displays 39% of the total variance and demonstrates the differences between the two communities (fig. 16-5). The first axis separates the South African faunas from the East African ones, while the Ibole and Hadar faunas have an important load along the second axis, and their original rodent composition has also a weight on this axis (*Millardia, Saidomys, Proceromys, Golunda*).

After 2.5 m.y., the Tanzanian faunas share between

Table 16-6. First appearance date (FAD) and last appearance date (LAD) occurrences and dispersal direction for all rodent genera found in fossil sites of East and South Africa.

Taxon	FAD (Ma)	Site of first occurrence	LAD (Ma)	Site of last occurrence	Direction of dispersal
Bathyergidae					
Bathyergus	5	Langebaanweg	0		None
Cryptomys	5	Langebaanweg	0		N
Georychus	3.3	Ngamiland	0		?
Gypsorychus	3.3	Makapansgat	2.5	Taung	N
Heterocephalus	4.3	Kakesio, Lower Laetoli Beds	0		N
Hystricidae					
Xenohystrix	3.7	Laetoli Beds	3.3	Makapansgat?	S
Hystrix	?				
Petromuridae					
Petromus	2.5?	Taung	0		S
Thryonomyidae					
Thryonomys	6–4?	Ibolé	0		?
Pedetidae					
Pedetes	3.7	Laetoli Beds	0		S
Sciuridae					
Xerus	3.7	Laetoli Beds	0		N
Paraxerus	3.7	Laetoli Beds	0		N
Cricetomyinae					
Saccostomus	10.5	Harasib 3A Namibie	0		N
Cricetomys	0.4	Kapthurin	0		?
Dendromurinae					
Dendromus	5	Langebaanweg	0		N
Steatomys	8.5	BA31 Namibie	0		N
Malacothrix	3.3	Makapansgat	0		None
Gerbillinae					
Tatera	?4	Kanapoi			S
Gerbillus	2.33	Omo F	0		S
Desmodillus	5	Langebaanweg	0		None
Gerbillurus	2.5	Jägersquelle-Nosib	0		?
Mystromyinae					
Mystromys	5	Langebaanweg	0		N
Proodontomys	3.7	Makapansgat	1	KA,KB, SK1	None
Delanymyinae					
Stenodontomys	5	Langebaanweg	1.5	BA54,Aig2,UIS	NW
Petromyscinae					
Petromyscus	10.5	Harasib 3a	0		None
Murinae					
Karnimata	10.5	Chorora, Harasib 3a	8.5	BA31	?
Saidomys	6–4	Ibolé	2	Omo G	N
Euryotomys	5	Langebaanweg	5	Langebaanweg	
Millardia	3	Ngamiland	3	Hadar	
Golunda	3.3	Hadar	2.5	Omo D	S
Acomys	5	Langebaanweg	0		N
Oenomys	3.3	Hadar	0		
Arvicanthis	3	Omo B	0		
Aethomys	5	Langebaanweg	0		N
Dasymys	3.3	Makapansgat	0		?N
Grammomys	3.3	Makapansgat	0		?N
Lemniscomis	3	Omo B	0		?
Mastomys	3.7	Laetoli Beds	0		N
Nannomys	3.3	Makapansgat	0		N

(continued)

Table 16-6. (*continued*)

Taxon	FAD (Ma)	Site of first occurrence	LAD (Ma)	Site of last occurrence	Direction of dispersal
Murinae					
Praomys	3.3	Hadar	0		S
Rhabdomys	?5	Langebaanweg	0		N
Thallomys	3.7	Laetoli Beds	0		N
Zelotomys	3	Ngamiland	0		N
Pelomys	3.3	Makapansgat	0		N
Otomys	3.3	Makapansgat	0		N
Myotomys	3.3	Makapansgat	0		None
Prototomys	3.3	Makapansgat	2.5	Taung	None
Dipodidae					
Jaculus	2.3	Omo F	0		S
Graphiuridae					
Graphiurus	5	Langebaanweg	0		N
Tachyoryctinae					
Tachyoryctes	3.3	Hadar	0		S
Anomaluridae					
Anomalurus	15.5	Kipsaramon	0		None

FAD, first appearance datum; LAD, last appearance datum.

five and eight common genera with the Transvaal faunas, while the East Turkana and Ethiopian Omo F and G share only two to four genera (table 16-9). The correspondence analysis diagram of the first two axes, displaying 42.6% of the total variance, shows a little less separation between eastern and southern faunas compared to the previous analysis. However, the first axis separates the Ethiopian Omo F and G faunas from the Tanzanian and the South African ones. The Humpata fauna is in between the South African Transvaal and the Tanzanian and Kenyan faunas. The Taung fauna is separated along the second axis with its extinct genera *Prototomys* and *Gypsorhychus* (fig. 16-6).

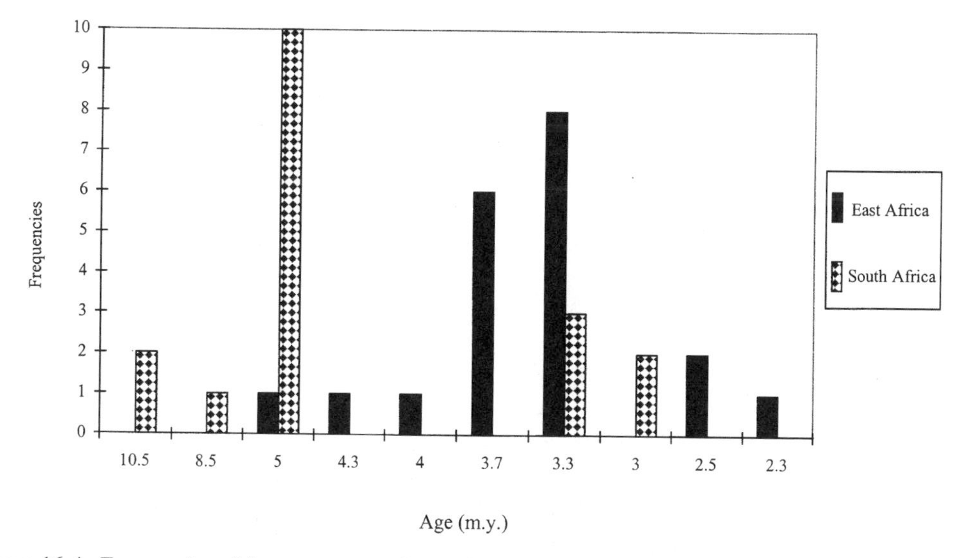

Figure 16-4. Frequencies of first appearance datum through time in East and South Africa (from table 16-14).

Table 16-7. Common genera among the modern biogeographic zones of Africa.

Faunas	SAH[a]	SEL	SUD	GUI	FOR	MFO	VIC	MAS	ZBZ	HIG	NAM	KAL
Sahelian savanna (SEL)	4	—	—	—	—	—	—	—	—	—	—	—
Sudanian savanna (SUD)	2	**9**	—	—	—	—	—	—	—	—	—	—
Guinean savanna (GUI)	2	6	**14**	—	—	—	—	—	—	—	—	—
Tropical forest (FOR)	1	4	7	12	—	—	—	—	—	—	—	—
Montane forest (MFO)	0	0	3	4	15	—	—	—	—	—	—	—
Victoria mosaic (VIC)	1	4	10	**14**	**17**	**14**	—	—	—	—	—	—
Somali—Masai region (MAS)	**5**	8	12	**14**	10	8	**18**	—	—	—	—	—
Zambezian savanna (ZBZ)	2	5	10	12	8	5	15	**21**	—	—	—	—
Temperate High Veld (HIG)	1	4	6	7	5	3	8	12	14	—	—	—
Namib desert (NAM)	1	2	2	3	1	1	5	15	8	7	—	—
South West arid (KAL)	1	4	5	7	4	3	10	8	**15**	**14**	**13**	—
Southwest Cape	2	3	6	8	4	3	9	10	11	10	7	**12**

[a]Sahara.

Table 16-8. Number of common genera between East and South African Early to middle Pliocene faunas.

Faunas	IBL[a]	UNB	LB	OMB	OMC	HAD	QSM	3AN	3AS	EXQRM	MRCIS	MLWD	NGA	JAG
Upper Ndolanya Beds (UNB)	2	—	—	—	—	—	—	—	—	—	—	—	—	—
Laetoli Beds (LB)	1	**6**	—	—	—	—	—	—	—	—	—	—	—	—
OMO Member B (OMB)	1	5	**5**	—	—	—	—	—	—	—	—	—	—	—
OMO Member C (OMC)	1	4	4	**7**	—	—	—	—	—	—	—	—	—	—
Hadar (HAD)	1	2	2	4	2	—	—	—	—	—	—	—	—	—
Langebaanweg QSM (QSM)	0	1	2	1	0	1	—	—	—	—	—	—	—	—
Langebaanweg PPM3AN (3AN)	0	1	2	1	0	1	**11**	—	—	—	—	—	—	—
Langebaanweg PPM3AS (3AS)	0	1	2	1	0	1	**11**	**11**	—	—	—	—	—	—
Makapansgat EXQRM (EXQRM)	0	0	2	1	0	2	6	6	6	—	—	—	—	—
Makapansgat MRCIS (MRCIS)	0	2	4	3	1	3	8	8	6	**15**	—	—	—	—
Makapansgat MLWD (MLWD)	0	1	3	2	1	2	6	6	6	14	**16**	—	—	—
Ngamiland (NGA)	0	1	2	2	0	2	2	2	2	2	6	6	—	—
Jägersquelle (JAG)	1	3	5	3	1	2	7	6	6	6	8	6	2	—
NOSIB	0	0	2	2	0	1	7	6	6	10	11	**10**	1	**9**

[a]Ibole (Narorga Valley).

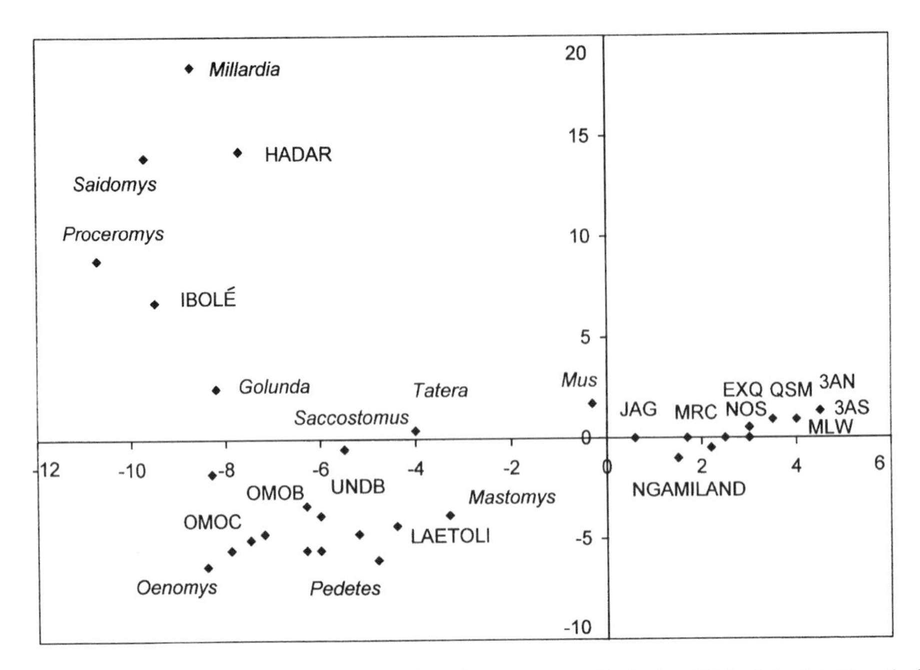

Figure 16-5. Correspondence analysis diagram of the first two axes displaying 39% of the total inertia for the Early Pliocene and Middle Pliocene fossil faunas.

The Angolan faunas share eight genera with Olduvai Bed I. From the Middle to the Late Pleistocene, the numbers of common taxa are low because most of these faunas have a low diversity due to sampling conditions, but the Holocene faunas of South Africa share four to seven genera with Olduvai Bed IV and Isenya, but only one with Masek Beds and two with Kapthurin (table 16-10). The correspondence analysis diagram of the first two axes displays 42.9% of the total variance (fig. 16-7). The first axis separates the two South African faunas from the Olduvai Masek Beds and Bir Tarfawi faunas. The Isenya and Olduvai Bed IV are intermediate, and the Melka Kunturé-Kapthurin assemblages have clearly a distinct influence on the second axis.

Discussion

Zoogeographical Affinities within the Fossil Faunas and Their Evolution

Miocene The earliest rodent faunas of East and South Africa in the Early Miocene deposits of Kenya, Uganda, and Namibia consist of well-diversified Pedetidae, Sciuridae, Bathyergidae, Anomaluridae, and Phiomyidae derived from the endemic Eocene stock of Africa. Many species in East Africa and Namibia are identical, but Lavocat (1978) describes Afrocricetodontinae only for the East African sites. Few Early-Middle Miocene rodent faunas are known in East Africa, but description of the Muruyur rodents of Kenya (15.5 m.y.) (Winkler, 1992) shows a relative faunal stability by that time, with many affinities with Namibia as well as Saudi Arabia. They differ from the well-documented contemporaneous sites of the Siwaliks (Pakistan) in lacking more advanced and diversified cricetid and muroid representatives. They also differ from the North African faunas of the same time in lacking ctenodactylids and graphiurids. Around 14 m.y., the African faunas underwent changes in composition. The most significant event is the appearance of Dendromurinae, Petromyscinae, and Myocricetodontinae. Some of these taxa are common to the Siwaliks and North Africa, and this confirms exchanges in a wide paleo-Saharo-Indian province. However, these exchanges were not complete because the East African faunas at that time do not comprise either the Asiatic *Potwarmus* or *Antemus,* both considered to be direct ancestors of true Murinae (Jacobs, 1978). At 9.5 m.y. (MN 9/10–11) a big turnover occurred in western Europe (Bernor et al., 1996a; Forlelius et al., 1996), but this event is diachronous between central and eastern Europe according to small mammals. At around 11.5–11 m.y. the differentiation of the first true murinae, *Progono-*

Table 16-9. Number of common genera between East and South Africa during Late Pliocene and Early Pleistocene times.

Faunas	TAUNG	STS	SE	SK	KA	KB	H2	BBI	SBI	BII	SII	NAT	ET	OMF
Sterkenfontein type site (STS)	**6**	—	—	—	—	—	—	—	—	—	—	—	—	—
Sterkenfontein extension (SE)	**6**	**14**	—	—	—	—	—	—	—	—	—	—	—	—
Swartkrans (SK1)	**6**	13	**15**	—	—	—	—	—	—	—	—	—	—	—
Kromdraai A (KA)	**6**	**14**	14	13	—	—	—	—	—	—	—	—	—	—
Kromdraai B (KB)	**6**	**14**	14	**14**	**15**	—	—	—	—	—	—	—	—	—
Humpata Level 2 (H2)	4	9	8	8	9	**9**	—	—	—	—	—	—	—	—
Olduvai Middle Bed I (BBI)	5	7	8	8	7	7	**8**	—	—	—	—	—	—	—
Olduvai Upper Bed I (SBI)	5	8	7	7	7	7	**8**	**12**	—	—	—	—	—	—
Olduvai Lower Bed II (BII)	5	6	6	6	6	6	6	10	**9**	—	—	—	—	—
Olduvai Upper Bed II (SII)	2	2	1	1	1	1	0	2	2	2	—	—	—	—
Natron	3	5	3	3	4	4	5	6	6	4	2	—	—	—
East Turkana (ET)	3	3	3	3	3	3	3	7	7	**5**	**3**	**5**	—	—
OMO Member F (OMF)	3	4	4	4	4	3	3	5	7	3	2	2	**6**	—
OMO Member G	3	3	2	2	2	2	2	5	2	4	2	3	**6**	**8**

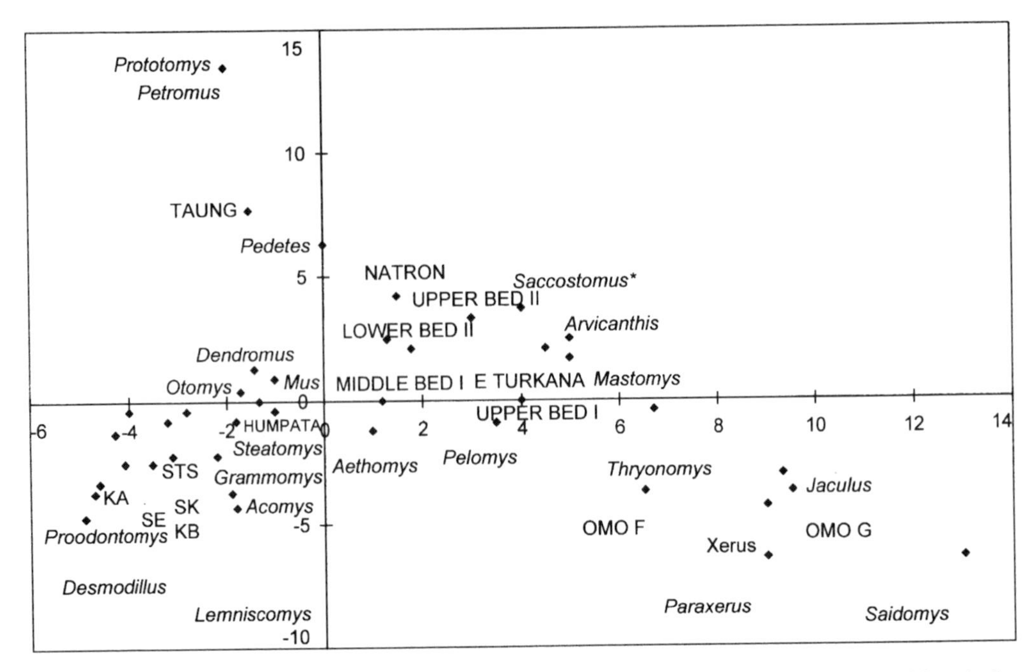

Figure 16-6. Correspondence analysis diagram of the first two axes displaying 42.6% of the total inertia for the Late Pliocene and Early Pleistocene fossil rodent faunas.

mys, took place in North Africa, Europe, and Pakistan. That genus is not recorded in contemporaneous faunas of East or South Africa. The first record of true Murinae is in Chorora (Ethiopia, about 10 m.y.) with an undetermined *Praomys*-like genus (Jaeger et al., 1980) and around 8–9 m.y. in Namibia (Senut et al., 1992) with *Karnimata,* also recorded in Afghanistan (Brandy et al., 1980) and Pakistan (Jacobs, 1978).

Late Miocene to Early Pliocene The analysis of the regional affinities of the fossil faunas are based on combining tables 16-1–16-4. The results are displayed in tables 16-11–16-14.

In East Africa, the Ibole fauna currently situated in the Somali-Masai region has yielded too few small mammals to be interpreted in biogeographical terms. However, the presence of *Saidomys* indicates an Asiatic component reflecting the existence of a wide paleo-province as described by Brandy *et al.* (1980), but with a larger extension than these authors suggested. The Sahabi fauna in Libya (Munthe, 1987) has yielded a mixture of rodent genera indicating a middle Miocene age rather than an Late Miocene one. All the rodents of Sahabi are of North African affinities and have (in contrast to large mammals) no connection with those at Langebaanweg. In Egypt, the Early

Table 16-10. Number of common genera between East and South Africa during Middle and Late Pleistocene times.

Faunas	BIV[a]	MSK	ISN	KAPTH	BTW	NGAB	MLK	KRM
Masek Beds (MSK)	3	—	—	—	—	—	—	—
Isenya (ISN)	5	2	—	—	—	—	—	—
Kapthurin (KAPTH)	1	0	2	—	—	—	—	—
Bir Tarfawi (BTW)	4	3	4	1	—	—	—	—
Ngaloba Beds (NGAB)	4	3	5	1	**9**	—	—	—
Melka Kunturé (MLK)	1	1	2	2	7	3	—	—
Klasie River Mouth (KRM)	4	1	4	2	6	3	3	—
Wondewerk Cave	**7**	1	**6**	2	1	3	3	**9**

[a]Olduvai Bed IV.

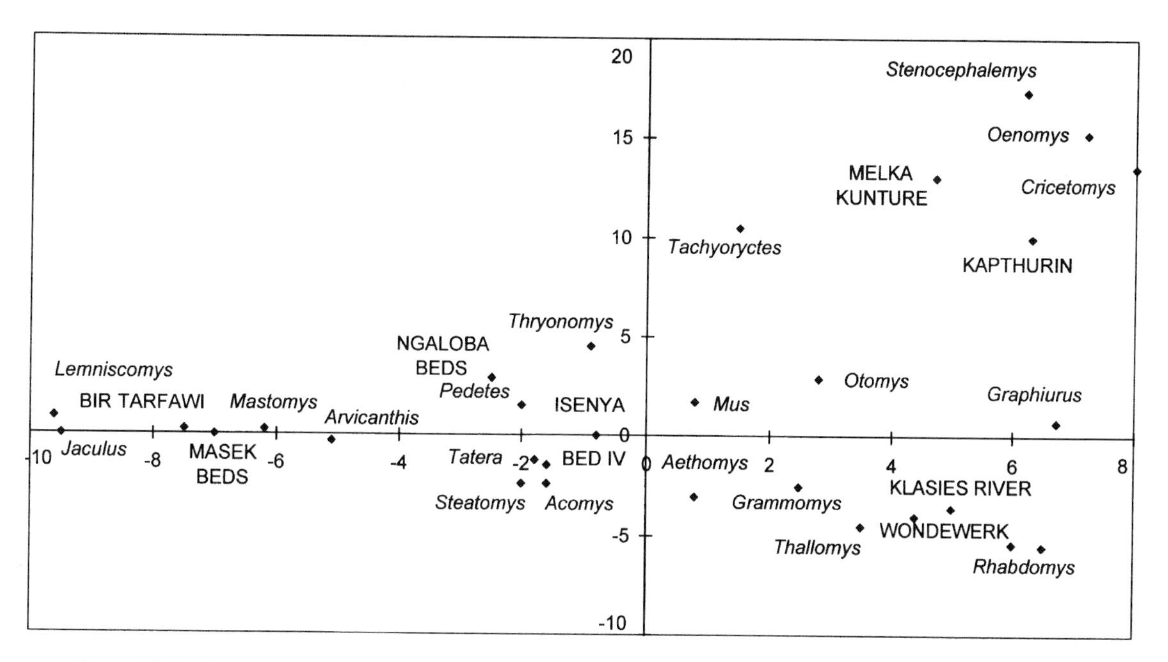

Figure 16-7. Correspondence analysis diagram of the first two axes displaying 42.9% of the total inertia for the Late Pliocene and Early Pleistocene fossil rodent faunas.

Table 16-11. Zoogeographical affinities of East and South Africa rodent faunas of the Early to Middle Pliocene.

Zone	UNB	LB	OMB	OMC	LBW	EXQRM	MRCIS	MLWD	NGA	HAD	JAG	NOS
Asiatic	0	0	1	1	0	0	0	0	0	3	0	0
Sahara	0	0	1	0	1	2	2	2	1	1	2	1
Sahelian	3	3	5	3	1	3	5	3	3	3	3	1
Sudanian	3	4	6	4	2	5	7	5	3	3	5	4
Guinean	4	4	7	5	3	6	8	6	3	4	6	5
Somali-Masai	**7**	**10**	**9**	8	6	9	**12**	11	5	**5**	10	**10**
Highveld	6	8	6	5	6	9	11	10	5	3	10	**10**
SW arid	5	8	5	4	**7**	7	9	8	**7**	3	**11**	**10**
Zambezian	6	8	8	**7**	6	9	**12**	**12**	6	3	9	**10**
Namib	3	4	3	2	2	3	4	3	3	2	7	4
SW Cape	2	4	3	1	7	**10**	11	10	5	2	9	**10**
Northern savanna	0	0	1	0	0	1	1	0	1	0	4	5
Southern savanna	2	5	2	0	3	4	5	6	4	0	6	5
Common savanna	4	4	5	4	4	5	7	6	2	3	0	0
Extinct genera	0	0	0	0	1	1	2	3	0	1	1	1

Key to site abbreviations

UNB	Upper Ndolanya Beds	MRCIS	Makapansgat rodent corner in situ
LB	Laetoli Beds	MLWD	Makapansgat Limework Dumps
OMB	Omo B	NGA	Ngamiland
OMC	Omo C	HAD	Hadar
LBW	Langebaanweg	JAG	Jägersquelle
EXQRM	Makapansgat exit quarry red mud	NOS	Nosib

Table 16-12. Zoogeographical affinities of East and South African rodent faunas of the Middle Pliocene to Early Pleistocene.

Zone	TAUNG	STS	SE	SK	KA	KB	H2	BBI	SBI	BII	SII	NAT	ET	OMF	OMG
Sahara	0	2	2	2	2	1	1	1	2	1	0	1	2	3	2
Sahelian	2	3	3	3	3	2	0	3	5	3	1	1	4	6	5
Sudanian	2	6	6	6	7	6	3	5	7	5	2	3	4	6	5
Guinean	2	7	7	7	8	7	4	6	7	5	3	4	6	7	5
Somali-Masai	7	13	**13**	**12**	**14**	**12**	9	**13**	**13**	**9**	**4**	**8**	**8**	**11**	**8**
Highveld	**9**	13	10	9	12	11	7	8	8	7	3	6	6	6	5
SW arid	**9**	12	10	9	12	11	8	10	10	7	2	7	6	5	5
Zambezian	7	13	**13**	**12**	**14**	**12**	**10**	12	11	7	3	7	7	10	5
Namib	5	7	5	4	6	5	3	6	5	4	2	5	5	3	3
SW Cape	5	**14**	12	11	12	**12**	7	7	6	5	1	5	4	3	2
Northern savanna	0	0	0	0	0	0	0	1	1	1	1	1	1	1	1
Southern savanna	5	6	6	5	7	5	6	8	5	3	1	4	2	2	1
Common savanna	2	7	7	7	7	8	5	4	6	4	2	3	5	7	4

Key to site abbreviations

TAUNG	Taung	SBI	Olduvai Upper Bed I
STS	Sterkfontein type site	BII	Olduvai Lower Bed II
SE	Sterkfontein extension	SII	Olduvai Upper Bed II
SK	Swartkrans	NAT	Natron Peninj site
KA	Kromdraai A	ET	East Turkana
KB	Kromdraai B	OMF	Omo F
H2	Humpata level 2	OMG	Omo G
BBI	Olduvai Middle Bed I		

Pliocene fauna of Wadi Natrun comprises *Saidomys* (James & Slaughter, 1974). Chaimanee et al. (1996) reported a cf.? *Saidomys* from undated Plio-Pleistocene fissure fillings in Thailand. The Langebaanweg fauna situated in the southwest Cape today shows southwest Cape and southwest arid affinities with a Zambezian and Somali-Masai component, indicating an early differentiation of these regions as early as the Early Pliocene. The Langebaanweg locality has no northern savanna rodents, but the assemblage indicates a mixture of grassland, swamps, riverine bush, and Fynbos habitats (Denys, 1990b, 1994, 1998).

Between 4 and 3 m.y., the East African faunas had already reached their maximum affinities with the Somali-Masai region, where they are currently situated (table 16-11), indicating a differentiation of this province as early as 3.7 m.y. The Laetoli Bed fauna has a high Zambezian proportion of taxa compared with Omo B and Hadar, which have a high Guinean proportion (northern savanna elements). The proportions of the rodent taxa are different from those observed today (Denys, 1985), but *Acacia* savanna environment is found at Laetolil beds (Denys, 1987a). The Omo B and C faunas represent, according to Wesselman (1984), a forest–savanna mosaic with *Acacia* woodland and mesic savanna. Hadar shows a mosaic of forest–moist savanna environments (Sabatier, 1982) and is characterized by Asiatic taxa (*Millardia, Golunda*) as well as by *Saidomys*. This confirms the persistence of a rodent community habitat in Ethiopia between Africa, Arabia, and Pakistan from the Late Miocene to the lower Pliocene, while in the south of Kenya and Tanzania, the Zambezian faunas were dominant. Differences between Ethiopian and Tanzanian faunas have been pointed out by Denys (1985, Denys et al., 1987a) and attributed to the Rift Valley structure and formation.

In South Africa, the Makapansgat assemblages do not correspond to the modern Transvaal Highveld habitats (Table 16-11; Denys, 1992b) and show a mixture of different communities probably reflecting mosaic environments, although their affinities are with the Zambezian savannas and even the Somali-Masai, indicating a shift toward the south of the Zambezian region. The Ngamiland fauna now situated in the South West Arid Zone has a southwest arid component in its faunas. It has been interpreted as dry savanna environment with bush and grassland (Denys, 1990a).

Table 16-13. Zoogeographical affinities of East and South African rodent faunas of the Middle to Late Pleistocene.

Zone	BIV	MSK	ISN	KAPTH	BTW	NGAB	MLK	KRM	WWK
Sahara	0	1	0	0	4	1	1	2	1
Sahelian	3	4	1	0	7	4	1	2	2
Sudanian	5	3	2	2	10	4	1	4	4
Guinean	5	3	3	3	10	4	1	4	5
Somali-Masai	**9**	**4**	**10**	**5**	**11**	**7**	**5**	10	12
Highveld	7	2	7	3	6	5	2	8	11
SW arid	7	2	7	2	5	5	2	8	**16**
Zambezian	9	2	9	4	8	5	2	9	13
Namib	4	1	6	1	3	4	2	6	10
SW Cape	5	1	5	2	4	3	2	**12**	12
Northern savanna	1	1	0	0	4	1	0	0	0
Southern savanna	4	0	6	1	1	2	1	5	7
Common savanna	5	2	3	3	8	3	1	4	5

Key to site abbreviations

BIV	Olduvai Bed IV	NGAB	Ngaloba Beds
MSK	Olduvai Masek Beds	MLK	Melka Knuturé
ISN	Isenya	KRM	Klasies River Mouth
KAPTH	Kapthurin	WWK	Wondewerk
BTW	Bir Tarfawi		

Late Pliocene-Early Pleistocene The Namibian faunas of 2.5 m.y. share little similarity with the modern Namib region but with the southwest arid region at Jägersquelle, southwest arid and southwest Cape at Nosib, indicating a less diversified fauna. At around 2.5 m.y., the Transvaal faunas situated today in the highveld region, such as the Sterkfontein type site, have higher similarities with the southwest Cape region but also retain a Zambezian savanna component (table 16-12). The more recent sites of Kromdraai, Swartkrans, and Sterkfontein extension have the same Somali-Masai and Zambezian dominance, indicating a shift toward the south of the savanna environment. The Omo C fauna already shows a Somali-Masai component at that time, but, contrary to the Omo Member B assemblage, does not show an important Guinean component. Instead, they have a Zambezian component confirming the retractation of the Saharo-Sahelian zone in the Northern Hemisphere. The Upper Ndolanya Beds at Laetoli show a Somali-Masai affinity.

In East Africa, between 2 and 1 m.y. the Somali-Masai element dominates in all the assemblages (table 16-12). There are few northern savanna elements and rather high percentages of southern savanna elements, except in the Omo F fauna, where the Guinean component is important. The Olduvai Beds I and II and the Natron rodent faunas appear close to modern Somali-Masai *Acacia* savanna mosaic. Wesselman (1984) has demonstrated an increase in aridity in the Omo Members F and G compared to Omo B and C. The Omo F and G members correspond, according to Wesselman, to open, dry savanna woodland and semi-arid steppe. At East Turkana, the rodent fauna suggests an arid environment with *Acacia* scrub and some riverine forest along intermittent stream channels (Black & Krishtalka, 1986). This site comprises the earliest occurrence of the gerboa *Jaculus,* 2000 km farther south than its modern distribution (Denys, 1990a). The Angola fauna at Humpata 2 situated in the Zambezian region has a Zambezian assemblage, with some elements of the Somali-Masai region. This indicates a Zambezian savanna zone as wide as today around 1.8 m.y. The Taung fauna, now in the South West Arid Zone, shows a maximum of Highveld and southwest arid elements.

Middle to Late Pleistocene There is a gap in the Middle to Late Pleistocene fossil-record in South Africa. In East Africa, there still is a predominance of Somali-Masai components in the faunas from Tanzania to Ethiopia, with few northern savanna ele-

Table 16-14. Zoogeographical affinities of the fossil faunas through time showing their changes compared with the modern situation (data from tables 11–13).

Time (Ma)	Site	Modern Region	Dominant assemblage	Other components
6–5	Ibolé	Somali-Masai	?	
	Langebaanweg	SW Cape	SW Cape + SW arid	Zambezian+Somali-Masai
4–3	Laetoli Beds	Somali-Masai	Somali-Masai	Zambezian
	Makapansgat			
	exit quarry red mud	Highveld	Zambezian	Somali-Masai+SW Cape
	rodent corner in situ	Highveld	Zambezian + Somali-Masai	SW Cape+SW arid
	Limework Dumps	Highveld	Zambezian	Highveld + Somali-Masai
3	Hadar	Somali-Masai	Somali Masai	Guinean
	Omo B	Somali-Masai	Somali-Masai	Zambezian-Guinean
	Ngamiland	SW arid	SW arid	Zambezian
2.5	Omo C	Somali-Masai	Somali-Masai	Zambezian
	Upper Ndolanya Beds	Somali-Masai	Somali-Masai	Highveld + Zambezian
	Jägerquelle	Namib	SW arid	Highveld + Somali-Masai
	Nosib	Namib	SW arid	Highveld + Somali-Masai
	Taung	SW arid	Highveld + SW arid	Zambezian + Somali-Masai
	Sterkfontein type site	Highveld	SW Cape	Zambezian + Highveld + Somali-Masai
2–1.5	Omo F	Somali-Masai	Somali-Masai	Zambezian + Guinean
	Omo G	Somali-Masai	Somali-Masai	?
	Sterkfontein extension	Highveld	Somali-Masai + Zambezian	SW Cape
	Swartrans member 1	Highveld	Somali-Masai + Zambezian	SW Cape
	Kromdraai A	Highveld	Somali-Masai + Zambezian	SW Cape
	Kromdraai B	Highveld	Somali-Masai + Zambezian	SW Cape + Highveld + SW arid
	Olduvai Middle Bed I	Somali-Masai	Somali-Masai	Zambezian
	Olduvai Upper Bed I	Somali-Masai	Somali-Masai	Zambezian
	East Turkana	Somali-Masai	Somali-Masai	Zambezian
	Natron	Somali-Masai	Somali-Masai	Zambezian + SW arid
1.5–1	Lower Bed II Olduvai	Somali-Masai	Somali-Masai	Zambezian + SW arid + Highveld
	Upper Bed II Olduvai	Somali-Masai	Somali-Masai	?
	Humpata 2	Zambezian	Zambezian	Somali-Masai
0.8–0.6	Melka Kunturé	Somali-Masai	Somali-Masai	?
	Olduvai Bed IV	Somali-Masai	Somali-Masai + Zambezian	Highveld + SW arid
	Isenya	Somali-Masai	Somali-Masai	Zambezian
0.6–0.4	Masek Beds	Somali-Masai	Somali-Masai + Sahelian	?
	Kapthurin	Somali-Masai	Somali-Masai	Zambezian
0.4–0	Ngaloba	Somali-Masai	Somali-Masai	Zambezian+Highveld + SW arid
	Bir Tarfawi	Sahara	Somali-Masai	Sudano-Guinean
	Klasies River mouth	SW Cape	SW Cape	Somali-Masai
	Wondewerk	SW arid	Zambezian	SW Cape Somali-Masai

The most abundant component is given in the third column and the second or third dominant one in the fourth column. A question mark means there are insufficient data for precision.

ments, but a good proportion of Zambezian ones (table 16-13). The Masek Beds fauna at Olduvai has a high proportion of Sahelian elements, indicating that the maximum expansion toward the south of the Sahara-Sahel zone was reached at around 0.6–0.4 m.y. At 120,000 years, in Bir Tarfawi (Egypt), today situated in the Sahara, one finds a dominance of Somali-Masai elements with a component of Sudano-Guinean ones. At Laetoli, the Ngaloba fauna (0.12 m.y.) shows only Somali-Masai and Zambezian elements. In South Africa, the Holocene fauna at Klasies River mouth today belongs to the Cape Province, and its fossil fauna also corresponds to this. In contrast, the Wondewerk cave assemblage situated in the southwest arid region has a dominant Zambezian component with a Cape and Somali-Masai component, indicating a change in the vegetation zones at that time.

Timing of the Regional Differentiations and Biome Shifts

Between 6 and 4 m.y., the differentiation of the southwest Cape Province and of the southwest arid one takes place in South Africa, while in East Africa there is a wide paleo-province composed of taxa found contemporaneously in Afghanistan and Arabia. Brandy et al. (1980) first described this province, and we know now that around 6–4 m.y. it included Tanzania. This "paleo-Somali-Masai province" may be a heritage of the Late Miocene faunal province widespread all over Africa and may correspond to an exchange phase between Afro and Arabian plates as seen for bovids (Thomas, 1979) and hipparion horses (Bernor & Armour-Chelu, this volume). But this also coincides with full-scale glaciation of Greenland, occurring around 7 m.y. ago (Larsen et al., 1994; Denton, this volume). The Early Pliocene East African fauna does not show any affinities with the Early Pliocene faunas of the Moroccan or the Algerian coasts. This could be due to an early differentiation of the Sahara as a barrier, preventing contacts between North and East Africa. There is evidence of Sahara differentiation since the Early Pliocene (Maley, 1996).

The paleoenvironment is indicated by species of this paleo-Somali-Masai fauna such as *Saidomys* with a grazing-like dental morphology and *Saccostomus* with a granivorous one. The presence of an *Acomys*-like dental pattern in the extinct genus *Proceromys* is in favor of a relatively arid savanna grassland environment for this paleo-Masai province, which is in agreement with the Middle Miocene development of grassland recorded from isotopic soil composition (Retallack, 1992). During that time, the uplift of the Rift Valley had started, but not the faulting, and although there are indications of an opening of the Red Sea, the passage across the Bab el Mandeb strait probably was still possible (Denys et al., 1987a). The "African genera" today found in Arabia, such as *Praomys, Acomys,* and *Arvicanthis* (Harrison, 1972), are phylogenetically related with *Saidomys* and *Proceromys* and could be the relicts of this old, relatively arid province. In the south, the southwest arid region was also differentiated, as attested by the presence of the Benguela current as early as 14 m.y. According to Diester-Haas (1987), it reached its present situation between 6.6 and 5.2 m.y. It shifted toward the equator during glacial periods and toward the pole during interglacials, but also some displacements to the west are recorded.

The Cape province differentiates during a wet, hot episode in Early Pliocene times and is characterized by a phase of modern Murinae radiation (including the Otomyinae). This last warm episode observed in the Southern Hemisphere could be the reason that the modern Murinae are more diversified in the tropical regions today (Misonne, 1969). But the murine radiation observed in the Early Pliocene (at 5 m.y.; Chevret, 1994) could only result from a faunal turnover after the severe arid conditions observed in the Late Miocene event. This radiation is not synchronous with extinctions, which is contra Vrba's theory. In fact, few extinctions are observed at this time, and they take place at least after 3 m.y. (Makapansgat) in South Africa, which could explain the higher diversity of the South African rodent faunas versus the eastern ones between 3 and 3.5 m.y. The persistence and cohabitation of Early Pliocene extinct forms and modern genera can have two explanations: first, this warm climatic episode provided sufficient niches for all the genera to persist; second, and more likely, a strong reelaboration process (which seems to be commonly found in cave formation; Denys et al., 1997) led to a mixture of diachronic populations in Makapansgat, Taung, Sterkfontein, and other caves.

During the Middle Pliocene (4–3 m.y.), the differentiation of the modern Somali-Masai region in Tanzania and of the Zambezian savanna in South Africa takes place. If the Laetoli beds faunas have fewer Asiatic elements than the older Ibole faunas and the younger and the more northerly situated Omo B, C and Hadar ones, this could indicate a change in the distribution of the regional zones between 4 and 3.7 m.y., with a shift toward the north of the old paleo-Masai province during global warming as well as the occurrence of Guinean savanna elements in the Ethiopian faunas [perhaps due to a general shift of the monsoon winds eastward; moonson winds being attested since Oligocene times (Maley, 1996)]. In East

Africa, this period corresponds to a high frequency of FAD of modern genera, especially Murinae but also Gerbillinae. At that time, the Rift Valley was composed of a succession of small basins developing lakes, separated by thresholds of altitude. These topographic conditions favor mosaic habitats, while volcanoes also play an important role in providing carbonatite lava, allowing the development of the Serengeti grassland ecosystem (Owen-Smith, this volume) and allowing speciation by isolation on the top of the mountains by producing altitudinal vegetation belts (Kingdon, 1990). In South Africa, the temperate highveld region of the Transvaal is not yet differentiated, and the Zambezian savanna is found farther south than today, indicating a more tropical climate in that region. This confirms the hypothesis of Rayner et al. (1993) that the uplift of the Transvaal Plateau (900 m uplift between 5 and 3 m.y.) was not yet achieved and that the climate was wetter than today. The fact that the Laetoli and Makapansgat faunas belong to different biogeographic zones confirms the difference of rodent species and faunal assemblages described by Denys (1990a), but fundamentally they both belong to a savanna system, and the Somali-Masai region for the rodents has clearly more Zambezian affinities than Guinean ones, and they could have differentiated from each other. According to Grubb et al. (this volume), this corresponds to the southern savanna zone of various authors, based mainly on large mammals.

Before or around 2.5 m.y., there is in the Namib the southwest arid and the Cape component, while in the Transvaal, the Sterkfontein type site fauna has a Cape component, indicating a strong shift toward the north of the arid region and winter rainfall regime. This is not accompanied by extinctions of genera in that region. In East Africa, the Upper Ndolanya Beds do not show such an important change, while at the Omo C the Zambezian element is important. More faunal records of this time are needed to improve the understanding of this event. The Lusso Beds in Zaire give evidence of an *Otomys* shift toward the north, which might confirm the general shift toward the north in all regions (southwest arid and Zambezian), probably in relation with the cold event observed at around 2.5 m.y. (Bonnefille, 1983; Denton, this volume). This was also suggested by Bromage et al. (1995) and seems to conform with Vrba's habitat theory (Vrba, 1992, this volume).

In the Early Pleistocene, 2–1.7 m.y., there is evidence of a strong shift toward the south of the Guinean savanna accompanied by Sudano-Sahelian arrival in the Rift Valley, but only in Ethiopia (Omo F) and North Kenya (East Turkana) (north of the equator), and there may have been an increase of arid open environments at low altitude only. In Tanzania, the faunas retain the typical Somali-Masai pattern and are close to their modern equivalents, and in some cases the modern species appear as early as 1.7 m.y. (Denys, 1992a), despite local paleoenvironmental fluctuations (Fernandez-Jalvo et al., 1998).

The last time period is less well documented, but there is an important extinction event in South Africa (Denys, 1990a), where between 1.5 and 0.3 m.y. now-extinct species were replaced by modern ones (the timing of these speciations is not known, and nothing can be said about their causes and eventual synchronism). According to Sikes (this volume), after 0.9 m.y., the modern types of soils existed in East Africa. The Holocene fauna at Klasies River mouth has a southwest Cape affinitiy corresponding with its present situation, but the Wondewerk fauna situated in the South West Arid Zone has a Zambezian dominant assemblage. In East Africa, the fossil record is better documented and if the Bed IV, Isenya and Melka Kunturé faunas, despite differences related to their geographical distribution, belong to the Somali-Masai region as today, this is not the case for the Masek beds faunas at Olduvai (0.6–0.4 m.y.), which have a Somali-Masai component but also a Sahelian one, compared to the others which have a Zambezian element.

According to paleoclimatic data, there is a strong global cooling at 0.6 m.y. (Denton, this volume), and this could have provoked a shift of the Sahara-Sahelian zone (or from a relict *Jaculus* population of the East Turkana region). This corresponds to a shift toward the south of about 700 km. The *Jaculus* dispersal is observed today in Senegal, where the gerboa as well as two other *Gerbillus* species have advanced 50 km in 20 years (Duplantier & Granjon, 1992). The dispersal is thought to be due to the present extension of the Sahara in response to the increase of severity of dry seasons in that part of the world. These shifts can be rapid and may be too fast to be recorded by fossilization. Taphonomic problems of underepresentation of some taxa also complicate this scenario.

There is much evidence for such vegetational shifts in the palynological record overall in tropical Africa (e.g., Neumann & Schulz, 1987; Ritchie & Haynes, 1987; Lezine, 1988; Dupont & Agwu, 1992). The existence of a latitudinal zonation of vegetation in North and West Africa is attested by Ritchie and Haynes (1987), who described a shift of the Sahelo-Sudanian zone of more than 450 km toward the north between 9500 and 4500 BP during a wet phase in Sudan. The Bir Tarfawi fauna illustrates the same tendency with a shift toward the north of the Somali-Masai region of more than 1000 km at 0.12 m.y. This shift in distribution was possible because of the persistence of woodland and more humid habitats along the Nile River (Kowalksi et al., 1989). The last 600,000 years are marked by an impressive shifts of biomes, and this could be related directly to the increasing intensity of glacial events seen from 0.9 m.y. (Denton, this volume) and may explain the phase of faunal extinction seen in South Africa, which is less well documented in East Africa (Denys, 1990a).

Resulting Speciation Patterns

It has been shown that the Early-Middle Pliocene time, with its wet, hot climate, is the time of the African murine radiation. Extinctions are not related to major biome shifts, but allopatric speciation occurs due to topographic or edaphic differences, such as in East Africa versus South Africa, and due to isolations either in the southwest arid province or in the Ethiopian Highlands. Low evolutionary rates are expected in the center of these biotic regions, except at the time of major shifts (2.5 m.y., 1 m.y., 0.6 m.y.), but in contrast, higher speciation rates must be found in contact zones between regions (e.g., Somali-Masai; parapatric speciation), and this could explain why the diversity is much higher in that region of Africa and greater than that seen in forest regions. The increasing intensity of Late Pleistocene glacial episodes may have favored the extinction of species and chromosomal speciation, which is found in most of the modern African rodent genera.

Conclusions

Mice and Men in the Same Trees?

Can we use the method discussed here to help infer the biogeographic specificity of early hominids? Are some hominid species related to one or the other vegetation zones like some rodents are? From the results based on mice, it can be concluded that *A. anamensis* and *A. afarensis* were species typical of the paleo-Somali-Masai province (dry savanna province) described for rodent assemblages in the deposits containing these hominids. For *A. anamensis,* dry, possibly open, wooded conditions are recorded by M. G. Leakey et al. (1995) in Kanapoi, but rodent faunas are not yet described, and at Allia Bay there is a gallery forest habitat. Based on the whole community and on analyses of the local habitat environment, Andrews & Humphries (this volume) find evidence of dry forest conditions for *Ardipithecus ramidus* at Aramis in Ethiopia, wooded savanna in Laetoli, and rich wooded environment in Hadar, but different environments from any existing today. Sikes (this volume) found grassy woodland at Laetoli. But there are signs of a north–south differentiation of this paleo-Masai province as early as 4 m.y. Leakey et al. (1995) concluded there is a contrast between the paleoenvironments of Kanapoi, Allia Bay, and Aramis sites. Similarly, the rodents of Laetoli indicate that in the south of this paleo-Masai province, there is an early differentiation of the modern Somali-Masai province with a strong Zambezian component (and no northern savanna elements compared to Hadar) as seen for small mammals between Ibolé and Laetoli. Consequently, the persistence of *A. afarensis* in Hadar could result from isolation in the Afar region of this taxon. The paleo-Somali-Masai province persists in the northern sites of Ethiopia until 3 m.y. and may be supportive of a northern shift of all vegetation zones north of the equator. The new Chad *A. bahrelgazali* (Brunet et al., 1996) now situated south of the Sahara region has yielded few small mammals, among them a new *Xerus* species (Denys et al., in preparation) and some Murinae. The presence of an australopithecine representative suggests that it either belongs to the paleo-Somali-Masai province or even indicates an early differentiation of the northern savannas (primitive Sudano-Guinean savanna region?) extending farther northward than today under warmer climatic conditions of the Middle Pliocene. Probable continuity of communications between the northern savannas and the paleo-Somali-Masai region during Pliocene is attested by the existence of modern rodents of Saharian-Sahelo-Sudanian regions still living in the Somali zone (like *Pectinator, Jaculus,* and *Gerbillus*). *Australopithecus africanus* is clearly a Zambezian savanna element [with a different rainfall and potential evapotranspiration and different wooded vegetation, and perhaps more edible plants (O'Brien, this volume) than north of the equator].

At 2.3 m.y. *Homo rudolfensis* is found first in Malawi. This region is today at the contact zone between the Somali-Masai and the Zambezian domains and at the crossroads of the Rift Valleys. No small mammals have been found despite intensive screening, but for the large mammals, Schrenk et al. (1995) found Zambezian affinities. Speciation could have occurred in favor of a shift toward the north of the Zambezian savanna domain and might explain a farther dispersal of this taxon to the north of the Rift Valley. In Tanzania, *H. habilis* belongs to Somali-Masai dominant faunas with a strong Zambezian component. By that time, at Olduvai Bed I, there is the first occurence in Tanzania of South African genera (such as *Otomys, Aethomys,* and *Zelotomys;* see Denys, 1990a) following a shift northward of the Zambezian element, probably the one dated at 2.5–2.3 m.y. The *Paranthropus* lineage differentiates in Ethiopia with the first species, *P. aethiopicus,* whose FAD is of 2.7 m.y. according to Kimbel (1995) at Omo Member C, a site where the Somali-Masai element dominates with a Zambezian component (fig. 16-14). At Omo Member G, which has a Somali-Masai dominant with one northern savanna element and less Zambezian influence (table 16-14), *P. aethiopicus* and *P. boisei* are found. In Peninj (West Natron site), which has yielded *P. boisei,* the rodent assemblage has a Somali-Masai component and Zambezian savanna affinities like at Olduvai Bed I, but all these faunas have one exclusive northern savanna element, contrary to the Laetoli or Hadar sites. In South Africa at around 2–1.7 m.y., we see a balance of the Zambezian and Somali-Masai elements but no northern savanna elements, and an important southwest Cape contribution (table 16-14), and *P. robustus* is clearly of Zambezian affinities. The West Turkana faunas have not yielded any rodents. The East Turkana fauna from Okoté formation indicates also a Somali-Masai and Zambezian component with two Saharian elements and eight Sudano-Sahelian ones, which suggests a more northern influence and arid environment and has yielded *P. boisei* and *H. erectus.*

What of the Forest Province?

During the Late Miocene and Early Pliocene, Africa was the theater of an important event in the form of the African murine radiation (and perhaps the hominine radiation), with the emergence of modern genera and the development of the modern vegetation zonation. These zones correspond broadly to vegetation types as well as rainfall patterns and therefore follow global climatic patterns. This work gives evidence of latitudinal zonation of the vegetation existing since at least the Early Pliocene and indicates that the Somali-Masai and Zambezian provinces are more similar to each other than they are to the Sudano-Guinean zone. The forest region has separated these two savanna domains. No fossil small mammals faunas are recorded in the forest environment, but there is indirect evidence of forest expansion toward the east (Williamson, 1985), and Maley (1987) described a scenario where the montane forest belt extended at a lower altitude during cold periods. It is necessary to keep in mind that climatic shifts are probably blocked by the meteorological equator, produced by the convergence of the northern and southern tradewinds, and that the latter are mainly controlled by the pole–equator thermal gradients in both hemispheres (Maley, 1996). We should also keep in mind that Ethiopian and North Kenya faunas are situated in the Northern Hemisphere, while the Tanzanian faunas and, of course, the South African faunas are in the Southern Hemisphere. The monsoon winds have prevented the development of forest in the Somali-Masai region at the level of the equator, but the variations in their intensity are not well known.

Further discoveries of rodent faunas of different ages might allow us to better understand the pattern and processes of rodent speciation in Africa. This approach shows that the vegetation shifts do not drastically change the community structure, but, at the contact zones, these frequent changes of environment may have been one of the motors of evolution of mice (and men) in tropical Africa.

Acknowledgments I am grateful to Friedemann Schrenk and Tim Bromage who gave me the opportunity to participate in this conference. This paper was originally prepared for the Wenner-Gren conference "African biogeography, climate change and early hominid evolution," held October 25–November 2, 1995, in Salima, Malawi. Ray Bernor, Alan Turner, and Michel Tranier carefully reviewed this chapter. K. Klaver corrected the English.

17

Relationships between Eastern and Southern African Mammal Fauna

Peter Grubb
Oliver Sandrock
Ottmar Kullmer
Thomas M. Kaiser
Friedemann Schrenk

Malawi lies between the early-hominid sites of eastern and southern Africa. Remains of the Plio-Pleistocene fauna recovered from the Malawi Rift (Bromage et al., 1995a; Schrenk et al., 1995) include both existing and extinct taxa otherwise known from only eastern or southern Africa, or from both. This chapter provides a background to these implications of endemism or cosmopolitanism by summarizing the geography of the extant mammal fauna of High Africa—eastern and southern Africa, encompassing the Great Rift Valley and most of the African mountains (fig. 17-1)—with emphasis on larger mammals (Grubb, this volume). Faunas have been reviewed in regional works (sources of biogeographical information in this chapter) covering Ethiopia (Largen et al., 1974; Yalden et al., 1976–1986), Somalia (Azzaroli & Simonetta, 1966; Funaioli & Simonetta, 1966), East Africa (Roosevelt & Heller, 1915; Hollister, 1918–1924; Kingdon, 1971–1982), Tanzania (Swynnerton & Hayman, 1951), Congo (Schouteden, 1944–1946), Zambia (Ansell, 1978), Malawi (Ansell & Dowsett, 1988), Angola (Hill & Carter, 1941), Mozambique (Smithers & Lobao Tello, 1976), Zimbabwe (Smithers & Wilson, 1979), Botswana (Smithers, 1971), Transvaal (Rautenbach, 1982), Orange Free State (Lynch, 1983), Namibia (Shortridge, 1934), Kwazulu-Natal (Pringle, 1974, 1977; Rowe-Rowe, 1978, 1991), Cape Province (Stuart, 1981; Herselman & Norton, 1985), and the Southern African Subregion (Stuart & Stuart, 1988; Skinner & Smithers, 1990). Systematics follows Wilson and Reeder (1993), except that infraspecific taxa near to full-species status are listed binomially, with the senior synonym in parentheses [e.g., *Papio (hamadryas) cynocephalus*].

The spectrum of mammalian dispersion from the Horn to the Cape is one of relatively discrete but overlapping regional faunas occupying centers of endemism (figs. 17-1, 17-2), accompanied by more widespread species. This pattern is related to latitudinal variation in climate, substrate and vegetation, to nutritional requirements, habitat preferences, and predation (Owen-Smith, this volume) and to constraints on dispersal and other historical contingencies. Ranges of major faunas correspond with biome-sectors or biotic zones, defined (Davis, 1962) by simplifying a vegetation map of Africa (Keay, 1959). In the light of the new map (F. White, 1983) their limits are here redrawn. Subzones are identified, and names of constituent phytogeographic units are given in figures 17-1 and 17-2, and by Denys (this volume).

Subregions and provinces in Africa determined independently by Turpie and Crowe (1994) from a study of large-mammal distributions correspond quite closely with the zones recognized here. The most extensive of these in High Africa is the Southern Savanna which separates the Somali and South West Arid Zones (figure 17-1). The two latter zones are connected climatically if not phytogeographically by the

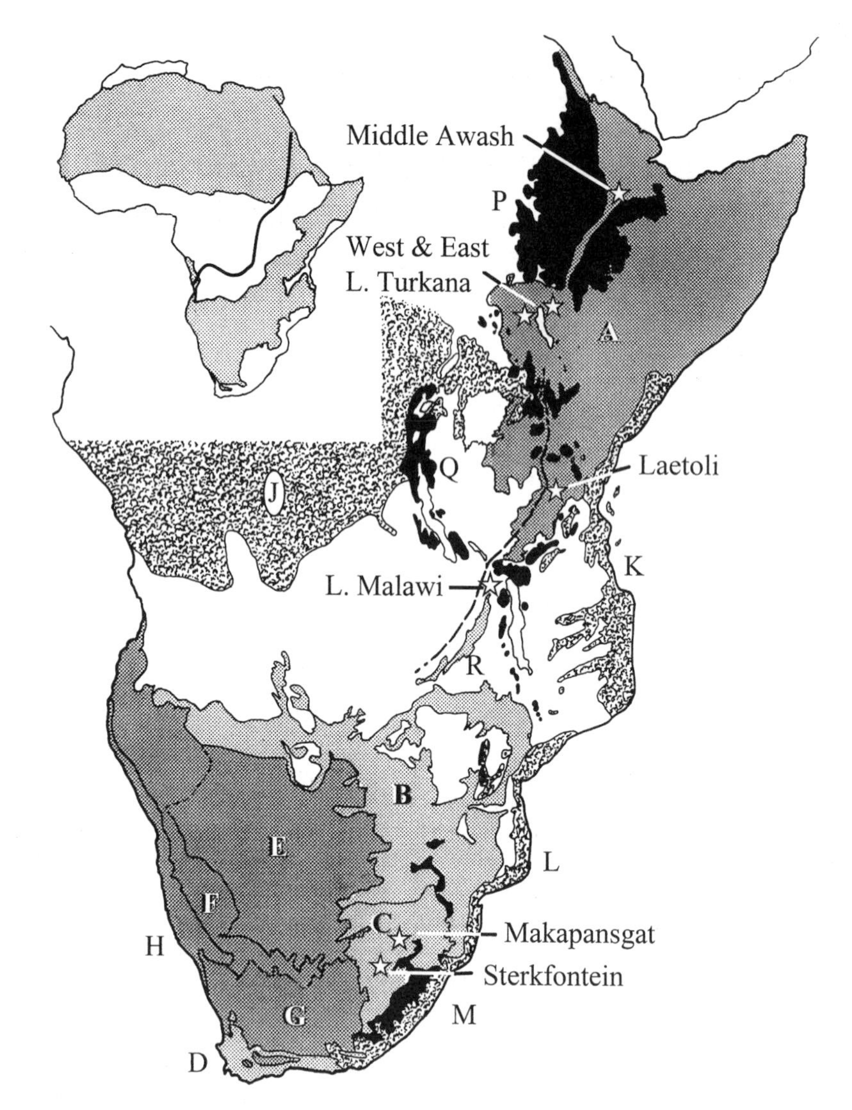

Figure 17-1. Map of contrasting arid and mesic biotic zones in High Africa with borderlines taken from F. White (1983), with his nomenclature for regional centers of endemism (RCE), transition zones (RTZ), or regional mosaics (RM) indicated in parentheses. Arid zones: Somali Arid Zone (Somalia-Masai RCE) (heavy stippling), A; Greater South West Arid Zone, transitional (light stippling): Dry Savanna (Zambezian RCE in part), B; Highveld (Highveld grassland, part of Kalahari-Highveld RTZ), C; Fynbos (Cape RCE), D; South West Arid Zone proper (heavy stippling): Kalahari (*Acacia* wooded grassland and deciduous bushland; and transitional vegetation types: part of Kalahari-Highveld RTZ), E; North Nama-Karoo (bushy Karoo-Namib shrubland; *Colophospermum mopane* woodland and scrub woodland; and transitional vegetation types: parts of Karoo-Namib and Zambezian RCEs), F; South Nama-Karoo (dwarf Karoo shrubland, with bushy Karoo-Namib shrubland, succulent Karoo shrubland, and transitional vegetation types, part of Karoo-Namib RCE), G; Namib (Namib Desert, part of Karoo-Namib RCE), H. Forest and forest/savanna transitional zones (shaded): Central Forest and transition with Southern Savanna (Guineo-Congolian RCE, Congolian/Zambesian RTZ, Lake Victoria RM), J; Eastern Forest Zone lowlands (Zanzibar, Inhambane, and Tongaland-Pondoland RM), K, L, M. Highlands, including highland forest of Central and Eastern Forest Zones (incorporating Afromontane RCE and Afroalpine RCE) (solid black), including Ethiopian Highlands, P, and Albertine Highlands, Q. Luangwa Valley, R. (Dotted line) Kingdon's line between Central and Eastern Forest Zones. (Inset: bold line separates Low Africa to northwest from High Africa. Stippling covers area in which monthly rainfall is <10 mm in at least 3 consecutive months and includes arid corridor in East Africa).

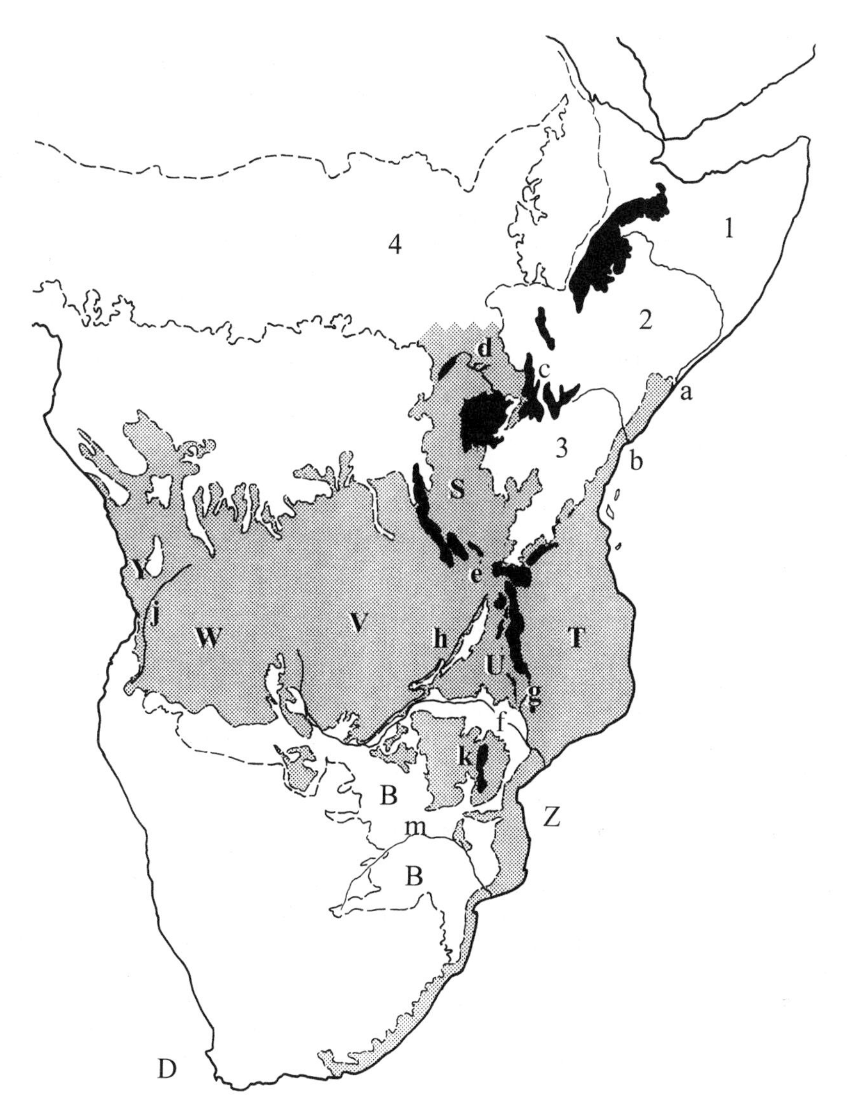

Figure 17-2. Map of Southern Savanna and surrounding biotic zones, together with topographic features that may contribute to endemism. Informal geographic divisions of Somali Arid Zone: Ogaden and Somali Plateau, 1; northern Kenya and Jubaland, 2; Masailand, 3; Northern Savanna, 4. Informal geographic divisions of Southern Savanna: Uganda to west Tanzania, S; Juba River to Zambezi River, T; Muchinga/Zambezi/Malawi triangle, U; Zambia/Katanga Plateau, V; Angola Plateau, W; Angola escarpment zone, Y; Mozambique Lowlands, east of Zimbabwe Highlands, Z. Dry Savanna zone, B. Topographic features: lakes and selected highlands (black); Webe Shebeli River, a; Tana River, b; Kenya Highlands partitioned by Gregory Rift, c; Victoria Nile, d; 'gate' between Ufipa Highlands/Lake Rukwa and Mbeya-Poroto Highlands, e; Zambezi River, f; Shire River, g; Zambezi-Muchinga Escarpment, h; Angola escarpment, j; Zimbabwe Highlands, k; Limpopo River, m.

drought corridor, a band of country crossing eastern Africa where monthly rainfall is <10 mm in at least 3 consecutive months (fig. 17-1; Balinsky, 1962). "Islands" of other biomes (Afroalpine, Afromontane, and lowland rainforest) are scattered through High Africa, rarely forming extensive blocks.

Faunas of Biotic Zones

Forest Faunas

The scattered Afromontane and lowland forests of High Africa can be partitioned by an analogue of

Wallace's line, "Kingdon's line" (fig. 17-1), crossing eastern Africa and running between Lakes Tanganyika and Malawi, through Tanzania to the floor of the Gregory Rift in Kenya. This division represents the optimal separation of overlapping central and eastern forest faunal elements.

The Albertine Rift massifs on the edge of High Africa have been sufficiently separated from each other for provincialism in their upland central-forest mammal fauna to evolve. Lowland central-forest species have ranged between these mountains into Uganda or Rwanda to the western Kenya Highlands (Mau Escarpment), and the eastern shore of Lake Tanganyika, passing through High Africa along a "northern route" (Kingdon, 1971–1982): of 88 nonvolant, strictly forest species in the Ituri Valley and neighboring lowland forest in eastern Congo, 33 reach no farther east than the Bwamba forest, Semliki Valley, on the Congo-Uganda border; 23 extend farther into Uganda but not east of the Victoria Nile; another 23 occur east of the Victoria Nile but no farther than the western border of the relatively dry Gregory (Kenya) Rift Valley; but only 9 species have breached this barrier (part of Kingdon's line) and occur still farther east, in the Aberdares, Mount Kenya, and even Mount Kilimanjaro. Rodgers et al. (1982) attribute this distribution pattern to postglacial dispersal. Similar patterns could nevertheless have existed repeatedly during previous climatic cycles.

The eastern forest fauna occupies montane forests east of Kingdon's Line and lowland forests of the Zanzibar-Inhambane and Tongoland-Pondoland mosaics (F. White, 1983) along the eastern African coast (Kingdon, 1971–1982; Rodgers et al., 1982; Grubb, 1983). The fauna is distributed across some 36° of latitude from the Juba Valley in southern Somalia to Knysna in Cape Province (fig. 17-1). The forest fauna is not merely depauperate, but also selectively deficient, lacking paludinous and aquatic taxa (*Potamogale, Scutisorex, Deomys, Malacomys, Colomys, Osbornictis, Hyemoschus, Aonyx congicus, Tragelaphus spekii*), many insectivorous bats, lowland forest shrews, most monkeys of the genus *Cercopithecus,* pangolins (*Manis* species), brush-tailed porcupines (*Atherurus africanus*), yellow-backed duiker (*Cephalophus silvicultor*), and, indeed, many otherwise widespread, even eurytopic, lowland forest mammals. Chimpanzees (*Pan troglodytes*), too, have not crossed the drought corridor or reached the Eastern Arc Mountains of Tanzania or the coastal forest. About 50 species are endemic to the eastern forest including among larger mammals, *Galagoides, Otolemur, Procolobus, Cercocebus, Cercopithecus, Dendrohyrax, Neotragus,* and *Cephalophus* species. Some endemics are in genera absent from the main forest blocks (*Otolemur, Beamys*). Others are allospecies of central forest taxa, or only subspecifically distinct. The eastern forest fauna therefore may have begun to differentiate at an early date, with the rise of the Albertine Rift highlands (Pickford, 1993) but has been episodically augmented by incursions from outside, at least in part along Kingdon's (1971–1982) "southern route." Compared with their nearest equivalents in Central Africa, mammals of these forests experience a more discontinuous habitat, interspersed through the savanna, and have wider habitat tolerances. This applies to species of *Nycteris, Galagoides, Otolemur, Colobus, Cercopithecus, Heliosciurus, Paraxerus, Rhynchocyon, Petrodromus, Bdeogale, Dendrohyrax, Neotragus,* and *Cephalophus.* The Eastern African Forest Zone is a center for colonization of the savanna separate from, and very much secondary to, the extensive Central Forest Zone.

Nonforest Faunas

About 102 zoogeographic species of large mammals are present outside rainforest in High Africa. Forty-one occur in relatively mesic habitats, 31 in relatively dry habitats only, and 30 in both. Of these last 30, 12 zoogeographic species have strongly differentiated arid-zone and savanna taxa (at the conventional species level, or close to it); 18 do not. A wide range of mammalian forms occurs in both savanna and arid zones, including pangolins, shrews, hedgehogs, microchiropteran bats (*Nycteris, Hipposideros, Rhinolophus, Pipistrellus, Laephotis, Tadarida*), galagos, baboons, mustelids, canids, mongooses, genets, hyaenas, felids, aardvarks, hyraxes, elephants, rhinoceroses, equids, warthogs, antelopes, hares, ground-squirrels, dormice, gerbils, mice, and porcupines.

Ethiopia

The Northern Savanna of Low Africa continues into the foothills of the northwestern Ethiopian Highlands (fig. 17-2), bringing with it some species that are not widespread in the Southern Savanna, that do not range south of Tanzania, and are replaced in most of the Southern Savanna by allotaxa (table 17-1). The floor of the Ethiopian Rift Valley allows lowland species, including those preferring relatively dry conditions, to disperse between the northwestern and southeastern Ethiopian Highland blocks. The southeastern block grades sharply into surrounding dry savanna including *Acacia* woodland, the Somali Arid Zone, which cuts off the Ethiopian Highlands from the Southern Savanna so that most High African savanna endemics

Table 17-1. Distributions of Southern Savanna mammals indicating endemism, vicariance, or limitation of distribution by topographic features.

Feature	Species
Northern/Southern Savanna vicariance, with contact zone in High Africa, the Northern Savannah taxa mostly present also in Ethiopia	*Galago senegalensis/G. moholi, C. (a.) tantalus/C. (a.) pygerythrus* [replaced by *C. (a.) aethiops* in Ethiopia], *Papio (h.) anubis/P. (h.) cynocephalus, Redunca redunca/ R. arundinum*
Affected by occurrence of western and eastern moieties of moist savanna in East Africa (Fig. 2):	
Endemic to east in Kenya:	*Epomophorus minimus* (also ranges into Somalia), *Taphozous hildegardiae, Rhinolophus deckeni, Miniopterus minor minor*
Populations in west and east separated by Taru Desert in Kenya	*Epomophorus wahlbergi, Potamochoerus larvatus, Ourebia ourebi, Redunca redunca, Damaliscus (lunatus) korrigum, Tragelaphus scriptus*
Confined to east within Tanzania	*Bdeogale crassicauda, Connochaetes taurinus nyasae*
Partitioned between west and east in Tanzania	*Rhynchocyon cirnei/R. petersi* (reputedly with hybrid zone; Kingdon, 1974), *Kobus (ellipsiprymnus) defassa/ K. (e.) ellipsiprymnus* (with hybrid zone; Peters, 1986), *Tatera (valida) taborae, T. inclusa*
Vicariance between Tanganyika and Zambia	*Tatera (valida) taborae* and *T. valida* sensu stricto; subspecies *cynocephalus* and *kindae* of yellow baboon, *Papio (hamadryas) cynocephalus*
Eastern distributions in southeast Tanzania and Mozambique east of Lake Malawi and Zimbabwe Highlands	*Tatera inclusa* (an allospecies of *T. valida*), *Genetta angolensis mossambica* (a replacement of the nominate race), *Neotragus moschatus* (different semispecies north and south of the Zambezi), *Bdeogale crassicauda* (ranges farther west up to Muchinga/Zambezi scarp), *Tragelaphus angasi* (south of Zambezi only)
Affected by Muchinga/Zambezi Escarpment	
Ranges from east up to scarp but not beyond (nor east of Shire River and Zimbabwe highlands)	*Paracynictis selousi*
Ranges from west across scarp but then contained farther east and south by Lake Malawi, Shire River, Zambezi River	*Epomops dobsoni*
Replaced by related taxa at or below Scarp	*Heliosciurus (gambiensis) rhodesiae* and *H. mutabilis; Kobus (ellipsiprymnus) defassa* and *K. (e.) ellipsiprymnus; Cercopithecus (mitis) mitis* and *C. (m.) albogularis*
Range from west, do not reach immediately beyond scarp, yet reappear some distance farther east	*Aethomys nyikae, Cryptomys mechowi, Genetta angolensis, Lutra maculicollis, Cephalophus monticola*
Range from west, do not reach beyond scarp, found nowhere farther east	*Crocidura mariquensis, Cryptomys hottentotus, Tatera valida, Cephalophus silvicultor, Kobus leche*
Absent or virtually absent from Angola Plateau	*Rhynchogale melleri, Raphicerus sharpei, Oreotragus oreotragus, Kobus vardoni, K. leche, Aepyceros melampus, Alcelaphus lichtensteinii, Hippotragus niger* (very localized isolate *H. n. variani* is present)
Confined to Angola-Zambia-Katanga Plateau	*Plerotes anchietae, Graphiurus monardi, Lemniscomys griselda, Rhabdomys (pumilio) angolae, Mastomys angolensis, Mus callewaerti*
Confined to escarpment zone of Angola	*Funisciurus congicus, Pelomys campanae, Aethomys bocagei, Heterohyrax brucei bocagei*

are absent from Ethiopia. Distinctive Ethiopian highland endemics (Yalden & Largen, 1992) are mostly small, nonforest mammals, but include a few larger species—Simien fox (*Canis simensis*), gelada (*Theropithecus gelada*), and mountain nyala (*Tragelaphus buxtoni*).

The Somali Arid Zone

East–west separation of forest mammals in High Africa is paralleled by north–south separation of South West Arid and Somali Arid faunas. The Somalia-Masai regional center of endemism defines a vegetation zone of *Acacia–Commiphora* deciduous bushland and thicket and of semidesert grassland and shrubland (F. White, 1983) and is the Somali Arid District or Biotic Zone of zoologists (fig. 17-1; Chapin, 1932; Davis, 1962). The boundaries set by White correspond closely with the ranges of a Somali Arid avifauna (Hall & Moreau, 1970; Snow, 1978) and with the distribution of such mammals as the beisa [*Oryx (gazella) beisa*] or dikdik (*Madoqua* spp.). The distinctive mammal fauna associated with the zone includes at least 60 endemic species together with well-marked and/or geographically isolated populations of at least 20 more-widespread taxa. The fauna has never been reviewed as a whole, and its geography and systematics are not well known: 16 of its endemic species were not listed by Honacki et al. (1982). A few Eremian (Palearctic desert) mammals are present: *Hemiechinus aethiopicus, Crocidura somalica, Asellia tridens, Gerbillus* spp., *Jaculus jaculus,* and others (table 17-2). *Rhinopoma, Asellia,* and *Gerbillus* are predominantly Eremian genera with Somali Arid endemic allospecies. Also occurring in the Somali Arid are some of the taxa widespread in the Sudanese Arid Zone, such as the gerbil *Tatera robusta* (Bates, 1988).

The Serengeti and adjacent Masailand Plains (fig. 17-2), like the Dry Savanna of southern Africa, are transitional with the Southern Savanna. They were excluded from the Somali Arid and placed by Turpie and Crowe (1994) in their large-mammal Savanna Subregion. The Masailand ungulate fauna includes endemic subspecies with South West or Sudanese Arid affinities as well as nonendemics, and there are more widespread Somali Arid endemics ranging through Masailand into Somalia and the Ogaden (table 17-2). The highlands of Kenya and northeastern Tanzania contribute to a faunal boundary (fig. 17-2; Roosevelt & Heller, 1915; Stewart & Stewart, 1963), which marks a shift from Masailand to Somalian faunas and is reinforced by the degeneration of the country north of the highlands into a semidesert grass- and shrubland. North of Masailand in Ethiopia and Somalia are additional endemics which may range farther, north of the Webi Shebeli River, whereas others occur only beyond this river [some included in Turpie and Crowe's (1994) ungulate arid province]. This enumeration of Somali Arid endemicity (table 17-2) reflects an environmental gradient of decreasing aridity, north to south, along which species of equids (*Equus* spp.), for instance, replace each other (Bauer et al., 1994). Vicariance of antelopes (*Madoqua* spp., *Gazella* spp., *Alcelaphus buselaphus* sspp.) and giraffes (*Giraffa camelopardalis* sspp.) (Funaioli & Simonetta, 1966; Yalden et al., 1976–1986) indicate that rivers and highland massifs have also been factors isolating taxa. Of 29 large-mammal species endemic to the Somali Arid or with endemic subspecies there, 17 are antelopes (table 17-2; Turpie & Crowe, 1994).

Southern Savanna

The Southern (moist) Savanna (the Zambezian Ecozone of Klein, 1984) is delimited by the "Sclater line" (Davis, 1962) running along the southern and eastern margins of the central forest and through the upper Nile swamps and Kenya highlands at about 1–2° N and by the edge of the Somali Arid Zone in Kenya, Tanzania, and Somalia. In terms of area, it is dominated by dystrophic woodland (Owen-Smith, this volume) or miombo (*Brachystegia, Julbernardia, Isoberlinia*), in the south meeting mopane (*Colophospermum, Combretum*) and other forms of eutrophic dry woodland of the Greater South West Arid Zone. The miombo is also dissected by the mopane—for instance in the valleys of the Zambezi and the Luangwa in Zambia and Zimbabwe (fig. 17-2)—allowing certain dry-country species to penetrate northward.

Approximately 120 mammal species are either confined to the Southern Savanna or range beyond it into Somali and South West Arid Zones but do not occur in the Northern Savanna (table 17-3), while another 115 species are shared with the Northern Savanna. Various topographical features divide up the Southern Savanna (fig. 17-2) and contribute to regional endemism, disjunct distributions, and introgression (table 17-1). Taxa which on a pan-African scale are predominantly Northern and Southern Savanna vicars may make contact within High African limits. In Tanzania and Kenya the savanna fauna is divided to some degree by the southward intrusion of the Somali Arid Zone, involving local endemism as well (Stewart & Stewart, 1963; Davis, 1966; Kingdon, 1974; Aggundey & Schlitter, 1984; Grubb, 1985; Bergmans, 1988; Claesson & De Vree, 1991; Juste & Ibanez, 1992). Faunal dispersal between Lakes Tanganyika and Malawi is restricted by the Ufipa and

Table 17-2. Distribution of some larger mammals in the Somali Arid zone.[a]

	Eremian species	Endemics
1 only		
	Vulpes rueppelli	*Genetta abyssinica*
	Equus asinus	*Madoqua piacentinii*
	Gazella dorcas pelzelni (endemic subspecies)	*Dorcatragus megalotis*
		Ammodorcas clarkei
		Gazella spekei
		Alcelaphus buselaphus swaynei
1 + 2		
		Galago gallarum
		Papio hamadryas
		Helogale hirtula
		Galerella (sanguinea) ochracea
		Equus grevyi
		Phacochoerus aethiopicus delamerei
		Madoqua saltiana
		M. guentheri
		Gazella soemmerringii.
1 + 2 + 3		
	Canis aureus	*Otocyon megalotis* sspp.
	Hyaena hyaena	*Canis mesomelas schmidti*
		Genetta (genetta) felina neumanni
		Proteles cristatus septentrionalis
		Litocranius walleri
		Oryx (gazella) beisa
		Tragelaphus imberbis
2 only		
		Giraffa camelopardalis reticulata
		Madoqua kirkii kirkii
		Gazella granti brighti
		Beatragus hunteri
2 + 3		
		Non-endemics, such as
		Equus burchellii
		Taurotragus oryx
		Aepyceros melampus
3 only		
		Giraffa camelopardalis tippleskirki
		Raphicerus campestris neumanni
		Madoqua kirkii sspp.
		Gazella granti sspp.
		Gazella thomsoni sspp.
		Alcelaphus buselaphus cokii
		Connochaetes taurinus sspp.

[a]1, north of the Webi Shebeli in Somali Plateau and Ogaden, Ethiopia; 2, between Webi Shebeli and Kenya Highlands/Tana River in northern Kenya and Jubaland; 3, Masailand, south of the Tana River and Kenya Highlands.

Mbeya-Poroto Highlands and Lake Rukwa, leading to some vicariance. The Muchinga/Zambezi Escarpment is a more significant boundary to the dispersion of mammals (Ansell, 1978). Forest mammals or close relatives of forest species (*Cephalophus, Cercopithecus, Heliosciurus, Bdeogale*) as well as savanna taxa are affected by the scarp, which therefore seems to mark a convergence-front between Central and East African faunal elements, whatever their ecological predilections, continuing the east–west partition of savanna in Tanzania. Endemism in the Angola and Zambia-Katanga plateaus or absence from the Angola Plateau may relate to Chapin's (1932) avifaunal "Rhodesian Highland District" (Hill & Carter, 1941)

Table 17-3. Larger-mammal species endemic to the Southern Savanna or ranging beyond its limits only into Somali and/or South West Arid zones (not into the Northern Savanna), or as specified.

Species
Galago moholi
Otolemur crassicaudatus
Papio (hamadryas) cynocephalus
Poecilogale albinucha
Bdeogale crassicauda
Rhynchogale melleri
Paracynictis selousi
Helogale parvula
Genetta angolensis
Dendrohyrax arboreus
Heterohyrax brucei (also in Ethiopian Highlands)
Equus burchellii
Potamochoerus larvatus (also in Ethiopian Highlands and Madagascar)
Raphicerus sharpei
Redunca arundinum
Kobus vardoni
K. leche
K. (e.) ellipsiprymnus
Hippotragus niger
Alcelaphus lichtensteinii
Connochaetes taurinus
Aepyceros melampus
Tragelaphus angasi
Taurotragus oryx (extends marginally into northern savanna east of the Nile in southern Sudan)

or Turpie and Crowe's (1994) southern province of their ungulate central forest subregion. An escarpment zone running north to south in western Angola divides the coastal desert and semidesert from the miombo of the inner plateau (Hall, 1960), influencing mammalian distribution (De Meneses Cabral, 1966; Crawford-Cabral, 1983, 1989a,b). Vicariance in baboons, waterbuck, gerbils, elephant shrews, and even fruit bats, among others (table 17-4), may have been facilitated by any factor that disrupted the Southern Savanna, including perhaps a former extension of arid vegetation through the zone.

The Greater South West Arid Zone

Native to political southern Africa are 284 land-mammal species (Meester et al., 1986), a number now raised to 300 (Wilson & Reeder, 1993). The zones into which this fauna can be partitioned have been determined by Davis (1962), Meester (1965), Rautenbach (1978), Coetzee (1983), Klein (1984), Turpie and Crowe (1994), and Gelderblom et al. (1995). In the light of their studies and the work of ornithologists and botanists (Moreau, 1966; Winterbottom, 1978; F. White, 1983), it is useful to recognize a South West Arid Zone proper including Namib, Nama-Karoo, and Kalahari sections, which together with Dry Savanna, Highveld, and Fynbos transitional subzones constitute a "Greater South West Arid Zone" (fig. 17-1, table 17-4). The Nama-Karoo is divided by the lower Orange River gorge into a southern section (Little Namaqualand and the Great Karoo) and a northern area (Namibian and Angolan mountainous scarp country and mopane, of Great Namaqualand, Damaraland, and the Kaokoveld). Southern (moist) Savanna and lowland and montane forest of the Eastern Forest Zone also extend into southern Africa. Mammalian diversity declines across the region (Rautenbach, 1978; Crowe, 1990) from 209 species in Savanna and 136 in the South West Arid to only 41 in the Namib Desert, strongly differentiated because of its faunal deficiency.

One hundred eight species and isolates (some of which can be regarded as semispecies) are more or less confined to the Greater South West Arid Zone. Though about a third of these are widespread, many are endemic to a single subzone (table 4). No less than 33 are restricted to the Namib and Nama-Karoo, where the lower Orange defines centers of endemism, but only 2 taxa, species of gerbils (*Gerbillurus*), are exclusive to

Table 17-4. Partition of endemic species and geographically isolated populations (isolates) among subzones of the Greater South West Arid zone. Names of Klein's (1984) ecozones are enclosed in brackets.

Zones and subzones	Numbers of endemic species and isolates	Endemic species and isolates of large mammals
Greater South West Arid zone	**87 endemic species; 21 isolates (total 108)** 36 (out of 108) are widespread	*Papio (hamadryas) ursinus* *Vulpes chama* *Galerella pulverulentus* *Cynictis penicillata* *Suricata suricatta* *Parahyaena brunnea* *Felis nigripes* *Procavia (capensis) capensis* *Equus quagga* (extinct) *Antidorcas marsupialis* *Pelea capreolus* *Oryx (gazella) gazella* *Alcelaphus (buselaphus) caama* *Otocyon megalotis megalotis* *Canis mesomelas mesomelas* *Genetta (genetta) felina* subspecies *Proteles cristatus cristatus* *Ceratotherium simum simum* *Giraffa camelopardalis giraffa* *Raphicerus campestris* subspecies *Redunca fulvorufula fulvorufula* *Connochaetes taurinus taurinus*
Transitional subzones	**32**	
Dry Savanna [Transvaalian]	14	*Damaliscus lunatus lunatus*
Highveld [Basutolian] plus Fynbos [Cape]	18	*Damaliscus pygargus*
Highveld only	4	*Damaliscus pygargus phillipsi* *Connochaetes gnou.*
Fynbos only	11	*Raphicerus melanotis* *Damaliscus pygargus pygargus* *Hippotragus leucophaeus* (extinct).
South West Arid zone proper	**40 (33 in Namib and Nama-Karoo)**	
Kalahari [Kalaharian]	7	none
Nama-Karoo [Karoo-Namaqualian in part], both north and south of Orange River	8	*Equus zebra*
Nama-Karoo, only south of Orange	12	*Phacochoerus aethiopicus aethiopicus* (extinct)
Nama-Karoo, only north of Orange	11	*Galerella flavescens* (= *nigratus*) *G. swalius* *Procavia (capensis) welwitschii* *Madoqua kirkii damarensis.*
Namib [Karoo-Namaqualian in part]	2	none

the Namib. No more than 7 taxa are confined to the Kalahari, where 34 more-widespread South West endemics are also found, some breaching the phytogeographic limits of the Greater South West Arid Zone (table 17-5). At least 26 species not confined to southern Africa occur here as well. Absentees from the Kalahari include golden moles (Chrysochloridae), the hedgehog *Atelerix frontalis,* the shrew *Crocidura cyanea,* and a suite of rock-living mammals: flat-headed bats (*Mormopterus petrophilus*), rock dormice (*Graphiurus* spp.), spiny mice (*Acomys* spp.), rock rabbits (*Pronolagus* spp.), rock elephant shrews (*Elephantulus* spp.), rock hyrax (*Procavia capensis*), and klipspringer (*Oreotragus oreotragus*).

Mammals confined to the South West Arid proper are by definition absent from the Highveld, where few species are strictly endemic and some are shared only with the Fynbos, a sclerophyllous shrub formation dominating southwest Cape Province. In spite of its tremendous phytodiversity (some 7000 species of vascular plants, of which more than half are endemic), the Fynbos Subzone has no more than 11 endemic species of mammals and some geographically isolated endemic subspecies. Ecogeographic relationships of the Fynbos fauna are diverse, but of 17 endemics or isolates, only two have affinity with the adjacent Karoo, where they are replaced by allospecies. The four taxa discontinuously distributed between Fynbos and Highveld ensure that the faunal resemblance factor between the two zones is higher than between any other zone and the Fynbos (Rautenbach, 1978).

Each major habitat of the Greater South West Arid Zone therefore has endemic mammalian taxa, and some have particular faunal deficiencies. The array of habitats has contributed to faunal diversity.

Distributional Discontinuities and Dispersal, Especially Involving Arid Zones

Eurytopic species, such as large carnivores, range from the Cape to Somalia, but other mammals face constraints upon dispersal. The miombo and adjacent vegetation formations have represented a more or less complete barrier to some species preferring quite mesic conditions including floodplains or other open habitats; kob (*Kobus kob*) and puku (*K. vardoni*) are examples. Though allotaxa occurring in miombo may fill the gap (table 17-5), a number of other species or species groups are discontinuously distributed between the South West Arid Zone and the Somali Arid or some other zone to the north. These striking discontinuities have attracted attention and emphasis (Meester, 1965) and have been interpreted in connection with the drought corridor (Balinsky, 1962; Benson et al., 1962; Cooke, 1964; Moreau, 1966; Bigalke, 1972; Klein, 1984; Turpie & Crowe, 1994). Along the corridor, a discontinuous channel of relatively dry vegetation exists today, which, because of the uplands between Lakes Tanganyika and Malawi, just fails to connect the South West Arid Zone with the Somali Arid Zone in southwest Tanzania, by way of the mopane woodland of the Zambezi and Luangwa valleys (fig. 17-1). In periods of extreme aridity, the dry vegetation might have broken through the miombo woodland completely, allowing faunal traffic between the arid zones.

But it must be borne in mind that the two zones have overall very different mammal faunas. Though they share 13 genera which include taxa endemic to one or other zone, these endemics are independently related to species or subspecies from nonarid habitats (e.g., in the genera *Atelerix, Crocidura, Elephantulus, Papio, Xerus, Galerella,* and *Alcelaphus*). The Somali Arid fauna has 29 genera lacking endemic species in the South West or not found there at all, while in the South West, the number of corresponding genera is 31. Only *Crocidura* and *Elephantulus* have more than one endemic species in each zone. Genera with more than one endemic species in the Somali Arid only are *Gerbillus, Tatera, Arvicanthis, Acomys, Madoqua,* and *Gazella;* genera in an equivalent position in the South West are *Chrysochloris, Amblysomus, Pronolagus, Xerus, Graphiurus, Gerbilliscus, Petromyscus, Otomys, Parotomys, Aethomys, Thallomys, Mus, Gallerella,* and *Procavia,* indicating contrasting trends in cladogenesis between the two zones which, with the exception of *Thallomys* spp., have no particularly close affinity in their endemic small, nonvolant mammals (Denys, this volume).

Thus mammals have colonized the two arid zones independently from more mesic habitats and have evolved independently as well. Although even at present there is a remnant of an arid-vegetation pathway, the faunal evidence suggests that relatively few arid-zone species (mostly larger mammals) were able to travel through it. At times there have been increases in aridity, but not a wholesale dispersal freeway during the Quaternary and earlier. Occurrence of *Ceratotherium simum, Giraffa camelopardalis,* and *Madoqua* sp. in the Chiwondo Beds of the Malawi Rift (Bromage et al., 1995a; Schrenk et al., 1995)—taxa which within historical times have been absent from Malawi (Ansell & Dowsett, 1988)—indicate the former more extensive ranges expected of these currently discontinuously distributed taxa when they were able to disperse through the arid-vegetation pathway. The Chiwondo record of *Camelus* sp. suggests the same phenomenon,

Table 17-5. Some mammals with populations confined to southern Africa, including those ranging north of the South West Arid zone at least as far as Zambia and those discontinuously distributed between Somali and South West Arid zones.

South West Arid zone taxa	In Southern Savanna? (or replaced)	In Somali Arid? (or replaced)
Tadarida aegyptiaca	Ranges north into Zambia	Widespread
Papio (h.) ursinus	Ranges north into Zambia; replaced by *P. (h.) cynocephalus, P. (h.) anubis*	Replaced by *P. (h.) hamadryas* north of Webi Shebeli only
Tatera brantsii	Ranges north into Zambia	
Thallomys nigricauda[a]	Ranges north into Zambia	Replaced by *T. loringi* in Masailand only
Aethomys namaquensis	Ranges north into Zambia	
Pedetes capensis	Ranges north into Katanga	Masailand only
Lepus capensis	Ranges north into Zambia	Widespread
Genetta (genetta) felina	Ranges north into Zambia; replaced by *G. angolensis*	Widespread
Proteles cristatus		Widespread
Parahyaena brunnea[b]		Replaced by *Hyaena hyaena*; widespread
Otocyon megalotis		Widespread
Canis mesomelas		Widespread
Ceratotherium simum	Formerly ranged north into Zambia; also in Northern Savanna	
Equus zebra, E. quagga	Replaced by *E. burchellii*	Replaced by *E. grevyi* north of Kenya Highlands
Phacochoerus aethiopicus in Karoo only[c]	Replaced by *P. africanus*	North of Kenya Highlands
Giraffa camelopardalis	Ranges north into Zambia; isolate in Luangwa Valley	South of Webi Shebeli only
Raphicerus campestris	Ranges north into Zambia	Masailand only
Madoqua kirkii in Damaraland only		South of Webi Shebeli only
Oryx gazella		Widespread
Damaliscus lunatus	Ranges north into Zambia with isolate in Chambeshi Valley; also Tanzania northward	
Connochaetes taurinus	Ranges north into Zambia; isolate in Luangwa Valley	Masailand, ranging from Mozambique in south

[a]For systematics, see Musser and Carlton (1993).
[b]For Plio-Pleistocene distributions, see Werdelin and Turner (1995).
[c]For systematics, see Grubb (1993).

though camels are not known to have reached the present South West Arid limits.

The Chiwondo Beds of northern Malawi yield faunal remains of 4.5 to 1.5 m.y. of age (Bromage et al., 1995). About 50% of the sample is made up by bovid fragments. The bovid fauna of Member 3A, aged at between 3 and 2 m.y., comprises 12 bovid genera (fig. 17-3). To evaluate the biogeographic role and relationships of the Chiwondo fauna in the Pliocene, the occurrence of these genera has been compared to 3–2 m.y. deposits of northeastern, eastern, and southern African sites (fig. 17-3). The overall bovid distributional pattern shows an overlap of 11 genera shared with eastern African sites (West and East Turkana and Ndolanya Beds) and of 9 genera shared with northeastern African sites (Omo and Hadar). The overlap

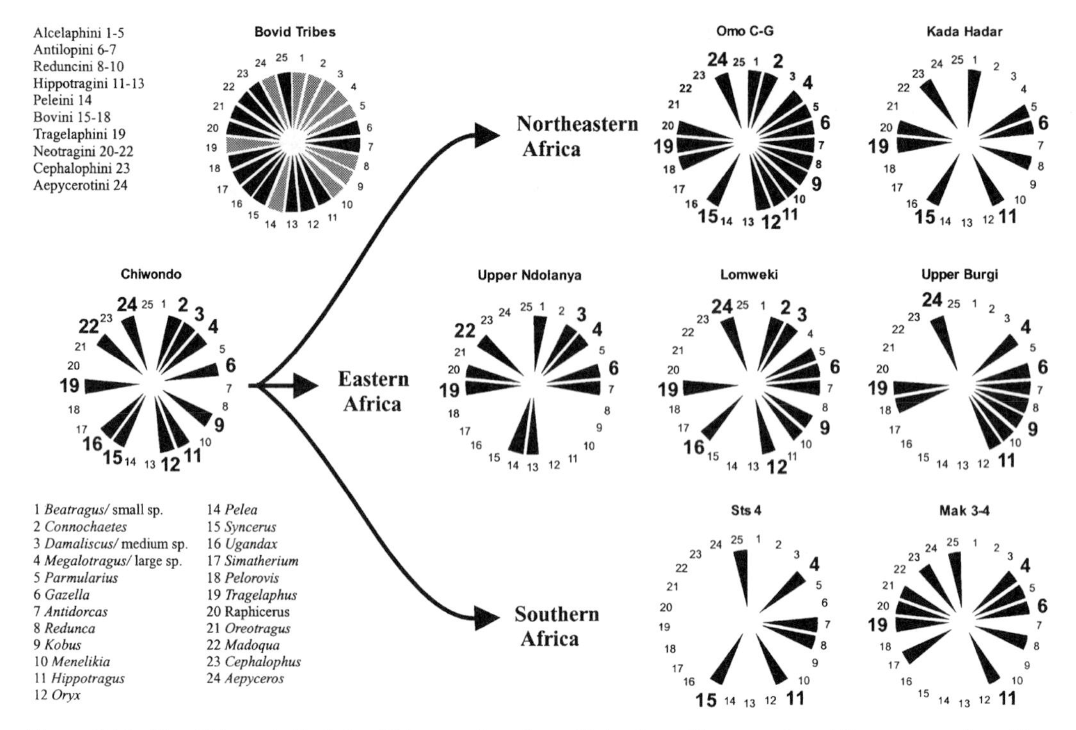

Figure 17-3. The Chiwondo Beds bovid fauna of northern Malawi aged between 3 and 2 m.y. comprises 12 bovid genera. The occurrence of these genera is compared to 3–2 m.y. deposits of northeastern, eastern, and southern African sites. The distributional pattern shows an overlap of 11 genera shared with eastern African sites (West and East Turkana and Ndolanya Beds) and of 9 genera shared with northeastern African sites (Omo and Hadar) and only 5 genera with southern African sites.

with southern African sites comprises only five genera (fig. 17-3).

The faunal assemblage of the Late Pliocene in Africa is marked by a rather fast radiation of suid taxa in eastern and southern Africa (Harris & White, 1979; Cooke, 1985). Often three or more species are represented in one stratigraphic unit, often of different evolutionary lineages (*Notochoerus, Metridiochoerus, Kolpochoerus,* and *Potamochoerus* lineages). Most of them are extinct. There are only five living species in three different genera (*Phacochoerus aethiopicus, P. africanus, Potamochoerus porcus, P. larvatus,* and *Hylochoerus meinertzhageni*). From northeastern African sites to the Chiwondo Beds in northern Malawi, the *Notochoerus* evolution is well documented, whereas in southern Africa (Vaal River gravels and Makapansgat), only one stage of the *Notochoerus* lineage occurs as a rare element (Kullmer, 1997). The appearance of the early metridiochoerine *Potamochoeroides shawi* at southern African sites demonstrates an endemic situation of the early metridiochoerine evolution. The more advanced *Metridiochoerus andrewsi* is known from northeastern, eastern, and with progressive types from southern African localities (Swartkrans, Kromdraai, and Bolt's Farm). The suid assemblage of the Chiwondo Beds consists of early to very progressive examples of *Metridiochoerus andrewsi* with all intermediate evolutionary stages, which illustrate the overlap with eastern African sites (Kullmer, in preparation).

The Chiwondo fauna therefore clearly is a mixed assemblage comprising faunal elements of the South West Arid Zone as well as of the Somali Arid Zone, pointing to a certain degree of faunal dispersal through the Pliocene arid-vegetation pathways. Even though the biotic zones are much larger than the area sampled by the Chiwondo Beds, it is striking that the number of shared taxa with northeastern and eastern African sites is double that shared with southern African sites. We may therefore conclude that the Pliocene Chiwondo Beds fauna predominantly was part of the eastern African faunal domain, yet, through some influence of southern African faunas made up a unique biogeographic area on its own in the Late Pliocene.

When dispersal between the arid zones was possible, taxa with a continuous miombo distribution may have been partitioned by unsuitable vegetation formations, as suggested above, or when forest was more widely distributed across eastern Africa. In keeping with the periodic model (Grubb, this volume) and the topography of High Africa (figs. 17-1, 17-2), species of different biomes are likely to have dispersed through the subcontinent at different times (Balinsky, 1962). Under the wettest conditions, continuity of forests would have facilitated dispersal between them, the ranges of savanna species would have been disrupted, and arid-zone communities would have been widely separated. With intermediate conditions, ranges of forest taxa were disrupted, dispersal for savanna species were facilitated, and arid zone taxa were not so widely separated. Finally, under dry conditions, dispersal between arid zones became a possibility, vicariance events occurred among savanna species, and separation of forest taxa was enhanced. Such cycles may not have been completed under long-term climatic amelioration or deterioration. But since the present disposition of biomes in High Africa is of considerable age (Axelrod & Raven, 1978), it is possible to envisage endemism, vicariance, and cladogenesis occurring in circumstances somewhat analogous to those of the present-day as far back as the Pliocene. Savanna ecosystems had differentiated and continued to expand; the Cape flora had long been established, though never to develop a very distinctive mammal fauna; eastern forests became separated from the central forest block with the rise of the Great Rift Valley system, engendering regional endemism yet without preventing continuing faunal influx from the richer central fauna; and Somali and South West Arid biota had evolved independently, judging by the history and endemism of the vegetation (see also F. White, 1983), the Somali Arid Zone having differentiated from a Paleo-Masai Province about 4 m.y. (Denys, this volume). Faunal interchange between these arid zones seems always to have been limited except for strongly eurytopic species and larger mammals of eutrophic savannas, not confined to the most xeric habitats; their dispersion would perhaps have been promoted in the Late Pliocene when wooded savannas gave way to more open, dry conditions (Owen-Smith, this volume).

Factors Affecting Distributions and Vicariance in Early Hominids

Human precursors emerged from the rainforest, and there are views as to where and how this happened. The Albertine Rift has been invoked as a major faunal barrier (Kortland, 1972; Brunet et al., 1995) generating vicariance events between forest and savanna faunas including Panidae and ancestral australopithecines (Coppens, 1994). Nowadays the rift highlands do not restrict the dispersion of 63% of nonvolant central-forest mammals, and Kingdon's line (or the drought corridor) is a better indicator of this fauna's present eastern limits. Whatever their former condition, the rift highlands could never have completely segregated forest and nonforest biota which would have been so widely in contact elsewhere in Africa. Huge swaths of interdigitating forest and savanna in the Central African Republic or in southern Zaire represent ideal Low African country for contact between forest and savanna faunas. Conceivably, colonization of the savanna from the forest by apes and other mammals occurred in one of these areas, calling into question Coppens's (1994) "East Side Story": in view of its more human propensities, the bonobo's (*Pan paniscus*) occurrence in southern Zaire may be suggestive. Landscapes of forest dissected by savanna have proportionally more ecotonal habitat than a sharp frontier between the biomes, especially perhaps under the impoverishment of rainforest in the Early Pliocene (Axelrod & Raven, 1978), and so could have facilitated adaptation to a savanna existence. The Eastern Forest Zone also provides a suitable area for savanna origins, apparent from the various eurytopic tendencies of its fauna, but once these forests became separated from the Central Zone, apes were not there to be a source of nonforest hominids.

Even when still confined to the forest biome, the lineage from which *Homo* was to evolve had probably come to acquire fewer allotaxa with broader ranges in each zoogeographic species relative to other forest primates. As part of a superspecies that was becoming amphibiomic, a stenotopic forest taxon could give rise to a stenotopic savanna taxon (ecological translation; Grubb, this volume), the lineage later evolving in nonforest environments. Mammal distribution today shows how nonforest topography can inhibit dispersion, offering opportunities for endemism and vicariance. We could expect lineages of early hominids to have been similarly affected. The range of habitats they occupied would have contributed to the extent of the distribution they acquired, but by comparison with other large, open-country mammals there would not have been more than five or six allopatric taxa at any one time, quite probably less, perhaps forming a superspecies. This is also in keeping with the spatiotemporal deployment of *Australopithecus* and *Paranthropus* taxa (Wood, 1994) known so far. But understanding the geography of proliferation by Pliocene hominid taxa (Foley, this volume) does not extend to a knowledge of their dispersion in different nonforest biotic zones, and various scenarios can be envisaged to account for the allopatry of eastern and southern African taxa of *Australopithecus* or *Paranthropus* (Bishop, this volume) or the sympatry of *Paranthropus* and *Homo*.

Australopithecines could have occurred in relatively moist, wooded savanna (Sikes, this volume) or could also have ranged into drier categories of woodland (Denys, this volume), such as mopane or acacia, or into open grasslands. If they indeed had such eurytopic tolerance, then the appearance of an arid-vegetation pathway in eastern Africa would not have been enough to cause vicariance, nor would it have been in the appropriate location to account for east/south allopatry. Extending the known range of australopithecines to Chad (Brunet et al., 1995) raises the possibility that their vicariance within High Africa, assumed to be of High African endemics, may really have been between widespread but predominantly Northern Savanna (Low African) taxa ranging into High Africa and endemics of the Southern Savanna, as in the primate cases cited in table 17-1.

There are other possibilities. Treeless grasslands (tending to fall within lower-rainfall regions today) would be secondary habitats, colonized from wooded savanna, and it may be questioned on grounds discussed by Sikes (this volume) whether any early-hominid taxon became wholly restricted to such environments. But some australopithecine taxa may have become confined to dry eutrophic zones in a broad sense (Owen-Smith, this volume), having differentiated from those in moister woodlands, even if association of the earlier australopithecines with the Paleo-Masai Province (Denys, this volume) may be in terms of geography rather than ecology. A great variety of mammalian lifestyles, including diets, is evidently possible in both mesic savanna and arid zones. Baboons, for example, occur in both. Interbiome vicariance, although relatively rare for large-mammal zoogeographic species (12 out of 102 in nonforested High Africa), is implicated for australopithecines by paleofaunal data (Andrews & Humphrey, this volume) and could account for the east/south allopatry.

Sympatry would have arisen between robust australopithecines (*Paranthropus*) and *Homo* by completion of speciation through a ring-species deployment (Grubb, this volume). The extent of the sympatric zone on a pan-African scale remains unknown. Sympatry would have been preceded by an allopatric state, but the region and ecological setting of this allopatry is also obscure. Possibly it lay along an ecotone between different vegetation formations. Or perhaps the taxa inhabited similar habitats but were separated by an unfavorable environment. Evolution in *Homo,* from occupation of a narrow range of habitats to a broader spectrum, presumably occurred through colonization of what had formerly been marginal habitat. This propensity might have facilitated access to the previously inaccessible range of *Paranthropus* (instead of *Paranthropus* invading the range of *Homo*), in accord with Rosensweig's suggestions (this volume), but was the dispersal ability acquired sufficiently early in the history of *Homo*? The several vicariance events presumably associated with the diversity of earliest *Homo* taxa might imply otherwise—that these taxa were relatively stenotopic and did not range widely, else they would not be subject to vicariance. Further speculation is hardly justified at the moment. Eventually, of course, extending the diversity of habitats humans occupied would be associated both with a lesser tendency to form allotaxa, in parallel with large carnivores or some very large ungulates, and with a greater potential for dispersal.

Modern mammalian distributions in High Africa can thus suggest scenarios involving vicariance and sympatry among early hominids, though paleoenvironmental sources are unlikely to be able to corroborate particular details of a cladogenetic event. Conversely, neontology, lacking in time depth, is necessarily silent on the evolutionary impact of long-term Neogene environmental changes in the High African context. These have led to the emergence of new mammalian communities (Owen-Smith, this volume) by channeling anagenetic trends occurring incrementally in cladogenetic events.

Summary

High Africa, the repository of important early hominid fossils, is dominated by nonforest biomes, though forest habitats are present and there is evidence for dispersal by forest mammals through the subcontinent. High African biomes are partitioned into major blocks or biotic zones: Central and Eastern Forest, Southern Savanna, Somali Arid and (Greater) South West Arid Zones, the last of which at least can be divided into well-defined subzones. Each biotic zone or subzone is characterized by endemic mammalian taxa, but there is also much overlap, particularly between dry- and moist-savanna species in the Dry Savanna or Masailand. There is evidence for long-term isolation of the arid biotic zones, dispersal between them having been restricted mainly to large mammals or to eurytopic species. Vicariance and endemism within the Southern Savanna and Somali Arid Zone are influenced by scarps and other physiographic features, while the particularly well-marked South West centers of endemism are delimited also by phytogeographic criteria.

Hominid cladogenesis took place among analogues of today's biotic zones and was affected by similar factors to those facilitating the process in other mammals. It involved dispersal from moister to drier biomes, but entry into the savanna from the forest was probably not made across the Albertine Highlands. Arid zones may then have been colonized so that *Australopithecus* or *Paranthropus* taxa occurred in both moist and dry savannas. Vicariance between eastern and southern African populations could have resulted from habitat segregation or perhaps from geographical partition of Northern Savanna and Southern Savanna faunas. But a well-supported scenario for the ecogeographic location of *Paranthropus* and *Homo* prior to their sympatry is still awaited.

Acknowledgments This chapter was developed from a paper originally prepared for the Wenner-Gren conference. We are grateful to the Wenner-Gren Foundation for the opportunity to attend this meeting, and to Norman Owen-Smith and Christiane Denys for critical advice concerning an earlier draft. Thanks to Tim Bromage and Rainer Abel for their help with the figures.

IV

HOMINID EVOLUTION

18

Introduction

Meave Leakey

In the early part of this century, studies of human evolution were traditionally based on morphological descriptions and comparisons of the few known hominid fossils. Scant attention was paid to context. Even as late as 1960, the fossil evidence was so sparse that Louis Leakey was able to pen a note to *Nature* describing the discovery of OH 5, for publication less than 1 month after its initial discovery (Morrell, 1995). Subsequent studies in the 1960s at Olduvai Gorge, the Omo Valley, and elsewhere began the current increasing trend to view hominid discoveries in the broader paleoenvironmental and geological context. Today's investigations into human prehistory benefit from a wealth of data accumulated since that time and originate from a variety of disciplines.

A common focus is now provided to participants in widely divergent areas by the relevance of their research to human evolution. This was the case at the Malawi Conference, and it is well illustrated by the fact that three-quarters of this volume are dedicated to the discussion of diverse topics with relatively little mention of human ancestors. The three earlier sections hardly touch on the specifics of the hominid fossil record but instead concentrate on broader issues relating to past climates and faunas, climatic perturbations, and the response of mammalian lineages and faunal communities to changing environments. In-depth studies of a few selected mammalian taxa, Cercopithecidae, Suidae, Equidae, and small mammals, illustrate evolutionary responses to specific environmental stimuli as shown by dietary adaptations, habitat preferences, migration and dispersion patterns, and biotic zonation. Only this final part of this volume concentrates more directly on specific aspects of the hominid fossil record, and here too, the emphasis is on placing the fossil hominids within the broader context of evolutionary theory, climatic change, and adaptive complexes.

Our view of human evolution and hominid habitats is necessarily prejudiced by the nature of the extraordinarily limited evidence. The task is equivalent to constructing the diversity of modern African habitats and faunas with knowledge of only a few selected tiny areas almost exclusively confined to the eastern and southern part of the continent. In the modern situation this would exclude the majority of Africa's diverse habitats, such as the central tropical rainforests, the Sahara and Namibian deserts, the Okavango Swamp, and the Sudd. Our knowledge of the past, glimpsed through the few preserved fossil sites that we have, is necessarily restricted to a minute proportion of the total past African landscape and has the added limitation of being spread through the temporal dimension as well as the spatial.

Because sedimentation is largely associated with hydrographic systems, the habitats sampled in the sedimentary record are largely those associated with water. Fortunately, remains of human ancestors are frequently encountered at these sites due to the water dependence of the human family so that we can at least say with certainty that at particular times our ancestors were using the habitats represented at these sites. But we cannot predict whether hominids were more commonly encountered elsewhere. Nor can we say from which habitats they were excluded. When hominids are present in the fossil record, they are consistently rare. The collections from the unusually rich deposits of the Turkana Basin show that the proportion of hominids recovered is very small. The West Turkana collections from the 1980s include only 15 specimens in a total collection of 1130 large mammals, representing 1.3% of the collection. At Kanapoi (4.17–4.07 m.y.) and Allia Bay, an excavation at site 261–1 (~3.9 m.y.), which both document shorter intervals of time, the percentages of hominids in the collection are 4.5% and 1.1%, respectively. Other sites, such as Manonga Valley in Tanzania (Harrison, 1997), either represent hominids even more rarely or do not yield them owing to their absence from these particular ecotypes.

Negative information should not be ignored, especially at sites where good collections are recovered. At Lothagam, for example, more than 1000 fossils have been collected from the Nawata Formation (~8–5.5), and these include excellently preserved cranial and postcranial specimens of a variety of mammals as well as monkeys. But in this entire collection, there is no evidence of hominoids until the uppermost deposits in the Nawata Formation and the base of the overlying Apak Member (~5.5 m.y.), where three fragmentary specimens have been recovered (M. G. Leakey et al., 1996). Although the absence of hominoids from the early part of the section could be attributed to preservation biases, such as the total destruction of entire hominoid carcasses by overzealous and efficient carnivores, it is tempting to interpret this as a true reflection of their absence. In this case it could be indicative of their avoidance at this time of fairly well-vegetated riparian woodland environments characteristic of the lower and middle Nawata Formation. If hominoids really were absent, does their appearance at ~5.5 m.y. indicate that hominoids migrated into the Lothagam environment from elsewhere because the progressive opening of the habitat, seen from other lines of evidence, had reached a point suitable for hominid exploitation? Or, at 5.5 m.y., had progressive evolutionary adaptive responses of ancestral hominoids in alternative favored habitats resulted in the development of adaptations suited to the Lothagam environment? More Late Miocene sites sampling alternative ecologies are needed to answer these questions, which are important in determining the preferred ancestral ecotypes of our earliest ancestors.

While recognizing the limitations of the fossil and geological evidence, it is all we have to work with. Researchers have designed a variety of innovative methods, many of which are discussed in this volume, to extract the maximum information from the infinitesimally small evidence that we have. But it is essential to define and recognize these limitations, and Feibel's chapter on the importance of basins provides a valuable contribution in this respect. Basins, the fundamental units of analysis in geology, are the source of nearly all data comprising both fossil and geological records and are also functioning ecosystems. As such, basins are important in demonstrating responses by local, specific habitats and organisms to large-scale phenomena such as global climate change. Feibel stresses that basins respond differently to external influences depending on their characteristics, and these must be evaluated. The Turkana Basin of Kenya and Ethiopia is one of the better documented examples of a Plio-Pleistocene ecosystem that has accumulated sediments episodically over a considerable length of time. This large basin (146,000 km^2) has been subjected to the stabilizing effects of perennial rivers providing fresh water throughout the year and thus is less likely to be affected by local climatic fluctuations than are smaller basins such as Olduvai (3,600 km^2) and Baringo (5,500 km^2), which lacked large perennial rivers. The sedimentary records of these smaller basins, in contrast to Turkana, are strongly influenced by climatically driven lake-level cycles and are more sensitive to changes of climate experienced within the depositional basin. Sites have to be viewed in the broader context of basinal and regional developments before responses can be evaluated in terms of global patterns.

Geographical and temporal variations in modern habitats illustrate the difficulties of interpreting the rather coarse resolution in the fossil record. The arid country that typifies Kanapoi today is almost without vegetation save for a few shrubs and the prevalent "wait-a-bit" thorns, *Acacia reficiens.* Yet only a few kilometers away the Kerio River supports a riverine forest in which elephants are found. Relatively short temporal changes in habitat are illustrated by von Höhnel's descriptions of Count Teleki's exploration of East Turkana in 1988 (von Höhnel, 1891), when the area must have been significantly more vegetated than today. Elephants and rhinoceroses, species unknown in the contemporary semiarid environment, were shot daily. Significant changes of habitat have been recorded in recent decades in many of East Africa's parks such as Amboseli, the Masai Mara, and Tsavo. These modern examples of lateral and temporal environmental variation illustrate the shortcomings of the geological record, which can only provide an averaged chronicle of long-term environmental signals but not a precise indication of the environment from which a specific fossil or fossil assemblage is derived.

Fossil soils are good indicators of past environments. Paleosols, if unaltered, record the average composition of soil carbonates over long periods of time and accurately reflect floral microhabitats in the immediate vicinity of the sample. Sikes analyzes the stable carbon and oxygen isotope composition of paleosol organic matter and pedogenic carbonates to estimate the vegetation at archaeological sites at Olduvai. She finds that C_4 plants were not a major component of early hominid ecosystems until after about 1.7 m.y. Sites in middle Bed I and early Bed II indicate that early hominids preferred woody C_3 habitats, perhaps riparian woodlands, rather than C_4 grasslands. This conclusion is supported by a recent study of rodents from middle Bed I, which indicates a rich, closed woodland environment, richer than any part of the present day savanna biome in Africa (Fernandez-Jalvo et al., 1998).

The interpretation of paleoenvironments is frequently based on fossil faunal assemblages. The conventional method was, for many years, confined to an assessment of the habitat preferences of a few dominant taxa in the faunal assemblage, and even today this method is used as a preliminary indicator. Because we cannot be sure that the preferred habitats of the modern taxa accurately reflect those of ancestral taxa, this method can be misleading. For example, the Aramis fauna is dominated by ancestral tragelaphines and colobines and is interpreted on this evidence as a closed, wooded environment. Tragelaphines include the eland and species of the genus *Tragelaphus* (the greater and lesser kudu, the bushbuck, bongo, and sitatunga) and are described by Kingdon (1997, p. 350) as "gleaners of green foliage in a wide variety of mostly unstable habitats." Only the bongo is a strictly forest-living species, and modern kudus inhabit bush and woodland. The lesser kudu is frequently encountered today in the semiarid bush and scrub east of Lake Turkana. Tragelaphines are therefore not good paleoecological indicators. Colobines today are specialized arboreal folivores, mostly confined to forest. But evidence suggests that the few remaining modern species are the survivors of a relatively recent adaptive radiation, and that in the past colobine diets may have included more seeds than leaves (Lucas & Teaford, 1994; Benefit, this volume). The majority of modern African cercopithecines, too, tend to be largely forest living as a result of the recent radiation of the guenons. The diversity of forest-living modern *Cercopithecus* and colobine monkeys has led to the commonly held view that monkeys are good forest indicators, whereas fossil records indicate that, for much of their evolutionary history, both colobines and cercopithecines were more open-country animals.

Analyses of faunal communities provide more meaningful, broader based comparisons. Andrews and Humphrey here analyze community structure of mammalian faunas in recognition that past ecosystems, like modern ones, may be defined on the basis of their constituent organism's interaction with the environment. These authors recognize two major aspects of the environment as the primary factors relating to the survival of individuals: the distribution of cover and shelter in the spatial niche and the distribution of food in the trophic niche. Selected characters reflecting these two factors identify community structure in 15 modern ecosystems and, by comparison, in past faunas. The method requires an extensive modern database, which Andrews has compiled over many years. Using this approach, the habitats of the Miocene apes and early hominids, for those sites where the published records are sufficiently detailed, are predicted. The limitation of the method primarily stems from taphonomic biases which leave fossil accumulations incomplete. In particular, fossil assemblages are often the result of predator preferences and so will not accurately reflect a modern faunal assemblage associated with a particular habitat. Microfauna, in particular, is usually accumulated by raptors such as owls or eagles, and larger bodied mammals by hyaenas or leopards. Many fossil specimens show carnivore tooth marks attesting to their previous role as carnivore prey.

Recently much attention has been given to extinction and speciation in faunal successions in an attempt to verify or falsify Vrba's Turnover Pulse Hypothesis (Vrba, 1985d). When originally formulated this theory was based on the Plio-Pleistocene evolution of the Bovidae, which indicated a major turnover pulse at 2.5 m.y., a time of known climatic change (deMenocal, 1995; deMenocal & Bloemendal, 1995; Shackleton et al., 1995b; Denton, this volume), which has been linked to the evolution of *Homo*. The theory has met considerable skepticism, and, in recent years, more independent studies have been generated to test this hypothesis than almost any other single theory. Foley uses the hominid record to detect whether climate is the primary and direct driving force for hominid evolutionary change, but, like other studies of mammalian lineages, finds that his data do not support this—an ironic conclusion given that the theory was originally formulated to explain the origins of a species of hominid.

Although the climatic forcing model predicts that first appearance data (FADS) will appear in association with climatic change, Foley argues that last appearance data (LADS) or extinctions are more immediately responsive. He finds some evidence for this in the hominid record. Foley stresses the importance of interpreting evolutionary patterns geographically. Extinctions seen in one small area fortuitously preserved in the geological record are unlikely to reflect a real absence from a larger geographical area. Thus a species can be recorded as extinct in the fossil record but still survive today. Similarly, FADS may record species that evolved in alternative habitats long before they are documented in the fossil record at known fossil sites. The sudden appearance in a fossil sequence of new, apparently well-adapted species tells us nothing of the earlier evolution of a particular adaptive radiation. The FAD for *Homo erectus* illustrates this problem. *Homo erectus* is first recorded in the Turkana Basin at about 1.8 m.y. with no obvious predecessor. Is this a result of immigration or punctuation? Did *H. erectus* evolve from *H. rudolfensis,* or is it an immigrant evolved from unknown predecessors not yet discovered in the fossil record?

At another level, species are recorded as FADS when part of a continuously evolving lineage. Many species are initially named when only the beginning and end of a such a lineage is known, but when the intermediate forms are recovered the lineage is seen to show a slowly changing set of characters, and there are no obvious points at which to draw lines. These FADS are then misleading. The early evolution of the australopithecines may well prove to be such a case. At present there are good collections of *A. afarensis* dating between 3.5 and 3.0 m.y. and sufficient evidence of *A. anamensis* at 4.1–3.9 m.y. to provide diagnostic specific distinctions (M. G. Leakey et al., 1995, 1998). But the evidence suggests that *A. anamensis* is ancestral to *A. afarensis.* When a good sample of fossils is eventually recovered from the 3.9–3.6 m.y. interval, will it be possible to securely assign them to one or other of the recognized species? Because species are genetically continuous, their temporal continuity becomes a paleontologist's recurring nightmare.

Collard and Wood's concept of grouping species according to grades or adaptive types avoids the problem of defining species divisions in continuous lineages. In applying this method to fossil hominid species, Collard and Wood identify adaptive types by considering features associated with diet, locomotion, body shape, and encephalization. These characters define two recognizable grades: one includes the australopithecines, paranthropines, *Homo habilis,* and *H. rudolfensis,* and the other comprises *H. ergaster.* There is an apparent shift of grade with the appearance of *H. ergaster,* although the mosaic nature of evolutionary progression prevents distinct divisions. This new and instructive method of looking at evolutionary change has the advantage that it groups ecomorphological characters irrespective of taxonomic affinities. Indeed, in this scheme the early species of *Homo* do not form a functionally coherent group: two species are included in the partially arboreal, small-brained, omnivorous australopithecine grade, whereas *H. ergaster,* although remaining relatively unencephalized, appears to have been a fully committed biped adapted to living on the open savanna. This raises the question of the validity of the generic attribution of the early species assigned to *Homo,* as logically all species within a genus should belong to one grade.

Teeth are particularly suitable for analyses of morphological changes in taxonomic lineages. Details of microwear and changing dental structure are frequently analyzed in this context (Teaford, 1984, 1994; Ungar, 1992, 1998). Recent studies have concentrated on the microstructure of dental enamel because this preserves a record of the tooth's developmental history (Macho et al., 1996). Ramirez Rozzi et al. apply this method to a sample of hominids from the Shungura Formation in the Turkana Basin with the expectation that dental developmental change should be observed over the course of early hominid evolution as it is in other mammals. A sample of 66 teeth dated between 3.36 and 2.1 m.y. are grouped into time intervals of 400,000 years irrespective of taxon. Each tooth is scored using carefully chosen characters of the enamel structure and thus assigned to morphs. This method shows that temporal changes in enamel microanatomy are related to environmental change and that the temporal trends that could be identified are most likely explained as adaptations to more abrasive diets. Further studies are required to confirm that these are not biotic responses to selective competition improving the adaptation to a particular food source. If hominids were indeed flexible, tolerant species, these dental adaptations may merely indicate selective improvement in the ability of teeth to process favored dietary resources.

In the final chapter of this section, Bromage evaluates hominid ecomorphological characters in an attempt to reveal the extent to which early hominids complied with habitat theory. A series of characters are chosen for their adaptive significance, including those of the skull and jaws reflecting locomotion and those of the dentition related to diet. These characters are interpreted as ecological indicators and in this way used to assess ecomorphology. *Australopithecus ramidus* is chosen as a test case because all the information is readily accessible in one publication (T. D. White et al., 1994) together with a second paper describing the paleoecology (WoldeGabriel et al., 1994). The shortcoming of this choice of hominid is that, with the limited evidence available, only provisional conclusions can be drawn as to the merits of this approach.

In spite of the many uncertainties, the increasing number of innovative methods of analysis designed to reveal the course of human evolution will undoubtedly gradually lead closer to the truth. Of all the hurdles that we encounter in unraveling the history of the past, perhaps the most difficult to overcome is the tendency that humans have to be influenced by preconceived ideas. As a consequence, the same data are too often seen as supporting opposing interpretations. The contrasting views about the locomotor patterns of *Australopithecus afarensis* as either a fully committed biped or an occasional biped that frequently climbed trees is perhaps the most vexing current example. Similarly disturbing and opposing are interpretations of the hominid fossils from Kanapoi and Allia Bay. These were assigned to a single sexually dimorphic species, *A. anamensis* (M. G. Leakey et al., 1995), in recognition of the primitive australopithecine characters and the lack of evidence for more than one species. Yet Senut (1996) proposes an alternative taxonomy in which she sees two major types of bipedal locomotion as far back as 4 m.y. Senut includes *A. anamensis* with the Laetoli hominids and some from Hadar as *Preanthropus africanus,* an early fully bipedal ancestor of *Homo.* The second locomotor type, bipeds that also climbed trees, comprises the early australopithecines *Australopithecus antiquus* and *A. africanus.* Although he uses different nomenclature, Coppens, in his introduction to part 2 of this volume, uses this interpretation in his scenario of human evolution, where he posits two forms of hominid at 4 m.y., one *A. afarensis,* a biped that still climbed trees, and the other *A. anamensis,* exclusively bipedal. As attractive as this hypothesis may be, however one looks at the data, the evidence for two locomotor types at 4 m.y. is simply not there (M. G. Leakey et al., 1998). Data must be examined objectively and conclusions drawn from the available facts. The data we have may be limited and often inadequate, but only rigorous scientific analyses will reveal the truth to us.

Kingdon ends his introduction to part 2 with an appeal for the conservation of Africa's Pleistocene habitats which form the basis of modern ecosystems. I will end my introduction with a plea for the preservation of Africa's rich record of the past. Increasingly, prehistoric sites are disappearing under agriculture and concrete because they lack the necessary protection or concern. Field work produces the primary data from which we interpret all that we know of the past. Continuing field work is essential in providing new evidence and broadening our perspectives. When there are no longer sites to investigate, future generations will not only criticize us for the loss of modern habitats but also for their loss of opportunities to explore those from the past.

19

Basin Evolution, Sedimentary Dynamics, and Hominid Habitats in East Africa

An Ecosystem Approach

Craig S. Feibel

Much has been made of the interaction between global-scale climatic phenomena and patterns of environmental change, habitat shift, and biotic evolution in Africa. What has been missing from the discussion is an understanding of how large-scale phenomena such as global climate change interact with specific habitats and organisms. The later Cenozoic presents an often detailed record of climatic history and biotic events, but linking these records is less than straightforward. Geographic separation, differing scales of analysis, and problems with numerical age control and correlation all tend to confuse the correspondence. Although a strong theoretical basis exists for a close action–response relationship between global-scale phenomena and local biotic events, the hard evidence has not yet demonstrated such a link.

A major concern in the integration of data pertaining to environmental change and biotic evolution is the connection from demonstrable global-scale phenomena to records of local habitat and its biotic components. Associated with the spatial scale problem is a parallel temporal difference, with large-scale phenomena often tracked over longer intervals, while the local record typically comprises short temporal windows.

Basin Analysis: The Middle Ground

One of the fundamental units of analysis in geology is the basin. The term "basin" can be used either in the limited sense of a depositional basin, actively accumulating sediments, or in the broader context of a drainage basin, delimited by an integrated hydrographic network. Both aspects are of significance in evaluating the hard evidence of habitats, change, and ultimate controls in the African record. Sedimentary basins are the source of nearly all data, comprising both the geological and fossil records. But the fundamental unit controlling the character of that data set is the drainage basin. The basin, including both the active depositional component and is contributing source area, is the fundamental unit of the "system" as discussed below.

A basin in a sense can be considered as a functioning ecosystem. It is generally an open system with respect to many components. For example, water enters the system as rainfall and may leave the system as an exiting river. Animals may migrate into and out of the system. Although it may be open to many components and is not delimited by biological boundaries demarking ecosystems as defined by dominant communities, the basin system is a basic unit for evaluation of environmental character and change. This has become an important approach in analysis of present-day ecological pattern and process (e.g., Bormann and Likens, 1979; Likens, 1985) and is a useful level of analysis for bridging the middle ground in paleoenvironmental studies.

Investigating ecosystem response at the basin level is particularly appropriate in analysis of paleosystems, as there is a close correspondence between the factors affecting the system and the geological and fossil evidence by which we reconstruct the past. For larger scale effects, it is harder to show a direct correspondence. For example, the recent analysis of African climate based on marine dust records (deMenocal, 1995) shows that some part of the continent exhibited fairly synchronous response to global-scale climatic fluctuations, but it cannot determine which parts of the terrestrial system were affected and to what extent. It is impossible to directly relate this record to the pattern of hominid evolution, as it is impossible to show that hominid habitats were in any way affected. Similarly, at a smaller scale than the basin, it is difficult to ascribe individual components of the database to particular environmental controls. A local change from forest to grassland might reflect global cooling, or it might simply be the result of a switch in channel pattern of a local drainage.

Response, Sensitivity, and Thresholds

Basins display a broad range of sizes, shapes, and features, and the characteristics of an individual basin will in large part determine how it responds to an external influence. Such responses are complex and not simply a function of latitude, elevation, and other primary controls. For example, lake level is a simple measure of response to climate change in a closed basin. Although early studies noted roughly synchronic lake-level response in the Holocene of East Africa (e.g., Butzer et al., 1972), subsequent studies have highlighted the independence of reactions in adjacent basins for all but extreme climatic swings. Basin size, and particularly catchment area at higher elevations, seems to be one of the major factors controlling how quickly and to what extent a basin responds to climatic forcing. The sensitivity of a basin is a measure of this response to perturbations such as climatic shifts. For highly sensitive basins, an external stimulus should elicit a rapid response, such as a rapid lake-level change. Less sensitive basins might not display any obvious response.

Compounding the problem of different sensitivities at the basin or system level is the reality that only parts of the system are recorded in the geological or fossil record (fig. 19-1), and thus these particular parts of the system must be sensitive enough to show a response. Otherwise a response might occur but never be recorded. From the biological perspective, then, we are left with a suite of marker habitats—those that will have a chance of marking in the geological or fossil record a response to environmental change.

Most environmental changes occur along a continuum rather than in discrete quantum units. Response may also occur as a gradual adjustment, but both physical and biotic systems tend to have critical thresholds that mark changes in character or state. Thus a river may gradually decrease in flow as a response to decreased rainfall, but there is a threshold that marks the transition from a truly perennial state to an intermittent one. Similarly, trees that depend on a year-round water supply may be able to tolerate shortages for a period of time, but at some stage this will lead to the death of the trees and a shift in vegetation structure. In detail the difference may seem subtle, but particularly in biotic systems there is an important difference. The nonthreshold loss of one tree out of a hundred may be insignificant to the character of a habitat over time, but the loss of the last tree is a different matter.

Controls

What, then, are the major controls on both the physical and biotic characteristics of a basin? This question has two answers. In a real sense, we can answer with one word: water. For a more elaborate response, though, the major factors would include climate, geology, topography, organisms, and time. This list is effectively the same as the pedologists' five soil-forming factors (Jenny, 1941), but to make a soil or an ecosystem, the same factors are at play. Several of these factors can be subdivided to introduce some important distinctions at this stage.

In a complex basin, several climatic regimes may affect different parts of the system, notably the higher elevation headwaters region(s) versus the lower depositional basin (Feibel, 1995). Geology operates at several levels. Tectonics determines subsidence rates and basinal configuration, which determines where events happen. The complexities of rift faulting in East Africa can introduce tremendous variety in the local landscape, controlling drainage patterns, topography, and distribution of volcanic centers (fig. 19-2). Explosive volcanic activity may disrupt a basin by choking drainage networks and mantling landscapes. It may even result in dramatic ecological displacement similar to the effects of climatic shifts (Harris & Van Couvering, 1995). Bedrock geology and/or soils form the substrate at the active interface between lithosphere, atmosphere, hydrosphere, and biosphere. Topography is particularly important in the sense of elevation, as both temperature and rainfall in Africa are strongly tied to altitude, and thus strong overall basinal relief is reflected in an apparent climate range. And, finally, time is a consideration both in terms of duration and of the temporal context, which relates

Figure 19-1. Generalized features of erosional versus depositional landscapes. Erosional landscapes (a) do not present the opportunity for burial and fossilization and are thus absent from the ancient record. Variations in bedrock lithology (volcanic at left, metamorphic on the right) climate and topography result in a diversity of habitats in this catchment portion of the basin. Depositional landscapes (b) provide opportunities for fossilization and form the basis for reconstruction of past habitats. Vegetation patterns are primarily influenced by water availability and substrate type, whereas topography and climate vary little within this portion of the basin.

conditions and events of the basin to Earth history.

For a single controlling factor, though, much of the character and history of ecosystems can be directly related to water. Rainfall is probably not the best measure of control by water in many habitats, however, particularly those of interest in the Plio-Pleistocene record of Africa. Water availability is the critical control on organisms, and this can limit occurrence as much from overavailability as from underavailability. A focal example is the distribution of trees on a landscape. Although in the broadest sense there is a correspondence of trees with latitudinal and elevational precipitation belts, in detail the correlation breaks down. Forests can exist in the midst of deserts given a groundwater source, and trees can be excluded from an otherwise optimal rainfall regime by an overly high water table. Water availability is often mediated by substrate, with porous, sandy soils supporting trees, and the less permeable clays mantled with grasses (Tinley, 1982). Both of these examples have direct relevance to the primary study areas of concern here. Water has an additional significance in this discussion, as nearly all of the fossiliferous deposits of Africa owe their existence to deposition by water. With few exceptions, this strong bias for sampling wet habitats pervades our perception of these past ecosystems.

The Turkana Basin and Ecosystem

One of the better documented examples of a Plio-Pleistocene ecosystem is the Turkana Basin of Kenya and Ethiopia. The basin accumulated sediments episodically from the late Oligocene, but this discussion will focus on the Plio-Pleistocene portion of the sequence, which reflects a large, integrated drainage and sedimentary basin (Brown & Feibel, 1991; Feibel et al., 1991). Although a closed basin today, it was an open system reaching to the Indian Ocean in the

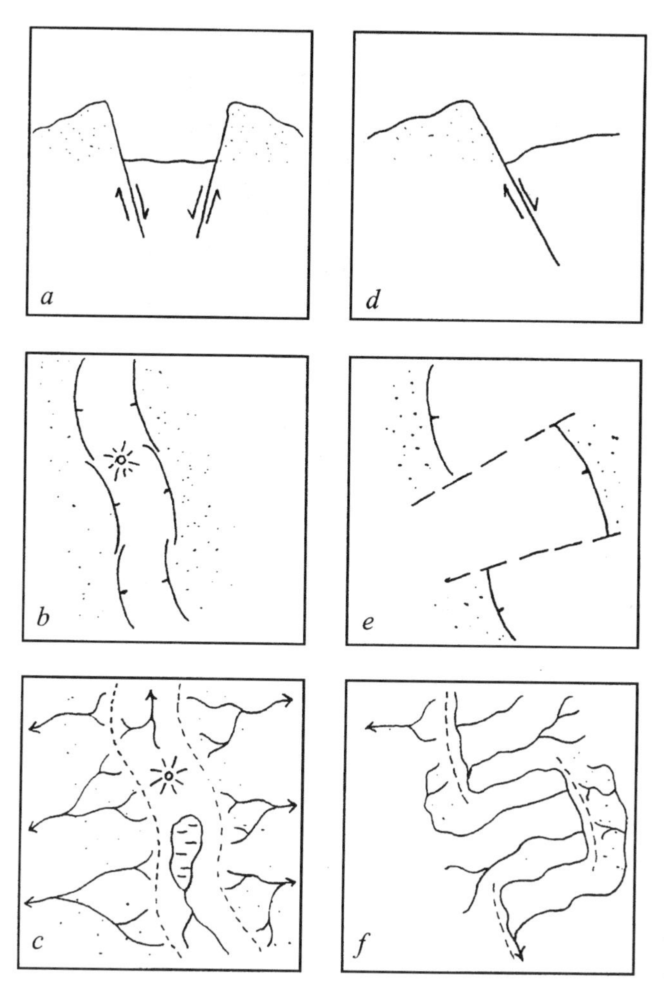

Figure 19-2. Schematic representations of the effects of differing tectonic character in a rift setting, with a full graben situation on the left, and a half-graben setting on the right. The two upper panels (a, d) show the cross-sectional difference between a full graben, with two antithetic normal faults, and a half-graben formed by a single normal fault. Areas uplifted above the valley floor are stippled. The second row (b, e) shows a map-view illustration of fault relations along the rift. The full graben (b) forms a relatively narrow valley bordered by higher terrain (stippled). The half-graben results in alternating centers of uplift (stippled). In the bottom row (c, f), the hydrography resulting from this pattern is illustrated. A full graben (c) directs much of the local precipitation away from the valley, with only rainfall within the valley able to flow toward the valley center and along the rift. Volcanic edifices may block the along-rift drainage, resulting in lakes. The half-graben setting (f) may premit extensive along-rift drainages to develop and tends to recapture much of the moisture falling on the rift flanks as well.

Pleistocene, and likely for much of the period of interest here (Feibel, 1994a).

Several salient points of the Turkana Basin record, reflecting environment–habitat interactions, can be summarized here. The basin covers some 146,000 km^2 (Butzer, 1971). Lake Turkana today has a surface elevation of about 375 m, and the highest point in the headwater region is 4203 m. Present-day climatic control is dramatically partitioned, with a headwater region climate that is warm temperate (Köppen Cfb and Cwb) or tropical (Am), while the lower sedimentary basin is hot arid or semi-arid (BWh and BSh). As this contrast results from the large-scale geographic context and relationships with the monsoons and elevation, it is unlikely that it has changed significantly in the past 4 m.y. As part of the East African Rift System, the basin is currently undergoing extension and is segmented into a series of half-grabens of alternating polarity. This has a strong influence on habitat distribution today, and similar patterns are discernible from the geologic past (Feibel, 1994b). The major drainage systems in the basin, the Omo, Kerio, and Turkwell rivers, all have extensive records, including critical markers of perennial flow in the form of the Nile oyster *Etheria elliptica.* The oysters indicate that the currently ephemeral Kerio River was a perennial stream at about 3.4 and 1.6 m.y. Volcanism has been an important, short-term influence on environments. Local volcanism in the form of basaltic lava flows may be responsible for initiation of lake phases in the basin. Explosive volcanism in the headwaters region mantled the basin with tephra, choking rivers and causing episodic aggradation in the sedimentary basin. Soils are extensively preserved in the stratigraphic record, but have only begun to yield data on environmental character. The database is still extremely small. The biotic record of the basin is unusually good and includes a broad temporal as well as geographic sample.

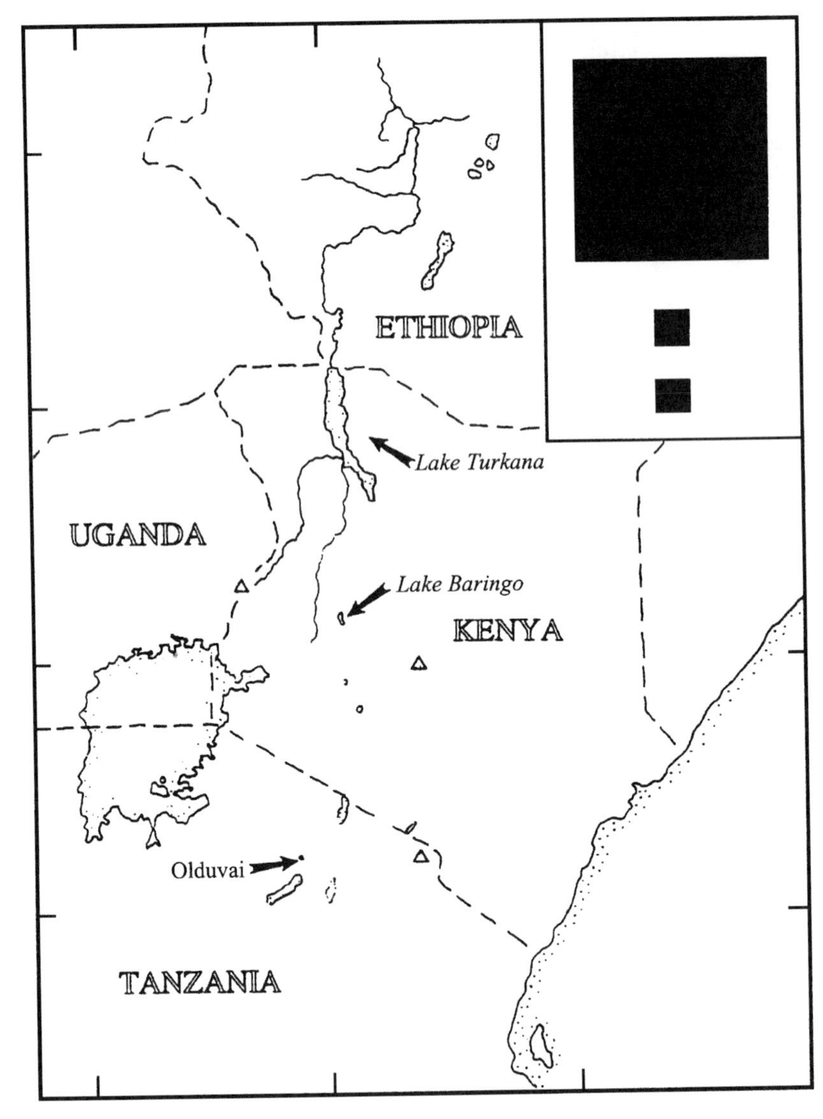

Figure 19-3. Map of the eastern rift showing the Turkana Basin, Baringo, and Olduvai. Inset illustrates the proportionate differences in area of the present-day Turkana (top) and Baringo (middle) basins, and the paleo-Olduvai Basin (bottom).

That sample is, however, limited by the usual taphonomic biases. A variety of different accumulation processes are responsible for the fossil record, and thus the assemblages reflect differing controls.

A critical aspect of the Turkana Basin complexity is the moderating effect that large, perennial rivers exert on local habitats within the depositional portions of the basin. In spite of the arid to semiarid local climate there, fresh water was available throughout the year at the rivers, and seasonal flooding, which may have persisted for periods of months, would have strongly tipped the moisture balance. The sedimentary record of the basin reflects the stabilizing effects of the perennial rivers. Base levels were tectonically controlled or episodically overtaken by volcanic signals, but climatically-driven base-level fluctuations (closed-basin lake-level fluctuations) are largely absent until a major tectonic reorganization of the basin in the Middle Pleistocene. In addition, the dominance of fluvial regimes along the basin axis instead of closed-basin lakes would have presented available moisture in a much different form.

But shouldn't major global-scale perturbations like the 2.8 m.y. signal (deMenocal, 1995) affect a fluvial source region as well as any other part of the system? The answer to this question may lie in the nature of thresholds. The overall effect of a large, high-altitude source area may have ameliorated climatic effects to the point where thresholds critical to habitat composition in the lower basin were not crossed. In this case, then, overall basin configuration and complexity are important to the stability of an ecosystem. This would have important implications for patterns of evolution and movement in association with climatic events.

Comparisons with Baringo and Olduvai

Although reflecting different temporal spans, the Baringo and Olduvai Basins present some useful points for comparison (fig. 19-3). Both have been extensively studied, and they have produced important contributions to the biotic evidence for the African Plio-Pleistocene.

The Baringo Basin lies within the Gregory Rift, and with the present-day elevation of Lake Baringo at about 967 m and catchment elevations in excess of 2700 m, the basin today covers an area of about 5500 km^2. Baringo is a closed basin today, and there is no indication that it was ever open or more extensive than at present. Like the Turkana Basin, Baringo covers several climatic zones, with a semiarid (BSh) basin floor and a subhumid temperate regime along the flanking rift walls. Because of the basins's physiography, water supply to the basin center is by way of numerous small, marginal streams. This dispersal of water sources leads to greater unreliability of their ephemeral flow.

The Olduvai Basin (Peters & Blumenschine, 1995) covers an area of about 3600 km^2, with an elevation of 1450 m at the gorge, and the Crater Highlands to the southeast reaching 2200 m. It is somewhat cooler due to its elevation, but still has a semiarid (BSh) climate. The Olduvai Basin contained one or more terminal lakes in the Early Pleistocene, but subsequent tectonic activity has changed the configuration of the basin and disrupted drainage patterns. The sedimentary record of Olduvai (Hay, 1976) is strongly influenced by volcanic activity, with much of Bed I composed of tuff, and significant components of tephra in the overlying units as well. The Olduvai sequence was strongly influenced by the closed-basin lakes, which occupy the center of the basin through much of the Plio-Pleistocene.

There are two obvious differences between these basins and the Turkana system. The first is size. Baringo and Olduvai are roughly comparable, wereas the Turkana Basin is some 32 times larger. For comparison, the present-day Serengeti ecosystem covers about 25,000 km^2 and falls in between (Sinclair, 1995). The major difference between these basins and the Turkana setting, however, is the lack of a large, perennial river dominating the system. The sedimentary record in both cases shows a more episodic character, influenced by the climatically driven lake-level cycles, and both basins can be expected to be far more sensitive to climate fluctuations experienced within the depositional basin.

Summary

It is highly unlikely that we will ever have the hard data necessary to reconstruct an entire ecosystem from the ancient record. That record does contain clues, however, to aspects of a system far from the actual site of investigation. In contrast, some individual sites and localities present such a wealth of information that we can make a credible reconstruction of habitats and the communities that existed there. Evolutionary and biogeographic processes are going on somewhere between these two scales, and it is dangerous to try to characterize an entire system as an extended expression of an individual site. Many of the environmental signals inferred from a local site record may change dramatically as the site is integrated into a larger picture at the ecosystem or basin level and as a component of a larger scale system.

The accumulating database shows that significant global-scale perturbations in climate are recorded in many settings and that even the low-latitude record shows a detailed pattern of climatic response to orbital periodicities. The trick now is to integrate this record with the history of evolution in Africa and to ascertain how these influences affected individual ecosystems and habitats. To understand how large-scale phenomena have influenced the pattern of biotic evolution and distribution, the record from individual localities and sites must be evaluated in the broader context of basinal and regional developments and be fitted into the global pattern.

Acknowledgments This work was supported by grants from the National Science Foundation and the Leakey Foundation. I thank Meave Leakey for her enthusiastic support of investigations into the environmental context of hominid evolution and Timothy Bromage and Friedemann Schrenk for organizing this symposium.

20

African Miocene Environments and the Transition to Early Hominines

Peter Andrews
Louise Humphrey

We are proposing a new paradigm for the study of paleoecology based on ecological first principles. It is our contention that the factors underlying modern ecosystems must first be understood, as far as possible, and then the same factors must be applied to evolutionary change in the fossil record. It is not enough to demonstrate taxonomic similarity between recent and fossil faunas, for there is no guarantee that ancient ecosystems were the same as present-day ones. Similarly, ecological equivalence in morphological structures of limited biota (ecomorphology) provides only a microcosm of the total ecosystem. The interplay between evolution and ecology therefore forms the basis for the new paradigm.

Evolutionary transitions at the species level range from adaptive to random, and there is little indication from the fossil record which of these might be operative for any one event. These transitions may be unrelated to specific environmental or temporal factors, and it follows from this that speciation within biotic communities does not necessarily provide useful information about environmental change or the evolution of ecosystems. In contrast to this, the structure of animal and plant communities and changes in structure are highly pertinent to environmental change, for they are not normally the result of random events. These changes can be identified and measured in the fossil record and are useful as a means of interpreting both the evolution of ecosystems (or lack of it) and the changing position of species within them (Andrews, 1992b).

It is our purpose here to reconstruct Neogene environments to compare Miocene hominoid environments with those of early hominines. Relationships between taxa are based on a consensus phylogeny, and the small differences in the numerous phylogenies that have been published in the last 5 years for hominoids are not relevant to the general trends discussed in this paper (Pilbeam et al., 1990; Andrews, 1992a; Begun, 1992, 1994; de Bonis & Koufos, 1993; Moya-Sola & Kohler, 1993) and for hominines (Wood, 1991; Kimbel et al., 1994; T. D. White et al., 1994).

Methods and Definitions

Environmental reconstructions will be attempted by analysis of the community structure of mammalian faunas. Past ecosystems may be defined on the basis of inferred interactions between fossil organisms and their environment, with the environment distinguished into abiotic and biotic elements (Odum, 1983). The biotic element comprises living communities of plants and animals, and a community of a particular type may exist at one time and place within broad abiotic limits of temperature, rainfall, and atmospheric composition. Communities form in response to changes in these parameters, and in extreme cases will be eliminated from that time and place, accompanied by extinction or emigration of parts of the community. Changes may be at local or global level, and it is not always possible to distinguish between these at sites that are restricted geographically.

In practice, it is rarely possible to reconstruct whole biotic communities from the fossil record. The term "community" is applied here solely to mammalian faunas (i.e., assemblages of mammals and their distributions within the multidimensional niche). The application of community analysis to fossil faunas is based on the recognition that mammal species occupying similar habitats across the world tend to show parallel or convergent adaptations. Convergent evolution results in the production of similar adaptations in phylogenetically unrelated organisms subject to similar agents of natural selection (Cody & Mooney, 1978), although convergence is constrained by availability of genetic variability in the converging lineages and sufficient time for it to act. For example, in lowland tropical forests in Asia, America, and Africa, gliding (flying) adaptations are present in unrelated families of rodents. Extending this to whole communities, mammalian communities in similar habitats also tend to have similar distributions of adaptations even where no species are held in common between them (Cody & Mooney, 1978). This results in similarities in the way niches are filled and, by extension, similarity in community structure.

The interface between animals and the ecosystem of which they are a part is through two major aspects of the environment: distribution of cover or shelter in the spatial niche, and the distribution of food in the trophic niche. All other factors relating to survival of individuals are secondary to these two features.

Spatial Niche

Cover can be in the form of vegetation or physical objects such as cliffs or caves, but whatever its form, its effect is to buffer the effects of climate, giving protection from climatic extremes. Cover is also a relative concept, for what is cover for a small animal, such as a clump of grass, is not so for a large animal, and while a tree may provide cover for a large animal, it may be the total habitat of a small one. As a consequence, to provide an adequate reconstruction of the degree of cover in a habitat, the full range of body sizes of animals occupying the habitat must be examined.

The nature and extent of cover present in a habitat can be reconstructed by the examination of the spatial distribution of certain groups of animals. This has been done for many groups of aquatic and marine fossil faunas, and in theory it should be possible to apply to terrestrial environments using groups such as birds, reptiles, and many groups of invertebrates, but in practice it is only for mammals that the first attempts have been made. Mammals lend themselves particularly well to spatial analysis, for their adaptations to the extremes of spatial distribution may be easily identified from their bones, extremes such as fossorial adaptations, and arboreal, aerial, and terrestrial specializations. There are ambiguities, of course, but some of these can be resolved by including an element of size in the analysis.

Faunas from different habitats have been found to have different patterns of spatial/size distribution, with, for example, differences between tropical habitats being distinguished at a high level of statistical probability. Fossil species with associated postcrania can also provide evidence of spatial distributions and can similarly be distinguished from each other and from living faunal distributions at high levels of probability. The spatial distribution alone does not indicate what kind of habitat was present, but it does provide a clue as to the kind of vegetation. For example, a small number of arboreal species indicates the presence of some trees but probably not a closed woodland habitat, whereas a large number of arboreal and scansorial species is strongly indicative of closed woodland.

Dietary Niche

The other main interface of animals with their habitat is in the food that they eat. Food is provided for primary consumers by the vegetation in the habitat, and the dietary analysis of animals making up present-day faunas provides information on the distribution of food and its availability in the habitat. Distribution of food has both a geographic element in its distribution in space, and also a temporal aspect, particularly in its seasonal distribution. There are also variations evident during the development and life of individual animals, with very young and very old animals often having reduced access to optimum foods, but with individual variations as well. Most animals are able to adapt their diets to a certain extent to take account of seasonal variations, but these different sources of variability make it difficult to identify the full range of diets on morphological evidence alone. In practice, only the major dietary elements can be identified by such means (e.g., dental morphology or microwear).

Secondary consumers can also provide clues as to the nature of the habitat in terms of their relative diversity. For example, high species richness of predators may be related to high density or biomass of prey rather than to high species richness of prey. Habitats with such a combination are generally rather impoverished in some way and usually consist of open grassland, steppe, or tundra. Habitats with high species richness of primary consumers and relatively lower richness of predators (although actual species numbers may still be greater) generally have rather rich forest vegetation with more complex niche structure.

There is also a relationship between body size and dietary distribution, but it differs from that seen in spatial distribution. In general, mammals with insectivorous diets are small; most weigh <1 kg. Larger species tend to specialize on particular groups of insects, such as ants or wood-boring grubs. Species with frugivorous diets are small to medium size, usually <10 kg but with some species reaching 70–80 kg. Herbivorous species tend to be medium sized to very large, with greatest diversity in the 10–100 kg size class but with many smaller and larger species as well. In terms of habitat reconstruction and community analysis, these correlations can provide some additional information.

What is being reconstructed by these means is not habitats as they are known today, but the degree and structure of the plant cover and the temporal and geographic availability of food. Comparisons must be made with modern habitats in order to understand the nature of the relationship, but it is not necessary to make direct parallels between past and present ecosystems. For example, in the case of the Miocene fauna from Songhor, it is clear from the large number of small arboreal, scansorial, and aerial mammal species that there was a structurally complex ecosystem, including an abundance of related but distinct niches in a complex forest canopy, and from this parallels may be drawn with present-day ecosystems, but the two levels of inference should be kept distinct. Similarly, the presence of major differences in proportions of browsing compared with grazing herbivores or frugivorous species could point to some general aspects of the environment—for example, in terms of the vegetation, such as the relative amounts of trees, bushes and herbs producing browse and fruit compared with grasses producing only graze. This can also be extended to climatic inferences, with equable climates being indicated by greater diversity (species richness) of frugivores requiring year-round abundance of fruit, or seasonality being indicated by greater diversity of herbivores relative to frugivores or of carnivores relative to their prey species.

Body Size

Body size is an integral part of the spatial and trophic niches, and it also makes up a third important aspect of the ecosystem through the multidimensional niche (Odum, 1983). There is also a more direct influence in the extent to which animals modify the environment. Elephants, for example, open up the environment, changing its nature at least locally from wooded to less wooded habitats, and some smaller animals such as beavers also have a major impact through their dam-building activities. There is also a major size element in grazing cycles, whereby large grazing mammals mow down coarse herbage to expose more delicate plant forms to more selective and smaller herbivores.

The contribution of body size analyses will be realized when attempts are made to understand the interrelationships of animals within communities. Past analyses of body size distributions, however, have been directed at finding a direct relationship with habitat, and these have not been successful. Most present-day habitats have mammalian faunas with similar size distributions, and faunas from habitats as distinct as open grassland and evergreen forest have been shown to have size distributions with no significant difference between them when analysed statistically (Andrews et al., 1979). Despite this, persistent attempts are still being made to relate body-size distributions to habitat, probably because size is the easiest ecological variable to estimate, but this only has ecological meaning when analyzed in conjunction with and in the context of spatial and dietary niche distributions.

Methods of Estimation

New data on ecomorphology are providing correlations between morphology and habitat, and where these data are available they can be used in the functional interpretation of morphological structures. In the absence of ecomorphological data, body size can be estimated by regressions of tooth or limb size on body weights of known individuals. The spatial niche is estimated on the basis of limb proportions and function, which readily distinguish between fossorial, terrestrial, arboreal, and flying adaptations. The trophic niche is based on correlations between diet and jaw and tooth anatomy, including data on tooth wear where available. These enable broad divisions to be recognized between insectivorous, frugivorous, and herbivorous adaptations.

Present-Day Environments

A survey is presented here of tropical and subtropical ecosystems (Richards, 1952; Eyre, 1963; H. Walter, 1979), recognizing 15 ecosystems that have potential relevance to hominoid evolution. The mammalian community structure of these ecosystems will be described based on an analysis of 44 present-day mammalian communities (table 20-1, figs. 20-1–20-3).

Table 20-1. List of recent localities that have provided the comparative mammalian faunas averages in figures 1–3.

Ecosystem					
Locality	Country	*N*	Latitude	Longitude	Altitude (M)
		African lowland tropical evergreen forest			
Irangi	Zaire	76	1°55′ S	28°28′ E	1000
Semliki	Uganda	72	0°4′ N	30°0′ E	750
Seredou	Guinea	76	7°35′ N	9°20′ W	700
		African montane evergreen forest			
Mt. Kenya	Kenya	67	0°15′ S	37°12′ E	2700
Semliki	Uganda	54	0°4′ N	30°0′ E	1800
Lemera	Zaire	64	2°8′ S	28°50′ E	2150
		African seasonal tropical forest			
Rwenzori	Uganda	59	0°30′ S	29°55′ E	1000
Amani	Tanzania	60	5°6′ S	38°38′ E	880
Kakamega	Kenya	62	0°20′ N	34°52′ E	1520
		Malayan tropical evergreen forest			
Bukit Lagong	Malaya	76	3°15′ N	101°40′ E	500
Ulu Gombak	Malaya	69	3°21′ N	101°47′ E	515
Ulu Langat	Malaya	67	3°05′ N	101°50′ E	500
		Indian tropical evergreen forest			
North Kenara	India	40	15° N	75° E	800
Coorg	India	52	12° N	76° E	900
Mokokchung	India	52	26°20′ N	94°30′ E	1500
		Central American tropical evergreen forest			
Balboa	Panama	38	8°57 N	80° W	50
Cristobal	Panama	38	9° N	80° W	50
Santa Rosa	Costa Rica	44	9°28′ N	85°09′ W	45
		Chinese paratropical forest			
E. Hsi Yunnan	China	50	22° N	101° E	1400
W. Hsi Yunnan	China	47	22° N	99° E	1400
Kwangsi	China	60	26° N	95°40′ E	1500
		Burmese paratropical forest			
Hkamti	Burma	56	12° N	99° E	600
S.Tenasse Rim	Burma	56	12° N	99° E	Sea level
Toungoo	Burma	56	19° N	96° E	Sea level
		Indian subtropical forest			
Kanha	India	39	22°45′ N	80°45′ E	500
Lyallapur	Pakistan	28	31°25′ N	73°10′ E	100
Imphal Valley	India	42	25° N	93°55′ E	1000
		Miombo woodland			
Nyika Plateau	Zambia	32	10°35′ S	33°41′ E	2180
Zinare	Mozambique	55	21° S	33° E	1000
Tambarara	Mozambique	41	18°40′ S	34°20′ E	300
		South African summer-rainfall woodland			
Kruger	South Africa	82	25° S	31°30′ E	700
Ingwaruma	South Africa	64	27°08′ S	32°01′ E	400
Ubombo	South Africa	48	27°34′ S	32°05′ E	400

(*continued*)

Table 20-1. (*continued*)

Ecosystem					
Locality	Country	*N*	Latitude	Longitude	Altitude (M)
		African tropical savanna			
Serengeti	Tanzania	75	2°15′ S	34°49′ E	1480
Serengeti	Tanzania	58	3°0′ S	35°02′ E	1520
Rwenzori	Uganda	54	0°5′ S	29°50′ E	1000
		African tropical grassland			
Serengeti	Tanzania	39	2°55′ S	35°10′ E	1530
Kafue	Zambia	49	15°40′ S	27° E	1000
		African tropical arid savanna			
Karamoja	Uganda	73	3° S	34°20′ E	1000
Tsavo	Kenya	67	3°2′ S	38°20′ E	700
Kapiti	Kenya	65	1°35′ S	36°54′ E	1600
		African mixed wood and grassland			
Tana floodplain	Kenya	48	2°15′ S	40°12′ E	40
Kalahari	SW Africa	60	25–26° S	20–21° E	100–500
Kagera	Rwanda	49	1°33′ S	30°23′ E	1500

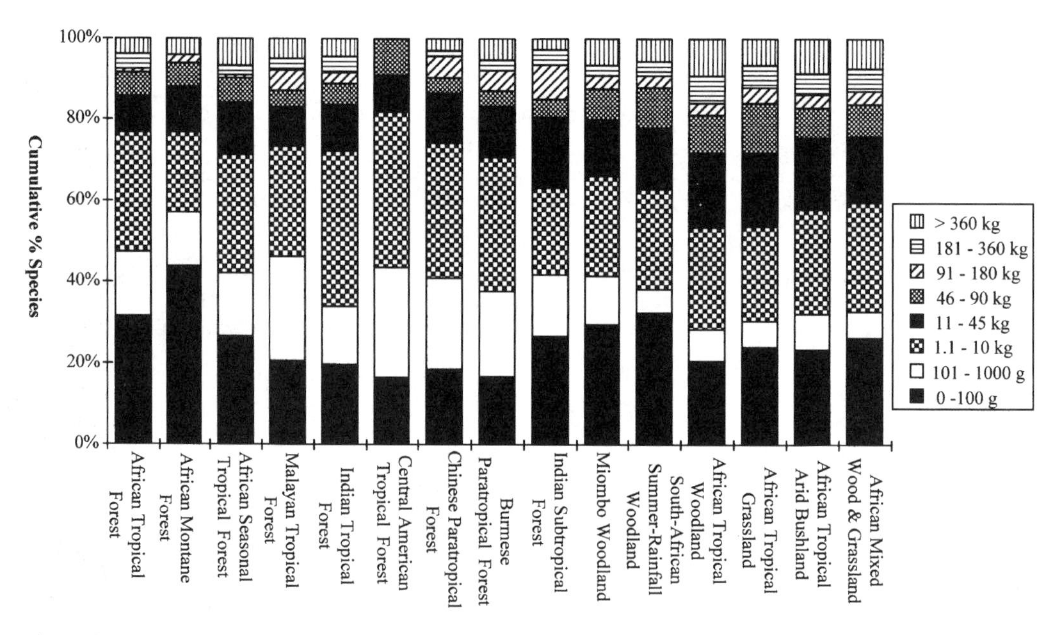

Figure 20-1. Proportions of species distributed by body size for 44 present-day mammalian faunas averaged for 15 vegetation associations. Each association is represented by three sites except for tropical grassland, for which only two present-day sites could be found (Serengeti short-grass plains and Kafue flood plain). Body weights in kilogram classes shown on the right.

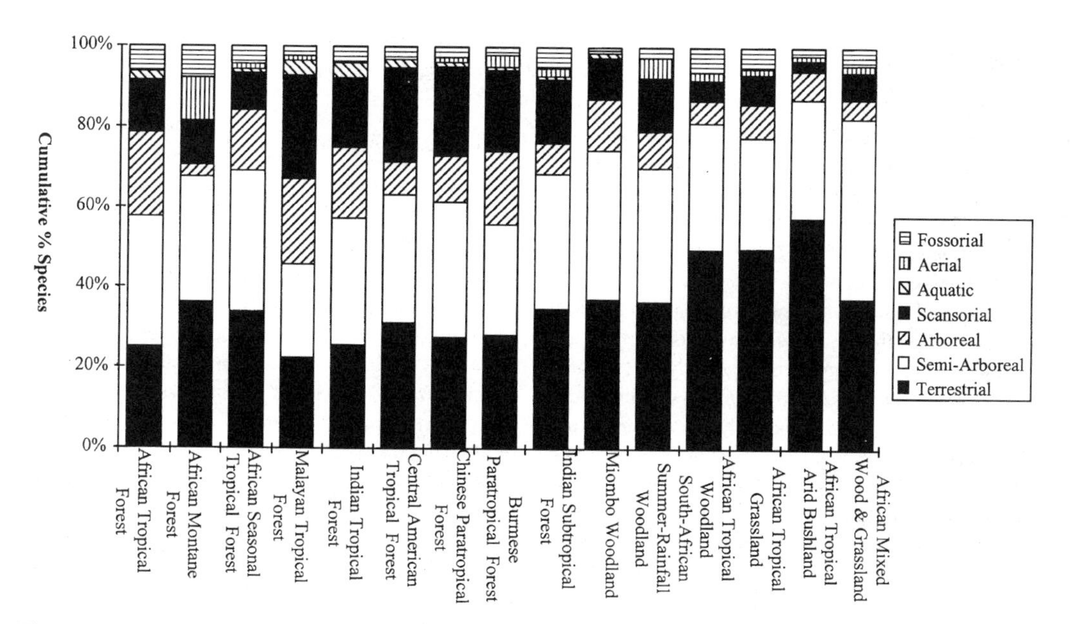

Figure 20-2. Proportions of species distributed by locomotor/spatial categories for 44 present-day mammalian faunas averaged for 15 vegetation associations. Each association is represented by three sites except for tropical grassland. Categories shown on the right.

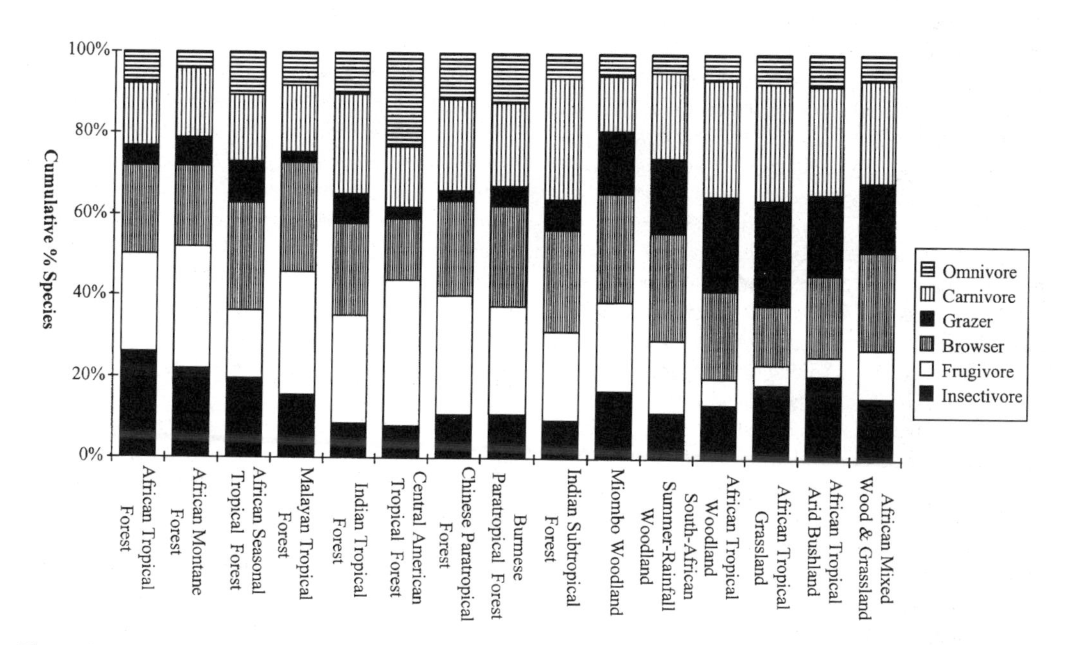

Figure 20-3. Proportions of species distributed by dietary guild categories for 44 present day mammalian faunas averaged for 15 vegetation associations. Each association is represented by three sites except for tropical grassland. Categories shown on the right.

Tropical Rainforest

Tropical rainforests today are characterized by an abundance of trees, lianas, and plants supported by the trees but not parasitic on them. The complexity of the tree canopies is enhanced by the abundance of epiphytes, which add to the species richness, form interconnections between and within the canopies, and form microcommunities within the main canopy structure. Flowering and fruiting are close to continuous on evergrowing trees, with both flowers and fruits from the previous flowering occurring simultaneously on the same trees. Fruiting becomes more strongly periodic as climatic seasonality increases and as the tree canopy structure becomes less complex. Numbers of deciduous tree species increase, although only in extreme cases does the forest as a whole become deciduous.

The mammalian communities inhabiting the various grades of tropical forest reflect the seasonality of climate and vegetation structure. Mammal biomass is low, contrasting with the high species numbers and the high productivity of the ecosystems. Small mammals are highly diverse, especially rodents, but no one species has high density or biomass (fig. 20-1). Spatial distributions of mammals in tropical forest ecosystems (i.e., the space occupied in the environment) are dominated by species with some degree of arboreality (fig. 20-2). Semiarboreal, arboreal, and scansorial species together make up 60–80% of the species. The most equable African forest ecosystems have high proportions of small- to medium-sized arboreal species and small-bodied semiarboreal species, but there is a marked reduction in the proportion of arboreal and scansorial species in more seasonal habitats. This is a reflection of the less complex and more open canopy structure in seasonal forests. The Asian and American forest ecosystems have generally higher proportions of scansorial species than African forests, particularly relative to small-sized semiarboreal species.

The distribution of dietary guilds in tropical forest ecosystems is dominated by small- to medium-sized frugivorous and browsing herbivores (fig. 20-3). Proportions are similar in African, Asian, and American ecosystems, with reductions in frugivory in more marginal African forests, notably montane forest and seasonal lowland ecosystems. In these ecosystems, the proportions of insectivorous species and grazing herbivores are higher as a result of the more open forest canopies evident in these habitats. The American and Asian forest ecosystems have lower proportions of insectivorous species compared with African forests, and while this makes an interesting parallel with the observed difference in the spatial niche, it does not appear to be directly related. Proportions of carnivorous species are low in all tropical rainforests, with little difference evident in more seasonal environments.

Paratropical Forest

The term "paratropical forest" was first introduced by Wolfe (1979) as a substitute for Richard's (1952) subtropical rainforest. It is mainly composed of evergreen tree species, but deciduous species and conifers may be present. Although temperate elements are present in paratropical rainforest, the majority of species are the same as in tropical rainforest. Paratropical rainforest is characterized by an equable climate, with no prolonged dry period. Tree physiognomy is similar to tropical rainforests, the main difference being the disappearance of the emergent tree canopy (Richards, 1952) and the break up of the lower canopy, which becomes discontinuous and effectively forms the emergent upper story of the paratropical forest. Temperate species such as *Acer, Alnus, Rhamnus, Salix,* and *Pinus* become more common toward the climatic limits of paratropical forest, together with *Glyptostrobus* and other conifers, which make up a minor element of the flora.

The mammalian faunas in paratropical forests of China (Yunnan) and Burma have similar community structures to those of tropical forests, with the following differences (figs. 20-1–20-3). There are fewer medium-sized species, particularly in the 1–10 kg size class, and a corresponding increase in large mammals. There is no significant difference in the spatial distribution, but there is a considerable reduction in the proportion of frugivorous species with corresponding increases in browsing herbivores and carnivorous species and a slight increase in grazing herbivory. These differences clearly reflect greater climatic seasonality and the loss of habitat complexity.

Subtropical Rainforest

Subtropical summer-wet forests may have canopy structures nearly as complex as tropical and paratropical forests, but they lack the epiphyte abundance that contributes so greatly to the species richness and adds an extra dimension to the structural complexity of the tropical forests. They have lower species diversity and more seasonal flower and fruit production. They occur in regions where convective summer rainfall climatic patterns predominate, close to the two tropics, but there is a much greater degree of seasonal variation in climate than found in tropical and paratropical forest regions. Subtropical forests may also be found in areas affected by monsoon conditions. Semideciduous associations occur in regions with a 3- to 4-month dry season, and deciduous forest associations in regions with at least 6 months dry. Asian forests have abundant bamboo replacing the sparse shrub layer, and climbing plants are abundant. In Africa, subtropical forests are greatly impoverished through recent human activities, but they would once have covered large parts of Malawi and Mozambique as well as other parts of the summer-rainfall zone of southern Africa. Grasses are rare, becoming more abundant under drier conditions as the trees become more widely spaced.

The mammalian faunas in subtropical forests represent a further stage of the trend seen in the transition from tropical to paratropical forest communities. There is a further decrease in medium-sized species, together with an increase in the proportion of large species, although in subtropical forests there is also an increase in proportions of small (<100 g) species (fig. 20-1). There is an increase in the proportion of terrestrial species and a reduction in arboreal species, although scansorial proportions remain high (fig. 20-2). The decline of frugivores and increases in browsing and grazing herbivores and carnivores also continues the trend between tropical and paratropical forest communities (fig. 20-3). Once again, these differences can be attributed to the more open environments and greater climatic seasonality.

Temperate/Subtropical Evergreen Forest

Several evergreen forest types have been recognized with differing floristic and physiognomic composition, and these must be briefly mentioned because of their potential importance to primate distributions. Microphyllous broadleaved evergreen forests cover extensive upland areas of Burma and central and southern China. Although the dominant species of Lauraceae and Fagaceae have tropical affinities, there is a considerable proportion of species with temperate affinities and a gradual transition into mixed mesophytic forests composed of deciduous and evergreen elements characteristic of temperate zones. A special association is present in areas typical of the Mediterranean region, in which the rainfall is uniquely confined to the winter months, and where the vegetation developed independently but convergently (Cody & Mooney, 1978). Species diversity of both flora and fauna are high, but productivity is low, with most plant growth occurring during the wet winter months.

Tropical Woodland, Bushland, and Grasslands

The term "savanna" has had a long and often acrimonious usage in African ecology. We avoid its usage here as far as possible because its ubiquitous usage to cover all nonforest African ecosystems, from the subtropical to temperate grasslands of South Africa to the tropical degraded or edaphic grasslands of East Africa, can lead to more confusion than understanding (Owen-Smith, this volume). We adopt the traditional division of nonforest ecosystems in Africa and recognize three distinct types: the Guinea/Zambezian, Sudan/bushveld, and Sahel/semiarid woodlands and bushlands (Werger, 1983). The Guinea woodlands occur in relatively wet climates, with up to 1500 mm of annual rainfall. The degree of tree cover is determined by fire (Werger & Coetzee, 1978). Where fire is excluded, closed-canopy forest occurs, with many species typical of tropical rainforests. With moderate and early fires, the vegetation changes to mixed trees and tall grass, while with intense and frequent fires open grassland may occur. The southern African equivalent of Guinea woodland/forests are the Zambezian Miombo woodlands that cover much of southern Tanzania and Mozambique in the east across to southern Zaire and Angola in the west. They have a diverse community of mammals, with high numbers of species and high equitability, but they lack the high biomass of large mammal species typical of drier ecosystems (see below).

The community structure of the mammalian faunas in the Miombo woodland ecosystem has similarities with that of the subtropical forests described above. Size distributions are similar, as are the relative proportions of terrestrial versus arboreal and scansorial species (figs. 20-1–20-2). The proportions of frugivores are lower than those of subtropical forest communities, and the proportions of grazing herbivores are higher (fig. 20-3).

The Sudan woodland and wooded grassland ecosystems have reduced rainfall and a prolonged dry season compared with the Guinea ecosystems. The southern African equivalent is the bushveld, summer-rainfall regions of woodland and bush intermediate between the Miombo woodlands to the north and the drier, arid zones to the south. Large mammals are less diverse than in the wetter climatic zones, and single species are often abundant. Small mammals (<100 g) may make up about 40% of the species numbers. Terrestrial species are slightly more abundant than in the Miombo woodlands, and there is a marked drop in the proportions of arboreal and scansorial species. There is a small increase in frugivorous species, as well as increases in grazing and browsing herbivores (figs. 20-1–20-3).

The East African equivalent to the Sudan woodland is rather different because it occurs within the equatorial zone. As a result, rainfall is distributed across two wet and two dry seasons rather than single seasons as in the higher latitudes. The high level of complexity in vegetation and habitat in East Africa reflects this pattern of rainfall together with more extreme topographic variation. Perennial grasses make up the ground vegetation, and a varying amount of tree cover is partly controlled by rainfall, fire, and soil conditions. Various classifications have been proposed to describe the varying amount of tree cover (Pratt et al., 1966; Grunblatt et al., 1989) based on percentages of canopy cover. Species diversities in the mammalian communities are higher in these tropical woodland, bushland, and grasslands than in the southern African ones, particularly for species >10 kg. There is a marked increase in the proportion of terrestrial species at the expense of arboreal and scansorial species, particularly in open grassland. There is a large reduction in the proportion of frugivores and correspondingly large increases in grazing and carnivorous species, again seen at its most extreme in open grassland environments (figs. 20-1–20-3).

The driest type of wooded grassland ecosystem is the Sahel, with rainfall <250–600 mm annually and prolonged dry periods. Most of the grasses that make up the ground flora are annuals, and open-canopy bush or small trees make up the dominant plant cover. As a result of the dry conditions, the ground is often bare between the bushes and trees during much of the dry season. Large mammals are relatively more abundant in this environment than are small- to medium-sized species, and all size classes are dominated by terrestrial species, which make up over 50% of the mammalian community. The distribution of dietary guilds is dominated by herbivores and carnivores, although the proportion of grazing herbivores is slightly less than is found in open grassland (figs. 20-1–20-3).

Miocene Apes and Their Environments

Fossil apes are well represented in Africa during the Early and Middle Miocene, from about 25–12 m.y. In the later part of the Miocene and until more recent times, almost no fossil apes are known from Africa, although they were abundant in the mid- to Late Miocene in Europe (15–8 m.y.) and Asia (12–7 m.y.). The fragmentary fossils from African Late Miocene deposits at Lukeino or Samburu Hills are not associated with faunas or floras sufficiently large to be diagnostic of any particular environment, and so in attempting to indicate the range of environments occupied by fossil apes, recourse has to be made to evidence outside Africa.

Recent great apes are generally forest-adapted species, arboreal to a greater or lesser extent, and except for the mountain gorilla, mainly frugivorous. Their distribution pattern closely follows that of tropical rainforests, and so, although they may venture outside the forests on occasion, this would not appear to be their optimum environment.

Miocene Environments

The environment of fossil apes is considered for five time periods and geographic regions. These are based on the sites listed in table 20-2, and they cover a period of climate change leading to cooler global temperatures (Denton, this volume), although these changes are unlikely to have had much effect within the equatorial region of Africa (O'Brien & Peters, this volume). The changing patterns in community structure of each is considered in general terms, without detailed descriptions of specific sites. The three ecological criteria of body size, spatial distribution, and dietary guilds is discussed in turn (figs. 20-4–20-6).

The distribution of body size (fig. 20-4) shows that the mammalian communities in the Early Miocene African sites are similar to present-day tropical forest communities. Just under 70% of species are small to medium sized (up to 10 kg), whereas in the African Middle Miocene communities this percentage is <50%, and the other faunas vary between 50% and 60%. In this respect, the Later Miocene communities are similar to the moist (Sudan-type) woodlands of tropical Africa and the summer-rainfall Miombo woodland ecosystems of southern Africa.

In terms of spatial distribution, the Early Miocene pattern is similar to those of present-day tropical African forest communities (fig. 20-5), particularly the tropical montane forests. Proportions of terrestrial species are low, and arboreal species are more common than scansorial species, in contrast to the Asian forest communities of today. There is a considerable change in the mammal communities from the African Middle Miocene, with an expansion of terrestrial species and reduction in arboreal and semiarboreal species. The African Middle Miocene mammalian communities show similarities with southern African Miombo woodland and East African equivalents. Monkeys become abundant for the first time (Benefit, this volume). The European Middle and Late Miocene mammal communities have lower terrestrial but also lower arboreal proportions, indicating forest conditions most similar to the subtropical forest communities of India and the seasonal tropical forests of Africa. Similarly, the Asian Late Miocene communities resemble present-day forest communities, particularly the Asian paratropical forest communities, with which they share a high proportion of scansorial species.

The distribution of dietary guilds in the fossil

Table 20-2. List of all fossil localities examined.

Locality	Country	Age	Hominoid	Data Sources
			Miocene	
Songhor	Kenya	Early	*Proconsul africanus, P. major, Rangwapithecus gordoni*	Andrews at al. (1979), Evans et al. (1981), Van Couvering (1980)
Koru	Kenya	Early	*P. africanus, P. major*	Evans et al. (1981)
Rusinga	Kenya	Early	*P. heseloni, P. nyanzae*	Walker et al. (1993)
Kalodirr	Kenya	Early	*Afropithecus turkanensis*	Leakey & Leakey (1986)
Maboko Island	Kenya	Middle	*Kenyapithecus africanus, Proconsul* sp.	Evans et al. (1981)
Fort Ternan	Kenya	Middle	*Kenyapithecus wickeri, Proconsul* sp.	Shipman (1986)
Ngorora	Kenya	Middle	*Hominoid,* indet.	Hill et al. (1985)
Neudorf	Slovakia	Middle	*Griphopithecus darwini*	L. Martin & Andrews (1993)
Pasalar	Turkey	Middle	*G. alpani*	Andrews (1990)
Can Ponsic	Spain	Mid-Late	*Dryopithecus laietanus*	Begun (1994)
Can Llobateres	Spain	Mid-Late	*D. crusafonti*	Begun (1994)
Rudabanya	Hungary	Mid-Late	*D. hungaricus*	Begun (1994)
Lufeng	China	Late	*Lufengpithecus lufengensis*	Guo-Qin (1993)
Ravin de la Pluie	Greece	Late	*Graecopithecus freybergi*	de Bonis & Koufos (1993)
			Plio-Pleistocene	
Aramis	Ethiopia	4.50 m.y.	*Ardipithecus ramidus*	WoldeGabriel et al. (1994)
Kanapoi	Kenya	4.10 m.y.	*Australopithecus anamensis*	M.G. Leakey et al. (1995)
Laetoli	Tanzania	3.76 m.y.	*Australopithecus afarensis*	M.D. Leakey & Harris (1987)
Omo Member B	Ethiopia	2.94 m.y.	*Australopithecus* sp.	Coppens & Howell (1976), Wesselman (1982),
Olduvai FLKNN 3–1 FLKN 6–1	Tanzania	1.76 m.y. 1.75 m.y.	*Paranthropus boisei, Homo habilis*	Fernandez-Jalvo et al. (1998)
Makapansgat	South Africa	3.00 m.y.	*Australopithecus africanus*	Delson (1984), Pocock (1987), Vrba (1987a)
Sterkfontein Member 4	South Africa	2.50 m.y.	*A. africanus*	Brain (1981b), Pocock (1987)
Swartkrans 1	South Africa	1.65 m.y.	*Paranthropus robustus*	Brain (1981b)

communities follows much the same pattern (fig. 20-6). The Early Miocene African communities are similar to African tropical forest communities (Andrews, 1992b) in all classes of dietary distribution. The African Middle Miocene communities show a higher proportion of grazing and browsing herbivores and fewer frugivorous species and are similar in pattern to present-day summer-rainfall miombo woodland and subtropical forest communities. This pattern is maintained throughout the European and Asian Middle to Late Miocene, apart from a decrease in grazers and a marked increase in browsers, which shift the similarity away from the African miombo woodlands more toward the Asian subtropical forests.

Miocene Ape Distributions

Relating these faunal patterns to actual groups of fossil ape associated with them, certain conclusions emerge. Species of the genus *Proconsul* in the Early Miocene appear, in general, to be associated with faunas having greatest similarities with tropical environments that were ever-wet, nonseasonal, and equable in terms of climate, with vegetation typical of evergreen and complex canopy tropical rainforest. Some sites, such as Songhor in Kenya, provide some indication of greater similarity to recent montane forest faunas (Evans et al., 1981). On the other hand, some species of *Proconsul,* for example, on Rusinga Island, appear

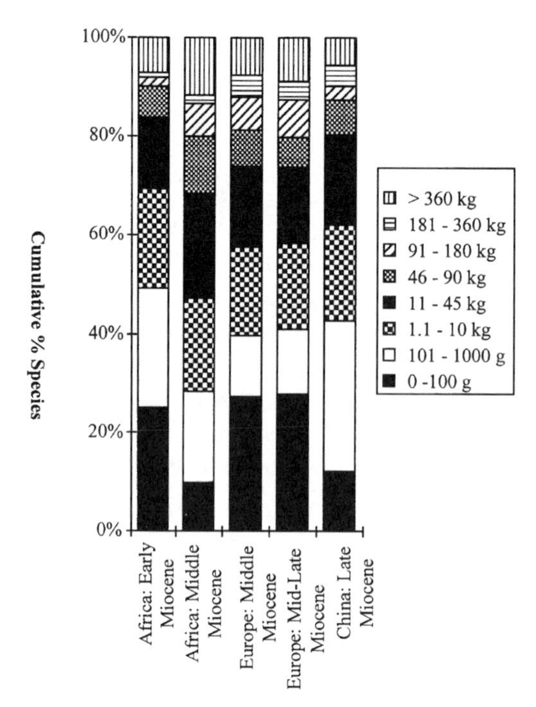

Figure 20-4. Proportions of species distributed by body size for the faunas from the Miocene sites listed in table 20-2. Body weights in kilogram classes shown on the right.

to be associated with less rich and less complex forests than are depicted here. The greatest similarity is with faunas from the seasonal *Chlorophora* forests of western Kenya (destroyed early this century) and with extant faunas from the Kenya coastal forests such as the Sokoke and Shimba Hills forests.

Even in the Early Miocene, therefore, there was some variability in ape environments, ranging from wet to dry tropical forests. These differences can be related to slight differences in adaptation of the species of *Proconsul* and other Early Miocene genera. Most species were arboreal frugivores, with at least two folivorous species (e.g., *Rangwapithecus*). They were all above-branch climbers like present-day cercopithecine monkeys, but the wet-forest species had adaptations for greater agility in trees than were present in the dry-forest species (references in Andrews et al., 1996). It would appear that the wet-forest species were better able to exploit the small branch niche in the tree canopies, although some species, like *Proconsul major,* were of a size that would have required some terrestrial activity. The dry-forest species were probably more generalized arboreal species like some of the more terrestrial cercopithecine monkeys today. One of the consequences of this functional and environmental variability is seen in the high species diversity of sympatric ape species during the Early Miocene (Andrews, 1981).

Present evidence suggests less habitat variability in the middle Miocene compared with the Early Miocene, and certainly habitat richness was reduced, but species numbers of hominoids remained high. Middle Miocene apes of the genus *Kenyapithecus* are associated with faunas that have community structures characteristic of more seasonal environments. Possible comparisons are with the miombo woodlands of southern Africa (Guinea woodland described above) and with the subtropical forest environments seen today in India. Both of these associations occur in summer-rainfall regions close to the southern and northern tropics, respectively, and so the climatic regime must have been different from Miocene sites such as Fort Ternan, situated almost on the equator. On the other hand, there is evidence of a flora at Ngorora indicating tropical rainforest at about 12.5 m.y. (Jacobs & Kabuye, 1987), but the fragmentary hominoid remains from this site have not yet been identified.

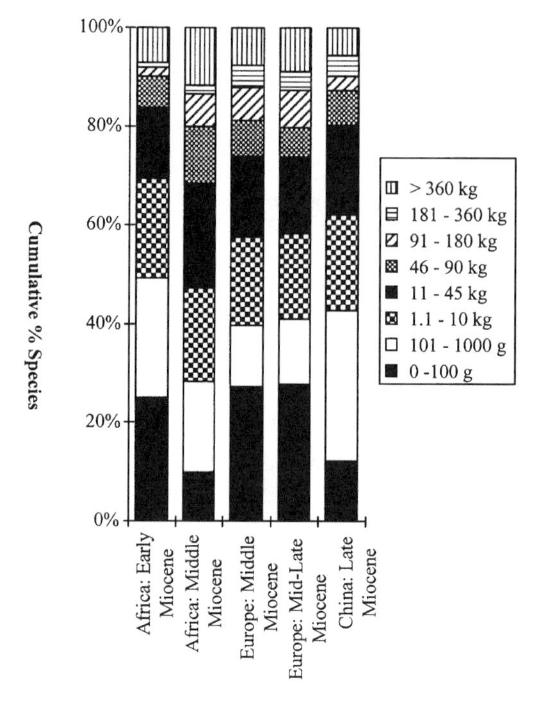

Figure 20-5. Proportions of species distributed by locomotor/spatial categories for the faunas from the Miocene sites listed in table 20-2. Categories are shown on the right.

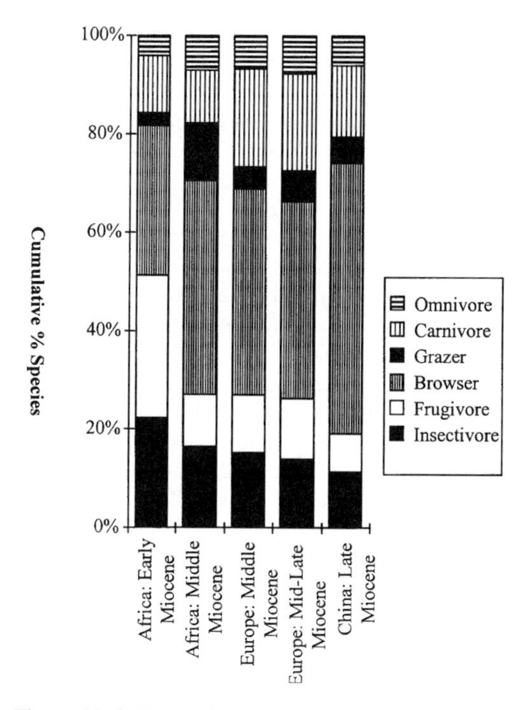

Figure 20-6. Proportions of species distributed by dietary guild categories for the faunas from the Miocene sites listed in table 20-2. Categories are shown on the right.

The few fossil remains of *Kenyapithecus* indicate postcranial adaptations little changed from those of *Proconsul* (which is also known at the same site). There is some evidence for terrestrial adaptation in both genera, particularly for *Kenyapithecus* (Benefit & McCrossin, 1995). In terms of diet, *Kenyapithecus* shows the development of a form of thick-enameled frugivory associated in living primates with drier, more seasonal habitats (references in Andrews et al., 1996). The forest-living baboons and mangabeys provide the most appropriate analogy for the role of these fossil apes in the Middle Miocene seasonal forests. Faunas including the fossil ape genus *Afropithecus* are unfortunately rather poorly known and mostly not suitable for ecological analysis. It has been suggested that the affinities of the fossil ape known from Maboko Island in Kenya are with this genus rather than with *Kenyapithecus* (Andrews, 1992a), and the environment at this site has been interpreted as dry seasonal and perhaps deciduous forest (Evans et al., 1981). Postcranial and dental adaptations of both the Maboko Island hominoid and *Afropithecus* were similar to those of *Kenyapithecus,* and this is consistent with their association with dry seasonal forest.

The European Middle and Late Miocene faunas include species of the genera *Dryopithecus* and *Griphopithecus.* Fewer species of hominoid were present, and it is rare to find more than one hominoid species in these higher latitude sites. The community structure of the faunas is similar to that of subtropical Indian forests with summer-rainfall climates and a prolonged winter dry season. Although there are differences between the sites at which these two fossil apes occur, these are variants on a theme, and the similarities are greater than the differences. *Dryopithecus* is associated with faunas having relatively more equable community structures and *Griphopithecus* with communities indicating seasonal forest in drier and more strongly seasonal conditions with summer rainfall and prolonged dry seasons. The adaptations of *Griphopithecus* appear to have been similar to those of *Kenyapithecus,* although there is no evidence of terrestriality, and this has been recognized both taxonomically and ecologically (Andrews, 1992a). *Dryopithecus,* in contrast, may show an early shift toward the below-branch arboreal adaptation seen in extant apes (Begun, 1994), indicating greater specialization to arboreal life, perhaps like the saki monkeys of South America.

The faunas associated with *Graecopithecus* are extremely poor, and taphonomic and biogeographic factors account for most aspects of the faunal composition (de Bonis et al., 1992), although it is possible to infer that Ravin de la Pluie, the site from which most hominoids have come, had "an open environment" (de Bonis et al., 1992).

The Asian Late Miocene sites at which *Sivapithecus* and *Lufengpithecus* are found have community structures most similar to modern subtropical forest communities. The size distributions of the faunas indicate an underrepresentation of small mammals (<100 g), which is probably a taphonomic effect, and a high proportion of medium-sized species. In terms of spatial distribution, terrestrial species are slightly more common than in present-day subtropical forest communities. This could be due to the underrepresentation of small species, which would tend to inflate other categories. Scansorial species, however, are well represented. In terms of dietary guild, these communities are dominated by browsing herbivores, with grazers and carnivores poorly represented. This pattern is not exactly comparable to any present-day community, but it is taken to indicate affinities with subtropical forest conditions at the wet end of the range. Like most other Miocene apes, the *Sivapithecus* species have thick tooth enamel and postcranial adaptations little changed from the Early Miocene. Indeed, the humeri described from Pakistan (Pilbeam et al.,

1990) are remarkably little changed from those of *Proconsul,* and it might be argued that they provide evidence of some degree of terrestrial behavior.

Pliocene Hominines and Their Environments

It has been shown in the previous section that Miocene apes occupied a variety of forested habitats throughout much of their evolutionary history. Caution is needed in interpreting these conclusions, however, for what has been presented is restricted by the fossil evidence available, and conclusions reached here do not exclude the possibility that fossil apes were also living in environments so far not represented in the fossil record. In addition, fossil apes almost certainly were living in environments additional to those indicated by the present analysis, either because they were mixed habitats or because they have no exact modern equivalent. For example, the available evidence indicates that Middle Miocene apes in Africa were associated with seasonal forest or more open woodland conditions, but there are no fossil sites from the Middle to Late Miocene of Africa containing wet forest faunas with hominoids in direct association. Such forests must surely have existed, if not in East Africa at least in Central or West Africa, and when a fossil site derived from such a habitat is found, it is likely that fossil apes will be found in the fauna. The same proviso also applies to the discussion in the present section on hominine environments.

Evidence is accumulating of relatively warm and humid conditions over much of Africa during the early and middle parts of the Pliocene. Scott (1995) has documented change in vegetation patterns in South Africa from subtropical to tropical forests and woodlands during the Late Miocene to Fynbos and subtropical woodlands in the Pliocene. Global temperatures were generally higher and conditions more stable, particularly during the period 3.6 to 3.3 m.y., as indicated by the oxygen isotope deep-sea record (Opdyke, 1995; Shackleton, 1995). Vegetation records indicate the presence of forest at the lower levels of the Hadar sequence (Bonnefille, 1995) and the considerable expansion northward of West African forests (Dupont & Leroy, 1995), and there are some indications also of warmer and more humid conditions from the fossil record of mammals (Wesselman, 1985c; Reumer, 1995) and molluscs (Williamson, 1985).

The sites providing evidence of early hominine environments are listed in table 20-2. They range from Pliocene to Early Pleistocene, encompassing the early stages of hominine evolution. The earliest record of hominines comes currently from two sites for which there is little palaeoecological information—namely, Tabarin and Lothagam—and the earliest fauna that we have been able to analyze comes from the Pliocene deposits at Aramis in Ethiopia.

Aramis

The site of Aramis on the Awash River in Ethiopia contains the recently described species *Ardipithecus ramidus* (T. D. White et al., 1994, 1995). Primates account for a large proportion of the mammalian fauna, with colobine monkeys alone accounting for 30% of the fauna (WoldeGabriel et al., 1994). The mammalian fauna has a community structure most similar to the most heavily wooded habitats in woodland ecosystems, particularly the miombo woodlands. The terrestrial component of the Aramis fauna is high, but so are the proportions of arboreal and scansorial species (fig. 20-7). Semiarboreal species (mainly small mammals clearly under represented in the Aramis collection; see fig. 20-8) have low proportions. In the distribution of dietary guilds, browsing herbivores have highest proportions, grazers low proportions, and frugivorous species are relatively well represented (fig. 20-9).

In addition to the evidence of the mammals, there are also fossilized plant remains that include large numbers of *Canthium* seeds. This is a genus common in present-day African woodlands and forest (WoldeGabriel et al., 1994), with 27 species known today in Kenya alone. Many are small trees or shrubs occuring in the dry forests of the Kenya coast and the miombo woodlands. *A. ramidus* is therefore associated with faunal and floral conditions indicating woodland or dry forest growing in a strongly seasonal environment.

Kanapoi

The community structure of the fauna from Kanapoi is biased, as is the Aramis fauna, by absence of small species (fig. 20-8). The fauna is dominated by medium- to large-sized terrestrial species (figs. 20-7, 20-8), and both browsers and grazers are well represented (fig. 20-9). The high terrestrial component in the fauna places it with the drier end of the present-day woodland and bushland ecosystems (fig. 20-10), suggesting open woodland with abundant grass.

Laetoli

Slightly later in time are the *Australopithecus afarensis* sites of Ethiopia and Tanzania, dating between 3 and 4 m.y. These are represented in the present analy-

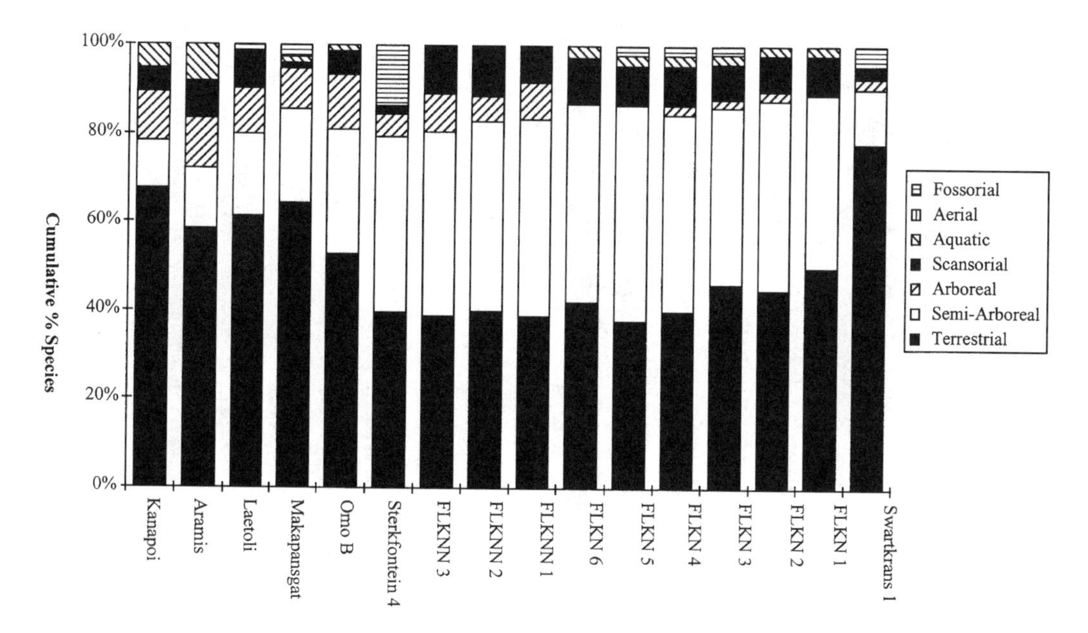

Figure 20-7. Proportions of species distributed by locomotor/spatial categories for nine faunas from the Plio-Pleistocene sites listed in table 20-2. Categories are shown on the right.

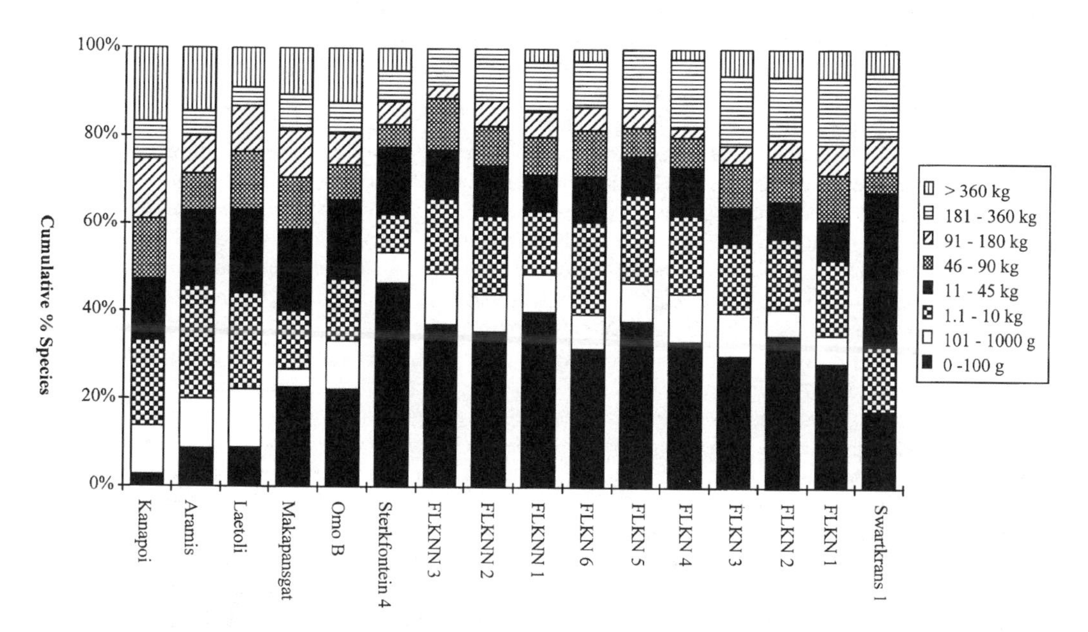

Figure 20-8. Proportions of species distributed by body size for faunas from the Plio-Pleistocene sites listed in table 20-2. Body weights in kilogram classes shown on the right.

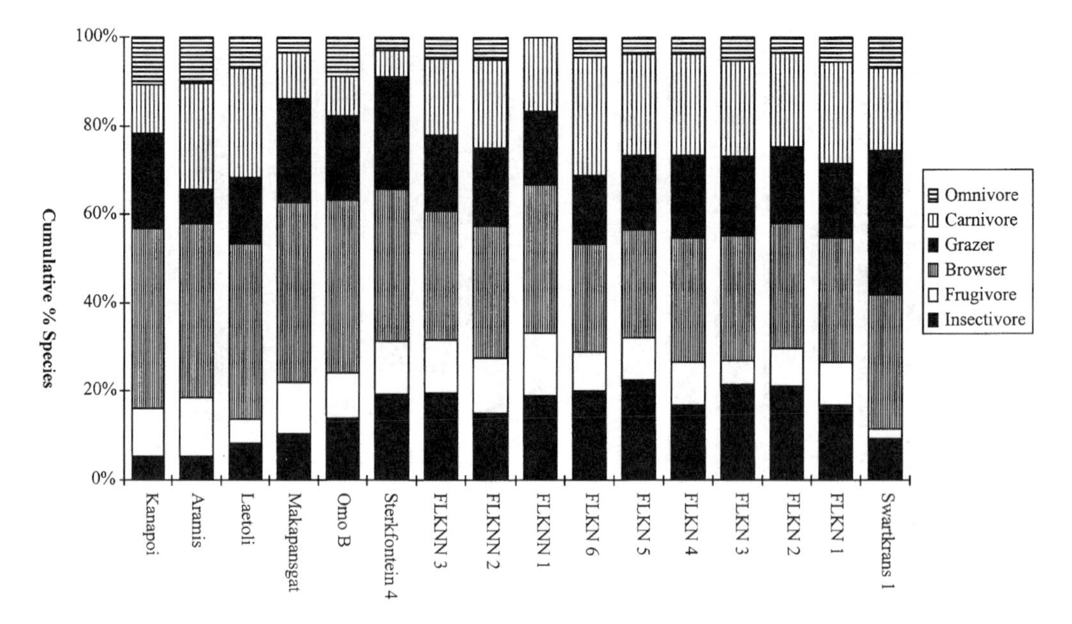

Figure 20-9. Proportions of species distributed by dietary guild categories for faunas from the Plio-Pleistocene sites listed in table 20-2. Categories are shown on the right.

sis by just one site, Laetoli in Tanzania, and the composite fauna from this site has a community structure consistent with the ecological variation within the Serengeti ecosystem today (figs. 20-7–20-9), but it tends more toward the wooded end of the range rather than toward the open grasslands (Andrews, 1989). The fauna is dominated by terrestrial species (fig. 20-7), but browsing herbivores greatly outnumber grazers (fig. 20-9). The relative abundance of grazers is similar to that of other dietary guilds (fig. 20-11), and so it is the abundance of browsers that has no exact equivalent present-day fauna. In both the Sudan woodlands (including the South African bushveld) and the Sahel bushlands, grazers are almost as diverse as browsers (fig. 20-12). It is a general feature of all other habitats with varying degrees of tree cover that browsers are more diverse than grazers, but in these cases other dietary classes are also more common, for example the frugivorous species proportions in the miombo woodlands, which are higher than in the Laetoli fauna (fig. 20-10). It would appear, therefore, that some form of wooded habitat was present at Laetoli, perhaps with canopy structure more closed and more complex than any present today in the Serengeti area but still continuing the trend in the Serengeti ecosystem.

Similarly anomalous conditions may have existed in the *A. afarensis* localities in Ethiopia. The coexistence of five proboscidean species in the Hadar Formation (Kalb, 1995) and similarly large proportions of browsing as opposed to grazing herbivores suggests some form of rich, wooded environment different from any existing today.

Olduvai

The faunas from Bed I Olduvai, in which *Homo habilis* and *Paranthropus boisei* are present, show a range of community structures intermediate between the Sudan/Sahel wooded grassland types in East Africa today and the dry seasonal end of the range of tropical African rainforest communities. In terms of proportions of terrestrial, frugivorous, and grazing species, the three middle Bed-I faunas from FLKNN levels 1–3 are closest to the present-day dry seasonal forest community patterns, and the faunas from FLKN levels 1–6 of upper Bed-I range from woodland affinities in the lower part to more open wooded bushland affinities at the top of Bed-I. The apparent forest affinities of the FLKNN faunas may be misleading, for in fact all nine Bed-I faunas appear to be structurally related to the East African seasonal wooded grassland ecosystems (Fernandez-Jalvo et al., 1998). The combinations of ecological categories in present-day woodland ecosystems show a trend from arid bush-

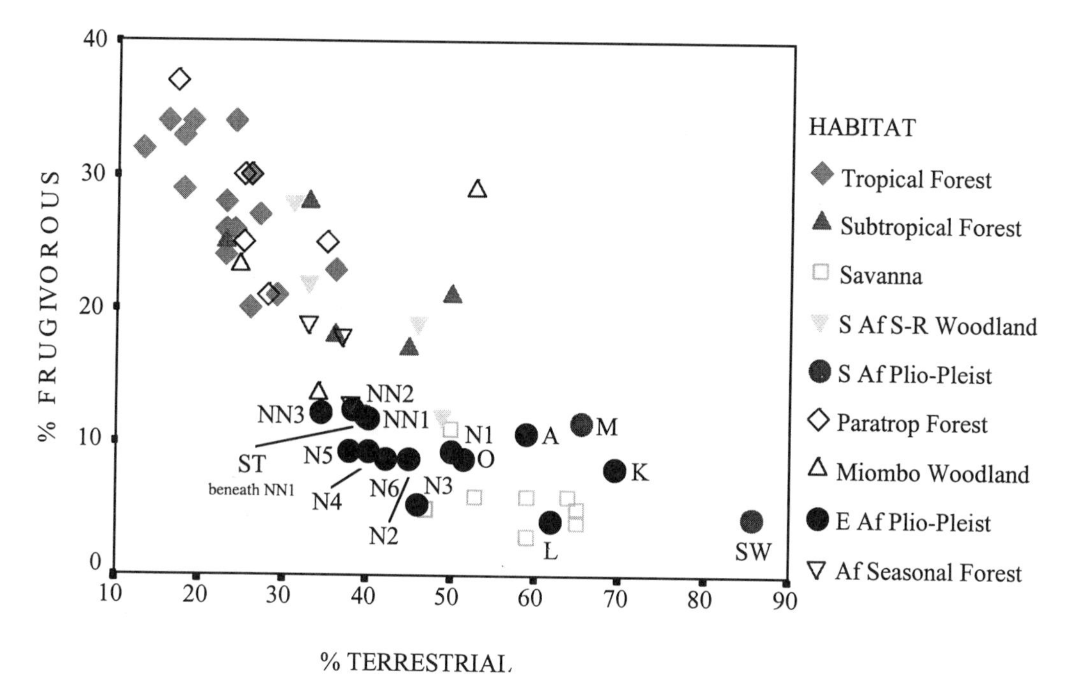

Figure 20-10. Bivariate plot of percentages of frugivorous mammal species in the dietary guild analysis against percentages of terrestrial species in the spatial analysis. Symbols represent individual faunas in every case, with fossil faunas shown in circles and recent faunas by other symbols as listed in the key. Tropical forest includes faunas from Africa, Asia, and America; subtropical forest includes faunas from India to China; "savanna" is used here as shorthand for tropical woodland and bushland faunas from East Africa; S Af S-R woodland is South African summer-rainfall woodland-forest; Paratrop is paratropical forests from Burma and China; Miombo woodland includes southern African deciduous woodlands dominated by species of *Brachystegia, Pterocarpus,* and so on; Af seasonal forest includes seasonal deciduous forests in East Africa such as the Sokoke forest. S Af Plio-Pleist has the following fossil faunas: M, Makapansgat; ST, Sterkfontein; SW, Swartkrans. E Af Plio-Pleist has the following East African fossil faunas: A, Aramis; K, Kanapoi; L, Laetoli; O, Omo; NN, Olduvai FLKNN-I; N, Olduvai FLKN-I (see table 20-1 for locality information).

land toward more wooded and closed conditions, and when the Olduvai faunas are plotted with the present-day communities, they continue the trend toward more wooded conditions from the top of Bed-I to the middle. This is particularly apparent when two of the most diagnostic ecological categories are plotted against each other (fig. 20-10), where the proportions of frugivorous against terrestrial species shows both the trend in recent wooded grassland faunas, with the closed woodland faunas falling on the left of the distribution, and the open grassland faunas falling on the right, and the trend in the Olduvai Bed I faunas, with the FLKN1–3 faunas on the right and FLKNN1–3 falling on the upper left. This trend does not lead directly to forest, however, for there is a major structural difference in the mammalian communities in forest and nonforest in that, for instance, a forest fauna is not simply a derivation of a fauna from a more closed environment. Although the Olduvai Bed I communities overlap the forest communities, they are off the forest line. In other words, we interpret the Olduvai Bed I pattern as indicating an extension of the present-day woodland ecosystems toward more closed and complex-canopied woodlands similar to but distinct from present day seasonal tropical forests (Fernandez-Jalvo et al., 1998). A similar trend is also apparent in figure 20-11, where proportions of frugivores are plotted against proportions of grazers, and when grazers and browsers are compared (fig. 20-12), the trend from FLKNN to FLKN is striking.

Such a range of environments no longer exist today, but nonetheless it is possible to reconstruct this

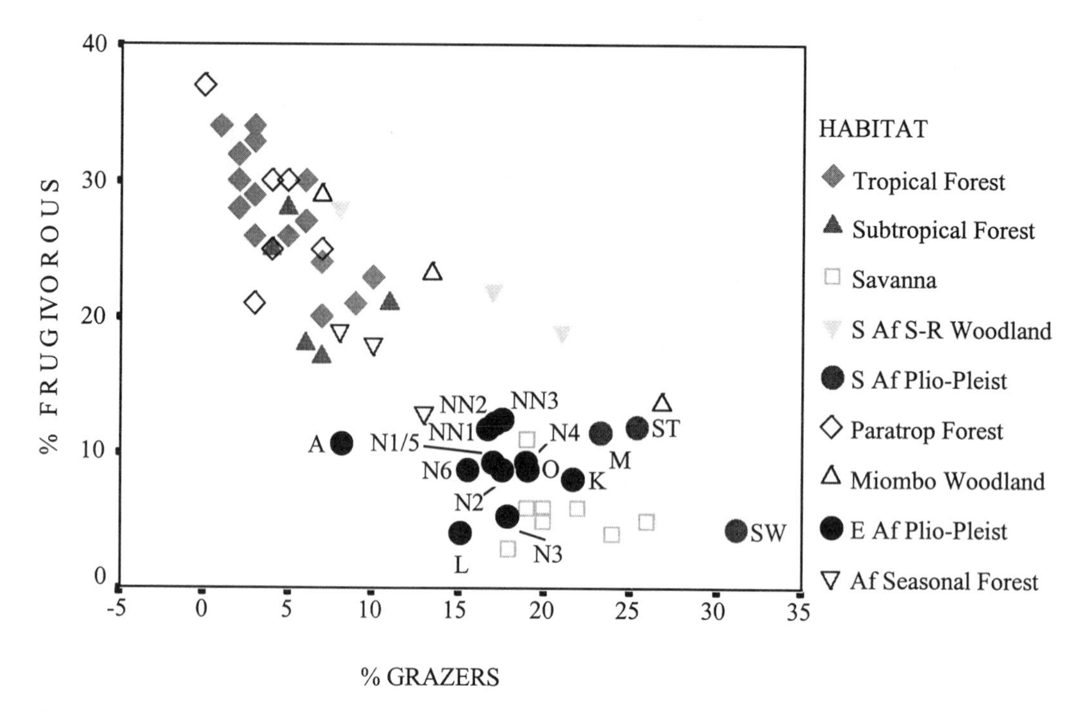

Figure 20-11. Bivariate plot of percentages of frugivorous against grazing adaptations in the dietary guild analysis. All notation as in figure 20-10.

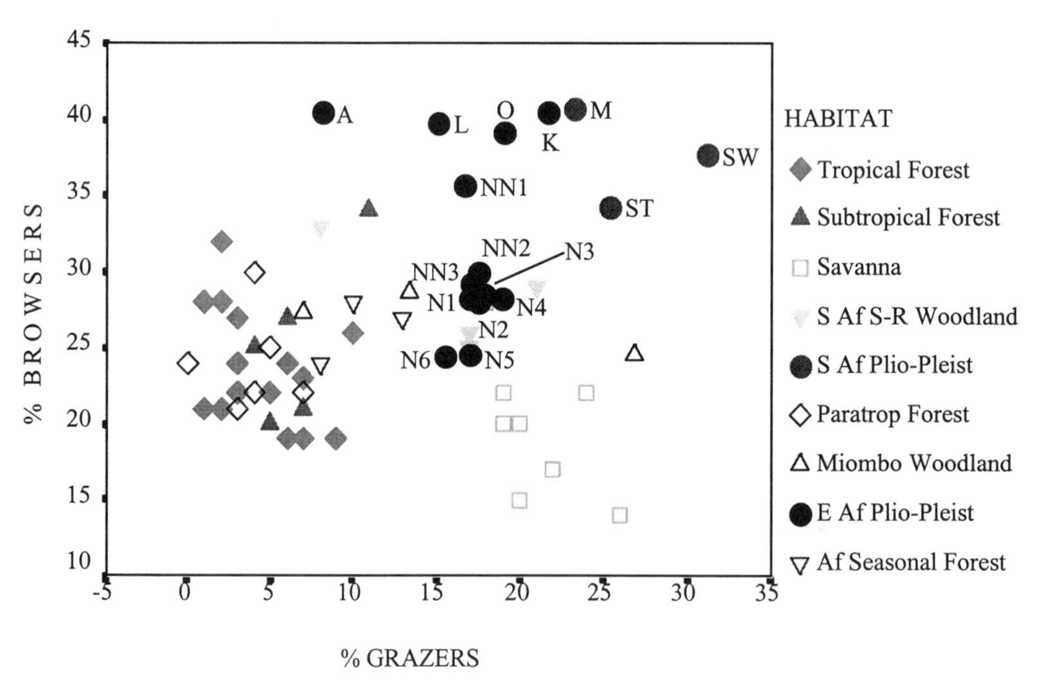

Figure 20-12. Bivariate plot of percentages of browsing against grazing adaptations in the dietary guild analysis. All notation as in figure 20-10.

range to a considerable extent. The FLKNN levels of middle Bed I would have had closed canopy woodland, probably with some evergreen species and with a dense and partly evergreen shrub layer and ground vegetation. Grasses would have been present, but not in great abundance. The lower three levels of upper Bed I (FLKN4–6) would have had less complex plant associations, but closed-canopy woodland was probably still present, together with the dense shrub and ground vegetation. Grasses would have been more abundant. At the top of Bed I, in levels FLKN1–3, conditions were more open, verging on open woodland with more scattered ground vegetation and much more abundant grass.

Omo

We have also attempted analyses on the Omo sequence faunas, but we have found them to be taxonomically biased to such a degree that they are difficult to interpret. Proportions of large mammal species are high throughout, and this could be because the small mammals have been well described (Wesselman, 1982), whereas the large mammals have been greatly oversplit taxonomically. It is interesting, however, that the Omo faunas all show the predominance of browsing herbivores mentioned above for the Laetoli fauna (fig. 20-12), and it may be that these faunas all come from communities that have no exact parallel in present-day habitats.

South Africa

Finally, analyses have been made of some of the South African cave faunas which include *Australopithecus africanus* (Sterkfontein Member 4 and Makapansgat) and *Paranthropus robustus* (Swartkrans 1). In all cases, terrestrial species dominate the pattern of community structure, more so in the case of the *P. robustus* site, whereas in the case of the *A. africanus* sites there are higher proportions of semiarboreal and fossorial species (fig. 20-7). Sterkfontein in particular has high proportions of fossorial species. In the distribution of dietary guilds, the *P. robustus* site has a fauna with higher proportions of grazing herbivores than any other hominine site, whereas the *A. africanus* sites have faunas with higher proportions of browsing and frugivorous herbivores (fig. 20-9). These point to more open environments in association with *P. robustus,* probably close to Sahel-type semiarid bushlands, and more closed environments with *A. africanus*. In all the South African sites, however, browsing herbivores greatly outnumber grazers (fig. 20-12).

These differences between gracile and robust australopithecine communities indicate a substantial environmental difference between them. Both South African hominines are associated with seasonal environments, and the robust australopithecines are linked with more open environments than any of the East African Plio-Pleistocene communities, essentially open grassland. The gracile anustralopithecines, however, are associated with closed-canopy woodlands perhaps similar in structure to those described above for the FLKNN faunas from Olduvai (fig. 20-11), although it is also similar to the pattern of community structure present today in the South African summer-rainfall woodlands. This divergence between environments associated with robust and gracile australopithecines suggests some degree of adaptive difference between them.

The general pattern of hominine diversity in the Pleistocene appears to be either one species at each site or one australopithecine and one species of the genus *Homo*. These appear to have occupied different niche space, but it is worth asking if this pattern was universal during hominine evolution. During much of hominoid evolution during the Miocene, multiple species were present in many African tropical communities, as at the present time with populations of cercopithecine monkeys. Miocene apes were medium-sized primates occupying the arboreal frugivorous niche in a range of environments from wet tropical forest to dry, highly seasonal, forest-woodlands; early hominines were also medium-sized primates, also frugivorous probably, and to some extent they may also have been partly arboreal (for Sterkfontein *A. africanus,* see Vrba, 1979, Ciochon & Corruccini, 1976; for Hadar *A. afarensis,* see Stern & Susman, 1983). From this perspective, and considering the degree of overlap between Miocene ape and Pliocene hominine environments, it may be that they occupied similar positions within the ecosystem, and the pattern of hominine diversity may be more similar to that of Miocene apes than to Later Pleistocene hominines.

Summary

Fossil apes show considerable diversity throughout the Miocene. This is greatest in the Early Miocene when they are associated with tropical forests ranging from multicanopy rainforests to more open and less complex forest and woodland types. The apes occupied an above-branch arboreal niche with frugivorous diets for the most part and ranged in size from smaller than gibbons to larger than female gorillas.

In the Middle Miocene, diversity was still high, but the known species of fossil ape are associated with either more seasonal, impoverished forest environments or with woodland environments. Apes were present both in tropical regions of Africa and in higher latitudes of Eurasia. The fossil apes had developed thick enamel on their teeth, probably associated with a tougher diet of fruit, and they still maintained similar locomotor strategies to Early Miocene forms, with some evidence of terrestriality.

In the Late Miocene hominoids are poorly represented in collections from Africa, and there is little information on either their morphology or environment. In Eurasia, some Middle Miocene taxa persist into the Late Miocene, particularly the thick-enameled forms linked phylogenetically with the orangutan. They are associated with subtropical and paratropical forest environments.

The earliest hominines in the Pliocene are associated with closed woodland ecosystems, in some cases approaching forest ecosystems in terms of community richness but not actually overlapping with them. The earliest hominines appear to have been medium-sized frugivores that were still at least partly arboreal, so that the niche space occupied by them overlapped the range of habitats associated with Miocene apes. The change to complete terrestriality, more varied diets, and from woodland environments to more open ones would then have occurred at a later stage of hominine evolution.

In the Early Pleistocene, the range of environments changed from middle Bed I Olduvai, when associated environments were dense, closed woodland verging on forest, to more open wooded grasslands at the top of Bed I Olduvai. This range of environments overlaps but goes beyond that present in comparable biomes today. No difference between environments occupied by australopithecines and early *Homo* were detected in this analysis.

In South Africa, the environments are more similar to those present in the same region today. The gracile australopithecines appeared to occupy dense summer-rainfall woodland/forests, whereas the robust australopithecines are associated with more open grassland environments.

Acknowledgments We are grateful to Timothy Bromage and Friedemann Schrenk for the invitation to contribute to this meeting and to the Wenner-Gren Foundation for financial support. We are also grateful to Barbara West for help with the ecological analyses, to Mikael Fortelius for access to his NOW database for the European faunas, to Nikos Solounias, Alan Gentry, and Ray Bernor for comments on the ecological categories, and to Norman Owen-Smith, Ray Bernor, and Libby Andrews for comments on the manuscript.

21

Plio-Pleistocene Floral Context and Habitat Preferences of Sympatric Hominid Species in East Africa

Nancy E. Sikes

The natural relationships between the carbon in terrestrial plants, soil organic matter, and pedogenic carbonates and between the oxygen in rainfall and pedogenic carbonates are sensitive indicators of the isotopic paleoecology of early hominid sites. Although the history of stable carbon and oxygen isotopic research on buried soils (paleosols) from hominoid and early hominid fossil localities in East Africa is only 20 years old, this method is well established in paleoanthropology as an independent means for reconstructing local and basinwide paleoenvironments. The available isotopic data, reviewed herein, from East African fossil localities do not show a dramatic change in the proportion of tropical C_4 grasses to woody C_3 vegetation coinciding with hominid speciation events about 5.5 or 2.5 m.y. (e.g., Brain, 1981a; Vrba, 1985c), or with the beginning of the archaeological record near 2.5 m.y. (e.g., Kibunjia, 1994). Likewise, the isotopic evidence suggests that early hominid floral habitats were more closed than previously assumed, and along with recent research in other fields, questions whether there was a dominant association between Plio-Pleistocene hominids and open C_4 savanna grasslands.

The majority of stable carbon and oxygen isotopic research has concentrated on diachronic change in East African early hominid paleoenvironmental contexts. Recent research has used this technique to study the synchronic variation in floral biomass present in stable Early Pleistocene land surfaces (Sikes, 1994, 1995, 1996). The reconstructed, site-specific floral microhabitat contexts of excavated archaeological traces at Olduvai Gorge (Tanzania) indicate that Oldowan hominids repeatedly exploited woody C_3 habitats. The range of stable carbon isotopic values from the 1.74 m.y. basal Bed II and the 1.77 m.y. middle Bed I *Zinjanthropus* paleosols indicates near–closed-canopy woodlands or riparian forest to grassy woodlands were present on the original land surfaces. These data suggest contemporary Plio-Pleistocene *Homo habilis* (sensu lato) and *Paranthropus boisei* may have preferred relatively closed woodland habitats that would have provided food resources, shade, and predator and sleeping refuge.

Although they may use the resources differently, the co-occurrence of two closely related species in the same microhabitats is common among mammals, including extant nonhuman primates (e.g., Waser, 1987; Kuroda, 1992). It is suggested here that contemporary early *Homo* and *P. boisei* may have exploited the same floral microhabitats in a manner similar to that exhibited by extant sympatric lowland gorillas and chimpanzees. Western and eastern lowland gorillas (*Gorilla g. gorilla* and *G. g. graueri*) are sympatric with chimpanzees (*Pan troglodytes*) in the tropical forests of central Africa, and both species eat many of the same foods (e.g., Tutin & Fernandez, 1985; Kuroda, 1992; Rogers et al., 1992; Tutin et al., 1992; Yamagiwa et al., 1992). Although both of these primate species prefer succulent fruits, they exhibit different habitat range and preference patterns. A brief review of early hominid morphology and physiology, as well as patterns in the Oldowan archaeological record, in the Discussion section of this chapter lends support to the argument for a preference by Early Pleistocene hominids for more closed wooded habitats.

Isotopic Background

Modern Plants and Soils

During photosynthesis, plants assimilate carbon from the atmosphere, which results in depletion of the heavier stable isotope (^{13}C) in terrestrial plant organic matter relative to atmospheric carbon dioxide. Plants may be separated into three major photosynthetic types, each with a characteristic stable carbon isotopic pattern. The three photosynthetic types are commonly called C_3, C_4, and CAM, and represent adaptations to different atmospheric and climatic conditions.

In tropical savanna ecosystems, C_4 species are better adapted than C_3 plants to conditions associated with higher temperatures and irradiance at lower latitudes and altitudes. C_3 plants outcompete tropical C_4 plants in winter-rainfall localities, shaded areas such as forests, and at high latitudes and altitudes. C_3 plants include nearly all trees, shrubs and herbaceous vegetation, some sedges, *Typha,* and grasses adapted to cool growing seasons and shade (including temperate and montane grasses). The majority of C_4 plants are nonwoody tropical grasses and sedges. C_4 plants may also be distinguished from C_3 plants by the concentric arrangement of their bundle sheath cells (also known as Kranz anatomy). In the tropics, there is an altitudinal transition from C_4 to C_3 grasses between 2000 and 3000 m (Tieszen et al., 1979). In forested settings at elevations below this transition zone, C_4 grasses are found only where the canopy is broken. CAM plants, which use both C_3 and C_4 pathways, include agaves, euphorbias, and cacti. These are rarely considered significant in nonarid land ecosystem studies and are not discussed further.

There is a distinct non-overlapping, bimodal distribution of the $\delta^{13}C$ values of C_3 versus C_4 plants (Smith & Epstein, 1971; Smith, 1972; Deines, 1980).[1] This distinction is the basis for this type of ecological research on plant stable carbon isotopes. C_3 plants discriminate greatly against atmospheric $^{13}CO_2$, which results in $\delta^{13}C$ values between $-35‰$ and $-22‰$. C_3 plants have a wide range of $\delta^{13}C$ values primarily due to two environmental effects. In closed-canopy forests, recycling of CO_2 near the forest floor results in more negative plant $\delta^{13}C$ values, typically ranging from -35 to $-28‰$ (Vogel, 1978). Second, under water-stressed conditions induced by low soil moisture availability, C_3 plant $\delta^{13}C$ values range from -26 to $-22‰$ (Ehleringer & Cooper, 1988). Discrimination against atmospheric $^{13}CO_2$ is far less for C_4 than for C_3 plants. The $\delta^{13}C$ values for C_4 plants thus range between -16 and $-8‰$. Atmospheric CO_2 currently has a $\delta^{13}C$ value of $-7.8‰$.

The carbon isotopic composition of soil organic matter and pedogenic carbonate is derived from the local plant biomass (Cerling, 1984; Cerling et al., 1989; Nadelhoffer & Fry, 1988; A. Martin et al., 1990; Ambrose & Sikes, 1991; Sikes, 1995). Soil organic matter $\delta^{13}C$ values may be slightly enriched ($\leq +1$ to $+2‰$) relative to those of plants, and range between -30 and $-23‰$ in habitats dominated by C_3 plants and between -15 and $-10‰$ in C_4 floral habitats. During pedogenic carbonate formation, there is a systematic enrichment of ^{13}C by 14‰ to 17‰ over the $\delta^{13}C$ values of the parent flora. Enrichment is due mainly to two isotopic effects, diffusion of soil CO_2 and temperature of calcite precipitation at 25°C to 0°C. As a result, the net difference between cogenetic soil organic matter and pedogenic carbonate $\delta^{13}C$ values also ranges from 14‰ to 17‰. Pedogenic carbonate $\delta^{13}C$ values in 100% C_3 and 100% C_4 floral habitats range from $-12‰$ to $-9‰$ and $+1‰$ to $+4‰$, respectively. Ecosystems with a mixture of C_3 and C_4 plants have intermediate soil organic and pedogenic carbonate $\delta^{13}C$ values.

Recent research on modern soil organic matter and pedogenic carbonate has demonstrated that $\delta^{13}C$ values may be used not only as a proxy for the proportion of C_3 versus C_4 floral biomass growing on the land surface, but they may also be used to differentiate standard physiognomic types of tropical plant communities, including those with mixtures of C_3 and C_4 plants (Cerling, 1984, 1992a; Ambrose & Sikes, 1991; Cerling et al., 1991; Sikes, 1994, 1995). Lateral changes in soil $\delta^{13}C$ values can be detected that mirror the complexity of the local vegetation mosaic. For example, soil organic $\delta^{13}C$ values of $-25‰$ and $-11‰$ reflect parent C_3 *Olea* forest and C_4 grassland, respectively, separated by only 20 m (Sikes, 1995). Likewise, lateral movement of soil CO_2 is generally <10 m (T. E. Cerling, personal communication, 1992). As a result, a fine scale of floral microhabitat resolution is possible with this technique.

In East Africa, where the tropical vegetation mosaic within the savanna biome can be complex, soil carbon isotopic ratios can thus indicate whether the vegetation community growing on the land surface is a closed-canopy forest or near–closed-canopy woodland (which includes riverine or riparian forest) where C_3 herbs and shrubs dominate the discontinuous ground cover, or one of three structural types of lowland vegetational savanna that have varying amounts of grass and woody plants. The three structural types are identified by percent canopy cover as (1) grassy woodland (including bushland and shrubland categories) with an open canopy of >20%, (2) wooded grassland (including bush grassland and shrub grassland categories) with a canopy cover of <20%, and (3) low and intermediate elevation treeless or sparsely treed grassland (including semidesert dwarf shrub grassland) with a canopy cover of <2% (Pratt et al., 1966, Pratt & Gwynne, 1977). In addition to an open canopy, each of the latter three physiognomic categories has a continuous herbaceous layer dominated by tropical C_4 grasses and a clear seasonality of development related to water stress.

The oxygen isotopic composition of soil carbonate is ultimately related to that of local meteoric water, which is well correlated with air temperature (e.g., Cerling, 1984; Cerling & Quade, 1993). In general, hot and dry areas today usually have less negative meteoric water and carbonate $\delta^{18}O$ values than cooler or moister regions, a trend that most likely occurred in the past (Cerling, 1992a). As a sensitive paleoclimatic indicator, the $\delta^{18}O$ values of soil carbonate can thus be used to compare local or regional change in climate over time, which in some cases has been correlated with shifts in the local percentage of C_4 biomass (e.g., Cerling & Hay, 1986; Cerling et al., 1988; Quade et al., 1989). In the tropical lowlands, for example, a warm area with low annual rainfall generally has a greater proportion of C_4 relative to C_3 plant biomass and thus more enriched soil carbonate $\delta^{18}O$ and $\delta^{13}C$ values. Cooler or moister conditions, on the other hand, benefit C_3 plants and result in more negative $\delta^{18}O$ and $\delta^{13}C$ values, respectively. This method of describing climate change thus differs from that used by other authors in this volume, comparing more recent "cooler and drier" conditions associated with "climatic deterioration" with the warm and wet, equitable climate prevailing before the global cooling trend of the last 50 m.y. (see, e.g., Denton, this volume).

The occurrence of pedogenic carbonate is dependent on rainfall patterns (e.g., Birkeland, 1984; Cerling, 1984). An inorganic mineral precipitated within the soil column, soil calcium carbonate ($CaCO_3$), is typical of areas receiving <750–850 mm of annual rainfall, and of soils developed under seasonal wet–dry climates. Massive layers of calcrete are common in arid regions with little or no vegetation cover. Under forested conditions with high soil CO_2 respiration rates and low pH and in soils receiving >1000 mm of annual rainfall where the calcium ions may be leached from the soil system, pedogenic carbonate is rare.

Paleosols

The same methods may be used to interpret the stable isotopic signatures of soil organic matter and pedogenic carbonate preserved *in situ* within buried, fossilized soils (paleosols). The C_4 photosynthetic pathway evolved from the C_3 pathway as an adaptation to lower atmospheric CO_2 concentrations (Ehleringer et al., 1991, 1997). Related to the effects of volcanism, burial of carbon in the oceans, photosynthesis, and continental weathering, CO_2 in the atmosphere has been generally declining during the Tertiary (e.g., Berner, 1990; Cerling, 1991, 1992b; Raymo & Ruddiman, 1992; Cerling et al., 1998). Perhaps since the beginning of the Cenozoic 65 m.y. and by at least 20 m.y., the concentration of atmospheric CO_2 was low enough for C_4 plant evolution. Although fossil evidence is rare, it is most likely that pre-Miocene ecosystems had little, if any, C_4 plant biomass (Cerling, 1991). The earliest known C_4 plant fossils with Kranz anatomy and enriched $\delta^{13}C$ values are from the Middle Miocene (12.5 m.y.) Ricardo Formation in North America (Nambudiri et al., 1978; Tidwell & Nambudiri, 1989; Whistler & Burbank, 1992). Indirect evidence from stable isotopic data on paleosol carbonates and fossil teeth indicate C_4 plants were present by at least 15 m.y. in East Africa (Kingston et al., 1994; Morgan et al., 1994a).

A global expansion of C_4 biomass between 8 and 6 m.y., as evidenced in extensive fossil tooth and limited paleosol carbonate isotopic data from Africa, North America, South America, and Asia, has been linked to a further reduction in atmospheric CO_2 levels to near current pressure (Cerling et al., 1993, 1994, 1997b, 1998; Quade et al., 1995; also see Kingston et al., 1994; Morgan et al., 1994a, 1994b). At this CO_2 "crossover point" of 500 ppmv, C_4 plants are favored over C_3 plants, with the global change in C_4 biomass occurring slightly earlier at lower latitudes with warmer temperatures (Cerling et al., 1997b; Ehleringer et al., 1997). Lowered atmospheric CO_2 concentrations imply that more negative $\delta^{13}C$ values, with little or no C_4 biomass, from low-altitude, nonwetlands paleosols are thus unlikely to have been C_3 grasslands, particularly after 8 m.y. (Cerling, 1992a).

During glacial periods, when atmospheric CO_2 concentrations are lowest, modeling predicts C_4 plants may be favored even at colder temperatures because quantum yields in C_3 plants are substantially reduced (Ehleringer et al., 1997; Cerling et al., 1998). In some tropical ecosystems in central Africa, an increase in grass pollen and $\delta^{13}C$ values of lake sediments and bogs shows this predicted expansion of C_4 plants during glacial periods (e.g., Talbot & Johannessen, 1992; Giresse et al., 1994; also see Ehleringer et al., 1997; Cerling et al., 1998). Other fossil pollen sequences from equatorial Africa show that during the last glacial period the maximum recorded descent for montane C_3 grassland zones is an altitude of 2230 m (e.g., Bonnefille et al., 1990; Taylor, 1992). This is well above the present or estimated-past altitude of the East African fossil localities discussed below. Soil $\delta^{13}C$ values may therefore be used as a signature for dryland woody C_3 versus grassy C_4 vegetation in tropical African lowlands. Seasonally flooded valley bottom grasslands (dambos) also have a significant C_4 grass component (Tieszen et al., 1979; Sikes, 1995).

Since pre-industrial CO_2 (before 1850) is slightly enriched in ^{13}C(~1‰) compared to modern atmospheric CO_2 (Friedli et al., 1986), this difference should be considered when interpreting some isotopic data (e.g., Cerling et al., 1997b, 1998). Any effect on paleoecologic interpretations from paleosol carbonate data is relatively minor, however, since the $\delta^{13}C$ values of our modern analogues represent the lifetime of a soil (>150 years). Paleosol organic and carbonate $\delta^{13}C$ values can thus be compared directly to modern soil values and used to quantify the density of plant cover and map the past abundance of tropical C_4 grasses to woody C_3 vegetation in intermediate and lowland paleohabitats within the African Rift. Where soil development may be only weakly expressed in lake margin zones, the presence of pedogenic carbonates, as well as their $\delta^{13}C$ and $\delta^{18}O$ values, can be used to separate non-wetland from wetland settings (Cerling & Hay, 1986; Sikes, 1994, 1995). Likewise, the 14‰ to 17‰ range of difference between co-existing soil organic and carbonate $\delta^{13}C$ values may be reliably used to indicate a lack of diagenesis in the paleosol components and thus suitability for paleoecologic reconstruction (Cerling et al., 1989).

Buried in place, paleosols may have a less complex taphonomic history compared with other fossilized remains, and, if unaltered, buried soils represent the vegetation in the immediate vicinity of the sample site. Fossil fauna and pollen, for example, both usually represent a much broader range of habitats. Further, the stable carbon isotopic signatures retained in an unaltered paleosol represent the annual fixation of carbon by C_3 and C_4 plants and hence the average composition of soil carbonate and organic matter over a longer span of time (i.e., the lifetime of the soil) compared with other fossilized remains.

Floral Biomass at Early Hominid Localities in East Africa

Carbon Isotopic Evidence from Paleosols

Stable carbon isotopic values on rare pedogenic carbonates from Middle Miocene sites in East Africa indicate local ecosystems were dominated by C_3 plants. At Lothidok in west Turkana $\delta^{13}C$ values indicate C_4 biomass ranged from only 5–20% at three points in time (17, 10, and 8.5 m.y.) (Cerling, 1992a). Likewise, isotopic analyses of paleosols (Cerling et al., 1991, 1992) and fossil tooth enamel (Cerling et al., 1997a) from the 14 m.y. hominoid site of Fort Ternan indicate C_4 plants, if present, were relatively rare (<5% of the flora) in the open canopy woodland inhabited by *Kenyapithecus wickeri.* Fossil grass specimens from Fort Ternan have not been conclusively identified as C_4 by either Kranz anatomy (Retallack et al., 1990; Dugas & Retallack, 1993) or $\delta^{13}C$ values (Cerling et al., 1997a). In the 15.3 m.y. Muruyur Formation in the Tugen Hills succession (Baringo Basin), $\delta^{13}C$ values from paleosols (Kingston et al., 1994) indicate C_4 plants composed ≤40% of the floral biomass in a grassy woodland setting. Fossil tooth isotopic data support this Middle Miocene presence of C_4 grasses at Baringo (Morgan et al., 1994a).

Some expansion of C_4 plant biomass is recorded during the Late Miocene at three fossil localities in Kenya. $\delta^{13}C$ values on paleosol carbonates from the Lothagam sequence (7 to 5 m.y.) (M. G. Leakey et al., 1996) and four soils in the Kanam Formation (6.1 m.y.) (Potts et al., 1997) represent grassy woodland and wooded grasslands with 30–80% C_4 biomass. Carbon isotopic data of fossil herbivore tooth enamel in the west Turkana sequence at Lothagam indicate the vegetation mosaic supported a wide range of dietary regimes with exclusive C_3 feeders and C_4 grazers, as well as various mixtures of C_3 and C_4 plants (M. G. Leakey et al., 1996). Between 7 and 5 m.y. at Baringo the paleosol data hint at a possible expansion of C_4 biomass (Kingston, 1992; Kingston et al., 1994). The tooth data show that in addition to mixed feeders, exclusive C_4 grazers appear by 6 m.y. in the Tugen Hills sequence (Morgan et al., 1994a).

Although limited, these data tend to support other geochemical evidence discussed above for global expansion of C_4 biomass between 8 and 6 m.y. None of the paleosol isotopic data indicate open C_4 savanna grasslands were present in East Africa during this period. Although the fossil tooth data indicate C_4 plants became the dominant food source of some African mammals between 8 and 6 m.y. (Morgan et al., 1994a; Leakey et al., 1996; Cerling et al., 1997b, 1998), C_4 grasses would have been available in the open canopy grassy woodlands and wooded grasslands. After 5 m.y. the available paleosol isotopic data from East African fossil localities indicate C_4 biomass is generally <50% under woodlands and grassy woodlands until about 1.8 to 1.7 m.y.

The 15.3 m.y. Tugen Hills sequence is the longest paleosol succession isotopically analyzed to date (Kingston, 1992; Kingston et al., 1994). The paleosol $\delta^{13}C$ values reflect an apparent expansion of C_4 plants between 7 and 5 m.y., as noted, and a possible decrease in the presence of C_3 flora at about 1.5 m.y. In contrast to other East African hominid localities with lower annual rainfall, however, there is no carbon isotopic evidence from paleosols or modern soils for open C_4 savanna grasslands at any time in the Tugen Hills succession. The 15 m.y. Baringo Basin sequence is described as a "persistent mosaic" of C_3 and C_4 plants that falls generally within the grassy woodland physiognomic category (Kingston et al., 1994); an environment that supported a wide range of herbivore diets (Morgan et al., 1994a). A significant change in the proportion of C_3/C_4 floral biomass coincident with hominid speciation events around 5.5 or 2.5 m.y. is not recorded in this sequence. The persistent mosaic of C_3/C_4 flora suggests that early hominids exploited a variety of wooded environments in the Baringo Basin.

In a summary of pedogenic carbonate stable isotopic research at a number of fossil localities in East Africa, dating from about 17 m.y. to the present, Cerling (1992a) argues that C_4 plants composed <50% of the floral biomass until about 1.7 m.y., and that relatively pure open C_4 grasslands did not appear until about 1 m.y. The data set includes diachronic research at Baringo (~9.6–4.9 m.y.), Laetoli (~3.7–2.5 m.y.), Olduvai Gorge (~2–0.02 m.y.), and Turkana (~3.7–0.6 m.y.), as well as points in time at Lothidok (West Turkana, ~17, 10, and 8.5 m.y.), Fort Ternan (14 m.y.), and Olorgesailie (0.7 m.y.) (Cerling, 1992a, personal communication, 1993). New carbonate values from Lothagam (M. G. Leakey et al., 1996) and Kanam (Potts et al., 1997) indicate C_4 plants composed up to 80% of the floral biomass between 7 and 5 m.y., but available data suggest this expansion was short-lived. Until about 1.8 m.y., the majority of pedogenic carbonate $\delta^{13}C$ values discussed below represent habitats with 15–50% C_4 plants.

The stable carbon isotopic evidence from pedogenic carbonates in the Koobi Fora Formation (east Turkana Basin, northern Kenya) indicates C_4 plants comprised only 20–40% of the vegetation from about 4 to 1.8 m.y. (Cerling et al., 1988; Cerling, 1992a). Although C_4 grazers are present in the Omo Basin by 2 m.y. (Ericson et al., 1981), an increase in the proportion of C_4 flora (up to 65%) at east Turkana does not occur until about 1.8 m.y. A further increase in the percentage of C_4 plants is seen around 0.7 m.y. A sparse dwarf shrub grassland dotted by various species of *Acacia* trees covers this arid part of northern Kenya today.

At the Kanam locality in western Kenya, carbonate $\delta^{13}C$ ratios from a 3.5 m.y. paleosol in the Homa Formation represent <45% C_4 plants, or a grassy woodland (Potts et al., 1997). New pedogenic carbonate $\delta^{13}C$ values from Kanjera South indicate relatively open habitats (>75% C_4) were present in western Kenya near 2.2 m.y. (Plummer et al., 1998). Grassy woodlands with 40% C_4 biomass are present later in the Kanjera South sequence, followed by open grassland at the top of the section. As noted above, there is a possible decrease in the presence of C_3 flora near 1.5 m.y. at Baringo, but the percentage of C_4 biomass still remains relatively low (about 50%) throughout the sequence in this west-central Kenyan basin (Kingston et al., 1994).

In northern Tanzania, an increase in $\delta^{13}C$ values at Laetoli indicates a brief expansion in C_4 biomass near 2.5 m.y. (Cerling, 1992a). This shift is not duplicated around 2 m.y. at nearby Olduvai Gorge (Cerling & Hay, 1986; Cerling, 1992a). The sequence at Olduvai has the best paleosol carbonate isotopic evidence for a 1.7 m.y. shift toward a greater proportion of C_4 floral biomass, beginning with the middle Bed II Lemuta Member (1.67 m.y.). Like the Koobi Fora record, forests and grassy woodlands were briefly replaced by wooded grasslands and some grassland with 60–80% C_4 biomass. Preliminary results on paleosol carbonate $\delta^{13}C$ values from synchronic study of lowermost Bed II between Tuffs IF and IIA (1.75–1.7 m.y.) suggest this expansion of C_4 biomass may have occurred slightly earlier in the Olduvai sequence (Sikes, in preparation). The percentage of C_4 grasses was then slightly reduced from about 1.6 to 1.4 m.y. (Cerling & Hay, 1986). After 1.2 m.y., the carbonate data indicate Beds III and IV, with greater than 70% C_4 biomass, are warmer and drier than at any time previously. $\delta^{13}C$ values from the Olduvai sequence further demonstrate the dominance of C_4 vegetation after 1 m.y., and a gradual approach to modern conditions with about 90% C_4 flora in this semiarid area in the midst of the Serengeti Plain.

By 1 m.y. at Olorgesailie, in southern Kenya only 170 km north of Olduvai Gorge, a C_4-dominant flora was also present. Pedogenic carbonate $\delta^{13}C$ values from Olorgesailie (Sikes, 1995; Sikes et al., in preparation), analyzed in collaboration with archaeological excavation of a 0.99 m.y. upper Member 1 volcanic ash-rich paleosol (Potts, 1989, 1994), indicate the 4 km paleolandscape near a shallow, freshwater lake supported a local biomass of 75–100% C_4 plants. Although the $\delta^{13}C$ values span the range from modern wooded grasslands to grassland, the majority (70%) represent an open C_4 grassland with >90% C_4 plants. Upper Member 1 stands out as a brief period in the Olorgesailie Basin when vegetation most approached present-day conditions. In later members of the Olorgesailie Formation, dating from 0.78 to 0.49 m.y., the majority of pedogenic carbonate $\delta^{13}C$ values represent open canopied grassy woodlands and wooded grasslands with 65–85% C_4 plants (Sikes et al., 1997, in preparation). Preliminary carbon isotopic values show that, when compared to upper Member 1, a greater proportion of C_3 plants was also present during formation of lower Member 1 paleosol carbonates around 1.2 m.y. (Sikes et al., in preparation). Modern pedogenic carbonate $\delta^{13}C$ values reflect the magnitude of C_4 grass cover (85–100%) in this semiarid basin, and a similarity to upper Member 1 time (Sikes, 1995; Sikes et al. in preparation).

A diversity of habitats is represented by pedogenic carbonate $\delta^{13}C$ values from the 90 k.y. Upper Semliki sequence in the western branch of the Rift Valley (Sikes et al., in preparation). Vegetation structure varied from riverine forest to grassy woodland and wooded grassland with 20–70% C_4 biomass, much like that present in the area today. In the eastern branch of the Rift at this time, open C_4 grassland was present at Olduvai (Cerling & Hay, 1986) and grassy woodland composed of 40–55% C_4 biomass about 50 k.y. in the Oltepesi Beds at Olorgesailie (Sikes et al., in preparation).

The majority of $\delta^{13}C$ values for modern pedogenic carbonates from Baringo, Laetoli, and Turkana represent 75%, 60%, and 80% C_4 biomass, respectively, which can be described as wooded grassland at Baringo and Laetoli and dwarf shrub grassland at Turkana. Like that described at Olduvai and Olorgesailie above, the fossil flora at Baringo, Laetoli, and Turkana comprised an equal or lesser percentage of C_4 biomass. For the five East African early hominid localities for which data are available (Baringo, Laetoli, Olduvai, Olorgesailie, Turkana), all the modern carbonates have $\delta^{13}C$ values the same as or more enriched in ^{13}C than their paleosol counterparts.

Despite the presence of C_4 plants by the Middle Miocene and the occurrence of exclusive C_4 grazers between 8 and 6 m.y., stable isotopic evidence from paleosols indicates C_4 grasses generally composed <50% of the vegetation mosaic at fossil localities in East Africa until after 1.8–1.7 m.y. Between 7 and 5 m.y. the expansion of C_4 biomass beyond 50% is relatively brief, and still not representative of open C_4 grassland. Only after 1 m.y., when C_3-dominant woodlands and grassy woodlands were generally replaced by C_4-dominant wooded grasslands and grasslands, is greater than 50% C_4 floral biomass consistently present at the majority of localities analyzed to date. Open C_4 grasslands, however, remain the least common vegetation community in the tropical savanna biome, a point reiterated below.

In sum, the isotopic evidence from East Africa does not support a sudden, dramatic shift from woody C_3 habitats to open C_4 savanna grasslands at any time. As summarized in table 21-1 by age, physiognomic category, and locality, the paleosol isotopic data record a more gradual approach to open vegetation communities with >50% C_4 biomass (wooded grasslands and grasslands). This general trend complements information gained by other means (e.g., Bonnefille, 1984; Bishop, 1996; Denys, 1996). Differences in interpretation of individual sites may be attributed to sampling scale, as well as to quantifying the proportion of grass to woody vegetation.

Oxygen Isotopic and Rainfall Evidence

Stable oxygen isotopic values from pedogenic carbonates are linked to local climatic conditions (i.e., rainfall and temperature). To date, such ratios from the East African hominid locality paleosol carbonates indicate that climatic conditions remained relatively stable from at least 5 to 1.8 m.y. at Baringo, Laetoli, Olduvai, Olorgesailie, and Turkana (Cerling et al., 1977, 1988; Cerling & Hay, 1986; Cerling, 1992a; Kingston, 1992; Sikes, 1995, in preparation). Fluctuations in carbonate $\delta^{18}O$ values at Turkana between 3.4 and 3.1 m.y. most likely reflect variation in the source of meteoric waters along the Omo River drainage system; no change is seen in the carbon isotopic ratios (Cerling et al., 1988). Oxygen isotopic ratios do show a slight enrichment in ^{18}O between 9 and 7 m.y. at Baringo, the only diachronic study covering this period (Kingston, 1992). This is followed by the apparent expansion of C_4 biomass between 7 and 5 m.y., but also by a depletion in ^{18}O from 7 to 1.5 m.y., suggesting a return to cooler and moister conditions. From 1.5 m.y. to the present, both oxygen and carbon isotopic ratios show a trend toward present-day con-

Table 21-1. The range of tropical plant communities, as indicated by paleosol and modern soil stable carbon isotopic values, present through time at fossil hominoid and hominid localities in East Africa.

Age (m.y.)	Closed canopy forest	Near-closed woodland or riparian forest	Grassy woodland	Wooded grassland	Grassland
	$<60\%$ C_4 plants			$>60\%$ C_4 plants	
0		Baringo Semliki	Baringo Semliki	Baringo Laetoli Semliki	Olduvai Olorgesailie Turkana
0.7			Baringo Turkana	Baringo Olduvai Olorgesailie Turkana	Olduvai Olorgesailie Turkana
1			Baringo Turkana	Baringo Olduvai Olorgesailie Turkana	Olduvai Olorgesailie
1.2			Baringo Turkana	Baringo Olduvai Olorgesailie	
1.6			Baringo Olduvai Turkana	Baringo Olduvai Turkana	
1.8–1.7			Baringo Olduvai Turkana	Baringo Olduvai Turkana	Olduvai
2		Baringo Turkana	Baringo Olduvai Turkana		
2.2		Baringo Turkana	Baringo Kanjera S. Olduvai Turkana	Kanjera S.	Kanjera S.
2.5		Baringo Laetoli Turkana	Baringo Laetoli Turkana	Laetoli	
3.5		Baringo Laetoli Turkana	Baringo Kanam Laetoli Turkana		
4		Baringo Laetoli Turkana	Baringo Laetoli Turkana		
7–5		Baringo	Baringo Kanam Lothagam	Kanam Lothagam	
9		Baringo Lothidok	Baringo		
14		Fort Ternan			
15	Baringo	Baringo			
17	Lothidok				

Dates were chosen to correspond with either the beginning of the isotopic record in each locality or shifts in the proportion of C_4 grasses to woody C_3 vegetation. A combination of analogue data from modern soil organic and carbonate stable carbon isotopic values (Ambrose & Sikes, 1991; Cerling, 1992a; Sikes, 1994, 1995) was used to categorize the fossil locality vegetation communities. In reality, because of isotopic effects during pedogenic carbonate formation discussed in the text (i.e., diffusion and temperature) as well as theoretical C_3 endmember assignment, there is not a precise one-to-one correlation between distinct physiognomic categories and C_4 percentages between the two soil components (Cerling, 1992b; Sikes, 1995). Nor is there an exact division between grassy woodland and wooded grassland biomass at 60% C_4 plants. For this table, in instances where paleocarbon isotopic ratios are reliable for both components, the soil and paleosol organic values were used to fine-tune interpretations (Sikes, 1995). Standard physiognomic and physiographic vegetation types are defined in the text. Sources for fossil locality isotopic data used to compile the table are Cerling and Hay (1986), Cerling et al. (1988, 1991, 1997a), Cerling (1992a), Kingston (1992), Kingston et al. (1994), Sikes (1994, 1995, in preparation), M. G. Leakey et al. (1996), Potts et al. (1997), Sikes et al. (1997, in preparation), and Plummer et al. (1998).

ditions in the Tugen Hills succession (Kingston, 1992; Kingston et al., 1994).

A shift in pedogenic carbonate oxygen isotopic values indicates that rainfall decreased at about 1.8 m.y. at Turkana, just before the change in carbon isotopic values toward a greater C_4 component (Cerling et al., 1988). A similar shift occurred at Olduvai at about 1.7 m.y., synchronous with an increase in the relative proportion of C_4 floral biomass in Bed II (Cerling et al., 1977; Cerling & Hay, 1986; Sikes, in preparation). An additional shift at about 1 m.y. toward permanent enrichment in $\delta^{18}O$ values is particularly evident in the diachronic study at Olduvai. This shift accompanies the dominance of C_4 vegetation in the basin at this time and the approach to modern-day conditions. The diachronic studies also indicate the modern ecosystems at Olduvai Gorge and Turkana, for example, are hotter and drier today than in the past (Cerling et al., 1977, 1988; Cerling & Hay, 1986).

At Olorgesailie, $\delta^{18}O$ values from diachronic study of Members 7, 8, 9, 11 and 13 indicate it was slightly cooler or moister in the basin between 0.78 and 0.49 m.y. than in either upper Member 1 (0.99 m.y.) or today (Sikes, 1995; Sikes et al., 1997, in preparation). These data support evidence for the slightly greater abundance of woody C_3 plants in the younger members compared to the open C_4 grassland of upper Member 1. Likewise, in the grassy woodlands of the Oltepesi Beds (50 k.y.), $\delta^{18}O$ values are comparatively less enriched.

Fossil evidence from research on pollen and molluscs also suggests that rainfall during the Plio-Pleistocene in the East African Rift was, in general, greater than at present (e.g., Hay, 1976; Bonnefille, 1984; Bonnefille & Vincens, 1985). For example, mean annual rainfall at Olduvai Gorge during middle Bed I at FLK *Zinjanthropus* and lower Bed II at HWK-E is estimated at 800–900 mm (Verdcourt, 1963; Hay, 1976; Bonnefille & Vincens, 1985; Cerling & Hay, 1986). This is substantially greater than rainfall in the Olduvai Basin today, which averages 566 mm. Likewise, the scarcity of paleosol carbonates in Bed I suggests that rainfall was higher in the past (>750 mm) (Cerling & Hay, 1986).

The presence of *Brachystegia* in the macrofossil assemblage from the Omo-Shungura Formation (Dechamps & Maes, 1985), a tree that is now exotic to the area and characteristic of miombo woodland, argues for greater rainfall 2 m.y. ago in the Turkana Basin (Feibel et al., 1991). Rainfall in *Brachystegia* miombo averages 500–1200 mm, whereas rainfall in the Turkana Basin today is <300–200 mm. Numerous lines of evidence (e.g., pollen, fauna, paleosols) suggest to Feibel et al. (1991) that the climate and vegetation during the period 1.9–1.6 m.y. in the Turkana Basin approached that of today, as also indicated by the stable isotopic evidence discussed above.

Terrestrial versus Marine Isotopic Records

This section briefly considers why the East African paleosol isotopic record does not mirror major global climatic events documented in the marine oxygen isotopic record. The terrestrial data, for example, do not record dramatic shifts in carbon or oxygen isotopic ratios around 5.5 or 2.5 m.y., dates that correspond to the Messisian salinity crisis and the beginning of glaciation in the Northern Hemisphere, respectively (reviewed by Denton, this volume). On the other hand, the terrestrial data do record a significant shift in both carbon and oxygen isotopic signatures at about 1.7 m.y. and again at about 1 m.y. that corresponds with an increase in aeolian dust from Africa and Arabia deposited in the oceans (deMenocal, 1995).

First, as suggested by deMenocal (1995:57), "remote changes" in climate in the high latitudes may have had less effect on conditions in Africa before 2.8 m.y. Marine dust records off West Africa and in the Indian Ocean from 7.3 m.y. do not show an increase in aeolian detritus from the Sahara and northeast Africa until after expansion of the northern latitude ice sheets near 2.8 m.y. and again at 1.7 and 1 m.y.

Second, the degree of sampling between the marine and terrestrial sedimentary records is not equivalent. The terrestrial sedimentary record is discontinuous and is also biased by concentration on sampling areas with hominoid or hominid fossils in East Africa. Hominid fossil and paleosol preservation does not coincide in South Africa. Paleosol sampling also depends on initial soil development (including precipitation of pedogenic carbonate), as well as burial, preservation, and exposure. Sampled paleosols from the Tugen Hills, for example, make up <1% of the 15.3 m.y. sedimentary sequence (Kingston et al., 1994). Further, at the present time only one long (Tugen Hills) and one short (Lothagam) paleosol isotopic sequence spans the Late Miocene event that ended in the 5.5 m.y. Messinian salinity crisis. On the other hand, paleosol isotopic signatures from three regions (Baringo, Olduvai-Laetoli, and Turkana) (Cerling & Hay, 1986; Cerling et al., 1988; Cerling 1992a; Kingston et al., 1994) cover the 2.5 m.y. Northern Hemisphere climatic event.

Third, there is a difference in scale between the terrestrial and marine records. The paleosol stable isotopic ratios reflect local habitats and climatic variables within a paleobasin, as noted above. Local environments, particularly in East Africa, cannot be expected to faithfully track global climatic changes because of the effects of nonglobal events, such as rifting and volcanism. In turn, changes in topography or relief resulting from tectonic events over the last 17 m.y. or so in Africa have affected regional climatic patterns. Examples of the complexity of subcontinental climatic variation are rainshadow and large-lake effects, as well as the influence of monsoonal patterns and the intertropical convergence zone (ITCZ) on the duration and intensity of seasonality including bimodal rainfall patterns in equatorial East Africa. Such complexity produces variation in both the floral community and the oxygen isotopic composition of meteoric water. Where feasible, however, lateral sampling of same-aged horizons within the same paleobasin has decreased the likelihood of site-specific bias in the studies of diachronic change discussed here. Throughout the extensive Tugen Hills sequence, for example, lateral samples were taken from 1 to 8 km away (Kingston et al., 1994).

Fourth, regional or local differences in vegetation communities in Africa are affected by altitude and temperature, as well as by the amount and pattern of annual rainfall and the geologic substrate (e.g., Bell, 1982; Tinley, 1982; Owen-Smith, this volume). In general, high-nutrient soils on volcanic parent material support a lesser diversity of woody C_3 plant species than low-nutrient soils (derived from basement rock regardless of yearly rainfall amount or leached of nutrients in well-drained areas with high-rainfall). C_4 grasses are thus promoted over C_3 woody taxa on high nutrient volcanic-derived and clay-rich soils where annual rainfall is between 400 mm and 1400 mm. Treeless grassland to sparsely wooded grassland usually occur where rainfall is <400 mm; forest is present when rainfall is >1400 mm. As a result of this relationship between soil nutrient status and moisture (including soil drainage), open C_4 grasslands such as the Serengeti with its underlying carbonatite ash are unusual and cover only a small proportion of East Africa (Pratt et al., 1966; F. White, 1983). Woodlands, on the other hand, cover approximately one-third of East Africa, and the most extensive vegetation type in East Africa today is wooded grasslands (Kingdon, 1974).

In sum, the effect of major global climatic events as seen in the terrestrial isotopic record is ameliorated by insensitivity to remote latitudinal changes, local and regional conditions, as well as sampling differences. Although changes in paleosol carbon or oxygen isotopic values at about 5.5 m.y. are not dramatic in the Tugen Hills record, for example, $\delta^{13}C$ values do reflect an apparent expansion of C_4 plants between 7 and 5 m.y., with exclusive C_4 grazers appearing at about 6 m.y. (Kingston et al., 1994; Morgan et al., 1994a). Any intensity of change expected from global events may have been lessened by local faulting and uplift of the proto-Tugen Hills at about 7 m.y. Further, global expansion of C_4 biomass between 8 and 6 m.y. has been linked, not to climatic change, but to a decrease in atmospheric CO_2 levels (Cerling et al., 1993, 1994, 1997b, 1998; Quade et al., 1995). Although this expansion is evidenced in the East African mammalian and paleosol isotopic records, the carbonate data indicate C_3-dominant vegetation communities are also present during this period. After 5 m.y. C_4 biomass is again generally reduced to $<50\%$ until expansion around 1.8 to 1.7 m.y. Likewise, dramatic change around 2.5 m.y. in the Baringo, Laetoli-Olduvai, and Turkana sequences is not indicated (Cerling & Hay, 1986; Cerling et al., 1988; Cerling, 1992a; Kingston, 1992; Kingston et al., 1994). There is only a brief increase in $\delta^{13}C$ values in the upper Ndolanya Beds at Laetoli about 2.5 m.y. (Cerling, 1992a) and in the excavations at Kanjera South at 2.2 m.y. (Plummer et al., 1998). In fact, the paleosol oxygen isotopic data cluster together near 2.5 m.y., and as noted above also suggest climatic conditions in this portion of East Africa remained relatively stable from at least 5 to 1.8 m.y.

An important regional shift is seen, however, from 1.8 to 1.5 m.y. at Baringo, Olduvai, and Turkana, with increased $\delta^{13}C$ and $\delta^{18}O$ values (Cerling et al., 1977, 1988; Cerling & Hay, 1986; Cerling, 1992a; Kingston, 1992). The change in paleosol isotopic ratios corresponds to a low-latitude climatic shift evidenced by an increase in aeolian detritus (deMenocal, 1995). The isotopic shift also occurs roughly coincident with regional tectonic and volcanic activity—namely, final uplift of the Tugen Hills (Kingston, 1992), widespread faulting and folding at Olduvai (Hay, 1976), and blockage of the axial outflow of the Omo River at Turkana (Brown & Feibel, 1988). These data suggest that local and regional climatic conditions in East Africa may be more responsive to change resulting from nonglobal events and cannot be expected to parallel global events.

Further, although records indicate expansion in northern latitude ice sheets, dominance of the 100 k.y. astronomical cycle, increase in aeolian detritus (deMenocal, 1995; Denton, this volume), and a change to warmer and drier conditions in East Africa (Cerling et al., 1977, 1988; Cerling & Hay, 1986; Cerling, 1992a; Kingston, 1992; Sikes et al., 1997, in preparation) near 1 m.y., regional or local influence over vegetation structure remains apparent. Open C_4 savanna grasslands, for example, are not present around

1 m.y. in the Turkana or Tugen Hills sequences (Cerling et al., 1988; Kingston et al., 1994), nor is this type of vegetation community reconstructed from modern pedogenic carbonate $\delta^{13}C$ values from Baringo, Laetoli, or Upper Semliki (Cerling, 1992a; Kingston et al., 1994; Sikes et al., in preparation).

Regardless of the relationship between global events recorded in the marine record and the terrestrial record, it is valuable and appropriate to combine local, intrabasinal paleosol isotopic data to examine regional and interbasinal trends or patterns through time. The East African paleosol stable isotopic data referenced here demonstrate an increase in aridity and percentage of C_4 floral biomass through time and an apparent association between early hominids and more wooded C_3 habitats before 1.7 m.y.

Early Hominid Habitat Preference

Habitat Selection by Contemporary Hominids

Stable carbon isotopic analyses from diachronic studies of East African hominid locality paleosols and carbonates (Cerling & Hay, 1986; Cerling et al., 1988; Cerling, 1992a; Kingston, 1992; Kingston et al., 1994) suggest that Plio-Pleistocene hominids may have exploited wooded areas with >50% C_3 plants, such as near–closed-canopy woodlands, riparian forests, and grassy woodlands. Recent research on the synchronic variation present in stable isotopic signatures from a stable land surface at Olduvai Gorge supports a hypothesis that early hominids may have preferred woody C_3 habitats (Sikes, 1994, 1995, 1996), although C_4 grass-dominant vegetation communities were also present in East Africa before 1.7 m.y.

Numerous lines of evidence, including pollen spectra (e.g., Bonnefille, 1984; Bonneville & Vincens, 1985) and faunal analyses (e.g., Bunn et al., 1980; Kappelman, 1984; Plummer & Bishop, 1994; Bishop, this volume), suggest that the vegetation near East African hominid localities included C_4 grasses. Further support for the early presence of C_4 flora is found in the lowering of global atmospheric CO_2 concentrations, coupled with an expansion of C_4 biomass between 8 and 6 m.y. (Cerling et al., 1993, 1994, 1997b, 1998) and the appearance of obligate C_4 grazers by about 6 m.y. in Lothagam and Baringo (Morgan et al., 1994a; M. G. Leakey et al., 1996) and by about 2 m.y. in the Omo (Ericson et al., 1981). Higher rainfall in the past on volcanic parent material most likely further promoted C_4 grasses (e.g., Bell, 1982; Tinley, 1982; Owen-Smith, this volume). Any C_4 savanna grasslands, however, may have been a transient feature and thus not preserved in the East African paleosol isotopic record prior to 1 m.y. except in rare cases, such as Kanjera South (Plummer et al., 1998). Alternatively, C_4-dominant wooded grasslands and open grasslands may have occurred more distal to known hominid localities or may have composed a smaller proportion of the tropical savanna vegetation mosaic than today.

In conjunction with a landscape archaeology project at Olduvai Gorge (Blumenschine & Masao, 1991), I used the stable isotopic composition of paleosol organic and pedogenic carbonate carbon to estimate the original proportions of tropical C_4 grasses to woody C_3 vegetation in a local habitat along a paleolandscape within basal Bed II (Sikes, 1994, 1995, 1996). The root-marked paleosol directly overlies Tuff IF, which marks the top of Bed I (Hay, 1976). Tuff IF is dated to 1.75 ± 0.01 m.y. (R. C. Walter et al., 1991). The paleosol is also equivalent to the HWK-E level 1 "occupation floor" excavated by M.D. Leakey (1971). During the time of soil formation, the 1-km^2 pilot study area, encompassing FLK-N to HWK-E in the gorge junction today, was within the eastern lake margin zone of the Olduvai paleobasin (Hay, 1976). New geological evidence (Ashley, 1996) indicates that the alkaline lake waters in this area were most likely freshened by a spring system, rather than by stream flow westward from the mountains (Hay, 1976).

The stable carbon isotopic values from the Olduvai basal Bed II paleosol organics and carbonates indicate the 1-km^2 study area supported a local floral biomass of 20–45% C_4 plants. By analogy with modern East African floral microhabitats, the narrow range of $\delta^{13}C$ values indicates that a relatively closed C_3 woodland to grassy woodland was present in the eastern paleolake margin about 1.74 m.y. Although this area, including HWK-E level 1, was repeatedly visited by Oldowan hominids (M. D. Leakey, 1971; Hay, 1976; Blumenschine & Masao, 1991), excavated coverage represents less than one-fifth of the lateral extent of the paleosol, and consequently only a small portion of the Olduvai paleobasin. Likewise, only a small proportion of available paleohabitats has been reported to date. Extensive isotopic analyses of lowermost Bed II paleosol and carbonate samples collected in four consecutive field seasons (1994–1997) throughout the Olduvai paleobasin between Tuffs IF and IIA (1.75–1.7 m.y.) in connection with the landscape archaeology project (Blumenschine & Masao, 1991) will be reported in the near future (Sikes et al., in preparation).

Results from stable carbon isotopic analyses (Cerling & Hay, 1986; Sikes, 1994, 1995) of the middle Bed I, FLK level 22, *Zinjanthropus* paleosol are similar to that from the basal Bed II study area. The data from the 1.77 m.y. *Zinjanthropus* level indicate a near–closed-canopy woodland or riparian forest to grassy woodland was present in the Olduvai eastern paleolake margin during middle Bed I.

Although synchronic analyses of the floral microhabitat context of early hominid behavior and land-use patterns are limited, such research may indicate a preference for woody C_3 habitats by Oldowan toolmakers during the Plio-Pleistocene in East Africa (Sikes, 1994, 1995, 1996). This view is supported by a review of modern ecological studies on the abundance and distribution of plant and animal resources in sub-Saharan Africa. Much of this modern actualistic research (referenced in Sikes, 1994, 1995) has been performed in settings that may be analogous to Plio-Pleistocene habitats. The studies indicate woodlands, particularly in riverine or riparian settings, may have offered hominids the best opportunity for access to a varied, dependable range of edible plant parts. Presumably plant foods were an essential and major ingredient in an early hominid diet. Dry season aquatic resources and scavengeable, abandoned carcasses may have also been more abundant and accessible in such settings. In contrast, areas dominated by C_4 grasses (wooded grasslands and treeless grasslands) may have been less successful settings for opportunistic hominid foragers. Not only are edible plant foods, including underground storage organs, scattered in diffuse patches in more open habitats, but competition for carcasses is more intense and the remains subject to rapid decay under the tropical sun. Finally, foraging by sweating hominids may have been riskier away from water or the arboreal refuge, for predator escape and sleeping accommodations, offered by wooded habitats.

These ecologically based studies were integrated into behavioral hypotheses of hominid foraging and land-use patterns (Sikes, 1994, 1995) as follows. If early hominids preferred to forage in woody C_3 habitats, one would expect the number and density of archaeological sites to vary with woody vegetation biomass. Archaeological traces (stones or bones) attest to a life association between hominids and the paleocommunity and represent the degree of use by early hominid toolmakers of certain places on the original land surface. If wooded habitats were preferred, one would expect surfaces across the paleolandscape with scant or no evidence of hominid activity to be in nonwoodland settings and, conversely, the highest density of archaeological materials and sites to be in more wooded areas.

Because the pedogenic carbonate and organic carbon isotopic ratios obtained to date within the basal Bed II study area at Olduvai Gorge are fairly homogeneous, no correlation between density of the archaeological traces and a preference for grassy C_4 versus more woody C_3 floral microhabitats could be demonstrated (Sikes, 1994, 1995, 1996). Nevertheless, a positive relationship between flaked stone and hammerstone-modified bone densities in the trenches excavated in this preliminary study suggests to Blumenschine and Masao (1991) that hominids may have preferentially discarded stone, possibly related to marrow extraction, at animal carcass concentrations across the paleolandscape. Such a conclusion suggests that food resources were the primary determinants of habitat exploitation or preference by Oldowan hominids. As noted, opportunistic foraging by hominids may have been more successful in woodlands.

Although carbon isotopic evidence for the floral microhabitat context of Plio-Pleistocene archaeological traces is sparse at present, and a demonstrated preference remains unknown, the synchronic study (Sikes, 1994, 1995, 1996) raises questions about a dominant association between early hominids and open savanna C_4 grasslands. The paleosol studies reviewed above do not contradict this assessment. In East Africa C_4 plants generally contribute $<50\%$ to hominid locality floral biomass until after about 1.7 m.y.

Range Use by Sympatric Nonhuman Primates

Skeletal material attributed to *Homo habilis* (sensu lato) and *Paranthropus boisei* has been found either singly or together at East African Plio-Pleistocene fossil localities. Both species were recovered, for example, from the middle Bed I FLK *Zinjanthropus* "living floor" at Olduvai Gorge (M.D. Leakey, 1971). This section considers how these sympatric species might partition the available food resources and whether we can use the ecology and behavior patterns of extant free-ranging nonhuman primates as appropriate analogues to explore that of extinct hominids.

Although nonhuman primates are not exact analogues for extinct hominid species (see, e.g., Kinzey, 1987), some aspects of their behavior may nevertheless be relevant for modeling early hominid behavior and land-use patterns. In particular, recent studies of sympatric lowland gorillas and chimpanzees in tropical deciduous forests offer insight into the behavior and ranging patterns, as well as dietary niche overlap, of *P. boisei* and contemporary early *Homo*. I have chosen these nonhuman primate species as an analogue because they exemplify the contrasting dietary patterns paleoanthropologists commonly associate with "robust" australopithecines versus early *Homo*—namely, a low- versus high-quality diet. In addition, recent research has demonstrated that lowland gorillas exploit a range of food types (including fruits, beans, leaves, stems, bark, pith, roots, and invertebrates; e.g., Tutin & Fernandez, 1985; Kuroda, 1992; Mwanza et al., 1992; Rogers et al., 1992; Tutin et al., 1992; Yamagiwa et al., 1992) and "robust" australopithecines may be more omnivorous than previously thought, exploiting either C_4 grass-eating vertebrates or invertebrates (Lee-Thorp, 1989; Lee-Thorp & van der Merwe, 1993).

Obviously gorillas are much larger than "petite-bodied" australopithecines or early *Homo* (McHenry, 1991, 1994b), and both gorillas and chimpanzees have dissimilar morphological and physiological traits (e.g., dentition, relative brain size, sexual dimorphism) as well as different life-history variables. However, these sympatric primates, like contemporary early *Homo* and *P. boisei,* are closely related species and are at least potential competitors. Likewise, lowland gorillas and chimpanzees exploit many of the same microhabitats, although they exhibit different habitat range and preference patterns, discussed further below.

In the tropical forests of central Africa, both western and eastern lowland gorillas (*Gorilla g. gorilla* and *G. g. graueri*) are more fruigivorous than mountain gorillas (*G. g. beringei*) (e.g., Tutin & Fernandez, 1985; Kuroda, 1992; Rogers et al., 1992; Tutin et al., 1992; Yamagiwa et al., 1992). As a result, the choice of foods by the two lowland gorilla subspecies overlaps with that of sympatric chimpanzees (*Pan troglodytes*). Lowland gorillas and chimpanzees each consume fruit, leaves, bark, pith, and roots, although the proportions consumed vary among species, groups, and habitats.

Studies indicate that ripe, succulent, sweet fruits (low in protein and fat) are the preferred foods of both subspecies of lowland gorilla. Most of the fruits consumed by eastern lowland gorillas in the Lopé Reserve (Gabon) are from trees >10 m in height (Rogers et al., 1992). Both young and female gorillas have even been observed to climb trees up to a height of 30 m to eat fruits, as well as beans (Kuroda, 1992; also see Yamagiwa et al., 1992). The observations raise the probability that if fruits were more available in the montane forests (which are not seasonal), mountain gorillas would also incorporate fruit into their diets more often. During the dry season, when fruits are less available in the tropical deciduous forests, lowland gorillas consume a greater variety of fibrous plants (Mwanza et al., 1992; Yamagiwa et al., 1992).

Chimpanzees and lowland gorillas supplement their diet with invertebrates (i.e., insects, ants, and termites) (Kuroda, 1992; Yamagiwa et al., 1992). Occurrences of meat eating by chimpanzees is now well known; vertebrate consumption by gorillas has not been reported. Overall, in terms of frequency, sympatric gorillas tend to eat more vegetable fiber than chimpanzees, whereas chimpanzees consume insects more frequently (Kuroda, 1992). Chimpanzees consume more types of fruits, whereas sympatric lowland gorillas consume more kinds of fibrous plants (Yamagiwa et al., 1992). Gorillas avoid high-fat fruits consumed by sympatric chimpanzees, as well as high-protein/high-fat seeds eaten by black colobus monkeys (*Colobus satanas*) (Rogers et al., 1992). (This observation is interesting from the perspective of dietary differences between larger-brained *Homo* and contemporary "robust" australopithecines.) When consuming fibrous matter, gorillas are highly selective feeders, extracting only the tenderest and most protein-rich plant parts (Rogers et al., 1992, and references therein). Like chimpanzees (e.g., Boesch & Boesch, 1983; Boesch-Achermann & Boesch, 1994; also see Peters, 1987a), lowland gorillas are also occasional seed predators (Kuroda, 1992; Rogers et al., 1992).

Tool use by gorillas has not been reported. Although the tool-use and tool-making behavior of common chimpanzees varies among research stations within Africa, observations of tool use by free-ranging chimpanzees are now routine (e.g., Boesch & Boesch, 1983; McGrew, 1992; Boesch-Achermann & Boesch, 1994).

The group size and population density of sympatric eastern lowland gorillas and chimpanzees are both relatively small. Lowland gorilla group size ranges from 1 to 30, with a population density of 0.3–1.3 animals/km^2 (Kuroda, 1992; Yamagiwa et al., 1992). Group size and density of sympatric chimpanzees are similar, with 1–25 individuals per community, and about 0.3–1.1 animals/km^2. In allopatry, gorilla group size (ranging from 9 to 17) is about the same as in sympatry with chimpanzees, with a population density of 0.5–1.4 animals/km^2 (Clutton-Brock & Harvey, 1977; Wrangham et al., 1993). Chimpanzee group size (ranging from 4 to 27) in allopatry is about the same, but population density (2.5 animals/km^2) is greater than in sympatry with gorillas. The relatively low density of sympatric lowland gorillas and chimpanzees, particularly that of *Pan,* may be related to ecological competition (Mwanza et al., 1992).

The equatorial floristic communities inhabited by sympatric lowland gorillas and chimpanzees are a mixture of primary and secondary forest and swamps. Most fruits are distributed in the primary and secondary forests; herbaceous plants are more common in secondary forest and swamps. Within the areas of sympatry, the lowland gorillas and chimpanzees use various vegetation communities differently depending on the seasonal availability of preferred foods (Mwanza et al., 1992). Lowland gorillas tend to range most frequently in the primary forest, but their use of secondary forest increases during the dry season. Likewise, sympatric chimpanzees increase their use of secondary forest during the dry season as fruit availability in the primary forest decreases. Lowland gorillas in the Kahuzi region (Congo) travel up the Rift Valley escarpment to nearby montane bamboo forests during the rainy season when bamboo shoots are available and forest fruits are patchily distributed in small clusters. Chimpanzees, however, never exploit swamps or bamboo forest (Mwanza et al., 1992), and they rarely range into secondary regenerating forest, where gorillas frequently exploit abundant herbaceous plants (Yamagiwa et al., 1992).

As a result of their preference for succulent fruits, both western and eastern lowland gorillas tend to range farther than mountain gorillas in search of preferred food (succulent fruit) (e.g., Tutin et al., 1992; Yamagiwa et al., 1992). Mountain gorillas have recorded home range areas of 4–10 km^2 and day ranges of only 350–1035 m (Tutin et al., 1992). Home range area for three groups of western lowland gorillas in Gabon is 4–14 km^2; eastern lowland gorillas in the Congo have home ranges of 16–34 km^2 (Tutin et al., 1992). Day range for the Gabon groups is 320–2600 m, averaging 1172 m, with an average for the eastern lowland group of 900 m. No information is currently available on the chimpanzees sympatric with the lowland gorillas, but the home and day range areas of common chimpanzees is typically much larger than for gorillas. In densely wooded regions, chimpanzee home range area is 5–38 km^2; this increases to 25–560 km^2 in sparsely wooded areas (Nishida & Hiraiwa-Hasegawa, 1987). At Gombe Stream Reserve (Tanzania), chimpanzee day range is 3–4 km.

Overlap has been recorded in the vertical distribution of nests between eastern lowland gorillas and chimpanzees (Rogers et al., 1992). Although the large lowland gorillas (averaging 118 kg for *G. g. gorilla;* McHenry, 1988) construct more than half their nests on the ground (56–77% depending on season), a comparatively large percentage of nests are built at heights comparable to chimpanzees' beds (>6 m above the ground). This information, plus the observation of gorillas climbing trees up to a height of 30 m to eat fruits and beans (Kuroda, 1992), has significance for how we interpret the tree-climbing abilities of Plio-Pleistocene hominids.

Differences in the chimp–gorilla frugivorous diet, such as greater consumption by gorillas of vegetable fiber and by chimpanzees of high-fat fruits and insects, as opposed to the traditional folivore-frugivore dichotomy, separate the dietary niches of sympatric lowland gorillas and chimpanzees. Both species prefer succulent fruits; seasonal differences in range use reflect the availability of this preferred food. What differentiates the two closely related species is the ability of gorillas to consume, and survive on if necessary, vegetal foods other than succulent fruits, such as foliage (leaves and stems) (Rogers et al., 1992). Likewise, dietary overlap undoubtedly occurred between contemporary early *Homo* and the "robust" australopithecines. The ability of "robust" australopithecines to survive on foods that require longer mastication, such as hard-cased fruits and seeds or underground storage organs (Grine, 1981; Walker, 1981; Kay & Grine, 1988; also see Hatley & Kappelman, 1980) is one feature that may have separated them from their contemporaries. Whether only one or both early hominid lineages supplemented their diets by procuring or processing food resources with tools during the Late Pliocene and Early Pleistocene is still an open question.

Discussion

The isotopic, ecologic, and actualistic evidence summarized here suggests that Plio-Pleistocene hominids in East Africa may have preferred more woody C_3 habitats, namely, near-closed woodland, riparian forest, or grassy woodland. Given the constraints against living in the open African savanna, such as susceptibility to predation, heat stress, and rarity of suitable plant foods, smaller-bodied Oldowan hominids in particular may have remained close to more wooded areas. The anatomical traits of contemporary "robust" australopithecines and early *Homo,* such as "apelike" postcranial elements (e.g., Susman & Stern, 1982; Wood, 1992a, 1993), as well as the body size and shape of these hominids (Ruff, 1991, 1993), support a more arboreal or tree-centered dependence in a relatively closed, wet environment. The small body size, coupled with a short stride length (Jungers, 1990) and less reliance on bipedality (Spoor et al., 1994), also suggest these hominids may not have foraged far from the trees. Likewise, the predominant use of local lithic raw material by Oldowan toolmakers, generally obtainable within 9 km (see, e.g., Hay, 1976; Rogers et al., 1994), supports the anatomical evidence for short foraging trips, particularly when compared with lithic transport distance during the succeeding Acheulean tradition (e.g., Hay, 1976; Potts, 1994) associated with large bodied, fully bipedal (A. L. Walker, 1993) *Homo erectus* (*H. ergaster*).

The relatively small size and low density of Oldowan archaeological assemblages compared with those in the later Acheulean (see, e.g., M. D. Leakey, 1971; Rogers et al., 1994) may be related to a greater abundance and distribution of food resources. Although relatively dense concentrations of Oldowan artifacts have been uncovered, the frequency of repeated activity at a given location on the ancient landscape may have been less than in the Acheulean. Because climatic conditions were more favorable to a greater proportion of woody C_3 vegetation during the Plio-Pleistocene in East Africa, as reviewed here, the relative abundance of food resources may have been greater and the distance between edible patches shorter throughout the Oldowan than later in the Acheulean. Also, with a small population density, perhaps comparable to that of extant sympatric lowland gorillas and chimpanzees in tropical deciduous forests, a smaller number of concentrated patches of stones or bones (now archaeological sites) would have been created by Oldowan hominids. Alternatively, larger, denser archaeological sites may still await discovery, although this seems unlikely in view of the effort already expended in surface survey and excavation in Africa over the past 50 years.

It seems reasonable that the East African Oldowan contemporaries, *H. habilis* (sensu lato) and *P. boisei,* could have had overlapping, relatively large home range areas with small day ranges in a manner analogous to sympatric extant lowland gorillas and chimpanzees. The potentially longer day ranges and larger home ranges of one sympatric species, whether early *Homo* or *P. boisei,* may have increased the quantity and diversity of ripe fruit available to each group. Dietary differences between the two hominid species may have allowed them to forage differently when the availability of fruits and other resources decreased seasonally. Within a home range, favored resting spots may have been used again and again, much like the pattern followed by extant chimpanzees who also repeatedly visit known plant and animal resources (e.g., nutting trees with associated anvils and hammers, or termite mounds) (Boesch & Boesch, 1983; McGrew, 1992; Boesch-Achermann & Boesch, 1994).

It may be argued that the environment inhabited by Oldowan hominids was unlike today's central African tropical deciduous forest exploited by sympatric lowland gorillas and chimpanzees. However, the stable isotopic evidence discussed here suggests that during the Plio-Pleistocene climatic conditions were cooler and moister than now, with more wooded C_3 vegetation communities. Based on the above discussion, C_4-dominant wooded grasslands and grasslands were present during the Plio-Pleistocene, but the overall proportions were much less than in today's hotter and drier East African landscape. Like closely related species as distinct as gorillas and chimpanzees, given adequate resource distribution and availability in a seasonal ecosystem, sympatric hominids may have also effectively exploited the same floral microhabitats, although the percentage of time and activities in each microhabitat most likely varied between species.

Summary

Stable isotopic analysis of paleosols is a powerful tool for floral microhabitat and climatic reconstruction. The stable carbon isotopic composition of paleosol organic matter and pedogenic carbonate can be used to estimate the original proportion of tropical C_4 grasses to woody C_3 vegetation growing on the ancient land surface in a variety of ecosystems. Standard physiognomic types of tropical plant communities, including those with mixtures of C_3 and C_4 plants, may also be differentiated using $\delta^{13}C$ values. Relative temperature and moisture may be estimated from pedogenic carbonate oxygen isotopic ratios.

The stable isotopic data reviewed here from diachronic and synchronic analyses of fossil locality paleosols in East Africa allow us to question whether Plio-Pleistocene hominids were predominantly associated with open C_4 Serengeti-type grasslands or wooded grasslands. C_4 plants were not a major component of early hominid ecosystems until after about 1.7 m.y. More directly, the isotopic data support actualistic and ecologic studies on the abundance and distribution of modern plant and animal resources in sub-Saharan Africa. These studies suggest that early hominids would have preferred woody C_3 habitats, perhaps C_3 riparian woodlands in particular, because of abundant plant foods, scavenging opportunities, and trees for shade, predator refuge, and sleeping platforms. In contrast, open C_4 grasslands and wooded grasslands may have been a riskier and less successful setting for opportunistic scavenging, plant food gathering, or sleeping.

It is suggested that contemporary *Homo habilis* (sensu lato) and *Paranthropus boisei* may have partitioned the foraging and feeding opportunities across the ancient East African landscape in a manner similar to extant sympatric lowland gorillas and chimpanzees. These sympatric nonhuman primates have different ranging patterns, although the preferred food of both species is ripe, succulent fruits. When fruit availability and distribution declines seasonally, the lowland gorillas consume a greater variety of vegetable fiber than the chimpanzees. The body size and shape, stride length, and postcranial and ear canal morphology of early *Homo* and *P. boisei* also suggest a more arboreal or tree-centered lifestyle. The apparent preferential use of locally available lithic raw material suggests that the tool-assisted foraging distances of Oldowan hominids may have been relatively limited. Likewise, the small size and low density of Oldowan archaeological assemblages suggest that food resources may have been relatively abundant for small groups of hominid foragers during the Plio-Pleistocene in East Africa.

Acknowledgments This chapter was originally prepared for the Wenner-Gren conference "African Biogeography, Climate Change, and Early Hominid Evolution," held October 25–November 2, 1995, in Salima, Malawi. I am grateful to Timothy Bromage and Friedemann Schrenk for their invitation and to Wenner-Gren for financial support of this meeting. The manuscript benefited from comments by Peter Andrews, Tim Bromage, Thure Cerling, Rick Potts, and Alan Turner. Stable isotope laboratory research was funded by the Wenner-Gren Foundation, Sigma Xi, and the Smithsonian Institution's Scholarly Studies Program. Fieldwork was supported by the Boise Fund, Leakey Foundation, National Science Foundation (BNS 9014160), and University of Illinois at Urbana. Isotope analyses were performed in the Smithsonian's Smithson Light Isotope Laboratory and the Department of Anthropology's Stable Isotope Laboratory in Urbana. This chapter was prepared while I was a member of NMNH's Human Origins Program.

Note

1. Stable isotopic ratios are expressed using the delta (δ) notation, as parts per thousand (‰, or per mil) relative to a standard, where $\delta‰ = (R_{sample}/R_{PDB} - 1) \times 1000$. R is the absolute $^{13}C/^{12}C$ or $^{18}C/^{16}O$ ratio, and PDB is the isotopic reference standard.

22

Grades among the African Early Hominids

Mark Collard
Bernard Wood

Palaeoanthropological systematics is principally concerned with the identification and formal recognition of natural groups among the fossil specimens that belong to the human lineage. It aims, in other words, to divide the fossil hominids into taxa that are the result of biological processes rather than abstractions of the human mind.

The groupings most commonly sought by hominid paleontologists are species and monophyletic clades. Although there is some debate over the theoretical basis of species in palaeontology (e.g., Martin & Kimbel, 1993), in practice the search for species usually involves an assessment of the extent and pattern of variation in a fossil assemblage in relation to that seen in appropriate extant comparator species (e.g., Lieberman et al., 1988; Wood et al., 1991; Wood, 1992a). The search for monophyletic clades, on the other hand, relies on the techniques of cladistics, which aims to reconstruct sister-group relationships on the basis of shared-derived character states (e.g., Skelton et al., 1986; Wood & Chamberlain, 1986; Chamberlain & Wood, 1987; Skelton & McHenry, 1992; Lieberman et al., 1996).

Because information about the "alpha taxonomy" of hominids and the pattern of their relationships are prerequisites for the successful interpretation of many other aspects of their biology, such as the evolution of function and adaptation, it is perhaps not surprising that paleoanthropologists have been preoccupied with the identification of species and monophyletic clades. Unfortunately, however, this emphasis on the delineation of species and the recovery of phylogenetic history of the hominids has resulted in the relative neglect of a third natural group, the grade.

In this chapter we aim to go some way toward rectifying this situation. We begin by discussing the concept of the grade, paying particular attention to its adaptive basis. Next, we outline criteria by which grades may be recognized among extant and fossil primate taxa, and then use these criteria to generate a grade classification of the better-represented African early hominid species. Following this, we briefly consider the timing and possible environmental causes of the grade shifts we identify.

Grades and Their Recognition

The Grade Concept

According to Huxley (1958), a grade classification attempts to identify the adaptive types that have appeared in a morphological trend. An adaptive type is a taxon with a more derived phenotypic pattern or organizational plan that is seen in the fossil record to replace an older taxon with a less derived organizational plan. In some cases the replacement is straightforward, involving just two taxa. In others the old organizational plan is replaced by an array of new organizational plans, which are then reduced in number by extinction, until finally only one is left. Regardless of the mode of replacement, the new taxon is called an "adaptive type" because, according to neo-Darwinian principles, it must have been more efficient than the taxa it superseded. The rise and success of a new organizational plan is evidence that it was better adapted than the older one and also better adapted than any potential competitor.

Like clades, grades are relative. They can only be properly delimited in relation to the particular trend being considered. For example, grades of the general organization of all animals will be different from those of the general organization of all vertebrates, which in turn will be different from the grades of all mammals. Similarly, each of these grades will be different from the grades for separate trends of specialization within a larger group-radiation such as those of the carnivores or the primates.

On the other hand, unlike clades, grades do not have to be monophyletic. They may also be polyphyletic, for convergent evolution can cause species from a number of distantly related lineages to arrive at the same adaptive type. *Aves* is an example of a monophyletic grade, whereas the homeotherms (birds and mammals) and the monkeys are examples of polyphyletic grades.

Although classifying by grades is a paleontological activity, Rosenzweig and McCord (1991) have recently argued that the grade has a neontological equivalent: the G-function of J. S. Brown and Vincent (1987). A G-function or "fitness generating function" is an equation used to calculate the fitness of a phenotype (Rosenzweig et al., 1987). It takes into account all the evolutionary factors that affect the success of an organism (e.g., densities and frequencies of other phenotypes) and "contains all the fitness trade-offs in terms of the costs and the benefits an organism receives for doing its business a certain way in a particular time and place" (Rosenzweig & McCord, 1991: 204). Because a G-function shows which phenotypes are possible and what fitness reward an individual gets for emphasizing a particular trait, it implies the design rules that govern an organizational plan (Rosenzweig et al., 1987). An adaptive type is thus a G-function with a less severe fitness trade-off than the G-function, or G-functions, it supersedes.

Rosenzweig et al. (1987) discuss a case of replacement in the evolution of the viper which illustrates these concepts quite well. Pit vipers (e.g., rattlesnakes, copperheads, and cottonmouths) have replaced true vipers in the Americas and are currently replacing them across the Old World. The success of the pit vipers appears to be due to their ability to detect both infrared and visible light. Because the focal length of electromagnetic radiation varies with its wavelength, vertebrates like the true viper must trade off sharpness of vision against the breadth of the spectrum they can see: they cannot focus sharply on both infrared and visible light. Pit vipers have overcome this problem by decoupling the ability to sense infrared from the ability to detect visible light. They have developed what amounts to a second pair of eyes, their loreal pits, which unlike their true eyes are sensitive to infrared. By avoiding the compromise between wavelength and the sharpness of the image, the pit vipers have reduced the severity of their fitness trade-off relative to that of the true viper. They have become, in Huxley's (1958) terminology, more efficient, and are consequently in the process of forming a new grade in the evolution of the viper.

Recognizing Grades

Huxley's (1958) criteria for recognizing a taxon as a grade are that it has to "emerge and persist"; emergence is proof of adaptive change, while persistence is evidence that it is a successful adaptive type. Unfortunately, these criteria are problematic for paleoanthropologists because they cannot easily be applied to recently evolved taxa. For taxa with long fossil records they work reasonably well, but persistence is difficult to apply to taxa with shorter evolutionary histories. Anatomically modern humans, for example, have probably existed as a distinct species for 150–200 k.y. Two hundred thousand years is a mere instant in geological time, so can *Homo sapiens* be said to have persisted? Humans have certainly emerged, but have they been around long enough to be called an adaptive type?

In this chapter we use criteria to recognize grades that are not time dependent and are, therefore, applicable to both recently and more distantly evolved taxa. For a primate taxon to emerge and persist, the individual animals that belong to it have to flourish in the face of the challenges posed by their environment to the extent that they can produce sufficient fertile offspring to repeat the process. To accomplish this they must meet three basic requirements. They must be able to maintain themselves in homeostasis—to sustain what Bernard has called their "mileu interieur"—despite fluctuations in the ambient levels of temperature and humidity and in spite of any restrictions in the availability of water. They also have to procure and process sufficient food to meet at least their minimum requirements for energy and for the amino acids and trace elements that are essential for continued function. Finally, they must be able to convince a member of the opposite sex to accept them as a mate in order to produce offspring.

The ways in which a species meets these fundamental requirements are clearly dependent on its adaptive organization. Thus, one method of assessing how many grades are represented in a sample of species is to look for major differences in the way in which they go about maintaining homeostasis, acquiring food, and producing offspring. There are, of course, many aspects of a primate's phenotype that help it carry out these three tasks, but some are clearly more important than others. For a hominid, the most significant are probably its mode of locomotion, dietary choices, brain size, and the shape and size of its body.

Although the importance of locomotion and diet is obvious, the significance of brain size and, especially, body shape and size requires some explanation. Brain size appears to determine the principal social interactions involved in reproduction (Dunbar, 1992b, 1995; Aiello & Dunbar, 1993). Primates with large neocortices tend to live in large social groups, whereas those with small neocortices tend to live in small groups. Dunbar (1992b, 1995) argues that this relationship arises from the role of the neocortex in processing information about social relationships. A larger neocortex allows a greater number of relationships to be tracked and maintained, and hence a larger social group to be formed. Body shape is closely linked to temperature regulation, water balance, and habitat (Ruff, 1991, 1993, 1994; Wheeler, 1991a, 1992; Ruff & Walker, 1993). Wheeler's (1991a, 1992, 1993) modeling work suggests that at low latitudes, a tall, linear body is advantageous for a hominid moving about in the open during the day. Relative to its mass, such a body leads to less heat gain from the sun, particularly near mid-day, and greater convective heat loss from the body, particularly in the morning and late afternoon. Ruff (1993) notes that in closed, forested environments with limited direct sunlight and little air movement, a tall, linear physique loses its advantages. Moreover, humid environments decrease the usefulness of a relatively large surface area for evaporative cooling by sweating. Because heat production is related to body size, the best way to avoid overheating under such conditions is to limit body size.

In the next four sections we examine data on locomotion, diet, neocortex size, and body shape for a maximum of seven African early hominid species: *Australopithecus afarensis, Australopithecus africanus, Paranthropus robustus, Paranthropus boisei, Homo habilis, Homo rudolfensis,* and *Homo ergaster.* We do not consider other early hominid species, such as *Ardipithecus ramidus* (T. D. White et al., 1994), *Australopithecus anamensis* (M. G. Leakey et al., 1995), and *Paranthropus aethiopicus* (A. C. Walker et al., 1986) because at the time of writing their fossil records are too sparse. Unless otherwise stated, we follow the taxonomy and specimen allocations for the early hominid species outlined by Wood (1991, 1992a). As inferences about fossil taxa can be made only by analogy with extant species, we also consider data on the four phenotypic parameters for *H. sapiens* and *Pan troglodytes.*

Locomotion

Locomotion in H. sapiens *and* P. troglodytes

Few would dispute that *H. sapiens* is best described as an obligate terrestrial biped (Prost, 1980; Rose, 1984). There may be some doubt about the extent to which eurocentrism has colored our perception of the efficiency with which modern humans can operate in an arboreal setting, but when compared to other anthropoids it is clear that the ability of adult *H. sapiens* to climb and travel through trees without the aid of technology is very limited. *H. sapiens* is basically adapted for a life of walking and running on the ground.

In contrast, the locomotor behavior of chimpanzees cannot be so readily categorized. Long thought to be an obligate terrestrial knuckle-walking quadruped which employs suspensory locomotion when in trees, it is now apparent that the range and flexibility of the locomotor repertoire of *P. troglodytes* has been underestimated. Work carried out by Hunt (1992, 1993) and Doran (1993a,b), for example, shows that in addition to knuckle-walking, vertical climbing, and underbranch swinging, chimpanzees employ arboreal quadrupedalism, terrestrial tripedalism, and terrestrial bipedalism as they move about their home ranges. Of equal importance is Hunt's (1992, 1993) observation that the different locomotor modes adopted by a chimpanzee are deployed strategically in response to factors like habitat, the availability of food, and even that individual's position in the dominance hierarchy. Thus, while the common chimpanzee seems to be principally a terrestrial quadruped and arboreal suspensor, there must be some doubt about the appropriateness of describing any one component of its locomotor behavior as obligate.

Early Hominid Locomotion

Evidence about the locomotor repertoire and capabilities of early hominids can come from a variety of sources. The most direct evidence comprises traces of locomotor behavior in the form of footprints, but in the event of hominids being both sympatric and synchronic, there is no way of being certain which species made the prints. For example, at one time it was held to be certain that the famous tracks at Laetoli were made by *A. afarensis,* but now that there is evidence of more than one taxon of australopithecine in that broad time range (M. G. Leakey et al., 1995), this conclusion no longer looks so convincing. In practice, most inferences about locomotion have to be drawn from skeletal evidence. For obvious reasons the postcranial skeleton has provided the bulk of the data, but recently some researchers have begun to obtain information about the posture and movement of the early hominids using novel evidence from the cranium.

Reconstructing the locomotor repertoire of *A. afarensis* is not a straightforward exercise. One group of traits has been interpreted as suggesting that *A. afarensis* was an obligate biped, exhibiting an "adaptation to full bipedality characteristic of more recent Plio-Pleistocene hominids" (Lovejoy, 1979:460; see also Lovejoy, 1981, 1988). These include the short, broad, backwardly extended iliac blades, the mechanically advantageous position of the anterior elements of the gluteal muscles, the valgus position of the knees, the nearly perpendicular orientation of the articular surface of the distal tibia relative to long axis of the tibia shaft, the nonopposable big toes, and the forward placement and downward orientation of its foramen magnum (Johanson & Coppens, 1976; Lovejoy, 1979; Johanson et al., 1982).

Other characters are thought by some authors to indicate that the gait of *A. afarensis* was different from that of modern humans. Stern and Susman (1983), for example, argue that because the form of the patella notch of the femur of *A. afarensis* is intermediate between those of modern humans and great apes, *A. afarensis* is likely to have walked in a more bent-kneed fashion than modern humans. Additionally, they suggest that because the iliac blades of *A. afarensis* are posteriorly oriented, whereas those of modern humans are laterally oriented, the gait of *A. afarensis* was probably also somewhat bent-hipped. Both of these hypotheses, however, have been challenged. Crompton and Li (1997), for instance, argue that it would have been extremely difficult, if not impossible, for *A. afarensis* to have walked with bent knees. Their computer simulations show that the inertial properties of its limbs simply would not have allowed it to do so. Likewise, Tague and Lovejoy (1986) reject the hypothesis of bent-hipped walking. They and others (e.g., Abitbol, 1995) argue that the dissimilarities between the pelvises of *A. afarensis* and *H. sapiens* reflect obstetric rather than locomotor differences.

Yet another group of traits points to *A. afarensis* having spent a considerable amount of time in trees. For example, its relatively long and curved proximal phalanges have been interpreted as adaptations for suspensory and climbing activities, as have its highly mobile hip, shoulder, and wrist joints, and its high humero-femoral index (Johanson & Taieb, 1976; Stern & Susman, 1983; Senut & Tardieu, 1985; Susman et al., 1984). Likewise, Schmid's (1991) reconstruction of the thoracic cage of AL 288–1 suggests that it was funnel-shaped, a trait associated in the pongids with the powerful muscle complex of the pectoral girdle used during arboreal locomotion.

On balance, there seems to be good reason to believe that *A. afarensis* combined a form of terrestrial bipedalism with an ability to move about efficiently and effectively in trees. *A. afarensis* had, in other words, a mixed locomotor repertoire, one that is not seen in extant primates.

McHenry (1986) has recently emphasized how similar in its postcranium *A. africanus* was to *A. afarensis.* He suggests that both were agile tree climbers as well as capable bipeds. The hypothesis of a mixed locomotor repertoire for *A. africanus* is also supported by Clarke and Tobias (1995), who describe four articulating bones from the left foot of an *A. africanus* individual (Stw 573). Found in deposits estimated to date between 3.0 m.y. and 3.5 m.y. at Sterkfontein, South Africa, these bones (the talus, navicular, medial cuneiform, and first metatarsal) suggest that the individual to which they belonged was capable of both bipedal locomotion and climbing. The foot has what Clarke and Tobias (1995) call a "compromise morphology," with the proximal end, especially the talus, displaying a suite of humanlike traits and the distal end recalling the divergent, highly mobile hallux of the common chimpanzee, *P. troglodytes.* It suggests that *A. africanus* was a facultative biped and climber, rather than an obligate terrestrial biped.

The postcranial skeleton of *P. robustus* is poorly known (Fleagle, 1988), and opinions differ about functional interpretation. Some authors suggest that *P. robustus* was more modern humanlike in both its hands and its feet than *A. afarensis.* Susman (1988), for example, argues that *P. robustus* hand bones show evidence of *Homo*-like manipulative abilities and that its foot bones point to a more humanlike form of locomotion than *A. afarensis.* On the other hand, the upper limbs of the type specimen (TM 1517) seem to have been longer in relation to its lower limbs than is the case in *H. sapiens,* which suggests that *P. robustus* was adapted to some extent for climbing (Aiello & Dean, 1990). Overall, it would appear that, even if *P. robustus* was not as arboreal as *A. afarensis,* it is likely that it spent a substantial proportion of its time in trees.

As with *P. robustus,* there are few limb bones that can be definitely attributed to *P. boisei.* However, several large forelimb bones from East African sites are often assigned to this species (Fleagle, 1988). These bones suggest that, like the other early hominids examined so far, it too could move about in trees with ease (McHenry, 1973; Howell & Wood, 1974; Howell, 1978). Similarly, various indices taken on the reasonably complete skeleton KNM-ER 1500, which some assign to *P. boisei* (e.g., Grausz et al., 1988, but see Wood, 1991), show that this fossil falls midway between modern humans and the great apes in its upper limb and lower limb proportions and in many ways is similar in these proportions to *A. afarensis* (Aiello & Dean, 1990). *P. boisei,* therefore, is also likely to have combined bipedal locomotion with an ability to climb effectively.

The hand bones associated with the type specimen of *H. habilis,* OH 7, have been interpreted by Susman and Stern (1979; 1982) as implying an apelike ability for under-branch suspension. Likewise, the relatively long arms of OH 62 suggest that *H. habilis* retained the tree-climbing ability of the australopithecines (Susman et al., 1984; Aiello & Dean, 1990). Although most of the postcranial material lacks epiphyseal ends (all except for the proximal ulna), comparisons with AL 288–1 indicate that the humerus of *H. habilis* was longer than that of *A. afarensis,* while its femur was either shorter or of equal size (Aiello & Dean, 1990; Hartwig-Scherer & Martin, 1991). Together, these data suggest that *H. habilis* was, like the other early hominid species considered so far, capable of both terrestrial bipedalism and arboreal locomotion.

At the moment there is no evidence for the locomotor behavior of *H. rudolfensis* because there is currently no postcranial material that can be reliably linked to this species. Some specimens have been tentatively suggested to be from *H. rudolfensis* (e.g., Wood, 1992a) but, as the date of the earliest *H. ergaster* specimens are close to those for *H. habilis* and *H. rudolfensis,* it is sensible to wait for evidence from associated skeletal evidence before making an assessment of the latter's locomotor habits.

In contrast to the other early hominids for which locomotor behavior can be inferred, *H. ergaster* seems to have been an obligate terrestrial biped much like *H. sapiens.* Its lower limb bones and pelvis suggest that it had a commitment to bipedal locomotion equivalent to that seen in modern humans, and there is no evidence in the upper limb bones for the sort of climbing abilities possessed by *Australopithecus, Paranthropus,* and *H. habilis* (Walker & Leakey, 1993). Furthermore, it is likely that the barrel-shaped thoracic cage and narrow waist of *H. ergaster* were adaptations to efficient bipedal walking and running. In modern humans, a barrel-shaped chest facilitates high levels of sustained activity, since it permits the upper part of the rib cage to be raised during inspiration (Aiello & Wheeler, 1995). This enlarges the thorax and consequently increases the efficiency of the respiratory system (Aiello & Wheeler, 1995). A relatively narrow waist helps stabilize the upper body during bipedal running, for it enables the arms to swing free in the lowered position and allows greater torsion in the abdominal region (Schmid, 1991).

The Bony Labyrinth and Early Hominid Locomotion

The hypothesized contrast between the locomotor repertoires of *Australopithecus, Paranthropus,* and *H. habilis* and that of *H. ergaster* is supported by recent computer tomography of the inner ear (Spoor et al., 1994, 1996). Spoor and colleagues (1994) argue that because the proportions of the vestibular apparatus of *Australopithecus* and *Paranthropus* are similar to those of the great apes, it is unlikely, given the relationship between inner ear morphology and locomotion, that either hominid species was a fully committed biped.

Spoor et al. (1994) also suggest that the vestibular dimensions of the early *Homo* specimen SK 847 are such that its locomotor behavior was probably much the same as *H. sapiens.* Some authors have likened SK 847 to *H. ergaster* (e.g., Wood, 1991), whereas others prefer to assign it, together with Stw 53, to *H. habilis* (e.g., Grine et al., 1993, 1996). If the latter hypothesis, were to be accepted, then we would need to account for the substantial differences between the inner ear morphologies of Stw 53 and SK 847. As such it seems preferable for the moment to accept the first hypothesis as more plausible and consider SK 847 to belong to *H. ergaster.* If this taxonomy is accepted as a working hypothesis. then Spoor et al.'s (1994) results are in line with the postcranial data in suggesting that *H. ergaster* was an obligate biped.

Surprisingly, Spoor et al. (1994) find the vestibular dimensions of the other early *Homo* specimen in their sample, Stw 53, to be most similar to those of large ground-dwelling quadrupedal primates like *Papio.* The exact meaning of this finding is unclear, but it does suggest that Stw 53 is unlikely to have been an obligate biped. Given that Stw 53 is usually assigned to *H. habilis,* Spoor and colleagues' study provides support for Hartwig-Scherer and Martin's (1991) observations about the arboreal orientation of that species.

Diet

The Diet of H. sapiens *and* P. troglodytes

While Fleagle (1988:222) is undoubtedly correct to suggest that "the 'natural' human diet is probably something that exists only in television commercials and on billboards," a working model for the diet of *H. sapiens* is clearly needed—one that is both environmentally and historically relevant. The most commonly used modern human diets on which to base such a model are those of the African mobile hunter-gatherers of the historical period, especially the !Kung of Botswana and the Hadza of Tanzania. Contrary to popular perception, these groups do not depend heavily on meat for their calories. The bulk of the diet of adults is composed of plant products, especially tubers, berries, and nuts. Lee (1965, 1972), for example, finds that during the early 1960s hunting provided only about 35% of the diet by weight of the Dobe !Kung, with the remainder coming primarily from gathered resources, especially, in order of declining dietary importance, the mongongo nut, the baobab nut, and the sour plum. Similarly, the Hadza only hunted about 20% by weight of their food (Hayden, 1981). Based on these data, *H. sapiens* is best described as an omnivore with a diet based principally on nuts, fruit, and meat.

For many years *P. troglodytes* was thought to be a vegetarian reliant on fruit and leaf matter. However, it is now clear that some common chimpanzees also incorporate significant quantities of meat in their diets. Hunting has been reported throughout the range of *P. troglodytes,* from Ugalla, Tanzania, in the extreme east of their present-day distribution to Mt. Assirik, Senegal, in the extreme west, and in every major habitat type they are known to occupy—primary forest, open forest–savanna, and savanna (Hladik, 1977, 1981; McGrew et al., 1979; Nishida et al., 1979; Teleki, 1981; Goodall, 1986; Boesch & Boesch, 1989; Wrangham & Van Zinnicq Bergmann Riss, 1990). For some chimpanzee communities (e.g., Kibale forest, Uganda) hunting is an incidental activity and predation rates are low (Uehara, 1986; Boesch & Boesch, 1989), but for others it is an important foraging strategy, supplying individuals with up to 25 kg of meat per year (Wrangham & Van Zinnicq Bergmann Riss, 1990). Thus, although *P. troglodytes* relies heavily on fruit (60%) and leaf matter (21%) (Fleagle, 1988), it is nevertheless best described as an opportunistic omnivore that combines the consumption of fruit, stems, and leaves with some meat eating.

Reconstructing Early Hominid Diets

In the absence of observational data for the dietary practices of the early hominids, paleoanthropologists are forced to reconstruct what they can from the dental and skeletal remains in the fossil record. Here we use an approach to dietary inference that assumes that if a species expends more energy developing a large masticatory apparatus than another species of the same body size, it is likely to have done so for functional reasons.

We examine three size-adjusted variables: the size of the crowns of the M_1 and M_3, and the cross-sectional area of the corpus of the mandible. All these variables are directly linked to the effectiveness with which the food items an animal consumes are converted into a form that can be dealt with by the chemicals in its digestive system. The relative size of the contact area or occlusal surface of the cheek teeth determines (all other things being equal) how efficiently a given quantity of food will be broken down. Molars with a relatively large occlusal surface are able to crush food more efficiently than molars with a small occlusal surface.

The cross-sectional area of the body of the mandible, on the other hand, is linked to the amount of force an individual can apply to an item of food. During mastication the opposite side of the mandible to the one on which the food item is being crushed (the balancing side) is bent in the sagittal plane (Aiello and Dean, 1990). As food is crushed between the teeth of the working side of the mandible, the balancing side is subject to three forces: the downward-acting condylar reaction force, the force transmitted from the balancing side to the working side via the symphysis, which is also downward acting, and the adductor muscle force on the balancing side, which acts in an upward direction. These forces cause a buildup of tensile stress at the alveolar margin of the balancing side and of compressive stress at its lower margin. The balancing side is thus bent in much the same way as a stick bends if its ends are forced toward one another. Just as the thickness of a stick determines how easily it can be bent, the thickness of the mandibular body determines the size of the bending forces it can withstand. A mandible body with a large cross-sectional area is able to withstand the stresses it is subjected to during chewing much better than one with a small cross-sectional area. Providing all other factors are equal, an individual with a robust mandible can, therefore, either break down tougher food items or process larger quantities of less resistant food than one with a gracile mandible.

Precision and Accuracy in Early Hominid Body Mass Estimates

Before we examine the size-adjusted data, it is appropriate to discuss how body mass estimates for the early hominid species are generated and, more importantly, how those estimates should be interpreted. Body masses for the early hominids can be estimated only by using surrogates from the skeleton. In practice, this usually involves the creation of a predictive model based on data from extant taxa. First, a skeletal variable is selected that is available on the fossil material for which body masses are required. This variable may be from the postcranial skeleton, such as the circumference of the femoral shaft (e.g., Jungers, 1988; McHenry, 1988, 1992), or from the cranium, such as orbital height or area (e.g., Aiello & Wood, 1994; Kappelman, 1996). Next, the variable is measured on the hominid material and on a representative sample of modern animals for which body mass data are available. The latter may be drawn from one species (e.g., McHenry, 1974) or from a number of species (e.g., Aiello & Wood, 1994). If the extant sample is composed of just one species, the variable data for the extant animals are then regressed directly against their body masses. Alternatively, if individuals from more than one extant species are measured, a mean is calculated for each extant species for the variable and for body mass. These means are then regressed against one another. Finally, the equation derived from the regression analysis is used to predict body masses for the hominid species. This is done by resolving the equation with either the individual values for the fossil specimens, if the equation is an intraspecific one, or with species means for the fossils, in the case of an interspecific equation.

However, the body weight estimates generated with this method cannot be interpreted in a straightforward manner because there are a number of reasons to doubt their precision and accuracy. For example, there are few data sets that can be used to verify the accuracy of the predictive regression formulae. This is a particular problem with intraspecific analyses based on relatively rare primate species, such as *P. troglodytes* or *Pan paniscus.* For a variety of reasons, the collectors responsible for acquiring most of the skeletal specimens of these species held in the major museums generally did not record the premaceration weights of the individuals. Consequently, the sample of specimens for which body weight data is available is quite small. To maximize the size of the sample used to create the predictive equations, all the specimens of known weight are usually used to generate the predictive equations, so the accuracy of the equations cannot easily be checked.

A second problem is that, although intraspecific regression is an intuitively satisfactory method for estimating the body masses of fossil specimens, it is questionable because no regression carried out using an extant species can be an entirely satisfactory substitute for determining the different body mass/variable relationship in a fossil species. Interspecific regression is often used to avoid the criticism of inappropriateness by determining a more robust regression line based on several species, but it does not, in fact, overcome the problem. It merely presents the problem in a different guise—namely, that of having to choose which of the residuals is the appropriate one for the fossil species.

Another difficulty is that the figures quoted as individual or species mean body mass estimates rarely indicate the size of the confidence intervals associated with them (Smith, 1996). Authors tend to give a single estimate for an individual or species, when in fact they should give a range of estimates. As Smith (1996) graphically demonstrates, this often leads to indefensible conclusions being drawn, especially where the body mass estimates are then used in a second regression analysis.

Counterbalancing these problems is a recent study by McHenry (1991), which suggests that the accuracy of the body weight estimates derived from the regression method may, in fact, be quite good. McHenry (1991) adopts a common-sense approach to the problem and compares elements of the postcrania of the southern African robust australopithecines to their homologues from human skeletons of three weight groups. He finds that the body weight estimates for the hominids produced by this method are similar to those derived from the regression technique. The congruence between the results of McHenry's (1991) analysis and those of the regression method suggests that, providing the precision of the regression-based body weight estimates is not overstated, for they are only indicative, "ball-park" figures; they can be used in other studies with some justification.

Dietary Inferences from the Size of the Early Hominid Masticatory System

It is evident from the values presented in table 22-1 that the two extant species in the sample, *H. sapiens* and *P. troglodytes,* have cheek teeth and mandibles of a similar relative size. Given the previously discussed differences in the diets of the two species, this similarity is perhaps somewhat surprising. If, as common sense would suggest, there are dissimilarities in the mechanical properties of the two diets, they are apparently not sufficient to be reflected in the area available for processing the food, nor in the forces applied by the masticatory muscles to the mandible. Whatever the differences in their diets, the food ingested by the taxa can be processed using a broadly similar-sized, though not necessarily similar-shaped, apparatus for crushing and grinding the food prior to its chemical digestion.

It is also apparent from the data that *H. ergaster* seems to have been the only early hominid species

Table 22-1. Means of body weight estimates, absolute and relative molar crown areas, and mandibular cross-sectional area (data from Wood, 1995).

Taxon	Body weight (Kg)	Absolute M_1 crown area (mm^2)	Relative M_1 crown area	Absolute M_3 crown area (mm^2)	Relative M_3 crown area	Absolute mandibular cross-sectional area (mm^2)	Relative mandibular cross-sectional area
H. sapiens	53	113	2.8	113	2.8	297	4.6
P. troglodytes	47	106	2.9	110	2.9	337	5.1
A. afarensis	38	166	3.8	193	4.1	488	6.6
A. africanus	35	179	4.1	218	4.5	568	7.3
P. robustus	36	207	4.4	254	4.8	786	8.5
P. boisei	41	239	4.5	327	5.2	960	9.0
H. habilis	31	166	4.1	201	4.5	421	6.5
H. rudolfensis	55	187	3.6	250	4.2	667	6.8
H. ergaster	56	144	3.1	170	3.4	455	5.6

able to survive with a mandible and chewing teeth that were in the *H. sapiens* and *P. troglodytes* size range. This suggests that the diet of *H. ergaster* was similar in terms of its mechanical properties to those of *H. sapiens* and *P. troglodytes*. The other six early hominid species have markedly larger relative tooth crown areas and mandibular bodies than do *H. sapiens* and *P. troglodytes*, which implies that their diets required considerably more bite force and/or processing than those of *H. sapiens* and *P. troglodytes*. The diets of *P. robustus* and *P. boisei* appear to have been particularly demanding, for their molars and mandibular corpora are consistently larger than those of *A. afarensis, A. africanus, H. habilis,* and *H. rudolfensis*.

These dietary inferences are supported by the results of recent dental microwear analyses, which suggest that the *Paranthropus* relied more heavily on difficult-to-process food than did *Australopithecus* (Teaford, 1995). Kay and Grine (1988), for example, find that the scratches on the teeth of *Paranthropus* specimens resemble those seen on the teeth of primates that eat hard food items, whereas the teeth of *Australopithecus* specimens tend to be damaged in a way that is reminiscent of primates that live on leaves and fleshy fruit.

Likewise, Aiello and Wheeler (1995) provide support for the idea that the diet of *H. ergaster* was less mechanically demanding than those of the other early hominids. Their analysis of the functional interrelationships between rib cage shape, gut size, metabolic rate, brain size, and dietary quality suggests that *H. ergaster* may have eaten considerably more meat than *A. afarensis*. Because meat is both calorie-rich and easily processed, a high level of meat consumption would have allowed *H. ergaster* to reduce its investment in its masticatory equipment. The role of meat in the diet of early *Homo* has also been highlighted in recent analyses of strontium/calcium stable isotope ratios (Sillen et al., 1995; but see Thackeray, 1995). These suggest that the diet of SK 847, which has been assigned to *H. ergaster* by some authors (e.g., Groves & Mazak, 1975), included a substantial contribution from animals and plant materials with a high Sr/Ca ratio, such as hyraxes and tubers.

Brain Size

At present it is not possible to determine the sizes of the neocortices of the early hominid species with any certainty (Smith, 1996). Therefore, we use overall size of the brain as a proxy measure of neocortex size (Passingham & Ettlinger, 1974).

Brain size, expressed in terms of endocranial capacity, can be determined from many early hominid crania (Holloway, 1978). Table 22-2 presents species estimates of brain size in both absolute and in relative terms, the latter being in the form of the encephalization quotient (EQ), which expresses relative brain size in relation to the estimated brain volume of a generalized placental mammal of the same body mass. The formula used to calculate EQ here is:

$$\text{EQ} = \text{observed endocranial volume}/0.0589(\text{body weight/g})^{0.76}$$

(Martin, 1981). The pattern of brain size differences is rather different from that of the masticatory variables. Although there are twofold differences in the mean absolute brain size of the early hominids, these differences are almost certainly not significant when body mass is taken into account (see table 22-2). A notable effect of body-mass correction is that the absolutely larger brain of *H. ergaster* is "cancelled out" by its substantial estimated body mass. The excellent preservation of KNM-WT 15000 means that this is a specimen for which such data are reliable.

Table 22-2. Means of body weights, cranial capacities, and encephalization quotients (EQs) (data from Tobias, 1987, and Aiello and Dean, 1990).

Taxon	Body weight (kg)	Cranial capacity (cc)	EQ
H. sapiens	53	1350	5.9
P. troglodytes	47	410	2
A. afarensis	38	410	2.3
A. africanus	35	440	2.6
P. robustus	36	530	3.1
P. boisei	41	515	2.7
H. habilis	31	610	4.0
H. rudolfensis	55	750	3.2
H. ergaster	56	850	3.6

Taken together with the gnathic evidence, these data suggest that there was a disjunction between food processing and diet, for in *H. sapiens* a reduced food-processing apparatus is combined with a relatively large brain, yet in *H. ergaster* it is not. If the larger brain of *H. sapiens* is related to its diet, then these data imply either that *H. ergaster* was eating different foods from *H. sapiens* or that contemporary *H. sapiens* manages to extract more energy from a similar diet. Systematic extra-oral food preparation by cooking is an obvious example of how this might be achieved.

Body Shape and Homeostasis

Body Shape in H. sapiens *and* P. troglodytes

The results of a wide survey of modern human data by Ruff (1993, 1994) support the relationship between body shape and habitats predicted from physiological principles by Wheeler (1991a, 1992, 1993) and Ruff (1993, 1994). For example, Ruff (1993, 1994) finds that all present-day populations that exhibit an extreme linearity of body build, like the Nilotics of East Africa, inhabit hot, dry, and relatively open environments, such as grasslands, whereas pygmies universally live in rainforest environments.

Because *P. troglodytes* is mainly a quadruped when on the ground, it is difficult to investigate its body shape in relation to Wheeler's and Ruff's predictions. However, as Wheeler (1991a, 1992) shows, when compared to the upright posture of a similarly proportioned model hominid, the generally quadrupedal stance of the chimpanzee has significantly poorer thermoregulatory properties under savanna conditions.

Early Hominid Body Shape

Body shape, in the form of limb proportions, can be deduced from a series of isolated fossils, but this is only justified if the taxonomic allocation of these fossils is reliable. This is generally not the case for isolated early hominid limb bones, so reliable data about body shape can only be gleaned from associated skeletons.

The best known associated hominid skeletons are those of AL 288–1, the *A. afarensis* specimen from Hadar, and KNM-WT 15000, the nearly complete juvenile male *H. ergaster* skeleton from West Lake Turkana. Ruff's (1993; 1994) comparison of these specimens with each other and with other less complete specimens (e.g., Sts 14) indicates that although KNM-WT 15000, when mature, would have been considerably taller than the gracile australopithecines, its body breadth would have been only marginally greater. The Turkana Boy was thus relatively tall and slender, while Lucy was relatively short and squat. Ruff argues that this difference in body form cannot be explained on the basis of obstetric or biomechanical factors; rather it is consistent with the constraints that theory suggests thermoregulation places on body shape. It is likely, therefore, Ruff asserts, that *H. ergaster* was limited in distribution to open, semiarid environments, for these are where its physique would have been adaptive. Smaller hominids like *A. afarensis* and *A. africanus,* on the other hand, probably spent most of their time in more closed environments.

If OH 62 is properly attributed to *H. habilis,* then that taxon also appears to have been short and relatively squat (Johanson et al., 1987). This suggests that like *A. afarensis* and *A. africanus, H. habilis* was principally an inhabitant of closed environments. Recent reconstructions of Olduvai Bed I habitats are congruent with this hypothesis (e.g., Plummer & Bishop, 1994).

Unfortunately, no reliable data on body shape are currently available for *P. robustus, P. boisei,* and *H. rudolfensis.*

Grades among the Early Hominids of Africa

Knowledge of locomotion, diet, encephalization, and body shape in the African early hominid species is frustratingly sketchy. What is known suggests that these species can be divided into two grades. One of these is characterized by a combination of terrestrial bipedalism and an ability to move effectively in trees; a diet considerably more mechanically demanding than those of *H. sapiens* and *P. troglodytes;* a low to moderate EQ; and a body shape that in terms of thermoregulation was best-suited to a relatively wooded environment. The other grade is characterized by a form of locomotion similar to that practiced by modern humans (i.e., terrestrial bipedalism with, in adults, a limited ability to climb and travel in trees); a diet that had similar mechanical properties to those of *H. sapiens* and *P. troglodytes;* a moderate EQ; and a physique adaptive on the open savanna. With varying degrees of certainty, *A. afarensis, A. africanus, P. boisei, P. robustus, H. habilis,* and *H. rudolfensis* can all be assigned to the first group, and *H. ergaster* can be assigned to the second.

When did this grade shift occur, and what caused it? Currently the first appearance date for *H. ergaster* is either approximately 1.9 m.y. (the mandible KNM-ER 1812 and the cranial fragment KNM-ER 2598) or approximately 1.85 m.y. (the cranial fragment KNM-ER 1648) (Feibel et al., 1989). The nature of the stra-

Table 22-3. Summary of African early hominid grades and their attributions (dates from Wood, 1992b, in press).

Grade	Characteristics	Member species (time range)
1	Locomotion: terrestrial bipedalism with climbing ability Diet: mechanically more demanding than those of *H. sapiens* and *P. troglodytes* EQ: low to medium Body shape: relatively short and broad	*A. afarensis* (>4.0 to 2.5 m.y.) *A. africanus* (3.0 to <2.5 m.y.) *H. habilis* (1.9 to 1.6 m.y.) *H. rudolfensis* (2.5 to 1.6 m.y.) *P. robustus* (2.6 to 1.2 m.y.) *P. boisei* (2.0 to 1.0 m.y.)
2	Locomotion: terrestrial bipedalism Diet: mechanically similar to those of *H. sapiens* and *P. troglodytes* EQ: medium to high Body shape: Relatively tall and narrow	*H. ergaster* (2.6–2.0 to 1.5 myr)

tigraphy at Koobi Fora, however, is such that both these dates are likely to be underestimates. There is a substantial period of time (in excess of half a million years) missing in the sedimentary sequence before 1.9 m.y. It is likely, therefore, that the first appearance of *H. ergaster* was between 2.6 and 2.0 m.y.

If we provisionally accept this date, it is clear from table 22-3 that the shift to the *H. ergaster* grade coincided with the appearance and disappearance of a number of hominid species. Because many other African large-mammal groups also experienced a period of intense cladogenetic and anagenetic evolutionary activity at this time (Turner & Wood, 1993a,b), it seems probable that the changes in the hominid lineage, including the grade shift we have identified, were caused by a widespread phenomenon. At the moment, the most likely candidate for this is the savanna expansion which followed the acceleration in the aridification of subtropical Africa around 2.8 m.y. (Vrba, 1988; deMenocal, 1995).

Clearly the hypothesis of the emergence of *H. ergaster* being driven by the aridification event that occurred some time after 2.8. m.y. and which intensified thereafter can be tested. For example, it would be falsified if those characters that we have linked to savanna life appear in a species before the aridification event. If *A. anamensis* has an *ergaster*-like postcranial skeleton and is linked to closed habitats, then our scenario for the emergence of such an adaptation is invalidated. Likewise, our present grade allocations would predict that an associated skeleton of *H. rudolfensis* would be morphologically more like those of the *Australopithecus* and *Paranthropus* than the skeletons of *H. ergaster* and *H. sapiens*.

Conclusions

The list of functions that it would be desirable to investigate in fossil hominids (Pilbeam, 1984) is a good deal longer than the list of those for which there is, or for which there is ever likely to be, reliable fossil evidence. Nevertheless, the data we have reviewed here suggest that *A. afarensis, A. africanus, P. robustus, P. boisei, H. habilis,* and *H. rudolfensis* were, to use Andrews's (1995) phrase, "bipedal apes". They spent much of their time moving about in trees, were equipped with a brain that was little bigger in relative terms than that of *P. troglodytes,* had an omnivorous diet that included a greater proportion of difficult-to-process items, such as seeds, than that of *P. troglodytes,* and would have found it easier to live in relatively wooded habitats than in the open.

The data also suggest that *H. ergaster* should be recognized as member of a different grade from *A. afarensis, A. africanus, P. robustus, P. boisei, H. habilis,* and *H. rudolfensis.* Although still relatively unencephalized, *H. ergaster* appears to have been a fully committed biped that was adapted to life on the open savanna and to a diet that was about as mechanically demanding as those of *H. sapiens* and *P. troglodytes.*

Finally, there is no formal taxonomic device for recognizing a grade, but it has become conventional for all the species within a genus to be belong to the same grade. We have seen that grade and genus are probably coextensive in the case of *Australopithecus* and *Paranthropus,* but is there the same degree of functional consistency within the genus *Homo* with respect to locomotion, relative tooth, jaw and brain size, and body shape? To judge from the evidence reviewed above, the answer must be no: the species assigned to *Homo* do not form a functionally coherent group.

Acknowledgments We thank the following people for their help in the preparation of this paper: Sally Gibbs, Patrick Quinney, Iain Spears, and Jon Tayler. We also thank our reviewers, Laura Bishop, Rob Foley, and Elisabeth Vrba, for their valuable comments and suggestions. Mark Collard is funded by The Wellcome Trust. Bernard Wood is supported by The Leverhulme Trust and the Henry Luce Foundation.

23

Evolutionary Geography of Pliocene African Hominids

Robert Foley

Palaeoanthropology, like other scientific disciplines, has seen marked change over the course of this century. New perspectives and interpretations can be ascribed to both classical empirical progress and to changes in the broader social and intellectual environment. There can be little doubt that the expansion of the fossil record from a few Neanderthal fragments to the thousands of specimens now known has been a significant driving force in the way human evolution is described and explained. Equally, though, the decline of notions of racial typology and the rejection of evolution as a progression up a *scala naturae* have played their part (Reader, 1988; Lewin, 1990).

In a discipline where the fiercest debates have been over empirical and chronological issues, it is perhaps not surprising that the impact of changes in evolutionary theory have sometimes been neglected when considering the history of the subject. However, it is clear that the development of the modern synthesis and the direct contribution of one of its architects (Mayr, 1950) steered postwar paleoanthropology into its essentially anagenetic position, resulting in major pruning of the hominid evolutionary tree and the basis for most of the arguments concerning the reluctance to accept new taxa of hominid (Smith & Spencer, 1982). The debate over macroevolutionary patterns and processes in the 1970s and 1980s (see Szalay, this volume), depending as they did on species level analysis, reversed this trend and played a part in the reintroduction of species and cladogenesis in hominid evolution. The development of behavioral ecology and sociobiology led to a greater emphasis on the role of behavior in human evolution during the 1970s, while evolutionary ecology provided one of the first coherent theoretical frameworks for linking the long-term evolutionary patterns of hominid evolution to the mechanisms of adaptation and selection (Foley, 1987). However, most paleoanthropologists have remained agnostic about detailed distinctions in evolutionary theory, and the current theoretical flavor is a well-established mix of functional adaptationism and behavioral and evolutionary ecology, placed in the context of a moderate level of speciation operating through various macroevolutionary mechanisms. This mix provides a powerful explanatory framework for the general pattern of hominid evolution: the Late Miocene diversification of the African apes, the development of bipedalism in Pliocene australopithecines on the eastern side of the continent, an adaptive radiation of African hominids between 4 and 1.5 m.y., and subsequent reduction in species richness, geographical expansion and divergence, with the survival of one *Homo* lineage only.

It may be asked what can usefully be added to this theoretical cocktail. Here I want to highlight two linked elements of evolutionary biology: the geographical basis for evolutionary change and the primary role of extinction in shaping evolutionary patterns. These two elements are particularly crucial because the main input of ideas concerning evolutionary mechanisms in hominid evolution have come from macroevolutionary perspectives—an emphasis on patterns above the species level, speciation as the primary evolutionary event, punctuated equilibrium as the mode, and an emphasis on external, nonbiotic causes of change, occurring in a manner associated with punctuated equilibrium or pulse-turnover hypotheses, at the expense of the role of natural selection operating at the level of populations. In other words, theory development in early hominid evolution has been primarily macroevolutionary and non-Darwinian. Here I develop a more microevolutionary approach that can account for the macroevolutionary patterns.

Evolutionary change occurs first as a result of spatial distribution—population dispersals and contractions, fragmentation and isolation in response to vicariance, shifting habitat boundaries, and so on. Changes in distribution result in altered selective pressures, different demographic parameters, new levels of gene flow or discontinuities, and as such evolutionary change through time is the end product of geographical processes. My conclusion is that from an evolutionary perspective there are three distinct types of evolutionary event: isolation, dispersal, and contraction. Intra- and intertaxon competition and external environmental factors determine population distributions and are the underlying conditions for speciation and extinction. Geographical patterns thus supply the microevolutionary underpinnings of macroevolutionary patterns and so form an intermediate and often confounding variable in simple links between climate and evolution.

The second novel element is a greater emphasis on the role of extinction in evolutionary theory. Where the tempo and mode of evolution has been studied, and in particular linked to the question of whether external factors such as climate are the primary determinants, speciation has been the main focus. Mechanisms of speciation are clearly crucial to the evolution of diversity and novelty, and it has therefore been assumed that speciation is the critical fulcrum around which evolutionary patterns pivot. Extinction has generally been thought of as an unfortunate by-product of other evolutionary processes. However, it is clear that the distribution of extinction patterns is nonrandom. Extinctions also have a major effect on habitat structure and thus have major ecological and evolutionary consequences. As hominid evolution, particularly in the early stages, consisted of multiple taxa, it is worth examining the role of extinction in the overall pattern of hominid evolution.

In this paper, therefore, I want to pursue an explanation for the overall pattern of early hominid evolution (rather than the more restricted problem of the evolution of the *Homo* lineage) in the context of evolutionary geography. The conclusion I shall be working toward is that evolutionary patterns are fundamentally a product of geographical factors. In looking at the overall pattern of early hominid evolution in Africa, we should identify these specific events in the fossil record and seek distinct explanations for each of them. In general, however, the evidence seems to suggest that extinction of local, regional, and subcontinental populations may be one of the driving forces of evolutionary change, and an extinction-based model of evolutionary dynamics is developed.

The level at which this investigation occurs is that of the species. Perhaps the most significant development in recent years is that, partly due to new fossil discoveries and partly due to new techniques of classification, the number of recognized hominid species has increased markedly. Hominid evolution is clearly no longer an anagenetic event. The number of hominid species recognized allows us to do something quite novel in paleoanthropology: to attempt quantitative comparative analyses of hominid taxa in an attempt to understand both pattern and process in hominid evolution. Species are the unit of analysis used here, and the analyses discussed examine hominid evolution as changes in species' distributions. In the first instance, therefore, it is necessary to consider briefly both what is meant by a hominid species and what are the species of early hominid evolution.

Species and Early Hominid Evolution

The Species Concept

The species concept has been a central problem for evolutionary biology since the publication of the *Origin of Species*. Indeed, Darwin (Darwin, 1859) established the basic framework for subsequent debate. The title *Origin of Species* showed that Darwin recognized that a key problem was the formation of new types of organisms. However, he also described the process of evolution (or perhaps from Darwin's perspective, more accurately the outcome of natural selection) as "descent with modification." Descent implies continuity, with all organisms linked through a vast tree of common ancestors, whereas types or species implies something different and distinct, with breaks in distribution. The reason the species concept is so problematic, and indeed is probably theoretically irresolvable, is that it must satisfy both elements—the continuity of descent, or change through time, and the discontinuities of the end products. Emphasis on the discontinuities is more likely to lead to a perspective in which species are discrete units, have distinct points of origin, are analytically equivalent to individuals in microevolution, and are more likely to be the product of punctuational cladogenesis. In contrast, the more the temporal dimension and continuity of descent is incorporated, the harder it becomes to treat species as discrete units, and as a mechanism punctuational cladogenesis becomes less clear cut, lost in the complexities of gene flow. Macroevolutionary perspectives on species become reducible to microevolutionary ones, and species concepts more fragile, once temporal dimensions are incorporated (see Turner et al., this volume).

The various definitions that have been developed lie somewhere along this continuum. Mayr's (1963) biological species concept, for example, emphasizes the discontinuities represented by separate gene pools and reproductive barriers. Other definitions, such as the specific mate-recognition concept (SMRS) (Paterson, 1985) or the phylogenetic species concept (PSC) (Cracraft, 1989b) also stress the mechanisms for isolation and separation and the endpoints of that separation, the terminal twigs of the evolutionary tree. Only Simpson's (1951, 1961) evolutionary species concept places more stress on the temporal dimensions and the problem of continuity of descent.

Such definitions obviously run into serious difficulties when paleobiological aspects are taken into account. Variation through time blurs the isolating mechanisms, as well as making it virtually impossible to test them. Furthermore, samples of biological diversity taken at any point in time will pick up different degrees of separation and isolation, degrees that are continuously changing.

To take chronological trends into account, however, makes the species concept even more slippery, as decisions become increasingly arbitrary. *Homo ergaster* must exist in some sort of ancestor-descendent relationship with *Homo habilis* and *Homo erectus.* To run them all into one species may make some sense, but why stop there, given that there must be a continuity of descent back to the African apes and through to *Homo sapiens.* In addition, with primates and hominids the whole nature of speciation mechanisms and the degree of isolation that does occur between closely related species is largely unknown for the present, let alone the past, such that defining a species becomes particularly hazardous.

Subspecies in Evolutionary Theory

Given that there can be no sound theoretical solution to this problem, a key criterion for defining species will be utility, and this in turn will hinge on the questions being asked. For the situation of the Plio-Pleistocene African hominids, the issue is variation in the relative intensity of evolutionary activity—essentially, is more going on at one period rather than another, or in one place rather than another? This does not necessarily depend on the question of reproductive isolation or isolating mechanisms, and it could be argued that much of what is of evolutionary interest (from the point of view of adaptive changes) will be completed by the time speciation occurs. The development of subspecies may be equally important and more interesting. From this point of view the most useful definition of species will be the one that identifies relatively small distinctions within the hominids. The most appropriate here is the evolutionary species; defined as any lineage that has an independent evolutionary trajectory or has the potential for an independent evolutionary trajectory (Simpson, 1953).

This definition of species has the disadvantage that it is not very precise, nor firmly tied to genetic mechanisms. A species in this sense may come into existence before full reproductive barriers are in place and thus may divide up what would be placed into a single species in the biological species concept. Furthermore, species are not discrete units in the sense proposed by Eldredge (1985) and others (Gould, 1980; Cracraft, 1989b), but merely relatively stable biological entities occurring at particular times and places. Such "species" may have a relatively brief existence before being absorbed back into the parent population, or in effect becoming extinct. Species are less watertight biological concepts than populations that may have the evolutionary potential to develop unique (apomorphic) traits, but the fate of which will be dependent on external, environmental, competitive, and stochastic factors. Some may simply die out, others may be swamped by other populations, while others may give rise to entirely new and distinctive taxa.

It could be argued, probably correctly, that this definition is closer to that of a subspecies than a species. In practice it may well be that this is the level at which, from the point of view of the ecological and competitive factors underlying evolutionary change, the main driving force of evolution occurs. In calling his book *The Origin of Species,* Darwin focused subsequent debate on this level of biological organization. However, Darwin was unaware of the genetic mechanisms involved in the erection of reproductive barriers, and was concerned more with a general understanding of the factors leading to diversity and its maintenance. Genetics is just part of that, and probably that part most associated with speciation, the isolation of fertilization, occurs relatively late in the process of divergence, especially where allopatry is involved. Thus the origin of species problem of Darwin has been monopolized increasingly by geneticists and the problem of reproductive isolating mechanisms and levels. In contrast, the species concept used here, which is closer in practice to subspecies, is not that distant from Darwin's own concept but is less concerned with the mechanisms of reproductive isolation and its genetic consequences. Ironically, where the comparative method is focusing more and more on the genus as the best level of analysis (Harvey & Pagel, 1992), and the proposal here that subspecies are the unit of evolutionary change, the species itself, the most argued about concept in evolutionary biology, is probably the least useful.

African Pliocene and Pleistocene Hominid Species

In practical terms, the definition of evolutionary species means that species in the hominid fossil record discussed here may not be full species in the BSC sense, nor will they necessarily have a separate species mate-recognition system. A consequence is that a relatively large number of species will be recognized (i.e., considerable splitting). Table 23-1 sets out the species used in this chapter.

A total of eleven species are recognized. These may be distributed across four genera, *Ardipithecus, Australopithecus, Paranthropus,* and *Homo,* but only two are used here, *Australopithecus* and *Homo.* This is in reasonably close accordance with theoretical expectations (Foley, 1991; Fleagle, 1995).

A number of questions and possibilities arise as a result of recognizing these taxa. The first is the observation that the hominid family, during the Pliocene and Early Pleistocene, is relatively speciose, a pattern consistent with an adaptive radiation and with evolutionary patterns in the catarrhines more generally (Foley, 1991). More interesting is to ask where and when this evolutionary activity is taking place, and thus to explore the mechanisms underlying macroevolutionary patterns in early hominid evolution.

Table 23-1. Dates of first and last appearances of fossil hominid taxa used in these analyses.[a]

Taxon	First appearance (m.y.)	Last appearance (m.y.)
A. ramidus[b]	5.1	4.4
A. anamensis[b]	4.2	3.9
A. afarensis[b]	3.7	2.89
A. africanus[c]	3.55	2.05
A. aethiopicus[d]	2.69	2.29
A. boisei[e]	2.11	0.81
A. crassidens[f]	1.55	0.65
A. robustus[f]	2.11	1.91
H. habilis[g]	2.35	1.61
H. rudolfensis[g]	1.91	1.61
H. ergaster[h]	1.61	1.15

[a]The level of accuracy shown is spurious but is used to force the taxa into the chronological units used.

[b]*A. afarensis* is taken here to include the material from Hadar and Laetoli. Earlier and more fragmentary material from Lothagam, Tabarin, Koobi Algi, Kanapoi, and middle Awash (Hill et al., 1992a) is assumed to belong to *Ardipithecus ramidus,* along with the material reported recently from Ethiopia (T. D. White et al., 1993.).

[c]*A. africanus* comprises specimens from Sterfontein and Makapansgat. Dates from Delson (1988). More than one species may be included in this taxon.

[d]*A. aethiopicus* is included by some authorities in *A. boisei* and is represented by specimens from West Turkana and Omo (see Grine, 1988).

[e]*A. boisei* (Grine, 1988).

[f]*A. crassidens* is recognized by some authors (Grine, 1989) as a distinct taxon. *A. robustus* consists of material from Kromdraai (Delson, 1988).

[g]Early *Homo* has been reported from Baringo by Hill et al. (1992b). Following Wood (1991), early *Homo* is divided into two groups, referred to here as *H. habilis* and *H. rudolfensis*, the former being the Olduvai material plus specimens from Koobi Fora such as KNM-ER 1813, the latter including the larger Koobi Fora material such as KNM-ER 1470.

[h]For the purposes of the analyses here, a distinction is made between the early African material that has been ascribed to *H. erectus* (e.g., KNM-ER 3733, 3883, WT-15000) (= *H. ergaster*) and the later African material (e.g., OH9) and the Asian material.

Macroevolutionary Explanations of Early Hominid Evolutionary Patterns

Interpretations of the pattern of early hominid evolution (or indeed any other evolutionary phenomena) exist on a continuum between two extremes. At one extreme, changes in hominid evolution can be seen as large-scale responses to a continental level of environmental trends and changes. At the other extreme, these large-scale trends are more apparent than real, and the total pattern of change is nothing more than the sum of small-scale changes occurring in particular times and places, in response to local competitive conditions. In a way these two positions almost amount to different evolutionary epistemologies, articulating to some extent with the debate over macroevolutionary processes. Here we can tackle the question of which end of this axis is appropriate by starting at the "macro" end and pursuing the problem down to the local and the particular. As indicated in the introduction of this chapter, by thinking about problems in a spatial context we can help resolve the issue of what factors underlie early hominid evolutionary trends. In the first instance I look at the relationship between early hominid evolution and climate before moving on to consider co-evolutionary patterns and ecological biogeography.

Climate and Early Hominid Evolution in Africa

There is little doubt that climatic and environmental change underlies much of evolution in general and hominid evolution in particular. The emergence of the hominid clade coincides with a global decline in temperatures, the establishment of polar and montane glaciation, increased mountain building and tectonic activity, and in Africa the break up of forested regions and the expansion of woodland, bushland, and grassland habitats. Many hominid traits may be adaptively linked to increased thermoregulatory stress and to movement in more open and terrestrial environments (Rodman & McHenry, 1980; Wheeler, 1991a,b; Foley & Elton, 1998). At this level there can be little disagreement about the role of climate change as a factor. Vrba (1985a,c) has developed this argument more strongly, proposing that climate change is a necessary condition for significant evolutionary change. This theory has taken a number of forms and has developed over the years (Vrba, 1992, 1996a; see Vrba, this volume, for a current formulation). At its strongest is the argument that major evolutionary change, particularly speciation, should be confined to periods of climate or environmental change, resulting in broad synchronicity across biomes—the Turnover Pulse Hypothesis. Other formulations, tied more to habitat theory, predict that speciation can also occur at other times. These, quite correctly, are general propositions about evolution in all lineages, and about the mechanisms of evolutionary change, not specific ones for humans. The specific claim was that the separation of the two major hominid lineages, *Homo* and *Australopithecus,* was a response to major climate change around 2.4 m.y. (see Denton, this volume; Vrba, this volume).

In an earlier paper I presented a statistical analysis of the relationship between hominid evolution as measured by first and last appearances of taxa and a number of climatic variables (Foley, 1994). This showed that there was no relationship between the first appearance of hominid taxa and any climatic variable. Significant relationships were found between the last appearances of species and climate change. Furthermore, no clear-cut relationship could be found between other large mammalian groups and climate change. Since that analysis, two new species of hominid, *A. ramidus* and *A. anamensis,* have been named (T. D. White et al., 1993; Leakey, 1995). In addition, it could have been argued that the earlier analysis conflated African patterns with those from high latitudes and weakened the effect by including later hominid taxa where more buffering of hominids from the climate could be expected.

Using the same methods, a new analysis is presented. *A. anamensis* and *A. ramidus* have been included and the age range of *A. afarensis* attenuated (table 23-2). Once again, no positive correlations were found with first appearances and temperature. This was the case where first appearances were measured in absolute terms or in relation to overall hominid diversity (number of first appearances/number of species). A relationship was found between the number of first appearances and the magnitude of climate change from the preceding period. Last appearances showed a marginally stronger relationship, being positively correlated with maximum δ-18 values (= minimum temperature), and climatic variability (both lagged effects). Extinction rate (number of extinctions/number of species) was positively correlated with minimum temperature, modal temperature, and climatic variability (table 23-2). When the number of new species is calculated relative to the number of species in the previous period or to the number of extinctions in the previous period, there is no significant relationship; there is, however, a significant relationship between maximum temperature and modal temperature for the number of new species relative to the number going extinct; in other words, periods with high minimum or modal temperatures have high rates of turnover. Using partial correlations to look at interactive effects, the

Table 23-2. Statistically significant (one tailed, $p < 0.01$) correlation co-efficients between climatic variables and the first and last appearances of hominids.

	Number of climatic cycles	Max. δ-18	Min. δ-18	Modal δ-18	Climatic variation	Magnitude of climatic change
Number of hominid species		.672			.705	.722
Number of hominid first appearances						.681(L)
Number of hominid last appearances		.741(L)			.731	
Speciation rate						
Extinction rate		.784(L)		.668	.693(L)	
Turnover			.810	−.762		

Time unit used is 500 k.y. L = correlations based on lagging evolutionary events after climatic change. See Foley (1994) for a fuller discussion of methods.

strongest link was found to be a relationship between modal temperature and the rate of extinction, controlling for the number of new species ($r = .814, p < .01$). Partial correlations can also be used to look independently at the interactions between a number of variables. In figure 23-1 it can be seen that there are strong interaction effects between temperature, extinction, and diversity, but weaker ones between speciation, diversity, and temperature. The relationship between extinction and speciation is intermediate.

These data do not support the view that climate is the primary and direct driving mechanism for hominid evolutionary change (see also Bishop, this volume). Although there was both climatic and evolutionary activity during the period 2–2.5 m.y., there are other periods that show species turnover and speciation without significant climate change. Furthermore, the period 2–2.5 m.y., were it taken in isolation, shows that there is a stronger tendency for the relationship between climate change and extinction rather than speciation.

A number of implications should be noted. The first is the general one: because human evolution is a one-off event and is a product of historical conditions that have occurred at a particular time and place, it is especially hard to demonstrate causal factors. As with most evolutionary processes, it is a matter of probability rather than deterministic rules, and therefore statistics must play a central role in any discussion. However, given the small sample size and the nature of the database, such statistics may well be inconclusive. The climatic forcing model argues that there is a strong statistical tendency for taxa to appear in association with climate change, not that all species do. The primary conclusion that should perhaps be drawn from the analysis presented here is that the evidence for the climatic driving model is far from convincing. The issue is not, though, to set up climate versus other factors, but to examine the interaction between different variables and to partition the amount of variation that can be explained between the factors that might be responsible. In this context it is interesting (but statistically insignificant) to note that a number of multivariate techniques failed to demonstrate a strong deterministic relationship with either climate or with the pattern of hominid diversity.

There are, however, reasons that climate change may not be expected to have a direct effect on evolution, even if it remains an important underlying factor. Species respond adaptively to changes in their relationships with other species (and, of course, to changes in intraspecific competition). Each species will be buffered from the direct impact of climate change because it will experience such change through the effect on other species, be they food sources, competitors, or predators. As such the effect will be both highly variable from one taxon to another and also subject to a series of time lags. The result will be that evolutionary change will be strung more or less continuously through time, even if rates of evolutionary change are far from constant. Empirically, therefore, any link between climate change and the appearance of any new species should appear as only a weak signal. The data presented above, therefore, are consistent with these theoretical expectations and the limited direct effect of any climatic forcing model.

To take things one stage farther, as stronger effects were noted when extinction and diversity patterns were considered, we can consider why these interactions should be more direct. With the number of species or species richness, the effect is best considered to be an ecological one rather than an evolution-

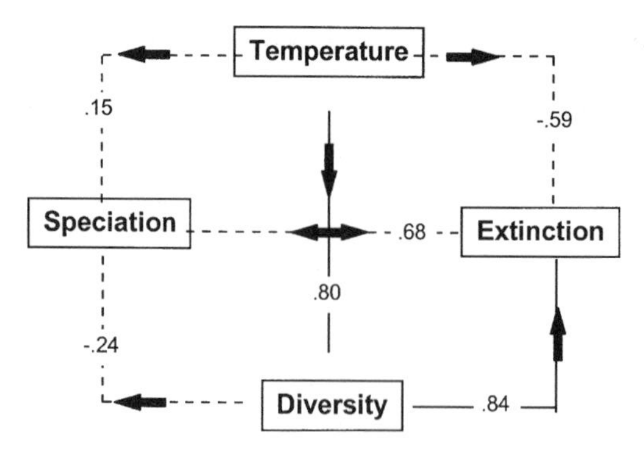

Figure 23-1. Partial correlations of the interactive effects of climate and speciation (first appearances), extinction (last appearances) and species diversity for the African hominids. Values indicate the r^2 between each pair of variables. Solid lines indicate statistically significant effects at the .01 level. The arrows indicate hypothesized direction of the interaction.

ary one. Certain climates and environments support more species and a higher level of complexity than others. Assuming, therefore, that the fossil data reflect some sort of temporary equilibrium for each time period, then species diversity will track climatic patterns. Although the data are not presented here, *Theropithecus* diversity was shown to closely reflect changes in temperature (Foley, 1993), and Dunbar (1993) has shown that the extant theropithecine at least (*T. gelada*) is highly sensitive in distribution to temperature. The rate at which these species evolve to reach some sort of ecological equilibrium, however, may either be below the resolution of the data or be too strongly influenced by nonclimatic factors for any relationship to be measurable.

With extinction, the effect of climate change can be expected to be more direct. This can best be seen by considering the geographical processes involved. If a climate change has a deleterious effect on a species reproductive rate, then it is likely to contract in its distribution. If this process continues, extinction will eventually occur. This, in effect, is a single process. For speciation to occur there must be both a change in distribution (which can be either expansion or contraction) and isolation of populations. As there are two conditions to satisfy rather than one, the probability of speciation occurring in relation to climate change must be that much lower than extinction. This interpretation, of course, has not taken into account any adaptive aspects that might be involved, but this would effectively be adding a further condition to both extinction and speciation.

In summary, empirically there seems to be little support for a climatic forcing model and several theoretical reasons that extinction and diversity rather than speciation should be more strongly influenced by climatic change. As climate change will act as a selective pressure through the secondary effects of changing distributions of other taxa, the effects will be dampened. Although extinction can result, the probability of new species evolving in response either to the new conditions or to fill vacant niches will be more dependent on local conditions. It should be stressed that this in no way implies that the appearance is a random process and that evolutionary patterns are anarchic—far from it. Rather, evolutionary theory indicates that there will be strongly deterministic elements to evolutionary change but that the level of such deterministic interaction will be the local population. The cumulative effect of many local changes may result in a strong periodic signal to evolutionary change, possibly in relation to climate, but this is not inevitable because minor local variations and asynchronies could blur the observable pattern considerably.

Community Evolution among Plio-Pleistocene Mammals

If climate is only a partial explanation because it operates through the intermediary of ecological interactions with other species, then perhaps it would be possible to explore briefly the relationship between hominid evolutionary patterns and those of other taxa. This has been considered earlier (Foley, 1984, 1987) in the context of the Red Queen hypothesis and community evolution in general. The most general hypothesis might be that, although climate does not directly influence the pattern and rate of evolutionary change, there is a synchronized pattern of change among species that are sympatric and contiguous. A generalized pattern of community evolution should therefore be observable. The Red Queen would constitute a more specific hypothesis (Van Valen, 1973). Due to the essentially deleterious effects of evolutionary change on competitive interactions for any species, the Red Queen predicts constant and rapid change across multiple taxa. However, such change is not necessarily synchronous (indeed, it is unlikely to be), but should consist of a complex web of interactive effects. Furthermore, although the engine of change in this model is competition, the diversity of competitive interactions is marked. Competition here is best thought of as the ecological conditions that result in selection favoring particular strategies, but these strategies may be co-operative or mutualistic.

Obviously there is a mass of potential relationships

possible here, and this is not the place to explore them in detail (see table 23-3). In terms of simple synchronized evolution, out of the hundreds of potential relationships, only a small handful are statistically significant. This in itself should be a warning, as by chance a small number of relationships are bound to be significant. There is, for example, a positive relationship between the diversity of baboons and the number of new species of bovid; there is also one between new bovid species and new hominid species; carnivore speciation is related to total species diversity in southern Africa, and there is a positive relationship between overall species diversity in East and South Africa.

These two-way relationships might allow us to track a series of events across biota and thus to discover the Red Queen's ecological network. When baboons are diverse (as a function of environmental distributions, perhaps), then there tends to be a high rate of speciation in both bovids and hominids; when overall species diversity is high, then carnivores are more prone to speciation. In this way the network of knock-on effects that provide the dynamism to community evolution could perhaps be untangled.

Tempting as it might be, there are two reasons for expressing caution. The first is that, given the interdependence of the data, a number of relationships are bound to occur, and it would be easy to construct stories to accommodate these. For example, bovids make up the bulk of species, so when species diversity is high, so too will be bovid diversity, and this may prompt an evolutionary arms race between carnivores and herbivores, thus producing a web of relationships. The difficulty lies in determining whether these are real relationships or simply plausible hypotheses. The second reason is that while the fossil record is, in general terms, good in quality, it is patchy geographically and temporally, and this makes it unreasonable, from a methodological point of view, to treat the samples of different lineages as if they were independent. Although the patterns may be a true reflection of evolutionary events, the possibility of taphonomical bias cannot be excluded. Although this might allow the exploration of robust and general patterns, detailed interactions may be nebulous.

We can perhaps conclude that there is no simple synchronous community evolution observable in the fossil record. Some signals indicating a network of Red Queen interactions can be observed, but these should not be overinterpreted. In particular, the logic of the Red Queen hypothesis is that these effects should be local and therefore that considering them at a continental or subcontinental scale is likely to introduce noise into the system. Although it would be premature to reject the Red Queen hypothesis (and, indeed, there are good reasons to think that it is correct), it will be necessary to find other means of testing it, the most promising of which would be within lineage rates of change and consist of detailed studies of particularly lineages such as pigs (see Bishop, this volume), carnivores (see Turner, this volume), cercopithecoids (see Benefit, this volume) or horses (see Bernor, this volume).

Ecological Rules: Latitude and Species Richness

As with the analyses in relation to climate change, there are reasons that when we consider overall species richness rather than issues of evolutionary dynamism we might expect a slightly different pattern. A large number of studies have shown that ecological factors underlie the patterns of biodiversity at a continental scale. The next level at which we might approach the problem of early hominid evolutionary patterns is to ask whether they conform to any general ecological rules.

One of the best documented relationships is between species diversity and latitude (for a general discussion, see Rosenzweig, 1995, this volume). Species richness declines with distance from the equator. Explanations for this phenomena have included energy availability, climatic stability, land surface area, and temperature. It has also been shown that species richness covaries inversely with species area (Stevens, 1989; Smith et al., 1994). In other words, close to the equator we can expect more species with relatively small ranges, whereas in high latitudes we can expect a few geographically extensive species. This relationship has been substantiated for African catarrhine primates (H. Eeley & R. A. Foley manuscript).

One simple hypothesis is that fossil patterns should reflect these basic biogeographical rules. Fossil data are geographically very limited, but it is possible to explore contrasts between equatorial eastern Africa and subtropical southern Africa. Table 23-4 presents some data on primate and hominid species numbers for eastern and southern Africa, and table 23-5 summarizes some of the more interesting elements. Although the trends remain the same for contemporary and paleodata, the contrast has not remained constant. In the past the differences between eastern and southern Africa have not been as marked. Furthermore, taking the fossil data at face value, there have been times when southern Africa has been more species rich than eastern Africa. We can also note that peak diversity is reached at different times in the two subcontinental regions.

Despite its taphonomical limitations, such data do

Table 23-3. Data on African Plio-Pleistocene climate and large mammal evolution.

	Time (m.y.)									
Climatic variables	0–0.5	0.5–1.0	1.0–1.5	1.5–2.0	2.0–2.5	2.5–3.0	3.0–3.5	3.5–4.0	4.0–4.5	4.5–5.0
Number of climatic cycles	6	7	8	9	8	11	7	8	4	5
Max. δ-18	4.40	4.50	4.60	3.70	4.70	3.10	3.30	3.00	2.65	2.95
Min. δ-18	2.40	2.70	2.20	2.80	2.10	2.40	2.00	1.70	1.90	2.30
Modal δ-18	3.40	3.50	3.25	3.25	3.00	2.80	2.40	2.30	2.40	2.50
Climatic variation	2.00	1.80	2.40	.90	2.60	.70	1.30	1.30	.75	.65
Magnitude of climatic change	.10	.25	.00	.25	.20	.40	.10	.10	.10	.10
Hominid species										
Hominid first appearances	3	0	1	3	3	1	0	2	2	0
Hominid last appearances	2	2	1	3	2	1	0	1	1	0
Number of hominid species	3	3	3	3	5	3	2	3	2	1
Papionine data										
Number of *Papio* spp.	6	1	4	6	4	8	2	4	0	2
Papio first appearances	1	0	1	3	1	2	0	0	0	1
Papio last appearances	0	3	2	1	3	0	0	0	0	0
Number of *Theropithecus* spp.	1	1	1	2	2	3	1	2	0	1
Theropithecus first appearances	0	0	0	0	1	0	1	2	0	0
Theropithecus last appearances	2	1	0	1	1	0	2	0	0	0
Papionine first appearances	1	0	1	3	2	2	1	2	0	1
Papionine last appearances	2	4	2	2	4	0	2	0	0	0
Number of papionine species	7	2	5	8	6	11	3	6	0	3
Other mammals										
Bovid first appearances (%)	75	5	40	55	100	50	15	50	50	50
South African carnivore first appearances	0	1	0	2	0	5	10	0		
East African carnivore first appearances	0	2	0	0	2	0	2	6		
South African carnivore last appearances	0	0	3	1	1	2	0	0		
East African carnivore last appearances	0	3	4	0	1	0	0	0		
South African suid first appearances	2	0	0	2	0	0	2	0		
East African suid first appearances	0	2	0	1	3	1	4	0		
South African suid last appearances	0	5	1	1	1	2	0	0		
East African suid last appearances	0	3	0	0	0	1	0	0		
Total East African diversity	20	36	44	49	51	50	50	24		
Total South African diversity	20	20	37	37	30	34	28	4		

See Foley (1994) for sources of data, modified here to take into account recent hominid discoveries (T. D. White et al., 1994; M. G. Leakey et al., 1995).

Table 23-4. Number of primate and hominid species in eastern and southern Africa, 4–0.5 m.y.

	Time (m.y.)							
Number of species	4	3.5	3	2.5	2	1.5	1	0.5
Primates								
South	0	0	6	7	4	6	2	1
East	1	1	9	6	9	5	6	1
Hominids								
South	1	1	1	1	1	3	2	1
East	2	3	1	2	2	4	2	1

Data from Turner and Wood (1993a,b).

serve to remind us that hominid evolution will be underlain by rules that apply more broadly. The variation in primate, including hominid, species diversity is generally consistent with a general rule derived from extant species, and certainly the pattern observed is broadly consistent. Furthermore, it has been shown that species range varies with latitude as well, such that species of primates extant today in equatorial regions have smaller geographical ranges than those in higher latitudes (Eeley, 1994). This reflects a generally observed biogeographical pattern (fig. 23-2). It can be observed that the current early hominid localities are spread across a marked latitudinal gradient. Although it is clear that range size varies with grade and adaptation (Rosenzweig, 1995; Foley, 1991), nonetheless it would be expected that the southern African hominids would have greater range area relative to those in equatorial regions, and the endemism of the southern African australopithecines may be consistent with this general rule. The most important conclusion is that when the temporal dimension not normally available to biogeographers is considered, the observable pattern is far more fluid. Evolutionary dynamism and changing environments together are likely to influence strongly the situation at any particular point in time.

It is clear from the absence of any simple climatic determinants, the complexity of the community structure and interactions, and the variability in broad biogeographical patterns that occurs over time that an alternative approach is necessary to understand the pattern of evolutionary change among hominids. This can be done by returning to the general theoretical points made at the outset—that evolutionary change starts as local shifts in species' distributions and that evolutionary events such as speciation and extinction are the outcome of patterns of dispersal and contraction.

Evolutionary Geography

The Geographical Basis for Evolutionary Processes: Dispersal, Contraction and Isolation

The limited explanatory value of global or continental-level contexts should not be interpreted as indicating that the large scale is insignificant or that there are no general underlying rules. Large-scale patterns are mediated through local events, and the specific geographical, phylogenetic, and ecological context are important. The key general principles are thus those derived from ecology, operating over short time

Table 23-5. Patterns of primate species richness in eastern and southern Africa between 4 and 0.5 m.y.

	Eastern Africa	Southern Africa
Current average primate diversity	10	2
Peak Paleo-primate diversity	9	7
Minimum Paleo-primate diversity	1	0
Time of peak primate diversity	3 m.y., 2 m.y.	2.5 m.y.
Time of minimum Paleo-diversity	4 m.y., 3.5 m.y.	4 m.y., 3.5 m.y.
Peak hominid diversity	2 m.y.	2 m.y.

Data on fossils from Turner and Wood (1993a,b). Modern data from Eeley and Foley (manuscript).

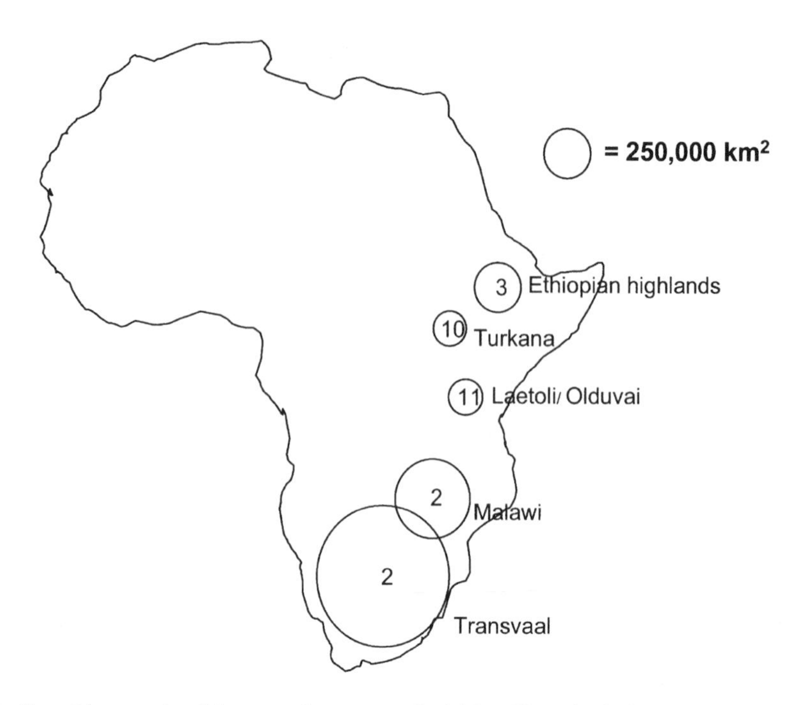

Figure 23-2. Catarrhine species richness and range area in Africa. The principal early hominid localities are marked on the map. The size of the circle shown indicates the average catarrhine species range for that latitude among extant groups. The number inside the circle shows the average species richness for that latitude among extant catarrhines. It can be seen that species richness and range area are inversely related to each other and that equatorial zones should have more species with smaller ranges, whereas southern and northern Africa will have fewer species ranged over larger areas. Extant data from Eeley and Foley (manuscript).

scales but repetitively built up over geological time. This is essentially a view that evolution is the product of classic microevolutionary mechanisms and that macroevolutionary patterns, which undoubtedly do occur and can be observed, should be explained in these terms.

As indicated at the beginning of this paper, considering evolutionary processes as changes in distribution is one option that may take us closer to the underlying processes that shaped early hominid evolutionary patterns. Evolutionary changes are the outcome of the demographic events, with populations expanding where reproduction exceeds mortality, remaining stable where they are in balance, or declining where mortality exceeds reproductive rate. Such patterns of reproduction feed directly into ecological and spatial patterns. Where populations expand, they will either increase locally in density or else disperse into new regions. Conversely, there will be a strong probability that declines in population will be associated with reduced or more fragmented geographical ranges. As such, demographic patterns are also the basis for selection, and macroevolutionary patterns will emerge from microevolutionary processes. The key elements are:

Dispersal: as populations increase (increased reproductive rate, decreased mortality, etc.), surplus population will disperse into appropriate adjacent habitats until either physical barriers occur or the effect of dispersal is to decrease reproductive output. Species range will increase through dispersal.

Contraction: populations in marginal habitats will become locally extinct when they fail to reproduce at replacement rates and are not replaced by other dispersing populations. Contraction of species range will occur.

Isolation: populations left isolated by either species range contraction, vicariance events, or changes in habitat distribution will be subject to particular evolutionary processes associated with small populations (local adaptation, founder effect and genetic drift, accelerated rates of genetic change, speciation, and extinction).

All evolutionary change is the result of these events. When a new species arises, it does so either through dispersal from a small core area and reduced gene flow over the entire range, or through isolation following contraction or vicariance events that occur through environmental change. When a species becomes extinct, it does so through a process of a failure of reproductive strategies to match mortality, leading to population reduction and range contraction (fig. 23-3; Tchernov, 1992). Such a distributional model can apply equally to the early hominids, and treating the genus *Australopithecus* (sensu lato) or *Homo* as such an evolving lineage can show that there is a potential relationship between geographical patterns and phylogeny.

A Microevolutionary Theory of Macroevolution

If the central point of the previous section is that climatic inputs are likely to be stronger in relation to the basic geographical distributions of populations, then two questions follow: Where does speciation fit into the broad sequence of causality? What is the role of biotic factors, and in particular, competition (including the competitive conditions that may produce more symbiotic, cooperative, and mutualistic associations) in producing evolutionary change?

In answer to the first question, the empirical evidence provided above suggests that the link with speciation is weak; it is likely to occur only stochastically as a result of changes in distribution, probably some time removed from any climate change. Speciation will depend on whether changes in distribution lead to isolation and hence the formation of genetic barriers. This will not be a function of the intensity of the climate change but of local effects and conditions. Speciation is thus the weakest link in the causal chain of evolutionary explanation.

Extinction, on the other hand, can be expected to be more tightly tied in. Limited habitats and reduced ranges will decrease population sizes, and in turn increase the probability of extinction. It will be expected that major climate changes leading to population and range contractions will in turn result in higher rates of extinction. There is some weak evidence that this is the case. This observation provides the basis for one answer to the second question. The extinction of populations and species from specific communities will change the dynamics of competition and equally will provide new and different initial conditions should the climate reverse and old habitats re-establish themselves. In this sense extinction, both local and general, will be a significant motor for evolutionary change. Extinction provides a key link between the general effects of external environmental change and the local community interactions, or, in other words, the link between macroevolutionary patterns and microevolutionary processes.

By placing particular evolutionary situations into this broad classification of geographical events and a microevolutionary basis for macroevolution, or the movement to novel competitive conditions and ecological equilibria, it will be possible to disentangle some general rules in what might otherwise be the conflation of several processes. Global climate changes are one cause of change, but they are neither necessary nor sufficient to explain the overall pattern. If the Turnover Pulse Hypothesis is a model for evolutionary change driven entirely by external factors and the Red Queen is one that demands an endless selective race, then perhaps we need a model that integrates the two (a less demanding Pink Queen?)—a model that recognizes that competition is the basic mechanism, but that it operates in a natural world in which there are both contingent local factors and global parameters. Such a model would not predict evolutionary change confined to particular periods, but would predict that different events are spread across broader periods. It would not predict constant and continuous change, as both local and global factors will accelerate rates of change at certain times, nor would it predict simply random change—patterns should be recognizable in the details of the strengths of relationships between different types of evolutionary event. In particular, geographical variation will be central to unraveling these relationships and thus central to evolutionary causality.

The microevolutionary framework set out here attempts to show how local demographic processes underly the macroevolutionary patterns and to show that these operate to induce extinction in populations as a basis for other types of evolutionary change. This conjunction of the small scale and the large scale in evolution brings into sharp focus the problems of integrating the short-term ecological with the longer term evolutionary (see Grubb, this volume; Owen Smith, this volume). The demographic processes discussed here can operate on time scales of 10^4 years or less, and it is unlikely that these processes will persist. A population declining for a thousand years will become extinct, and a population expanding for ten thousand years will swamp the world. These are not time scales that are visible under most conditions seen in the fossil record. It is possible, therefore, that many of the extinction processes discussed here will occur rapidly and will appear virtually instantaneous in the fossil record. The same is likely to be the case with major dispersals. The result will be that the microevolution-

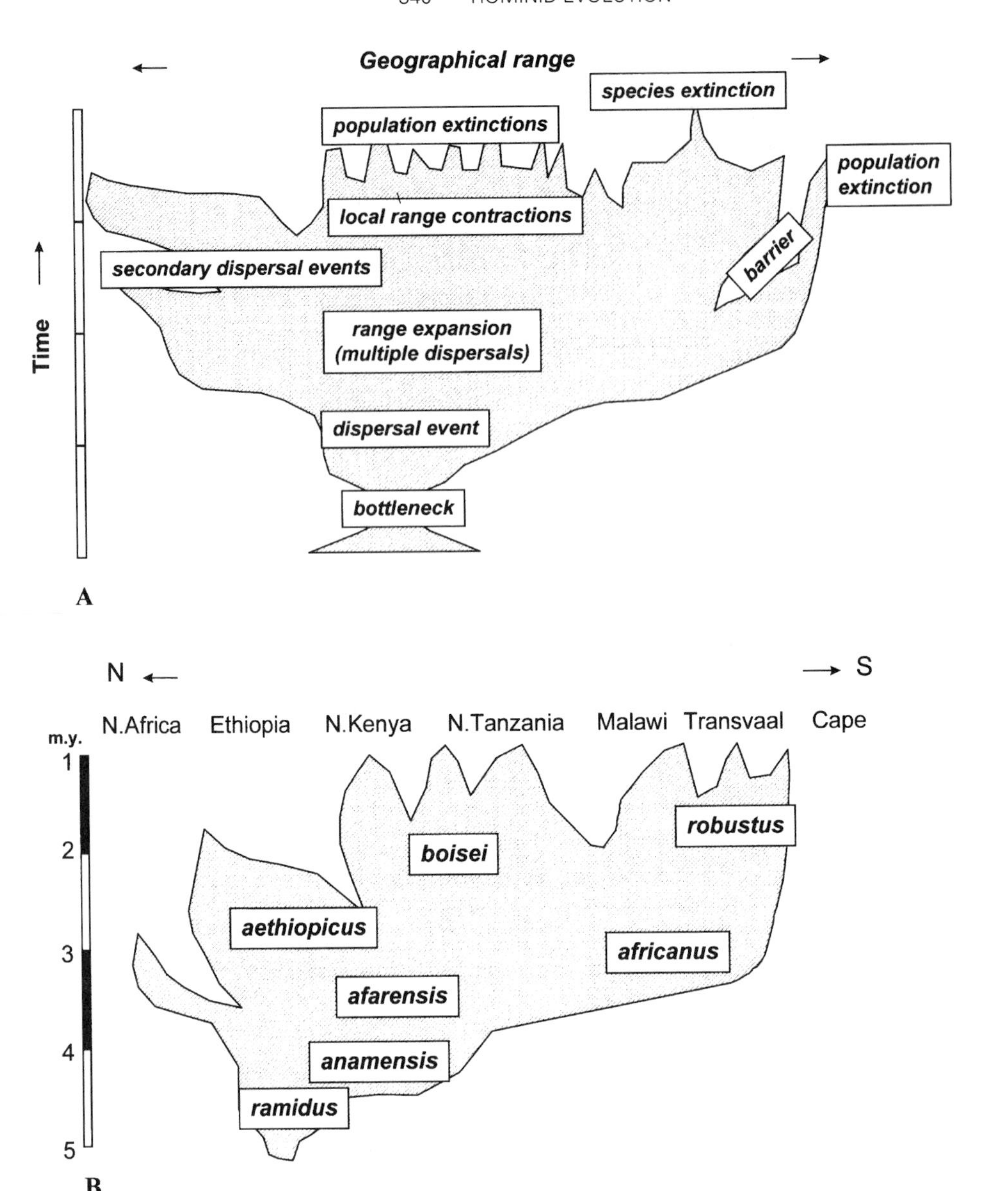

Figure 23-3. (A) A generalized model of the evolutionary geography of a species (modified from Tchernov, 1992); (B) applied to the evolution of the australopithecines (incorporating *Ardipithecus* and *Paranthropus*); and (C) applied to *Homo*.

ary processes underlying ecological change (and evolutionary change as well) will often lead to rapid changes in ecosystems and a punctuated appearance to evolutionary change. The final section of this chapter therefore, outlines the major geographical patterns of African early hominid evolution and evaluates them in the light of the proposed classification.

Early Hominid Evolutionary Geography

African Biogeographical Patterns

The biogeography of Africa has been extensively described and discussed (see Kingdon, 1989; Grubb, this

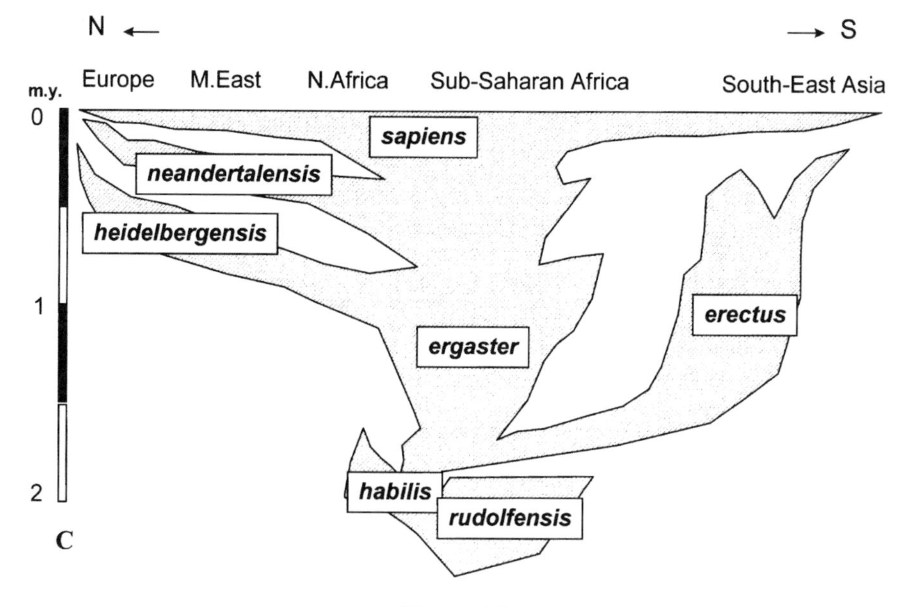

Figure 23-3. *(continued)*

volume). Paleontological approaches are now revealing patterns of long-term development (see Maglio & Cooke, 1978; Andrews et al., this volume). In setting the context for considering early hominid evolutionary geography, two factors are central: the development of the Rift Valley and the climate changes of the late Neogene.

The evolutionary consequences of the development of a major rifting system across the eastern side of Africa are massive. Among the key consequences are increased altitude and resulting habitats due to rifting and faulting; altered habitats produced by rich volcanic soils (see Owen Smith, this volume); and highly variable and unstable topography, with major lake basins, geographical barriers, and migratory corridors. The end results seen in today's environments are the rich and variable savannas and woodland and forest mosaics, cut by mountains, lakes, and volcanic debris into a series of relatively clearly demarcated ecosystems (see Feibel, this volume). To this general picture should be added the temporal element: this system did not appear instantaneously, but developed over the late Neogene, primarily from the north (the Ethiopian Highlands) in its early phases to the south (the Okavambo swamp) in its latest one. Uplift such as occurred in the Rift Valley cut sub-Saharan Africa into a low center, and a high eastern/southern portion.

While the Rift Valley developed, climate changes occurred (see Vrba, 1996; Denton, this volume). The last 6 m.y. have seen an overall decline in global temperatures. The broad effect on African environments was increased aridity and seasonality with lower temperatures, especially along the eastern side of the continent. The major trajectory of environmental change is therefore toward savanna regions in the eastern and southern parts, with forests in the center. This general picture should be qualified by pointing out that within the broad trends there are many reversals as well as periods of accelerated and more extreme change.

When these two factors are combined to provide the geographical context for hominid evolution, a useful model can be developed (fig. 23-4). When conditions are relatively wet and warm, there will be a greater distribution of forest, essentially expanding from a more westerly point. This will have the effect of both reducing savanna regions and also occasionally cutting the eastern biome into equatorial and southern parts. When conditions are drier, forests are reduced, the savanna corridors would be more continuous, and forests fragmented east to west. For presumed savanna species such as hominids, dispersals would be associated with the presence of the savanna corridor along the Rift Valley and contractions and isolation with the wetter periods. Figure 23-4 shows that up to 3.5 m.y.

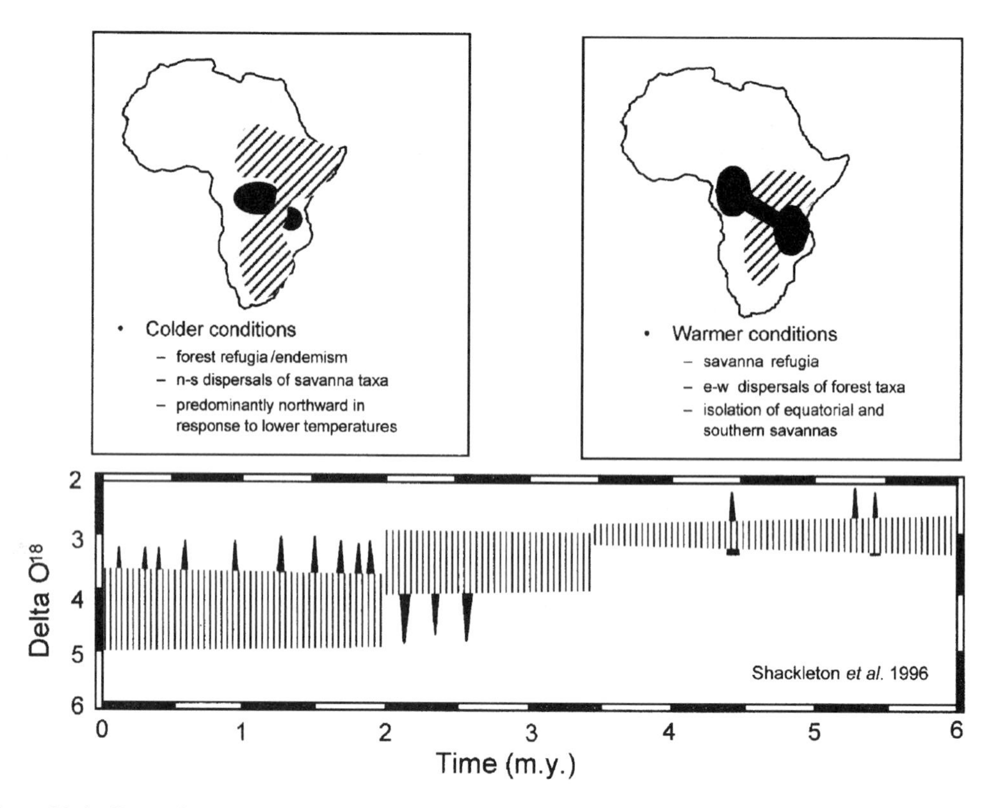

Figure 23-4. Generalized biogeographical pattern of Africa in relation to the pattern of paleoclimatic change. Under warmer conditions, such as those that occurred in the later Miocene and during Pleistocene interglacials, forests expanded along an east–west gradient, cutting the more open biomes of High (eastern) Africa into more isolated zones. During colder conditions the more widespread savanna zones tended to restrict the forests and divide them into eastern and western/central zones. The deep-sea oxygen isotope record shows that the later Miocene and early Pliocene (6–3.5 m.y.) was relatively warm, with a few especially warm spikes occurring. There was a major shift to colder climates at 3.5 m.y., resulting in more seasonal and arid environments.

the climate was wetter than later, and major fluctuations, when they occurred, were toward even warmer and moister conditions (see fig. 8-2 in Denton, this volume; Shackleton, 1996). The period 3.5–2.0 m.y. saw a shift toward relatively greater aridity, and at the end of this period there was a series of major perturbations in the direction of yet cooler and drier conditions. After 2.0 m.y. these extreme oscillations became the norm, with the establishment of an essentially dry climate. During the course of the Pleistocene the major perturbations were short periods that were much warmer and wetter. The spikes shown on figure 23-4 indicate when, in relative terms, dispersals and contraction/isolation for more savanna-prone species are likely to occur. In particular, the savannas would have been at their most restricted during short phases around 5.5 and 4.5 m.y.; during the Pleistocene, break up of the savannas probably occurred repetitively, possibly as many as 10 times, or approximately every 100,000 yr (see Sikes, this volume). Apart from the overall trend toward savannas, the period between about 2.8 and 2.0 m.y. saw at least three periods of, for the Pliocene, extreme dry spikes.

The whole of the later Neogene shows a complex and continuous pattern of climate change, with major long-term shifts occurring at 3.5 and 2.0 m.y., but with numerous shorter oscillations of even greater magnitude, and in the direction of both greater aridity and greater moisture. According to the general model of evolutionary change shown here, hominids and other mammals would have responded demographically and spatially to these changes. Extinctions would also occur probabilistically. Speciation, however, would be buffered from these direct effects and would show only a weak signal. However, the overall pattern of hominid diversity and phylogeny ought to be consistent with this paleoenvironmental context, a context not of pulses and turnovers, but more repeated and continuous change

Community Evolution among Plio-Pleistocene Mammals

Figure 23-5 shows maps of inferred hominid distribution for various time periods between 5.0 m.y. and 1.0 m.y. The phylogenetic relationships implied by these distributions are shown next to each map. These are a modified version of those presented in an earlier work (Foley, 1987:253). The description of the sequence of events is provided in the caption to figure 23-5. The central point they express is that phylogenetic reconstructions are two-dimensional representations of what is essentially a geographical process. In considering this reconstruction, two notes of caution should be made. First, we are clearly dependent on the two areas that have yielded fossils—the Eastern Rift Valley and the Transvaal. The recent work in Malawi shows that inferred geographical distributions of hominids can be significantly changed by opening up new research areas (see Bromage et al., 1995a,b; Grubb et al., this volume). Assumptions about other regions must be speculative. I am assuming here a continuity between East and South Africa with no significant penetration into central or West Africa. Parts of the Sahel were probably northward extensions of the eastern African savannas, and may well have been inhabited by hominids (see Brunet et al., 1995).

Second, is there is no consensus phylogeny, and a number of alternatives may be equally supported by the evidence. The one employed here is put as one among many possibilities for the purpose of exploring evolutionary hypotheses, and others should be tested. Indeed, it could be argued that exploring the geographical implications of a phylogeny is a good way of assessing its plausibility.

Evolutionary Events in Early Hominid Evolution

The history of all taxa consists of the same three events—isolation, dispersal, contraction. These events may be repeated, and their significance will vary between taxa. Some species may always be small, local populations. Some may be geographically extensive, but, because they are a product of vicariance, dispersal has preceded isolation (and therefore occurred before true speciation). The evolutionary pattern of the early hominids is thus the cumulative effect of these processes repeated and overlain over several million years. Phylogeny is also the outcome of them. Explanation of the patterns of early hominid evolution should therefore consist of relating these events to the context. By recognizing three types of event, we can raise the possibility identifying the conditions that yield different outcomes. Isolation, for example, is almost by definition an event that will occur locally and may be subject to more stochastic processes, which may in turn explain why we can see little by way of a direct relationship with climate change. Dispersal, on the other hand, is an event that will occur over a larger geographical area (or at least it is these that are observable in the fossil record), and thus may be more easily linked to the continental or regional scale of data that we have in the fossil and paleoenvironmental record. Dispersal, we can perhaps expect, will be more closely related to climate change and to the related changes in the distribution of other species and their competitive interactions. Contraction may also occur in the same way, although it is likely that refuge populations may survive for considerable periods of time after major reductions in overall species range have occurred. Ultimate extinction may be less linked to climatic changes or interspecific dynamics than to range contraction into refugia (fig. 23-6).

Looking at the three types of evolutionary event (table 23-6), speciations (the product of geographical isolation) occur in the earlier periods probably around 4.5 m.y., and then are essentially continuously between 3.0 and 1.5 m.y. Extinctions, or at least range contractions, of the early bipeds occur between 4.0 and 2.5 m.y., and then again from around 2.0 m.y. Dispersals occur early (5.0–3.0 m.y.) and late (after 2.0 m.y., or more probably somewhat earlier).

Although at a more restricted scale than the initial analyses of the relationship with climate, the geographical scale here is still relatively large and in some cases unspecified. It is unclear, for example, whether isolation of a population refers to a demographic scale of fewer than 500 individuals, fewer than 5000, or indeed even fewer than 50,000. The difference would become significant should we want to look at the mechanisms more precisely. The same would apply to dispersals and contractions. Furthermore, the geographical scale treated here has been simply in terms of eastern and southern Africa, but this is clearly too coarse grained and places insufficient attention to habitat distributions, relief and drainage patterns, and moisture regimes.

Nonetheless, looking at the hominids in this geographical perspective a number of insights can be gained. In particular, the geography of evolution shows that there are benefits to be derived from distinguishing between different types of evolutionary event and

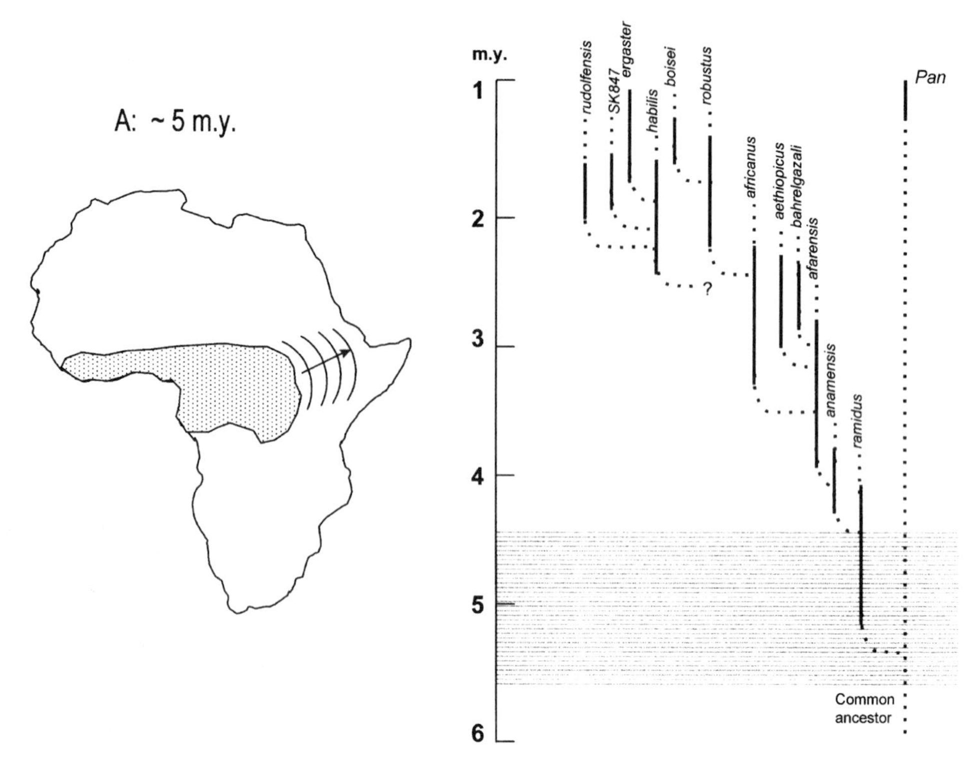

Figure 23-5. The evolutionary geography of the Pliocene and early African hominids. Each map shows schematically the distribution of the African hominid taxa for successive time periods and alongside each map is a phylogeny of the evolutionary relationships expressed by the geographical patterning. The dark shaded area indicates the current distribution of chimpanzees. The distribution of the various hominids are shown by radiating lines or dots. Following is a gloss on the maps. In reading these it should be stressed that this is interpreting the fossil record as if it is a complete record, whereas in reality it is likely that what is recorded is a minimal number of events. (A) Around 5 m.y. the ancestors of later hominids diverged from the ancestors of *Pan.* At some stage this probably involved an eastward dispersal of apes into the more seasonal and open environments, probably with some greater level of bipedalism than occurs in chimpanzees. *A. ramidus* is the current best contender for at least one of the taxa involved in this dispersal. (B) Between 4.5 and 3.5 m.y. these bipedal apes in eastern Africa underwent range expansion. The ultimate evolutionary consequence of this was the southern African clade represented by *A.africanus.* More local within eastern Africa dispersals probably also occurred, as did isolation leading to the occurrence in eastern Africa of several species—at least *A. ramidus, A. anamensis,* and *A. afarensis.* (C) The period 3.5 to >2.0 m.y. must have seen major range contractions and ultimately extinctions of the more primitive hominids (represented by isolated dots of populations). Within the overall range there were also new evolutionary events—in particular a probable trend toward megadonty in both eastern and southern Africa. One can speculate that *Homo,* which was probably around by the later part of this period, was a much more isolated and insignificant trend. (D) By 2.0 m.y. the fossil record suggests that all the primitive hominids were extinct, although it is possible that small refuge populations continued to survive. The primary events of this period appear to be the range expansion southward of at least one *Homo,* and the expansion of robust australopithecines. This is represented here as occurring from south to north, but alternative interpretations are possible. (E) After 1.5 m.y. there was a range contraction of the robust australopithecines into refuge areas, with their extinction in place by 1.0 m.y. Real or pseudo extinctions of *Homo* populations would have occurred in this period, as would the dispersal out of its core area (and ultimately out of Africa) of *H. ergaster.*

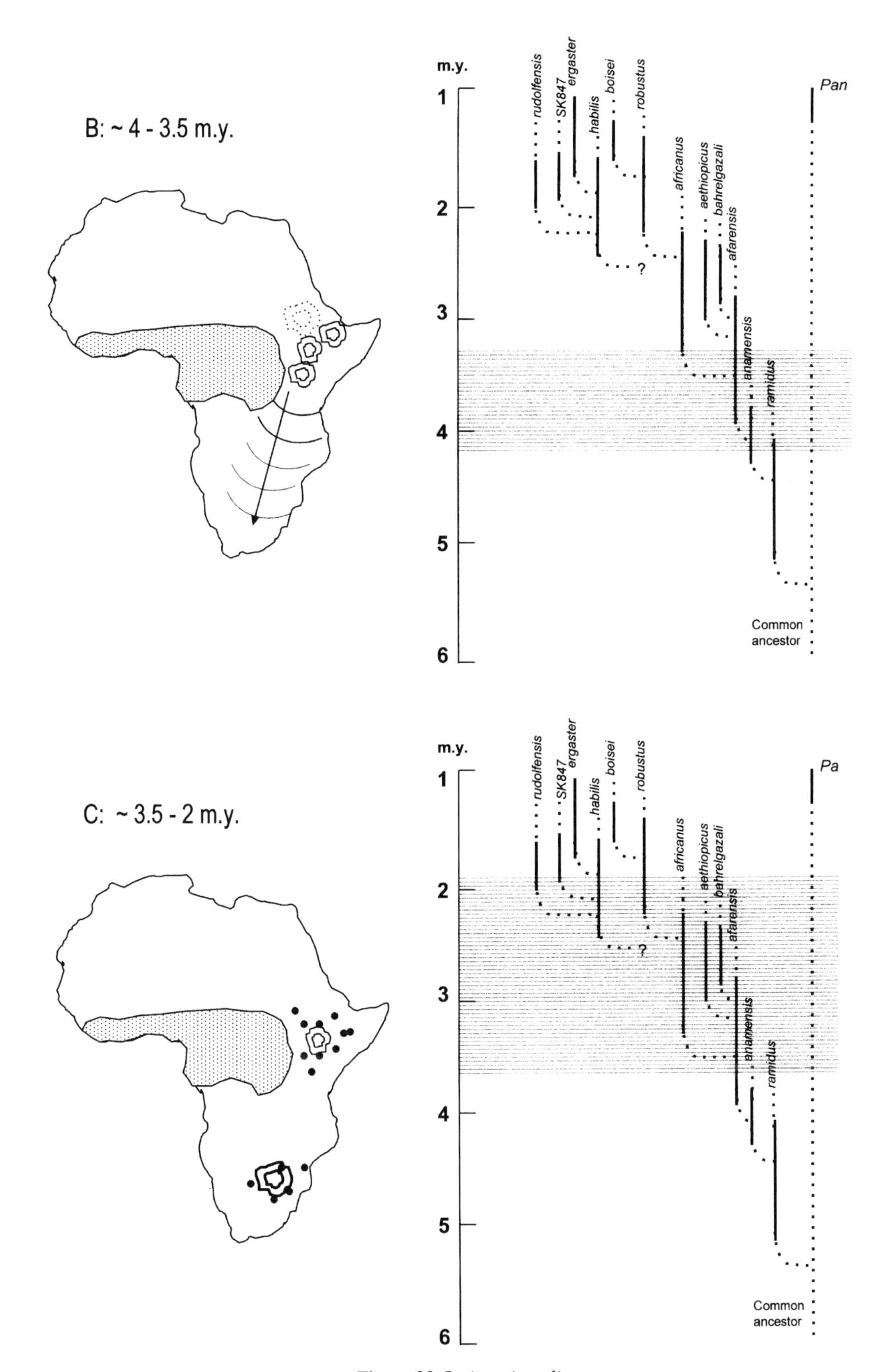

Figure 23-5. (*continued*)

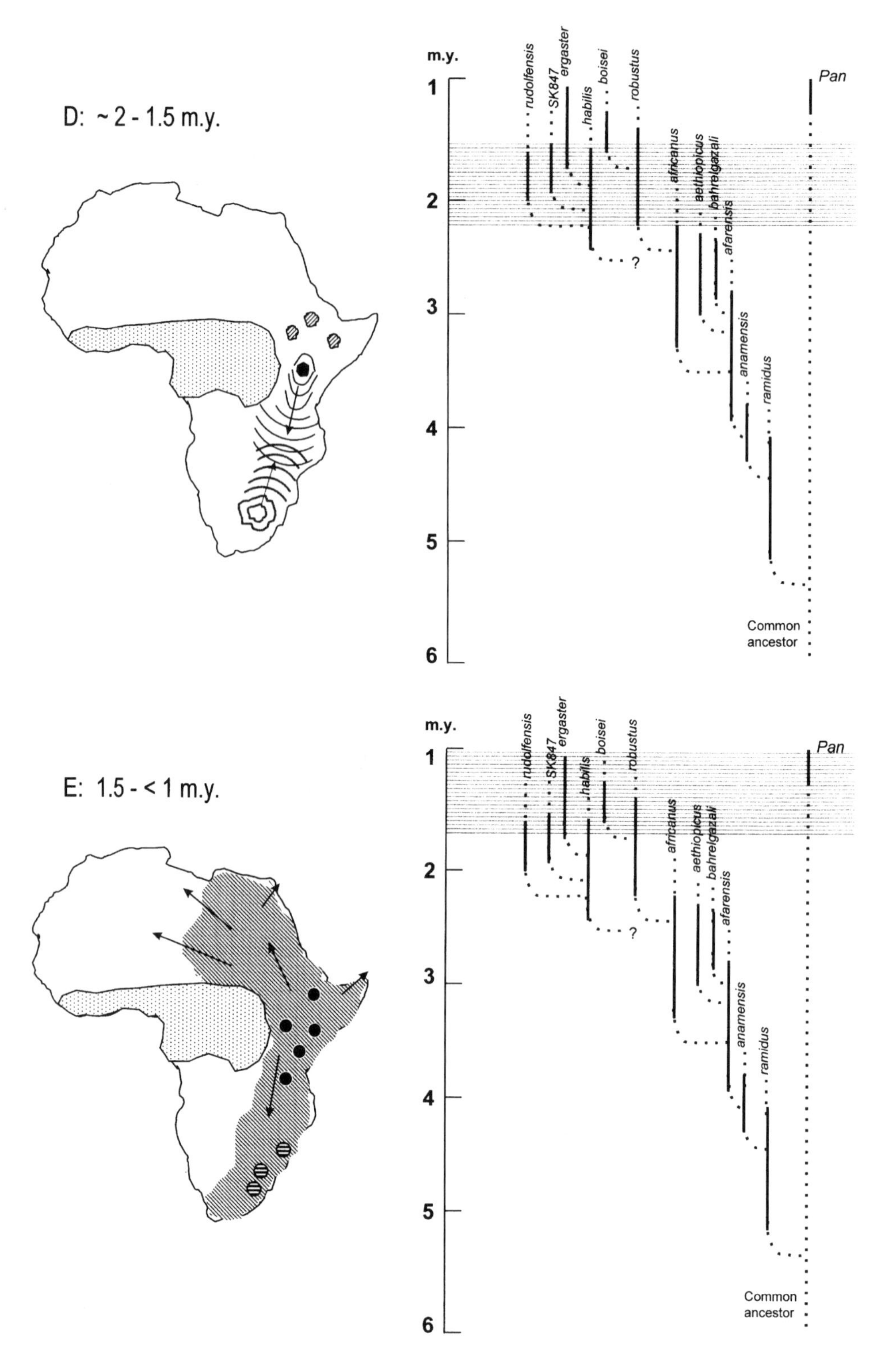

Figure 23-5. (*continued*)

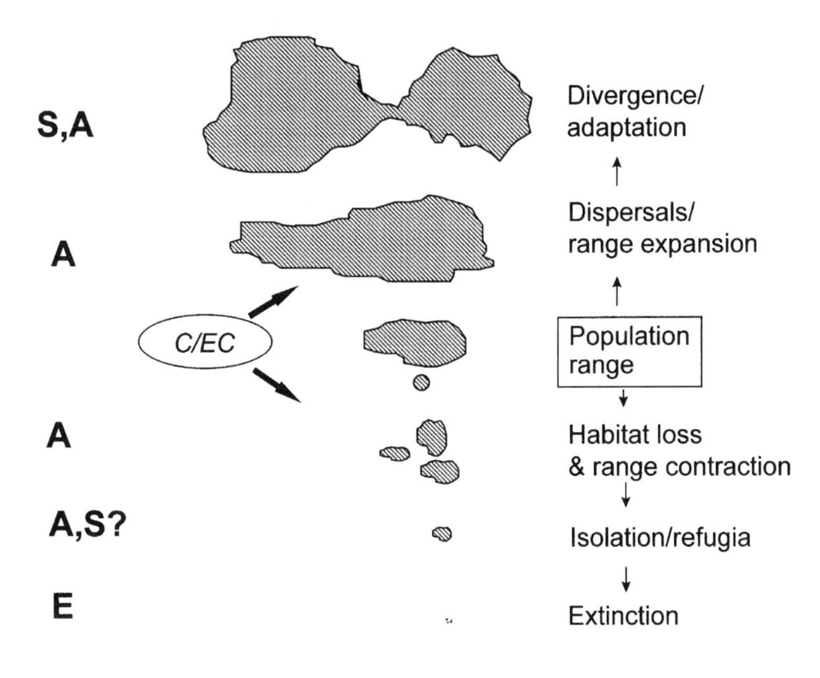

Figure 23-6. The geography of evolutionary change. Species exist in terms of a spatial distribution, which may either expand or contract. Both may lead to speciation through cladogenesis (S), resulting from isolation or reduced gene flow. Environmental or climatic change (C/EC), however, is more likely to be associated with the actual changes in distribution or anagenesis (A) rather than the end products of speciation or extinction (E).

from distancing them from macroevolutionary concepts. Although speciation and extinction are the evolutionary endpoints, by looking at the proximate demographic patterns we are in a better position to explain evolutionary patterns. Geography and ecology can take us closer to the actual dynamics of evolution.

More specifically, the analyses presented show two things. The first is that there can be a concordance between geographical patterns and phylogenetic ones. This should remind us that as well as locating species in time, we should equally locate them firmly in space. Although different interpretations from the one presented here are possible, the conceptual framework proposed may remain useful. Second, evolutionary events are spread throughout the period considered here (5–1 m.y.). By discriminating between different events occurring at different times and in different places, we are in a better position to relate the patterns to their environmental context. And third, Pliocene hominid evolution is clearly made up of multiple events (of which we know only a few). It is the totality of these events that must be used to test hypotheses rather than just selected ones. In particular, the split between australopithecines and other hominines does not, in the work presented here, take on any great significance in the overall evolutionary patterns. The significance in the genus *Homo* lies in what happened after, but we should be wary of backwardly imposing significance to the original split. In particular, it may be that what distinguishes *Homo* from other hominids lies in the extent of its dispersal potential, which in turn will relate to its social structure, its foraging behavior, reproductive rate, and, as Vrba (this volume) has argued, its habitat tolerance.

Summary

The central assumptions of this paper are that Pliocene hominid evolution is sufficiently speciose for the patterns to be explored using species as the unit of analysis and that evolutionary patterns should be interpreted geographically as much as chronologically or phylogenetically. It was also asserted that, although evolutionary explanations of species patterns can lie along a continuum from the macro (global/continental) to the micro (local population effects), geographical patterns provide a means of accounting for macroevolutionary patterns in terms of microevolutionary mechanisms. The following conclusions can be drawn:

1. Pliocene hominid evolution was examined using available data at these various scales. It was found that the most global scale (climate change) could not account for the observed pattern of the appearance of new species and novelty in hominid evolution.
2. Climate change could be shown to have an effect on extinctions and patterns of dispersals. It

was argued that this was consistent with expectations that, although speciation was the product of a number of mechanisms, there was likely to be a more direct relationship between diversity and environments and extinction and changing habitats.

3. Theoretically, it was argued that interspecific competition (sensu lato) would impinge more directly on evolutionary patterns than climate and that the logic of the Red Queen hypothesis was such that evolutionary change would not be synchronous across an ecological community, and nor would it be confined to particular periods.
4. The available data on community level patterns supported these propositions, but the extent to which detailed statistical relationships could be explained was questionable.
5. The biogeographic patterns of species richness in African primates today provide another level of explanation, although the fossil data indicate that the precise pattern has shifted through time.
6. The implication of a Red Queen hypothesis of interspecific interactive effects led to the proposal that evolutionary patterns should be considered more locally and with greater emphasis on geographical patterns. Community-level interactions are not necessarily arms races but can vary in intensity according to local conditions. In particular, when species become extinct, which may occur in relation to climate change, competitive conditions will change and may lead to evolutionary shifts.
7. Three categories of evolutionary geographical event were identified (dispersal, isolation, and contraction), and maps of hominid phylogenetic geography were used to uncover the pattern for hominid evolution. The evidence can be tentatively interpreted to suggest that Pliocene hominid phylogeny is strongly geographical and that isolation and dispersal events do not occur at the same time. Contraction of hominid populations was probably widespread, leading to periodic endemism, refugia formation, and local and continental extinction. Such events occur throughout the Pliocene and Early Pleistocene.

In considering these conclusions, two final points can be emphasized. The first is that perhaps the most important development in paleoanthropology in recent years has been the greater emphasis placed on interpreting hominid patterns in the context of broader evolutionary theory. Ideas such as the climatic forcing model or the Red Queen hypothesis are useful because they apply more generally and link hominids to the broader evolutionary context. Differences between the various hypotheses are, in this more historical context, less significant than what they have in common. The second is that this chapter has argued for the importance of viewing macroevolutionary patterns as the product of geographically local ecological and demographic conditions. This is where the working mechanisms of differential reproductive success must lie. The empirical elements of this paper have moved some way toward locating explanations of Pliocene hominid evolution in the more specific (in this case, the geographical elements). However, there is still a considerable way to go, particularly in terms of taking precise habitat distributions into account, reconstructing biogeographic regimes and, most of all, taking into account the adaptive traits of the hominids themselves.

24

Early Hominid Dental Development and Climate Change

Fernando Ramirez Rozzi
Christopher Walker
Timothy G. Bromage

Plio-Pleistocene global climate change resulted in terrestrial biotic shifts toward more open environments at low latitudes (Coppens, 1989; Vrba, 1992). Such environmental change has been proposed to explain the origins of *Homo* and robust australopithecines around 2.4 m.y. (Coppens, 1975b, 1989; Vrba, 1985c, 1988, 1995b). Little is known, however, about the relationship between environmental changes and specific anatomical modifications in Hominidae. As Rightmire (1993:45) has stated, "It will be a challenge [to paleoanthropology] to measure the impact of climatic change on the evolution of this family."

The scenario typically advanced suggests that the spread of more open environments during the Pliocene engendered more fibrous and abrasive vegetative foods (e.g., Bromage & Schrenk, 1995). Mammals adapted to this new condition by developing either more hypsodont teeth or teeth with more complex occlusal faces (Turner & Wood, 1993a). Therefore, environmental change may exercise selective pressure on dental development and tooth structure, the morphologies of which can be used as indicators of habitat specificity.

Fortunately, patterns of dental development are naturally "fossilized" in enamel, and any change in the process of enamel formation is observable as a modification in one or more microanatomical characteristics. Analyses of enamel microstructure enable us to reconstruct changes in dental development through time. It is possible, if not expected, therefore, that, as for other Pliocene mammal groups, dental development change observed over the course of early hominid evolution may be related to environmental change.

The three microanatomical structures relevant to this investigation are the cross-striation, striae of Retzius and the Hunter-Schreger band. Their characteristics enable us to define how teeth were formed (fig. 24-1) (for description of these structures, see Boyde, 1965, 1976; Rose, 1979; Gohdo, 1982; Shellis, 1984; Risnes, 1985, 1986; Boyde & Fortelius, 1986). Cross-striations are incremental growth markers produced by a circadian rhythm (Bromage, 1991), the striae of Retzius is an increment reflecting a circaseptan rhythm ranging from 6 to 12 days between individuals (Schour & Poncher, 1937; Massler & Schour, 1946; Fitzgerald, 1998; see Kimura, 1978; Dean, 1987, 1989; Ramirez Rozzi, 1992, for a large bibliography). Hunter Schreger bands are microanatomical indicators of the trajectory of enamel development through the thickness of enamel. Measures of cross-striations, striae of Retzius, and Hunter Schreger bands represent both the rate and pattern of crown formation processes. Therefore, analyses of enamel microstructure reveal modifications in dental development over evolutionary time frames that may reflect adaptations to changing environmental circumstances and hence habitat specificities.

Taxonomic value has been given to enamel microanatomical structures for teeth attributed to Plio-Pleistocene hominid taxa identified on the basis of gross cranial and dental features (Beynon & Wood, 1986, 1987; Grine & Martin, 1988a,b). In this study,

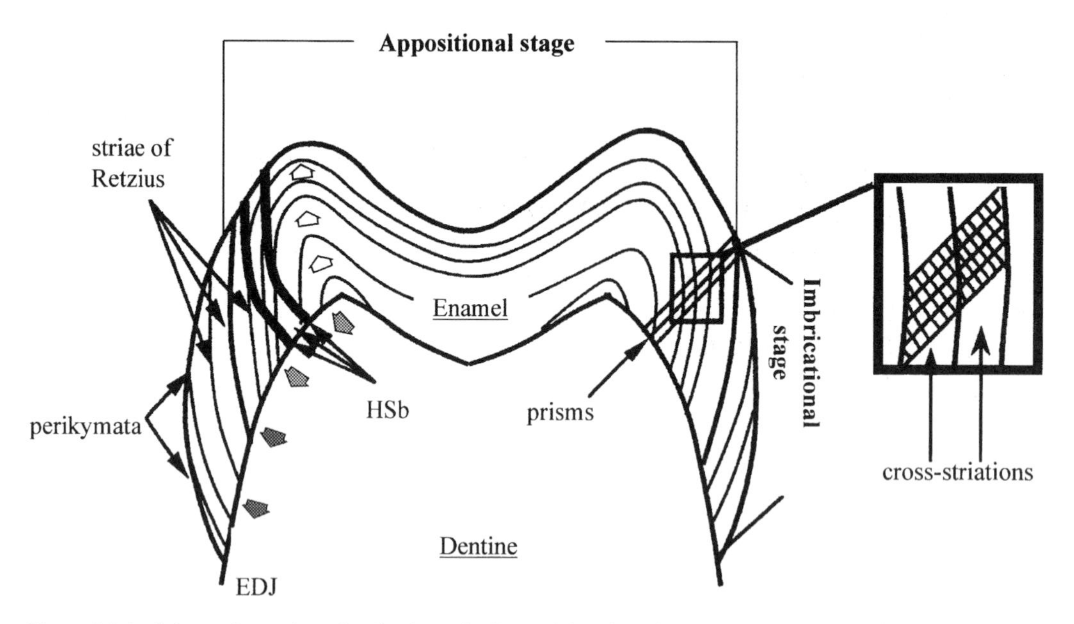

Figure 24-1. Schematic section of a cheek tooth. Enamel develops in an outward direction from the enamel–dentine junction (EDJ) to the enamel surface (open blunt arrows), and in a longitudinal direction from the dentine horn to the cervix (closed blunt arrows). The first stria of Retzius arriving at the enamel surface determines the limit between the appositional stage, where striae do not reach the enamel surface, and the imbricational stage which presents the enamel layers in an imbricational position along the sides of the tooth. Where striae of Retzius reach the enamel surface, a slight depression, the perikymata, is produced (Beynon and Wood, 1987; Risnes, 1985). Hunter Schreger bands (HSb) are shown as heavy lines through the thickness of enamel.

however, the hominid sample is derived mainly from isolated teeth, and taxonomic attribution is provisional (e.g., Howell & Coppens, 1976). For this reason, although enamel microstructural characteristics may reflect taxonomic differences, our general aim is, instead, to examine any legitimate correspondence between these characteristics and climate/environmental change.

The specific aims of this chapter are to (1) characterize enamel microstructural features of early hominid postcanine teeth; (2) group teeth sharing enamel microstructural characteristics into "morphs"; (3) observe the relation between the temporal distribution of morphs and tooth size with the record of global environmental change over the Pliocene; and (4) examine whether dental trends exhibited by early hominids follow those reported for other mammal groups said to be due to climate change.

Materials and Methods

The sedimentary sequence of the Omo Group, Ethiopia, is continuous from 3.5 to 1.05 m.y. Deltaic, lacustrine, and fluvial sediments are intercalated with layers of volcanic ash, which has allowed workers to establish a stratigraphic sequence of Members and Submembers named from A (oldest) to L (youngest). An accurate chronology, a well-known lithostratigraphy and biostratigraphy, a well-recorded evolution of fauna and flora, and an abundance of hominid fossils make the Omo Group a site of reference for the Plio-Pleistocene (e.g., Coppens, 1972; Brown et al., 1985; Brown & Feibel, 1988; Harris et al., 1988; Feibel et al., 1989) and make possible the study of modifications in dental development patterns over time.

The enamel microstructure of 21 naturally broken premolars and 45 molars of the Omo Group were analyzed. We also examined measures of the crown and cusp areas of 34 mandibular molars. The teeth date between 3.36 m.y. (the base of Member B) and 2.1 m.y. (the base of Submember G9) of the Usno and Shungura Formations (Feibel et al., 1989). Teeth were grouped by time spans (periods) of approximately 0.4 m.y: period 1, 3.36–2.85 m.y.; period 2, 2.85–2.4 m.y., and period 3, 2.4–2.1 m.y., corresponding, respectively, to Member B, Members C and D, and Members E, F, and G.

Enamel Microstructure

Enamel microstructural details were observed from the surfaces of naturally occurring occluso-cervical fractures. Specimens were examined with a zoom stereomicroscope (Leica M8) and photographed with a Nikon 2020 AF camera. Images were projected onto transparencies using a Prestinox Diasystem Lant 2200 GTS. Angles were determined with a 100 × 1 mm reticle (see Ramirez Rozzi, 1992, for details). The following characteristics were recorded: the number of imbricational striae (Isr), the number of observed and calculated appositional striae (Asrc and Asr, respectively), the total number of striae (Sr), lateral enamel thickness (ELT; Beynon & Wood, 1986), the height of the apparent dentine cap (Hdc), and curvature of the Hunter Schreger bands (curv) (Beynon & Wood, 1986) which gives an estimation of the number of changes in the orientation of the prisms (Beynon & Wood, 1986; Ramirez Rozzi, 1992, 1993a,b, 1997). Curvature is measured as the number of parazones and diazones crossed by a line established parallel to the bands in the outer enamel as it then proceeds toward the EDJ (Beynon & Wood, 1986). The values for these characteristics are listed in table 24-1. The contour of the striae of Retzius and the angle between the striae of Retzius and the enamel–dentine junction (EDJ) were also assessed. Estimates of the time of crown formation for these teeth were given in Ramirez Rozzi (1995, 1997).

In a few premolars and molars one can enumerate the number of striae at different distances from the cusp tip (Ramirez Rozzi, 1992). For the most part, however, fractures do not cross the cusp tip in the teeth studied. Rather, they cut the occlusal line at varying distances from the cusp tip, so they do not include the first striae formed. The number of cuspal striae not observed is related to the distance between the broken surface and the cusp tip and can thus be estimated. We have assumed this relationship to be similar for all teeth. In addition, the last period of crown formation was not always represented on fractured cervical margins. The perikymata (an enamel surface manifestation of a stria of Retzius) corresponding to the last stria observed on such a margin was identified and its course followed around the tooth until intact cervical enamel was reached. Perikymata placed cervical to this point were counted and added to the number of striae. Thus, the number of imbricational striae reported does not necessarily correspond to the lateral face where the count was carried out, but was corrected for the crown. The total number of striae was calculated by adding the number of appositional and imbricational striae.

Some characteristics do not show apparent intratooth variation to date (e.g., the strial contour, the modification of angles along the EDJ; Ramirez Rozzi, 1992, 1998), but others may vary depending on the tooth face considered (i.e., enamel lateral thickness). For instance, crown formation begins and ends at different times around the tooth. The number of appositional striae, then, corresponds to the closest cusp to the fracture plane. Thus, if the number of appositional striae is obtained in the cusp where enamel formation first starts, the time of appositional stage formation will be well estimated, while in another cusp the time of appositional stage formation will be underestimated. Contrary to estimates of appositional striae, the number of imbricational striae was obtained for the crown and not for any particular tooth face because last periods of imbricational stage formation not present in the broken surface were added from the greatest count of perikymata cervical to the last striae observed in the fracture plane. Therefore, the number of imbricational striae, the stria's contour, and the modification of angles along the EDJ were analyzed separately for premolars and molars without regard to tooth face. The other characteristics studied were examined separately for guiding (lower lingual and upper buccal faces) and supporting (lower buccal and upper lingual faces) cusps. Lower and upper mesial faces were considered together, as were lower and upper distal faces. The number of appositional striae was compared separately for each cusp. Tests of differences between groups of teeth from different periods were carried out (*t*-test). Correlations between time and characteristics were also calculated for the molar sample to examine the correspondence between dental modification and time period.

Morphs were determined and teeth assigned to them based on examination of the most variable enamel microstructural characteristics. The first step in the determination of morphs was to group premolars and molars within each period on the basis of those enamel microstructural characteristics that best explained the observed variation. This was achieved by using principal components analysis (PCA). Angles were not included in the PCA because more than one measure was obtained for each broken surface and attention was focused on their modifications along the EDJ and not on the real values. PCA was not calculated for Hunter Schreger band curvature because the result for each tooth corresponds, in general, to a range. Neither was it calculated for the strial contour because this is a qualitative characteristic. The characteristic that best explained the variation according to the PCA and the characteristic with the strongest variability among those not included in the PCA were used to regroup

Table 24-1. Dental sample pertaining to the study of enamel microanatomical characteristics.

Specimen	Member	Tooth	Face	Cusp	ISr	ASrc	ASr	No. Sr	ELT	Hdc	Curv
B8–14b	B	ul P4	D	—	43	—	—	—	—	—	—
W8–750	B	ul P4	M	Protocone	71	50	53	124	2130	5190	3
W8–751	B	ll P4	L	Metaconid	41	55	55	96	1590	5740	5
L1–468	B	ur P4	L	Protocone	58	53	56	114	2290	5390	4–5
Omo 33e–3495	E	ul P4	D	Paracone	79	72	72	151	1820	5470	3–4
Omo 44–1410	E	ur P4	M	Protocone	85	63	70	155	1600	4400	2
Omo 57.4–147	E	ll P4	B	Protoconid	93	65	68	161	1690	4490	2
Omo 57.4–148	E	ur P3	D	Paracone	63	55	55	118	2100	5590	0
Omo 57.5–123	E	ur P	M	Paracone	—	53	57	—	—	—	0–1
L338x–33	E	ur P4	M	Paracone	76	54	54	130	1350	5660	2
Omo 33–31	F	ur P4	L	Protocone	40	73	77	117	2200	5400	2
Omo 33–62	F	ul P4	M-D	Paracone	77	62	69	146	1600	4340	5–6
Omo 33–507	F	lr P4	B	Protoconid	75	76	76	151	2630	6280	3
Omo 33–740	F	ur P3	B	Paracone	—	77	77	—	2520	5600	6–7
L398–1223	F	lr P4	M	Protoconid	49	64	64	113	1810	6140	6
L420–15	F	ll P4	M	Protoconid	69	64	66	135	1730	5270	5
L465–113	F	ul P	M	Protocone	79	62	71	150	2260	5560	7–8
Omo 75i–1255	G	lr P3	B	Protoconid	54	65	73	127	1950	5700	4
L628–5	G	lr P4	M	Protoconid	58	71	71	129	1630	5620	1
L726–11	G	ur P4	M	Paracone	—	79	79	—	2600	7620	4
L797–1	G	ll P4	L	Protoconid	52	68	68	120	1970	4980	1
B7–39c	B	ul M1	L	Hypocone	47	46	51	98	1630	4280	0–1
B8–14a	B	ul M3	L	Protocone	67	55	61	128	2540	7060	0–1
B8–20p	B	ur M1	D	Metacone	—	42	52	—	1400	—	4
B8–28m	B	ll M3/2	L	Entoconid	—	52	52	—	1490	—	—
W7–559i	B	ur M	L	Protocone	72	52	58	130	1750	4920	6
W8–578	B	ll M2/1	L	Metaconid	50	51	59	109	1470	4045	3
W8–752	B	lr M1	B	—	60	—	—	—	—	—	—
L2–79	B	ll M2/3	L	Entoconid	46	47	47	93	1240	4720	0–1
Omo 84–100	C	lr M2	M	Protoconid	62	68	68	130	1960	5410	4
L183–11	C	ul M1	B	Paracone	54	50	53	107	2240	3990	2
L849–1	C	lr M2/1	L	Metaconid	52	57	57	109	2040	4180	0–1
L9–11	D	lr M3	L	Entoconid	68	49	58	126	1570	3680	2
L9–92	D	ul M3	M	Protocone	76	48	60	136	1770	4430	4
L296–1	D	lr M3	L	Protoconid	60	48	53	113	1500	4000	3
L10–21	E	lr M3	L	Entoconid	—	63	63	—	2100	5320	4
L26–59	E	ul M2/3	D	Paracone	56	70	70	126	2880	6460	3
Omo 57.5–320	E	lr M2/1	B	Hypoconid	58	64	64	122	2410	7320	2–3
L338x–32	E	ul M1	D	Metacone	39	61	70	109	1500	4230	3–4
L209–17	E-F	lr M2/1	D	Entoconid	46	73	80	126	2190	4280	5
L209–18	E-F	ll M	B	Hypoconid	56	67	69	125	2580	5320	6
F22–1a	F	lr M2	L	Entoconid	87	73	81	168	2340	5590	3–4
F22–1b	F	lr M3	L	Entoconid	88	66	68	156	2480	5180	5
Omo 33–65	F	ul M3	L	Protocone	59	96	96	155	2730	6420	4–5
Omo 33–509	F	ur M	D	Metacone	66	66	73	139	1730	5730	3
Omo 33–3325	F	ul M3	B	Metacone	28	76	81	109	2160	3820	2
Omo 33–3721	F	ur M3	B	Metacone	54	76	82	136	1840	3770	1
Omo 33–5497	F	ur M1	M	Metacone	86	87	90	176	2020	5050	2
Omo 33–6172	F	lr M3	M	Metaconid	43	77	89	132	1650	3780	4
Omo 57.6–244	F	ur M2/1	B	Hypocone	38	67	71	109	1680	5410	0–1
L398–14	F	lr M	L	Entoconid	34	67	67	101	1340	6320	3
L398–264	F	ur M1/2	M	Protocone	98	74	74	172	1630	5720	—
L398–266	F	lr M3	L	Entoconid	95	71	73	168	1930	4830	7–8
L398–847	F	ll M3	M	Protoconid	54	85	85	139	2350	6230	3

(*continued*)

Table 24-1. *(continued)*

Specimen	Member	Tooth	Face	Cusp	ISr	ASrc	ASr	No. Sr	ELT	Hdc	Curv
L465–112	F	ul M2	B	Entoconid	57	75	75	132	1670	4630	3
Omo 263–5479	F-G	ur M	M	Protocone	60	62	76	136	—	4970	2–3
Omo 35–4022	G	u M	M	Protocone	55	66	80	135	1870	4220	5
Omo 35–4023	G	ll M2/3	L	Metaconid	62	80	80	142	—	—	7
Omo 35–4024	G	ur M2/1	B	Metacone	60	60	71	131	2560	5620	7
Omo 47–20	G	u M	D	Metacone	48	77	91	139	2040	4140	5–6
Omo 76–37	G	ll M3	M	Hypoconulid	95	80	83	178	2410	5860	5
Omo 136–1	G	ll M3	M	Protoconid	80	70	74	154	2520	5950	3–4
Omo 136–2	G	ll M2	B	Metaconid	63	73	73	136	1960	5520	4
Omo 136–3	G	lr M2	B	Hypoconid	64	71	73	137	2420	6100	—
Omo 141–1	G	ur M3	B	Protocone	53	91	91	144	2020	5000	1
Omo 141–2	G	ul M2	D	Metacone	68	82	96	164	2150	4440	1

The first column lists the specimens for premolars and molars analyzed; the second column lists the geological Member to which the teeth belong (Feibel et al., 1989); the third lists the tooth type (ul, upper left; ur, upper right; ll, lower left; lr, lower right); the fourth lists the lateral tooth face analyzed; the fifth lists the cusp within which the number of appositional striae was assessed. ISr, number of imbricational striae; ASrc, number of appositional striae counted on the broken surface; ASr, number of appositional striae corrected by adding the calculated first striae for the cusp; No Sr, total number of striae (ISr + ASr); ELT, enamel lateral thickness (μm); Hdc: height of the apparent dentine cap (μm); curv, number of Hunter Schreger bands cut by the virtual line.

teeth and to define the morphs (Ramirez & Rozzi, 1998). It should be kept in mind that morphs do not necessarily correspond to species. Instead, enamel microstructural characteristics were compared between morphs to observe if they might potentially be related to different environmental circumstances.

Crown and Cusp Areas

We examined 34 mandibular molars from Members B–G that were complete enough to calculate base crown area (CA) and relative cusp areas (table 24-2). This sample complements recent studies of premolar crown and cusp areas from the lower Omo Basin (Suwa, 1990; Suwa et al., 1996). The occlusal plane of each tooth was oriented so that three points on the cervical margin created a plane parallel with the focal plane. Because only complete, or nearly complete, teeth were measured, there is little overlap of specimens between this sample and that employed in the study of enamel microstructure (which are necessarily fractured). Nonetheless, these teeth derive from the same sample and may be presumed to represent the same taxic assemblage.

Calibrated occlusal images were first digitally acquired with a Hitachi CCD camera mounted to a macro lens and then transferred to a Quantimet 500+ image analysis system. Image processing was performed to detect the crown area from background and to improve the visual contrast of occlusal fissures. Digital editing of detected images was carried out to correct for interstitial wear and/or missing enamel according to natural anatomical outlines. Fissure patterns were used to digitally dissect cusps from one another to calculate their separate areas. Relative cusp areas were calculated as a percentage of the CA.

Results

Quantitative results for all micro- and macrostructural characteristics (excepting the striae of Retzius–EDJ angle; see below) are presented in tables 24-1 and 24-2. In addition to these results, certain qualitative observations were made. The angle at which the striae of Retzius meets the EDJ was observed to vary along the EDJ in three general configurations. The angle increases linearly toward the cervix, shows a marked increase centrally, becoming less marked cervically, and shows an initial increase centrally and then decreases cervically. Five different strial contours were observed (for a detailed description, see Ramirez Rozzi, 1992, 1993d, 1997). The striae present either (1) a long, almost straight contour, (2) a slightly concave contour relative to the EDJ, (3) a concave contour with or without an occlusal-cervical orientation, (4) a very concave contour, or (5) a boomeranglike contour.

Curvature of the Hunter Schreger bands was quite variable in the study sample. The results vary between 0 and 8 in both premolars and molars.

Trends Over Time

Premolars Configuration 1 of striae of Retzius-EDS angle is found in premolars from all periods. Config-

Table 24-2. Dental sample pertaining to the study of crown and cusp areas.

Specimen	Member	Tooth	BCA	Proto		Meta		Ento		Hypcd		Hycld	
				Abs	%	Abs	%	Abs	%	Abs	%	Abs	%
L1–294	B	lr M3	186.2	51.0	27.7	43.8	23.5	24.4	15.3	30.5	16.4	34.4	18.5
L1–398	B	lr M2/3	150.5	38.3	25.8	46.2	30.7	17.4	11.5	29.1	19.3	17.5	11.6
Omo 28–19	B	lr M3	128.3	27.7	21.8	36.4	28.4	16.8	13.1	29.2	22.7	16.8	16.8
Omo 28s–30	B	lr M3	214.8	52.1	24.6	42.5	19.8	37.3	17.4	37.0	17.2	42.9	20.0
Omo 212–1950	B	ll M1	114.8	30.8	26.8	23.6	20.6	11.6	10.1	26.6	26.6	16.9	14.7
W7–508	B	lr M1	134.8	31.6	23.4	28.6	21.2	20.6	15.3	28.9	21.4	23.4	17.4
L795–1	B-C	lr M3/2	229.5	63.5	28.0	53.3	23.2	35.3	15.4	48.5	21.1	26.3	11.5
L45–2	C	lr M1	130.0	33.3	25.6	28.6	22.0	17.6	13.5	30.7	23.6	18.4	14.2
L51–1	C	ll M1/2	152.7	37.5	24.6	28.8	18.8	19.3	12.6	37.8	24.7	27.7	18.1
L62–17	C	lr M2	193.3	44.0	22.7	42.8	22.2	34.2	17.7	39.5	20.4	30.6	15.8
Omo 18s–34	C	ll M1/2	193.1	41.9	21.7	35.4	18.4	36.9	19.1	39.3	20.3	37.3	19.3
L26–1g	E	ll M2/1	160.3	30.0	18.7	35.8	22.3	28.9	18.1	31.3	19.5	32.4	20.2
L338x–39	E	ll M3	235.1	54.6	23.3	43.1	18.3	44.7	19.0	42.6	18.1	47.4	20.2
F22–1a	F	lr M2	261.1	57.7	22.1	63.0	24.5	42.4	16.3	51.4	19.7	42.6	16.3
F22–1b	F	lr M3	275.0	67.1	24.7	60.6	22.0	53.2	19.3	42.1	15.3	48.9	17.8
L28–30	F	lr M3	168.2	39.6	23.8	33.6	20.0	28.6	17.0	29.3	17.4	35.2	20.9
L28–31	F	lr M2	156.9	34.9	22.3	35.4	22.6	29.8	19.0	33.1	21.1	21.9	14.0
L157–35	F	ll M2	270.3	61.0	22.6	59.6	22.1	53.4	19.7	52.2	19.3	41.3	15.3
L398–630	F	lr M3	199.3	51.3	26.0	39.5	19.8	35.6	17.9	33.8	17.0	37.1	18.6
L398–2608	F	ll M2	162.2	41.1	25.6	31.0	19.1	17.7	10.9	50.2	30.9	20.4	12.6
Omo 33–9	F	lr M3	226.6	45.5	20.3	56.7	25.0	38.9	17.2	59.8	26.4	23.2	10.2
Omo 33–6172	F	lr M3	211.3	43.3	20.7	43.4	20.5	50.0	23.6	35.4	16.8	37.0	17.5
L427–7	G	lr M2	194.9	48.5	25.0	39.7	20.4	38.7	19.9	38.0	19.5	27.6	14.1
L628–3	G	ll M3	229.3	56.4	24.9	65.0	28.3	26.5	11.5	44.9	19.6	34.1	14.9
L628–9	G	ll M1	174.2	35.6	20.4	43.6	25.0	23.6	13.6	36.8	21.1	32.5	18.7
L628–10	G	ll M1/2	165.2	44.8	27.1	31.5	19.1	27.3	16.5	31.4	19.0	28.3	17.1
Omo 47–46	G	lr M2	237.2	60.0	25.3	48.9	20.6	26.5	11.2	44.4	18.7	54.7	23.1
Omo 47–1500	G	lr M2	184.9	45.8	24.8	36.9	19.9	30.2	16.3	38.8	21.0	31.2	16.9
Omo 75–14a	G	ll M2	180.3	44.0	24.4	35.8	19.9	22.7	12.6	34.4	19.1	41.1	22.8
Omo 75–14b	G	lr M2	214.7	48.0	22.3	46.4	21.6	33.2	15.5	45.9	21.4	38.6	18.0
Omo 75s–15	G	ll M1	153.4	39.0	25.4	33.2	21.6	23.8	15.5	38.2	24.9	17.7	11.5
Omo 75s–16	G	lr M3	136.9	31.7	23.4	36.1	26.3	26.0	19.0	26.9	19.7	14.8	10.8
Omo 136–1	G	ll M3	211.8	61.3	29.3	39.8	18.8	41.5	19.6	41.8	19.7	25.0	11.8
Omo 136–2	G	ll M2	193.5	47.4	24.5	39.2	20.3	34.9	18.0	38.3	19.8	31.5	16.3

The first column lists the specimens for premolars and molars analyzed; the second column lists the geological Member to which the teeth belong (Feibel et al., 1989); the third lists the tooth type (ul, upper left; ur, upper right; ll, lower left; lr, lower right); the fourth lists the base crown area (BCA); and in subsequent columns are the absolute area (Abs) and relative, or, area percentage (%), of BCA for the protoconid (Proto), metaconid (Meta), entoconid (Ento), hypoconid (Hypcd), and hypoconulid (Hycld).

uration 2 is found only in some premolars from period 1 (Member B). Configuration 3 is present in some premolars from period 3 (Members E, F, and G), but is not found in premolars from period 1. The absence of configuration 3 in premolars from period 1 may be due to the small sample size, but the absence of configuration 2 in the more recent premolars indicates that it is no longer present in premolars from 2.4 to 2.1 m.y.

The premolars from different periods show no significant difference in number of imbricational striae. The number does vary, however, between Members: those from Member E have a high number of imbricational striae, significantly higher than those from Members B and G.

It was possible to study the number of appositional striae for certain cusps in the upper premolars (in the lower premolars, this analysis was not possible). For the paracone, teeth from period 3 (from Members F and G) show a higher number of appositional striae than those from Member E ($p = .0395$). In the protocone, the number of appositional striae is higher in premolars from period 3 than in those from period 1 ($p = .0093$), suggesting an increase through time (fig. 24-2).

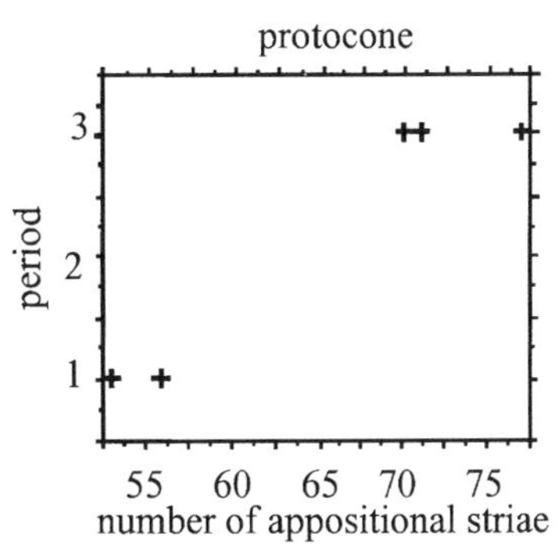

Figure 24-2. Change in microanatomical characteristics through time in premolars. Period 1 comprises teeth from Member B (3.36–2.85 m.y.), period 2 from Members C and D (2.85–2.4 m.y.), and period 3 from Members E, F, and G (2.4–2.1 m.y.). The numbers of appositional striae were analyzed only in the protocone: the number is higher in teeth from period 3 than from period 1.

We did not investigate premolar base crown area (CA), but Suwa (1990) presented data on the CA for 64 maxillary and mandibular premolars from the lower Omo Basin. We have culled the lower premolar data for comparison to our study of the lower molars from the lower Omo Basin. The average CA for the third and fourth lower premolars, plotted together with that of the lower molars, have been grouped by period (fig. 24-3). There is an increase from period 1 through periods 2 and 3.

Molars Although all three angular configurations between the striae of Retzius and the EDJ are found in molars from all periods, configuration 3 occurs in 33% of the molars from period 1 and in 67% of molars from period 3.

The number of appositional striae increases with time. In the protocone, molars from period 3 present a higher number of appositional striae than those from period 1 ($p = .0216$) and period 2 ($p = .0909$; fig. 24-4A). On the metacone, the number of appositional striae in molars from period 3 is higher than in those from period 2 ($p = .0254$). In the hypocone and the paracone, represented by a small sample ($n = 2$), the lowest number of appositional striae is found in molars from period 1 and period 2, respectively.

On the metaconid, molars from period 3 have a higher number of appositional striae than those from periods 1 and 2 ($p = .0328$; fig. 24-4B). In the entoconid, period 3 appositional striae increase from period 1 ($p = .0029$) and period 2 ($p = .0936$; fig. 24-4C). In the protoconid ($n = 4$), the lowest values are found in molars from period 2. The hypoconid was not studied in any tooth from periods 1 and 2. Cumulatively these results suggest that the number of appositional striae and, thus, appositional enamel formation time, increased around 2.4 m.y.

A higher total number of striae are also found on molars from period 3. There is no difference between groups when molars are analyzed separately according to the cusp, except on the metaconid, where molars from period 3 have a higher number of total striae than those from period 1 and period 2 ($p = .0414$) or from periods 1 and 2 together ($p = .0052$). However, molars from period 1 and period 2 have the lowest number of total striae in all tooth quadrants investigated: the protocone, paracone, hypocone, protoconid, and entoconid. When all molars are analyzed together, using whichever cusp is available, molars from period 3 have a higher number of total striae than those from periods 1 and 2 ($p = .0012$). An increase in the total number of striae through time, which would demonstrate an increase in the time of crown formation, cannot be ruled out (fig. 24-4D).

Modifications through time were investigated separately for guiding and supporting cusps. On the lateral faces of guiding cusps, enamel thickness is significantly increased in the comparison between period 1 and period 3 ($p = .0211$; fig. 24-4E). The *t*-test remains significant when molars from period 3 are compared with those from both periods 1 and 2 ($p = .0727$). Enamel thickness does not change significantly in other

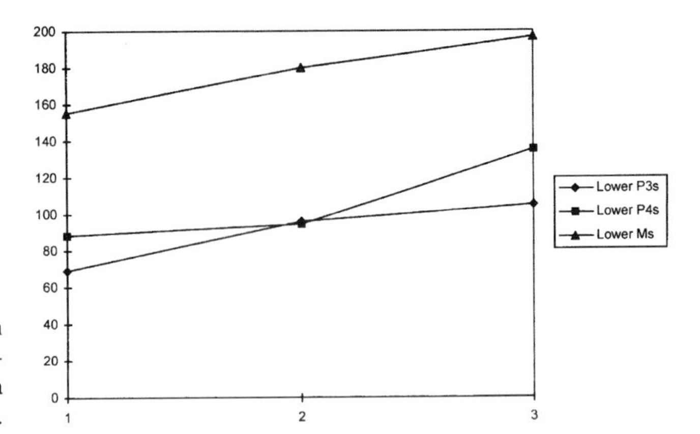

Figure 24-3. Changes in mean crown areas for mandibular molars and premolars over periods 1 through 3. Data for premolars from Suwa (1990).

lateral tooth faces, though the distal face of one molar (specimen B8–20p) from period 1 shows the thinnest lateral enamel (table 24-1).

The lateral faces of supporting cusps from three period 1 molars were investigated. Though levels of statistically significant results are not obtained for mesial cusps, perhaps due to the small sample size, the lateral enamel in teeth studied is nonetheless thicker in distal and lateral faces of supporting cusps from period 3 than in those from period 1.

The apparent height of the dentine cap increases in the lateral faces of guiding cusps from period 2 (Members C and D) to period 3 (p = .0357; fig. 24-4F). When molars from period 3 are compared with those from periods 1 and 2, the difference remains significant (p = .0299).

Figure 24-3 shows the changes in mean CA for the mandibular molars and premolars for periods 1–3. The mean molar CA increased from period 1 to period 2 by 16% (p = .05), the CA from periods 1 plus 2 to period 3 increased by 18% (p = .0429), and the increase from period 2 to period 3 is not statistically significant. Figure 24-5 illustrates trends for the protoconid, metaconid, and entoconid cusps through time. The relative entoconid area increased by 31% (p = .089) from period 1 to period 2; the other period-to-period changes are not statistically significant. However, when the data from periods 1 and 2 are pooled and compared with the data from period 3, the absolute area of the protoconid increases by 15% (p = .2026), that of the metaconid increases by 16% (p = .0827), and that of the entoconid increases by 37% (p = .0095). It follows that the relative cusp areas changed differentially. The relative protoconid area decreased by 4% (p = .0161), the relative metaconid area decreased by 1% (p = .3416), and the relative entoconid area increased by 16% (p = .0062).

Dental Sets and Morphs

Principal components analysis of the quantitative microstructural data has been presented elsewhere (Ramirez Rozzi, 1992, 1997, 1998). The number of imbricational striae most affects the first factor in premolars and molars. Among the characteristics not included in the PCA analysis, strial contour is that with the greatest variability in both premolars and molars. Therefore, the number of imbricational striae and the striae's contour, characteristics without intratooth variation, were used to group teeth.

Number of Imbricational Striae Premolars and molars were ranked according to the number of imbricational striae in each period. The groups were delineated by the largest intervals between specimens. This method is arbitrary, but it seems to be a reasonable means for separating a sample into groups based on characteristics present in a continuum. The coefficient of variation of the number of imbricational striae is 26.4 and 22.7 for premolars from periods 1 and 3, and 19.3, 14.43 and 30.1 for molars from periods 1, 2, and 3, respectively. A division between groups was established when a minimum of seven striae existed between adjacently ranked premolars or molars. Premolars and molars do not differ in the number of imbricational striae, so premolars and molars were grouped together (Ramirez Rozzi, 1998). Three groups were established in the first period, two in the second, and four in the third. The groups from all periods were then compared and found to form four sets: set ISr1 com-

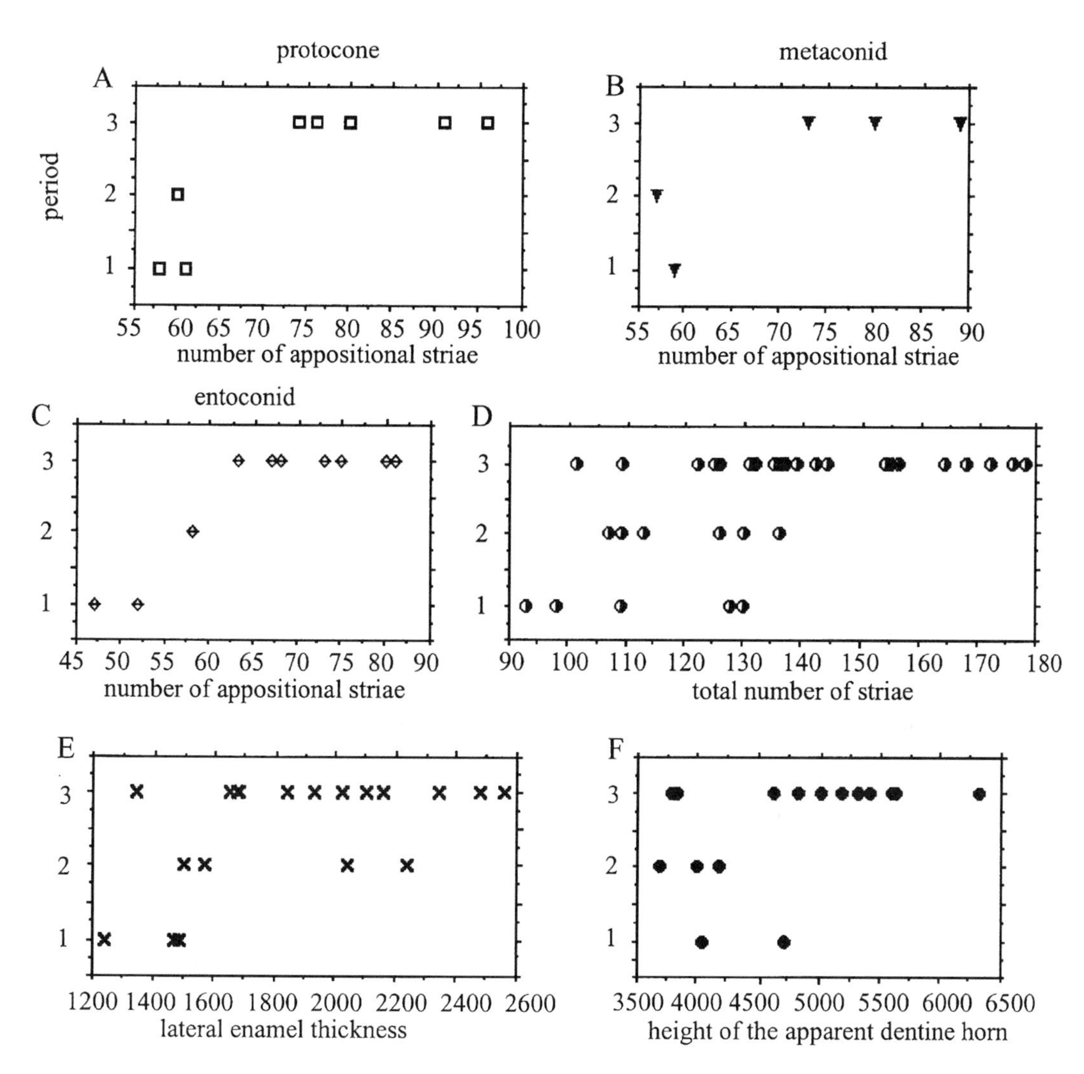

Figure 24-4. Change in microanatomical characteristics through time in molars: the numbers of appositional striae increase in molars. The numbers become higher in the protocone (A), metaconid (B), and entoconid (C) from 2.4 m.y. The total number of striae increases from 2.4 m.y. (D). The lateral enamel is significantly thicker in teeth from period 3 than in those from period 1 (E). The height of the apparent dentine cap increases from period 2 to period 3 (F).

prises teeth with 28–50 imbricational striae, set ISr2 teeth with 49–69, set ISr3 teeth with 67–88, and set ISr4 teeth with 93–98. The overlap in the number of imbricational striae between sets stems from the fact that sets do not have the same chronological distribution. Set ISr1 is found in periods 1 and 3 (3.36–2.85 and 2.4–2.1 m.y.), set ISr4 is confined to period 3, and sets ISr2 and ISr3 are represented in all periods.

Strial Contour The grouping of teeth according to the strial contour was qualitatively assessed. Sets were determined principally by the contour of the last appositional stria and first imbricational stria because they correspond to the longest stria in a tooth's lateral face. Comparison between groups from all periods led us to establish three sets, grouping teeth with the same strial contour. Teeth of set SC1 are characterized by

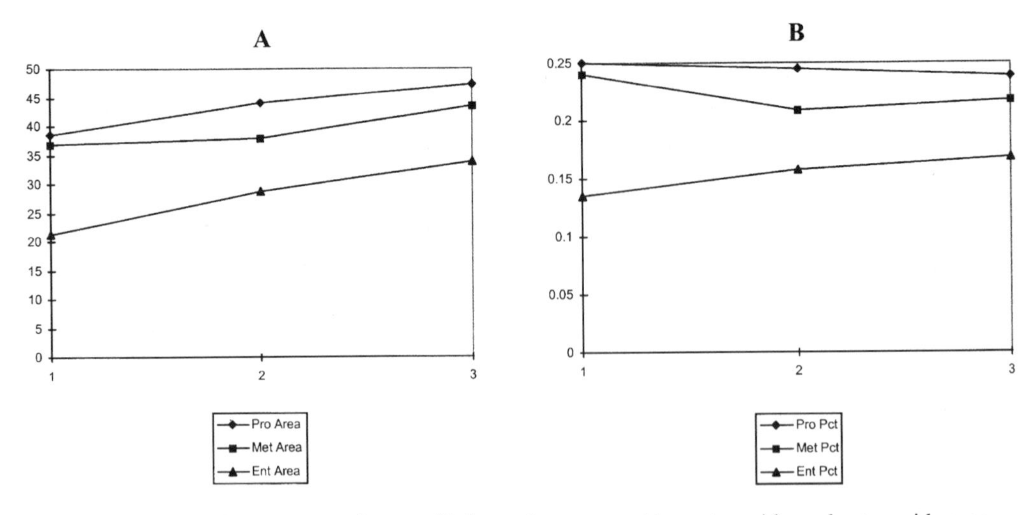

Figure 24-5. Changes in mean areas for mandibular molar protoconids, metaconids, and entoconids over periods 1 through 3. (A) changes in absolute areas; (B) changes in relative areas.

concave striae (different degrees in the concavity are found). These striae can show a noticeable occluso-cervical orientation, and they may run parallel to the enamel surface. The striae with a boomeranglike contour characterize teeth of set SC2 (Ramirez Rozzi, 1992, 1997). The striae of teeth from set SC3 are long, with an occluso-cervical orientation, and their contour is almost straight. Set SC1 is represented in all periods, set SC2 is present in periods 2 and 3, and set SC3 is found in only period 3.

Comparison of Sets and Determination of Morphs Membership in morphs is determined by combinations of sets defined on the basis of imbricational striae and strial contour described above. Set SC1 includes principally teeth of set ISr2, but it also includes some teeth from set ISr1. Set SC2 is made up of teeth from sets ISr2 and ISr3. Set SC3 includes only teeth from set ISr1. Thus, five morphs were recognized from the comparison of sets (table 24-3). The teeth of morph I are characterized by a low number of imbricational striae and a concave strial contour. Morph II consists of teeth with the same concave strial contour but with a lower number of imbricational striae than morph I. A high number of imbricational striae and a boomeranglike strial contour characterize teeth from morph III. Morph IV is composed of teeth with the boomerang contour but with a lower number of imbricational striae. Teeth from Morph V are characterized by a low number of imbricational striae and a rather straight strial contour. The chronological distribution of Morphs is complex. Morphs I and II are present in all periods, morph III is found in periods 2 (only one tooth from Member D) and 3, and teeth from morphs IV and V come from period 3 only.

Discussion

Enamel thickness is related to two main characteristics of dental formation: the degree of prism decussation and the length of prisms which, in turn, are influenced by the active life of ameloblasts and the appositional rate. The higher the degree of prism decussation, the thinner the enamel; the longer the prism, the thicker the enamel. A high appositional rate or prolonged ameloblast activity produces a longer prism and therefore thicker enamel (Ramirez Rozzi, 1992, 1993b,c, 1996, 1997). The number of appositional striae corresponds to the active life of ameloblasts during the first phase of enamel formation. Thus, the number of appositional striae plays a major role in determining enamel thickness. In premolars, a higher number of appositional striae from 2.4 m.y. is accompanied by a stable enamel thickness. Similar changes are observed in molars, but somewhat differently from premolars: modification in the number of appositional striae produce an increase in the lateral enamel thickness around 2.4 m.y. The number of appositional striae is significantly correlated with enamel thickness ($p < .01$; Ramirez Rozzi, 1997). The increase in the number of appositional striae and the thicker enamel from 2.4 m.y. indicate that, in molars, the development of the

Table 24-3. Dental sample assigned to morphs on the basis of enamel microanatomical characteristics.

Morph	Set	Teeth that define the morph	Teeth that could be included in morph by virtue of having a similar	
			No. of imbricational striae	Strial contour
I	Isr2, SC1	W8–752, L1–468, L9–11, L183–11, L209–18, L296–1, L398–847, L398–1223, L465–112, L628–5, L797–1, L849–1, Omo 33–3721, Omo 35–4023, Omo 35–4024, Omo 57.4–148, Omo 57.5–320, Omo 75i–1255, Omo 84–100, Omo 136–2, Omo 136–3	L26–59, L420- 15, Omo33–509, Omo35–4022, Omo141–1, Omo141–2, Omo263–5479	B7–39c, B8–14a, B8–14b, W7–559i, W7–758, W8–750, W8–751, L2–79, L209–17, L398–266°, Omo33–31, Omo33–507, Omo33–3325, Omo33–6172, Omo136–1
II	Isr1, SC1	B7–39c, B8–14b, W7–578, W8–751, L2–79, L209–17, Omo33–31, Omo33–3325, Omo 33–6172	L338x–32, L398–14, Omo47–20, Omo57.6–244	B8–14a, W7–559i, W8–750, W8–752, L1–468, L9–11, L183–11, L209–18, L296–1, L398–266°, L398–847, L398–1223, L465–112, L628–5, L797–1, L849–1, Omo33–507, Omo33–3721, Omo35–4023, Omo35–4024, Omo57.4–148, Omo57.5–320, Omo75i–1225, Omo84–100, Omo136–1, Omo136–2, Omo136–3
III	Isr3, SC2	L9–92, L338X-33, L465–113, F22–1a, F22–1b, Omo33–62, Omo33–5497, Omo33e-3495, Omo44–1410	B8–14a, W7–559i, W8–750, Omo–507, Omo136–1	L26–59, L420–15, Omo33–509, Omo35–4022, Omo47–20, Omo57.4–147°, Omo76–37°, Omo141–1, Omo141–2, Omo263–5479
IV	Isr2, SC2	L26–59, L420–15, Omo33–509, Omo35–4022, Omo141–1, Omo141–2, Omo263–5479	W8–752, L1–468, L9–11, L183–11, L209–18, L296–1, L398–847, L398–1223, L465–112, L628–5, L797–1, L849–1, Omo33–3721, Omo35–4023, Omo35–4024, Omo57.4–148, Omo57.5–320, Omo75i–1255, Omo84–100, Omo136–2, Omo136–3	L9–92, L338X–33, L465–113, F22–1a, F22–1b, Omo33–62, Omo33–5497, Omo33e–3495, Omo44–1410, Omo47–20, Omo57.4 –147°, Omo76–37°
V	Isr1, SC3	L338x–32, L398–14, Omo57.6–244	B7–39c, B8–14b, W7–578, W8–751, L2–79, L209–17, Omo33–31, Omo33–3325, Omo33–6172, Omo47–20	

Teeth were assigned to a morph on the basis of their number of imbricational striae (ISr) and strial contour (SC). ISr and SC were categorized into four and three sets respectively, as described in the text. L398–264, L398–266, Omo57.4–147, and Omo76–37 show a high number of imbricational striae and, indeed, characterize set 4, but each tooth has a different strial course (except that of Omo398–264, which could not be determined). Consequently, this ISr4 cannot be used to define any morph.

appositional stage becomes more pronounced over time, whereas in premolars it remains stable. This could be a functional difference related to increased masticatory force and/or lifetime abrasive load in the posterior dentition. Nevertheless, thicker enamel is not always present in teeth with a longer functional occlusal life (Aiello et al., 1991).

The premolar and molar pattern 3 striae–EDJ angle configuration (initial increase centrally, and then decreasing cervically) becomes more common passing from period 1 to period 3. The striae–EDJ angle depends on the rate of ameloblasts becoming active. Thus, this angle gives an estimation of the extension rate of enamel: the more open the angle, the lower the extension rate. A modification of the angle along the EDJ of individual teeth enables us to reconstruct the change in extension rate during crown formation. A pattern 3 configuration indicates that the extension rate decreases during the first two-thirds but increases in the last third of crown formation. The number of appositional striae increases from 2.4 m.y. and reflects, therefore, a longer time of appositional stage formation. A high extension rate in the cervical one-third of the crown would balance the increase in appositional stage formation time and avoid a longer period devoted to dental formation.

The number of imbricational striae and strial contour were combined to recognize five morphs in the dental sample. Morphs III–V may be considered to traverse the time span associated with the markedly more cool and arid climes of the Late Pliocene. The boomeranglike (morphs III and IV) and almost straight (morph V) strial contours and higher numbers of imbricational striae (morph III) appear during this time. Teeth from morphs III and IV show a variable enamel lateral thickness; however, only a few teeth present thick enamel. Although teeth with a similar number of appositional striae may be found in other morphs, morphs III and IV mostly comprise teeth with a high number of appositional striae. In addition, teeth from morph III are characterized by a high number of total striae. Morph V comprises teeth with a low number of imbricational striae and thin enamel, similar to some of the teeth from morph II; however, the straight strial contour characterizes this morph.

Teeth from morph V not only differ from other teeth by the low number of imbricational striae and the almost straight strial contour, but also by a low number of appositional striae and thin lateral enamel. This pattern characterizes great ape teeth. Although we make no claims for species assignments based on morphs, it is possible that the particular pattern of enamel distribution in these teeth suggests the presence of some ape fossil species in Omo from 2.4 m.y. This would be consistent with the existence of mosaic habitats in the lower Omo Basin (Geraads & Coppens, 1995).

Earlier studies of the Omo dental sample based any provisional taxonomy on the loose dichotomy between robust versus nonrobust teeth. Research on this sample may now acknowledge species considered more recently (e.g., *Australopithecus aethiopicus, Homo rudolfensis*). Comparisons between morphs and taxonomic assignments indicates that robust teeth (*A. boisei* White, 1988; *A. aff aethiopicus:* Howell et al., 1987; Suwa, 1990) are found in all morphs, even those from morph V with thin enamel. Teeth attributed to *A. aethiopicus* in a previous study on enamel microstructure (Ramirez Rozzi, 1993d) are dispersed into three morphs, and two out of four teeth assigned to *A. boisei* are included in morph IV. Two teeth, one in morph I and the other in Morph III, were attributed to *aff. Homo* (Suwa, 1990; T. D. White, 1988). Two teeth reported to *A. aff. afarensis* (Suwa, 1990) are included in morphs II and V. Beynon and Wood (1986) have suggested that teeth from *Homo* sp. are differentiated from those of *A. boisei* by a strong curvature of the Hunter Schreger bands. Although this observation needs to be confirmed in light of the presence of *A. aethiopicus* and *H. rudolfensis,* no difference is observed in the curvature of the Hunter Schreger bands between teeth from different morphs. Therefore, morphs, as described in this study, do not seem to represent hominid species, *provided* that previous taxonomic attributions assigned to the specimens investigated are accepted (Ramirez Rozzi, 1998). Morphs would indicate a large overlap in those enamel microstructural characteristics analyzed in this study of Pliocene hominid species.

There are only five mandibular molars for which we have both the information necessary to assign them to morphs and also accurate size (CA) measurements. Omo 136–2, an M_2 with a CA of 193.5 mm^2, has been assigned to morph I. Omo 33–6172, an M_3 with a CA of 211.3 mm^2, has been assigned to morph II. F22–1a, an M_2 with a CA of 261.1 mm^2 and F22–1b, an M_3 with a CA of 275.0 mm^2, have been assigned to morph III. Omo 136–1, an M_3 with a CA of 211.8 mm^2, may belong to morph II (or possibly morph III). Although this subsample is very small, it does suggest a possible relation between morphs and CA: morph I CA $<$ morph II CA $<$ morph III CA.

The average CA of mandibular molars increased from 154.9 mm^2 in period 1 to 179.7 mm^2 in period 2 to 208.5 mm^2 in period 3. This pattern of change ties in with the changes seen in other mammalian teeth and also fits with the inferred climate changes about 2.5 m.y.

Also of interest are the size changes in the protoconid, metaconid, and entoconid and changes in the number of striae in these cusps. The absolute CA of the entoconid increased more than twice as much as the CA of the protoconid or metaconid, the relative areas of the protoconid and metaconid changed hardly at all, and the relative entoconid area increased by 16%. The number of appositional striae increased in all three cusps: by 31% in the protoconid, by 47% in the metaconid, and by 40% in the entoconid; the number of total striae also increased by 21%, 25%, and 37%, respectively (because of the small sample sizes, these numbers are meant to be suggestive rather than statistically significant). Thus, although the increase in molar CA involved an increase in the time of formation, the entoconid increased its proportion of the total CA without adding to the increase any more than the other cusps.

Teeth have a principal role in alimentation and are thus subjected to strong selective pressure. The dentitions of organisms will then have adaptive characteristics ensuring efficient mastication of preferred foods found in the organism's environment. This means that environmental change resulting in changes in the material properties of vegetative food items (e.g., texture, toughness, amount of grit adhering to surfaces) will in turn generate selective pressures on organisms depending on them for dental characteristics adapted to this change. Therefore, dental characteristics (and thus dental development) can be said to be linked to habitat specificities. Indeed, alteration of tooth structure and size in 2.5-m.y. African faunas appear to be temporally related to changes in the environment (Turner and Wood, 1993a). Consequently, the changes observed in premolars and molars of Pliocene hominids might be best explained by a change in climatic conditions.

Pliocene hominids from Omo are dated between 3.36 and 2.1 m.y. Paleoclimate data (see, e.g., Denton, this volume) show that global temperatures fell throughout this period, with the sharpest downturn at about 2.5 m.y. Also, previous work has suggested that the fauna and flora changed between 3 and 2 m.y. in East Africa. Three main lines of evidence are changes in floral communities, faunal communities, and dental morphology. Important floral evidence has been published by Bonnefille (1995) and co-workers (Bonnefille & Dechamps, 1983; Bonnefille & Vincens, 1985). They have proposed that the abundance of tree species decreases and the extension of meadowland increases from Member A to Member G, with the predominance of Gramineae pollen throughout the period indicating the existence of openings in forests and woodlands or of wooded grasslands and grasslands. *Olea* and *Typha,* good indicators of humidity, become less prominent in the spectrum of pollen during this time. At the end of period 2, the highlands became an estimated 5–6°C cooler. An acceleration in floral change has also been found, with the number of species of trees and the amount of tree cover becoming markedly less in Members F and G. Through period 3, there was a shift in the relative floral proportions from wettest broadleaved forest taxa to dry-forest taxa. Late in period 3, it was still dryer, although more humid than modern conditions.

A number of workers have documented changing faunal communities in East Africa during the Pliocene. For example, Wesselman (1995) has studied the changing composition of micromammal faunas. In period 1 (Member B), he found that species characteristic of tropical forests and mesic woodlands predominated, with only a few species of more open savanna-woodland type. Many of the former taxa resemble living taxa that inhabit the central and western African rainforests. The species from period 2 (Member C) suggest a decline of tropical high forest. These species suggest mesic woodlands and dense thickets. By period 3 (Member F), only about 7% of recovered species suggest mesic savannas, gallery forests, or riverine environments, while about 70% of the taxa suggest xeric conditions. Twenty-eight percent of the total number of species are strongly arid-adapted types; included is *Jaculus orientalis,* which lives today in open desert, independent of water. Coppens (1972, 1975a) has identified four faunistic associations in the Omo Group. These successive faunistic associations are increasingly adapted to open environments with low precipitation, indicating a shift in climatic conditions.

Hypsodonty develops as an adaptation to a more fibrous and silicaceous diet and accompanies, so we expect, those changes in tooth size and shape that ensure more effective chewing of harder food (Van Valen, 1960). In East Africa, between 3 and 2 m.y., species of proboscideans, equids, suids, and bovids developed increasingly hypsodont molars as an adaptation to dry grassland (Beden, 1983, 1987; Harris, 1983b, 1991d; Eisenmann, 1985; Hooijer and Churcher, 1985). Some taxa increased tooth size. For example, Cooke (1976) showed that the upper and lower third molars of species of *Mesochoerus* increased steadily in length and crown area from period 1 to period 3 (fig. 24-6). Their breadth and hypsodonty index also increased during this time. Environmental change was also commensurate with an increase of the occlusal area in teeth of some suid species at about 2.5 m.y. (Harris, 1983b), the development of hypsodonty in other suid species around 2.3 m.y. (Feibel et al., 1991), the first appearance of *Equus* around 2.3 m.y., and the acceleration of the evolution of hypsodonty in *Elephas recki atavus*

(Proboscidea) around 2.3 m.y. (Beden, 1987). Thus, studies of fauna and flora document a gradual change from 3 to 2 m.y. in East Africa, with a short period around 2.4 m.y. during which modifications seem to have been especially apparent (Coppens, 1975a,b, 1989; Turner & Wood, 1993a). These studies indicate that regional African environments became dryer to include more open savanna or steppe as a consequence of a shift in the climatic regime (Prentice & Denton, 1988; Denton, this volume).

Evolutionary changes in teeth, such as thicker enamel and greater occlusal area, are related, in a way similar to that of hypsodonty, to the adaptation for an increase in the resistance to abrasion (Van Valen, 1960; Peters, 1981; L. B. Martin, 1985; Grine & Martin, 1988a,b). The increase of the appositional stage in Plio-Pleistocene hominid molar development can be seen as an adaptation to changing occlusal demands. This change occurs around 2.4 m.y., at which time faunal and floral changes are also taking place (as described above), suggesting that they could be related to an adaptation to a more abrasive, more fibrous, harder, or more resistant diet as a result of dryer environmental conditions.

Summary

The microanatomical characteristics of Pliocene hominid premolar and molar teeth from Members of the Shungura Formation (Omo Basin, Ethiopia) were investigated. Teeth were assigned to morphs on the basis of their combination of characteristics. Crown and cusp areas were also measured. Temporal changes in enamel microanatomy and crown and cusp areas were related to environmental change and to changes in the dentitions of other African Pliocene mammals.

Teeth from Member B (3.36–2.85 m.y.) were placed in period 1, teeth from Members C and D (2.85–2.4 m.y.) were placed in period 2, and teeth from Members E, F, and G (2.4–2.1 m.y.) were placed in period 3. These periods were used to establish temporal trends in dental development.

In both premolars and molars, the number of appositional striae increased through time, with a significant change observed around 2.4 m.y. It is probable that the total number of striae increased at the same time. Also, the ameloblast extension rate changed at 2.4 m.y.: teeth with a high extension rate in the cervical one-third became more common in period 3.

In the lateral faces of the guiding cusps of molars, the height of the apparent dentine cap increased around 2.4 m.y., and the lateral enamel became thicker after 2.4 m.y. The increase in enamel thickness and of the number of appositional striae in molars indicates a greater importance for the development of the appositional stage through time. Total crown area increased through time in both premolars and molars; the most significant increase occurred at about 2.4 m.y.

The increases documented in this study of such molar features as enamel thickness, number of appositional striae, and base crown area parallel changes in other East African mammalian taxa from the same

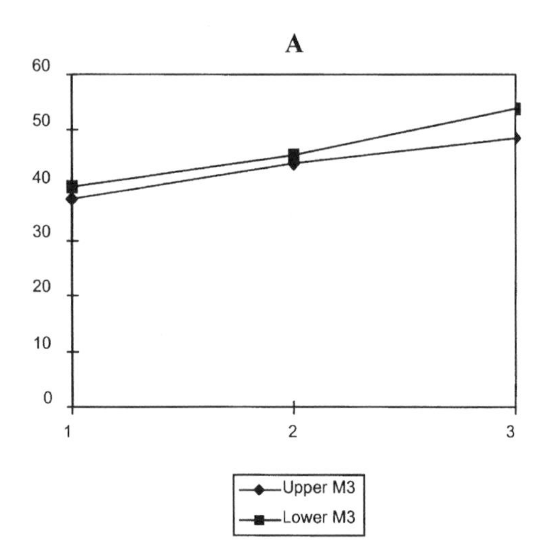

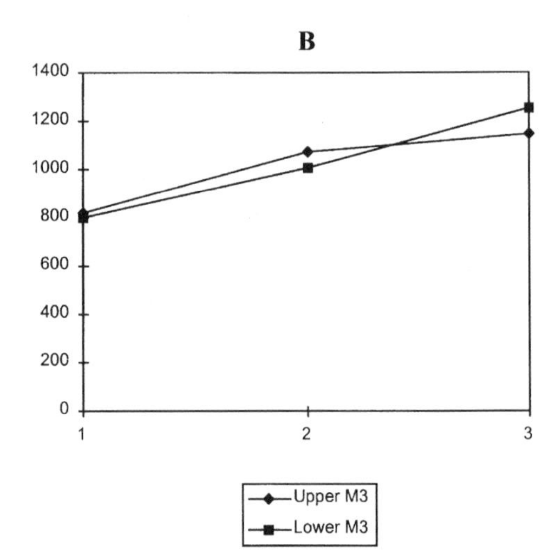

Figure 24-6. Changes in *Mesochoerus* species' upper and lower third molars over periods 1 through 3. (A) changes in mean length; (B) changes in mean crown area. (Data from Cooke, 1976.)

time periods. These changes coincide with evidence for an increasingly cool and dry climate and for increasingly tough and fibrous vegetation. Hence, we suggest that the changes in molars are best explained as an adaptation to dietary changes.

Results from this study suggest that the adaptation to more abrasive food was not a feature peculiar to robust australopithecines (Ramirez Rozzi, 1992, 1993d), but that it was a general phenomenon characterizing Pliocene hominids between 3 and 2 m.y. Therefore, given the amount of parallelism likely to have been associated with dental adaptations for heavy chewing, the dental characteristics studied here would appear, at present, to be of limited value to Pliocene hominid taxonomy.

Acknowledgments A version of this chapter was presented at the Wenner-Gren conference "African Biogeography, Climate Change, and Early Hominid Evolution," held October 25–November 2, 1995, in Salima, Malawi. We thank the Wenner-Gren Foundation for financial support. We are grateful to all participants of this symposium for interesting discussions on East African past and present ecology. We thank Y. Coppens for his support of this research. A portion of this research was made possible by grants provided by the National Science Foundation (SBR-9512373) and the Graduate Research Initiative to the Analytical Microscopy and Imaging Center in Anthropology (AMICA), Hunter College.

25

Habitat Specificity and Early Hominid Craniodental Ecomorphology

Timothy G. Bromage

Habitat theory (as promulgated in recent years by Vrba, e.g., 1992) derives from a set of hypotheses relating ecological conditions to faunal evolution and biogeography (see Vrba, this volume). Necessary features of this theory are that terrestrial vegetational habitat specificities are heritable and thus characteristic for clades, that terrestrial mammalian biomes may be distinguished by their gross vegetational physiognomy, and that distribution movements (drift) occur in the context of physical environmental change. One of our interests is whether an examination of hominoid ecomorphological characters may reveal the extent to which early hominids complied with conditions of habitat theory.

Habitat differences and climate change are important variables that influence faunal patterns through dispersion and extinction/speciation and that provide a deeper context for our understanding of character evolution. Indeed, one of the tenets of organismal biology is that character morphology and transformation occur deep within this ecological context. This elicits a certain ecocentric view regarding the nature of paleospecies.

Van der Klaauw (1948) coined the term "ecological morphology" and defined the term as the study of the relationship between the morphology of the organism and its environment. Subsequently, Bock and von Wahlert (1965) defined the function of structures in terms of their possible actions, or how they work, and made a distinction between function and biological role (biorole)—the latter referring to how the structure is used in the life of the organism. Perhaps because of the lack of concerted efforts by morphologists to conduct and rationalize behavioral research (in order to appreciate the biorole) with functional morphology, Szalay and Bock (1991) referred to any character differences associated with different environments as adaptations unless it could be demonstrated otherwise.

It is clear, however, that although research on the taxonomy, phylogeny, and functional morphology of early hominids has resulted in a plethora of characters and a fertile knowledge, a substantial research effort to understand the significance of the relations between early hominids and their environment, as revealed by their hard tissue anatomy, still lies ahead. I wish to help launch this research effort by providing a rationale for the selection of ecomorphologically significant characters (those causally related to a specific niche and operating within a specific habitat) to be employed in studies of early hominid systematics, biogeography, and paleoecology.

Ecomorphological Characters

Characters for this study were chosen largely through an inspection of anatomy, coupled with a review of the literature and an evaluation of the ability of the measurements to reveal function related to bioroles. Features providing evidence of the dietary niche (see Andrews and Humphrey, this volume) are considered here. Those features of the skull and jaws largely reflect head posture and musculoskeletal features concerned with masticatory function. Dental features concentrate on food procurement and preparation efficicacy and behaviors.

Measurements of the skull and jaws are listed and discussed below.

1. Nuchal breadth, relative to body size, partly determines the lever arm for the sternocleidomastoid muscle, a rotator of the head on the spinal column. Kunimatsu (1992) has suggested that the more terrestrial and quadrupedal the locomotor adaptation, the more narrow the nuchal breadth. This would emphasize sagittal movements of the head versus more rotational behaviors characteristic of other locomotor repertoires.
2. Presence and disposition of the sagittal crest, compound temporal/nuchal crest, and the asterionic notch as an indicator of temporalis muscle fiber orientation and function (Kimbel & Rak, 1985). This musculo-skeletal arrangement, together with the disposition of the other masticatory muscles, may reflect proportions of anterior and posterior food procurement and preparation behaviors.
3. Increases in posterior facial height (orbital apex-maxillary tuberosity relative to orbital midpoint-prosthion; see Bromage, 1992), combined with a tall vertical ramus and a deep posterior palate (cf. Bromage, 1989), constitute a masticatory adaptation for an efficient posterior bite and a reflection of the extent of tooth–tooth contacts (cf. Ward & Molnar, 1980). This functional aggregate may reflect diet-related functional morphology of the posterior facial complex.
4. Nasal morphology related to moisture conservation: angular transition of the nasal bones and the frontal processes of the maxillae onto the plane of the upper-mid face, nasal index, and maximum pyriform aperture width (see Franciscus & Trinkaus, 1988). These measures are suggested to covary with ambient humidity.
5. Subnasal conformation of the hard palate and alveolar process among hominoids (i.e., relative inclination and size of the alveolus; see Ward & Kimbel, 1983; and McCullum et al., 1993) may reflect incisal procurement and preparation behavior.
6. Maximum bite force may be estimated according to a measurement protocol developed by Demes and Creel (1988), comprising measured load arms of bite force, infratemporal fossa area and masseter origin length, and lever arms of the temporal and masseter muscles. Summed molar areas are isometrically related to bite force as well molar area and bite force relationships must be rationalized with evidence for a specific dietary adaptation.
7. Cross-sectional area and shape parameters of the postcanine mandibular corpus reflect interspecific differences in the resistance of the mandible to stress: a vertically deep corpus and parasagittal bending, a cylindrical corpus and torsion, a transversely wide corpus and transverse bending, and a large cross-sectional area and shear stress (Hylander, 1979, 1988). These measures give some indication of the stress regime that must corroborate with evidence for postulated dietary adaptation.

Measurements of the dentition are listed and discussed below.

1. Mesial-distal width of the upper central incisor. This dimension, relative to body size, reflects primate folivory/frugivory procurement and preparation behaviors (Hylander, 1975; Kay, 1985; Ungar & Grine, 1991).
2. Incisor versus premolar–molar size among frugivores (i.e., large or small in some mesiodistal, buccolingual, or occlusal surface measure) may relate to the manner by which seeds are processed (i.e., swallowed, cleaned, spit, or crushed) (Lucas, 1988).
3. Molar morphometrics relating to a stereophotogrammetric determination of location, size, and vertical height of cusps and a canonical variates analysis protocol have been suggested to account for a diet-related functional morphology of the hominoid dentition (measurement protocol of Hartman, 1989).
4. The combination of increased height of cusps, more acutely angled and relatively longer shearing blades, and larger crushing basins have been considered adaptations to a primarily folivorous diet (Kay & Hylander, 1978).
5. Molar relative cusp area variability has been postulated to account for possible dietary differences between hominoid taxa (Matsumura et al., 1992; measurement protocol of Wood & Abbott, 1983; Wood et al., 1983).
6. The buccolingual/mesial-distal (corrected for mesial and distal wear) crown shape index [(buccolingual/mesial-distal) × 100] describes the "squareness" of the molar teeth—relatively buccolingually wide molars in contrast to relatively narrower, mesiodistally elongated, molars. I sug-

gest that relative buccolingual breadth is related to an emphasis on masticatory function in the transverse plane. This character may contribute to a knowledge of the diet-related functional morphology of the dentition.

7. Premolar–molar root form and placement (cf. Wood et al., 1988) may reflect diet-related functional morphology of the dentition (e.g., the distribution of masticatory stresses on the dentition).
8. Dental microwear on the incisor and canine/premolar complex. Length, width, and orientation measures of fine-wear striae and linear gauges, the diameters of pits, microwear density, and microflaking of the enamel provide information on diet, feeding behavior, and dental function. These parameters have been described by Ryan and Johanson (1989) and are expected to vary according to the geometrical and mechanical properties of the food eaten (cf. Hylander & Johnson, 1994).

Ardipithecus (= *Australopithecus*) *ramidus*: A Case Study

The preceding lists of presumed ecomorphological characters were arrived at independently of any consideration for the published descriptions of the craniodental anatomy for any hominid species. Thus, it would be useful to evaluate such a description for whatever it might reveal in the way of habitat preferences, or ecomorphology, of some hominid. We might find that such descriptions are good enough, however unsystematic in respect to our paleoecological interests they may be. I half expected, however, to find little overt ecomorphological connections, as much anatomical description in paleoanthropology is concerned with taxonomic distinctions and not necessarily concerned with the broader ecological issues.

I chose *A. ramidus* for a case study because it is so tractable: the whole of its description of 17 fossils (13 dental, 2 cranial, and 2 postcranial) is available in one publication (T. D. White et al., 1994). The hominids derive from deposits dated between 4.5 and 4.3 m.y. at Aramis in the middle Awash region of Ethiopia. The small sample size is, of course, an obvious shortcoming, but that not withstanding, the whole of a sample is all there is, and we should endeavor to examine the material as deeply as the sample size will permit.

Further to the tractability of the *A. ramidus* sample, the publication by T. D. White et al. (1994) is followed in the same volume by another publication describing the paleoecology proposed for the hominid-bearing horizon (WoldeGabriel et al., 1994). This is extremely useful because, although paleoanthropologists may assert the importance of paleoecological reconstruction, there is little, if any, evidence that this makes any difference in the descriptive approach. Morphological approaches are normally concerned more with taxonomic differentiation, as mentioned above, than with describing specimens in a way that would intentionally make connections with the paleoenvironmental setting. Naturally, it is desirable to rationalize the two descriptive realms. Below, I have made an effort to extract ecomorphological information from the published description of *A. ramidus* by T. D. White et al. (1995) whether or not the characters match those described above from the literature. This will then be compared to the approach proposed in this chapter.

The Characters

The preserved base of the cranium, just anterior to the spinal cord, is said to be shortened relative to African apes. This would imply a more habitually erect posture (bipedality is tentatively proposed by T. D. White et al., 1994) than seen in African apes. (However, we do not know from the description whether the basicranium is shortened relative to, say, *A. afarensis).* This would suggest some semblance of a less obliquely disposed masticatory musculature and a relatively efficient posterior bite compared with an African ape.

On the other hand, this cranial architecture is combined with relatively blunt and uninflated mastoid processes which serve as attachment sites for the sternocleidomastoid muscles. These mucles assist the turning of the head from side to side, and they participate in flexion and extension of the head on the spinal column. A more erect posture is typically characterized by anteriorly positioned articulations of the skull with the spinal column that must be matched by similarly positioned insertions of the sternocleidomastoid muscles. Fully erect postures increase the requirement for anteriorly projecting bone processes to facilitate head positioning, unlike the condition described for *A. ramidus.*

Permanent central incisor proportions are described as small in relation to the postcanine dentition, unlike extant chimpanzees, but like the gorilla. However, the postcanine dentition is also described as being significantly smaller than that of *A. afarensis.* Calculation of the sum of the mesiodistal × buccolingual diameters of the *A. ramidus* P_4–M_2 yields an area of 390 mm^2. This compares with 460 mm^2 for *A. afarensis* and 227 mm^2, 294 mm^2, and 654 mm^2 for *Pan paniscus, Pan troglodytes,* and *Gorilla gorilla,* respectively (McHenry, 1994). The *A. ramidus* postcanine teeth thus appear to be rather large compared to extant chimpanzees, and it is for this reason, perhaps, that the central incisors are said to be relatively small. Furthermore, if the body size is somewhat more than 30 kg, as Wood (1994) has suggested, then *A. ramidus* has 1.5 times the tooth area expected for its body size (based on 35 kg) compared to values reported by McHenry (1994) of 1.7 for *A. afarensis* and 0.9, 0.9, and 1.0 for *Pan paniscus, Pan troglodytes,* and *Gorilla gorilla,* respectively. If actual tooth area is measured for the *A. ramidus* sample, yielding an area of 311 mm^2 (Bromage et al., 1995b), the postcanine teeth remain large, at 1.2 times the tooth area expected for its body size. If these results are substantiated, they would herald the trend for megadontia in the earliest (as yet) hominids and denote a fundamental shift in food procurement and processing by the turn of the Pliocene. This adaptation, incidently, would support the comments above that *A. ramidus* features a more efficient posterior bite than, for instance, the chimpanzee.

Although *A. ramidus* postcanine megadontia is likely derived, this feature is combined with thin enamel presumed to be like that of *Pan,* and most likely primitively retained. This combination of features may at first appear puzzling, but it is not inconsistent with our understanding of masticatory mechanics and mosaic evolution. Relatively larger postcanine teeth do suggest increased bite forces, but this increase is probably isometric (Demes & Creel, 1988), and so masticatory force per square millimeter of enamel surface remains constant.

It would seem that *A. ramidus* had feeding preferences for items requiring less incisal procurement and preparation than extant chimpanzees, but which nevertheless emulated in some way the consistency of fleshy fruits gnashed by chimpanzees on their postcanine dentition. Postcanine megadontia may be related to an increased number of chewing cycles required to sufficiently reduce a somewhat tougher fruit than commonly eaten by chimpanzees. It need not be an indication of an adaptation to hard-object feeding.

The Paleoecology

WoldeGabriel et al. (1994) cite the predominance of colobines and medium-sized kudu and the rarity of large mammals and aquatic fauna as evidence for a closed woodland setting at moderate altitude (~1600 m). They note an abundance of seeds of the genus *Canthium* within the vertebrate fossil-bearing horizon at Aramis. Species of this tree, such as *C. lactescens,* are associated with contemporary seasonal subhumid, subtropical African woodlands and are described by Peters & O'Brien (1994) as bearing edible fleshy fruits. Such fruits ripen during the early part of the dry season, or possibly later, and may have tougher mechanical properties than other fruits ripening during the wet season.

A. ramidus Ecomorphology and Habitat Preferences

Characterizations of the *A. ramidus* craniodental anatomy by T. D. White et al. (1994) and of the paleoecology by WoldeGabriel et al. (1994) enable us only to make first-order interpretations of their ecomorphology and habitat preferences. Ideally the postcranium must be used to determine *A. ramidus* positional behavior and, thus, its spatial niche, but from the cranium, the basicranial anatomy of this species is at least consistent with a vertical posture (though perhaps not fully erect terrestrially). This characteristic is not much help, but it may add to the ecomorphological significance of postcranial features to be characterized in larger forthcoming samples. Meanwhile, *A. ramidus* may be provisionally interpreted to feed principally on arboreal fleshy fruit in a seasonal woodland habitat. *A. ramidus* also had dental proportions and enamel thickness measures that suggest a certain concentration on seasonally tough fruiting resources.

No doubt new fossil discoveries will improve our understanding of the relationship between *A. ramidus* anatomy and habitat. However, we find that we are unable to create little more than a brief scenario concerning this relationship at this time. Although there is currently little cranial material on which to address the ecomorphological characters of the skull and jaws, the dental sample could be described further to include additional measurements, as described above, that confer more depth to the relationship between morphological traits and the environment.

Beyond what may be accomplished with the hominid material alone, we need to know something of the range of variation of the features in modern apes and how this variation is distributed. Then we could determine more confidently whether features are sensitive or not to environmental differences and whether variation in these features is associated with specific behaviors that are systematically distributed across an environmental spectrum.

Conclusion

There are several considerations that may lead to a sound knowledge of early hominid ecomorphology and responsiveness to habitat change. First, we require a knowledge about the ecosensitivity of skeletal characters (including all bone and tooth features). This requires cataloging presumed ecomorphological characters for analysis from not only the skull and dentition but from the postcranial skeleton as well (e.g., those related to postural activities and to feeding).

The most efficacious means by which we believe the ecomorphology of early hominids may be teased out is by detailing and comparing such characters obtained from different populations of the African great ape clade within known distributional, ecological, and behavioral contexts. I acknowledge, though, that however much an analysis of ecomorphological characters might document early hominid habitat preferences, it remains to be shown how one would distinguish between environmentally influenced subspecies- and species-level character differences. The objective at present, then, while using other means (e.g., genetics) of differentiating taxa at these low levels, would be to evaluate the currently accepted hominoid species and subspecies designations in an ecomorphological context. Such a discussion may shed light on the relationship between micro- and macroevolution in this group related to climate and environmental change.

There is a final and critical caveat: we must not forget to include ontogenetic considerations. We should distinguish the developmental plasticity of characters in different environments from historically constrained ecomorphological adaptations. Morphological variability over population sample ontogenies will need to be assessed to determine the extent to which characters are canalized and represent the functional needs of behaviors from juveniles throughout life histories.

It is expected that presumed patterns of early hominid endemism and distribution drift will convey additional information about the likelihood of trait transformations, and subsequently the polarity of various morphoclines. If so, then a biogeographic perspective on ecomorphological character evolution (see Collard and Wood, this volume), combined with more traditional character analyses, ought to provide an improved accounting of early hominid systematics in the larger context of faunal biogeography and climate change.

Acknowledgments I thank Fred Szalay and Berrian Hobby for their help in the preparation of this contribution. I am grateful to the Wenner-Gren Foundation for their invitation and support of my participation in the symposium "African Biogeography, Climate Change and Early Hominid Evolution."

Appendix

A Locality-based Listing of African Plio-Pleistocene Mammals

Alan Turner
Laura C. Bishop
Christiane Denys
Jeffrey K. McKee

In view of the evident interest in paleobiogeographic patterns across a wide range of disciplines concerned with the Neogene history of Africa, it was proposed at the meeting that a list of Plio-Pleistocene fossil mammalian occurrences at specific African localities be drawn up with the present authors acting as a small organizing group. Such a list, it was argued, would offer a valuable distillation from what is now an increasingly diversified literature, much enlarged to include new localities and discoveries since completion of the massive compilation of Maglio and Cooke (1978) and useful for the nonspecialist. Even for the specialist, a list of this kind may offer a good starting point for any wider scale investigations that seek to combine the theoretical and the empirical, the combination with "the ultimate power to convince us of scientific reality" (Rosenzweig, 1995:5). However, all such compilations are hazardous and provide many hostages to fortune given the vagaries of the fossil record and the changing picture of distributions that may occur with new discoveries. In these circumstances, it is as well to offer a clear statement of what this list is and what it is not.

The aim of this listing is to provide an up-to-date summary of presence and absence at localities of African Plio-Pleistocene Mammalia, taking account of the most recent taxonomic evaluations of the various groups. We have included Lukeino on the grounds that Plio-Pleistocene material is known from localities in the same area and that it may therefore provide a picture of longer-term continuity. We would have liked to have included material from Lothagam for the same reasons, but it is evident that much work remains to be completed and published on this locality (M. G. Leakey, et al., 1996) and that the taxonomic information now available is not in a form suitable for inclusion here.

The data presented builds on that underpinning the list of time ranges for East and South African large mammals provided by Turner and Wood (1993b), which was compiled from a combination of an extensive, locality-based literature review augmented by the more direct knowledge of the two authors about the Carnivora and the Hominidae. It is divided for the sake of convenience into two lists, one of larger (table 1) and one of smaller (table 2) taxa. The present lists are also based in part on the literature and on the specialist knowledge of the authors on Carnivora, Suidae, micromammals, and more recent South Africa discoveries, but with the addition of more detailed information and/or comment and corrections provided by authorities on various groups: E.S. Vrba (Bovidae); R. Bernor (Equidae), and B. Benefit (Primates).

On the other hand, we do not want the listing to become a fossil itself. For this reason, and unlike Turner and Wood (1993b), who were addressing the particular question of faunal similarities in eastern and southern Africa and the changing pattern over time, we have made no effort to deal with the chronological evidence and its problems in relation to some of the localities. The listings therefore do not offer a chronological summary of the time ranges of taxa based on the date of deposits nor have we attempted to list the sites in chronological order. The problems of constructing such a list of taxon time ranges were discussed in some detail by Maglio (1975), who sought a compromise solution to the difficulties thus posed by dealing in very broad time units of the African and Eurasian Pleistocene and provided an appropriately broad analysis of the patterns that emerged. What we provide is more akin to the site-based compilation for the European Pleistocene provided by Kahlke (1975) in the same volume as Maglio (although it is true that he, too, divided localities into geographic and chronological groupings) in being a list of what is currently known at each locality.

Of course, we recognize that a listing of this kind may be considered inadequate by readers who wish for a quick and convenient summary of distributions in both time and space. In our view, this shortcoming is minimal. We believe that most users of this list will be active researchers pursuing some aspect of faunal evolution and diversity, with some knowledge of and access to the literature relating to various localities. The well-dated localities are extensively reported, and the references listed for each in the site key include such detail. We also indicate those localities for which chronological control is poor and provide a separate listing of key references for major taxonomic groups. Those interested in doing so will thus be able to make their own estimations of first and last appearance datums. Because dating of sites changes but geographical position does not, we hope that listing sites from north to south will increase the useful lifetime of this compilation. To continue the comparison made above, we are therefore providing a summary of the data, like Kahlke, rather than analyzing them, as Maglio did.

Site Key

Northern Africa

Algeria

1. Ain Hanech: Arambourg, 1970, 1979.

Libya

2. Sahabi: Bernor & Pavlakis, 1987; Boaz et al., 1987; Munthe, 1987.

Egypt

3. Wadi Natrun: James & Slaughter, 1974.

Eastern Africa

Ethiopia

4. Hadar: Sabatier, 1978, 1982; Boaz et al., 1981, 1983; Kalb et al., 1982; Sarna-Wojcicki et al., 1985.
5. Awash, Matabaeitu Formation: Kalb et al., 1982; F. H. Brown et al., 1985.
6. Awash, Gamedah Formation
7. Aramis: T. D. White et al., 1994.
8. Omo Usno Formation: F. H. Brown et al., 1985; Feibel et al., 1989. Omo Shungura Formation
9. Member B10: Wesselman, 1984; F. H. Brown et al., 1985; Feibel et al., 1989.
10. Member C: As B10.
11. Member D: As B10.
12. Member E: As B10 and Howell et al., 1987.
13. Member F: As B10 and E.
14. Member G: As B10
15. Member H: As B10.
16. Member J: As B10.
17. Member L: As B10.

Kenya

East Turkana

18. Moiti Member: F. H. Brown & Feibel, 1991; F. H. Brown et al., 1985.
19. Lower Burgi Member: As Moiti.
20. Upper Burgi Member: As Moiti.
21. KBS Member: As Moiti.
22. Okote Member: As Moiti, and Black & Krishtalka, 1986.
23. Chari Member: As Moiti.

West Turkana

24. Kataboi Member: Harris et al., 1988; F. H. Brown & Feibel, 1991.
25. Lower Lomekwi Member: As Kataboi.
26. Middle Lomekwi Member: As Kataboi.
27. Upper Lomekwi Member: As Kataboi.
28. Lokalalei Member: As Kataboi
29. Kalochoro Member: As Kataboi.
30. Kaitio Member: As Kataboi.
31. Natoo Member: As Kataboi.
32. Nariokotome Member: As Kataboi.
33. Kanapoi: Behrensmeyer, 1976; M. G. Leakey et al., 1995.

Tugen Hills

34. Lukeino Formation: Hill et al., 1985, 1986; Hill, 1995.
35. Chemeron Formation: Hill, 1985, 1995; Hill et al., 1985, 1986, 1992b.

Tanzania

Olduvai Gorge: Hay, 1976; Butler, 1978a,b; Savage, 1978, Walker, 1978; Denys, 1990a.

36. Bed 1
37. Bed 2
38. Bed 3

Laetoli

39. Upper Laetolil Beds: A. C. Walker, 1978; Butler, 1987; Davies, 1987a,b; Denys, 1987;

Drake & Curtis, 1987; M. D. Leakey & Harris, 1987; Petter, 1987.

South Africa

Makapansgat
40. Member 3: Vrba 1982, 1987a, 1995c.
41. Member 4: As Member 3.
Sterkfontein
42. Member 2: Clarke, 1994; Kuman, 1994; Turner, 1987, 1997.
43. Member 4: Vrba, 1976, 1982, 1985, 1995c; Brain, 1981b; and as Member 2
Kromdraai
44. Member A: Denys, 1990a; Brain, 1981b.
Swartkrans
45. Member 1: Brain, 1981b, 1993, Denys, 1990; Avery, 1995.
46. Member 2: As Member 1
47. Member 3: As Member 1
Langebaanweg: Hendey, 1974, 1981, 1984; Denys, 1990a.
48. Quartzose Sand Member
49. Pelletal Phosphorite Member

Taxonomic Groups

Hominoidea
Howell et al., 1987; Tobias, 1991; Wood, 1991, 1992a; T. D. White et al., 1994, 1995; M. G. Leakey et al., 1995.
Cercopithecoidea
Delson, 1984, 1988.
Lorisidae
Walker, 1978, 1987; Wesselman 1985b.
Carnivora
Turner, 1990; Werdelin et al., 1994; Werdelin & Turner, 1996.
Small Carnivora
Savage, 1978; Hendey, 1981; Wesselman, 1985b; Petter, 1987.
Proboscidea
Beden, 1983, 1985; Kalb & Mebrate, 1993.
Equidae
Churcher & Richardson, 1978; Eisenmann, 1983, 1984.
Rhinocerotidae
Hooijer, 1978.
Chalicotheriidae
Butler, 1978.
Dicotylidae
Cooke & Wilkinson, 1978.
Suidae
Cooke & Wilkinson, 1978; Harris & White, 1979; Harris, 1983b; Cooke, 1985; Cooke & Hendey, 1992; Bishop, 1994; T. D. White, 1995.
Hippopotamidae
Geze, 1985; Harris, 1991a.
Giraffidae
Harris, 1991c.
Camelidae
Harris, 1991b.
Bovidae
Vrba, 1976, 1982, 1985a, 1987a, 1995c; Gentry, 1978, 1985a,b, 1987; Gentry & Gentry, 1978; Harris, 1991d.

The faunal lists of small mammals have been ordered under the classification followed by Wilson & Reeder (1993) for modern mammals.

Rodentia
Sabatier, 1978, 1982; Wesselman, 1985b, 1995; Black & Krishtalka, 1986; Davis, 1987a; Denys, 1987a, 1990a,b,c; Munthe, 1987; Pocock, 1987; Avery, 1995; McKee et al., 1995.
Insectivora and Chiroptera
Butler, 1978b, 1987; Wesselman, 1984, 1995; Black & Krishtalka, 1986; Pocock, 1987; Avery, 1995; McKee et al., 1995.
Lagomorpha
Davies, 1987b; McKee et al., 1995.
Hyracoidea
Meyer, 1978; Wesselman, 1985b, 1995; McKee et al., 1995.

Table A-1. Larger mammals. Table sorted geographically, north-south and then chronologically within localities.

	1	2	3	4	5	6	7	8	9	10	11	12	13	14	15	16	17	18	19	20	21	22	23
PRIMATES																							
HOMINOIDEA																							
Hominoidea indet																							
HOMINIDAE																							
Paranthropus boisei											■	■		■						■	■	■	
Paranthropus aethiopicus										■	■	■	■										
Paranthropus robustus																							
Ardipithecus ramidus							■																
Australopithecus anamensis																							
Australopithecus afarensis				■																			
Australopithecus africanus																							
Homo habilis																				■	■		
Homo rudolfensis																				■	■	■	
Homo erectus																							
Homo ergaster																				■	■	■	
Homo sp.																				■	■	■	
Hominidae indet																							
CERCOPITHECOIDEA																							
Cercopithecoidea indet																							
CERCOPITHECIDAE																							
Cercopithecidae indet																							
CERCOPITHECINAE																							
PAPIONINI																							
Theropithecus oswaldi					■					aff	?	?	■	■	■		cf			■	■	■	
Theropithecus darti				■																			
Theropithecus brumpti									■	■	■	■	■	■									
Theropithecus sp.								■															
Parapapio jonesi				cf																			
Parapapio ado																							
Parapapio whitei																							
Parapapio broomi																							
Parapapio sp.				■			cf							■									
Papio quadratirostris								■	■	■	■	■	■										
Papio hamadryas																							
Papio izodi																							
Papio ingens																							
Papio beringensis																							
Papio sp.										■	■	■	■	?			?			■	■		
Gorgopithecus major										■										■	■		
Cercocebus sp.										■		?	?	?						■	■	■	
Macaca sp.		■																					
CERCOPITHECINI																							
Cercopithecus aethiops																					cf.		
Cercopithecus sp.								■	■					■									
COLOBINAE																							
Cercopithecoides williamsi																							

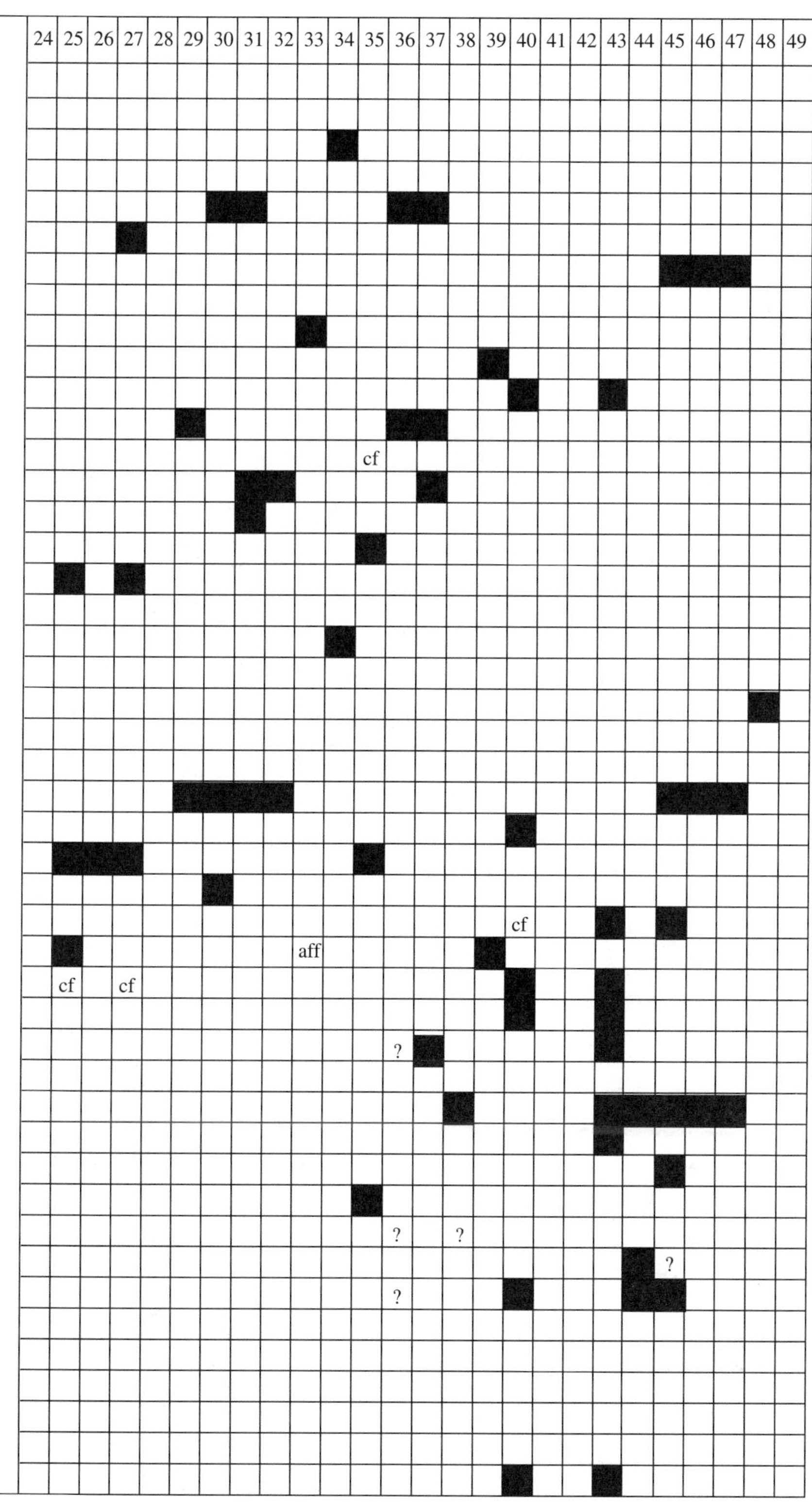
24 25 26 27 28 29 30 31 32 33 34 35 36 37 38 39 40 41 42 43 44 45 46 47 48 49
cf
cf
aff
cf
cf
?
?
?
?
?

(*continued*)

Table A-1. (*continued*)

	1	2	3	4	5	6	7	8	9	10	11	12	13	14	15	16	17	18	19	20	21	22	23
Cercopithecoides kimeui																							
Cercopithecoides sp.																				new			
Colobus sp.																	cf				■	■	
Rhinocolobus turkanaensis				cf				■	■	■	■	■	■	■			■						
Paracolobus mutiwa									■	■	■	■	■	■									
Paracolobus chemeroni					■																		
Paracolobus sp.					■		cf			■													
Colobinae indet.				■			A													new	new		
LORISIDAE																							
GALAGINAE																							
Galago sandimanensis																							
Galago howelli									■														
Galago demidovi									■														
Galago senegalensis																							
CARNIVORA																							
HERPESTIDAE																							
HERPESTINAE																							
Atilax sp.																							
Crossarchus transvaalensis																							
Crossarchus sp.																							
Cynictis penicillata																							
Helogale paleogracilis																							
Helogale hirtula													■	■									
Helogale kitafe									■														
Herpestes sp.																							
Herpestes paleoserengetensis																							
Herpestes ichneumon																							
Herpestes mesotes																							
Herpestes primitivus																							
Herpestes sanguineus																							
Herpestes delibis																							
Ichneumia albicauda																							
Mungus dietrichi																							
Mungus minutus																							
Suricatta suricatta																							
VIVERRIDAE																							
VIVERRINAE																							
Genetta sp.																							
Genetta tigrina																							
Viverra leakeyi																							
Viverra sp.																							
Pseudocivetta ingens																							
MUSTELIDAE																							
LUTRINAE																							
Aonyx capensis																							
Enhydriodon sp.				■																			

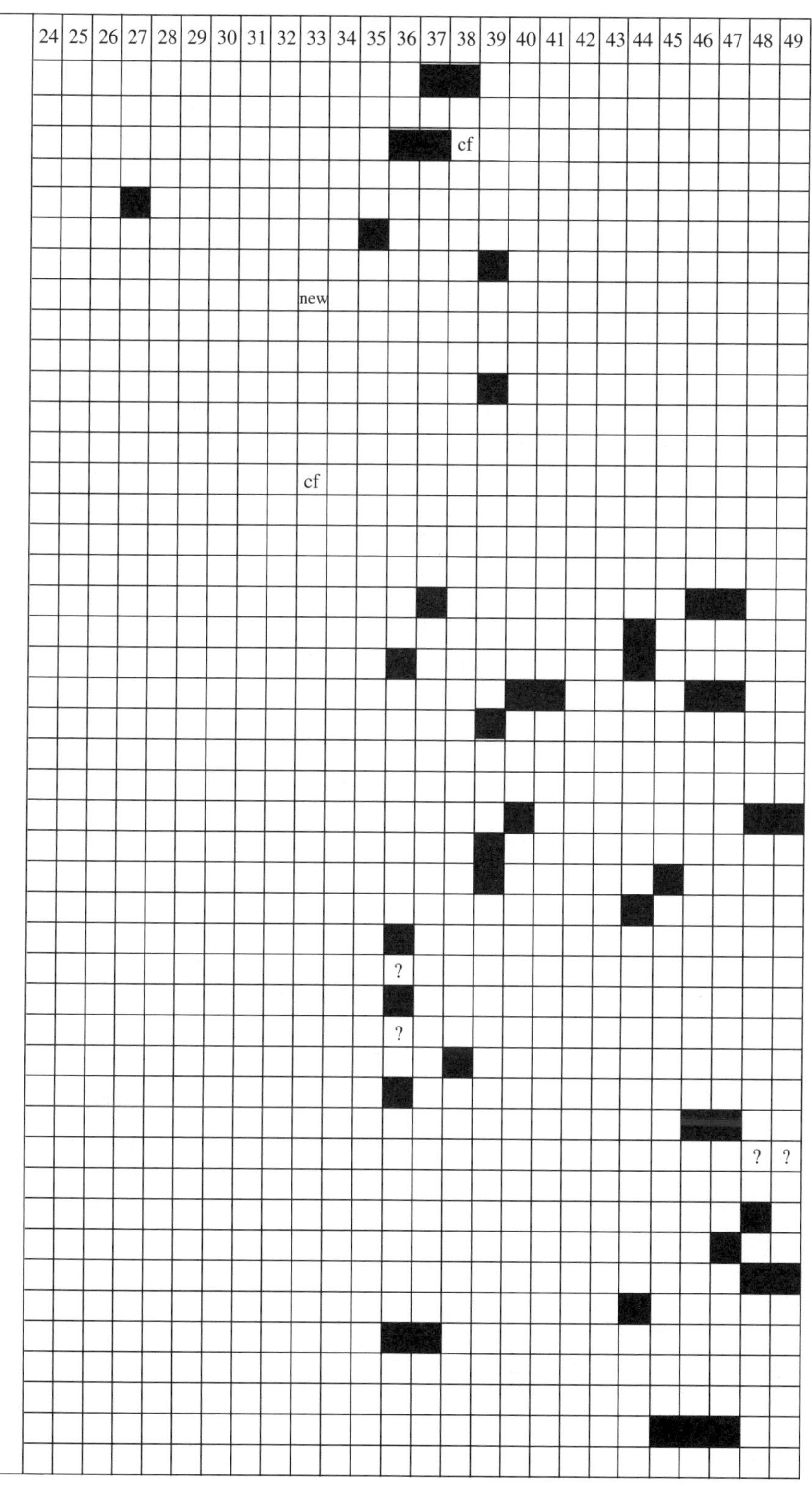
24 25 26 27 28 29 30 31 32 33 34 35 36 37 38 39 40 41 42 43 44 45 46 47 48 49
cf
new
cf
?
?
? ?

(*continued*)

Table A-1. (*continued*)

	1	2	3	4	5	6	7	8	9	10	11	12	13	14	15	16	17	18	19	20	21	22	23
Enhydriodon africanus																							
Enhydriodon pattersoni																							
Lutra lybica, hessica			■																				
Lutra maculicollis																							
MELLIVORINAE																							
Mellivora benfieldi																							
Mellivora capensis																							
MUSTELINAE																							
Plesiogulo monspessulanus																							
Propoecilogale bolti																							
FELIDAE																							
Adelphailurus sp.																							
Machairodus sp		■																					
Homotherium crenatidens								■	■	■	■	■		■						■	■	■	
Dinofelis barlowi									■											■			
Dinofelis piveteaui					■					■	■	■	■									cf	
Dinofelis sp.				cf				cf						?									
Megantereon cultridens				■	■		cf	■	?	■		■	■	■						cf.	■	■	
Panthera leo								?						■								cf	
Panthera pardus								■	■	■			■	■									
Panthera sp.				■																			
Acinonyx jubatus										■			■	■									
Felis serval																							
Felis caracal									■				■										
Felis sp.				■																			
PERCROCUTIDAE																							
Percrocuta senyureki		aff																					
HYAENIDAE																							
Adcrocuta eximia		■																					
Ikelohyaena abronia																							
Hyaenictitherium namaquensis		■																					
Hyaenictis hendeyi																							
Pachycrocuta brevirostris																							
Crocuta crocuta				■	■									■						■	■	■	
Hyaena hyaena								■	aff.	■		■	■	■						■	■	■	
Parahyaena brunnea								?		?													
Parahyaena sp																							
Chasmaporthetes silberbergi																							
Chasmaporthetes nitidula																							
Chasmaporthetes australis																							
Chasmaporthetes sp.				■					■				■										
Proteles sp																							
Hyaenidae indet							■																
CANIDAE																							
Nyctereutes sp							cf																
Canis brevirostris																							

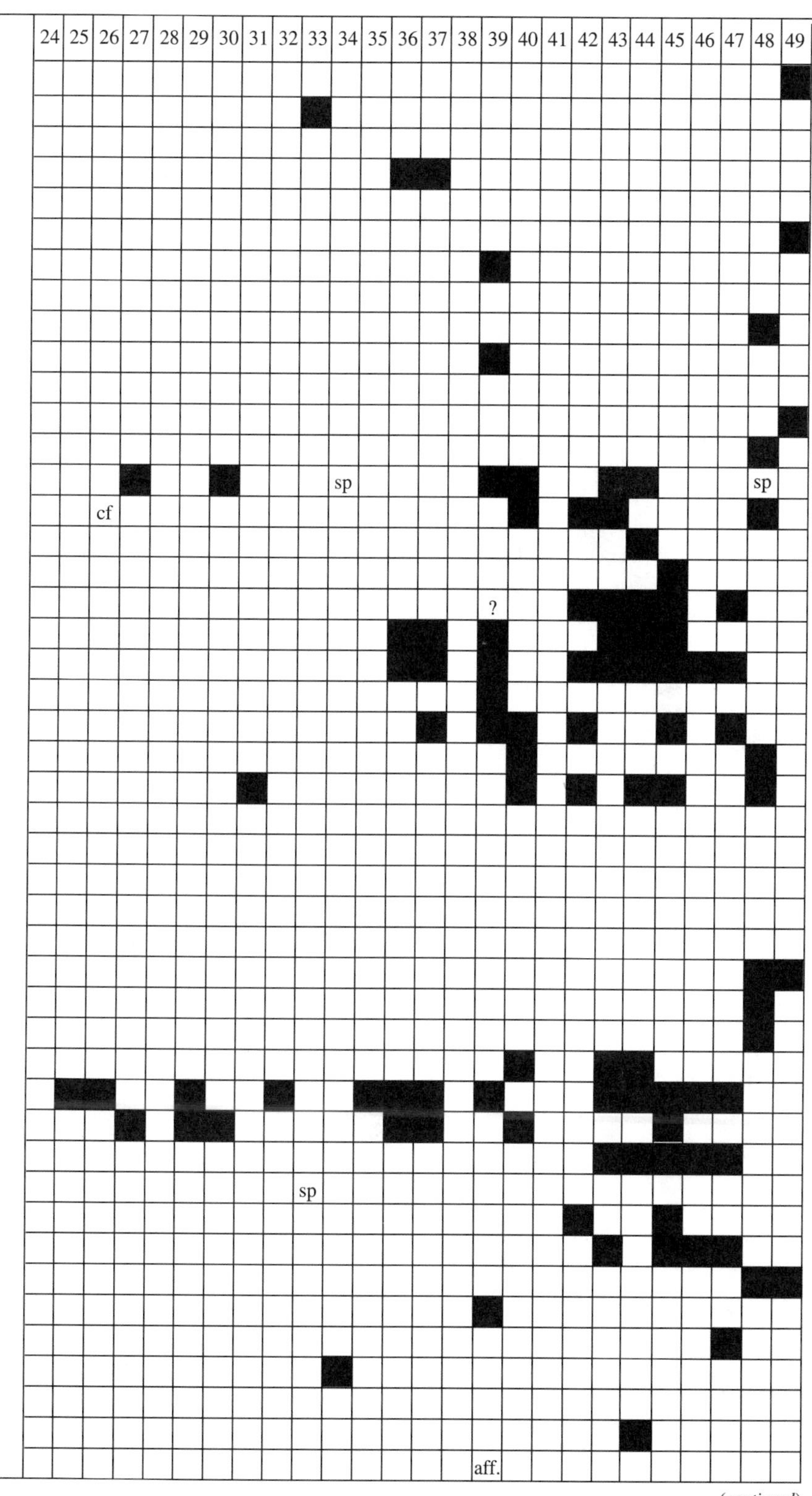
24 25 26 27 28 29 30 31 32 33 34 35 36 37 38 39 40 41 42 43 44 45 46 47 48 49
sp
sp
cf
?
sp
aff.

(continued)

Table A-1. (*continued*)

	1	2	3	4	5	6	7	8	9	10	11	12	13	14	15	16	17	18	19	20	21	22	23
Canis mesomelas				aff.																	■		
Canis adustus																							
Lycaon pictus																						?	
Canis sp.																							
Otocuon sp.																							
Megacyon sp.																							
Vulpes sp.																							
Vulpes pulcher																							
Vulpes chama																							
URSIDAE																							
Agriotherium africanum		cf					sp																
Indarctos atticus		■																					
PROBOSCIDEA							cf																
DEINOTHEROIDEA																							
DEINOTHERIIDAE																							
Deinotherium bozasi				■	■		sp	■	■	■	■		■							■	■		
Deinotheriidae indet																							
ELEPHANTOIDEA																							
GOMPHOTHERIIDAE																							
AMEBELODONTINAE																							
Amebelodon cyrenaicus		■																					
ANANCINAE																							
Anancus kenyensis							sp																
Anancus sp																							
ELEPHANTIDAE																							
Primelephas gomphotheroides																							
Stegotetrabelodon orbus																							
Stegotetrabelodon lybicus		■																					
Loxodonta adaurora				■						?	■									■			
Loxodonta exoptata																							
Loxodonta atlantica											cf.												
Loxodonta africana																							
Loxodonta sp.																							
Elephas ekorensis				■				■															
Elephas recki				■	■				■	■	■	■	■	■						■	■	■	
Elephas sp.																							
Mammuthus subplanifrons																							
Mammuthus sp.				■																			
Elephantidae indet.							■																
PERISSODACTYLA																							
EQUIDAE																							
Hipparion baardi																							
Hipparion hasumense																				■			
Hipparion ethiopicum										?	?			■						cf.	cf.	■	
Hipparion afarense				■	■																		

24	25	26	27	28	29	30	31	32	33	34	35	36	37	38	39	40	41	42	43	44	45	46	47	48	49
					cf.		■					■							■	■	cf	cf	cf		
																cf									
					cf.		?																		
															■					large		larg	large		
															cf.										
															?										
															■	■									■
																				■	■				
																■	■						■		
																									■
■	■	■	■		■	■			■		■														
										■															
									■	■															
																								■	■
										■															
										■															
■	■								■																
	■										cf				■										
						■																			
	■	■	■	■	■	■		■			■		■						■						
																cf				■	■				
																								■	■
	■																								
																								cf	cf
■	■	■	■																						
					■	■	■				cf														

(*continued*)

Table A-1. *(continued)*

	1	2	3	4	5	6	7	8	9	10	11	12	13	14	15	16	17	18	19	20	21	22	23
Hipparion sitifense		cf						■	■					cf.									
Hipparion lybicum																							
Hipparion albertense										cf.	cf.												
Hipparion cornelianum																				■			
Hipparion sp.		■			■		■													B	B		
Equus capensis																							
Equus koobiforensis																				■	■		
Equus tabeti																					cf.		
Equus burchelli																						cf	
Equus grevyi																						cf	
Equus sp.														■						■	■	■	
Equidae indet.																							
RHINOCEROTIDAE																							
Ceratotherium praecox				■			cf																
Ceratotherium simum				■	■			■	■	■	■			■						■	■	■	
Ceratotherium sp.																							
Diceros neumayri		■																					
Diceros bicornis				■				■	■		■			■						■	■	■	
Rhinocerotidae indet																							
CHALICOTHERIIDAE																							
Ancylotherium hennigi																							
ARTIODACTYLA																							
DICOTYLIDAE																							
Pecarichoerus africanus																							
SUIDAE																							
Nyanzachoerus devauxi		■																					
Nyanzachoerus syrticus		■																					
Nyanzachoerus jaegeri							■																
Nyanzachoerus kanamensis		■		■			■	■		■													
Notochoerus euilus				■	■			■	■	■													
Notochoerus scotti					■					■	■	■	■	■						■	■		
Notochoerus capensis																							
Notochoerus sp.														new									
Kolpochoerus afarensis				■	■																		
Kolpochoerus limnetes					■				■	■	■	■	■	■	■		■			■	■	■	
Kolpochoerus majus																	■						
Metridiochoerus andrewsi									■	■	■	■	■	■						■	■		
Metridiochoerus compactus																					■	■	
Metridiochoerus sp.					■																		
Metridiochoerus hopwoodi																							
Metridiochoerus modestus																							
Metridiochoerus sp.																							
Potamochoerus porcus																							
Potamochoerus sp.																							
Phacochoerus aethiopicus														cf.									
Hylochoerus meinertzhagen																							

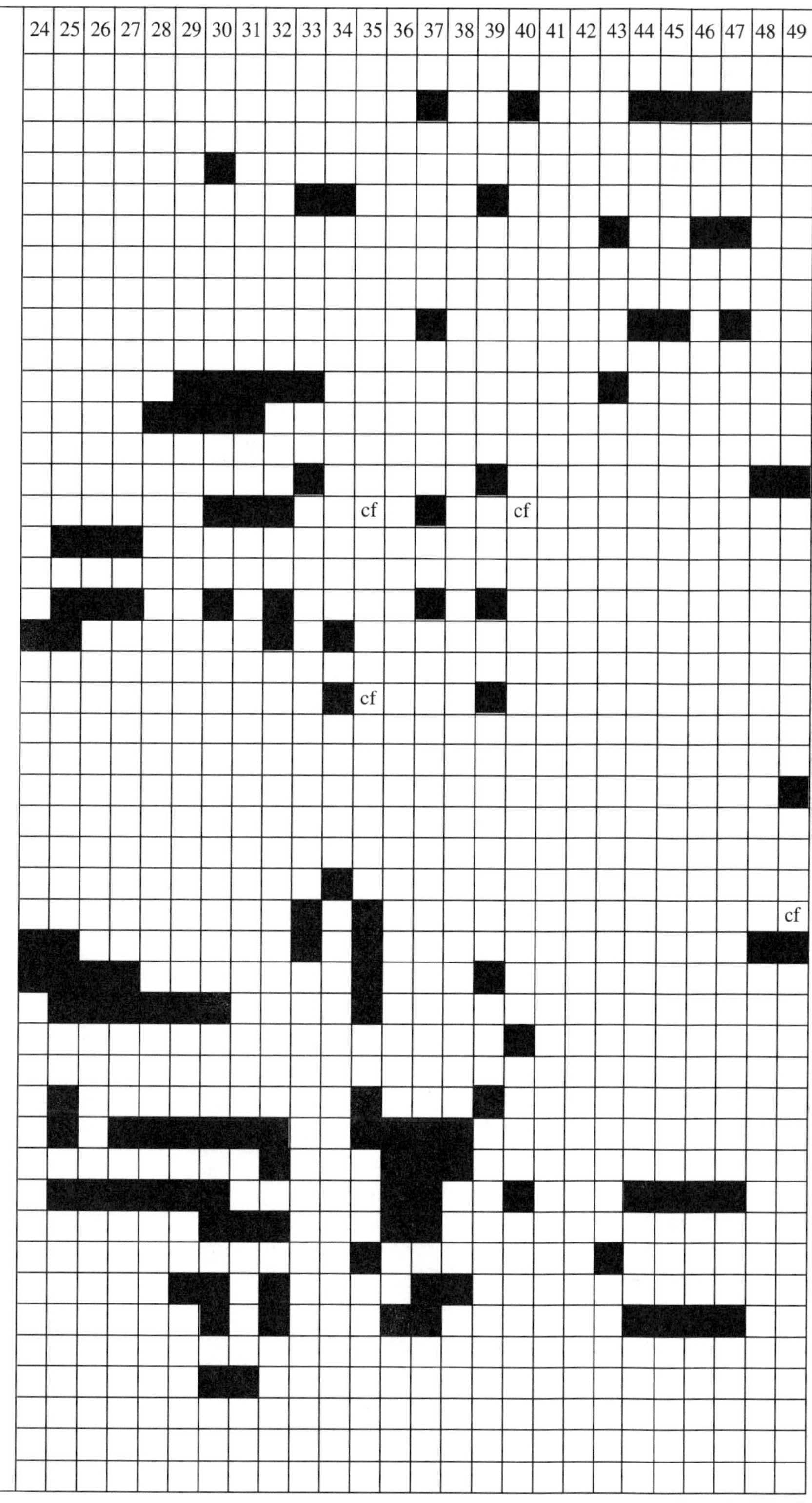
24
25
26
27
28
29
30
31
32
33
34
35
36
37
38
39
40
41
42
43
44
45
46
47
48
49
cf
cf
cf
cf

(continued)

Table A-1. (*continued*)

	1	2	3	4	5	6	7	8	9	10	11	12	13	14	15	16	17	18	19	20	21	22	23
Suidae indet																							
HIPPOPOTAMIDAE																							
Merycopotamus petrocchii		■																					
Hexaprotodon sahabiensis		■																					
Hexaprotodon imagunculus																							
Hippopotamus kaisensis																							
Hexaprotodon shungurensis									■														
Hexaprotodon protamphibius				aff.				aff.				?											
Hexaprotodon karumensis																				■	■		
Hexaprotodon sp.					■		■																
Hippopotamus gorgops														■						■	■	■	
Hippopotamus aethiopicus																				■	■	■	
Hippopotamus protamphibius										■	■	?		■									
Hippopotamus amphibius																							
Hippopotamus sp.					■						■			A+B									
Hippopotamidae indet																							
GIRAFFIDAE																							
Samotherium sp.		■																					
Sivatherium hendeyi																							
Sivatherium maurusium				■	■			■	■											■	■	■	
Sivatherium sp.										■	■			■									
Palaeotragus germaini																							
Giraffa stillei							aff																
Giraffa jumae				■			aff	?	■	cf.				cf.						■	■	■	
Giraffa gracilis				aff.				■	■	■	■			■						■	■	■	
Giraffa pygmaeus				cf.	cf.				■											■	■	■	
Giraffa camelopardalis																							
Giraffa sp.										new	new			new									
Giraffid indet																							
CAMELIDAE																							
Camelus sp.												■			■						■		
BOVIDAE																							
HIPPOTRAGINAE																							
ALCELAPHINI																							
Beatragus antiquus						■								■									
Beatragus hunteri																							■
Beatragus whitei					■																		
Beatragus sp.																							
Alcelaphus sp.																							
Sigmoceros sp.																							
Connochaetes taurinus																							
Connochaetes africanus																							
Connochaetes gentryi																							
Connochaetes sp.													■	■									
Damaliscus niro																							
Damaliscus sp.										?													

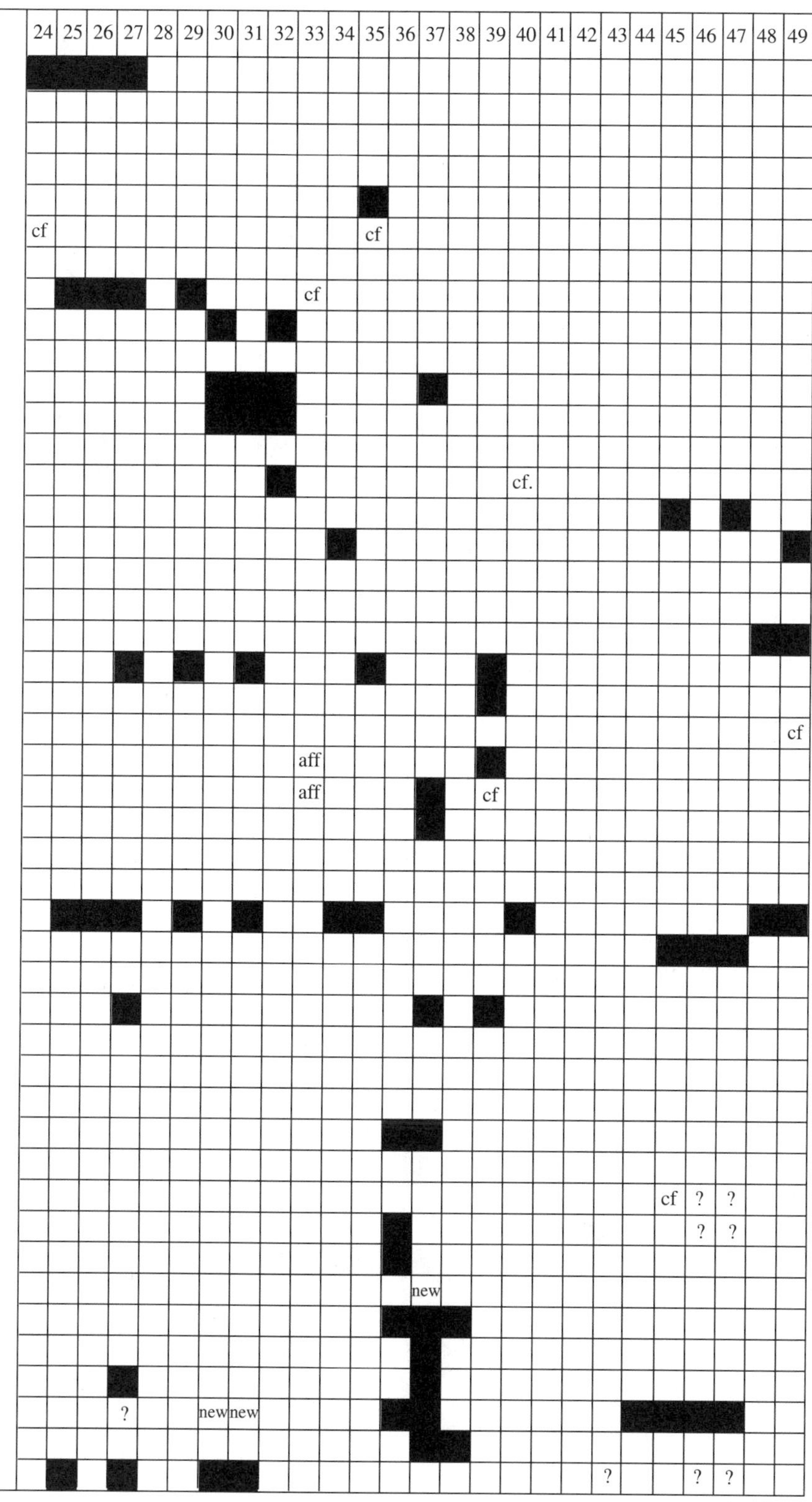
24 25 26 27 28 29 30 31 32 33 34 35 36 37 38 39 40 41 42 43 44 45 46 47 48 49
cf
cf
cf
cf.
cf
aff
aff
cf
cf ? ?
? ?
new
?
new new
? ? ?

(*continued*)

Table A-1. (*continued*)

	1	2	3	4	5	6	7	8	9	10	11	12	13	14	15	16	17	18	19	20	21	22	23
Damaliscus dorcas																							
Damaliscus lunatus																							
Damaliscus asfawi						■																	
Damaliscus agelaius																							
Awashia suwai					■																		
Damalops palaeindicus				?					aff.														
Megalotragus isaaci																				■		■	
Megalotragus kattwinke									cf.		?			■			?			■	■	■	
Megalotragus sp.																							
Parmularius braini										■													
Parmularius pandatus																							
Parmularius eppsi										■												■	
Parmularius altidens														?	■						■		
Parmularius angusticorn																					■		
Parmularius rugosus																							
Parmularius parvus																							
Parmularius sp.																					new	new	
Damalacra neanica																							
Damalacra acalla			■																				
Damalacra sp		cf																					
Rabaticeras arambourgi																							
Rabaticeras porrocornutus																							
Rabaticeras sp.																					■		
Numidocapra crassicornis	■																						
Alcelaphini indet.				new																			
HIPPOTRAGINI																							
Hippotragus equinus																							
Hippotragus gigas						■				■										■	■		
Hippotragus cookei																							
Hippotragus niger																							
Hippotragus sp.		new												■									
Wellsiana torticornuta																							
Praedamalis deturi									■														
Brabovus nanincisivus																							
Oryx sp.										■				■							■		
Hippotragini indet										■													
REDUNCINI																							
Dorcadoxa porrecticornis				■																			
Kobus sigmoidalis										■	■	■	■	■									
Kobus oricornis																		■					
Kobus kob				aff.					■	■	■	■		■			■					cf.	cf
Kobus ancystrocera									■	■		■		■		■						■	
Kobus ellipsiprymnus														■			?				■	■	■
Kobus leche																						cf.	cf
Kobus subdolus		■																					

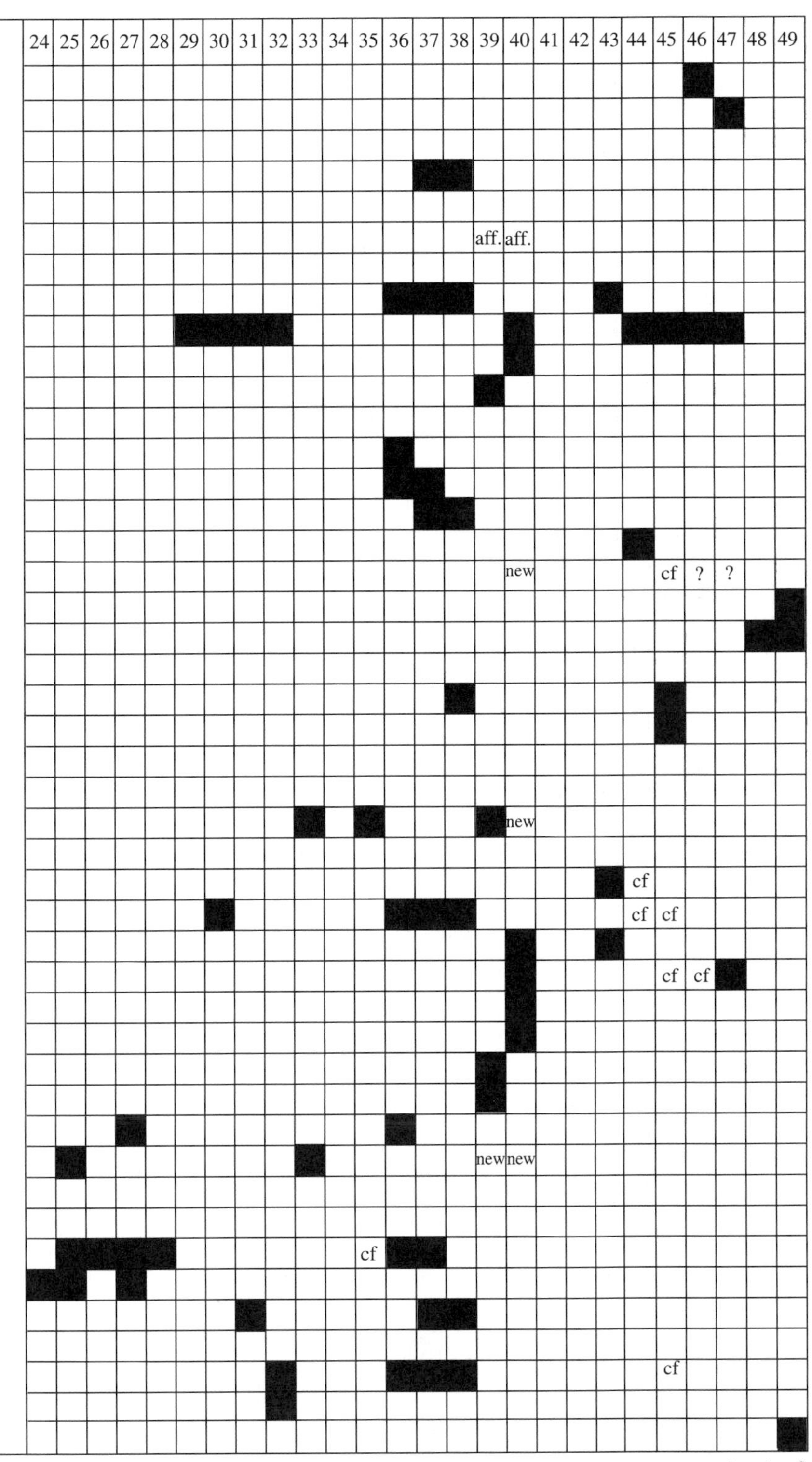
24 25 26 27 28 29 30 31 32 33 34 35 36 37 38 39 40 41 42 43 44 45 46 47 48 49
aff. aff.
new
cf ? ?
new
cf
cf cf
cf cf
new new
cf
cf

(*continued*)

Table A-1. (*continued*)

	1	2	3	4	5	6	7	8	9	10	11	12	13	14	15	16	17	18	19	20	21	22	23
Kobus sp.A				■																			
B				■																			
C				■																			
D																							
Kobus sp.																							
Menelikia lyrocera										■		■	■	■	■	■				■	■	■	
Menelikia leakeyi																		■					
Menelikia sp.										new	new												
Redunca darti			aff																				
Redunca arundinum										■													
Redunca fulvirufula																							
Redunca sp.									■	■		■		■	■		■				■		
Reduncini indet																							
PELEINI																							
Pelea capreolus																							
Pelea sp.																							
ANTILOPINAE																							
ANTILOPINI																							
Gazella janenschi																							
Gazella vanhoepeni																							
Gazella granti																							
Gazella praethomsoni													■	■	■								
Gazella pomeli	■																						
Gazella sp.		■		■	■			■												■	■	■	
Antilope subtorta										aff.													
Antidorcas recki									■				■	■	■					■	■	■	
Antidorcas australis																							
Antidorcas bondi																							
Antidorcas marsupialis																							
Antidorcas sp.																							
Prostrepsiceros libycus		■																					
Antilopini indet.																	■						
NEOTRAGINI																							
Raphicerus campestris																							
Raphicerus paralius																							
Raphicerus sp.		■		?						■		■		?									
Orebia ourebi																							
Oreotragus oreotragus																							
Oreotragus major																							
Oreotragus sp.																							
Madoqua aviflumnis				cf.																			
Madoqua sp.																					■		
Neotragini indet.							■						■	■									
BOVINAE																							
BOSELAHPHINI																							
Miotragoceros cyrenaicus		■																					

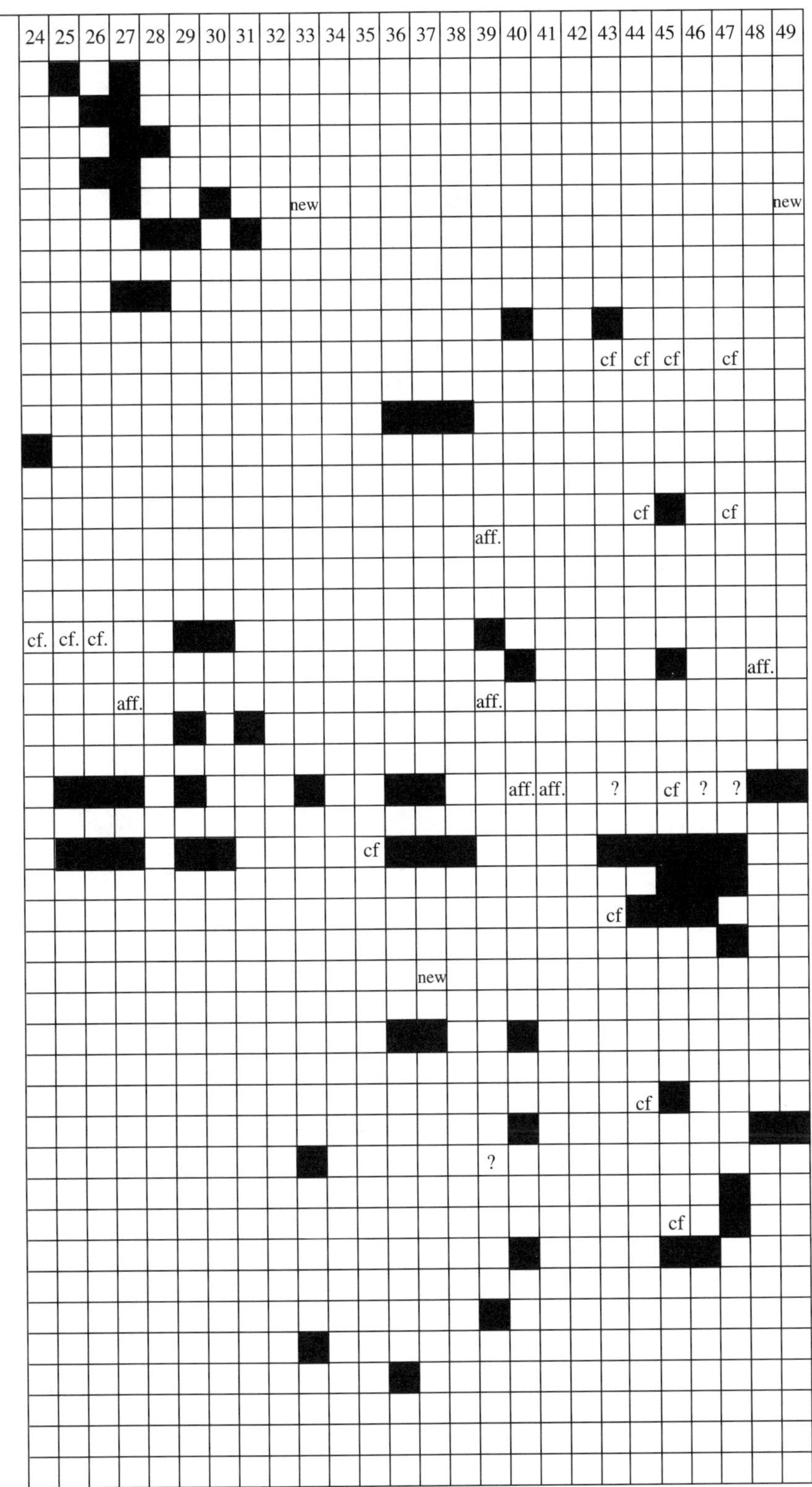
24 25 26 27 28 29 30 31 32 33 34 35 36 37 38 39 40 41 42 43 44 45 46 47 48 49
new
new
cf cf cf cf
cf cf
aff.
cf. cf. cf.
aff.
aff.
aff.
aff. aff. ? cf ? ?
cf
cf
new
cf
?
cf

(continued)

Table A-1. (*continued*)

	1	2	3	4	5	6	7	8	9	10	11	12	13	14	15	16	17	18	19	20	21	22	23
Mesembriportax acrae																							
Boselaphini sp.																							
TRAGELAPHINI																							
Tragelaphus kyaloae																		■					
Tragelaphus nakuae				■	cf.			■	■	■	■	■	■	■	■			■		■	■		
Tragelaphus gaudryi						■						■	■	■									
Tragelaphus scriptus																							
Tragelaphus angasii																							
Tragelaphus strepsiceros														■						■	■	■	
Tragelaphus pricei										■													
Tragelaphus sp.							■			■	■	■	■										
Taurotragus arkelli																							
Taurotragus oryx																							
Taurotragus sp.									■														
Tragelaphini sp.				new																			
BOVINI																							
Leptobos syrticus		■																					
Ugandax gautieri				cf																			
Ugandax sp.				new																			
Syncerus acoelotus										■		?		?									
Syncerus caffer																							
Syncerus sp.									■														
Pelorovis turkanensis																				■			■
Pelorovis oldowayensis										■												■	
Pelorovis sp.				?					■	■	■		■	■	■					new	new	new	
Simatherium demissum																							
Simatherium kohllarseni																							
Bovini indet.							■																
CEPHALOPHINAE																							
CEPHALOPHINI																							
Cephalophus sp.																							
Cephalophini indet																							
AEPYCEROTINAE																							
AEPYCEROTINI																							
Aepyceros shungurae										■	■	■	■	■	■		?	■					
Aepyceros melampus																	?	■					
Aepyceros sp.				■	■			new	new				new	new						■	■	sp	
CAPRINAE																							
OVIBOVINI																							
Damalavus makapaani																							
Makapania broomi																							
Makapania sp.				aff.																			
Ovibovini indet.										■	■			■									
CAPRINI																							
Caprini indet											1		1										

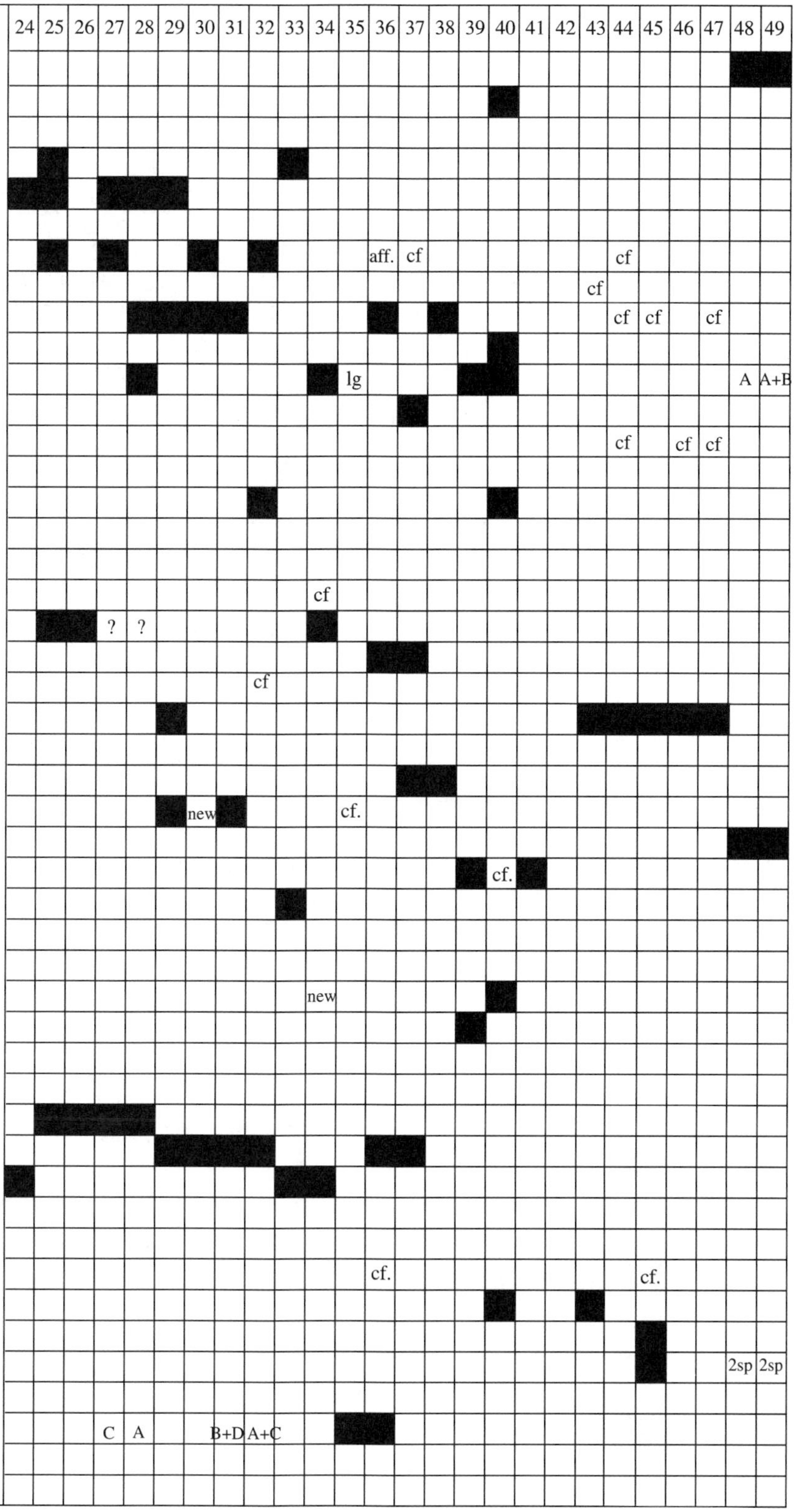

24	25	26	27	28	29	30	31	32	33	34	35	36	37	38	39	40	41	42	43	44	45	46	47	48	49
												aff.	cf							cf					
																			cf						
																				cf	cf		cf		
											lg													A	A+B
																				cf		cf	cf		
										cf															
			?	?																					
								cf																	
						new					cf.														
																cf.									
										new															
												cf.									cf.				
																								2sp	2sp
			C	A			B+D	A+C																	

Table A-2. Smaller mammals. Table sorted as for Table 1.

	1	2	3	4	5	6	7	8	9	10	11	12	13	14	15	16	17	18	19	20	21	22	23
SCIURIDAE																							
Sciuridae indet.																							
cf. Atlantoxerus getulus		■																					
?Xerus				■																			
Xerus sp.indet.																							
Xerus janenschi																							
Xerus erythropus									■	■													
Xerus cf. inauris																							
Paraxerus ochraceus									■	■													
Paraxerus sp.																							
CTENODACTYLIDAE																							
Saimys sp.		■																					
DIPODIDAE																							
Jaculus orientalis												■	■	■								■	
MURIDAE																							
Cricetomyinae																							
Saccostomus sp.																						■	
Saccostomus major																							
Saccostomus cf. mearnsi																							
DENDROMURINAE																							
Dendromus averyi																							
Dendromus darti																							
Dendromus sp.																							
Dendromus sp1																							
Dendromus sp.2																							
D. mesomelas																							
Steatomys sp.																							
Steatomys cf.opimus																							
Steatomys sp.2																							
Steatomys sp.1																							
Malacothrix typica																							
Malacothrix sp.																							
GERBILLINAE																							
Protatera		3 sp.																					
Gerbillus sp.																							
Tatera inclusa																							
Tatera brantsii																							
Tatera leucogaster																							
Tatera gentryi																							
Tatera sp.A				■	■	■	■	■	■													■	
Tatera sp.B																							
Desmodillus sp.																							
Taterillus																							
MYOCRICETODONTINAE																							
Myocricetodon cherifensis		aff.																					

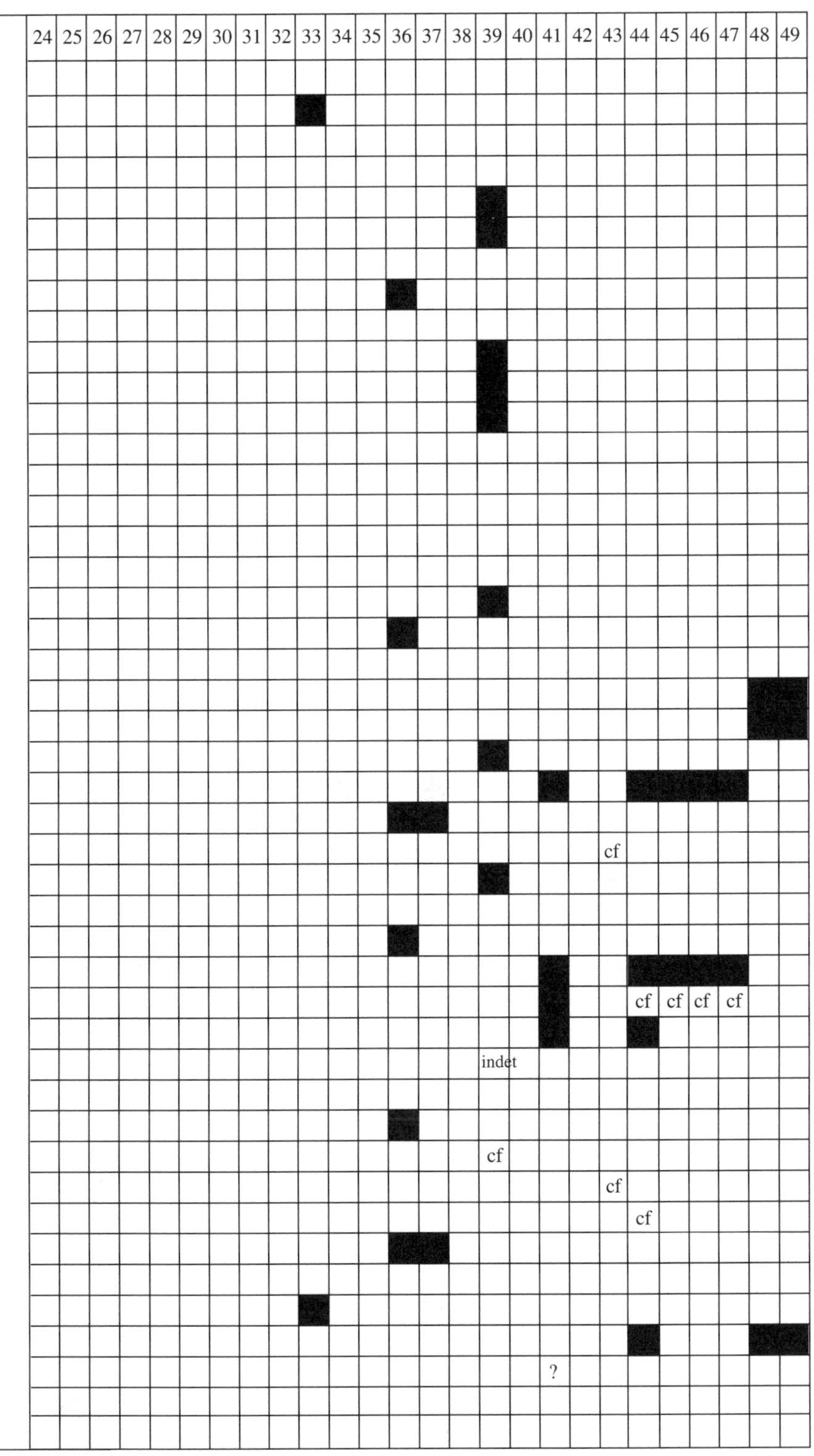
24 25 26 27 28 29 30 31 32 33 34 35 36 37 38 39 40 41 42 43 44 45 46 47 48 49
cf
cf cf cf cf
indet
cf
cf
cf
?

(continued)

Table A-2. (*continued*)

	1	2	3	4	5	6	7	8	9	10	11	12	13	14	15	16	17	18	19	20	21	22	23
MURINAE																							
Progonomys sp.		■																					
Arvicanthis sp.									■													■	
Arvicanthis primaevus																							
Arvicanthis cf.niloticus																							
Aethomys modernis																							
Aethomys adamanticola																							
A. cf.chrysophilus																							
A. cf. namaquensis																							
Aethomys sp.																						■	
Aethomys lavocati																							
A.deheinzelini												■	■	■									
Acomys mabele																							
Acomys coppensi				■																			
Acomys sp.													■										
Dasymys																							
Dasymys sp.nov.																							
Euryotomys pelomyoides																							
Grammomys sp.																							
Golunda gurai				■	■	■	■	■	■	■													
Millardia taiebi				■																			
Lemniscomys aff. striatus									■	■		■											
Lemniscomys sp.																							
Mastomys minor									■	■		■	■	■								■	
Mastomys cinereus																							
Mastomys sp.																							
Mus sp.				■																		■	
Mus aff. minutoides									■														
Mus petteri																							
Oenomys tiercelini				■																			
Oenomys sp.										■													
Oenomys olduvaiensis																							
Pelomys sp.													■										
Pelomys dietrichi																							
Myomyscus sp.																							
Praomys sp.				■																			
Rhabdomys sp.																							
Saidomys natrunensis			■																				
Saidomys afarensis				■																			
Saidomys sp.															■								
Thallomys jaegeri									■	■													
Thallomys laetolilensis																							
Thallomys sp.																							
Thallomys quadrilobatus												■	■	■								■	
Zelotomys sp.																							
Zelotomys leakeyi																							

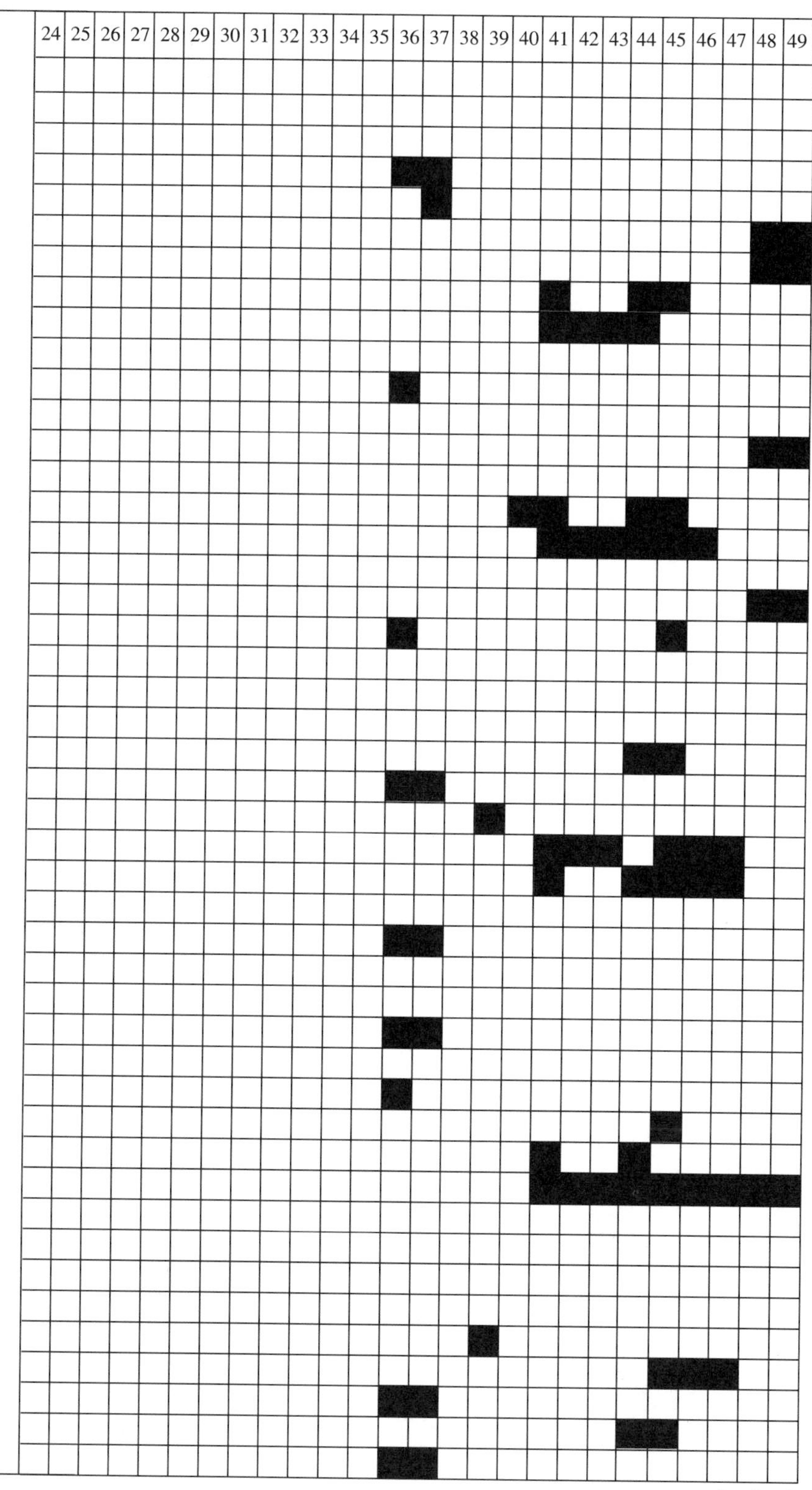
24 25 26 27 28 29 30 31 32 33 34 35 36 37 38 39 40 41 42 43 44 45 46 47 48 49

(continued)

Table A-2. (*continued*)

	1	2	3	4	5	6	7	8	9	10	11	12	13	14	15	16	17	18	19	20	21	22	23
MYSTROMYINAE																							
Mystromys hausleitneri																							
Mystromys pocockei																							
Mystromys antiquus																							
Proodontomys cookei																							
OTOMYINAE																							
Otomys gracilis																							
Otomys sp.																							
Otomys slogetti																							
Otomys petteri																							
PETROMYSCINAE																							
Stenodontomys darti																							
Stenodontomys saldanhae																							
RHIZOMYINAE																							
Tachyoryctes pliocaenicus				■																			
PEDETIDAE																							
Pedetes laetoliensis																							
Pedetes sp.																							
MYOXIDA																							
Graphiurus cf.monardi																							
Graphiurus sp.																							
BATHYERGIDAE																							
Cryptomys broomi																							
Cryptomys robertsi																							
Cryptomys hottentotus																							
Gypsorhychus makapani																							
Georychus capensis																							
Bathyergus hendeyi																							
Heterocephalus quenstedti																							
Heterocephalus jaegeri																							
H. atikoi												■	■	■									
HYSTRICIDAE																							
Xenohystrix crassidens				■	■	■	■	■	■														
Hystrix leakeyi																							
Hystrix makapanensis																							
Hystrix sp.																						■	■
H. africaeaustralis																							
Hystrix cristata				■																			
THRYONOMYIDAE																							
Thryonomys sp.																						■	
Thryonomys gregorianus									■	■	■	■	■	■									
Thryonomyidae swinderianus														■	■								

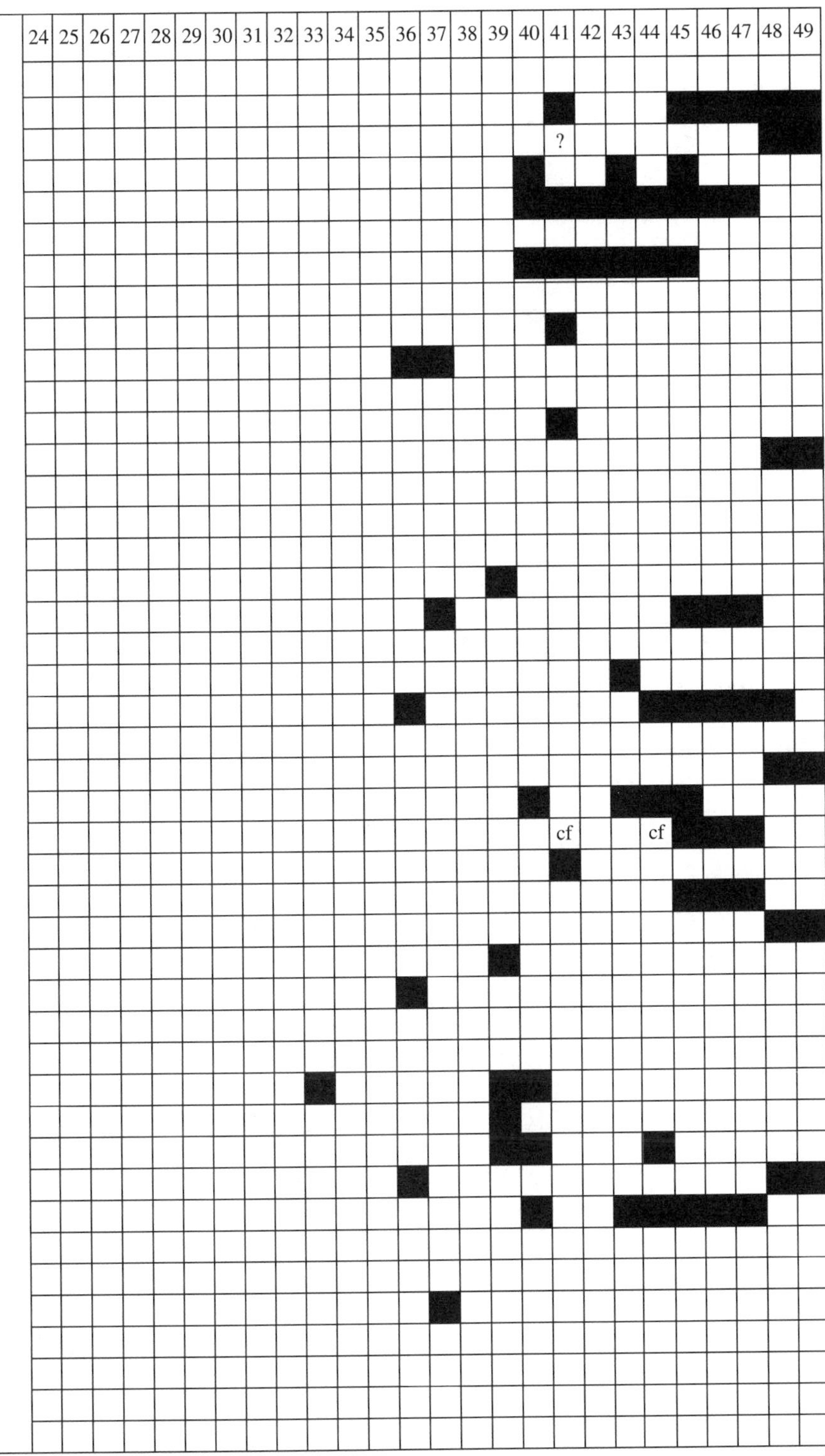
24 25 26 27 28 29 30 31 32 33 34 35 36 37 38 39 40 41 42 43 44 45 46 47 48 49
?
cf
cf

(continued)

Table A-2. (*continued*)

	1	2	3	4	5	6	7	8	9	10	11	12	13	14	15	16	17	18	19	20	21	22	23
INSECTIVORA																							
MACROSCELIDIDAE																							
Elephantulus antiquus																							
E. broomi																							
E. fuscus leakeyi																							
Elephantulus sp.																							
E. brachyrhynchus																							
Rhynchocyon pliocaenicus																							
Macroscelides proboscideus																							
SORICIDAE																							
Myosorex sp.																							
Myosorex varius																							
Myosorex cafer																							
Myosorex robinsoni									■														
Sylvisorex granti																							
Suncus varilla																							
S. infinitessimus																							
Suncus sp.																							
Suncus haesaertsi									■														
Suncus aff.lixus									■														
Suncus shungurensis									■														
Diplomesodon fossorius																							
Crocidura hindei																							
?Crocidura sp.																							
Crocidura aethiops									■	■	■	■	■										
Crocidura aff. dolichura									■													■	
Crocidura aff. nana																						■	
Crocidurinae indet.		■																					
ERINACEIDAE																							
Erinaceus broomi																							
CHRYSOCHLORIDAE																							
Chrysochloris sp.																							
Calcochloris hamiltoni																							
A. hamiltoni																							
A. gumingi																							
Chlorotalpa sp.																							
Chlorotalpa spelea																							
C. villosus																							
CHIROPTERA																							
RHINOLOPHIDAE																							
RHINOLOPHINAE																							
Rhinolophus darlingi																							
R. capensis																							
R. clivosus																							

	24	25	26	27	28	29	30	31	32	33	34	35	36	37	38	39	40	41	42	43	44	45	46	47	48	49
										cf.																
										■							■	■	■	■						■
										■																
																■					■				■	■
																	cf.									
														■												
																				■						■
																■										
																	■									
																				cf.	cf.	cf.				
										■								■	■	■	■					■
										cf																
										cf							cf	■			■	■	■			■
										cf							■	■								
																■										
										■																
																	■									
										cf																
														■												
										■																
																■									■	
																	■									
																	■									
																					cf.	cf.	cf.			
																				■						
																		■								
																					cf.	cf.	cf.			
																	cf.	cf.								
																					cf.	cf.	cf.			
																	cf.									

(*continued*)

Table A-2. (*continued*)

	1	2	3	4	5	6	7	8	9	10	11	12	13	14	15	16	17	18	19	20	21	22	23
HIPPOSIDERINAE																							
Hipposideros aff.cyclops									■														
H. kaumbului													■										
Hipposideros aff. camerunensis													■										
MYZOPODIDAE																							
Myzopoda sp.																							
VESPERTILIONIDAE																							
VESPERTILINAE																							
Eptesicus hottentotus																							
E. bottae																							
Eptesicus sp.																							
Myotis sp.																							
cf. Nycticeius schlieffeni																							
cf. Pipistrellus rueppelli																							
Scotophilus sp.																						■	
MINIOPTERINAE																							
Miniopterus sp.																							
M. schreibersi																							
EMBALLONURIDAE																							
Coleura muthokai													■										
Taphozous abitus									■	■	■	■	■										
NYCTERIDAE																							
Nycteris sp.																						■	
PTEROPODIDAE																						indet.	
Eidolon aff.helvum									■														
MEGADERMATIDAE																							
Cardioderma sp.																							
LAGOMORPHA																							
Pronolagus sp.																							
Serengetilagus praecapensis																							
Lepus capensis												■	■	■									
HYRACOIDEA																							
PROCAVIIDAE																							
Gigantohyrax maguirei									■	■													
Heterohyrax brucei									■														
Procavia antiqua																							
Procavia transvaalensis																							

24	25	26	27	28	29	30	31	32	33	34	35	36	37	38	39	40	41	42	43	44	45	46	47	48	49
									■																
									cf							cf.				cf.	cf.				
																■									
															■										
									■																
									■																
									■																
																■									
									cf																
									■																
															■	■								■	
													■												
															cf	■	■	■	■	■	■	■			■
																			■	■	■	■			

Glossary

abiotic nonliving (vs. biotic)

ablation weathering of rock by erosion, e.g., wind

Acheulean ca. 0.4–0.15 m.y.; paleolithic period following the Oldowan, characterized by stone tools with bifacial chipping

adaptation adjustment to change in the environment; any character contributing to a life form's evolutionary success

adaptationism a view of the evolutionary process that adaptation is the predominant force of evolutionary change

adaptive radiation diversification of an ancestral species into several descendant lineages

advection transfer of air with change of temperature

aeolian caused by the action of wind

albedo radiation reflected by surface

allele any of the variants of a gene coding for a variable character

Allen's Rule decrease in extremity size associated with increasing latitude or altitude (cf. Bergmann's Rule)

allochemistry exchange of chemical compounds between species (e.g., scents)

allometry change in body proportions with increasing body size

allopatry geographical separation of populations

allopatric speciation formation of new species following geographical separation that prevents populations from interbreeding

allospecies allopatric species; any taxon part of a superspecies

allotaxon allopatric taxon; any subspecies part of a polytypic species

alpha taxonomy designations of populations and species taxa based on all sorts of information

altricial relating to young born dependent and helpless (vs. precocial)

alveolus bony cavity enclosing a tooth

ameloblast dental cell involved in enamel formation

amino acid one of twenty basic constituents of proteins

amphibiomic able to live in two different biomes

anagenesis evolutionary change within a lineage (from population to species)

anthropocentric a perspective in terms of human values

anthropogenic caused or influenced by human action

Anthropoidea simian primates; comprising Platyrrhini (New World monkeys) and Catarrhini

anthropomorphic an assessment of animal behavior in terms of human attributes *apex/apical* toward the tip of a tooth root

apomorphy derived character

appositional enamel relating to the cuspal part of the tooth (cf. imbricational)

arboreal living in or relating to trees

Ardipithecus proposed extinct hominoid genus; suggested to be sister taxon of Hominidae (4.5 m.y., East Africa)

arid relating to continental regions with water deficit (<250 mm; evapotranspiration exceeding precipitation); supporting only sparse vegetation (vs. semiarid, humid)

aridification development of barrenness due to drought

articular relating to, or forming, a joint

Artiodactyla order of even-toed ungulates; including Bovidae, Camelidae, Cervidae (deer), Giraffidae, Hippopotamidae, Suidae (pigs)

aseasonal without significant change throughout the year

atmosphere sum of layers of air surrounding the Earth (e.g., troposphere)

atmospheric pressure weight of air mass, varying due to convection

atrophy reduced development (cf. hypertrophy)

australopithecus hominid genus of "gracile" australopithecine

autapomorphy derived character not shared with sister taxa

autecology ecology of an individual life form

autocatalytic evolution (sensu Kauffmann) evolutionary change as a result of the self-organizing ability of an organism independent of natural selection

avifauna bird wildlife

axial in direction of the predominant orientation of a geologic structure; relating to the North-South axis of the Earth; term commonly used also in anatomy as related to structural long axes

axiom established principle

bar low ridge

basalt dark, fine-grained volcanic rock

basicranial flexion relating to angulation of the base of the skull

basin geologic formation of rock beds dipping toward a center (vs. dome)

bauplan structural design, usually referring to ancestral constraints

bauxite sedimentary clay containing aluminum oxide

bedrock undisturbed underlying rock

benthic in bottom waters

Bergmann's Rule increase in body size associated with increasing latitude or altitude (*see also* Allen's Rule)

Beringia Bering Strait and neighboring regions when unflooded during cool periods

bilophodonty transverse ridges formed by pairs of aligned cusps; molar pattern characteristic of Cercopithecoidea Phalangeriform marsupials, and many other groups of mammals

bimodal having two prevailing peaks on a graph

biochronology dating method based on absolutely or relatively dated fossil evidence

biogeography geographical distribution of life forms

biological species concept (*BSC*) collective of individuals interbreeding among themselves and who are sympatrically either ecologically, reproductively, or genetically separated from other such entities (*see* reproductive isolation)

biomass total dry weight of living matter in a locality

biome community typical of a geographical region comprising specific life forms as determined by climate and substrate; flora and fauna at a geographic scale (e.g., savanna)

biosphere sum of localities with life forms

biostratigraphy science of geologic correlation based on the distribution of fossils in different strata

biota fauna, flora and microorganisms of a locality

biotic living (vs. abiotic)

biotic zone geographical region comprising a specific combination of fauna and flora; bioregion (e.g., Namib)

bivariate occurring jointly

boreal with northern (vs. austral) affinities; relating to the climate characteristic of semiarid (400–650 mm rainfall) cold-temperate regions, with short summers and long, cold winters, supporting conifer forest (cf. Alpine, polar, Mediterranean)

bottleneck bias in gene frequencies in populations that have suffered a severe reduction in the number of reproductive individuals

bottom water cold and dense water at the sea bottom (*see* thermohaline)

boundary current ocean current longitudinally flowing along continental margins

brachiation locomotion in trees by arm swinging; a plesiomorphy in Hominoidea

Brachiopoda phylum of marine invertebrates

breccia rock consisting of cemented rubble

BSC see biological species concept

buccal toward the cheek

buccinator cheek muscle

bushveld semiarid open bushland savanna in South Africa

C_3 plants with the Calvin photosynthetic pathway, with a three-carbon compound as the first product; including aquatic plants, most trees, shrubs, and herbaceaous plants, as well as some grasses and sedges; adapted to low temperature and low irradiation in temperate, montane, and wetland habitats, thus outcompeting C_4 plants in shaded localities such as forests, at high altitudes and latitudes; discrimination against atmospheric carbon-13 resulting in distinctly negative $\delta^{13}C$ values

C_4 plants with the Hatch-Slack photosynthetic pathway, with a four-carbon compound as the first product; including most nonwoody tropical grasses and sedges, as well as some shrubs; adapted to low atmospheric carbon dioxide concentrations, high temperature and high irradiation in water-stressed habitats, thus outcompeting C_3 plants in open localities such as savannas, at low altitudes and latitudes; less discrimination against atmospheric carbon-13 resulting in less negative $\delta^{13}C$ values

CAM Crassulacean acid metabolism; including mostly succulent plants with switching of the photosynthetic pathway in response to environmental factors, allowing temporal separation of carbon assimilation and fixation; use of both C_3 and C_4 photosynthetic pathways, resulting in intermediate $\delta^{13}C$ values

caballine horselike

calcareous of calcium carbonate

calcite crystalline calcium carbonate

calcrete layer containing calcium carbonate forming at the surface of semiarid regions

cambium growth tissue in roots and shoots

caniniform in the shape of a canine tooth

carbon element with two stable, ^{12}C (98.9%) and ^{13}C (1.1%), and one radioactive isotope, ^{14}C (used in radiometric dating); bimodal distribution of $\delta^{13}C$ values of terrestrial C_3 and C_4 plants allowing paleoecologic reconstruction of tropical ecosystems (*see* PDB)

carbon cycle circulation of carbon (photosynthesis, assimilation, decomposition, combustion)

carbonatite igneous rock rich in carbonate minerals

catalyst promoter (e.g., enzyme)

Catarrhini Old World Anthropoidea; comprises Cercopithecoidea and Hominoidea and several other basal lineages

causality relationship between an evolutionary event and its effect; possible relationship between correlated events

centrifugal speciation formation of new species at the periphery of the ancestral species

centripetal selection stabilizing selection

Cercopithecoidea catarrhine superfamily of Old World monkeys

cerebrum enlarged part of forebrain

cervix neck; necklike part

chlorite any chlorous silicate mineral

chorology biogeography

chronocline gradual variation in a lineage over geological time; fossils displaying chronologically successive variation in a character (*see* cline)

chronospecies any descendant sufficiently different from the ancestor to deserve a taxonomic recognition; designation of fossil species on grounds of morphological and temporal aspects; successive morphs in an evolutionary lineage (e.g., *Homo erectus*)

Cichlidae family of tropical freshwater fish

cirque steep-walled, semicircular hollow on a mountain due to glacial erosion

clade collective of populations or species (or higher taxa) that descended from a common ancestor; monophyletic group

cladism classification of life forms according to sister-group relationships

cladistics establishment of cladistic relationships on grounds of shared derived characters, indicative of a common ancestor

cladogenesis evolutionary branching of a lineage

cladogeny branching of a lineage from populations of the same species to different lineages

cladogram diagram depicting nodes of divergence, as indicated by shared derived characters

classification arrangement of life forms in hierarchical Linnaean categories of successively subordinate ranks (e.g., kingdom, phylum, class, order, family, genus, species)

climate prevailing weather of a locality or period (latitude, altitude, distance from the sea; irradiation, temperature, relative humidity, precipitation, atmospheric pressure, wind)

climatic forcing change in climate conditions initiated by some atmospheric agent arising from outside the climate system

cline gradual variation along a geographic gradient within a species (cf. chronocline, morphocline)

coeval of the same age

coevolution causally interrelated evolution of associated life forms; evolutionary change in one life form influencing evolutionary change in another

cogenetic of the same origin

colluvium sediment accumulated at the base of an incline

Colobidae/Colobinae cercopithecoid family/subfamily of leaf-eating monkeys

community assemblage of life forms interacting in a locality (sensu Andrews & Humphries)

community ecology interactions among species at small scales of time and space

competition struggle for access to limited resources

competitive coexistence continued existence of species in the same biogeographic province despite their competitive interaction

condyle bony process forming a joint

congener taxon of the same genus

conspecific any member of the same species

constraint limit to evolutionary change due to environmental, genomic, and historical contingencies (*see* initial and boundary conditions)

continental relating to the climate characteristic of continental interiors, with low and seasonal precipitation and much seasonal fluctuation in temperature; summer rain (vs. maritime)

continental shelf shallow sea between the continental coast and the deep ocean

convection ascension of warm air (condensation) and descension of cold air (precipitation) (e.g., warm equatorial air cooling over the subtropics); rising of light warm water and sinking of dense cold water

convergence the opposite of homology, evolution of similarity not based on common descent

conveyer-belt circulation Gulf stream having a warming influence on Western Europe climate, interrupted during Heinrich events

cordillera group of parallel mountain ranges

corolla whorl of flower petals

corollary proposition deduced from a principle

cosmopolitanism worldwide occurrence

craniofacial relating to the skull

crepuscular active in the twilight (cf. diurnal, nocturnal)

crista obliqua ridge on hominoid upper molars linking protocone and metacone; it incorporates the metaconule

cross-striation cross-barring of enamel prisms; circadian incremental markers in enamel

cryosphere sum of localities frozen throughout the year

crypt alveolus

cursorial running

cusp protuberance on the occlusal surface of a tooth

cuspule small cusp

dambo treeless wetland, waterlogged in wet and grassy in dry seasons (cf. grassland, marsh)

Dansgaard-Oeschger event short climatic cycles observable in Greenland ice cores

delta 1. (δ) ratio of two stable isotopes in relation to an international standard (*see* PDB); 2. triangular river mouth with accumulating sediments

deme local population of a species

demography distribution of life conditions such as age and mortality within a population

demographic sink any population with death rate exceeding birth rate

depositional accumulating sediments

desert habitat in arid regions with erratic or little rainfall (<250 mm)

detritus debris resulting from the decay of organic matter; any transported material originating from the disintegration of rocks

diachronic/diachronous nonidentical time periods (vs. isochronous); of an extended time period (vs. synchronic)

diagenesis process of physical and chemical changes in the sediments after deposition (compaction, cementation, etc.)

Diatoma class of unicellular alga

dichopatry division of populations by emerging geographical barriers

dichotomous branching into two branches

dicotyledon any plant with two embryonic seed leaves, even-numbered structures and branching venation, including all woody plants (e.g., *Nothofagus,* extinct beech; cf. monocotyledon)

dimorphism (*see* sexual dimorphism)

dispersal range expansion or shift; dispersion of life forms from their original range *distal* away from the center of any point of reference

distinct preference community (sensu Rosenzweig) mode of competitive coexistence, with each species outcompeting all others in a specific habitat (vs. shared preference/tolerance community)

distribution drift (sensu Vrba) any distribution change enforced by environmental change

diurnal active during the day (vs. crepuscular); daily

diversity species richness; variety and relative abundance of different life forms present in a locality; morphological or genetic differentiation of particular populations (sensu Szalay)

DNA/DNA hybridization technique for determining as a quantitative measure overall genetic similarity

dolerite basaltic rock

dome geologic formation of rock beds dipping away from the center

drainage erosion by flowing waters

drift random fluctuation in gene frequencies in a population, with the probability of fixation increasing with decreasing population size

drift sheet deposit of loose surface material

dumb-bell population separation with barriers closing in from two sides

dystrophic nutrient poor; relating to infertile soils subjected to much leaching (cf. eutrophic)

eccentricity in reference to 100 k.y.; 400 k.y.; periodic fluctuations of the Earth's orbit (*see* Milankovitch; cf. obliquity, precession)

ecocentric a perspective that emphasizes ecological aspects of organismal evolution and adaptation in the larger context of their habitat specificities

ecogeography geographical distribution of environmental conditions

ecology study of the sum of all interactions between life forms and their total environment

ecomorph variant resulting from particular environmental conditions

ecomorphology environmentally influenced morphology; adaptationally oriented study of morphology

ecophenotype variation in character expression resulting from particular environmental conditions

ecosensitivity relating to the disposition to be affected by changes in the environment

ecosystem sum of interactions between a community and its environment (physical, chemical, biological); interactions of a particular life form with its environment (sensu Vrba)

ecotone transitional zone between two biotic zones

ectostylid small buccal cusp of ungulate lower molars

edaphic relating to or influenced by the soil

eddy oceanic whirlpool

"*Edentata*" artificial assemblage of mammals including such independent groups as Xenarthra, Pholidota, Palaeanodonta, Afredentata, and Bibimalagasia.

effect hypothesis the specific influences of phylogenetic constraints on the evolutionary process of all species (sensu Szalay and Vrba)

Ekman relating to wind influencing ocean current flow

Ekman divergence zone where Ekman drifts from both hemispheres converge

Ekman drift water flowing at an angle of 45° to the wind direction, to the right in the Northern Hemisphere and to the left in the Southern Hemisphere

empirical based on observation and experiment testing in general, rather than on theory alone

encephalization relative brain enlargement compared to body size

endemism occurrence restricted to a particular region (vs. cosmopolitanism, provincialism)

endocranial within the skull

endocrine hormonal

endosperm edible pulp between the skin and germ of seeds

entire marginal leaves having a continuous edge

entoconid distolingual cusp of mammal lower molars

environment sum of biotic and abiotic components in a locality

enzyme any protein acting as catalyst in a specific chemical reaction

epeirogenesis deformations of the earth's crust affecting continental structures

epicondyle lateral protrusion of a femoral or humeral condyle

epicontinental on the continental shelf

epigamic relating to features involved in sexual selection

epigenetics development of the phenotype from the genotype due to interactive, as opposed to genetic, influence

epiphysis center of ossification separated from the

bone shaft by cartilage, the two unite at the completion of bone growth

epiphyte any plant growing on another without being a parasite

epistasis interaction of two or more genes

epistemic pertaining to the validation of knowledge (i.e., epistemology)

equatorial trough low atmospheric pressure in the intertropical convergence zone

Eremian living in paleoarctic deserts

erg sand desert

erosion breaking down of matter, followed by transportation and deposition

erratic rock transported by a glacier

estuary funnel-shaped river mouth formed by tidal action

ESS evolutionarily stable strategy; that adopted by members of a population and which cannot be improved in that context (Maynard Smith); optimal compromise possibly causing stasis in punctuated equilibria (sensu Rosenzweig)

ethology the study of the evolution of animal behavior

eukaryote any life form with a nucleated cell; comprises animals, plants, and fungi (cf. prokaryote)

eurybiomic/eurytopic living in a wide range of habitats; habitat generalist

eustasy, eustatic relating to global sea level changes

eutrophic nutrient rich; relating to fertile soils, subjected to little leaching (cf. dystrophic)

evaporite sedimentary rock resulting from the evaporation of salt water

evapotranspiration sum of water lost from soil evaporation and plant transpiration

evolutionary species concept not a testable concept but a descriptive reference to the evolutionary course of a lineage segment or a multidimensional species taxon (*see* chronospecies) (sensu Szalay)

exhumation exposure of a surface by erosion

exponential ob increasing at a growing rate

FAD first appearance datum

family (-idae) monophyletic taxonomic group of subfamilies and genera (*see* classification)

faulting rock fracture due to shift

fauna animal life of a locality

feeder dyke intrusion of rock across sedimentary strata from a magma chamber passage

fertilization system (*see* specific mate-recognition system)

fitness relative reproductive success; relative ability to leave viable offspring and to contribute to succeeding generations

Fitness-Generating Function equation predicting the theoretical fitness of a phenotype in a particular environment (sensu Rosenzweig)

fixation establishment of an allele in a population to the exclusion of all other alleles of that gene

floodplain river bed with unconsolidated detritus, periodically flooded

flora plant life of a locality

floristic relating to the geographic distribution of plants

fluvial/fluviatile relating to or produced by a river

foramen magnum opening at the base of the skull for the exit of the spinal chord

Foraminifera order of marine unicellular Protozoa, their shells constitute the main ingredient of chalk

fossa cavity or channel in anatomy

fossiliferous containing fossils

fossilization transformation of biotic remains into fossils

fossorial burrowing or living underground

founder effect the establishment of a colony from only a few individuals of a population

frugivorous feeding mainly or exclusively on fruit

fumarole ob volcanic vent releasing vapors and gases

Fynbos evergreen sclerophyllous scrub in the Mediterranean Cape region of South Africa

G-function (*see* fitness-generating function)

gallery fertile habitat along rivers in arid regions

game theory mathematical evaluation of strategies for participants in conflict

Gastropoda mollusck family of snails and slugs

Gauss/Matuyama boundary 2.43 m.y.; reversal in polarity between two paleomagnetic epochs

gene unit of heredity coding for a protein

gene flow exchange of genes between populations through interbreeding

gene frequency relative occurrence of a particular allele in a population

gene leakage integration of a gene from another species

gene pool sum of alleles present in a population

generic relating to the genus level

genetic drift (*see* drift)

genetic revolution hypothesis that posits bouts of genetic mutation leading to sudden developmental reorganization and evolutionary change

genome complete genetic material of a life form

genotype sum of genetic characteristics of a life form (vs. phenotype)

genus classificatory category comprising species with shared derived characters, several genera constitute a family

geographic speciation lineage splitting due to allopatry (vicariance)

geomorphology development of landform features

geophyte any plant with an underground storage organ (e.g., bulb, tuber)

geostrophic relating to currents flowing parallel to a pressure gradient, being deflected to the left (Southern Hemisphere) by the Earth's rotation

glacial interval of geologic time with worldwide cooling resulting in increasing glaciation; ice age; relating to glaciers

glacio-eustatic relating to global sea level changes due to the melting or growing of glaciers

glaciology dynamics and effects of glaciers

global climate change period of coupled but not necessarily phased alteration of climatic factors in more than one part of the world (e.g., by astronomical forcing) (sensu Vrba)

gluteus any of three muscles that originate on the innominate bone and sacrum

gnathic relating to the jaw

Gondwana supercontinent in the Southern Hemisphere, breaking into South America, Africa, India, Australia, and Antarctica at the end of the Jurassic (ca. 150 m.y.)

gradation leveling out of land

grade evolutionary achievement of similar level of organization by lineages

gradualism theory of evolution by Darwin (often miscast as steady-rate evolution) as a gradual and accumulative process but independent of notion of rate; anagenesis is the predominant mode of evolutionary change, even in cladogenetic events; macroevolution is an accumulation of microevolution (vs. punctuationism)

graminoid monocotyledon

graminivorous feeding mainly or exclusively on grains and seeds

granite coarse-grained igneous rock, derived from magma or continental crust

grassland habitat dominated by grasses and other nonwoody plants in semiarid to subhumid regions (250–1800 mm) with seasonal precipitation and periodic drought, mostly in the continental interior; in cooler and wetter regions than savanna (sensu O'Brien & Peters) any treeless vegetation form (cf. dambo, marsh)

group selection notion that evolution acts on groups rather than on individuals, selecting for characters benefiting the whole group rather than individuals (cf. species selection)

guild group of life forms exploiting and competing with each other for the same resource; coexisting competitors sharing a G-function (sensu Rosenzweig)

habitat environment of a particular kind (e.g., rainforest); kind of environment occupied by a particular life form; sum of localities supporting a particular life form (sensu Szalay); localities of natural occurrence (sensu Vrba)

habitat, fundamental localities including all the resources required by a life form

habitat, realized localities including a life form (sensu Vrba)

habitat specifity resource requirements of a life form (sensu Vrba)

habitat theory vicariance induced habitat change as the main force of speciation (sensu Vrba)

habitus adaptation of a lineage to its current (real-time) mode of life (sensu Szalay)

hallux big toe

halophyte any plant living in a saline habitat

hammada rocky desert of exposed bedrock and boulders or pebbles

headwater any stream being a source of a larger stream

Heinrich event melting of icebergs at intervals of 10 k.y., after North Atlantic cooling maximum

Hennigian pertaining to the phylogenetic philosophy of W. Hennig (*see* phylogenetic species concept)

herb any annual nonwoody terrestrial plant (*see* monocotyledon)

heterochrony evolutionary change in the rate or timing of character development

heterogenous dissimilar; diverse (vs. homogenous)

heteromorphy variation in morphology

hierarchy classificatory categorization; levels of organization; level specificity of evolution (*see* macroevolution) (cf. LIMA)

highveld temperate montane grassland in South Africa

hinterland region at the rear of a mountain range

holophyletic a taxon that includes all the descendants of a common ancestor (cf. paraphyletic)

homeostasis maintenance of a stable state by self-regulating processes

Hominidae hominoid family diagnosed by bipedal walking

Homininae Hominidae; *Homo*

Hominoidea catarrhine superfamily of apes and hominids

Homo hominid genus of modern humans and their extinct relatives

homogeneous similar; uniform (vs. heterogenous)

homology hypothesized origin of traits in two or more organisms inherited from their latest common ancestor

homoplasy a non-corroborated homology hypothesis (sensu Szalay)

horst uplifted block of rock

humero-femoral index relationship between the lengths of humerus and femur

Hunter Schreger bands alternating dark and light bands in the inner enamel as a result of prism decussation

hybridization interbreeding of different populations or species taxa

hydrography characteristics of water bodies

hydrogeography distribution and characteristics of water bodies

hydrology distribution and movement of water

hydrologic cycle circulation of water (evaporation, condensation, precipitation, glaciation)

hydromorphic with excess moisture

hydrosphere sum of localities with water

hyper- extensive

hypermasticatory having extensively developed chewing structures

hypermorphosis extended development of ancestral characters in the descendant

hypermorphosis, rate by increased descendant growth rates

hypermorphosis, sequential time by prolonged growth in several bouts

hypermorphosis, time by prolonged growth

hypertrophy increased development (vs. atrophy)

hypodigm collective of samples used in the description of a species taxon

hypothermal relating to periods of climatic cooling

hypsodonty molars having high or deep crowns; height of tooth crown greater than its breadth

igneous produced by volcanic or magmatic action (cf. sedimentary, metamorphic)

iliac relating to the ilium (one of three bones forming the pelvis)

illite common clay mineral

imbricational relating to the part of the crown with perikymata; comprising striae of Retzius having reached the enamel surface (producing perikymata), deposited in shinglelike, partially overlapping layers (cf. appositional)

individual single organism, not a class of individuals; unit subjected to natural selection (sensu Vrba)

infrared relating to wavelengths between visible red and radio waves

infraspecific relating to taxa below the level of species, or individuals of a species

initial and boundary conditions the inherited components and constraints of an organism and/or species and the environment that guides its ontogenetic and (for species) evolutionary change (sensu Szalay)

initiating causes physical and biotic events of the environment initiating either speciation or extinction of taxa (sensu Vrba)

insectivorous feeding mainly or exclusively on insects

inselberg isolated round hill in savanna lowland; kopje

interfluve elevation between adjacent streams

interglacial interval between glacial epochs with worldwide warming resulting in reduced glaciation

intergradation interbreeding of two subspecies or related species in range overlaps; intercalation of gene pools

intermembral correlating the lengths of the extremities, indicative of the mode of locomotion

interspecific between species; pertaining to different species

intertropical convergence zone (*ITCZ*) belt just north of the equator where air masses from the two hemispheres converge, resulting in bimodal rainfall

intraspecific within species; pertaining to individuals of a species

introgression (*see* gene leakage)

ischial callosities sitting pads on the ischium (one of three bones forming the pelvis); characteristic of Cercopithecoidea

isochronous of the same time period (vs. diachronous)

isostatic relating to balancing of earth crust dynamics

isotope variant of an element, differing in relative atomic mass and nuclear but not chemical properties (*see also* carbon, oxygen)

karstic relating to limy rocks whose dissolution results in underground cavities

key adaptation trait responsible for the success of lineages; new character relaxing a trade-off constraint, facilitating an improved ESS and improving versatility (sensu Rosenzweig)

keystone species any taxon crucial for the functioning of an ecosystem

kin selection evolution acting on all copies of a gene, including those present in related individuals, favoring those benefiting kin (Hamilton)

Köppen system of climate classification

LAD last appearance datum

laterite reddish clay

levator labii superioris muscle elevating the upper lip

LIMA level-interactive-modular-array; collective of complexes within which constituents impose constraints on each other without clear supersession (distinct from notions of hierarchy; sensu Szalay)

limestone sedimentary rock composed mainly from the calcium carbonate of marine microorganisms (e.g., Foraminifera)

lineage line of descent; continuous evolutionary sequence; evolutionary course of a species taxon; a segment or all previous and subsequent history of a realtime species (sensu Szalay)

lithology composition and characteristics of rock

lithosphere sum of crust and upper mantle of the earth

lithostratigraphy position and order of rocks

loess loam of windborne Pleistocene dust

macchia Mediterranean sclerophyllous scrub

machairodont sabre-toothed cat

macroevolution taxic evolution; large-scale evolutionary processes above the level of populations; selection on supraorganismic properties; large-scale evolutionary processes decoupled from microevolution (sensu e.g., Vrba), accumulation of microevolutionary processes (sensu e.g., Foley, Szalay) (*see* punctuationism)

macrofossil any fossilized remains large enough to be studied without microscope

macromolecule any large chemical compound

malar of the cheek or cheekbone

Malthusian relating to unsustainably growing populations

mantle plume mantle uplift by underlying magmatic activity, resulting in domes

maritime relating to climate characteristic of coastal regions, with high and aseasonal precipitation and little seasonal fluctuation in temperature; humid (cf. continental; Mediterranean)

marsh waterlogged, treeless wetland (cf. dambo, swamp, grassland)

mass balance balance of elements in ocean water

Mayrian pertaining to the biological species concept advocated by E. Mayr

Mediterranean relating to the climate characteristic of semiarid (500–900 mm) warm-temperate west coasts, with hot, dry summers and mild, wet winters, supporting sclerophyllous vegetation (cf. boreal)

megadontia dental enlargement

megafauna collective of mammals exceeding 10 kg in adult body mass (cf. microfauna)

megaherbivore herbivorous mammal exceeding 1000 kg in adult body mass

meridional with southern affinities (cf. boreal); longitudinal (cf. zonal)

mesa terrace; tableland

mesic habitat or climate with moderate moisture

mesophyte any plant requiring only moderate moisture

metamorphism transformation of rock by such means as heat and compression

metamorphosis significant change in form

meteoric originating in the atmosphere

microevolution generational evolution; evolutionary processes involving changes in phenotype or genotype frequencies within populations (*see* gradualism)

microfauna collective of invertebrates (e.g., insects, Gastropoda), small reptiles, micromammals (cf. megafauna)

micromammal any small mammal (e.g., rodents, bats)

micronutrient vitamins and minerals

microphyllous having reduced leaves

microwear microscopic marks on teeth resulting from use

Milankovitch combined effects of the eccentricity, obliquity, and precession cycles of the earth, proposed to lead to cyclical climatic changes during the Quaternary, including glacial periods; 23 k.y. precession cycle, 41 k.y. obliquity cycle, 100 k.y. eccentricity cycle

2.5 million year event period of significant cooling

miombo dystrophic evergreen open-canopy woodland in continental East and South Africa (cf. Fynbos, mopane)

modal most frequently observed

mode means of evolutionary change (anagenesis, cladogenesis)

Modern Synthesis theoretical merging of Darwinian natural selection and Mendelian and population genetics (Simpson, 1953); neoDarwinism

Mollusca phylum of invertebrates with shells (e.g., Gastropoda)

monocotyledon any plant with one embryonic seed leaf, three-parted structures, and parallel venation (e.g., grasses; cf. dicotyledon)

monogeneric comprising one genus

monophyletic a clade or taxon diagnosed as derived from a single ancestor

monospecific relating to a taxon comprising only one species

monotypic comprising of one representative taxon

monsoon seasonal change in jet stream direction causing winds and change in weather

montane of high altitudes (1000–2000 m) below the treeline, with a moist and cool climate and evergreen vegetation (cf. Alpine, submontane, temperate)

mopane eutrophic, open-canopy woodland in South Africa (cf. Fynbos, miombo)

moraine mound of glacially deposited material

morphocline the known diversity of a series of homologous traits in a group

morphogenesis developmental determination of form

morphometric relating to the measurement of physical proportions

morphotype the estimated ancestral condition of a group

mosaic evolution differential rates in the evolution of discrete characters within a lineage or clade

mtDNA mitrochondrial DNA; DNA in the mitochondria, the organelles of respiration and energy production in the cell

mudstone dark clay sediment

multiplicity number of lineages or taxa at a given time; number of descendants of a particular ancestor; number of lineages making up a clade (sensu Szalay)

multispecific comprising several species

multivariate sampling many variables simultaneously

mutation any change in the genome

mutualism mutually beneficient association between two life forms; symbiosis

natural selection as a cause, mechanism, or process resulting in differential survival and reproduction (hence fitness) of individuals (Darwin); the preferential propagation of heritable characters (and individuals expressing them) best adapted to their total environment (causally different from sexual selection; sensu Szalay)

nectarivorous feeding mainly or exclusively on nectar

neocortex evolutionarily youngest and most highly developed portion of the brain

neoDarwinism (*see* Modern Synthesis)

neural relating to the nervous system

neurocranium part of the skull encasing the brain

neutral relating to heritable mutations not influencing function and biological roles, thus not being subjected to natural selection

niche way of living of a life form; position of a life form in its environment, constrained by its tolerances and requirements (e.g., habitat, activity, interactions, and resource use); sum of interactions of an organism or species with its environment (sensu Szalay)

nocturnal active in the night (vs. diurnal)

node point of divergence

non-Darwinian pertaining to evolutionary processes based on random variation and independent of natural selection

nuchal crest bony ridge at the rear of the skull for muscle attachment

^{18}O (*see* oxygen)

obliquity 41 ky; periodic fluctuations in the plane of the Earth's orbit (*see* Milankovitch) (vs. eccentricity, precession)

omnivorous feeding on both plant and animal matter

ontogeny development of an organism (vs. phylogeny)

orogeny formation of mountains

orography distribution and characteristics of mountains

orthogenesis notion of a predestined evolutionary course as a result of inherent mechanisms rather than the accepted causes and mechanisms of evolutionary change

orthograde upright

Ostracoda subclass of small marine crustaceans

outcome (sensu Szalay) condition of an entity at the end of a cause-effected process; consequence

outgroup any taxon used for comparison in with the group studied

oxygen element with three stable isotopes, ^{16}O, ^{17}O, and ^{18}O; ratio of oxygen-16 and oxygen-18 isotopes ($\sim$500:1) indicating water temperature, $\delta^{18}O$ values increasing with decreasing water temperature; international standards SMOW (standard mean ocean water) and PDB

paedomorphosis retention of ancestral juvenile characters in later developmental stages of descendants

paleolatitude latitude in relation to the position of the equator during a particular epoch

paleosol undisturbed buried soil

paludinous waterlogged; marshy (cf. dambo, marsh)

palynology remains of flora and microorganisms inferring geological age and environmental conditions

Pangaea supercontinent breaking up during the Triassic

Pan African ape genus, purported closest living relative of *Homo*

Papionini cercopithecine tribe

Paranthropus hominid genus of "robust" australopithecines

parapatry adjacency of populations

parapatric speciation formation of new species adjacent to the ancestral population

paraphyletic monophyletic group including some or most but not all descendants of a common ancestor

Paratethys Central European sea in the geologic past extending from the Alps to the Aral Sea

paratropical characteristic of a subtropical equable climate without prolonged dry periods

parsimony a group of concepts used by parsimony-cladists based on the assumption that the fewest evolutionary changes lead to the correct cladistic reconstruction

PCA principal component analysis; statistical method for measuring the interdependence between three or more variables

PDB Pee Dee Formation belemnite; fossiliferous marine limestone rich in carbon-13, used as an international standard in the determination of (consequently negative) $\delta^{13}C$ and pedogenic $\delta^{18}O$ values

pediplain extensive plain at the base of a mountain in semiarid regions

pedogenic formed in the soil

pedology formation and characteristics of soil

peneplain extensive plain as an end product of erosion

perikymata horizontal furrows where striae of Retzius reach the enamel surface; circling the tooth crown

periodic model (sensu Grubb) repeated vicariance resulting in cladogenesis

peripatry barrier crossing of founder populations

peripatric speciation formation of new species at the periphery of the ancestral population

PET potential evapotranspiration

phenetic pertaining to overall similarity, both primitive and advanced traits combined

phenon any morphological cluster of organisms or attributes

phenotype sum of characteristics of a life form resulting from the interaction of genotype and environment

photosynthesis assimilation and fixation of carbon dioxide and subsequent conversion into organic compounds by plants

photosynthetic type C_3, C_4, CAM; adaptive variants to differing atmospheric and climatic conditions; characterized by specific stable carbon isotopic patterns, allowing paleoecologic reconstruction of tropical ecosystems

phyletic pertaining to a lineage; pertaining to an evolutionary trend

phylogenetic species concept (*PSC*) segment of a lineage between two speciation events; Hennigian/cladistic species concept

phylogeny evolutionary history of any lineage segment of life; consequence of sundry causes acting on organisms through time (sensu Szalay)

phylum (*see* classification)

phytochorion/phytogeography geographical distribution of plants

Pitheciinae group of New World primates

plankton collective of small aquatic life forms drifting in sea water

plate, tectonic any of the segments making up the earth's crust

playa frequently flooded basin between mountains

plesiomorph primitive character

plication dental folding

polar relating to climate characteristic of high latitudes and altitudes, with permanent underlying frost (cf. Alpine, boreal, temperate)

polarity periodic reversal of the earth's magnetic field (e.g., Olduvai event, 1.83 m.y.; Gauss/ Matuyama)

pollinator agent of pollen transfer (e.g., insects, birds, primates, bats, wind, water)

polyphyletic a taxon that does not include the common ancestor of its members; derived from several ancestors

polyploid having more than the general two sets of chromosomes

polytypic comprising several subordinate populations or taxa

Pongidae hominoid great apes

population group of interbreeding life forms sharing a common gene pool

precocial relating to young born mobile and relatively well developed (vs. altricial)

precession 23 k.y.; periodic fluctuations in the Earth's rotational axis due to planetary influences (*see* Milankovitch) (cf. eccentricity, obliquity)

precipitation sum of deposition of moisture after condensation (rain, dew, snow); depostion of a substance in a solution

premaceration predating the removal of flesh

pressure cell semipermanent zones of particular atmospheric pressure (e.g., subtropical high)

prism decussation groups of enamel prisms coursing in a sinusoidal fashion, at some angle to adjacent groups of prisms, from the enamel–dentine junction outward

proboscis trunk (elephants), feeding organ of nectarivorous insects; enlarged upper lip (*see also* Ungulata)

proceptivity female behavior of seeking initiation of copulation

process course of change effected by one or more causes acting on an entity with its associated constraints (sensu Szalay)

prognathy protruding jaws

progress pertaining to acquisition of any adaptation in a taxon; any superior adaptation in a taxon, resulting in improved competitive ability; evolutionary change in a lineage (*see also* Red Queen)

progressivist advocacy of evolution as resulting in increasingly superior adaptations in a given context

prokaryote any life form without a nucleus and a single DNA molecule per cell (bacteria; cf. eukaryote)

prosthion point between the base of the two middle upper incisors

proximate relating to underlying physiological and behavioral mechanisms; causal (vs. ultimate)

proxy replacement of ions by other elements

PSC (*see* phylogenetic species concept)

pulse (*see* turnover pulse)

punctuated equilibrium/punctuationism model of evolution of new lineages that involves rapid punctuational origin and prolonged stasis, with cladogenesis being the predominant mode of evolutionary changes; macroevolution heirarchically decoupled from microevolution (vs. "gradualism")

punctuation (without equilibrium) instance of rapid and pronounced evolutionary change, quantum evolution, resulting in new lineages (vs. stasis) or rapid change in same lineage (sensu Szalay)

quantum evolution rapid evolutionary change within a lineage as the origin of a new lineage due to sudden changes in the adaptive requirements; largely undetectable in the fossil record (sensu Simpson)

radiation (see adaptive radiation; heat emission)

radiocarbon/radiometric absolute dating by the proportion of stable carbon-12 and radioactive carbon-14 (half-life approximately 5730 years; usable up to 50,000 BP; cf. $^{39}Ar/^{40}Ar$)

rainshadow decreased precipitation on the sheltered lee side of a mountain range

rate measure of the relative or absolute evolutionary change of a feature

realtime lifetime of an organism or of a stable population; segment of a lineage subjected to causally driven evolutionary change (sensu Szalay)

receptivity female consent to copulation

recognition (species) concept collective of individuals recognizing each other as sharing a common fertilization system and constituting a genetic entity (*see* specific mate-recognition system; cf. biological species concept)

Red Queen pertaining to evolutionary competition; evolutionary progress not resulting in improved competitive ability due to effected coevolution in associated life forms

reproductive barrier/isolation prezygotic mechanisms inhibiting hybridization (spatial, temporal, ethological, mechanical, chemical, genetic)

resource any component of the environment a life form can make use of (e.g., temperature, humidity, pH, salinity, water flow velocity, substrate, light, shelter, nesting sites, inorganic and organic compounds, prey, mates)

reticulate evolution repeated interbreeding of previously separated populations

retrotime pertaining to the evolutionary past of an organism or species in realtime (sensu Szalay)

retroviral capable of reverse transcription (insertion of foreign DNA)

rhizome underground roots and shoots

rifting process of crustal separation between elongate parallel faults

ring species two sympatric, yet reproductively isolated, species at the distal ends of a linked circle of interbreeding populations

riparian of or on a river bank

Rodentia mammalian order, whose members have continuously growing, chisellike incisors

Ruminantia artiodactyl suborder of ruminant herbivores with complex stomachs

sagittal crest bony ridge along the top midline of the skull for temporalis muscle attachment; characteristic of males of *Gorilla* and *Paranthropus*

saltation mutation with major evolutionary effects

salt pan depression in semiarid regions accumulating salty deposits (e.g., evaporite)

sapropel unconsolidated organic sludge developing on shallow lake bottoms

savanna grassland habitat interspersed with trees and bushes in semiarid tropical and subtropical regions, situated between rainforest and desert, with seasonal (sutropical) or bimodal (tropical) rainfall (250–1200 mm) and periods of drought; any nonforest African ecosystem

Scala Naturae Aristotelian and pre-Darwinian notion of a graded series of ascending complexity, culminating in *H. sapiens*

scansorial adapted to climbing

sclerophyllous having thickened leaves; associated with semiarid habitats (*see also* Mediterranean)

secondary compound/metabolite any substance not ingested purposefully, including plant toxins (e.g., phenolics)

secondary succession gradual recovery of plant communities after perturbation

selection (*see* natural selection)

selection, stabilizing favoring a modal character resulting in its stasis

semispecies any component species of a superspecies; allospecies; subspecies

sexual dimorphism difference between the sexes in a character (e.g., body size, ornamentation)

sexual selection causally distinct from natural selection, favoring hereditary characters promoting mating success; differential fitness depending on success in the competition for or selection of mates

shale sedimentary rock of consolidated mud or clay

shared preference community mode of competitive coexistence, with each species preferring the same habitat but varying in the tolerance of other habitats; tolerance community organization (cf. distinct preference community) (sensu Rosenzweig)

shifting balance theory genetic drift in small demes facilitating selection and resulting in the spread of these in a population or species (sensu Wright)

shoal submerged accumulation of sediments in shallow water

sibling species species differentiated genetically, reproductively, or ecologically but difficult to distinguish morphologically

sill sheet of igneous rock intruding layers of other rock

simian shelf transverse buttressing of the catarrhine mandibular symphysis

sinkhole surface depression in karstic regions resulting from a collapsed cavity

sister group any lineage or taxon that is the most recent relative

sister taxon (nonancestral) taxon sharing the most recent common ancestor

sorting consequential evolutionary change and differential representation (survival) of lineages after divergence

speciation development of new closed lineages in the process of evolution; complete cladogenesis

species individuals sharing a gene pool resulting in fertile offspring, separated from other such entities by reproductive, genetic, or ecological barriers; relatively stable biological entities at a particular time and place (sensu Foley); basic concept in evolutionary biology, conceptualized in many ways by taxonomists (sensu Szalay); category in taxonomy

species concept (*see under* biological, evolutionary, phylogenetic, recognition)

species selection differential survival among species descended from a common ancestor; hypothesis of selection promoting the evolution and proliferation of particular supraorganismic, species level characters (*see also* group selection; macroevolution)

species taxon terminal or intermediate segments of a lineage defined in space and time

specific mate recognition concept (*SMRS*) features of the fertilization system encouraging only mating partners of the own species

stable nonradioactive (*see also* isotope)

stasipatric speciation rapid formation of new species by chromosomal mutation

stasis evolutionary equilibrium of traits or whole lineage; interval of relative evolutionary constancy in a character or taxon; adaptive stabilization of a character or taxon (cf. punctuation)

stenobiomic/stenotopic living in a narrow range of habitats; habitat specialist

steppe treeless grasslands in semiarid temperate regions (sensu Denys); semiarid grassland savanna

stratigraphy branch of geology that studies the formation, composition, sequence, and correlation of rocks

striae of Retzius near weekly incremental lines related to successive forming fronts during enamel development (cf. cross-striation)

subduction descent of one geologic plate beneath another by subsidence of the earth's crust

subfamily (-inae) monophyletic group of genera

subfossil any remains of a life form dating from the Holocene

subhumid relating to regions with occasional water deficit (1000–2000 mm), seasonal precipitation, and supporting decideous vegetation (cf. humid, semiarid)

submontane of middle altitudes (500–1000 m; cf. montane)

subsidence sinking of ground

subspecies taxonomic category below the species; geographically separated variant population

subtropical 35°–40° latitude; latidudinal belt between the tropical and the temperate zones with convective summer rainfall (cf. tropical)

supergene sequence of genes preserved by selection

superspecies monophyletic group of allopatric species or subspecies that cannot be tested for species designation through sympatry, frequently equivalent to a genus

swamp waterlogged wetland covered with bushes and trees (cf. marsh, forest)

sympatry geographical overlapping of populations

sympatric speciation formation of new species in overlapping geographical ranges

symplesiomorphy shared primitive character

synapomorphy homologously shared derived character

synergism mutualism; symbiosis

synspecies similar taxonomic species occurring at the same time in different places

Synthesis (*see* Modern Synthesis)

taphonomy process of burial and fossilization; probability of fossilization

taphonomic bias/effect differential probability of fossilization, increasing with size and compactness of structures (e.g., teeth)

taxogram cladogram that is directly translated into a classification and vice versa

taxon any taxonomic entity; any classified group of organisms

taxon effect differential evolution of taxa due to inherent contingencies

taxonomy classification of extant and extinct life forms

tectonics dynamics and features of the earth's crust

teleological assuming intentional and preordained rather than chance- and selection-based courses of evolution

temperate characteristic of mid latitudes and altitudes, with large seasonal fluctuations in temperature, including winter frost (e.g., montane; cf. polar, tropical)

tempo rate of evolutionary change in a character or taxon

teratology study of malformations and congenital defects

Tethys sea of the geological past between the northern (Laurasia) and southern (Gondwana) continents of the Old World

tetraconodont having molars with four cusps

thermohaline relating to the joint influence of salinity and temperature in oceans *thermohaline circulation* waters sinking after having cooled at the surface

thermoregulatory controlling internal temperature

till heterogeneous detritus deposited by glaciers

tolerance community organization shared preference community organization

top predator predator not preyed upon by any other predator

topography surface features of a region

topographic relief altitude and slope variations of a region

topology spatial distribution

tradewind tropical easterly wind flowing from subtropical high into equatorial low pressure zones and deflected by the Earth's rotation (*see also* intertropical convergence zone)

trade-off trade rules conflict among functions a life form can perform, necessitating a compromise of optimal strategies (sensu Rosenzweig)

transposition displacement of DNA sequences to another site in the genome

transposable elements DNA sequences prone to displacement

trellis parallel streams with tributaries set at a right angle

trend continual evolutionary change in a particular direction in a given lineage or lineages (e.g., megadontia, encephalization)

tributary any stream flowing into a larger river or lake

tropical characteristic of low latitudes, with small seasonal fluctuations in temperature and wet seasons; equatorial (cf. subtropical, temperate)

troposphere lowest layer of the atmosphere, up to 10–17 km (highest in the tropics)

tundra treeless biome characteristic of polar regions

turnover exchange or relay of species by speciation, extinction, or migration, altering the species composition in a locality (sensu Vrba)

turnover pulse statistically significant burst of turnover of species across several taxa, associated with physical change (sensu Vrba)

ultimate relating to past evolutionary, developmental or adaptational causes (vs. proximate)

ultra-Darwinian euphemism for those who reject the punctuated equilibrium model

Ungulata superorder at cohort of herbivorous, hoofed mammals

uniformitarianism theory that geological processes are due to the same continuously and uniformly operating forces in the past that operate in the present

upwelling winds (e.g., trade winds) bringing up (nutrient-rich) cold water, on converging with warmer surface water giving rise to abundant marine fauna (e.g., off western Africa, along the equator); deep waters rising in open oceans, replacing diverging surface currents

vascular having vessels for water and nutrient transport (all terrestrial plants except mosses, liverworts, hornworts)

vegetation zone geographic region comprising specific plants

vegetative relating to growth; relating to asexual reproduction; characteristic of plants

versatility evolvability; relaxation from trade-off constraints; ability to evolve in more than one way (sensu Rosenzweig)

vertisol clay soil of tropical regions with surface cracks in the dry season

vestibular towards the gums and lips

vicariance fragmentation of formerly continuous range into allopatric populations; allopatric speciation (sensu Grubb)

vicariance event any incident leading to allopatric speciation following vicariance

vleis open grassland savanna on limestone in South Africa

Wallace's Line boundary between Oriental and Australian faunas

weathering chemical decomposition and physical disintegration of rock by environmental agents (e.g., temperature, wind, rain)

woodland open-canopy forest habitat with little undergrowth and widely spaced trees; forest

xeric with little moisture; arid

xerophyte any plant requiring little moisture

zoogeography geographical distribution of animals

zygostructure patterns of breeding within and between populations of a taxon

zygote fertilized ovum

References Cited

Aagaard, K. (1988). The Arctic thermohaline circulation [Abst.]. *EOS Transactions of the American Geophysical Union* **69,** 1043.

Abitbol, M. M. (1995). Reconstruction of the Sts 14 (*A. africanus*) pelvis. *American Journal of Physical Anthropology* **96,** 143–158.

Acres, B. D., Rains, A. B., King, R. B., Lawton, R. M., Mitchell, A. J. B. & Rackham, L. J. (1985). African dambos: their distribution, characteristics and use. In M. F. Thomas & A. S. Goudie, eds., *Dambos: Small Channelless Valleys in the Tropics—Characteristics, Formation, Utilisation.* Zeitschrift fur Geomorphologie, Supplement band 52. Berlin: Gebruder Borntraeger.

Aggundey, I. R. & Schlitter, D. A. (1984). Annotated checklist of the mammals of Kenya. I. Chiroptera. *Annals of Carnegie Museum of Natural History* **53,** 119–161.

Aguilar, J. & Michaux, J. (1987). Essai d'estimation du pouvoir separateur de la méthode biostratigraphique des lignées évolutives chez les rongeurs neogènes. *Bulletin de la Societé Geologique Française* **8** (6), 1113–1124.

Aguirre, E. (1970). Identificacion de *'Paranthropus'* en Makapansgat. *Crónica del XI Congresso Nacional Arqueologia,* 98–124.

Aguirre, E. & Alberdi, M.-T. (1974). Hipparion remains from the Northern part of the Rift Valley (Kenya). *Proceedings, Koninklijke Nederlandse Akademie van Wetenschappen, Series B* **77,** 146–157.

Aharon, P., Goldstein, S. L., Wheeler, C. W. & Jacobson, G. (1993). Sea-level events in the South Pacific linked with the Messinian salinity crisis. *Geology* **26,** 771–775.

Aiello, L. C. & Dean, C. (1990). *An Introduction to Human Evolutionary Anatomy.* London: Academic Press.

Aiello, L. C. & Dunbar, R. I. M. (1993). Neocortex size, group size and the evolution of language. *Current Anthropology* **34,** 184–193.

Aiello, L. C., Montgomery, C. & Dean, M. C. (1991). The natural history of deciduous tooth attrition in hominoids. *Journal of Human Evolution* **21,** 397–412.

Aiello, L. C. & Wheeler, P. (1995). The expensive-tissue hypothesis: the brain and the digestive system in human and primate evolution. *Current Anthropology* **36,** 199–221.

Aiello, L. C. & Wood, B. A. (1994). Cranial variables as predictors of body mass. *American Journal of Physical Anthropology* **95,** 409–426.

Alberch, P. (1980). Ontogenesis and morphological diversification. *American Zoologist* **20,** 653–667.

Albrecht, G. H. (1978). The craniofacial morphology of the Sulawesi macaques: multivariate approaches to biological problems. *Contributions to Primatology* **13,** 1–151.

Allen, J. A. (1877). The influence of physical conditions in the genesis of species. *Radical Review* **1,** 108–140.

Allen, T. F. H. & Starr, T. B. (1982). *Hierarchy: Perspectives for Ecological Complexity.* Chicago: University of Chicago Press.

Ambrose, S. & Sikes, N. (1991). Soil carbon isotope evidence for Holocene habitat change in the Kenya Rift Valley. *Science* **253,** 1402–1405.

Anderson, J. B. & Bartek, L. R. (1992). Cenozoic glacial history of the Ross Sea revealed by intermediate resolution seismic reflection data combined with drill site information. *Antarctic Research Series* **56,** 61–73.

Andrews, P. (1981). Species diversity and diet in monkeys and apes during the Miocene. In C. B. Stringer, ed., *Aspects of Human Evolution,* pp. 25–61. London: Taylor and Francis.

Andrews, P. (1989). Palaeoecology of Laetoli. *Journal of Human Evolution* **18,** 173–181.

Andrews, P. (1990). *Owls, Caves and Fossils.* London: Natural History Museum Publications.

Andrews, P. (1992a). Evolution and environment in the Hominoidea. *Nature* **360,** 641–646.

Andrews, P. (1992b). Community evolution in forest habitats. *Journal of Human Evolution* **22,** 423–438.

Andrews, P. (1995). Ecological apes and ancestors. *Nature* **376,** 555–556.

Andrews, P., Begun, D. & Zylstra, M. (1996). Interrelationships between functional morphology and paleoenvironments in Miocene hominoids. In D. Begun, C. Ward & M. Rose, eds., *Function Phylogeny, and Fossils,* pp. 29–58. New York: Plenum.

Andrews, P., Lord, J. & Evans, E. M. N. (1979). Patterns of ecological diversity in fossil and modern mammalian faunas. *Biological Journal of the Linnean Society* **11,** 177–205.

Andrews, P., Meyer, G., Pilbeam, D. R., Van Couvering, J. A. & Van Couvering, J. A. H. (1981). The Miocene fossil beds of Maboko Island, Kenya: geology, age, taphonomy and paleontology. *Journal of Human Evolution* **10,** 35–48.

Andrews, P. A. & Van Couvering, J. A. H. (1975). Palaeoenvironments in the East African Miocene. In F. S. Szalay, ed., *Contributions to Primatology,* Vol. 5. *Approaches to Primate Paleobiology,* pp. 63–103. New York: Basel Karger.

Andrews, P. & Walker, A. (1976). The primate and other fauna from Fort Ternan, Kenya. In G. Ll. Isaac & E. R. McCown, eds., *Human Origins: Louis Leakey and the East African Evidence,* pp. 279–304. Menlo Park, CA: W. A. Benjamin.

Ansell, W. F. H. (1971). Artiodactyla. In J. Meester & H. Setzer, eds., *The Mammals of Africa. An Identification Manual,* Vol. 15, pp. 1–84. Washington, DC: Smithsonian Institution Press.

Ansell, W. F. H. (1972). Order Artiodactyla, part 15. In J. Meester & H. W. Setzer, eds., *The Mammals of Africa. An Identification Manual,* pp. 1–84. Washington, DC: Smithsonian Institution Press.

Ansell, W. F. H. (1978). *The Mammals of Zambia.* Chilanga, Zambia: The National Parks and Wildlife Service.

Ansell, W. F. H. & Dowsett, R. J. (1988). *Mammals of Malawi. An Annotated Check List and Atlas.* St Ives, England: Trendrine Press.

Arambourg, C. (1947). *Mission scientifique de l'Omo* (1932–1933), vol. I. Paléontol. fasc. 3. Muséum National d'Histoire Naturelle. Paris: Éditions du Muséum.

Arambourg, C. (1959). Vertébrés continentaux du Miocène supérieur de l'Afrique du nord. Publications Service de la Carte Geologique d'Algérie (nouvelle serie). *Paleontologie* **4,** 5–159.

Arambourg, C. (1970). Les vertébrés du Pleistocène de l'Afrique du Nord. *Archives de Museum Nationale d'Histoire Naturelle, Paris* **(7) X,** 1–128.

Arambourg, C. (1979). Les Vertébrés Villafranchiens d'Afrique du nord. Paris: Singer-Polignac.

Arambourg, C., Chavaillon, J. & Coppens, Y. (1967). Premiers résultats de la nouvelle mission de l'Omo (1967). *Comptes Rendus de l'Académie des Sciences Paris* **265,** 1891–1896.

Arambourg, C., Chavaillon, J. & Coppens, Y. (1969). Premiers résultats de la nouvelle mission de l'Omo (2c campagne 1968). *Comptes Rendus de l'Académie des Sciences Paris* **268,** 759–761.

Arambourg, C., Chavaillon, J. & Coppens, Y. (1972). Expédition internationale de recherches paléontologiques dansla vallée de l'Omo (Ethiopie) en 1967. Rapport de l'équipe française. *Actes du 6e Congrès panafricain de Préhistoire et d'Etudes du Quaternaire,* Dakar, 2–8 décembre 1967, pp. 135–140.

Abitbol, M. (1995). Lateral view of *Australopithecus afarensis:* primitive aspects of bipedal positional behaviour in the earliest hominids. *Journal of Human Evolution* **28,** 211–229.

Arnold, S. J., Alberch, P., Csanyi, V., Dawkins, R. C., Emerson, S. B., Fritzsch, B., Horder, T. J., Maynard-Smith, J., Starck, M., Vrba, E. S., Wagner, G. P. & Wake, D. B. (1989). How do complex organisms evolve? In D. B. Wake & G. Roth, eds., *Complex Organismal Functions: Integration and Evolution in Vertebrates,* pp. 403–433. Chichester, UK: John Wiley and Sons.

Ashley, G. M. (1996). Springs, pools and adjacent wetlands, a newly recognized habitat, lowermost Bed II, Olduvai Gorge, Tanzania. *Geological Society of America Abstracts* **28** (6), 28.

Askin, R. A. (1992). Late Cretaceous-early Tertiary Antarctic outcrop evidence for past vegetation and climates. *Antarctic Research Series* **56,** pp. 61–73.

Aulagnier, S. & Thévenot, M. (1987). Catalogue des

mammifères sauvages du Maroc. *Travaux de l'Institut des Sciences de Rabat, Zoologie* **41,** 1–164.

Avery, D. M. (1981). Holocene micromammalian faunas from the northern Cape Province, South Africa. *South African Journal of Science* **77,** 265–273.

Avery, D. M. (1982). Micromammals as palaeoenvironmental indicators and an interpretation of the late Quaternary in the southern Cape Province, South Africa. *Annals of the South African Museum* **85(2),** 183–374.

Avery, D. M. (1995). Southern savannas and Pleistocene hominid adaptations: the micromammalian perspective. In E. S. Vrba, G. H. Denton, T. C. Partridge & L. H. Burckle, eds., *Paleoclimate and Evolution with Emphasis on Human Origins,* pp. 459–478. New Haven, CT: Yale University Press.

Axelrod, D. I. & Raven, P. H. (1978). Late Cretaceous and Tertiary vegetation history of Africa. In M. J. A. Werger, ed., *Biogeography and Ecology of Southern Africa,* vol. 31, pp. 77–131. The Hague: W. Junk.

Azzaroli, A. (1995). The "elephant-*Equus*" and the "end-Villafranchian" events in Eurasia. In E. S. Vrba, G. H. Denton, T. C. Partridge & L. H. Burckle, eds., *Paleoclimate and Evolution with Emphasis on Hominid Origins,* pp. 311–318. New Haven, CT: Yale University Press.

Azzaroli, A., De Giuli, C., Ficcarelli, G. & Torre, D. (1988). Late Pliocene to early mid-Pleistocene mammals in Eurasia: faunal succession and dispersal events. *Palaeogeography, Palaeoclimatology, Palaeoecology* **66,** 77–100.

Azzaroli, L. & Simonetta, A. M. (1966). Carnivori della Somalia ex-Italiana. *Monitore Zoologico Italiana* **74** (supplement), 102–195.

Balinsky, B. I. (1962). Patterns of animal distribution on the African continent. *Annals of the Cape Province Museum* **2,** 299–310.

Barker, P. F. & Burrell, J. (1977). The opening of the Drake Passage. *Marine Geology* **25,** 15–34.

Barker, P. F. & Burrell, J. (1982). The influence upon Southern Ocean circulation, sedimentation, and climate of the opening of the Drake Passage. In C. Craddock, ed., *Antarctic Geoscience,* pp. 377–385. Madison, WI: University of Wisconsin Press.

Barnard, P. (1973). Mesozoic floras. In N. Hughes, ed., *Organisms and Continents through Time.* Special Paper of the Paleontological Society of London no. 12, pp. 175–188.

Barrett, P. J., ed. (1989). Antarctic Cenozoic history from the CIROS-1 drillhole, McMurdo Sound. *DSIR Bulletin* **245,** 1–254.

Barrett, P. J. (1987). Oligocene sequence cored at CIROS-1, western McMurdo Sound. *New Zealand Antarctic Record* **7,** 1–7.

Barron, E. J. (1983). A warm, equable Cretaceous: the nature of the problem. *Earth Science Reviews* **19,** 305–338.

Barron, E. J. (1985). Explanations of the Tertiary global cooling trend. *Palaeogeography, Palaeoclimatology, Palaeocecology* **50,** 45–61.

Barron, E. J., Sloan, J. L. & Harrison, C. G. A. (1980). Potential significance of land-sea distribution and surface albedo variations as a climatic forcing factor: 180 M.y. to the present. *Palaeogeography, Palaeoclimatology, Palaeoecology* **30,** 17–40.

Barron, J. A. (1996). Diatom constraints on the position of the Antarctic Polar Front in the middle part of the Pliocene. *Marine Micropaleontology* **27,** 195–213.

Barry, J. C. & Flynn, L. J. (1990). Key biostratigraphic events in the Siwaliks sequence. In E. L. Lindsay et al., eds., *European Neogene Mammal Chronology,* pp. 557–571. New York: Plenum.

Barry, J. C., Morgan, M. E., Flynn, L. J., Pilbeam, D., Jacobs, L. L., Lindsay, E. H., Raza, S. M. & Solounias, N. (1995). Patterns of faunal turnover and diversity in the Neogene Siwaliks of Northern Pakistan. *Palaeogeography, Palaeoclimatology, Palaeoecology* **115,** 209–226.

Bartek, L. R., Sloan, L. C., Anderson, J. B. & Ross, M. I. (1992). Evidence from the Antarctic continental margin of late Paleogene ice sheets: a manifestation of plate reorganization and synchronous changes in atmospheric circulation over the emerging Southern Ocean? In D. R. Protheno & W. A. Berggren, eds., *Eocene-Oligocene Climatic and Biotic Evolution,* pp. 131–159. Princeton, NJ: Princeton University Press.

Barton, N. H. (1988). Speciation. In A. A. Myers & P. S. Giller, eds., *Analytical Biogeography,* pp. 185–218. London: Chapman and Hall.

Bates, P. J. J. (1988). Systematics and zoogeography of *Tatera* (Rodentia: Gerbillinae) of north-east Africa and Asia. *Bonner zoologische Beiträge* **39,** 265–303.

Bauer, A. M. (1993). African-South American relationships: a perspective from the Reptilia. In P. Goldblat, ed., *Biological Relationships between Africa and South America,* pp. 244–288. New Haven, CT: Yale University Press.

Bauer, I. E., McMorrow, J. & Yalden, D. W. (1994). The historic ranges of three equid species in north-east Africa: a quantitative comparison of environmental tolerances. *Journal of Biogeography* **21,** 169–182.

Beadle, L. C. (1981). *The Inland Waters of Tropical Africa: An Introduction to Tropical Limnology,* 2nd ed. New York: Longman.

Beck, R. A., Burbank, D. W., Sercombe, W. J., Olson, T. L. & Khan, A. M. (1995). Organic carbon exhumation and global warming during the early Himalayan collision. *Geology* **23,** 387–390.

Beden, M. (1983). Family Elephantidae. In J. M. Harris, ed., *Koobi Fora Research Project,* vol. 2. *The Fossil Ungulates: Proboscidea, Perissodactyla and Suidae,* pp. 40–129. Oxford: Clarendon Press.

Beden, M. (1985). Les proboscidiens des grands gisements a hominidés Plio-Pléistocènes d'Afrique orientale. In Y. Coppens, ed., *L'Environnement des Hominidés au Plio-Pléistocène,* pp. 21–44. Paris: Masson.

Beden, M. (1987). Les éléphantidés (Mammalia–Proboscidea). In Y. Coppens & F. C. Howell, eds., *Les faunes plio-pleistocenes de la basse vallée de l'Omo (Ethiopie), tome 2,* pp. 1–162. Paris: C. N. R. S.

Begun, D. (1992). Miocene fossil hominids and the chimp-human clade. *Science* **257,** 1929–1933.

Begun, D. (1994). Relations among the great apes and humans: new interpretations based on the fossil great ape *Dryopithecus. Yearbook of Physical Anthropology* **37,** 11–63.

Behrensmeyer, A. K. (1976). Lothagham, Kanapoi and Ekora: a general summary of stratigraphy and fauna. In Y. Coppens, F. C. Howell, G. L. Isaac & R. E. F. Leakey, eds., *Earliest Man and Environments in the Omo Deposits,* pp. 163–170. Chicago: University of Chicago Press.

Behrensmeyer, A. K., Damuth, J. D., DiMichele, W. A., Potts, R., Sues, H.-D. & Wing, S. L. (1992). *Terrestrial Ecosystems through Time.* Chicago: University of Chicago Press.

Behrensmeyer, A. K., Todd, N. E., Potts, R. & McBrinn, G. E. (1997). Late Pliocene faunal turnover in the Turkana Basin, Kenya and Ethiopia. *Science* **278,** 1589–1594.

Bell, R. H. (1969). *The Use of the Herbaceous Layer by Grazing Ungulates in the Serengeti National Park, Tanzania* (Ph.D. thesis), Manchester: University of Manchester.

Bell, R. H. V. (1971). A grazing ecosystem in the Serengeti. *Scientific American* **225,** 86–93.

Bell, R. H. V. (1981). An outline of a management plan for Kasungu National Park, Malawi. In P. A. Jewell, S. Holt & D. Hart, eds., *Problems in Management of Locally Abundant Wild Mammals,* pp. 69–89. New York: Academic Press.

Bell, R. H. V. (1982). The effect of soil nutrient availability on community structure in African ecosystems. In B. J. Huntley & B. H. Walker, eds., *The Ecology of Tropical Savannas,* pp. 193–216. Berlin: Springer-Verlag.

Belsky, A. J. (1990). Tree/grass ratios in East African savannas: a comparison of existing models. *Journal of Biogeography* **17,** 483–489.

Benammi, M., Orth, B., Vianey-Liaud, M., Chaimanee, Y., Suteethorn, V., Feraud, G., Hernandez, J. & Jaeger, J. J. (1995). Micromammifères et biochronologie des formations néogènes du flanc sud du Haut-Atlas Marocain: implications biogéographiques, stratigraphiques et tectoniques. *Africa Geoscience Review* **2,** 279–310.

Bender, M., et al. (1994). Climate correlations between Greenland and Antarctica during the past 100,000 years. *Nature* **372,** 663–666.

Benefit, B. R. (1987). The molar morphology, natural history, and phylogenetic position of the middle Miocene monkey *Victoriapithecus* (Ph.D. dissertation). New York: New York University.

Benefit, B. R. (1990). Fossil evidence for the dietary evolution of Old World monkeys [abstract]. *American Journal of Physical Anthropology* **81,** 193.

Benefit, B. R. (1991). The taxonomic status of Maboko small apes [abstract]. *American Journal of Physical Anthropology* **12,** 50–51.

Benefit, B. R. (1993). The permanent dentition and phylogenetic position of *Victoriapithecus* from Maboko Island, Kenya. *Journal of Human Evolution* **25,** 83–172.

Benefit, B. R. (1994). Phylogenetic, paleodemographic, and taphonomic implications of *Victoriapithecus* deciduous teeth. *American Journal of Physical Anthropology* **95,** 277–331.

Benefit, B. R. (1999). *Victoriapithecus,* the key to Old World monkey and catarrhine origins. *Evolutionary Anthropology* **7,** 155–174.

Benefit, B. R. (in press). Old World monkey origins and diversification: an evolutionary study of diet and dentition. In P. Whitehead & J. Jolly, eds., *Old World Monkeys.* Cambridge: Cambridge University Press.

Benefit, B. R. & McCrossin, M. L. (1990). Diet, species diversity, and distribution of African fossil baboons. *Kroeber Anthropological Society Papers* **72,** 77–93.

Benefit, B. R. & McCrossin, M. L. (1991). Ancestral facial morphology of Old World higher primates. *Proceedings of the National Academy of Science USA* **88,** 5267–5271.

Benefit, B. R. & McCrossin, M. L. (1992). *Kenyapithecus* from Maboko Island. In J. A. Van Couvering, ed., *Apes or Ancestors?* pp. 11–12. New York: American Museum of Natural History.

Benefit, B. R. & McCrossin, M. L. (1993a). New *Kenyapithecus* postcrania and other primate fossils from Maboko Island, Kenya [abstract]. *American Journal of Physical Anthropology Supplement* **16,** 55–56.

Benefit, B. R. & McCrossin, M. L. (1993b). The facial anatomy of *Victoriapithecus* and its relevance to the ancestral cranial morphology of

Old World monkeys and apes. *American Journal of Physical Anthropology* **92,** 329–370.

Benefit, B. R. & McCrossin, M. L. (1995). Miocene hominoids and hominid origins. *Annual Review of Anthropology* **24,** 237–256.

Benefit, B. R. & McCrossin, M. L. (1997). Earliest known Old World monkey skull. *Nature* **388,** 368–371.

Benefit, B. R. & Pickford, M. (1986). Miocene fossil cercopithecoids from Kenya. *American Journal of Physical Anthropology* **69,** 441–464.

Benson, C. W., Stuart Irving, M. P. & White, C. M. N. (1962). The significance of valleys as avian zoogeographical barriers. *Annals of the Cape Province Museum* **2,** 155–189.

Benton, M. J. (1987). Progress and competition in macroevolution. *Biological Reviews* **62,** 305–338.

Berge, C. (1994). How did the australopithecines walk? A biomechanical study of the hip and thigh of *Australopithecus afarensis. Journal of Human Evolution* **26,** 259–273.

Berger, L. R. (1994). Functional morphology of the hominoid shoulder, past and present (Ph.D. thesis). Johannesburg: University of the Witwatersrand.

Berggren, W. A., Kent, D. V., Swisher, C. C. & Aubry, M. P. (1995). A revised Cenozoic geochronology and chronostratigraphy. In W. A. Berggren, D. V. Kent & J. Hardenbol, eds., *Geochronology, Time Scales, and Global Stratigraphic Correlations: A Unified Temporal Framework for A Historical Geology.* Society of Economic Paleontology and Mineralogy, Special Publication no. 54, pp. 129–212. Tulsa: Society of Economic Paleontology and Mineralogy.

Berggren, W. A., Kent, D. V. & Van Couvering, J. A. (1985). Neogene geochronology and chronostratigraphy. *Geological Society of London Memoir* **10,** 211–260.

Berggren, W. A. & Van Couvering, J. A. (1974). The late Neogene biostratigraphy, geochronology and paleoclimatology of the last 15 million years in marine and continental sequences. *Palaeogeography, Palaeoclimatology, Palaeoecology* **16(1/2),** 1–216.

Bergmann, C. (1846). Uber die Verhaltnisse der Warmeokonomie der Thiere zu ihzer Grosse. *Gottinger Studien* **31,** 595–708.

Bergmans, W. (1988). Taxonomy and biogeography of African fruit bats (Mammalia, Megachiroptera), 1, General introduction; material and methods; results: the genus *Epomophorus* Bennett, 1836. *Beaufortia* **38,** 75–146.

Berner, R. A. (1990). Atmospheric carbon dioxide levels over Phanerozoic time. *Science* **249,** 1382–1386.

Berner, R. A., Lasaga, A. C. & Garrels, R. M. (1983). The carbonate-silicate geochemical cycle and its effect on atmospheric carbon dioxide over the past 100 million years. *American Journal of Science* **283,** 641–683.

Bernor, R. L. (1983). Geochronology and zoogeographic relationships of Miocene Hominoidea. In R. L. Ciochon & R. S. Corruccini, eds., *New Interpretations of Ape and Human Ancestry,* pp. 21–64. New York: Plenum.

Bernor, R. L. (1984). A zoogeographic theater and biochronologic play: the time/biofacies phenomena of Eurasian and African Miocene mammal provinces. *Paléobiologie Continentale* **14,** 121–142.

Bernor, R. L. (1985). Systematic and evolutionary relationships of the hipparionine horses from Maragheh, Iran (late Miocene, Turolian age). *Palaeovertebrata* **15(4),** 173–269.

Bernor, R. L. (1993). Evolution of terrestrial ecosystems through time [book review]. *American Journal of Physical Anthropology* **91(2),** 258–260.

Bernor, R. L., Andrews, P., Solounias N. & Van Couvering, J. A. H. (1979). The evolution of "Pontian" mammal faunas: some zoogeographic, paleoecologic and chronostratigraphic considerations. *Annales Geologica Pays Hellenica* **1,** 81–90.

Bernor, R. L. & Armour-Chelu, M. (1997). Later Neogene hipparions from the Manonga Valley, Tanzania. In T. Harrison, ed, *Neogene Paleontology of the Manonga Valley, Tanzania,* pp. 219–264. New York: Plenum.

Bernor, R. L., Brunet, M., Ginsburg, L., Mein, P., Pickford, M., Rögl, F., Sen, S., Steininger, F. & Thomas, H. (1987a). A consideration of some major topics concerning old world Miocene mammalian chronology, migrations and paleogeography. *Geobios* **20,** 431–439.

Bernor, R. L., Fahlbusch, V., Andrews, P., de Bruijn, H. Fortelius, M., Rögl, F., Steininger, F. F. & Werdelin, L. (1996a). The evolution of Western Eurasian later Neogene faunas: a chronologic, systematic, biogeographic and paleoenvironmental synthesis. In R. L. Bernor, V. Fahlbusch & H.-W. Mittman, eds., *The Evolution of Western Eurasian Neogene Faunas,* pp. 449–469. New York: Columbia University Press.

Bernor, R. L., Fahlbusch, V. & Mittman, H. W. (1996b). *The Evolution of Western Eurasian Neogene Mammal Faunas.* New York: Columbia University Press.

Bernor, R. L., Fahlbusch, V., Mittmann, H.-W. & Rietschel, S. (1996c). The evolution of western Eurasian Neogene mammal faunas: The 1992 Schloss Reisensburg Workshop Concept. In R. L. Bernor, V. Fahlbusch & H.-W. Mittmann, eds., *The Evolution of Western Eurasian Later*

Neogene Faunas, pp. 1–4. New York: Columbia University Press.

Bernor, R. L. & Franzen, J. (1997). The equids (Mammalia, Perissodactyla) from the late Miocene (Early Turolian) of Dorn-Dürkheim 1 (Germany, Rheinhessen). *Courier Forschungs-Institut Senckenberg* **197,** 117–186.

Bernor, R. L., Heissig, K. & Tobien, H. (1987b). Early Pliocene Perissodactyla from Sahabi, Libya. In N. T. Boaz, A. El-Arnauti, A. W. Gaziry, J. de Heinzelin & D. D. Boaz, eds., *Neogene Paleontology and Geology of Sahabi,* pp. 233–254. New York: Alan R. Liss.

Bernor, R. L. & Hussain, S. T. (1985). An assessment of the systematic, phylogenetic and biogeographic relationships of Siwalik hipparionine horses. *Journal of Vertebrate Paleontology* **5,** 32–87.

Bernor, R. L., Koufos, G., Woodburne, M. O. & Fortelius, M. (1996d). The evolutionary history and biochronology of European and Southwestern Asian late Miocene and Pliocene hipparionine horses. In R. L. Bernor, V. Fahlbusch & H.-W. Mittmann, eds., *The Evolution of Western Eurasian Later Neogene Faunas,* pp. 307–338. New York: Columbia University Press.

Bernor, R. L., Kovar-Eder, J., Lipscomb, D., Rögl, F., Sen, S. & Tobien, H. (1988). Systematics, stratigraphic and paleoenvironmental contexts of first-appearing Hipparion in the Vienna Basin, Austria. *Journal Vertebrate Paleontology* **8,** 427–452.

Bernor, R. L., Kovar-Eder, J., Suc, J.-P. & Tobien, H. (1990). A contribution to the evolutionary history of European late Miocene age hipparionines (Mammalia: Equidae). *Paleobiologie Continentale* **XVII,** 291–309.

Bernor, R. L., Kretzoi, M., Mittmann, H.-W. & Tobien, H. (1993a). A preliminary systematic assessment of the Rudabánya hipparions. *Mitteilungen der Bayerischen Staatssammlung fur Palaontologie und historisches Geologie* **33,** 1–20.

Bernor, R. L. & Lipscomb, D. (1991). The systematic position of *Plesiohipparion aff. huangheense* (Equidae, Hipparionini) from Gülyazi, Turkey. *Mitteilung der Bayerischen Staatsslammlung für Paläontologie und historische Geologie* **31,** 107–123.

Bernor, R. L. & Lipscomb, D. (1995). A consideration of Old World hipparionine horse phylogeny and global abiotic processes. In E. Vrba et al., eds., *Paleoclimate and Evolution, with Emphasis on Human Origins,* pp. 164–177. New Haven, CT: Yale University Press.

Bernor, R. L., Mittmann, H.-W. & Rögl, F. (1993b). The Gözendorf *Hipparion. Annales Naturhistoriche Museum, Wien* **95A,** 101–120.

Bernor, R. L. & Pavlakis, P. P. (1987). Zoogeographic relationships of the Sahabi large mammal fauna (Early Pliocene Libya). In N. T. Boaz, A. El-Arnati, A. W. Gaziry, J. de Heinzelin & D. D. Boaz, eds., *Neogene Paleontology and Geology of Sahabi,* pp. 349–383. New York: Alan R. Liss.

Bernor, R. L. & Tobien, H. (1989). Two small species of Cremohipparion (Equidae, Mammalia) from Samos, Greece. *Mitteilungen der Bayerischen Staatssammlung fur Palaontologie und historisches Geologie* **29,** 207–226.

Bernor, R. L., Tobien, H., Hayek, L.-A. & Mittmann, H.-W. (1997). *Hippotherium primigenium* (Equidae, Mammalia) from the late Miocene of Howenegg (Hegau, Germany). *Andrias* **10,** 1–230.

Bernor, R. L., Tobien, H. & Woodburne, M. O. (1989). Patterns of Old World hipparionine evolutionary diversification and biogeographic extension. In E. H. Lindsay, V. Fahlbusch & P. Mein, eds., *Topics on European Mammalian Chronology,* pp. 263–319. New York: Plenum.

Bernor, R. L., Woodburne, M. O. & Van Couvering, J. A. (1980). A contribution to the chronology of some Old World Miocene faunas based on hipparionine horses. *Géobios* **13,** 25–59.

Beynon, A. D. & Wood, B. A. (1986). Variations in enamel thickness and structure in east African hominids. *American Journal of Physical Anthropology* **70,** 177–195.

Beynon, A. D. & Wood, B. A. (1987). Patterns and rates of molar crown formation times in east African fossil hominids. *Nature* **326,** 493–496.

Bielicki, T. (1969). Deviation-amplifying cybernetic systems and hominid evolution. *Materialy i Prace Anthropology* **77,** 57–40.

Bigalke, R. C. (1972). The contemporary mammal fauna of Africa. In A. Keast, F. C. Erk & B. Glass, eds., *Evolution, Mammals, and Southern Continents,* pp. 141–194. Albany: University of New York Press.

Birchette, M. G. (1981). Postcranial remains of *Cercopithecoides* [abstract]. *American Journal of Physical Anthropology* **54,** 201.

Birchette, M. G. (1982). The postcranial skeleton of *Paracolobus chemeroni* (Ph.D. dissertation). Cambridge, MA: Harvard University.

Birkeland, P. W. (1984). *Soils and Geomorphology.* New York: Oxford University Press.

Bischof, J. F. (1994). The decay of the Barents ice sheet as documented in Nordic seas ice-rafted debris. *Marine Geology* **117,** 35–55.

Bishop, L. C. (1993). Hominids of the East African Rift Valley in a macroevolutionary context. *American Journal of Physical Anthropology* Supplement **16,** 57.

Bishop, L. C. (1994). Pigs and the ancestors: hominids, suids and environments during the Plio-Pleistocene and East Africa (Ph.D. dissertation). Yale University. Ann Arbor, MI: University Microforms International.

Bishop, L. C. (1995). Habitat availability at East African Pliocene and Pleistocene hominid localities as indicated by suid remains [abstract]. *American Journal of Physical Anthropology Supplement* **20,** 65–66.

Bishop, L. C. (1997). Suidae from the Manonga Valley, Tanzania. In T. Harrison, ed., *Neogene Paleontology and Geology of the Manonga Valley,* pp. 191–217. New York: Plenum Press.

Bishop, W. W. (1968). The evolution of fossil environments in East Africa. *Transactions of the Leicester Literary and Philosophical Society* **62,** 22–44.

Bishop, W. W. (1980). Paleogeomorphology and continental taphonomy. In A. K. Behrensmeyer & A. P. Hill, eds., *Fossils in the Making,* pp. 20–37. Chicago: University of Chicago Press.

Bishop, W. W. & Clarke, J. D., eds. (1967). *Background to Evolution in Africa.* Chicago: University of Chicago Press.

Black, C. G. & Krishtalka, L. (1986). Rodents, bats, and insectivores from the Plio-Pleistocene sediments to the east of Lake Turkana, Kenya. *Contribution in Science* **372,** 1–15.

Blumenschine, R. J. (1987). Characteristics of an early hominid scavenging niche. *Current Anthropology* **28,** 383–407.

Blumenschine, R. J. & Masao, F. T. (1991). Living sites at Olduvai Gorge, Tanzania? Preliminary landscape archaeology results in the basal Bed II lake margin zone. *Journal of Human Evolution* **21,** 451–462.

Boaz, N. (1977). Palaeoecology of early Hominidae in Africa. *Kroeber Anthropological Society Papers* **50,** 37–62.

Boaz, N. T., Bernor, R. L., Brooks, A. S., Cooke, H. B. S., de Heinzelin, J., Dechamps, R., Delson, E., Gentry, A. W., Harris, J. W. K., Meylan, P., Pavlakis, P. P., Sanders, W. J., Stewart, K. M., Verniers, J., Williamson, P. G. & Winkler, A. J. (1992). A new evaluation of the significance of the Late Neogene Lusso beds, Upper Semliki Valley, Zaïre. *Journal of Human Evolution* **22,** 505–517.

Boaz, N. T., Howell, F. C. & McCrossin, M. L. (1982). Faunal age of the Usno, Shungura B and Hadar Formations, Ethiopia. *Nature* **200,** 633–635.

Boaz, N. T., El-Arnauti, A., Gaziry, A. W., de Heinzelin, J. & Boaz, D. D. eds. (1987). *Neogene Paleontology and Geology of Sahabi.* New York: Alan R. Liss.

Bock, C. E. (1987). Distribution-abundance relationships of some Arizona landbirds: a matter of scale? *Ecology* **68,** 124–129.

Bock, C. E. & Ricklefs, R. E. (1983). Range size and local abundance of some North American songbirds: a positive correlation. *American Naturalist* **122,** 295–299.

Bock, W. J. (1972). Species interactions and macroevolution. In T. Dobzhansky, M. K. Hecht & W. C. Steere, eds., *Evolutionary Biology,* pp. 1–24. New York: Appleton-Century-Crofts.

Bock, W. J. (1979). The synthetic explanation of macroevolutionary change—a reductionist approach. *Bulletin of the Carnegie Museum of Natural History* **13,** 20–69.

Bock, W. J. (1981). Functional-adaptive analysis in evolutionary classification. *American Zoologist* **21,** 5–20.

Bock, W. J. (1986). Species concepts, speciations, and macroevolution. In K. Iwatsuki, P. H. Raven & W. J. Bock, eds., *Modern Aspects of Species,* pp. 31–57. Tokyo: University of Tokyo Press.

Bock, W. J. (1991). Levels of complexity and organismic organization. In G. Lanzavecchia & R. Valvassori, eds., *Form and Function in Zoology;* 181–212. Modena: Mucchi.

Bock, W. J. (1993). Selection and fitness: Definitions and uses: 1859 and now. *Proceedings of Zoological Society, Calcutta* **Haldane Commemorative Volume,** 7–26.

Bock, W. J. (1995). The species concept versus the species taxon: their roles in biodiversity analyses and conservation. In R. Arai, M. Kato & Y. Doi, eds., *Biodiversity and Evolution,* 47–72. Tokyo: The National Science Museum Foundation.

Bock, W. J. & Farrand, J. (1980). The number of species and genera of recent birds: a contribution to comparative systematics. *American Museum Novitates* **2703,** 1–29.

Bock, W. J. & von Whalert, G. (1965). Adaptation and the form-function complex. *Evolution* **19,** 269–299.

Boellstorff, J. (1978). A need for redefinition of North American Pleistocene stages. *Transactions of the Gulf Coast Association Geological Society* **28,** 65–74.

Boesch, C. & Boesch, H. (1983). Optimisation of nut-cracking with natural hammers by wild chimpanzees. *Behavior* **83,** 265–285.

Boesch, C. & Boesch, H. (1989). Hunting behaviour of wild chimpanzees in the Tai National Park. *American Journal of Physical Anthropology* **78,** 547–573.

Boesch-Achermann, H. & Boesch, C. (1994). Hominization in the rainforest: the chimpanzee's piece of the puzzle. *Evolutionary Anthropology* **3(1),** 9–16.

Bond, G., Broecker, W., Johnson, S., McManus, J., McManus, J., Labeyrie, L., Jouzel, J. & Bonani, G. (1993). Correlations between climate records from North Atlantic sediments and Greenland ice. *Nature* **365,** 143–147.

Bond, G., Heinrich, H., Broecker, W., Labeyrie, L., McManus, J., Andrews, J., Huon, S., Jantschik, R., Clasen, S., Simet, C., Tedesco, K., Klas, M., Bonani, G. & Ivy, S. (1992). Evidence for massive discharges of icebergs into the glacial northern Atlantic. *Nature* **360,** 245–249.

Bond, G. C. & Lotti, R. (1995). Iceberg discharges into the North Atlantic on millennial time scales during the last glaciation. *Science* **267,** 1005–1010.

Boné, E. L. & Singer, R. (1965). *Hipparion* from Langebaanweg, Cape Province and a revision of the genus in Africa. *Annals of the South African Museum* **48,** 273–397.

Bonnefille, R. (1976). Palynological evidence for an important change in the vegetation of the Omo Basin between 2.5 and 2.0 million years ago. In Y. Coppens, F. C. Howell, G. L. Isaac & R. E. F. Leakey, eds., *Earliest Man and Environments in the Lake Rudolf Basin,* pp. 421–431. Chicago: University of Chicago Press.

Bonnefille, R. (1983). Evidence for a cooler and drier climate in Ethiopia uplands towards 2.5 Myr ago. *Nature* **303,** 487–491.

Bonnefille, R. (1984). Cenozoic vegetation and environments of early hominids in East Africa. In R. O. Whyte, ed., *The Evolution of the East Asia Environment,* pp. 579–612. Hong Kong: Centre of Asia Studies, University of Hong Kong.

Bonnefille, R. (1995). A reassessment of the Plio-Pleistocene pollen record of East Africa. In E. S. Vrba, G. H. Denton, T. C. Partridge & L. H. Burckle, eds., *Paleoclimate and Evolution with Emphasis on Human Origins,* pp. 299–310. New Haven, CT: Yale University Press.

Bonnefille, R. & Dechamps, R. (1983). Data on fossil flora. In J. de Heinzelin, ed., *The Omo Group: Archives of the International Omo Research Expedition,* Annales du Musée Royal de l'Afrique Centrale, Sciences Géologiques, vol. 85, pp. 191–207. Tervuren: Musée Royal de l'Afrique Centrale.

Bonnefille, R., Hamilton, A. C., Linder, H. P. & Riollet, G. (1990). 30,000-Year-old fossil Restionaceae pollen from Central Equatorial Africa and its biogeographical significance. *Journal of Biogeography* **17,** 307–314.

Bonnefille, R. & Hamilton, A. (1986). Quaternary and late tertiary history of Ethiopian vegetation. *Symb. Bot. Ups.* **26(2),** 48–63.

Bonnefille, R., Roeland, J. C. & Guiot, J. (1990). Temperature and rainfall estimates for the past 40,000 years in equatorial Africa. *Nature* **346,** 347–349.

Bonnefille, R. & Vincens, A. (1985). Apport de la palynologie a l'environnement des hominidés d'Afrique orientale. In Y. Coppens, ed., *L'Environnement des Hominidès au Plio-Plèistocene,* pp. 237–278. Paris: Masson.

Bonnefille, R., Vincens, A. & Buchet, G. (1987). Palynology, stratigraphy, and palaeoenvironment of a Pliocene hominid site (2.9-3.3 M.Y.) at Hadar, Ethiopia. *Palaeogeography, Palaeoclimatology, Palaeoecology* **60,** 249–281.

Booth, A. H. (1958). The Niger, the Volta and the Dahomey Gap as geographic barriers. *Evolution* **12,** 48–62.

Bormann, F. H. & Likens, G. E. (1979). *Pattern and Process in a Forested Ecosystem.* New York: Springer-Verlag.

Boucot, A. J. (1975). *Evolution and Extinction Rate Controls.* Amsterdam: Elsevier.

Boucot, A. J. (1978). Community evolution and rates of cladogenesis. In M. K. Hecht, W. C. Steere & B. Wallace, eds., *Evolutionary Biology,* pp. 545–657. New York: Plenum.

Boucot, A. J. (1982). Ecostratigraphic framework for the Lower Devonian of the North American Apomihimchi subprovince. *Neues Jahrbuch Geologische Paleontologische Abhandlung* **163,** 81–121.

Boucot, A. J. (1983). Does evolution take place in an ecological vacuum? *Journal of Paleontology* **57,** 1–30.

Boucot, A. J. (1990). *Evolutionary Paleobiology of Behavior and Coevolution.* New York: Elsevier.

Bovbjerg, R. V. (1970). Ecological isolation and competitive exclusion in two crayfish (*Orconectes virilis* and *Orconectes immunis*). *Ecology* **51,** 225–236.

Boyde, A. (1965). The structure of developing mammalian dental enamel. In M. V. Stack & R. W. Fearnhead, eds., *Tooth Enamel,* pp. 163–167. Bristol: J. Wright.

Boyde, A. (1976). Amelogenesis and the development of teeth. In B. Cohen & I. R. H. Kramer, eds., *Scientific Foundations of Dentistry,* pp. 335–352. London: Heinemann.

Boyde, A. & Fortelius, M. (1986). Development, structure and function of rhinoceros enamel. *Zoological Journal of the Linnean Society* **87,** 181–214.

Bradley, R., Bard, E., Farguhar, G., Joussaume, S., Lautenschlager, M., Molfino, B., Rascheke, E., Shackleton, N. J., Sirocko, F., Stauffer, B. & White, J. (1993). Group report: evaluating strategies for reconstructing past global changes—what and where are the gaps? In J. A. Eddy & H. Oeschger, eds., *Global Changes in the Perspective of the Past,* pp. 145–171. New York: John Wiley & Sons.

Brain, C. K. (1981a). The evolution of man in Africa. Was it a consequence of Cainozoic cooling? *Annex to the Transactions of the Geological Society of South Africa* **84,** 19.

Brain, C. K. (1981b). *The Hunters or the Hunted? An Introduction to African Cave Taphonomy.* Chicago: University of Chicago Press.

Brain, C. K., ed. (1993). *Swartkrans: A Cave's Chronicle of Early Man.* Pretoria: Transvaal Museum Monograph no. 8.

Brandy, L. D., Sabatier, M. & Jaeger, J. J. (1980). Implications phylogénétiques et biogéographiques des dernières découvertes de Muridae en Afghanistan, au Pakistan et en Ethiopie. *Géobios* **13(4),** 639–643.

Brass, G. W., Saltzman, E., Sloan, J. L., II, Southam, J. R., Hay, W. W., Holser, W. T. & Peterson, W. H. (1982). Ocean circulation, plate tectonics, and climate. 83–89.

Brigham-Grette, J. & Carter, L. D. (1992). Pliocene marine transgressions of northern Alaska: circumarctic correlations and paleoclimatic interpretations. *Arctic* **45,** 74–89.

Britton-Davidian, J., Catalan, J., Granjon, L. & Duplantier, M. (1995). Chromosomal phylogeny and evolution in the genus *Mastomys* (Mammalia, Rodentia). *Journal of Mammalogy* **76(1),** 248–262.

Broecker, W. S. (1994). Massive iceberg discharges as triggers for global climate change. *Nature* **372,** 421–424.

Broecker, W. S. (1995a). *The Glacial World According to Wally.* Palisades, NY: Eldigio Press.

Broecker, W. S. (1995b). Chaotic climate. *Scientific American* **273,** 62–68.

Broecker, W. S., Bond, G., Klas, M., Bonani, G. & Wolfli, W. (1990). A salt oscillator in the glacial Atlantic? 1. The concept. *Paleoceanography* **5,** 469–477.

Broecker, W. S., Bond, G., Klas, M., Clark, E. & McManus, J. (1992). Origin of the northern Atlantic's Henrich events. *Climate Dynamics* **6,** 265–273.

Broecker, W. S. & Denton, G. H. (1989). The role of ocean-atmosphere reorganizations in glacial cycles. *Geochimica et Cosmochimica Acta* **53,** 2465–2501.

Broecker, W. S. & Denton, G. H. (1990). What drives glacial cycles? *Scientific American* **262,** 48–56.

Bromage, T. G. (1989). Ontogeny of the early hominid face. *Journal of Human Evolution* **18,** 751–773.

Bromage, T. G. (1990). Early hominid development and life history. In C. J. DeRousseau, ed., *Primate Life History and Evolution,* pp. 105–113. New York: Wiley-Liss.

Bromage, T. G. (1991). Enamel incremental periodicity in the pigtailed macaque: a polychrome fluorescent labeling study of dental hard tissues. *American Journal of Physical Anthropology* **86,** 205–214.

Bromage, T. G. (1992). Ontogeny of *Pan troglodytes* craniofacial architectural relationships and implications for early hominids. *Journal of Human Evolution* **23,** 235–251.

Bromage, T. G. & Schrenk, F. (1995). Biogeographic and climatic basis for a narrative of early hominid evolution. *Journal of Human Evolution* **28,** 109–114.

Bromage, T. G., Schrenk, F. & Juwayeyi, Y. M. (1995a). Paleobiogeography of the Malawi Rift: age and vertebrate paleontology of the Chiwondo Beds, northern Malawi. *Journal of Human Evolution* **28,** 37–57.

Bromage, T. G., Schrenk, F. & Zonneveld, F. W. (1995b). Paleoanthropology of the Malawi Rift: an early hominid mandible from the Chiwondo Beds, northern Malawi. *Journal of Human Evolution* **28,** 71–108.

Broom, R. (1934). On the fossil remains associated with *Australopithecus africanus. South African Journal of Science* **31,** 471–480.

Broom, R. (1937). Notes on new fossil mammals from the Transvaal caves. *Annals and Magazine of Natural History* **20,** 509–514.

Broom, R. & Robinson, J. T. (1950). Further evidence of the structure of the Sterkfontein apeman *Plesianthropus. Transvaal Museum Memoir* **4,** 11–83.

Brown, F. H. (1995). The potential of the Turkana Basin for paleoclimatic reconstruction in East Africa. In E. S. Vrba, G. H. Denton, T. C. Partridge & L. H. Burckle, eds., *Paleoclimate and Evolution with Emphasis on Human Origins,* pp. 319–331. New Haven, CT: Yale University Press.

Brown, F. H. & Nash, W. P. (1976). Radiometric dating and tuff mineralogy of Omo Group deposits. In Y. Coppens, F. C. Howell, G. L. Isaac & R. E. F. Leakey, eds., *Earliest Man and Environments in the Lake Rudolf Basin,* pp. 50–63. Chicago: University of Chicago Press.

Brown, F. H. & Feibel, C. S. (1988). 'Robust' hominids and Plio-Pleistocene paleogeography of the Turkana basin, Kenya and Ethiopia. In F. E. Grine, ed., *Evolutionary History of the "Robust" Australopithecines,* pp. 325–341. New York: Aldine de Gruyter.

Brown, F. H. & Feibel, C. S. (1991). Stratigraphy, depositional environments, and palaeogeography of the Koobi Fora Formation. In J. M. Harris, ed., *Koobi Fora Research Project,* vol. 3. *Stratigraphy, Artiodactyls and Palaeoenvironments,* pp. 1–30. Oxford: Clarendon Press.

Brown, F. H., Harris, J., Leakey, R. E. F. & Walker, A. (1985a). Early *Homo erectus* skeleton from West Turkana, Kenya. *Nature* **316,** 788–792.

Brown, F. H., McDougall, L., Davies, I. & Maier, R. (1985). An integrated Plio-Pleistocene chronology for the Turkana basin. In E. Delson, ed., *Ancestors: The Hard Evidence,* pp. 82–90. New York: Alan R. Liss.

Brown, J. H. (1984). On the relationship between abundance and distribution of species. *American Naturalist* **124,** 225–279.

Brown, J. H., Davidson, D. W. & Reichman, O. J. (1979). An experimental study of competition between seed eating desert rodents and ants. *American Zoologist* **19,** 1129–1143.

Brown, J. H., Kodric-Brown, A., Whitham, T. G. & Bond, H. W. (1981). Competition between hummingbirds and insects for the nectar of two species of shrubs. *Southwestern Naturalist* **26,** 133–145.

Brown, J. S. (1989a). Coexistence on a seasonal resource. *American Naturalist* **133,** 168–182.

Brown, J. S. (1989b). Desert rodent community structure: a test of four mechanisms of coexistence. *Ecological Monographs* **59,** 1–20.

Brown, J. S. & Vincent, T. L. (1987). A theory for the evolutionary game. *Theoretical Population Biology* **31,** 140–166.

Brugal, J. P. & Denys, C. (1989). Vertébrés du site acheuléen d'Isenya (Kenya, district de Kajiado). Implications paléoécologiques et paléobiogéographiques. *Comptes Rendus de l'Académie des Sciences de Paris,* séries II **308,** 1503–1508.

Brunet, M., Beauvilain, A., Coppens, Y., Heintz, E., Moutaye, A. H. E. & Pilbeam, D. (1995). The first australopithecine 2500 kilometers west of the Rift Valley (Chad). *Nature* **378,** 273–275.

Brunet, M., Beauvilain, A., Coppens, Y., Heintz, E., Moutaye, A. H. E. & Pilbeam, D. (1996). *Australopithecus bahrelghazali,* une nouvelle espèce d'Hominidé ancien de la région de Koro Toro (Tchad). *Comptes Rendus de l'Académie des Sciences Paris* séries IIa **322,** 907–913.

Bunn, H. T., Harris, J. W. K., Isaac, G., Kaufulu, Z., Kroll, E., Schick, K., Toth, N. & Behrensmeyer, A. K. (1980). FxJj50: an early Pleistocene site in northern Kenya. *World Archaeology* **12,** 109–139.

Burckle, L. H. (1995). Current issues in Pliocene paleoclimatology. In E. S. Vrba, G. H. Denton, T. C. Partridge & L. H. Burckle, eds., *Paleoclimate and Evolution, with Emphasis on Human Origins,* pp. 319–331. New Haven: Yale University Press.

Burckle, L. H. & Potter, N., Jr. (1996). Pliocene-Pleistocene diatoms in Paleozoic and Mesozoic sedimentary and igneous rocks from Antarctica: a Sirius problem solved. *Geology* **24,** 235–238.

Burgh van der, J., Visscher, H., Dilcher, D. L. & Kkrschner, W. H. (1993). Paleoatmospheric signatures in Neogene fossil leaves. *Science* **260,** 1788–1790.

Bush, G. L. (1975). Modes of animal speciation. *Annual Review of Ecology and Systematics* **6,** 339–364.

Butler, P. M. (1978a). Chalicotheriidae. In V. J. Maglio & H. B. S. Cooke, eds., *Evolution of African Mammals,* pp. 368–370. Cambridge, MA: Harvard University Press.

Butler, P. M. (1978b). Insectivora and Chiroptera. In V. J. Maglio & H. B. S. Cooke, eds., *Evolution of African Mammals,* pp. 56–68. Cambridge, MA: Harvard University Press.

Butler, P. M. (1987). Fossil insectivores from Laetoli. In M. D. Leakey & J. M. Harris, eds., *Laetoli: A Pliocene Site in Northern Tanzania,* pp. 85–87. Oxford: Clarendon Press.

Butzer, K. W., Isaac, G. L., Richardson, J. L. & Washbourn-Kamau, C. (1972). Radiocarbon dating of East African lake levels. *Science* **175,** 1069–1076.

Cadman, A. & Rayner, R. J. (1989). Climatic change and the appearance of *Australopithecus africanus* in the Makapansgat sediments. *Journal of Human Evolution* **18,** 107–113.

Cahen, L. (1954). *Geologie du Congo Belge.* Liege: Vaillant-Carmanne.

Cain, A. (1988). Evolution. *Journal of Evolutionary Biology* **1,** 185–194.

Cande, S. C. & Kent, D. (1992). A new geomagnetic polarity time scale for the Late Cretaceous and Cenozoic. *Journal Geophysical Research* **97,** 13,917–13,951.

Carcasson, R. H. (1964). A preliminary survey of the zoogeography of African butterflies. *East African Wildlife Journal* **2,** 122–157.

Carson, H. L. (1975). The genetics of speciation at the diploid level. *The American Naturalist* **109 (965),** 83–92.

Carson, H. L. (1982). Speciation as a major reorganization of polygenic balances. In C. Barigozzi, ed., *Mechanisms of Speciation,* pp. 411–433. New York: Alan R. Liss.

Carson, H. L. (1987). The process whereby species originate. *BioScience* **37,** 715–720.

Carson, H. L. & Templeton, A. R. (1984). Genetic revolutions in relation to speciation phenomena: The founding of new populations. *Annual Review of Ecology and Systematics* **15,** 97–131.

Cartmill, M. (1981). Hypothesis testing and phylogenetic reconstruction. *Zeitschrift fur zoologische Systematik und Evolutionforschung* **19,** 73–96.

Cerling, T. E. (1984). The stable isotopic composition of modern soil carbonate and its relationship to climate. *Earth and Planetary Science Letters* **71,** 229–240.

Cerling, T. E. (1991). Carbon dioxide in the atmosphere: evidence from Cenozoic and Mesozoic paleosols. *American Journal of Science* **291,** 377–400.

Cerling, T. E. (1992a). Development of grasslands and savannas in East Africa during the

Neogene. *Palaeogeography, Palaeoclimatology, Palaeoecology* **97,** 241–247.

Cerling, T. E. (1992b). Use of carbon isotopes in paleosols as an indicator of the $P(CO_2)$ of the paleoatmosphere. *Global Biogeochemical Cycles* **6(3),** 307–314.

Cerling, T., Ehleringer, J. R. & Harris, J. M. (1998). Carbon dioxide starvation, the development of C_4 ecosystems, and mammalian evolution. *Philosophical Transactions of the Royal Society of London* B **353,** 159–171.

Cerling, T. E., Harris, J. M., Ambrose, S. H., Leakey, M. G. & Solounias, N. (1997a). Dietary and environmental reconstruction with stable isotope analyses of herbivore tooth enamel from the Miocene locality of Fort Ternan, Kenya. *Journal of Human Evolution* **33,** 635–650.

Cerling, T. E., Harris, J. M., MacFadden, B. J., Leakey, M. G., Quade, J., Eisenmann, V. & Ehleringer, J. R. (1997b). Global vegetation change through the Miocene/Pliocene boundary. *Nature* **389,** 153–158.

Cerling, T. E., Bowman, J. R. & O'Neil, J. R. (1988). An isotopic study of a fluvial-lacustrine sequence: the Plio-Pleistocene Koobi Fora sequence, East Africa. *Palaeogeography, Palaeoclimatology, Palaeoecology* **63,** 335–356.

Cerling, T. E. & Hay, R. L. (1986). An isotopic study of paleosol carbonates from Olduvai Gorge. *Quaternary Research* **25,** 63–78.

Cerling, T. E., Hay, R. L. & O'Neil, J. R. (1977). Isotopic evidence for dramatic climatic changes in East Africa during the Pleistocene. *Nature* **267,** 137–138.

Cerling, T. E., Kappelman, J., Quade, J., Ambrose, S. H., Sikes, N. E. & Andrews, P. (1992). Reply to comment on the paleoenvironment of *Kenyapithecus* at Fort Ternan. *Journal of Human Evolution* **23,** 371–377.

Cerling, T. E. & Quade, J. (1993). Stable carbon and oxygen isotopes in soil carbonates. In P. K. Swart, K. C. Lohmann, J. McKenzie & S. Savin, eds., *Climate Change in Continental Isotopic Records* Geophysical Monograph **78,** 217–231.

Cerling, T. E., Quade, J., Ambrose, S. H. & Sikes, N. E. (1991). Miocene fossil soils, grasses and carbon isotopes from Fort Ternan (Kenya): grassland or woodland? *Journal of Human Evolution* **21,** 295–306.

Cerling, T. E., Quade, J., Wang, Y. & Bowman, J. R. (1989). Carbon isotopes in soils and paleosols as ecology and paleoecology indicators. *Nature* **341,** 138–139.

Cerling, T. E., Quade, J. & Wang, Y. (1994). Expansion and emergence of C_4 plants. *Nature* **371,** 112.

Cerling, T. E., Wang, Y. & Quade, J. (1993). Expansion of C_4 ecosystems as an indicator of global ecological change in the late Miocene. *Nature* **361,** 344–345.

Chaimanee Y., Suteethorn, V., Triamwichanon, S. & Jaeger, J. J. (1996). A new stephanodont Murinae (Mammalia, Rodentia) from the early Pleistocene of Thailand and the age and place of the *Rattus* adaptive radiation in South East Asia. *Comptes Rendus de l'Académie des Sciences de Paris* séries IIa **322,** 155–162.

Chamberlain, A. T. & Wood, B. A. (1987). Early hominid phylogeny. *Journal of Human Evolution* **16,** 118–133.

Chamberlain, T. C. (1899). An attempt to frame a working hypothesis of the cause of glacial periods on an atmospheric basis. *Journal of Geology* **7,** 545–584, 667–685, 751–787.

Chamberlain, T. C. (1906). On a possible reversal of deep sea circulation and its influence on geological climates. *Journal of Geology* **14,** 363–373.

Chapin, J. P. (1932). The birds of the Belgian Congo, part 1. *Bulletin of the American Museum of Natural History* **65,** 1–756.

Charlesworth, B. (1995). Simpson 50 years on (review of "Tempo and mode in evolution: genetics and paleontology 50 years after Simpson"). *Trends in Ecology and Evolution* **10,** 507–508.

Charlesworth, B., Lande, R. & Slatkin, M. (1982). A neo-Darwinian commentary on macroevolution. *Evolution* **36,** 474–498.

Chesser, R. T. & R. M. Zink (1994). Modes of speciation in birds: a test of Lynch's method. *Evolution* **48,** 490–497.

Chesson, P. & Huntly, N. (1988). Community consequences of life-history traits in a variable environment. *Annales Zoologici Fennici* **25,** 5–16.

Chesson, P. & Huntly, N. (1993). Temporal hierarchies of variation and the maintenance of diversity. *Plant Species Biology* **8,** 195–206.

Chevret, P. (1994). Etude évolutive des Murinae (rongeurs: Mammiferès) africains par hybridations ADN/ADN. Comparaisons avec les approches morphologiques et paléontologiques Thèse Doctora. Montpellier: Université Montpellier II.

Chevret, P., Denys, C., Jaeger, J. J., Michaux, J. & Catzeflis, F. (1993a). Molecular and Paleontological aspects of the tempo and mode of evolution in *Otomys* (Otomyinae: Muridae: Mammalia). *Biochemical Systematics and Ecology* **21(1),** 123–131.

Chevret, P., Denys, C., Jaeger, J. J., Michaux, J. & Catzeflis, F. M. (1993b). Molecular evidence that the spiny mouse (*Acomys*) is more closely related to the gerbils (Gerbillinae) than to true mice (Murinae). *Proceedings of the National Academy of Sciences USA* **90,** 3433–3436.

Churcher, C. S. & Richardson, M. L. (1978). Equidae. In V. J. Maglio & H. B. S. Cooke, eds., *Evolution of African Mammals,* pp. 379–422. Cambridge, MA: Harvard University Press.

Cifelli, R. L., Ibui, A. K., Jacobs, L. L. & Thorington, R. W. (1986). A giant tree squirrel from the late Miocene of Kenya. *Journal of Mammalogy* **67,** 274–283.

Ciochon, R. L. (1986). The Cercopithecoid forelimb: anatomical implications for the evolution of African Plio-Pleistocene species (Ph.D. dissertation). Berkeley: University of California.

Ciochon, R. L. & Corruccini, R. S. (1976). Shoulder joint of Sterkfontein *Australopithecus. South African Journal of Science* **72,** 80–82.

Claessen, C. J. & De Vree, F. (1991). Systematic and taxonomic notes on the *Epomophorus anurus-labiatus-minor* complex with the description of a new species (Mammalia: Chiroptera: Pteropodidae). *Senckenbergiana Biologica* **71,** 209–238.

Clarke, R. J. (1985). *Australopithecus* and early *Homo* in southern Africa. In E. Delson, ed., *Ancestors: The Hard Evidence,* pp. 171–177. New York: Alan R. Liss.

Clarke, R. J. (1994). On some new interpretations of Sterkfontein stratigraphy. *South African Journal of Science* **90,** 211–214.

Clarke, R. & Tobias, P. V. (1995). Sterkfontein Member 2 foot bones of the oldest South African hominid. *Science* **269,** 521–524.

Clauzon, G., Suc, J., Gautier, F., Berger, A. & Loutre, M. (1996). Alternate interpretation of the Messinian salinity crisis: controversy resolved? *Geology* **24,** 363–366.

Clutton-Brock, T. H. & Harvey, P. H. (1977). Primate ecology and social organization. *Journal of Zoology* **183,** 1–39.

Cody, M. L. & Mooney, H. A. (1978). Convergence versus nonconvergence in Mediterranean-climate ecosystems. *Annual Reviews of Ecology and Systematics* **9,** 265–321.

Coetzee, C. G. (1983). An analysis of the distribution patterns of the Namibian terrestrial mammals (bats excluded). *Annales du Musée Royal de l'Afrique Centrale, Sciences Zoologiques* **237,** 63–73.

Coifaait, B. & Coiffait, P. E. (1981). Découverte d'un gisement de micromammifères d'âge pliocène dans le bassin de Constantine (Algérie). Présence d'un muridé nouveau *Paraethomys athmenia. Palaeovertebrata* **11(1),** 1–15.

Colbert, E. H. (1973). Continental drift and the distribution of fossil reptiles. In D. H. Tarling & S. K. Runcorn, eds., *Implications of Continental Drift for the Earth Sciences,* pp. 395–412. New York: Academic Press.

Cole, M. M. (1982). The influence of soils, geomorphology and geology on the distribution of plant communities in savanna ecosystems. In B. J. Huntley & B. H. Walker, eds., *The Ecology of Tropical Savannas,* pp. 145–174. Berlin: Springer-Verlag.

Cole, M. (1986). *The Savannahs. Biogeography and Geobotany.* London: Academic Press.

Collins, L. S., Coates, A. G., Berggren, W. A., Aubry, M.-P. & Zhang, J. (1996). The late Miocene Panama isthmian strait. *Geology* **24,** 687–690.

Colyn, M. (1991). *L'importance zoogéographique du basin du fleuve Zaire pour la spéciation: le cas des primates simiens. Annalen. Koninklijk Museum voor Midden-Afrika,* Tervuren 264, ix + 250 pp.

Colyn, M. & Van Rompaey, H. (1994). Morphometric evidence of the monotypic status of the African long-nosed mongoose *Xenogale naso* (Carnivora, Herpestidae). *Belgian Journal of Zoology* **124,** 175–192.

Contrafatto, G., Meester, J. A., Willan, K., Taylor, P. J., Roberts, M. A. & Baker, C. M. (1992). Genetic variation in the African rodent subfamily Otomyinae (Muridae). II. Chromosomal changes in some populations of *Otomys irroratus. Cytogenetics Cell Genetics* **59,** 293–299.

Contrafatto, G., David, D. & Goossens-Le Clerq, V. (1994). Genetic variation in the African rodent subfamily Otomyinae. V: Phylogeny inferred by immuno-electrotransfer analysis. *Durban Museum Novitates* **19,** 1–7.

Cooke, H. B. S. (1950). A critical revision of the Quaternary Perissodactyla of southern Africa. *Annals of the South African Museum* **31,** 393–479.

Cooke, H. B. S. (1962). The Pleistocene environment in Southern Africa—Hypothetical vegetation in Southern Africa during the Pleistocene. *Annals of the Cape Province Museum* **2,** 11–15.

Cooke, H. B. S. (1964). The Pleistocene environment in Southern Africa. In D. H. S. Davis, ed., *Ecological Studies in Southern Africa,* pp. 1–23. The Hague: W. Junk.

Cooke, H. B. S. (1968). Evolution of mammals on southern continents: II. The fossil mammal fauna of Africa. *Quarterly Review of Biology* **43,** 234–264.

Cooke, H. B. S. (1976). Suidae from Plio-Pleistocene strata of the Rudolf Basin. In Y. Coppens, F. C. Howell, G. Ll. Isaac & R. E. F. Leakey, eds., *Earliest Man and Environments in the Lake Rudolf Basin,* pp. 251–263. Chicago: University of Chicago Press.

Cooke, H. B. S. (1978). Suid evolution and correlation of African hominid localities: an alternate taxonomy. *Science* **201,** 460–463.

Cooke, H. B. S. (1984). Horse, elephants and pigs as clues in the African later Cainozoic. In J. C. Vogel, ed., *Late Cainozoic Palaeoclimates of the Southern Hemisphere,* pp. 473–482. Rotterdam: A. A. Balkema.

Cooke, H. B. S. (1985). Plio-Pleistocene Suidae in relation to African hominid deposits. In Y. Coppens, ed., *L'Environnement des Hominides au Plio-Pléistocène,* pp. 101–117. Paris: Masson.

Cooke, H. B. S. (1987). Fossil Suidae from Sahabi, Libya. In N. T. Boaz, A. El-Arnauti, A. W. Gaziry, J. de Heinzelin & D. D. Boaz, eds., *Neogene Paleontology and Geology of Sahabi,* pp. 255–266. New York: Alan R. Liss.

Cooke, H. B. S. & Hendey, Q. B. (1992). *Nyanzachoerus* (Mammalia: Suidae: Etraconodontinae) from Langebaanweg, South Africa. *Durban Museum Novitates* **17,** 1–20.

Cooke, H. B. S. & Wilkinson, A. F. (1978). Suidae and Tayassuidae. In V. J. Maglio & H. B. S. Cooke, eds., *Evolution of African Mammals,* pp. 435–482. Cambridge, MA: Harvard University Press.

Cooke, H. J. (1980). Landform evolution in the context of climatic change and neo-tectonism in the Middle Kalahari of north-central Botswana. *Transactions, Institute of British Geographers* **5(1),** 80–99.

Coope, C. R. (1979). Late Cenozoic fossil Coleoptera: evolution, biogeography, ecology. *Annual Review of Ecology and Systematics* **10,** 247–267.

Coppens, Y. (1972). Tentative de zonation du Pliocène et du Pléistocène ancien d'Afrique par les grands mammifères. *Comptes Rendus de l'Académie des Sciences Paris* **274,** 181–184.

Coppens, Y. (1974). Les faunes de Vertébrés du Pliocène et du Pléistocène ancien d'Afrique. Ve Congrès du Néogène méditerranéen, t. 1. *Mémoires du B.R.G.M.* **78,** 109–119.

Coppens, Y. (1975a). Evolution des Mammifères, de leurs fréquences et deleurs associations, au cours du Plio-Pleistocène dans la basse vallée de l'Omo en Ethiopie. *Comptes Rendus de l'Académie des Sciences Paris* **281,** 1571–1574.

Coppens, Y. (1975b). Evolution des Hominidés et de leur environnement au cours du Plio-Pléistocène dans la basse vallée de l'Omo en Ethiopie. *Comptes Rendus de L'Académie des Sciences Paris* **281,** 1693–1696.

Coppens, Y. (1976). Evolution de l'environnement des Hominidés dans la basse vallée de l'Omo en Ethiopie au cours du Plio-Pléistocène. *Palaeocology of Africa* **IX,** 1972-74, 1976, 103–105.

Coppens, Y. (1978a). Le Lothagamien et le Shungurien, étages continentaux du Pliocène est-africain. *Bulletin de la Societé Géologique de France* **7e, XX,** 1, 39–44.

Coppens, Y. (1978b). Les Hominidés du Pliocène et du Pléistocène d'Ethiopie, chronologie, systématique, environnement. In J. Pivetcau, ed., *Les Origines Humaines et les époques de l'Intelligence,* pp. 79–106. Paris: Fondation Singer-Polignac, Masson Ed.

Coppens, Y. (1978c). Evolution of the Hominids and of their environment during the Plio-Pleistocene in the lower Omo Valley, Ethiopia. In W. W. Bishop, ed., *Geological Background to Fossil Man,* 499–506. Edinburgh: Scottish Academic Press.

Coppens, Y. (1979). Les Hominidés du Pliocène et du Pléistocène de la Rift Valley. Séance à thème de la Société Géologique de France, Geologie et Paléontologie de la Rift Valley. *Bulletin de la Societé Géologique de France* **7, XXI,** 3, 313–320.

Coppens, Y. (1980). The differences between *Australopithecus* and *Homo:* preliminary conclusions from the Omo Research Expedition's studies. In L. Konigsson, ed., *Current Argument on Early Man, Report from a Nobel Symposium,* pp. 207–225. London: Pergamon.

Coppens, Y. (1983a). Les Hominidés du Pliocène et du Pléistocène d'Afrique orientale et leur environnement. In C.N.R.S., ed., *Table Ronde* C.N.R.S., *Morphologie évolutive, morphogénèse du crâne et anthropogénèse,* pp. 185–194. Paris: VIIIe Congrès de la Société de Primatologie Internationale.

Coppens, Y. (1983b). Systématique, phylogénie, environnement et culture des Australopithèques; hypothèses et synthèse. Séance à thème de la Société d'Anthropologie de Paris. "Les Australopithèques." *Bulletin et Mémores de la Societé d'Anthropologie de Paris* séries XIII **10(3),** 273–284.

Coppens, Y. (1983c). Les plus anciens Hominidés. In *Groupe de travail de l'Académie pontificale des Sciences, Récents progrés dans nos connaissances concernant l'évolution des Primates,* pp. 1–9. Rome: Pontificiae Academiae Scientarium Scripta Varia Vatican City, pp. 1–9.

Coppens, Y. (1983d). *Le Singe, l'Afrique et l'Homme.* Collection, "Le temps des Sciences," sous la direction d'Odile Jacob. Paris: Edition Fayard.

Coppens, Y. (1984). Hominoidés, Hominidés et Hommes. *La Vie des Sciences,* comptes rendus s.g. **1(5),** 459–486.

Coppens, Y., ed. (1985). *L'environnement des Hominidés au Plio-Pléistocène.* Paris: Masson.

Coppens, Y. (1986). L'évolution de l'Homme. *La Vie des Sciences,* comptes rendus, s.g. **3,** 227–243.

Coppens, Y. (1989). Hominid evolution and the evolution of the environment. *Ossa* **14,** 157–163.

Coppens, Y. (1994). East Side Story: the origin of humankind. *Scientific American* **270,** 62–69.

Coppens, Y. & Howell, F. C. (1976). Mammalian

faunas of the Omo group: distributional and biostratigraphical aspects. In Y. Coppens, F. C. Howell, G. L. Isaac & R. E. F. Leakey, eds., *Earliest Man and the Environments in the Lake Rudolf Basin,* pp. 177–192. Chicago: University of Chicago Press.

Corliss, B. H., et al. (1984). The Eocene/Oligocene boundary event in the deep sea. *Science* **226,** 806–810.

Covey, C. & Barron, E. (1988). The role of ocean heat transport in climatic change. *Earth Science Reviews* **24,** 429–445.

Covey, C. & Thompson, S. L. (1989). Testing the effects of ocean heat transport on climate. *Palaeogeography, Palaeoclimatology, Palaeoecology* **75,** 331–341.

Cowling, R., ed. (1992). *The Ecology of Fynbos: Nutrients, Fire and Diversity.* Cape Town: Oxford University Press.

Coyne, J. (1992). Genetics and speciation. *Nature* **355,** 511–515.

Coyne, J. (1996). Speciation in action. *Science* **272,** 700–701.

Cracraft, J. (1982). Geographic differentiation, cladistics, and vicariance biogeography: reconstructing the tempo and mode of evolution. *American Zoologist* **22,** 411–424.

Cracraft, J. (1983). Cladistic analysis and vicariance biogeography. *American Scientist* **71,** 273–281.

Cracraft, J. (1984). The terminology of allopatric speciation. *Systematic Zoology* **33,** 115–116.

Cracraft, J. (1986). Origin and evolution of continental biotas: speciation and historical congruence within the Australian avifauna. *Evolution* **40,** 977–996.

Cracraft, J. (1988). Deep-history biogeography: retrieving the historical pattern of evolving continental biotas. *Systematic Zoology* **37,** 221–236.

Cracraft, J. (1989a). Species as entities of biological theory. In M. Ruse, ed., *What the Philosophy of Biology Is,* pp. 31–52. Dordrecht: Kluwer.

Cracraft, J. (1989b). Speciation and its ontology: the empirical consequences of alternative species concepts for understanding patterns and processes of differentiation. In D. Otte & J. A. Endler, eds., *Speciation and Its Consequences,* pp. 28–59. Sunderland, MA: Sinauer Associates.

Crawford-Cabral, J. (1983). Patterns of allopatric speciation in some Angolan Muridae. *Annales Musée Royal de l'Afrique Centrale, Sciences Zoologiques* **237,** 153–157.

Crawford-Cabral, J. (1989a). A list of Angolan Chiroptera with notes on their distribution. *Garcia de Orta, Serie Zoologico* **13,** 7–48.

Crawford-Cabral, J. (1989b). Distributional data and notes on Angolan carnivores (Mammalia: Carnivora). 1—Small and median-sized species. *Garcia de Orta, Serie Zoologico* **14,** 3–27.

Crawford-Cabral, J. (1993). Parapatry as a secondary event. *Garcia de Orta, Serie Zoologico* **19,** 1–6.

Croizat, L., Nelson, G. & Rosen, D. E. (1974). Centers of origin and related concepts. *Systematic Zoology* **23,** 265–287.

Crompton, R. H. & Li, Y. (1997). Running before they could walk? Locomotor adaptation and bipedalism in early hominids. In A. Sinclair, E. Slater & J. A. J. Gowlett, eds., *Archaeological Sciences 1995.* Oxford Monograph 64. Oxford: Oxbow Books.

Cronin, T. M. (1991). Pliocene shallow water paleoceanography of the North Atlantic Ocean based on marine ostracodes. *Quaternary Science Reviews* **10,** 175–188.

Crowe, T. M. (1990). A quantitative analysis of patterns of distribution, species richness and endemism in southern African vertebrates. In G. Peters & R. Hutterer, eds., *Vertebrates in the Tropics,* pp. 161–175. Bonn: Alexander Koenig Zoological Research Institute and Zoological Museum.

Crowe, T. M. & Crowe, A. A. (1982). Patterns of distribution, diversity and endemism in Afrotropical birds. *Journal of Zoology* **198,** 417–442.

Crowley, T. J. (1991). Modeling Pliocene warmth. *Quaternary Science Reviews* **10,** 275–282.

Crowley, T. J. (1996). Pliocene climates: the nature of the problem. *Marine Micropaleontology* **27,** 3–12.

Crowley, T. J. & North, G. R. (1991). *Paleoclimatology.* New York: Oxford University Press.

Cumming, D. H. M. (1975). *A Field Study of the Ecology and Behaviour of Warthog.* Museum Memoir 7. Salisbury: Trustees of the National Museums and Monuments of Rhodesia.

Curtis, G. H. & Hay, R. L. (1972). Further geological studies and potassium-argon dating at Olduvai Gorge and Ngorongoro crater. In W. W. Bishop & J. A. Miller, eds., *Calibration of Hominoid Evolution,* pp. 289–310. Edinburg: Scottish Academic Press.

Damuth, J. (1985). Selection among "species": A formulation in terms of natural functional units. *Evolution* **39,** 1132–1146.

Darwin, C. (1859). *On The Origin of Species by Means of Natural Selection, or the Preservation of Favoured Races in the Struggle for Life.* London: John Murray.

Darwin, C. (1871). *The Descent of Man, and Selection in Relation to Sex.* London: John Murray.

Davies, A. G. & Oates, J. F., eds. (1994). *Colobine Monkeys: Their Ecology, Behavior and*

Evolution. Cambridge: Cambridge University Press.

Davies, C. (1987a). Fossil Pedetidae (Rodentia) from Laetoli. In M. D. Leakey & J. M. Harris, eds., *Laetoli: a Pliocene Site in Northern Tanzania,* pp. 171–189. Oxford: Clarendon Press.

Davies, C. (1987b). Notes on the fossil Lagomorph from Laetoli. In M. D. Leakey & J. M. Harris, eds., *Laetoli: a Pliocene Site in Northern Tanzania,* pp. 190–193. Oxford: Clarendon Press.

Davis, D. H. S. (1962). Distribution patterns of Southern African Muridae, with notes on some of their fossil antecedents. *Annals of the Cape Province Museum* **2,** 56–76.

Davis, D. H. S. (1966). Contribution to the revision of the genus *Tatera* in Africa. *Annales Musée Royal de l'Afrique Centrale, Serie in-8°, Sciences Zoologiques* **144,** 49–65.

Day, M. H., Leakey, M. D. & Magori, C. (1980). A new fossil skull (L.H. 18) from the Ngaloba Beds, Laetoli, northern Tanzania. *Nature* **284,** 55–56.

Deacon, H. J., Jury, M. R. & Ellis, F. (1992). Selective regime and time. In R. M. Cowling, ed., *The Ecology of Fynbos: Nutrients, Fire and Diversity,* pp. 6–22. Cape Town: Oxford University Press.

Dean, M. C. (1987). Growth layers and incremental markings in hard tissues; a review of the literature and some preliminary observations amount enamel structure in *Paranthropus boisei. Journal of Human Evolution* **16,** 157–172.

Dean, M. C. (1989). The developing dentition and tooth structure in hominids. *Folia Primatology* **53,** 160–176.

de Bonis, L., Bouvrain, G., Geraads, D. & Koufos, G. (1992). Diversity and palaeoecology of Greek late Miocene mammalian faunas. *Palaeogeography, Palaeoclimatology, Palaeoecology* **91,** 99–121.

de Bonis, L. & Koufos, G. (1993). The face and mandible of *Ouranopithecus macedoniensis:* description of new specimens and comparisons. *Journal of Human Evolution* **24,** 469–491.

Dechamps, R. & Maes, F. (1985). Essai de reconstitution des climats et des végétations de la basse vallée de l'Omo au Plio-Pléistocène à l'aide des bois fossiles. In Y. Coppens, ed., *L'Environnement des Hominidés au Plio-Pléistocène,* pp. 175–222. Paris: Masson.

de Heinzelin, J. (1963). Addendum to Transcript of Discussions. In F. C. Howell & F. Bourliere, eds., *African Ecology and Human Evolution,* pp. 648–654. Chicago: Aldine.

de Heinzelin, J. (1983). The Omo Group: archives of the International Omo Research Expedition. *Musée Royal de L'Afrique Centrale Annales Sciences Geologiques* **85,** 1–365.

Deines, P. (1980). The isotopic composition of reduced organic carbon. In P. Fritz & J. C. Fontes, eds., *Handbook of Environmental Isotope Geochemistry,* vol. 1. *The Terrestrial Environment,* pp. 329–406. Amsterdam: Elsevier.

De Jong, J. (1988). Climatic variability during the past three million years, as indicated by vegetational evolution in northwest Europe and with emphasis on data from the Netherlands. *Philosophical Transactions of the Royal Society of London* B **318,** 603–671.

Delson, E. (1973). Fossil colobine monkeys of the circum-Mediterranean region and the evolutionary history of the Cercopithecidae (Primates, Mammalia) (Ph.D. dissertation). New York: Columbia University.

Delson, E. (1975). Evolutionary history of the Cercopithecidae. In F. S. Szalay, ed., *Approaches to Primate Paleobiology,* vol. 5. *Contributions to Primatology,* pp. 167–217. Basel: Karger.

Delson, E. (1984). Cercopithecid biochronology of the African Plio-Pleistocene: Correlation among eastern and southern hominid-bearing localities. *Courier Forschungs Institute, Senckenberg* **69,** 199–218.

Delson, E., ed. (1985). *Ancestors: The Hard Evidence.* New York: Alan R. Liss.

Delson, E. (1988). Chronology of South African australopith site units. In F. E. Grine, ed., *The Evolutionary History of the "Robust" Australopithecines,* pp. 317–324. New York: Aldine de Gruyter.

Delson, E. (1993). *Theropithecus* fossils from Africa and India and the taxonomy of the genus. In N. G. Jablonski, ed., *Theropithecus—The Rise and Fall of a Primate Genus,* pp. 157–189. Cambridge: Cambridge University Press.

Delson, E., Eldredge, N. & Tattersall, I. (1977). Reconstruction of hominid phylogeny: a testable framework based on cladistic analysis. *Journal of Human Evolution* **6,** 263–278.

Demars, P. Y. & Hublin, J. J. (1989). La transition néandertaliens/hommes de type moderne en Europe occidentale: aspects paléontologiques et culturels. In B. Vandermeersch, ed., *L'homme de Néandertal 7—La transition,* pp. 29–42. Liège: ERAUL.

De Meneses Cabral, J. C. (1966). Some new data on Angolan Muridae. *Zoologica Africana* **2,** 193–203.

deMenocal, P. B. (1995). Plio-Pleistocene African climate. *Science* **270,** 53–59.

deMenocal, P. B. & Bloemendal, J. (1995). Plio-Pleistocene climatic variability in subtropical Africa and the paleoenvironment of hominid evolution: a combined data-model approach. In E. S. Vrba et al., ed., *Paleoclimate and*

Evolution with Emphasis on Human Origin, pp. 262–288. New Haven, CT: Yale University Press.

deMenocal, P. B. & Rind, D. (1993). Sensitivity of Asian and African Climate to Variations in Seasonal Insolation, Glacial Ice Cover, Sea Surface Temperature, and Asian Orography. *Journal of Geophysical Research* **98(D4),** 7265–7287.

Demes, B. & Creel, N. (1988). Bite force, diet, and cranial morphology of fossil hominids. *Journal of Human Evolution* **17,** 657–670.

Dennett, D. C. (1995). *Darwin's Dangerous Idea.* New York: Simon & Schuster.

Denton, G. H. (1985). Did the antarctic ice sheet influence Late Cainozoic climate and evolution in the Southern Hemisphere? *South African Journal of Science* **81,** 224–229.

Denton, G. H. (1995). The problem of Pliocene paleoclimate and ice-sheet evolution in Antarctica. In E. S. Vrba, et al., eds., *Paleoclimate and Evolution with Emphasis on Human Origin,* pp. 213–229. New Haven, CT: Yale University Press.

Denton, G. H. & Hughes, T. (1981). *The Last Great Ice Sheets.* New York: Wiley.

Denton, G. H., Sugden, D. E., Marchant, D. R., Hall, B. L. & Wilch, T. I. (1993). East Antarctic ice-sheet sensitivity to Pliocene climatic change from a Dry Valleys perspective. *Geografiska Annaler* **75A,** 155–204.

Denton, G. H. et al. (1999). Interhemispheric linkage of paleoclimatic during the last glaciation. *Geografiska Annaler* **81A,** 107–153.

Denys, C. (1985). Paleoenvironmental and paleobiogeographical significance of the fossil rodent assemblages of Laetoli (Pliocene, Tanzania). *Palaeogeography, Palaeoclimatology, Palaeoecology* **52,** 77–97.

Denys, C. (1987a). Rodentia and Lagomorpha. Fossil rodents (other than Pedetidae) from Laetoli. In M. D. Leakey & J. M. Harris, eds., *Laetoli: a Pliocene site in Tanzania,* pp. 118–170. Oxford: Clarendon.

Denys, C. (1987b). Micromammals from the West Natron Pleistocene deposits (Tanzania). Biostratigraphy and Paleoecology. *Sciences Géologiques Bulletin* **40(1–2),** 185–201.

Denys, C. (1989a). A new species of Bathyergid rodent from Olduvai Bed I (Tanzania, Lower Pleistocene). *Neues Jachbuch für Geologie und Paläontologie Monathefte* **5,** 257–264.

Denys, C. (1989b). Phylogenetic affinities of the oldest East African *Otomys* (Rodentia, Mammalia) from Olduvai bed I (Pleistocene, Tanzania). *Neues Jachbuch für Geologie und Paläontologie Monathefte* **12,** 705–25.

Denys, C. (1989c). Two new Gerbillids (Rodentia, Mammalia) from Oluvai bed I (Pleistocene, Tanzania). *Neues Jachbuch für Geologie und Paläontologie Abhandlungen* **178,** 243–265.

Denys, C. (1989d). Implications paléoécologiques et paléobiogéographiques de la présence d'une gerboise (Rodentia, mammalia) dans le rift est africain au Pléistocène moyen. *Comptes Rendus de l'Académie des Sciences de Paris* séries II **309,** 1261–1266.

Denys, C. (1990a). *Implications paléoécologiques et paléobiogéographiques de l'étude de rongeurs Plio-Pléistocènes d'Afrique orientale et australe* (Thèse doctorat d'Etat). Paris: Université Paris.

Denys, C. (1990b). Deux nouvelles espèces d'*Aethomys* (Rodentia, Muridae) à Langebaanweg (Pliocène, Afrique du Sud): implications phylogénétiques et paléoécologiques. *Annales de Paléontologie* **76,** fasc.1, 41–69.

Denys, C. (1990c). First occurrence of *Xerus* cf. *inauris* (Rodentia, Sciuridae) at Olduvai Bed I (Lower Pleistocene, Tanzania). *Paläontologische Zeitschrift* **64(3/4),** 359–365.

Denys, C. (1992a). Les analyses multivariées: une aide à la reconstitution des paléoenvironnements. L'exemple des rongeurs plio-pléistocènes d'Afrique australe. *Geobios* (Mémoire spécial) **14,** 209–217.

Denys, C. (1992b). Présence de *Saccostomus* (Rodentia, Mammalia) à Olduvai Bed I (Tanzanie, Pléistocène inférieur). Implications phylétiques et paléobiogéographiques. *Geobios* **25(1),** 145–154.

Denys, C. (1994). Deux nouvelles espèces de *Dendromus* (Rongeurs, Muroidea) à Langebaanweg (Pliocene, Afrique du Sud). Conséquences stratigraphiques et paléoécologiques. *Palaeovertebrata* **23(1–4),** 153–176.

Denys, C. (1997). Rodent faunal lists in karstic and open-air sites of Africa: an attempt to evaluate predation and fossilisation biases on paleodiversity. *Cuadernos de Geologia Ibérica* **23,** 73–94.

Denys, C. (1998). Phylogenetic implications of the existence of two modern genera of Bathyergidae genera (Rodentia, Mammalia) in the Pliocene site of Langebaanweg (South Africa). *Annals of the South African Museum* **105,** 265–286.

Denys, C., Chorowicz, J., Tiercelin, J.-J. (1987a). Tectonic and environmental control on rodent diversity in the Plio-Pleistocene sediments of the African rift system. In Frostick, L. E. et al., eds., *Sedimentation in the African rifts* Geological Society Special Papers Publication **25,** 363–372. London: Geological Society.

Denys, C., Dauphin, Y. & Fernandez-Jalvo, Y. (1997). Apports biostratigraphiques et paleoecologiques de l'étude taphonomique des assemblages de micromammifères. Bilan et perspectives. In P. R. Racheboeuf & M. Gayet, eds., *Actualités paléontologiques en l'honneur de C. Babin. Geobios* Mémoire special) **20,** 197–206.

Denys, C., Dauphin, Y., Rzebik-Kowalska, B., Kowalski, K. (1996). Taphonomical study of Algerian owl pellets assemblages and differential preservation of some rodents. Paleontological implications. *Acta zoologica cracoviensis* **39(1),** 103–116.

Denys, C., Gautun, J. C., Tranier, M. & Volobouev, V. (1994). Evolution of the genus *Acomys* (Rodentia, Muridae) from dental and chromosomal patterns. *Israelian Journal of Zoology* **40,** 215–246.

Denys, C. & Jaeger, J. J. (1986). A biostratigraphic problem: the case of the east African Plio-Pleistocene rodent faunas. *Modern Geology* **10,** 215–233.

Denys, C., Michaux, J., Hendey, Q. B. (1987b). Les rongeurs (Mammalia) *Euryotomys* et *Otomys:* un exemple d'évolution parallèle en Afrique tropicale? *Comptes Rendus de l'Académie des Sciences de Paris séries II* **305,** 1389–1395.

de Vries, H. (1906). *Species and Varieties: Their Origin by Mutation.* Chicago: The Open Court Publishing Co.

Diamond, J. M. (1987). Will grebes provide the answer? *Nature* **325,** 16–17.

Dieckmann, U., Marrow, P. & Law, R. (1995). Evolutionary cycling in predator-prey interactions: population dynamics and the Red Queen. *Journal of Theoretical Biology* **176,** 91–102.

Diester-Haas, L. (1987). History of the Benguela current off SW Africa (DSDP sites 362 and 532). *Palaeoecology of Africa* **18,** 55–70.

Dieterlen, F. (1989). Rodents. In H. Lieth & M. J. A. Werger, eds., *Tropical Rainforest Ecosystems,* pp. 383–400. Amsterdam: Elsevier.

Dingle, R. V., Siesser, W. G. & Newton, A. R. (1983). *Mesozoic and Tertiary Geology of Southern Africa.* Rotterdam: A. A. Balkema.

Doran, D. M. (1993a). Comparative locomotor behavior of chimpanzees and bonobos: the influence of morphology on locomotion. *American Journal of Physical Anthropology* **91,** 83–98.

Doran, D. M. (1993b). Sex differences in adult chimpanzee positional behavior: the influence of body size on locomotion and posture. *American Journal of Physical Anthropology* **91,** 99–115.

Dorst, J. & Dandelot, P. (1970). *A Field Guide to the Larger Mammals of Africa.* London: Collins.

Dowsett, H., Barron, J. & Poore, R. (1996). Middle Pliocene sea surface temperatures: a global reconstruction. *Marine Micropaleontology* **27,** 13–25.

Dowsett, H. J. & Cronin, T. M. (1990). High eustatic sea level during the middle Pliocene: evidence from the southeastern U.S. Atlantic Coastal Plain. *Geology* **18,** 435–438.

Dowsett, H. J., Cronin, T. M., Poore, R. Z., Thompson, R. S., Whatley, R. C. & Wood, A. M. (1992). Micropaleontological evidence for increased meridional heat transport in the North Atlantic Ocean during the Pliocene. *Science* **258,** 1133–1135.

Dowsett, H. J. & Poore, R. Z. (1991). Pliocene sea surface temperatures of the North Atlantic Ocean at 3.0 Ma. *Quaternary Science Reviews* **10,** 189–204.

Drake, R. & Curtis, G. H. (1987). K-Ar geochronology of the Laetoli fossil localities. In M. D. Leakey & J. M. Harris, eds., *Laetoli: a Pliocene Site in Northern Tanzania,* pp. 48–52. Oxford: Clarendon Press.

Dublin, H. (1995). Vegetation dynamics in the Serengeti-Mara ecosystem: the role of elephants, fire, and other factors. In A. R. E. Sinclair and P. Arcese, eds., *Serengeti II,* pp. 71–90. Chicago: University of Chicago Press.

Dugas, D. P. & Retallack, R. J. (1993). Middle Miocene fossil grasses from Fort Ternan, Kenya. *Journal of Paleontology* **67(1),** 113–128.

Dunbar, R. I. M. (1983). Theropithecines and hominids: contrasting solutions to the same ecological problem. *Journal of Human Evolution* **12,** 647–658.

Dunbar, R. I. M. (1988). *Primate Social Systems.* Beckenham, Kent: Croom Held.

Dunbar, R. I. M. (1992a). Behavioral ecology of the extinct papionines. *Journal of Human Evolution* **22,** 407–421.

Dunbar, R. I. M. (1992b). Neocortex size as a constraint on group size in primates. *Journal of Human Evolution* **22,** 469–493.

Dunbar, R. I. M. (1993). Socioecology of the extinct theropithecines: a modelling approach. In N. Jablonski, ed., *Theropithecus—the Rise and Fall of a Primate Genus,* pp. 465–486. Cambridge: Cambridge University Press.

Dunbar, R. I. M. (1995). Neocortex size and group size in primates: a test of the hypothesis. *Journal of Human Evolution* **28,** 287–296.

Dunbar, R. I. M. & Dunbar, P. (1974). Ecological relations and niche separation among sympatric terrestrial primates in Ethiopia. *Folia Primatologica* **21,** 36–60.

Dunne, T. (1978). Rates of chemical denudation of silicate rocks in tropical catchments. *Nature* **274,** 244–246.

Duplantier, J. M. & Granjon, L. (1992). Liste révisée des rongeurs du Sénégal. *Mammalia* **56, 3,** 425–431.

Duplessy, J. C., Shackleton, N. J., Fairbanks, R. G., Labeyrie, L., Oppo, D. & Kallel, N. (1988). Deepwater source variations during the last climatic cycle and their impact on the gloval deepwater circulation. *Paleoceanography* **3,** 343–360.

Dupont, L. M. & Agwu, C. O. C. (1992). Latitudinal shifts of forest and savannah in NW Africa during the Bruhnes chron further marine palynological results from site M16415 (9°N–19°W). *Vegetation History and Archaeobotany* **1,** 163–175.

Dupont, L. M. & Leroy, S. A. G. (1995). Steps towards drier climatic conditions in northwestern Africa during the upper Pliocene. In E. S. Vrba, G. H. Denton, T. C. Partridge & L. H. Burckle, eds., *Paleoclimate and Evolution,* pp. 289–298. Chicago: University of Chicago Press.

Dwyer, G. S., Cronin, T. M., Baker, P. A., Raymo, M. E., Buzos, J. S. & CorrJge, T. (1995). North Atlantic deepwater temperature change during late Pliocene and late Quaternary climatic cycles. *Science* **270,** 1347–1351.

Dyke, C. (1988). *The Evolutionary Dynamics of Complex Systems.* New York: Oxford University Press.

East, R. (1984). Rainfall, soil nutrient status and biomass of large African savanna animals. *African Journal of Ecology* **22,** 245–270.

Eck, G. G. (1976). Cercopithecoidea from Omo Group deposits. In Y. Coppens, F. C. Howell, C. Ll. Isaac & R. E. F. Leakey, eds., *Earliest Man and Environments in the Lake Rudolf Basin,* pp. 332–344. Chicago: University of Chicago Press.

Eddy, J. A. & Oeschger, H. (1993). *Global Changes in the Perspective of the Past.* New York: John Wiley & Sons.

Edel, A. (1988). Integrative levels: some reflections on a philosophical dimension. In G. Greenberg and E. Tobach, eds., *Evolution of Social Behavior and Integrative Level,* pp. 65–74. Hillsdale, NJ: Lawrence Erlbaum Associates.

Edmond, J. M., Palmer, M. R., Measures, C. I., Brown, E. T. & Huk, Y. (1996). Fluvial geochemistry of the eastern slope of the northeastern Andes and its foredeep in the drainage of the Orinocoin, Columbia and Venezuela. *Geochimica et Cosmochimica Acta* **60,** 2949–2976.

Edwards, A. R. (1968). The calcareous nannoplankton evidence for New Zealand Tertiary climate. *Tuatara* **16,** 26–31.

Eeley, H. (1994). Species range area in catarrhine primates (Ph.D. thesis). Cambridge: University of Cambridge.

Ehleringer, J. R. & Cooper, T. A. (1988). Correlation between carbon isotope ratio and microhabitat in desert plants. *Oecologia* **76,** 562–566.

Ehleringer, J. R., Sage, R. F., Flanagan, L. B. & Pearcy, R. W. (1991). Climate change and the evolution of C_4 photosynthesis. *Trends in Ecology and Evolution* **6,** 95–99.

Ehleringer, J. R., Cerling, T. E. & Helliker, B. R. (1997). C_4 photosynthesis, atmospheric CO_2, and climate. *Oecologia* **112,** 285–299.

Ehrlich, P. & Raven, P. (1969). Differentiation of populations. *Science* **165,** 1228–1232.

Ehrmann, W. & Mackensen, A. (1992). Sedimentological evidence for the formation of an East Antarctic ice sheet in Eocene/Oligocene time. *Palaeogeography, Palaeoclimatology, Palaeoecology* **93,** 85–112.

Ehrmann, W., Hambrey, M., et al. (1992). History of Antarctic glaciation: an Indian Ocean perspective. Synthesis of results from scientific drilling in the Indian Ocean. *American Geophysical Union Geophysical Monograph* **70,** 423–446.

Einarsson, T. & Albertsson, K. J. (1988). The glacial history of Iceland during the past three million years. *Philosophical Transactions of the Royal Society of London Series B* **318,** 637–644.

Eisenmann, V. (1976). Nouveaux crânes d'Hipparions (Mammalia, Perissodactyla) Plio-Pléistocènes d'Afrique Orientale (Ethiopie et Kenya): *Hipparion* sp., *Hipparion cf. ethiopicum,* et *Hipparion afarense nov. sp. Géobios* **9(5),** 577–605.

Eisenmann, V. (1977). Les Hipparions africains: valeur et signification de quelques caractères des jugales inférieures. *Bulletin Museum national d'Histoire naturelle. Paris 3e Series, Science de Terre* **60,** 69–87.

Eisenmann, V. (1979). Le genre Hipparion (Mammalia, Perissodactyla) et son intérê t stratigraphique en Afrique. *Bulletin Societé géologique de France,* 7e Série **21,** 277–281.

Eisenmann, V. (1980). Caractères spécifiques et problèmes taxinomiques relatifs à certains Hipparions africains. *Actes du 8e Congress Panafricain de Préhistoire,* 77–81.

Eisenmann, V. (1983). Family Equidae. In J. M. Harris, ed., *Koobi Fora Research Project,* vol. 2. *The Fossil Ungulates: Proboscidea, Perissodactyla, and Suidae,* pp. 156–214. Oxford: Clarendon Press.

Eisenmann, V. (1984). Sur quelques caractères adaptifs du squelette d'*Equus* (Mammalia, Perissodactyla) et leurs implications paléoécologiques. *Bulletin du Muséum National d'Histoire Naturelle* **6C(2),** 185–195.

Eisenmann, V. (1985). Les équidés des gisements de la vallée de l'Omo en Ethiopie. In Y. Coppens & F. C. Howell, eds., *Les Faunes Plio-Pléistocènes de la Basse Vallée de l'Omo (Ethiopie), tome 1, Périssodactyles, Artiodactyles (Bovidae),* pp. 13–56. Paris: C.N.R.S.

Eisenmann, V. (1994). Equidae of the Albertine Rift Valley, Uganda. *CIFEG Occasional Publication* **29,** pp. 289–307.

Eisenmann, V., Alberdi, M.-T., de Giuli, C. & Staesche, U. (1988). Studying fossil horses, vol. I. Methodology. In M. O. Woodburne & P. Y. Sondaar, eds., *Collected Papers after the "New York International Hipparion Conference, 1981,"* pp. 1–71. Leiden: Brill.

Eldredge, N. (1979). Alternative approaches to evolutionary theory. *Bulletin of the Carnegie Museum of Natural History* **13,** 7–19.

Eldredge, N. (1985). *Unfinished Synthesis.* New York: Oxford University Press.

Eldredge, N. (1989). *Macroevolutionary Dynamics. Species, Niches, and Adaptive Peaks.* New York: McGraw Hill.

Eldredge, N. (1994). Species, selection, and Paterson's concept of the specific-mate recognition system. In D. H. Lambert & H. G. Spencer, eds., *Speciation and the Recognition Concept—Theory and Application,* pp. 464–476. Baltimore, MD: The Johns Hopkins University Press.

Eldredge, N. (1995). *Reinventing Darwin: The Great Debate at the High Table of Evolutionary Theory.* New York: John Wiley & Sons.

Eldredge, N. & Cracraft, J. (1980). *Phylogenetic Patterns and the Evolutionary Process.* New York: Columbia University Press.

Eldredge, N. & Gould, S. J. (1972). Punctuated equilibria: an alternative to phyletic gradualism. In J. M. Schopf, ed., *Models in Paleobiology,* pp. 82–125. New York: Freeman, Cooper & Co.

Emlen, J. T. & Schaller, G. B. (1960). Distribution and status of the mountain gorilla (*Gorilla gorilla beringei*)—1959. *Zoologica* **45,** 41–52.

Endler, J. A. (1982). Pleistocene refuges: fact or fancy? In G. Prance, ed., *Biological Diversification in the Tropics,* pp. 641–657. New York: Columbia University Press.

Ericson, J. H., Sullivan, C. H. & Boaz, N. T. (1981). Diets of Pliocene mammals from Omo, Ethiopia, deduced from carbon isotope ratios in tooth apatite. *Palaeogeography, Palaeoclimatology, Palaeoecology* **36,** 69–73.

Estes, R. & Hutchinson, J. H. (1980). Eocene lower vertebrates from Ellesmere Island, Canadian Arctic Archipelago. *Palaeogeography, Palaeoclimatology, Palaeoecology* **30,** 325–347.

Evans, E. M. N., Van Couvering, J. A. H. & Andrews, P. (1981). Palaeoecology of Miocene sites in western Kenya. *Journal of Human Evolution* **10,** 99–116.

Ewer, R. F. (1954). Sabre-toothed tigers. *New Biology* **17,** 27–40.

Ewer, R. F. (1958). Adaptive features in the skulls of African Suidae. *Proceedings of the Zoological Society of London* **131,** 135–155.

Ewer, R. F. (1967). The fossil hyaenids of Africa—a reappraisal. In W. W. Bishop & J. D. Clark, eds., *Background to Evolution in Africa,* pp. 109–123. Chicago: University of Chicago Press.

Exon, N. F., Hill, P. J. & Yoyer, J. Y. (1995). New maps of crust off Tasmania expand research possibilities. *EOS, Transactions of the American Geophysical Union* **76(20),** 201, 206–207.

Eyre, S. R. (1963). *Vegetation and Soils: A World Picture.* London: Arnold.

Fabian, B. (1985). Ontogenetic explorations into the nature of evolutionary change. In E. S. Vrba, ed., *Species and Speciation: Transvaal Museum Monograph* 4, pp. 77–85. Pretoria: Transvaal Museum.

Falconer, D. S. (1992). Early selection experiments. *Annual Review of Genetics* **26,** 1–14.

Feibel, C. S. (1994a). Freshwater stingrays from the Plio-Pleistocene of the Turkana Basin, Kenya and Ethiopia. *Lethaia* **26,** 359–366.

Feibel, C. S. (1994b). Controls on sedimentation in a Plio-Pleistocene, fluvial-dominated rift basin, the Turkana Basin of East Africa. *AAPG Annual Meeting, Denver, Abstracts* **3,** 148.

Feibel, C. S. (1995). Geological context and the ecology of *Homo erectus* in East Africa. In J. R. F. Bower & S. Sartono, eds., *Evolution and Ecology of Homo erectus,* pp. 67–74. Leiden: Pithecanthropus Centennial Foundation.

Feibel, C. S. & Brown, F. H. (1991). Age of the primate-bearing deposits on Maboko Island, Kenya. *Journal of Human Evolution* **21,** 221–225.

Feibel, C. S., Brown, F. H. & McDougall, I. (1989). Stratigraphic context of fossil hominids from the Omo Group deposits: northern Turkana Basin and Ethiopia. *American Journal of Physical Anthropology* **78,** 595–622.

Feibel, C. S., Harris, J. M. & Brown, F. H. (1991). Palaeoenvironmental context for the late Neogene of the Turkana Basin. In J. M. Harris, ed., *Koobi Fora Research Project,* vol. 3, *The Fossil Ungulates: Geology, Fossil Artiodactyls, and Palaeoenvironments,* pp. 321–370. Oxford: Clarendon Press.

Felsenstein, J. (1971). On the biological significance of the cost of gene substitution. *The American Naturalist* **105,** 1–11.

Fernández-Jalvo, Y., Denys, C., Andrews, P., Williams, T., Dauphin, Y. & Humphrey, L. (1998). Taphonomy and palaeoecology of Olduvai Bed-I (Pleistocene, Tanzania). *Journal of Human Evolution* **34,** 137–172.

Fisher, R. A. (1958). *The Genetical Theory of Natural Selection.* New York: Dover Publications.

Fitch, W. M. & Ayala, F. J. (1995). *Tempo and Mode in Evolution. Genetics and Paleontology 50 Years after Simpson.* Washington, DC: National Academy Press.

Fitzgerald, C. M. (1998). Do enamel microstructures have regular time dependency? Conclusions from the literature and a large-scale study. *Journal of Human Evolution* **35,** 371–386.

Fitzpatrick, J. W. (1981). Search strategies of tyrant flycatchers. *Animal Behavior* **29,** 810–821.

Fleagle, J. G. (1975). A small gibbon-like hominoid from the Miocene of Uganda. *Folia Primatologica* **24,** 1–15.

Fleagle, J. G. (1988). *Primate Adaptation and Evolution.* London: Academic Press.

Fleagle, J. (1995). Too many species. *Evolutionary Anthropology* **4,** 37–38.

Flower, B. P. & Kennett, J. P. (1993a). Middle Miocene ocean-climate transition: high-resolution oxygen and carbon isotopic records from Deep Sea Drilling Project Site 588A, southwest Pacific. *Paleoceanography* **8,** 811–843.

Flower, B. P. & Kennett, J. P. (1993b). Relationships between Monterey Formation deposition and middle Micoene global cooling: Naples Beach section, California. *Geology* **21,** 877–880.

Flynn, L. J. & Sabatier, M. (1984). A muroid rodent of Asian affinity from the Miocene of Kenya. *Journal of Vertebrate Paleontology* **3,** 160–165.

Flynn, L. J., Tedford, R. H. & Qiu, Z. (1991). Enrichment and stability in the Pliocene mammalian fauna of North China. *Paleobiology* **17,** 246–265.

Foley, R. A. (1984). Early man and the Red Queen: tropical African community evolution and hominid adaptation. In R. Foley, ed., *Human Evolution and Community Ecology,* pp. 85–110. London: Academic Press.

Foley, R. A. (1987). *Another Unique Species: Patterns of Human Evolutionary Ecology.* Harlow: Longman.

Foley, R. A. (1991). How many hominid species should there be? *Journal of Human Evolution* **20,** 413–427.

Foley, R. A. (1993). African terrestrial primates: the comparative evolutionary biology of *Theropithecus* and hominids. In N. Jablonski, ed., *Theropithecus—the Rise and Fall of a Primate Genus,* pp. 254–270. Cambridge: Cambridge University Press.

Foley, R. A. (1994). Speciation, extinction and climatic change in hominid evolution. *Journal of Human Evolution* **26,** 275–289.

Foley, R. A. & Elton, S. (1998). Time and energy: the ecological context for the evolution of bipedalism. In E. Strasser, J. Fleagle, A. Rosenberger & H. McHenry, eds., *Primate Locomotion: Recent Advances,* pp. 419–434. New York: Plenum.

Foley, R. A. & Lee, P. C. (1989). Finite social space, evolutionary pathways, and reconstructing hominid behavior. *Science* **243,** 901–906.

Ford, S. M. (1994). Taxonomy and distribution of the owl monkey. In J. F. Baer, R. E. Weller & I. Kakoma, eds., *Aotus: The Owl Monkey,* pp. 1–57. New York: Academic Press.

Forsten, A. (1968). Revision of Palearctic hipparion. *Acta Zoologica Fennici* **119,** 1–134.

Fortelius, M., Werdelin, L., Andrews, P., Bernor, R. L., Gentry, A., Humphrey, L., Mittmann, W. & Viratana, S. (1996). Provinciality, diversity, turnover and paleoecology in land mammal faunas of the later Miocene of Western Eurasia. In R. L. Bernor, V. Fahlbusch & H.-W. Mittmann, eds., *The Evolution of Western Eurasion Neogene Faunas,* pp. 414–448. New York: Columbia University Press.

Fox, L. R. (1981). Defense and dynamics in plant-herbivore systems. *American Zoologist* **21,** 853–864.

Frakes, L. (1979). *Climate through Geologic Time.* New York: Elsevier.

Franciscus, R. G. & Trinkaus, E. (1988). Nasal morphology and the emergence of *Homo erectus. American Journal of Physical Anthropology* **75,** 517–527.

Freedman, L. (1957). The fossil Cercopithecoidea of South Africa. *Annals of the Transvaal Museum* **23,** 121–262.

Freedman, L. (1976). South African fossil Cercopithecoidea: a reassessment including a description of new material from Makapansgat, Sterkfontein, and Taung. *Journal of Human Evolution* **5,** 297–315.

Friedli, H., Lotscher, H., Oeschger, H., Siegenthaler, U. & Stauffer, B. (1986). Ice core record of the $^{13}C/^{12}C$ ratio of atmospheric CO_2 in the past two centuries. *Nature* **324,** 237–238.

Fritz, H. & Duncan, P. (1993). On the carrying capacity for large ungulates of African savanna ecosystems. *Proceedings of the Royal Society of London Series B* **256,** 77–82.

Frost, P. G., Menaut, J. C., Walker, B. H., Medina, E., Solbrig, O. T. & Swift, M. (1986). Responses of savannas to stress and disturbance. *Biology International* Special Issue no. 10.

Fryxell, J. M. & Sinclair, A. R. E. (1988). Causes and consequences of migration by large herbivores. *Trends in Ecology and Evolution* **3,** 237–241.

Funaioli, U. & Simonetta, A. M. (1966). The mammalian fauna of the Somali Republic: status and conservation problems. *Monitore Zoologico Italiano* **74** (supplement), 285–347.

Funder, S., Abrahamsen, N., Bennike, O. & Feyling-Hansen, R. W. (1985). Forested Arctic: evidence from North Greenland. *Geology* **13,** 542–546.

Futuyma, D. J. & Mayer, G. C. (1980). Non-allopatric speciation in animals. *Systematics and Zoology* **29,** 254–271.

Gaston, K. J. (1996). The multiple forms of the interspecific abundance-distribution relationship. *Oikos* **76,** 211–220.

Gatesy, J., Yelon, D., Desalle, R. & Vrba, E. S. (1992). Phylogeny of the Bovidae (Artiodactyla, Mammalia), based on mitochondrial ribosomal DNA sequences. *Molecular Biology and Evolution* **9,** 433–466.

Gautier, J. P., Moysan, F. & Feistner, A. T. C. (1992). The distribution of *Cercopithecus (lhoesti) solatus,* an endemic guenon of Gabon. *Terre et Vie* **47,** 367–381.

Gautun, J. C., Tranier, M. & Sicard, B. (1985). Liste préliminaire des rongeurs du Burkina Faso (ex Haute Volta). *Mammalia* **49,** 537–542.

Geist, V. (1971). The relation of social evolution and dispersal in ungulates during the Pleistocene, with emphasis on the Old World deer and the genus *Bison. Quaternary Research* **1,** 285–314.

Geist, V. (1987). On the evolution of optical signals in deer: a preliminary analysis. In C. M. Wemmer, ed., *Biology and Management of the Cervidae,* pp. 235–255. Washington, DC: Smithsonian Institution Press.

Gelderblom, C. M., Broner, G. N., Lombard, A. T. & Taylor, P. J. (1995). Patterns of distribution and current protection status of the Carnivora, Chiroptera and Insectivora in South Africa. *South African Journal of Zoology* **30,** 103–114.

Gentry, A. W. (1970). The Bovidae (Mammalia) of the Fort Ternan fossil fauna. In L. S. B. Leakey & R. J. G. Savage, eds., *Fossil Vertebrates of Africa,* pp. 234–324. London: Academic Press.

Gentry, A. W. (1978). Bovidae. In V. J. Maglio & H. B. S. Cooke, eds., *Evolution of African Mammals,* pp. 540–572. Cambridge, MA: Harvard University Press.

Gentry, A. W. (1985a). Pliocene and Pleistocene Bovidae in Africa. In Y. Coppens, eds., *L'Environnement des Hominidés au Plio-Pléistocène,* pp. 119–132. Paris: Masson.

Gentry, A. W. (1985b). The Bovidae of the Omo Group deposits. In Y. Coppens, ed., *Les Faunes Plio-Pléistocènes de la Basse Vallée de L'Omo (Éthiopie),* tome 1, pp. 119–191. Paris: Editions du CNRS.

Gentry, A. W. (1987). Pliocene Bovidae from Laetoli. In M. D. Leakey & J. M. Harris, eds., *Laetoli: A Pliocene Site in Northern Tanzania,* pp. 378–408. Oxford: Clarendon Press.

Gentry, A. W. (1994). The Miocene differentiation of Old World Pecora (Mammalia). *Historical Biology* **7,** 115–158.

Gentry, A. W. & Gentry, A. (1978a). Fossil Bovidae (Mammalia) of Olduvai Gorge, Tanzania. *Bulletin of the British Museum (Natural History)* **29,** 289–446.

Gentry, A. W. & Gentry, A. (1978b). Fossil Bovidae (Mammalia) of Olduvai Gorge, Tanzania. *Bulletin of the British Museum (Natural History)* **30,** 1–83.

Geraads, D. (1987). Dating the Northern African cercopithecid fossil record. *Human Evolution* **2(1),** 19–27.

Geraads, D. & Coppens, Y. (1995). Evolution des faunes de mammifères dans le plio-pléistocène de la basse vallée de l'Omo (Ethiopie): apports de l'analyse factorielle. *Comptes Rendus de l'Académie des Sciences Paris* **320,** 625–637.

Geze, R. (1985). Répartition paléoecologique et relations phylogénétiques des Hippopotamidae (Mammalia, Artiodactyla) du Néogène d'Afrique orientale. In Y. Coppens, ed., *L'Environnement des Hominidés au Plio-Pléistocène,* pp. 81–100. Paris: Masson.

Gibbs Russell, G. E. (1985). Analysis of the size and composition of the southern African flora. *Bothalia* **15,** 613–630.

Gibbs Russell, G. E. (1987). Preliminary floristic analysis of the major biomes in southern Africa. *Bothalia* **17(2),** 213–227.

Gill, A. E. & Bryan, K. (1971). Effects of geometry on the circulation of a three-dimensional southern hemisphere ocean model. *Deep Sea Research* **18,** 685–721.

Giresse, P., Maley, J. & Brenac, P. (1994). Late Quaternary palaeoenvironments in Lake Barombi Mbo (West Cameroon) deduced from pollen and carbon isotopes of organic matter. *Palaeogeography, Palaeoclimatology, Palaeoecology* **107,** 65–78.

Gladenkov, Y. B., Barinov, K. B., Basilian, A. E. & Cronin, T. M. (1991). Stratigraphy and paleoceanography of Pliocene deposits of Karaginsky Island, Eastern Kamchatka, U.S.S.R. *Quaternary Science Reviews* **10,** 239–246.

Gleick, J. (1987). *Chaos: Making a New Science.* New York: Viking.

Godfrey, L. & Jacobs, K. H. (1981). Gradual, autocatalytic and punctuational models of hominid brain evolution: a cautionary tale. *Journal of Human Evolution* **10,** 255–272.

Gohdo, F. (1982). Differential rates of enamel formation on human tooth surfaces deduced from the striae of Retzius. *Archives of Oral Biology* **27,** 289–296.

Goodall, J. (1986). *Chimpanzees of Gombe National Park: Patterns of Behavior.* Cambridge, MA: Harvard University Press.

Gordon, A. L. (1988). The Southern Ocean and global climate. *Oceanus* **31,** 39–46.

Gordon, D. H. (1986). Extensive chromosomal variation in the pouched mouse, *Saccostomus campestris* (Rodentia, cricetidae) from southern

Africa: a preliminary investigation of evolutionary status. *Cimbebasia* **8(5),** 38–47.

Gordon, I. J. & Illius, A. W. (1988). Incisor arcade structure and diet selection in ruminants. *Functional Ecology* **2,** 15–22.

Gould, S. J. (1977a). Eternal metaphors in paleontology. In A. Hallam, ed., *Patterns of Evolution as Illustrated by the Fossil Record,* pp. 1–26. Amsterdam: Elsevier.

Gould, S. J. (1977b). *Ontogeny and Phylogeny.* Cambridge, MA: Harvard University Press.

Gould, S. J. (1980). Is a new and general theory of evolution emerging? *Paleobiology* **6,** 119–130.

Gould, S. J. (1982a). Darwinism and the expansion of evolutionary theory. *Science* **216,** 380–387.

Gould, S. J. (1982b). The meaning of punctuated equilibrium and its role in validating a hierarchical approach to macroevolution. In R. Milkman, ed., *Perspectives on Evolution,* pp. 83–104. Sunderland, MA: Sinauer Associates Inc.

Gould, S. J. (1983). Irrelevance, submission, and partnership: the changing role of paleontology in Darwin's three centennials, and a modest proposal for macroevolution. In D. S. Bendall, ed., *Evolution from Molecules to Men,* pp. 347–366. Cambridge: Cambridge University Press.

Gould, S. J. (1985). The paradox of the first tier: an agenda for paleobiology. *Paleobiology* **11,** 2–12.

Gould, S. J. (1988). Trends as changes in variance: a new slant on progress and directionality in evolution. *Journal of Paleontology* **62,** 319–329.

Gould, S. J. (1990). *The Individual in Darwin's World,* Edinborough: Edinborough University Press.

Gould, S. J. (1993). The inexorable logic of the punctuational paradigm: Hugo de Vries on species selection. In D. R. Lees & D. Edwards, eds., *Evolutionary Patterns and Processes,* pp. 3–17. New York: Academic Press.

Gould, S. J. (1994). Tempo and mode in the macroevolutionary reconstruction of Darwinism. *Proceedings of the National Academy of Sciences* **91,** 6764–6771.

Gould, S. J. (1995a). Spin doctoring Darwin. *Natural History* **7,** 6–10.

Gould, S. J. (1995b). The Darwinian Body. *Neues Jahrbuch Geologie Paläontologie* **195,** 267–278.

Gould, S. J. (1995c). A task for paleobiology at the threshold of majority. *Paleobiology* **21,** 1–14.

Gould, S. J. & Eldredge, N. (1977). Punctuated equilibria: the tempo and mode of evolution reconsidered. *Paleobiology* **3,** 115–151.

Gould, S. J. & Eldredge, N. (1993). Punctuated equilibrium comes of age. *Nature* **366,** 223–227.

Grande, L. & Rieppel, O., eds. (1994). *Interpreting the Hierarchy of Nature. From Systematic Patterns to Evolutionary Process Theories.* San Diego: Academic Press.

Grant, P. R. (1994). Ecological character displacement. *Science* **266,** 746–747.

Grant, V. (1982). Punctuated equilibrium: a critique. *Biologisches Zentralblatt* **101,** 175–184.

Grausz, H. M., Leakey, R. E., Walker, A. C. & Ward, C. V. (1988). Associated cranial and post-cranial bones of *Australopithecus boisei.* In F. E. Grine, ed., *Evolutionary History of the 'Robust' Australopithecines,* pp. 127–132. New York: Aldine de Gruyter.

Greenacre, M. J. & Vrba, E. S. (1984). Graphical display and interpretation of antelope census data in African wildlife areas, using correspondence analysis. *Ecology* **65,** 984–997.

Griffiths, J. F. (1976). *Climate and the Environment.* Boulder: Westview Press.

Griffiths, T. A. (1994). Phylogenetic systematics of slit-faced bats (Chiroptera, Nycteridae), based on hyoid and other morphology. *American Museum Novitates* **3090.**

Grine, F. E. (1981). Trophic differences between "gracile" and "robust" australopithecines: a scanning electron microscope analysis of occlusal events. *South African Journal of Science* **77,** 203–230.

Grine, F. E., ed. (1988). *Evolutionary History of the "Robust" Australopithecines.* New York: Aldine.

Grine, F. E., Demes, B., Jungers, W. L. & Cole, T. M. (1993). Taxonomic affinity of early *Homo* cranium from Swartkrans, South Africa. *American Journal of Physical Anthropology* **92,** 411–426.

Grine, F. E. & Hendey, Q. B. (1981). Earliest primate remains from South Africa. *South African Journal of Science* **77,** 374–376.

Grine, F. E., Jungers, W. L. & Schultz, J. (1996). Phenetic affinities among early *Homo* from East and South Africa. *Journal of Human Evolution* **30,** 189–225.

Grine, F. E. & Martin, L. B. (1988a). Enamel thickness and development in *Australopithecus* and *Paranthropus.* In E. Grine, ed., *Evolutionary History of the "Robust" Australopithecines,* pp. 3–42. New York: Aldine de Gruyter.

Grine, F. E. & Martin, L. B. (1988b). New evidence for the distinctiveness of *Paranthropus. American Journal of Physical Anthropology* **75,** 217–218.

Gromova, V. (1952). Le Genre *Hipparion,* trans. S. Aubin. *Bureau de Recherche Mineralogie et Geologie C.E.D.P.* **12,** 1–473.

Groves, C. P. (1989). *A Theory of Human and Primate Evolution.* Oxford: Clarendon Press.

Groves, C. P. & Mazak, V. (1975). An approach to

the taxon of *Hominidae:* gracile Villfranchian hominids of Africa. *Casopsis Pro Mineralogii Geologii* **20,** 225–246.
Grubb, P. (1972). Variation and incipient speciation in the African buffalo. *Zeitschrift für Säugetierkunde* **37,** 121–144.
Grubb, P. (1973). Chance, change and diversity in the African fauna. *Universitas* (n.s.) **2(3),** 31–48.
Grubb, P. (1978). Patterns of speciation in African mammals. *Bulletin of the Carnegie Museum of Natural History* **6,** 152–167.
Grubb, P. (1982). Refuges and dispersal in the speciation of African forest mammals. In G. T. Prance, ed., *Biological Diversification in the Tropics,* pp. 537–553. New York: Columbia University Press.
Grubb, P. (1983). The biogeographic significance of forest mammals in Eastern Africa. *Annales, Musée royal de l'Afrique centrale, Sciences zoologiques* **237,** 75–86.
Grubb, P. (1985). Geographical variation in the bushbuck of eastern Africa (*Tragelaphus scriptus;* Bovidae). In L. Schuchmann, ed., *Proceedings of the International Symposium on African Vertebrates,* pp. 11–27. Bonn: Zoologisches Forschungsinstitut und Museum Alexander Koenig.
Grubb, P. (1990). Primate geography in the Afrotropical forest biome. In G. Peters & R. Hutterer, eds., *Vertebrates in the Tropics,* pp. 187–214. Bonn: Museum Alexander Koenig.
Grubb, P. (1993). The Afrotropical suids *Phacochoerus, Hylochoerus* and *Potamochoerus.* Taxonomy and description. In W. L. R. Oliver, ed., *Pigs, Peccaries, and Hippos. Status Survey and Action Plan,* pp. 66–75. Gland: IUCN.
Grube, F., Christensen, S. & Vollmer, T. (1986). Glaciations in northwest Germany. *Quaternary Science Reviews* **5,** 347–358.
Grunblatt, J., Ottichilo, W. K. & Sinange, R. K. (1989). A hierarchical approach to vegetation classification in Kenya. *African Journal of Ecology* **27,** 45–51.
Guo-Qin, Q. (1993). The environmental ecology of the Lufeng hominoids. *Journal of Human Evolution* **24,** 3–11.
Gwiazda, P. H., Hemming, S. R. & Broecker, W. S. (1996). Provenance of icebergs during Heinrich events and the contrast to their sources during other Heinrich episodes. *Paleoceanography* **11,** 371–378.
Habicht, J. K. A. (1979). *Paleoclimate, Paleomagnetism, and Continental Drift.* Studies in Geology no. 9. Tulsa, OK: American Association of Petroleum Geologists.
Haflidason, H., Sejrup, H. P., Kristensen, D. R. & Johnsen, S. (1995). Coupled response of the late glacial climate shifts of northwest Europe reflected in Greenland ice cores: Evidence from the northern North Sea. *Geology* **23,** 1059–1062.
Hafner, J. C. & Hafner, M. S. (1988). Heterochrony in rodents. In M. L. McKinney, ed., *Heterochrony in Evolution: A Multidisciplinary Approach,* pp. 217–235. New York: Plenum Press.
Hall, B. P. (1960). The faunistic importance of the Scarp of Angola. *Ibis* **102,** 420–442.
Hall, B. P. & Moreau, R. E. (1970). *An Atlas of Speciation in African Passerine Birds.* London: British Museum, Natural History.
Hambrey, M. J. & Barrett, P. J. (1993). Cenozoic sedimentary and climate record, Ross Sea Region, Antarctica. In J. P. Kennett & D. A. Warnke, eds., *American Geophysical Union Antarctic Research Series* **60,** 91–124.
Hambrey, M. J., Ehrmann, W. U. & Larsen, B. (1991). The Cenozoic glacial record from the Prydz Bay continental shelf, East Antarctica. *Proceedings of the Ocean Drilling Program Scientific Results* **119,** 77–132.
Hamilton, A. C. (1976). The significance of patterns of distribution shown by forest plants and animals in tropical Africa for the reconstruction of Upper Pleistocene palaeoenvironments: a review. *Palaeoecology of Africa* **9,** 63–97.
Hamilton, W. J., Buskirk, R. E. & Buskirk, W. H. (1978). Omnivory and utilization of food resources by chacma baboons, *Papio ursinus. American Naturalist* **112,** 110–120.
Hammen van der, T., Wijmstra, T. A. & Zagwijn, W. H. (1971). The floral record of the late Cenozoic in Europe. In K. K. Turekian, ed., *The Late Cenozoic Glacial Ages,* pp. 391–424. New Haven, CT: Yale University Press.
Hanski, I. (1982). Dynamics of regional distribution: the core and satellite species hypothesis. *Oikos* **38,** 210–221.
Hanski, I., Kouki, J. & Halkka, A. (1993). Three explanations of the positive relationship between distribution and abundance of species. In R. Ricklefs & D. Schluter, eds., *Species Diversity in Ecological Communities: Historical and Geographical Perspectives,* pp. 108–116. Chicago: University of Chicago Press.
Happold, D. C. D. (1987). *The Mammals of Nigeria.* Oxford: Clarendon Press.
Haq, B. U. (1980). Biogeographic history of Miocene calcareous nannoplankton and paleoceanography of the Atlantic Ocean. *Micropaleontology* **26,** 414–443.
Haq, B. U., Hardenbol, J. & Vail, P. R. (1987). Chronology of fluctuating sea levels since the Triassic. *Science* **235,** 1156–1167.

Hardin, G. (1961). The competitive exclusion principle. *Science* **131,** 1292–1297.
Harger, J. R. E. (1968). The role of behavioral traits in influencing the distribution of two species of sea mussel *Mytilus edulis* and *Mytilus californianus. Veliger* **11,** 45–49.
Harger, J. R. E. (1970a). The effect of wave impact on some aspects of the biology of sea mussels. *Veliger* **12,** 401–414.
Harger, J. R. E. (1970b). The effect of species composition on the survival of mixed populations of the sea mussels. *Mytilus edulis* and *Mytilus californianus. Veliger* **13,** 147–152.
Harger, J. R. E. (1972). Competitive co-existence: maintenance of interacting associations of the sea mussels *Mytilus edulis* and *Mytilus californianus. Veliger* **14,** 387–410.
Harrington, R., Owen-Smith, N., Viljoen, P. C., Biggs, H. C., Mason, D. R. & Funston, P. (in press). Establishing the causes of the roan antelope decline in the Kruger National Park, South Africa. *Biological Conservation.*
Harris, J. & Van Couvering, J. (1995). Mock aridity and the paleoecology of volcanically influenced ecosystems. *Geology* **23,** 593–596.
Harris, J. M. (1983a). Correlation of the Koobi Fora succession. In J. M. Harris, ed., *Koobi Fora Research Project,* vol. 2. *The Fossil Ungulates: Proboscidea, Perissodactyla and Suidae,* pp. 303–318. Oxford: Clarendon Press.
Harris, J. M. (1983b). Family Suidae. In J. M. Harris, ed., *Koobi Fora Research Project,* vol. 2. *The Fossil Ungulates: Proboscidea, Perissodactyla and Suidae,* pp. 215–302. Oxford: Clarendon Press.
Harris, J. M. (1985). Age and paleoecology of the Upper Laetolil Beds, Laetoli, Tanzania. In E. Delson, ed., *Ancestors: The Hard Evidence,* pp. 76–81. New York: Alan R. Liss.
Harris, J. M. (1987). Fossil Suidae from Laetoli. In M. D. Leakey & J. M. Harris, eds., *Laetoli: A Pliocene Site in Northern Tanzania,* pp. 349–358. Oxford: Clarendon Press.
Harris, J. M. (1991a). Family Hippopotamidae. In J. M. Harris, ed., *Koobi Fora Research Project,* vol. 3. *The Fossil Ungulates: Geology, Fossil Artiodactyls, and Palaeoenvironments,* pp. 31–85. Oxford: Clarendon Press.
Harris, J. M. (1991b). Family Camelidae. In J. M. Harris, ed., *Koobi Fora Research Project,* vol. 3. *The Fossil Ungulates: Geology, Fossil Artiodactyls, and Palaeoenvironments,* pp. 86–92. Oxford: Clarendon Press.
Harris, J. M. (1991c). Family Giraffidae. In J. M. Harris, ed., *Koobi Fora Research Project,* vol. 3. *The Fossil Ungulates: Geology, Fossil Artiodactyls, and Palaeoenvironments,* pp. 93–138. Oxford: Clarendon Press.
Harris, J. M. (1991d). Family Bovidae. In J. M. Harris, ed., *Koobi Fora Research Project,* vol. 3. *The Fossil Ungulates: Geology, Fossil Artiodactyls, and Palaeoenvironments,* pp. 139–320. Oxford: Clarendon Press.
Harris, J. M., Brown, F. H. & Leakey, M. G. (1988). Stratigraphy and paleontology of Pliocene and Pleistocene localities west of Lake Turkana, Kenya. *Natural History Museum of Los Angeles County. Contributions in Science* **399,** 1–128.
Harris, J. M. & White, T. D. (1979). Evolution of the Plio-Pleistocene African Suidae. *Transactions of the American Philosophical Society* **69(2),** 1–128.
Harrison, D. L. (1972). *The Mammals of Arabia,* vol. 3. *Lagomorpha, Rodentia.* London: Ernest Benn Limited Press.
Harrison, T. (1989). New postcranial remains of *Victoriapithecus* from the middle Miocene of Kenya. *Journal of Human Evolution* **18,** 3–54.
Harrison, T., ed. (1997). *Neogene Paleontology of the Manonga Valley, Tanzania. A Window into the Evolutionary History of East Africa.* New York: Plenum Press.
Hartman, S. E. (1989). Stereophotogrammetric analysis of occlusal morphology of extant hominoid molars: phenetics and function. *American Journal of Physical Anthropology* **80,** 145–166.
Hartwig-Scherer, S. & Martin, R. D. (1991). Was "Lucy" more human than her "child"? Observations on early hominid post-cranial skeletons. *Journal of Human Evolution* **21,** 439–449.
Harvey, P. H. & Clutton-Brock, T. H. (1985). Life history variation in primates. *Evolution* **39,** 559–581.
Harvey, P. & Pagel, M. (1992). *The Comparative Method in Evolutionary Biology.* Oxford: Oxford University Press.
Hatley, T. & Kappelman, J. (1980). Bears, pigs, and Plio-Pleistocene hominids: a case for the exploitation of belowground food resources. *Human Ecology* **8,** 371–387.
Haughton, S. H. (1932). The fossil Equidae of South Africa. *Annals of the South African Museum* **28,** 407–427.
Hay, R. L. (1976). *Geology of the Olduvai Gorge.* Berkeley: University of California Press.
Hayden, B. (1981). Subsistence and ecological adaptations of modern hunter/gatherers. In G. Teleki & R. Harding, eds., *Omnivorous Primates: Gathering and Hunting in Human Evolution,* pp. 344–422. New York: Columbia University Press.
Hayek, L.-A., Bernor, R. L., Solounias, N. & Steigerwald, P. (1991). Preliminary studies of hipparionine horse diet as measured by tooth microwear. In A. Forsten, M. Fortelius & L. Werdelin, eds., *Bjorn Kurten—a memorial vol-*

ume. Annales Zoologici Fennici **28(3-4),** 187–200.

Hays, J. D., Imbrie, J. & Shackleton, N. J. (1976). Variations in the Earth's orbit: pacemaker of the ice ages. *Science* **194,** 1121–1132.

Hecht, M. K. & Hoffman, A. (1986). Why not neo-Darwinism? A critique of paleobiological challenges. *Oxford Surveys of Evolutionary Biology* **3,** 1–46.

Heinrich, H. (1988). Origin and consequences of cyclic ice rating in the northeast Atlantic Ocean during the past 130,000 years. *Quaternary Research* **29,** 142–152.

Hendey, Q. B. (1974). The Late Cenozoic Carnivora of the south-western Cape Province. *Annals of the South African Museum* **63,** 1–369.

Hendey, Q. B. (1984). Southern African late Tertiary vertebrates. In R. G. Klein, ed., *South African Prehistory and Paleoenvironments,* pp. 121–145. Rotterdam: A. A. Balkema.

Hendey, Q. B. (1981). Geological succession at Laangebaanweg, Cape Province, and global events of the late Tertiary. *South African Journal of Science* **77,** 33–38.

Henneberg, M. (1987). Hominid cranial capacity change through time: a darwinian process. *Human Evolution* **2,** 213–220.

Henneberg, M. (1992). Continuing human evolution: bodies, brains and the role of variability. *Transactions of the Royal Society of South Africa* **48,** 159–182.

Hennig, W. (1966 [1950]). *Phylogenetic Systematics.* Urbana: University of Illinois.

Herbert, T. D. & Fischer, A. G. (1986). Milankovitch climatic origin of mid-Cretaceous black shale rhythms in central Italy. *Nature* **321,** 739–743.

Herlocker, D. J., Dirschl, H. J. & Frame, G. (1993). Grasslands of East Africa. In R. T. Coupland, ed., *Natural Grasslands: Eastern Hemisphere & Resume,* pp. 221–264. London: Elsevier Science Publishers.

Herselman, J. E. & Norton, P. M. (1985). The distribution and status of bats (Mammalia: Chiroptera) in the Cape Province. *Annals of the Cape Province Museum* **16,** 73–126.

Heusser, L. E. & van de Geer, G. (1994). Direct correlation of terrestrial and marine paleoclimatic records from four glacial-interglacial cycles—DSDP Site 594, southwest Pacific. *Quaternary Science Reviews* **13,** 273–282.

Hilgen, F. J. (1991a). Extension of the astronomically calibrated (polarity) time scale to the Miocene/Pliocene boundary. *Earth and Planetary Science Letters* **107,** 349–368.

Hilgen, F. J. (1991b). Astronomical calibration of Gauss to Matuyama sapropels in the Mediterranean and implications for the Geomagnetic Polarity Time Scale. *Earth and Planetary Science Letters* **104,** 226–244.

Hill, A. (1985). Early hominid from Baringo, Kenya. *Nature* **315,** 222–224.

Hill, A. (1987). Causes of perceived faunal change in the later Neogene of East Africa. *Journal of Human Evolution* **16,** 583–596.

Hill, A. (1995). Faunal and environmental change in the Neogene of East Africa: evidence from the Tugen Hills sequence, Baringo District, Kenya. In E. S. Vrba, G. H. Denton, T. C. Partridge & L. H. Burckle, eds., *Paleoclimate and Evolution with Emphasis on Human Origins,* pp. 178–193. New Haven, CT: Yale University Press.

Hill, A., Drake, R., Tauxe, L., Monaghan, M., Barry, J., Behrensmeyer, A. K., Curtis, G., Jacobs, L., Johnson, N. & Pilbeam, D. R. (1985). Neogene paleontology and geochronology of the Baringo Basin, Kenya. *Journal of Human Evolution* **14,** 759–773.

Hill, A., Curtis, G. & Drake, R. (1986). Sedimentary stratigraphy of the Tugen Hills, Baringo, Kenya. In L. E. Frostick, ed., *Sedimentation in the African Rifts,* pp. 285–295. Geological Society of London Special Publication No. 25. Oxford: Blackwell.

Hill, A. & Ward, S. (1988). Origin of the Hominidae: the record of African large hominoid evolution between 14 M.y. and 4 M.y. *Yearbook of Physical Anthropology* **31,** 49–83.

Hill, A., Ward, S. & Brown, B. (1992a). Anatomy and age of the Lothagam mandible. *Journal of Human Evolution* **22,** 439–451.

Hill, A., Ward, S., Delno, A., Curtis, G. & Drake, R. (1992b). Earliest *Homo* from Baringo, Kenya. *Nature* **355,** 719–722.

Hill, J. E. & Carter, T. D. (1941). The mammals of Angola, Africa. *Bulletin of the American Museum of Natural History* **83,** 1–211.

Hladik, C. M. (1977). Chimpanzees of Gabon and chimpanzees of Gombe: some comparative data on the diet. In T. H. Clutton-Brock, ed., *Primate Ecology: Studies of Feeding and Ranging Behaviour in Lemurs, Monkeys and Apes,* pp. 481–503. London: Academic Press.

Hladik, C. M. (1981). Diet and the evolution of feeding strategies among forest primates. In R. S. O. Harding & G. Teleki, eds., *Omnivorous Primates: Gathering and Hunting in Human Evolution,* pp. 215–254. New York: Columbia Press.

Hodell, D. A. & Venz, K. (1992). Toward a high-resolution stable isotopic record of the Southern Ocean during the Pliocene-Pleistocene (4.8 to 0.8 ma). In J. P. Kennett & D. A. Warnke, eds., *The Antarctic Paleoenvironment: A Perspective on Global Change, Antarctic Research Series* **56,** 61–73.

Hoffman, A. (1989). *Arguments on Evolution. A Paleontologist's Perspective.* New York: Oxford University Press.

Hofman, M. A. (1983). Energy metabolism, brain

size and longevity in mammals. *Quarterly Review of Biology* **58,** 495–512.

Hofmann, R. R. (1989). Evolutionary steps of ecophysiological adaptation and diversification of ruminants: a comparative view of their digestive system. *Oecologia* **78,** 443–457.

Hofmann, R. R. & Stewart, D. R. M. (1972). Grazer or browser: a classification based on the stomach structure and feeding habits of East African ruminants. *Mammalia* **36,** 226–240.

Hollister, N. (1918–1924). East African mammals in the United States National Museum. *Bulletin of the United States National Museum* **99,** 1–194, 1–184, 1–164.

Holloway, R. L. (1970). New endocranial values for the australopithecines. *Nature* **227,** 199–200.

Holloway, R. L. (1972a). Australopithecine endocasts, brain evolution in the Hominoidea, and a model of hominid evolution. In R. Tuttle, ed., *The Functional and Evolutionary Biology of Primates,* pp. 185–203. Chicago: Aldine-Atherton.

Holloway, R. L. (1972b). New australopithecine endocast SK 1585, from Swartkrans, South Africa. *American Journal of Physical Anthropology* **37,** 173–186.

Holloway, R. L. (1978). Problems of brain endocast interpretation and African hominid evolution. In C. J. Jolly, ed., *Early Hominids of Africa,* pp. 379–401. London: Duckworth.

Holloway, R. L. & Kimbel, W. H. (1986). Endocast morphology of Hadar hominid AL 162-68, *Nature* **321,** 536–538.

Holt, R. D. (1984). Spatial heterogeneity, indirect interactions, and the coexistence of prey species. *American Naturalist* **124,** 377–406.

Homewood, K. M. (1978). Feeding strategy of the Tana mangabey (*Cercocebus galeritus*) (Mammalia: Primates). *Journal of Zoology* **186,** 375–391.

Honacki, J. H., Kinman, K. E. & Koeppl, J. W. (1982). *Mammal Species of the World. A Taxonomic and Geographic Reference.* Lawrence, KS: Allen Press and the Association of Systematics Collections.

Hooghiemstra, H. (1984). Vegetational and climatic history of the high plain of Bogota, Columbia: a continuous record of the last 3.5 million years. *Dissertaciones Botanicae* **79.**

Hooghiemstra, H. (1995). Environmental and paleoclimatic evolution in Late Pliocene-Quaternary Columbia. In E. S. Vrba, ed. *Paleoclimate and Evolution with Emphasis on Human Origin,* pp. 249–261. New Haven, CT: Yale University Press.

Hooijer, D. A. (1974). Hipparions from the late Miocene and Pliocene of northwestern Kenya. *Zoologische Verhandelingen, Leiden* **134,** 1–34.

Hooijer, D. A. (1975). Miocene to Pleistocene Hipparions of Kenya, Tanzania and Ethiopia. *Zoologische Verhandelingen, Leiden* **142,** 1–80.

Hooijer, D. A. (1976). The late Pliocene Equidae of Langebaanweg, Cape Province, South Africa. *Zoologische Verhandelingen, Leiden* **148,** 1–39.

Hooijer, D. A. (1978). Rhinocerotidae. In V. J. Maglio & H. B. S. Cooke, eds., *Evolution of African Mammals,* pp. 371–378. Cambridge, MA: Harvard University Press.

Hooijer, D. A. & Churcher, C. S. (1985). Perissodactyla of the Omo Group deposits. In Y. Coppens & F. C. Howell, eds., *Les Faunes Plio-Pléistocènes de la Basse Vallée de l'Omo (Ethiopie), tome 1, Périssodactyles, Artiodactyles (Bovidae),* pp. 97–117. Paris: C.N.R.S.

Hooijer, D. A. & Maglio, V. J. (1974). Hipparions from the late Miocene and Pliocene of northwestern Kenya. *Zoologische Verhandelingen* **134,** 3–34.

Hopewood, A. T. (1937). Die fossilen Pferde von Oldoway. *Wissenschaftliche Ergebnisse der Olduvay Expedition 1913* **4,** 111–136.

Horgan, J. (1995a). From complexity to perplexity. *Scientific American* **272,** 104–109.

Horgan, J. (1995b). Profile: Stephen Jay Gould. Escaping in a cloud of ink. *Scientific American* **273,** 37–41.

Hornibrook, N. de B. (1992). New Zealand Cenozoic marine paleoclimates: a review based on the distribution of some shallow water and terrestrial biota. In R. Tsuch & J. C. Ingle, eds., *Pacific Neogene,* pp. 83–106. Tokyo: University of Tokyo Press.

House, M. R. (1995). Orbital forcing timescales: an introduction. In M. R. House & A. S. Gale, eds., *Orbital Forcing Timescales and Cyclostratigraphy,* pp. 1–18. London: The Geological Society of London.

Howard, D. J. (1993). Reinforcement: origin, dynamics, and fate of an evolutionary hypothesis. In R. G. Harrison, ed., *Hybrid Zones and the Evolutionary Process,* pp. 46–69. Oxford: Oxford University Press.

Howard, W. R. (1985). Late Quaternary Southern Indian Ocean circulation. *South African Journal of Science* **81,** 253–254.

Howell, F. C. (1978). Hominidae. In V. J. Maglio & H. B. S. Cooke, eds., *Evolution of African Mammals,* pp. 154–248. Cambridge, MA: Harvard University Press.

Howell, F. C. & Bourlière, F. (1963). *African Ecology and Human Evolution.* Chicago: Aldine.

Howell, F. C. & Coppens, Y. (1976). An overview of Hominidae from the Omo succession, Ethiopia. In Y. Coppens, F. C. Howell, G. L. Isaac & R. E. Leakey, eds., *Earliest Man and Environ-*

ments in the Lake Rudolf Basin, pp. 522–532. Chicago: University of Chicago Press.

Howell, F. C., Haesaerts, P., de Heinzelin, J. (1987). Depositional environments, archeological occurrences and hominids from Members E and F of the Shungura Formation (Omo basin, Ethiopia). *Journal of Human Evolution* **16,** 665–700.

Howell, F. C. & Wood, B. A. (1974). Early hominid ulna from the Omo basin, Ethiopia. *Nature* **249,** 174–176.

Hublin, J. J. (1990). Les peuplements paléolithiques de l'Europe: un point de vue paléobiogéographique. In C. Farizy, ed., *Paléolithique moyen récent et Paléolithique supérieur ancien en Europe,* pp. 29–37. Nemours: Association pour la Promotion de la Recherche Archeologique en Ile de France.

Hublin, J. J., Barroso Ruiz, C., Medina Lara, P., Fontagne, M. & Reyss, J.-L. (1995). The Mousterian site of Zafarraya (Andalucia, Spain): dating and implications on the palaeolithic peopling processes of Western Europe, *Comptes Rendus de l'Académie des Sciences Paris Séries IIa* **321,** 931–937.

Hublin, J. J., Spoor, F., Braun, M., Zonneveld, F. & Condemi, S. (1996). A late Neanderthal associated with upper Paleolithic artefacts. *Nature* **381,** 224–226.

Hull, D. C. (1979). The limits of cladism. *Systematic Zoology* **28,** 416–440.

Hulley, P. E., Walter, G. H. & Craig, A. J. F. K. (1988). Interspecific competition and community structure, I. Shortcomings of the competition paradigm. *Rivista di Biologia* **81,** 57–71.

Hunt, K. D. (1992). Social rank and body size as determinants of positional behaviour in *Pan troglodytes. Primates* **33,** 347–357.

Hunt, K. D. (1993). The evolution of human bipedality: ecology and functional morphology. *Journal of Human Evolution* **26,** 183–202.

Huntley, B. J. (1982). Southern African savannas. In B. J. Huntley & B. H. Walker, eds., *The Ecology of Tropical Savannas,* pp. 101–119. Berlin: Springer-Verlag.

Huntley, B. & Webb, T. (1989). Migration: species' response to climatic variations caused by changes in the Earth's orbit. *Journal of Biogeography* **16,** 5–19.

Hussain, S. T. (1975). Evolutionary and functional anatomy of the pelvic limb in fossil and recent *Equidae* (Perissodactyla, Mammalia). *Anatomy, Histology, Embryology* **4,** 179–222.

Huxley, J. (1958). Evolutionary processes and taxonomy with special reference to grades. *Uppsala. Univ. Arssks.* **6,** 21–38.

Hylander, W. L. (1975). Incisor size and diet in anthropoids with special reference to Cercopithecidae. *Science* **189,** 1095–1098.

Hylander, W. L. (1979). The functional significance of primate mandibular form. *Journal of Morphology* **160,** 223–240.

Hylander, W. L. (1988). Implications of *in vivo* experiments for interpreting the functional significance of "robust" australopithecine jaws. In F. E. Grine, ed., *Evolutionary History of the "Robust" Australopithecines,* pp. 55–83. New York: Aldine de Gruyter.

Hylander, W. L. & Johnson, K. R. (1994). Jaw muscle function and wishboning of the mandible during mastication in macaques and baboons. *American Journal of Physical Anthropology* **94,** 523–547.

Hynes,T. & Benefit, B. R. (1995). Phylogenetic relationships of the long-snouted fossil colobines. *American Journal of Physical Anthropology Supplement* **20,** 115.

Imbrie, J. (1985). A theoretical framework for the Pleistocene ice ages. *Journal of the Geological Society* **142,** 417–432.

Imbrie, J., Berger, A., Boyle, E. A., Clemens, S. C., Duffy, A., Howard, W. R., Kukla, G., Kutzbach, J., Martinson, D. G., McIntyre, A., Mix, A. C., Molfino, B., Morley, J. J., Peterson, L. C., Pisias, N. G., Prell, W. L., Raymo, M. E., Shackleton, N. J. & Toggweiler, J. R. (1993). On the structure and origin of major glaciation cycles. 2. The 100,000-year cycle. *Paleoceanography* **8,** 699–735.

Imbrie, J., Berger, A. & Shackleton, N. J. (1993). Role of orbital forcing: A two million year perspective. In J. A. Eddy & H. Oeschger, eds., *Global Changes in the Perspective of the Past,* pp. 263–277. New York: John Wiley & Sons.

Imbrie, J., Boyle, E. A., Clemens, S. C., Duffy, A., Howard, W. R., Kukla, G., Kutzbach, J., Martinson, D. G., McIntyre, A., Mix, A. C., Molfino, B., Morley, J. J., Peterson, L. C., Pisias, N. G., Prell, W. L., Raymo, M. E., Shackleton, J. J. & Toggweiler, J. R. (1992). On the structure and origin of major glaciation cycles. 1. Linear responses to Milankovitch forcing. *Paleoceanography* **7,** 701–738.

Imbrie, J., Hays, J. D., Martinson, D. G., McIntyre, A., Mix, A., Morley, J. J., Pisias, N. G., Prell, W. & Shackleton, N. J. (1984). The orbital theory of Pleistocene climate: support from a revised chronology of the marine $\delta 18_O$ record. In A. Berger, et al., eds., *Milankovitch and Climate,* pp. 269–305. Hingham, MA: D. Reidel.

Jackson, S. P. (1961). *Climatological Atlas of Africa.* Pretoria: CCTA/CSA, Government Printer.

Jacobs, B. F. & Kabuye, C. (1987). A middle Miocene (12.2 my old) forest in the East African Rift Valley, Kenya. *Journal of Human Evolution* **16,** 147–155.

Jacobs, L. L. (1978). Fossil rodents (Rhizomyidae

and Muridae) from Neogene Siwalik deposits, Pakistan. *Museum of Northern Arizona Press Bulletin* Series **52,** 1–103.

Jaeger, J. J. (1976). Les Ronguers (Mammalia, rodentia) du pléistocène inférieur d'Olduvai Bed I (Tanzanie). Iè partie: les Muridés. In R. J. G. Savage & S. C. Coryndon, eds., *Fossil vertebrates of Africa,* vol. 4, pp. 58–120. New York: Academic Press.

Jaeger, J. J. (1979). Les faunes de rongeurs et de lagomorphes du Pliocène et du Pléistocène d'Afrique orientale. *Bulletin de la Société Géologique de France* **21,** 301–308.

Jaeger, J. J., Michaux, J. & Sabatier, M. (1980). Premières données sur les rongeurs de la formation de Ch'orora (Ethiopie) d'âge Miocène supérieur. I.-Thryonomyidés. *Palaeovertebrata* (Mémoire Jubilaire en l'honneur de R. Lavocat) 365–374.

James, G. T. & Slaughter, B. H. (1974). A primitive new middle Pliocene murid form Wadi el Natrun, Egypt. *Annals of the Geological Survey of Egypt Cairo* **4,** 333–362.

James, M. (1986). *Classification Algorithms.* New York: John Wiley and Sons.

Janecek, T. R. & Ruddiman, W. F. (1987). Plio-Pleistocene sedimentation in the south equatorial Atlantic divergence. In. *XII INQUA Research, Ottawa, Canada, Program and Abstracts.*

Janis, C. (1982). Evolution of horns in ungulates: ecology and paleoecology. *Biological Reviews of the Cambridge Philosophical Society* **57,** 261–318.

Janis, C. & Erhardt, D. (1988). Correlation of relative muzzle width and relative incisor width with dietary preference in ungulates. *Zoological Journal of the Linnean Society* **92,** 267–284.

Jansen, E. & Sjoholm, J. (1991). Reconstruction of glaciation over the past 6 Myr from ice-borne deposits in the Norwegian Sea. *Nature* **349,** 600–603.

Jeffries, M. J. & Lawton, J. H. (1984). Enemy-free space and the structure of ecological communities. *Biological Journal of the Linnaean Society* **23,** 269–286.

Jenkins, D. G. (1973). Diversity changes in New Zealand Cenozoic foraminifera. *Journal of Foraminifera Research* **3,** 78–88.

Jenny, H. (1941). *Factors of Soil Formation.* New York: McGraw-Hill.

Johanson, D. C. & Coppens, Y. (1976). A preliminary anatomical diagnosis of the first Plio-Pleistocene hominid discoveries in the Central Afar, Ethiopia. *American Journal of Physical Anthropology* **45,** 217–234.

Johanson, D. C. & Edey, M. (1981). *Lucy: The Beginnings of Humankind.* New York: Simon and Schuster.

Johanson, D. C., Masao, F. T., Eck, G. G., White, T. D., Walter, R. C., Kimbel, W. H., Asfaw, B., Manega, P., Ndessokia, R. & Suwa, G. (1987). New partial skeleton of *Homo habilis* from Olduvai Gorge. *Nature* **327,** 205–209.

Johanson, D. C. & Taieb, M. (1976). Plio-Pleistocene hominid discoveries in Hadar, Ethiopia. *Nature* **260,** 293–297.

Johanson, D. C., Taieb, M. & Coppens, Y. (1982). Pliocene hominids from the Hadar formation, Ethiopia (1973–1977): stratigraphic, chronologic, and paleoenvironmental contexts, with notes on hominid morphology and systematics. *American Journal of Physical Anthropology* **57,** 373–402.

Johanson, D. C., White, T. D. & Coppens, Y. (1978). A new species of the genus *Australopithecus* (Primates: Hominidae) from the Pliocene eastern Africa. *Kirtlandia* **28,** 1–14.

Joleaud, L. (1933). Un nouveau genre d'Equidé quaternaire de l'Omo (Abyssinie): *Libyhipparion ethiopicum. Bulletin Societé géologique de France* **3,** 7–27.

Jolly, C. J. (1967). The evolution of baboons. In H. Vagtborg, ed., *The Baboon in Medical Research,* vol. 2, pp. 23–50. Austin: University of Texas Press.

Jolly, C. J. (1970). The large African monkeys as an adaptive array. In J. R. Napier & P. H. Napier, eds., *Old World Monkeys,* pp. 141–174. New York: Academic Press.

Jolly, C. J. (1972). The classification and natural history of *Theropithecus* (*Simopithecus*) (Andrews, 1916) baboons of the African Plio-Pleistocene. *Bulletin of the British Museum of Natural History, Geology Series* **22,** 1–123.

Jolly, C. J. (1993). Species, subspecies, and baboon systematics. In W. H.Kimbel & L. B. Martin, eds., *Species, Species Concepts, and Primate Evolution,* pp. 67–107. New York: Plenum Press.

Jungers, W. L. (1988). New estimates of body size in australopithecines. In F. E. Grine, ed., *Evolutionary History of the "Robust" Australopithecines,* pp. 115–125. New York: Aldine de Gruyter.

Jungers, W. L. (1990). Scaling of hominoid femoral head size and the evolution of hominid bipedalism. *American Journal of Physical Anthropology* **81,** 246.

Juste, J. & Ibanez, C. (1992). Taxonomic review of *Miniopterus minor* Peters, 1867 (Mammalia: Chiroptera) from western central Africa. *Bonner zoologische Beiträge* **43,** 355–365.

Kahlke, H. D. (1975). The macro-faunas of continental Europe during the Middle Pleistocene: stratigraphic sequence and problems of intercorrelation. In K. Butzer & G. L. Isaac, eds., *After the australopithecines,* pp. 309–374. The Hague: Mouton.

Kalb, J. E. (1995). Fossil elephantoids, Awash paleolake basins and the Afar triple junction, Ethiopia. *Palaeogeography, Palaeclimatology, Palaeoecology* **114,** 357–368.

Kalb, J. E., Jolly, C. J., Tebedge, S., Mebrate, A., Smart, C., Oswald, E. B., Whitehead, P. F., Wood, C. B., Adefri, T. & Rawn-Schatzinger, V. (1982). Vertebrate faunas from the Awash Group. *Journal of Vertebrate Paleontology* **2,** 237–258.

Kalb, J. E. & Mebrate, A. (1993). Fossil elephantoids from the hominid-bearing Awash Group, Middle Awash Valley, Afar Depression, Ethiopia. *Transactions of the American Philosophical Society* **83(1).**

Kaneps, A. G. (1979). Gulf Stream: velocity fluctuations during the latest Cenozoic. *Science* **204,** 297–301.

Kappelman, J. (1984). Plio-Pleistocene environments of Bed I and lower Bed II, Olduvai Gorge, Tanzania. *Palaeogeography, Palaeoclimatology, Palaeoecology* **48,** 171–196.

Kappelman, J. (1986). Plio-Pleistocene marine-continental correlation using habitat indicators from Olduvai Gorge, Tanzania. *Quarternary Research* **25,** 141–149.

Kappelman, J. (1988). Morphology and locomotor adaptations of the bovid femur in relation to habitat. *Journal of Morphology* **198,** 119–130.

Kappelman, J. (1991). The paleoenvironment of Kenyapithecus at Fort Ternan. *Journal of Human Evolution* **20,** 95–129.

Kappelman, J. (1996). The evolution of body mass and relative brain size in fossil hominids. *Journal of Human Evolution* **30,** 243–276.

Kappelman, J., Bishop, L., Plummer, T., Appleton, S. & Duncan, A. (1997). Bovids as indicators of Plio-Pleistocene paleoenvironments in East Africa. *Journal of Human Evolution* **32,** 229–256.

Kappelmann, J., Sen, S., Fortelius, M., Duncan, A., Alpagut, B., Crabaugh, J., Gentry, A., Lunkka, J. P., McDowell, F., Solounias, N., Viranta, S. & Werdelin, L. (1996). Chronology and biostratigraphy of the Miocene Sinap Formation of Central Turkey. In: R. L. Bernor, V. Fahlbusch & H.-W. Mittmann, eds., *The Evolution of Western Eurasian Neogene Faunas,* pp. 78–95. New York: Columbia University Press.

Karp, L. E. (1987). Allometric effects and habitat influences on the postcranial skeleton of suids and tayassuids (M.S. thesis). New Brunswick, NJ: Rutgers University.

Kastens, K. A. (1992). Did glacio-eustatic sea level drop trigger the Messinian salinity crisis? New evidence from Ocean Drilling Program site 654 in the Tyrrhenian Sea. *Paleoceanography* **7,** 333–356.

Kauffman, S. (1993). *The Origins of Order: Self-Organization and Selection in Evolution.* New York: Oxford University Press.

Kauffman, S. (1995). *At Home in the Universe.* New York: Oxford University Press.

Kay, R. F. (1975). The functional adaptations of primate molar teeth. *American Journal of Physical Anthropology* **43,** 195–215.

Kay, R. F. (1977). The evolution of molar occlusion in the Cercopithecidae and early catarrhines. *American Journal of Physical Anthropology* **46,** 327–352.

Kay, R. F. (1978). Molar structure and diet in extant Cercopithecidae. In P. M. Butler & K. Joysey, eds., *Studies in the Development, Function, and Evolution of Teeth,* pp. 309–339. London: Academic Press.

Kay, R. F. (1981). The nut-crackers—a new theory of the adaptations of the Ramapithecinae. *American Journal of Physical Anthropology* **55,** 141–152.

Kay, R. F. (1984). On the use of anatomical features to infer foraging behavior in extinct primates. In P. S. Rodman and J. G. H. Cant, eds., *Adaptations for Foraging in Nonhuman Primates,* pp. 21–53. New York: Columbia University Press.

Kay, R. F. (1985). Dental evidence for the diet of *Australopithecus. Annual Review of Anthropology* **14,** 315–341.

Kay, R. F. & Covert, H. H. (1984). Anatomy and behavior of extinct primates. In D. J. Chivers, B. A. Wood & A. Bilsborough, eds., *Food Acquisition and Processing in Primates,* pp. 467–508. New York: Plenum Press.

Kay, R. F. & Grine, F. E. (1988). Tooth morphology, wear and diet in *Australopithecus* and *Paranthropus* from southern Africa. In F. E. Grine, ed., *The Evolutionary History of the "Robust" Australopithecines,* pp. 427–447. New York: Aldine de Gruyter.

Kay, R. F. & Hylander, W. L. (1978). The dental structure of mammalian folivores with special reference to primates and phalangeroids (Marsupialia). In G. G. Montgomery, ed., *The Ecology of Arboreal Folivores,* pp. 173–192. Washington, DC: Smithsonian Institution Press.

Keay, R. W. J., ed. (1959). *Vegetation Map of Africa South of the Tropic of Cancer.* London: Oxford University Press.

Keigwin, L. D. Jr. (1982). Isotopic paleoceanography of the Caribbean and East Pacific: role of Panama uplift in Late Neogene time. *Science* **217,** 350–353.

Keigwin, L. D. Jr. (1987). Pliocene stable-isotope recod of Deep Sea Drilling Project Site 606: Sequential events in the $\delta^{18}O$ enrichment beginning at 3,1 Ma. In W. F. Ruddiman et al., eds., *Initial reports of the Deep Sea Drilling Project* no. 94, pp. 911–920. Washington, DC: National Science Foundation.

Kellogg, D. E. & Kellogg, T. B. (1996). Diatoms in South Pole ice: implications for eolian contami-

nation of Sirius Group deposits. *Geology* **24,** 115–118.
Kemp, E. M. & Barrett, J. P. (1975). Antarctic glaciation and early Tertiary vegetation. *Nature* **258,** 507–508.
Kemp, A. C. & Crowe, T. M. (1985). The systematics and zoogeography of Afrotropical hornbills (Aves: Bucerotidae). In K. L. Schuchmann, ed., *Proceedings of the International Symposium on African Vertebrates,* pp. 279–324. Bonn: Zoologisches Forschungsinstitut und Museum Alexander Koenig.
Kenagy, C. J. (1972). Saltbush leaves: excision of hyper-saline tissue by a kangaroo rat. *Science* **178,** 1094–1096.
Kennett, J. P. (1977). Cenozoic evolution of Antarctic glaciation, the Circum-Antarctic Ocean, and their impact on global paleoceanography. *Journal of Geophysical Research* **82,** 3843–3860.
Kennett, J. (1995). A review of polar climatic evolution during the Neogene based on the marine sediment record. In E. S. Vrba, G. H. Denton, T. C. Partridge & L. H. Burckle, eds., *Paleoclimate and Evolution with Emphasis on Human Origins,* pp. 49–64. New Haven, CT: Yale University Press.
Kennett, J. P. & Hodell, D. A. (1993). Evidence for relative climatic stability of Antarctica during the early Pliocene: a marine perspective. *Geografiska Annaler* **75A,** 205–220.
Kennett, J. P. & Shackleton, N. J. (1976). Oxygen isotope evidence for the development of the cryosphere 38 myr ago. *Nature* **260,** 513–515.
Kibunjia, M. (1994). Pliocene archaeological occurrences in the Lake Turkana basin. *Journal of Human Evolution* **27,** 159–171.
Kimbel, W. H. (1994). The first skull and other new discoveries of *Australopithecus afarensis* at Hadar, Ethiopia. *Nature* **368,** 449–451.
Kimbel, W. H. (1995). Hominid speciation and Pliocene climatic change. In E. S. Vrba, C. H. Denton, T. C. Partridge & L. H. Buckle, eds., *Paleoclimate and Evolution with Emphasis on Human Origins,* pp. 425–437. New Haven, CT: Yale University Press.
Kimbel, W. H., Johanson, D. C. & Rak, Y. (1994). The first skull and other new discoveries of *Australopithecus afarensis* at Hadar, Ethiopia. *Nature* **368,** 449–451.
Kimbel, W. H. & Martin, L. B., eds. (1993). *Species, Species Concepts and Primate Evolution.* New York: Plenum Press.
Kimbel, W. H. & Rak, Y. (1985). Functional morphology of the asterionic region in extant hominoids and fossil hominids. *American Journal of Physical Anthropology* **66,** 31–54.
Kimbel, W. H. & Rak, Y. (1993). The importance of species taxa in paleoanthropology and an argument for the phylogenetic concept of the species category. In W. H. Kimbel & L. B. Martin, eds., *Species, Species Concepts and Primate Evolution,* pp. 461–484. New York: Plenum Press.
Kimura, M. (1978). *A Bibliography of Age Determination of Mammals (with Special Emphasis on the Cetacea).* Available on request from National Oceanic and Atmospheric Administration National Marine Fisheries Service, Southwest Fisheries Center, La Jolla, CA.
King, L. (1978). The geomorphology of central and southern Africa. In M. J. A Wenger, ed., *Biogeography and Ecology of Southern Africa,* Vol. 31, pp. 1–17. The Hague: Dr. W. Junk.
Kingdon, J. (1971–1982). *East African Mammals: An Atlas of Evolution in Africa.* London: Academic Press.
Kingdon, J. (1974). *East African Mammals: An Atlas of Evolution in Africa,* vols. 1 and 2. Chicago: University of Chicago Press.
Kingdon, J. (1997). *The Kingdon Field Guide to African Mammals.* London: Academic Press.
Kingdon, J. (1990). *Island Africa.* London: William Collins Sons.
Kingston, J. D. (1992). Stable isotopic evidence for hominid paleoenvironments in East Africa (Ph.D. dissertation). Cambridge, MA: Harvard University.
Kingston, J. D., Marino, B. D. & Hill, A. P. (1994). Isotopic evidence for Neogene hominid paleoenvironments in the Kenya Rift Valley. *Science* **264,** 955–959.
Kinzey, W. G., ed. (1987). *The Evolution of Human Behavior: Primate Models.* Albany: State University of New York.
Kitts, D. B. (1977). Karl Popper, verifiability, and systematic zoology. *Systematic Zoology* **26,** 185–194.
Klein, R. G. (1984). The large mammals of southern Africa; late Pliocene to recent. In R. G. Klein, ed., *Southern African Prehistory and Paleoenvironments,* pp. 107–146. Rotterdam: A. A. Balkema.
Klein, R. G., Cruz-Uribe, K. & Beaumont, P. B. (1991). Environmental, ecological, and paleoanthropological implications of the Late Pleistocene mammalian fauna from Equus Cave, Northern Cape Province, South Africa. *Quaternary Research* **36,** 94–119.
Koch, D. (1978). Vergleichende Untersuchung einiger Säugetiere im südlichen Niger. *Senckenbergiana Biologica* **58, 3-4,** 113–136.
Köhler, M. (1993). *Skeleton and Habitat of Recent and Fossil Ruminants.* Munich: Müncher Geowissenschaftliche Abhandlungen.
Kominz, M. A. (1984). Oceanic ridge volumes and sea-level change—an error analysis. In J. S.

Schlee, ed., *Interregional Unconformities and Hydrocarbon Accumulation,* pp. 108–123. Tulsa, OK: American Association of Petroleum Geologists Memoir 36.

Koppen, W. (1931). *Grudiss der Klimkunde.* Berlin: Walter de Gruyter & Co.

Kortlandt, A. (1972). *New Perspectives on Ape and Human Evolution.* Amsterdam: Stichting voor Psychobiologie.

Kovar-Eder, J., Kvacek, Z., Zastawniak, E., Givulescu, R., Hably, L., Mihajlovic, D., Teslenko, J. & Walther, H. (1996). Floristic trends in the vegetation of the Paratethys surrounding areas during Neogene time. In R. L. Bernor, V. Fahlbusch & H.-W. Mittmann, eds., *The Evolution of Western Eurasian Neogene Faunas,* pp. 395–413. New York: Columbia University York.

Kowalski, K. & Rzebik-Kowalska, B. (1991). *Mammals of Algeria.* Warsaw: Ossolineum.

Kowalski, K., Van Neer, W., Bochenski, Z., Mlynarski, M., Rzebik-Kowalska, B., Szyndlar, Z., Gautier, A., Schild, R., Close, A. E. & Wendorf, F. (1989). A last interglacial fauna from the Eastern Sahara. *Quaternary Research* **32,** 335–341.

Krishtalka, L. & Stucky, R. K. (1985). Revision of the Wind River faunas, early Eocene of central Wyoming. Part 7: Revision of *Diacodexis* (Mammalia, Artiodactyla). *Annals of the Carnegie Museum of Natural History* **54,** 413–486.

Kuhn, H. J. (1965). A provisional check-list of the mammals of Liberia. *Senckenbergiana Biologica* **46,** 321–340.

Kukla, G. (1987). Loess stratigraphy in central China. *Quaternary Science Reviews* **6,** 191–219.

Kukla, G. & Cilek, V. (1996). Plio-Pleistocene megacycles: Record of climate and tectonics. *Palaeogeography, Palaeoclimatology, Palaeoecology* **120(1-2),** 171–194.

Kullmer, O. (1997). Die Evolution der Suiden im Plio-Pleistozän Afrikas und ihre biostratigraphische, paläobiogeographische und paläoökologische Bedeutung (Thesis). Mainz: University of Mainz.

Kuman, K. (1994). The archaeology of Sterkfontein—past and present. *Journal of Human Evolution* **27,** 471–495.

Kummer, H. (1968). *Social Organization of Hamadryas Baboons.* Basel: Karger.

Kunimatsu, Y. (1992). Allometry of the nuchal plane in hominoids, and its reduction in relation to bipedalism in hominids. In S. Matano R. H. Tuttle & H. Ishida, eds., *Topics in Primatology,* vol. 3. *Evolutionary Biology, Reproductive Endocrinology, and Virology,* pp. 209–220. Tokyo: University of Tokyo Press.

Kuroda, S. (1992). Ecological interspecies relationships between gorillas and chimpanzees in the Ndoki-Nouabale Reserve, northern Congo. In N. Itoigawa, Y. Sugiyama, G. P. Sackett & R. K. R. Thompson, eds., *Topics in Primatology,* vol. 2. *Behavior, Ecology, and Conservation,* pp. 385–394. Tokyo: University of Tokyo Press.

Kurschner, W. M., van der Burgh, J., Visscher, H. & Dilcher, D. L. (1996). Oak leaves as biosensors of late Neogene and early Pleistocene paleoatmospheric CO_2 concentrations. *Marine Micropaleontology* **27,** 299–312.

Kurtén, B. (1957). Mammal migrations. Cenozoic stratigraphy, and the age of Peking man and the australopithecines. *Journal of Paleontology* **31,** 215–227.

Kurtén, B. (1959a). On the longevity of mammalian species in the Tertiary. *Commentationes Biologicae Societas Scientarium Fennica* **21,** 1–14.

Kurtén, B. (1959b). Rates of evolution in fossil mammals. *Cold Spring Harbor Symposium on Quantitative Biology* **24,** 205–215.

Kvasov, D. D. & Verbitsky, M. Y. (1981). Causes of Antarctic glaciation in the Cenozoic. *Quaternary Research* **15,** 1–17.

Lagoe, M. B., Eyles, C. H., Eyles, N. & Hale, C. (1993). Timing of late Cenozoic tidewater glaciation in the far North Pacific. *Geological Society of America Bulletin* **105,** 1542–1560.

Lamarck, J. (1809). *Philosophie Zoologique.* Paris: Methuen.

Lande, R. (1980). Genetic variation and phenotypic evolution during allopatric speciation. *American Naturalist* **116,** 463–479.

Lande, R. (1985). Expected time for random genetic drift of a population between stable phenotypic rates. *Proceedings of the National Academy of Science USA* **82,** 7641–7645.

Lande, R. (1986). The dynamics of peak shifts and the pattern of morphological evolution. *Paleobiology* **12,** 343–354.

Lande, R. (1994). Risk of population extinction from fixation of new deleterious mutations. *Evolution* **48,** 1460–1469.

Largen, M. J., Kock, D. & Yalden, D. W. (1974). Catalogue of the mammals of Ethiopia. 1. Chiroptera. *Monitore zoologico Italiano* (n.s.) *Supplemento* **5,** 221–298.

Larsen, H. C., Saunders, A. D., Clift, P. D., Beget, J., Wei, W. & Spezzaferri, S. (1994). ODPLEg 152, Scientific party, seven million years of glaciation in Greenland. *Science* **264,** 952–955.

Lavocat, R. (1978). Rodentia and Lagomorpha. In V. J. Maglio & H. B. S. Cooke, eds., *Evolution of African Mammals,* pp. 69–89. Cambridge, MA: Harvard University Press.

Leuthold, W. (1981). Contact between formerly allopatric subspecies of Grant's gazelle (*Gazells granti* Brooke, 1872) owing to vegetation

changes in Tsavo National Park, Kenya. *Zeitschrift für Säugetierkunde* **46,** 48–55.
Leakey, L. S. B. (1958). *Some East African Pleistocene Suidae. Fossil Mammals of Africa,* no. 14. London: British Museum (Natural History).
Leakey, L. S. B. (1965). *Olduvai Gorge. 1951–1961.* Cambridge: Cambridge University Press.
Leakey, M. D. (1971). *Olduvai Gorge,* vol. 3. *Excavations in Beds I and II, 1960–1963.* Cambridge: Cambridge University Press.
Leakey, M. D. & Harris, J. M., eds. (1987). *Laetoli: A Pliocene Site in Northern Tanzania.* Oxford: Clarendon Press.
Leakey, M. D., Hay, R. L., Curtis, G. H., Drake, R. E., Jackes, M. K. & White, T. D. (1976). Fossil hominids from the Laetoli Beds. *Nature* **262,** 460–466.
Leakey, M. D. & Hay, R. (1982). Chronological position of the fossil hominids in Tanzania. *Proceedings of the 1st Congress of Human Paleontology* **2,** 253–265.
Leakey, M. G. (1976). Cercopithecoidea of the East Rudolf succession. In Y. Coppens, F. C. Howell, G. Ll. Isaac & R. E. F. Leakey, eds., *Earliest Man and Environments in the Lake Rudolf Basin,* pp. 345–350. Chicago: University of Chicago Press.
Leakey, M. G. (1982). Extinct large colobines from the Plio-Pleistocene of Africa. *American Journal of Physical Anthropology* **58,** 153–172.
Leakey, M. G. (1985). Early Miocene cercopithecids from Buluk, Northern Kenya. *Folia Primatologica* **44,** 1–14.
Leakey, M. G. (1993). Evolution of *Theropithecus* in the Turkana Basin. In N. G. Jablonski, ed., *Theropithecus—The Rise and Fall of a Primate Genus,* pp. 85–123. Cambridge: Cambridge University Press.
Leakey, M. G., Feibel, C. S., Bernor, R. L., Harris, J. M., Cerling, T. E., Stewart, K. M., Storrs, G. W., Walker, A., Werdelin, L. & Winkler, A. J. (1996). Lothagam: a record of faunal change in the late Miocene of East Africa. *Journal of Vertebrate Paleontology* **16(3),** 556–570.
Leakey, M. G., Feibel, C. S., McDougall, I. & Walker, A. (1995). New four-million-year-old hominid species from Kanapoi and Allia Bay, Kenya. *Nature* **376,** 565–571.
Leakey, M. G., Feibel, C. S., McDougall, I., Ward, C. & Walker, A. (1998). New specimens and confirmation of an early age for *Australopithecus anamensis. Nature* **393,** 62–66.
Leakey, R. E. & Leakey, M. G. (1986). A new Miocene hominoid from Kenya. *Nature* **324,** 143–145.
Lee, R. B. (1965). Subsistence ecology of !Kung bushmen (Ph.D. dissertation). University of California, Berkeley. Ann Arbor, MI: University Microfilms Inc.
Lee, R. B. (1972). The !Kung bushmen of Botswana. In M. Bicchieri, ed., *Hunters and Gatherers Today,* pp. 327–368. New York: Holt, Rhinehart & Winston.
Leroy, S. & Dupont, L. (1994). Development of vegetation and continental aridity in northwestern Africa during the Late Pliocene: the pollen record of ODP Site 658. *Palaeogeography, Palaeoclimatology, Palaeoecology* **109,** 295–316.
Lee-Thorp, J. A. (1989). Stable carbon isotopes in deep time: the diets of fossil fauna and hominids (Ph.D. dissertation). Cape Town, South Africa: University of Cape Town.
Lee-Thorp, J. A. & van der Merwe, N. J. (1993). Stable carbon isotope studies of Swartkrans fossils. In C. K. Brain, ed., *Swartkrans: A Cave's Chronicle of Early Man,* pp. 251–256. Pretoria: Transvaal Museum.
Lee-Thorp, J. A., van der Merwe, N. J. & Brain, C. K. (1989). Isotopic evidence for dietary differences between two extinct baboon species from Swartkrans. *Journal of Human Evolution* **18,** 183–190.
Lee-Thorp, J. A., van der Merwe, N. J. & Brain, C. K. (1994). Diet of *Australopithecus robustus* at Swartkrans from stable carbon isotopic analysis. *Journal of Human Evolution* **27,** 361–372.
Lehman, S. J. & Keigwin, L.D., Jr. (1992). Sudden changes in North Atlantic circulation during the last deglaciation. *Nature* **356,** 757–762.
Le Houerou, H. N. (1993). Grasslands of the Sahel. In R. T. Coupland, ed., *Natural Grasslands: Eastern Hemisphere & Resume,* pp. 197–220. London: Elsevier.
Levins, R. (1962). Theory of fitness in a heterogeneous environment. I. The fitness set and adaptive function. *American Naturalist* **96,** 361–373.
Levinton, J. (1988). *Genetics, Paleontology, and Macroevolution.* New York: Cambridge University Press.
Lewin, R. (1990). *Bones of Contention.* New York: Simon & Schuster.
Lewis, M. E. (1994). Carnivoran paleoguilds of Plio/Pleistocene Africa: implications for hominid food procurement strategies [abstract]. *American Journal of Physical Anthropology Supplement* **18,** 130.
Lezine, A. M. (1988). Les variations de la couverture forestiere mésophile d'Afrique occidentale au cours de l'Holocène. *Comptes Rendus de l'Académie des Sciences de Paris Séries II,* 439–445.
Li, T. & Etler, D. A. (1992). New middle Pleistocene hominid crania from Yunzian in China. *Nature* **357,** 404–407.
Lieberman, B. S., Brett, C. E. & Eldredge, N.

(1995). A study of stasis and change in two species lineages from the middle Devonian of New York state. *Paleobiology* **21,** 15–27.

Lieberman, D. E., Pilbeam, D. R. & Wood, B. A. (1988). A probablistic approach to the problem of sexual dimorphism in *H. habilis:* a comparison of KNM-ER 1470 and KNM-ER 1813. *Journal of Human Evolution* **17,** 503–511.

Lieberman, B. S. & Vrba, E. S. (1995). Hierarchy theory, selection, and sorting. *BioScience* **45,** 394–399.

Lieberman, D. E., Wood, B. A. & Pilbeam, D. R. (1996). Homoplasy and early *Homo:* an analysis of the evolutionary relationships of *H. habilis sensu stricto* and *H. rudolfensis. Journal of Human Evolution* **30,** 97–120.

Likens, G. E., ed. (1985). *An Ecosystem Approach to Aquatic Ecology: Mirror Lake and Its Environment.* New York: Springer-Verlag.

Lindsay, E. H., Fahlbusch, V. & Mein, P. (1989). *European Neogene Mammal Chronology.* New York: Plenum Press.

Lindsay, E. H., Opdyke, N. D. & Johnson, N. M. (1980). Pliocene dispersal of the horse, *Equus,* and late Cenozoic mammalian dispersal events. *Nature* **287,** 135–138.

Livingstone, D. A. (1967). Postglacial vegetation of the Ruwenzori Mountains of Equatorial Africa. *Ecological Monographs* **37,** 25–52.

Livingstone, D. A. (1993). Evolution of African climate. In P. Goldblatt, ed., *Biological Relationships between Africa and South America;* 455–472. New Haven, CT: Yale University Press.

Lloyd, E. A. (1988). *The Structure and Confirmation of Evolutionary Theory.* New York: Greenwood Press.

Lobeck, A. K. (1946). *Physiographic Diagram of Africa* [reprinted 1952]. New York: The Geographical Press.

Locker, S. & Martini, E. (1985). Phytoliths from the southwest Pacific, Site 591. In J. P. Kennett et al, eds., *Initial Reports of the Deep Sea Drilling Project* **90,** 1079–1084.

Lönnberg, E. (1918). Klimatväxlingars inflytande på Afrikas högre djurvarld. *Kungliga Svenska Vetenskapsakademiens Arsbok for 1918,* 246–288.

Lönnberg, E. (1929). The development and distribution of the African fauna in connection with and depending upon climatic changes. *Arkiv for Zoologi* **21A(4),** 133.

Loutit, T., Kennett, J. P. & Savin, S. M. (1983). Miocene equatorial and southwest Pacific paleoceanography from stable isotope evidence. *Marine Micropaleontology* **8,** 215–233.

Lovejoy, C. O. (1979). A reconstruction of the pelvis of AL-288 (Hadar Formation, Ethiopia) [abstract]. *American Journal of Physical Anthropology* **50,** 460.

Lovejoy, C. O. (1981). The origin of Man. *Science* **211,** 311–350.

Lovejoy, C. O. (1988). The evolution of human walking. *Scientific American* **259,** 118–125.

Lowell, T. V., Heusser, C. J., Andersen, B. G., Moreno, P. I., Hauser, A., Schlüchter, C., Marchant, D. R. & Denton, G. H. (1995). Interhemispheric correlations of late Pleistocene glacial events. *Science* **269,** 1541–1549.

Lucas, P. W. (1988). A new theory relating seed processing by primates to their relative tooth sizes. *Proceedings of the Australasian Society of Human Biology* **2,** 37–49.

Lucas, P. W. & Teaford, M. F. (1994). Functional morphology of colobine teeth. In A. G. Davies & J. F. Oates, eds., *Colobine Monkeys: Their Ecology, Behavior and Evolution,* pp. 173–204. Cambridge: Cambridge University Press.

Lynch, C. D. (1983). The mammals of the Orange Free State. *Memoirs van die Nasionale Museum Bloemfontein* **18,** 1–218.

Lynch, J. D. (1989). The gauge of speciation: on the frequences of modes of speciation. In D. Otte & J. A. Endler, eds., Speciation and Its Consequences, pp. 527–553. Sunderland, MA: Sinauer Associates.

Maasch, K. A. (1988). Statistical detection of the mid-Pleistocene transition. *Climate Dynamics* **2,** 133–143.

Maasch, K. A. & Saltzman, B. (1990). A low-order dynamical model of global climatic variability during the full Pleistocene. *Journal of Geophysical Research* **95,** 1955–1963.

Mabberley, D. J. (1987). *The Plant Book.* Cambridge: Cambridge University Press.

MacArthur, R. H. (1964). Environmental factors affecting bird species diversity. *American Naturalist* **98,** 387–397.

MacAyeal, D. R. (1993). Binge/purge oscillations of the Laurentide Ice Sheet as a cause of the North Atlantic's Heinrich events. *Paleoceanography* **8,** 775–784.

Macho, G. A. (1996). Climatic effects on dental development of *Theropithecus oswaldi* from Koobi Fora and Olorgesailie. *Journal of Human Evolution* **30,** 57–70.

Macho, G. A., Reid, D. J., Leakey, M. G., Jablonski, J. & Beynon, A. D. (1996). Climatic effects on dental development of *Theropithecus oswaldi* from Koobi Fora and Olorgesailie. *Journal of Human Evolution* **30,** 57–70.

Mackensen, A. & Ehrmann, W. (1992). Middle Eocene through early Oligocene climate history and paleoceanography in the Southern Ocean: stable oxygen and carbon isotopes from ODP site on Maud Rise and Kerguelen Plateau. *Marine Geology* **108,** 1–27.

Maglio, V. J. (1973). Origin and evolution of the

Elephantidae. *Transactions of the American Philosophical Society* **63(3).**
Maglio, V. J. (1975). Pleistocene faunal evolution in Africa and Eurasia. In K. Butzer & G. L. Issac, eds., *After the Australopithecines,* pp. 419–476. The Hague: Mouton.
Maglio, V. & Cooke, H. B. S. (1978). *Evolution of African Mammals.* Cambridge, MA: Harvard University Press.
Mahaney, W. C. (1990). *Ice on the Equator.* Sister Bay, WI: W. Caxton Press Ltd.
Maier, W. (1971). New fossil Cercopithecoidea from the Lower Pleistocene cave deposits of the Makapansgat Limeworks, South Africa. *Palaeontologia Africana* **13,** 69–108.
Maier, W. (1977). Die Evolution der bilophodonten Molaren der Cercopithecoidea. *Zeitschrift fur Morphologie und Anthropologie* **68,** 26–56.
Maier-Reimer, E., Mikolajewicz, U. & Crowley, T. J. (1990). Ocean general circulation model sensitivity experiment with an open Central American isthmus. *Paleoceanography* **5,** 349–366.
Maier-Reimer, E., Mikolajewicz, U. & Hasselman, K. (1993). Mean circulation of the Hamburg LGS OGCM and its sensitivity to the thermohaline surface forcing. *Journal of Physical Oceanography* **23,** 731–757.
Maley, J. (1980). Les changements climatiques de la fin due Tertiare en Afrique: leur consequence sur l'apparition du Sahara et de sa vegetation. In A. J. W. Martin & F. Hugues, eds., *The Sahara and the Nile,* pp. 63–86. Rotterdam: A. A. Balkema.
Maley, J. (1987) Fragmentation de la forêt dense humide africaine et extension des biotopes montagnards au quaternaire récent: nouvelles données polliniques et chronologiques. Implications paléoclimatiques et biogéographiques. *Palaeoecology of Africa* **18,** 307–334.
Maley, J. (1996). The African rain forest—main characteristics of changes in vegetation and climate from the Upper Cretaceous to the Quaternary. *Proceedings of the Royal Society of Edinburgh* **104B,** 31–73.
Manakbe, S. & Stouffer, R. J. (1993). Century-scale effects of increased atmospheric CO_2 on the ocean-atmospheric system. *Nature* **364,** 215–218.
Mandelbrot, B., Laff, A. & Hubbard, D. (1979). Fractals and the rebirth of iteration theory. In B. Mandelbrot, ed., *The Beauty of Fractals.* London: Academic Press, pp. 151–160.
Manighetti, B. & McCave, I. N. (1995). Depositional fluxes, paleoproductivity and ice rafting in the NE Atlantic over the past 30 ka. *Paleoceanography* **10,** 579–592.
Manighetti, B., McCave, I. N., Maslin, M. & Shackleton, N. J. (1995). Chronology for climate change: developing age models for t he Biogeochemical Ocean Flux Study cores. *Paleoceanography* **10,** 513–525.
Maples, W. R. (1972). Systematic reconsideration and a revision of the nomenclature of Kenya baboons. *American Journal of Physical Anthropology* **36,** 9–20.
Marchant, D. R., Denton, G. H., Sugden, D. E. & Swisher, C. C. III (1993a). Miocene glacial stratigraphy and landscape evolution of the western Asgard Range, Antarctica. *Geografiska Annaler* **74a,** 303–351.
Marchant, D. R., Denton, G. H. & Swisher, C. C. III (1993b). Miocene-Pliocene-Pleistocene glacial history of Arena Valley, Quartermain Mountains, Antarctica. *Geografiska Annaler* **75a,** 269–302.
Marchant, D. R., Swisher, C. C. III, Lux, D. R., West, D. P., Jr. & Denton, G. H. (1993c). Pliocene paleoclimate and East Antarctic Ice Sheet history from surficial ash deposits. *Science* **260,** 667–670.
Marshall, L. G., Webb, S. D., Sepkoski, J. J. & Raup, D. M. (1982). Mammalian evolution and the Great American Interchange. *Science* **215,** 1351–1357.
Martin, A., Mariotti, A., Balesdent, J., Lavelle, P. & Vauttoux, R. (1990). Estimate of organic matter turnover rate in a savanna soil by ^{13}C natural abundance measurements. *Soil Biology and Biochemistry* **22,** 517–523.
Martin, L. B. (1985). Significance of enamel thickness in hominid evolution. *Nature* **314,** 260–263.
Martin, L. & Andrews, P. (1993). Species recognition in middle Miocene hominoids. In W. H. Kimbel & L. Martin, eds., *Species, Species Concepts and Primate Evolution,* pp. 393–427. New York: Plenum Press.
Martin, L. B. & Kimbel, W. H., eds. (1993). *Species, Species Concepts and Primate Evolution.* New York: Plenum Press.
Martin, R. A. (1993). Patterns of variation and speciation in Quaternary rodents. In R. A. Martin & A. D. Barnosky, eds., *Morphological Change in Quaternary Mammals of North America,* pp. 226–280. Cambridge: Cambridge University Press.
Martin, R. D. (1981). Relative brain size and basal metabolic rate in terrestrial vertebrates. *Nature* **293,** 57–60.
Martin, R. D. (1983). *Human Brain Evolution in an Ecological Context. Second James Arthur Lecture on the Evolution of the Human Brain.* New York: American Museum of Natural History Publications.
Maslin, T. P. (1952). Morphological criteria of phyletic relationships. *Systematic Zoology* **1,** 49–70.

Massler, M. & Schour, I. (1946). The appositional life span of the enamel and dentin-forming cells. *Journal of Dental Research* **25,** 145–150.

Masters, J. C. & Rayner, R. J. (1993). Competition and macroevolution: the ghost of competition yet to come? *Biological Journal of the Linnean Society* **49,** 87–98.

Matsumura, H., Nakatsukasa, M. & Ishida, H. (1992). Comparative study of crown cusp areas in the upper and lower molars of African apes. *Bulletin of the National Science Museum, Tokyo, Series D* **18,** 1–15.

Matthews, J. V., Jr. (1990). New data on Pliocene floras/faunas from the Canadian Arctic and Greenland. In: *Pliocene Climates: Scenario for Global Warming. Abstracts from USGS Workshop, Denver Colorado, October 23–25, 1989.* U.S. Geological Survey Open File Report 90-64, pp. 29–33. Washington, DC: U.S. Geological Survey.

Maynard Smith, J. (1981). Sympatric speciation. *American Naturalist* **100,** 386–392.

Maynard Smith, J. (1989). *Evolutionary Genetics.* Oxford: Oxford University Press.

Maynard Smith, J. & Price, G. R. (1973). The logic of animal conflict. *Nature* **246,** 15–18.

Mayr, E. (1942). *Systematics and the Origin of Species.* New York: Columbia University Press.

Mayr, E. (1950). Taxonomic categories of fossil hominids. *Cold Spring Harbor Symposium on Quantitative Biology* **15,** 109–118.

Mayr, E. (1963). *Animal Species and Evolution.* Cambridge, MA: The Belknap Press.

Mayr, E. (1976). *Evolution and the Diversity of Life.* Cambridge, MA: The Belknap Press.

Mayr, E. (1989). Speciational evolution or punctuated equilibria. *Journal of Social Biological Structure* **12,** 137–158.

Mayr, E. (1992). Controversies in retrospect. *Oxford Surveys in Evolutionary Biology* **8,** 1–34.

Mayr, E. & O'Hara, R. J. (1986). The biogeographic evidence supporting the Pleistocene forest refuge hypothesis. *Evolution* **40,** 55–67.

Mayr, E. & Provine, W. B. (1980). *The Evolutionary Synthesis.* Cambridge: Harvard University Press.

McCann, K. & Yodzis, P. (1994). Nonlinear dynamics and population disappearances. *The American Naturalist* **144,** 873–879.

McCollum, M. A., Grine, F. E., Ward, S. C. & Kimbel, W. H. (1993). Subnasal morphological variation in extant hominoids and fossil hominids. *Journal of Human Evolution* **24,** 87–111.

McCrossin, M. L. (1983). Postcranial remains of *Nyanzachoerus* (suidae, artiodactyla) from the Sahabi Formation, Libya (M.A. thesis). New York: New York University.

McCrossin, M. L. (1987). Postcranial remains of fossil Suidae from the Sahabi Formation, Libya. In N. T. Boaz, A. el-Arnauti, A. W. Gaziry, J. de Heinzelin & D. Dechant Boaz, eds., *Neogene Paleontology and Geology of Sahabi,* pp. 267–286. New York. Alan R. Liss.

McCrossin, M. L. (1992). An oreopithecid proximal humerus from the middle Miocene of Maboko Island, Kenya. *International Journal of Primatology* **13,** 659–677.

McCrossin, M. L. (1994a). The phylogenetic relationships, adaptations, and ecology of *Kenyapithecus* (Ph.D. dissertation). University of California at Berkeley. Ann Arbor, MI: University Microfilms International.

McCrossin, M. L. (1994b). Semi-terrestrial adaptations of *Kenyapithecus* [abstract]. *American Journal of Physical Anthropology Supplement* **18,** 142–143.

McCrossin, M. L. (1995). New perspectives on the origins of terrestriality among Old World higher primates [abstract]. *American Journal of Physical Anthropology Supplement* **20,** 147.

McCrossin, M. L. (1996). A reassessment of forelimb evidence for the phylogenetic relationships of *Kenyapithecus* and other large-bodied hominoids of the middle-late Miocene [abstract]. *American Journal of Physical Anthropology Supplement* **22,** 161–162.

McCrossin, M. L. & Benefit, B. R. (1992a). Maboko Island and the evolutionary history of Old World monkeys and apes [abstract]. *American Anthropological Association,* 226.

McCrossin, M. L. & Benefit, B. R. (1992b). Comparative assessment of the ischial morphology of *Victoriapithecus macinnesi. American Journal of Physical Anthropology* **87,** 277–290.

McCrossin, M. L. & Benefit, B. R. (1993). Recently recovered *Kenyapithecus* mandible and its implications for great ape and human origins. *Proceedings of the National Academy of Sciences USA* **90,** 1962–1966.

McCrossin, M. L. & Benefit, B. R. (1994). Maboko Island and the evolutionary history of Old World monkeys and apes. In R. S. Corruccini & R. L. Ciochon, eds., *Integrative Paths to the Past: Paleoanthropological Advances in Honor of F. C. Howell,* pp. 95–122. New York: Prentice Hall.

McCrossin, M. L. & Benefit, B. R. (1997). On the relationships and adaptations of *Kenyapithecus,* a large-bodied hominoid from the middle Miocene of eastern Africa. In D. R. Begun, C. V. Ward & M. D. Rose, eds., *Miocene Hominoid Fossils: Functional and Phylogenetic Implication,* pp. 241–267. New York: Plenum.

McCrossin, M. L., Benefit, B. R., Gitau, S. N., Palmer, A. & Blue, K. (1998). Fossil evidence for the origins of terrestriality among Old World monkeys and apes. In E. Strasser, J. G. Fleagle,

H. M. McHenry, eds., *Primate Locomotion: Recent Advances,* pp. 353–396. New York: Plenum.

McDonald, J. F. (1990). Macroevolution and retroviral elements. *BioScience* **40,** 183–191.

McGrew, W. C. (1992). *Chimpanzee Material Culture: Implications for Human Evolution.* Cambridge: Cambridge University Press.

McGrew, W. C., Tutin, C. E. G. & Baldwin, P. J. (1979). New data on meat eating by wild chimpanzees. *Current Anthropology* **20,** 238–239.

McHenry, H. M. (1973). Early hominid humerus from East Rudolf, Kenya. *Science* **180,** 739–741.

McHenry, H. M. (1974). How large were the australopithecines? *American Journal of Physical Anthropology* **40,** 329–340.

McHenry, H. M. (1986). The first bipeds: a comparison of the *A. afarensis* and *A. africanus* postcranium and implications for the evolution of bipedalism. *Journal of Human Evolution* **15,** 177–191.

McHenry, H. M. (1988). New estimates of body weight in early hominids and their significance to encephalization and the megadontia in 'robust australopithecines'. In R. E. Grine, ed., *The Evolutionary History of the Robust Australopithecines,* pp. 133–148. New York: Aldine.

McHenry, H. M. (1991). Petite bodies of the "robust" australopithecines. *American Journal of Physical Anthropology* **86,** 445–454.

McHenry, H. M. (1992). Body size and proportions in early hominids. *American Journal of Physical Anthropology* **87,** 407–431.

McHenry, H. M. (1994a). Tempo and mode in human evolution. *Proceedings of the National Academy of Science USA* **91,** 6780–6786.

McHenry, H. M. (1994b). Behavioral ecological implications of early hominid body size. *Journal of Human Evolution* **27,** 77–87.

McIntyre, A. & Molfino, B. (1996). Forcing of Atlantic equatorial and subpolar millenial cycles by precession. *Science* **274,** 1867–1870.

McKee, J. K. (1984). A genetics model of dental reduction through the probable mutation effect. *American Journal of Physical Anthropology* **65,** 229–241.

McKee, J. K. (1989). Australopithecine anterior pillars: reassessment of the functional morphology and phylogenetic relevance. *American Journal of Physical Anthropology* **80,** 1–9.

McKee, J. K. (1991). Palaeo-ecology of the Sterkfontein hominids: a review and synthesis. *Palaeontologia Africana* **28,** 41–51.

McKee, J. K. (1993a). The faunal age of the Taung hominid deposit. *Journal of Human Evolution* **25,** 363–376.

McKee, J. K. (1993b). Formation and geomorphology of caves in calcareous tufas and implications for the study of the Taung fossil deposits. *Transactions of the Royal Society of South Africa* **48,** 307–322.

McKee, J. K. (1994). Catalogue of fossil sites at the Buxton Limeworks, Taung. *Palaeontologia Africana* **31,** 73–81.

McKee, J. K. (1995). Turnover patterns and species longevity of large mammals from the Late Pliocene and Pleistocene of southern Africa: a comparison of simulated and empirical data. *Journal of Theoretical Biology* **172,** 141–147.

McKee, J. K. (1996a). Faunal evidence and Sterkfontein Member 2 foot bones of early hominid. *Science* **271,** 1301–1302.

McKee, J. K. (1996b). Faunal turnover patterns of the Pliocene and Pleistocene of southern Africa. *South African Journal of Science* **92,** 111–113.

McKee, J. K., Thackeray, J. F. & Berger, L. R. (1995). Faunal assemblage seriation of southern African Pliocene and Pleistocene fossil deposits. *American Journal of Physical Anthropology* **96,** 235–250.

McKee, J. K. & Tobias, P. V. (1994). Taung stratigraphy and taphonomy: preliminary results based on the 1988–93 excavations. *South African Journal of Science* **90,** 233–235.

McKeon, G. M., Day, K. A., Howden, S. M., Mott, J. J., Orr, D. M., Scattini, W. J. & Weston, E. J. (1991). Northern Australia savannas: management for pastoral production. In P. A. Werner, ed., *Savanna Ecology and Management,* pp. 11–28. Oxford: Blackwell.

McKinney, M. L. & McNamara, K. J. (1991). *Heterochrony.* New York: Plenum Press.

McNaughton, S. J. (1988). Mineral nutrition and spatial concentrations of African ungulates. *Nature* **334,** 343–345.

McNaughton, S. J. (1990). Mineral nutrition and seasonal movements of African migratory ungulates. *Nature* **345,** 613–615.

Medina, E. & Silva, J. F. (1991). Savannas of northern South America: a steady state regulated by water-fire interactions on a background of low nutrient availability. In P. A. Werner, ed., *Savanna Ecology and Management,* pp. 59–69. Oxford: Blackwell.

Meester, J. (1965). The origins of the Southern African mammal fauna. *Zoologica Africana* **1,** 87–93.

Meester, J. (1988). Chromosomal speciation in Southern African mammals. *South African Journal of Science* **84,** 721–724.

Meester, J., Rautenbach, I. L. & Dippenaar, N. J. (1986). Classification of Southern African mammals. *Transvaal Museum Monograph* **5,** 1–359.

Meester, J. & Setzer, H. W., eds. (1971–1977). *The Mammals of Africa. An Identification Manual.* Washington, DC: Smithsonian Institution Press.

Meikle, W. E. (1987). Fossil Cercopithecidae from the Sahabi Formation. In N. t. Boaz, A. el-Arnauti, A. W. Gaziry, J. de Heinzelin & D. Dechant Boaz, eds., *Neogene Paleontology and Geology of Sahabi,* pp. 119–127. New York: Alan R. Liss.

Mein, P. (1975). Report on the activity of the R.C.M.N.S. working groups (1971–75). In J. Senes, ed., *Bratislava: Vertebrata. I.U.G.S. Commission on Stratigraphy, Subcommission on Neogene Stratigraphy,* pp. 77–81. Bratislava: I.U.G.S.

Mein, P. (1979). Rapport d'activitie du group de travail vertébrés mise ă jour de la biostratigraphie du Négène basée sur les mammifères. *Annales Geologica Pays Hellenica* **3,** 1367–1372.

Menaut, C. (1983). The vegetation of African savannas. In J. Bourlière, ed., *Ecosystems of the World, 13. Tropical Savannahs,* pp. 109–149. Amsterdam: Elsevier.

Mercier, N., Valladas, H., Joron, J.-L., Reyss, J. I., Levêque, F. & Vandermeersh, B. (1991). Thermoluminescence dating of the late Neandertal remains from Saint Césaire. *Nature* **351,** 737–739.

Metzler, D. E. (1977). *Biochemistry: The Chemical Reactions of Living Cells.* New York: Academic Press.

Meyer, G. E. (1978). Hyracoidea. In V. J. Maglio & H. B. S. Cooke, eds., *Evolution of African Mammals,* pp. 284–314. Cambridge, MA: Harvard University Press.

Michaelson, J. (1993). Biology of disease. *Laboratory Investigation* **69,** 136–149.

Michelmore, A. P. G. (1939). Observations on tropical African grasslands. *Journal of Ecology* **27,** 282–330.

Milankovitch, M. (1941). *Kanon der Erdbestrahlung und seine Anwendung auf das Eiszeitenproblem.* Königlich Serbische Akademie Beograd **132,** 1–484.

Mildenhall, D. C. (1989). Terrestrial palynology. In P. J. Barrett, ed., *Antarctic Cenozoic history from the CIROS-1 drill hole, McMurdo Sound, DSIR Bulletin* 245, pp. 119–127. Wellington: Deep Sea Drilling Project.

Miller, G. H., Manley, W., Duvall, M. & Kaufman, D. (1993). Ice-sheet/ocean interaction at the mouth of Hudson Strait, Arctic Canada. In *Abstracts for the Geological Society of American Fall meeting,* 1993.

Miller, K. G., Fairbanks, R. G. & Mountain, G. S. (1987). Tertiary oxygen isotope synthesis, sea level history, and continental margin erosion. *Paleoceanography* **2,** 1–19.

Miller, K. G., Mountain, G. S., et al. (1996). Drilling and dating New Jersey Oligocene-Miocene sequences: ice volume, global sea level, and Exxon records. *Science* **271,** 1092–1095.

Miller, K. G. & Sugarman, P. J. (1995). Correlating Miocene sequences in onshore New Jersey boreholes (ODP Leg 150X) with global $\delta^{18}O$ and Maryland outcrops. *Geology* **23,** 747–750.

Miller, K. G., Wright, J. D. & Fairbanks, R. G. (1991). Unlocking the ice house: Oligocene-Miocene oxygen isotopes, eustasy, and margin erosion. *Journal of Geophysical Research* **96,** 6829–6848.

Mills, M. G. L., Biggs, H. C. & Whyte, I. J. (1995). The relationship between lion predation, population trends in African herbivores and rainfall. *Wildlife Research* **22,** 75–87.

Misonne, X. (1969). African and Indo-Australian Muridae evolutionary trends. *Annales du Musée Royal d'Afrique Centrale de Tervuren, Sciences Zoologiques* **172,** 1–219.

Misonne, X. & Verschuren, J. (1966). Les rongeurs et lagomorphes de la région du parc national du Serengeti. *Mammalia* **30,** 517–537.

Mitchell, G. & Skinner, J. D. (1993). How giraffe adapt to their extraordinary shape. *Transactions of the Royal Society of South Africa* **48,** 207–218.

Molnar, P. & England, P. (1990). Late Cenozoic uplift of mountain ranges and global climate change: Chicken or egg? *Nature* **346,** 29–34.

Moreau, R. E. (1966). *The Bird Faunas of Africa and Its Islands.* New York: Academic Press.

Morgan, M. E., Kingston, J. D. & Marino, B. D. (1994a). Carbon isotopic evidence for the emergence of C_4 plants in the Neogene from Pakistan and Kenya. *Nature* **367,** 162–165.

Morgan, M. E., Kingston, J. D. & Marino, B. D. (1994b). Reply. *Nature* **371,** 112–113.

Morgan, T. H. (1910). Chance or purpose in the origin and evolution of adaptation. *Science* **31,** 201–210.

Morin, P. A., Moore, J. J., Chakraborty, R., Jin, L., Goodall, J. & Woodruff, D. S. (1994). Kin selection, social structure, gene flow, and the evolution of chimpanzees. *Science* **265,** 1193–1201.

Morrell, V. (1995). *Ancestral Passions. The Leakey Family and the Quest for Humankind's Beginnings.* New York: Simon & Schuster.

Morrison, D. F. (1976). *Multivariate Statistical Methods.* New York: McGraw-Hill.

Moya-Sola, S. & Kohler, M. (1993). Recent discoveries of *Dryopithecus* shed new light on evolution of great apes. *Nature* **365,** 543–545.

Munthe, J. (1987). Small-Mammal fossils from the Pliocene Sahabi formation of Lybia. In N. T. Boaz, A. El-Arnauti, A. W. Gaziry, J. de Heinzelin & D. Dechant Boaz, eds., *Neogene Palaeontology and Geology of Sahabi,* pp. 135–144. New York: Alan R. Liss.

Murray, M. G. & Brown, D. (1993). Niche separation of grazing ungulates in the Serengeti: an experimental test. *Journal of Animal Ecology* **62,** 380–389.

Musser, G. G. (1987). The mammals of Sulawesi. In T. C. Whitmore, ed., *Biogeographical Evolution of the Malay Archipelago,* pp. 73–93. Oxford: Oxford University Press.

Mwanza, N., Yamagiwa, J., Yumoto, T. & Maruhashi, T. (1992). Distribution and range utilization of eastern lowland gorillas. In N. Itoigawa, Y. Sugiyama, G. P. Sackett & R. K. R. Thompson, eds., *Topics in Primatology,* vol. 2. *Behavior, Ecology, and Conservation,* pp. 283–300. Tokyo: University of Tokyo Press.

Myers, A. A. & Giller, P. S., eds. (1988a). *Analytical Biogeography: An Integrated Approach to the Study of Animal and Plant Distributions.* London: Chapman and Hall.

Myers, A. A. & Giller, P. S., eds. (1988b). Process, pattern and scale in biogeography. In A. A. Myers & P. S. Giller, eds., *Analytical Biogeography,* pp. 3–12. London: Chapman and Hall.

Nadelhoffer, K. J. & Fry, B. (1988). Controls of natural nitrogen-15 and carbon-13 abundances in forest soil organic matter. *Soil Science Society of America Journal* **52,** 1633–1640.

Nakaya, H. (1993). Les Faunes de mammifères du Miocène supérieur de Samburu Hills, Kenya, Afrique de l'est et l'environnement des pré-hominidés. *L'Anthropologie* **97,** 9–16.

Nakaya, H. (in press). Mammalian fauna of late Miocene Samburu Hills, Kenya, East Africa and environments of pre-hominids. *L'Anthropologie* **93(4).**

Nakaya, H., Pickford, M., Nakano, Y. & Ishida, H. (1984). The late Miocene large mammalian fauna from the Namurungule Formation, Samburu Hills, Northern Kenya. *African Study Monograph Supplementary Issue* **2,** 87–131.

Nakaya, H., Pickford, M., Yasuis, K. & Nakano, Y. (in press). Additional large mammalian fauna from the Namurungule Formation, Samburu Hills, Northern Kenya. *Africa Study Monograph Supplementary Issue.*

Nambudiri, E. M. V., Tidwell, W. D., Smith, B. N. & Hebbert, N. P. (1978). A C_4 plant from the Pliocene. *Nature* **276,** 816–817.

Napier, J. R. & Davis, P. R. (1959). The forelimb skeleton and associated remains of *Proconsul africanus. Fossil Mammals of Africa* **16,** 1–69.

Napier, J. R. (1970). Paleoecology and catarrhine evolution. In J. R. Napier & P. H. Napier, eds., *Old World Monkeys,* pp. 53–96. New York: Academic Press.

Nelson, C. S., Hendy, C. H., Jarrett, G. R. & Cuthbertson, A. M. (1985). Near-synchroneity of New Zealand alpine glaciations and Northern Hemisphere continental glaciations during the past 750 kyr. *Nature* **318,** 361–363.

Nelson, G. (1974). Historical biogeography: an alternative formalization. *Systematic Zoology* **23,** 555–557.

Nelson, G. J. (1992). Why, after all, must it? *Cladistics* **8,** 139–146.

Nelson, G. & Platnick, N. (1981). *Systematics and biogeography: cladistics and vicariance.* New York: Columbia University Press.

Neumann, K. & Schulz, E. (1987). Middle Holocene savannah vegetation in the central Sahara. Preliminary report. *Paleoecology of Africa* **18,** 163–166.

Nevo, E. (1989). Modes of speciation: the nature and role of peripheral isolates in the origin of species. In V. L. Giddins, K. Y. Kaneshiro & W. W. Anderson, eds., *Genetics, Speciation, and the Founder Principle,* pp. 205–236. New York: Oxford University Press.

Nevo, E. (1991). Evolutionary theory and processes of speciation and adaptive radiation in subterranean mole rats *Spalax ehrenbergi* superspecies, in Israel. *Evolutionary Biology* **25,** 1–125.

Nevo, E. (1996). Evolutionary processes and theory: micro- and macroevolution. In S. P. Wasser, ed., *Botany and Mycology for the Next Millenium: Collection of Scientific Article Devoted to the 70th Anniversary of Academician K. M. Sytnik,* pp. 63–83. Kyeiv: N. G. Kolodny Institute of Botany, National Academy of Sciences of Ukraine.

Newman, C. M., Cohen, J. E. & Kipnis, C. (1985). Neo-Darwinian evolution implies punctuated equilibria. *Nature* **315,** 400–401.

Newnham, R. M., Lowe, D. J. & Green, J. D. (1989). Palynology, vegetation and climate in the Waikato lowlands, North Island, since c. 18,000 years ago. *Journal of the Royal Society of New Zealand* **19,** 127–150.

Nishida, T. & Hiraiwa-Hasegawa, M. (1987). Chimpanzees and bonobos: cooperative relationships among males. In B. B. Smuts, D. L. Cheney, R. M. Seyfarth, R. W. Wrangham & T. T. Struhsaker, eds., *Primate Societies,* pp. 165–177. Chicago: University of Chicago Press.

Nishida, T., Uehara, S. & Nyundo, R. (1979). Predatory behaviour among wild chimpanzees of the Mahale Mountains. *Primates* **20,** 1–20.

Nixon, K. C. & Wheeler, Q. D. (1992). Extinction and the origin of species. In M. J. Novacek and Q. D. Wheeler, eds., *Extinction and Phylogeny,* pp. 119–143. New York: Columbia University Press.

Oates, J. F. (1988). The distribution of *Cercopithecus* monkeys in West African forests. In A. Gautier-Hion, F. Bourlière & J.-P. Gautier, eds., *A Primate Radiation: Evolutionary Biology of the African Guenons,* pp. 79–103. Cambridge: Cambridge University Press.

Oates, J. F. (1981). Mapping the distribution of West African rain-forest monkeys: issues, methods, and preliminary results. *Annals of the New York Academy of Sciences* **376,** 53–64.

Oates, J. F. & Trocco, T. F. (1983). Taxonomy and phylogeny of black-and-white colobus monkeys. *Folia Primatologica* **40,** 83–13.

Oboussier, H. (1972). Morphologische und quantitative Neocortexuntersuchungen bei Boviden, ein Beitrag zur Phylogenie dieser Familie. III. Formen uber 75 kg Korpergewicht. *Mitt. Hamburg Zoologisches Museum und Institut* **68,** 271–292.

O'Brien, E. M. (1989). *Climate and Woody Plant Species Richness: Analyses Based Upon Southern Africa's Native Flora with Extrapolations to Subsaharan Africa* (Ph.D. diss.). Oxford: Oxford University.

O'Brien, E. M. (1993). Climatic gradients in plant species richness: towards an explanation based on an analysis of southern Africa's woody flora. *Journal of Biogeography* **20,** 181–198.

O'Brien, E. M. (1998a). Climate and woody plant diversity in southern Africa: relationships at species, genus and family levels. *Ecography* **21,** 495–509.

O'Brien, E. M. (1998b). Water-energy dynamics, climate, and prediction of woody plant species richness: An interim general model. *Journal of Biogeography* **25,** 379–398.

O'Brien, E. M. & Peters, C. R. (1991). Ecobotanical contexts for African hominids. In J. D. Clark, ed., *Cultural Beginnings: Approaches to Understanding Early Hominid Life-ways In the African Savanna,* pp. 1–15. Bonn: Rudolf Hablet GMBH.

O'Brien, E. M. & Peters, C. R. (1998). Wild fruit trees and shrubs of southern Africa: geographic distribution of species richness. *Economic Botany* **10,** 245–256.

O'Brien, E. M. & Peters, C. R. (1999). Climactic perspectives for neogene environmental reconstructions. In J. Agusti & P. J. Andrews, eds. *Hominoid Evolution and Climactic and Environmental Change in the Neogene of Europe,* pp. 53–78. Cambridge: Cambridge University Press.

Odum, E. P. (1983). *Basic Ecology.* Philadelphia: W. B. Saunders.

Olsen, P. E. (1986). A 40-million-year lake record of early Mesozoic orbital climatic forcing. *Science* **234,** 842–848.

Olsen, P. E. & Kent, D. V. (1996). Milankovitch climate forcing in the tropics of Pangaea during the late Triassic. *Palaeogeography, Palaeoclimatology, Palaeoecology* **122,** 1–26.

Opdyke, N. D. (1995). Mammalian migration and climate over the last seven million years. In E. S. Vrba, G. H. Denton, T. C. Partridge & L. H. Burckle, eds., *Paleoclimate and Evolution with emphasis on Human Origins,* pp. 109–114. New Haven, CT: Yale University Press.

Osborn, D. J. & Helmy, I. (1980). The contemporary land mammals of Egypt (including Sinai). *Fieldiana, Zoology* **5,** 1–579.

Oster, G. & Alberch, P. (1982). Evolution and bifurcation of developmental programs. *Evolution* **36,** 444–459.

Owen-Smith, N. (1982). Factors influencing the consumption of plant products by large herbivores. In B. J. Huntley & B. H. Walker, eds., *The Ecology of Tropical Savannas,* pp. 359–404. Berlin: Springer-Verlag.

Owen-Smith, N. (1985). Niche separation among African ungulates. In E. S. Vrba, ed., *Species and Speciation,* pp. 167–171. Transvaal Museum Monograph No. 4. Pretoria: Transvaal Museum.

Owen-Smith, N. (1987). Late Pleistocene extinctions: the pivotal role of megaherbivores. *Paleobiology* **13,** 351–362.

Owen-Smith, N. (1988a). *Megaherbivores. The Influence of Very Large Body Size on Ecology.* Cambridge: Cambridge University Press.

Owen-Smith, N. (1988b). Niche separation among African ungulates. In E. Vrba, ed., *Species and Speciation,* pp. 167–171. Pretoria: Transvaal Museum Monograph no. 4.

Owen-Smith, N. (1989a). Morphological factors and their consequences for resource partitioning among African savanna ungulates: a simulation modelling approach. In D. W. Morris, Z. Abramsky, B. J. Fox & M. L. Willig, eds., *Patterns in the Structure of Mammalian Communities,* pp. 155–165. Texas Tech Museum Special Publications Series. Lubbock, TX: Texas Tech Museum.

Owen-Smith, N. (1989b). Megafaunal extinctions: the conservation message from 11,000 years BP. *Conservation Biology* **3,** 405–412.

Owen-Smith, N. (1990). Demography of a large herbivore, the greater kudu, in relation to rainfall. *Journal of Animal Ecology* **59,** 893–913.

Owen-Smith, N. (1992). Grazers and browsers: ecological and social contrasts among African ruminants. In F. Spitz, G. Janeau & S. Aulagnier, eds., *Ongulés/Ungulates 91;* 175–182. Paris: SFEPM-IRGM.

Owen-Smith, N. (1993a). Woody plants, browsers and tannins in southern African savannas. *South African Journal of Science* **89,** 505–510.

Owen-Smith, N. (1993b). Comparative mortality rates of male and female kudus: the costs of sexual size dimorphism. *Journal of Animal Ecology* **62,** 428–440.

Owen-Smith, N. (1998). How high ambient temperature affects the daily activity and foraging time

of a subtropical ungulate, the greater Kudu. *Journal of Zoology*, London **246.**
Owen-Smith, N. & Cooper, S. M. (1987). Palatability of woody plants to browsing ruminants in a South African savanna. *Ecology* **68,** 319–331.
Owen-Smith, N. & Cooper, S. M. (1989). Nutritional ecology of a browsing ruminant, the kudu through the seasonal cycle. *Journal of Zoology* **219,** 29–43.
Owen-Smith, N. & Cumming, D. H. M. (1993). Comparative foraging strategies of grazing ungulates in African savanna grasslands. *Proceedings of the XVII International Grassland Congress,* pp. 691–698.
Parker, I. S. C. (1983). The Tsavo story: an ecological case history. In R. N. Owen-Smith, ed., *Management of Large Mammals in African Conservation Areas,* pp. 37–50. Pretoria: Haum.
Partridge, T. C., Bond, G., Hartnady, C. J., deMenocal, P. & Ruddiman, W. (1995a). Climatic effects of Late Neogene tectonism and volcanism. In E. S. Vrba, G. H. Denton, T. C. Partridge & L. H. Burckle, eds., *Paleoclimate and Evolution with Emphasis on Human Origins,* pp. 8–23. New Haven, CT: Yale University Press.
Partridge, T. C., Wood, B. A. & deMenocal, P. B. (1995b). The influence of global climatic change and regional uplift on large-mammalian evolution in East and southern Africa. In E. S. Vrba, G. H. Denton, T. C. Partridge & L. H. Burckle, eds., *Paleoclimate and Evolution with Emphasis on Human Origins,* pp. 331–355. New Haven, CT: Yale University Press.
Pascual, R. & Juareguizar, E. O. (1990). Evolving climates and mammal faunas in Cenozoic South America. *Journal of Human Evolution* **19,** 23–60.
Passingham, R. E. & Ettlinger, G. (1974). A comparison of cortical functions in man and the other primates. *International Review in Neurobiology* **16,** 233–299.
Paterson, H. E. H. (1978). More evidence against speciation by reinforcement. *South African Journal of Science* **74,** 369–371.
Paterson, H. E. H. (1981). The continuing search for the unknown and unknowable: a critique of contemporary ideas on speciation. *South African Journal of Science* **77,** 113–119.
Paterson, H. E. H. (1982). Perspective on speciation by reinforcement. *South African Journal of Science* **78,** 53–57.
Paterson, H. E. H. (1985). The recognition concept of species. In E. S. Vrba, ed., *Species and Speciation.* Pretoria: Transvaal Museum Monograph no. 4, pp. 21–34.
Paterson, H. E. H. (1986). Environment and species. *South African Journal of Science* **82,** 62–65.
Paterson, H. E. H. (1988). On defining species in terms of sterility: problems and alternatives. *Pacific Science* **42,** 65–71.
Paterson, H. E. H. (1993). *Evolution and the Recognition Concept of Species,* Shane F. McEvey, ed. Baltimore, MD: The Johns Hopkins University Press.
Peabody, F. E. (1954). Travertines and cave deposits of the Kaap escarpment of South Africa, and the type locality of *Australopithecus africanus* Dart. *Bulletin of the Geological Society of America* **65,** 671–706.
Perret, J. L. & Aellen, V. (1956). Mammifères du Cameroun de la collection J. L. Perret. *Revue Suisse de Zoologie* **26,** 395–450.
Peters, C. R. (1981). Robust vs. gracile early hominid masticatory capabilities: the natural competitive advantage of the megadonts. In L. L. Mai, E. Shanklin & R. W. Sussman, eds., *The Perceptions of Human Evolution,* pp. 161–181. Los Angeles: University of California Press.
Peters, C. R. (1987a). Nut-like oil seeds: food for monkeys, chimpanzees, humans, and probably ape-men. *American Journal of Physical Anthropology* **73,** 333–363.
Peters, C. R. (1987b). *Ricinodendron rautanenii* (Euphorbiaceae): Zambezian wild food plant for all seasons. *Economic Botany* **41(4),** 494–502.
Peters, C. R. (1988). Notes on the distribution and relative abundance of *Sclerocarya birrea* (A. Rich.) Hochst. (Anacardiaceae). *Monographs in Systematic Botany of the Missouri Botanic Garden* **25,** 403–410.
Peters, C. R. (1990). African wild plants with rootstocks reported to be eaten raw: the monocotyledons, part 1. *Mitteilungen aus dem Institut für Allgemeine Botanik Hamburg* **23,** 935–952.
Peters, C. R. (1993). Shell strength and primate seed predation of nontoxic species in eastern and southern Asia. *International Journal of Primatology* **14(2),** 315–344.
Peters, C. R. (1994). African wild plants with rootstocks reported to be eaten raw: the monocotyledons, part II. In J. H. Seyani & A. C. Chikuni, eds., *Proceedings of the XIIIth Plenary Meeting of AETFAT,* pp. 25–38. Zomba: National Herbarium & Botanic Gardens of Malawi.
Peters, C. R. (1996). African wild plants with rootstocks reported to be eaten raw: the monocotyledons, part III. In L J. G. van der Maesen, X. M. van der Burgt & J. M. van Medenbach de Rooy, eds., *Proceedings of the XIVth Plenary Meeting of AETFAT,* pp. 665–677. Dordrecht: Kluwer Academic Publishers.
Peters, C. R. & Blumenschine, R. J. (1995). Landscape perspectives on possible land use patterns for Early Pleistocene hominids in the

Olduvai Basin, Tanzania. *Journal of Human Evolution* **29,** 321–362.

Peters, C. R. & Blumenschine, R. J. (1996). Landscape perspectives on possible land use patterns for Early Pleistocene hominids in the Olduvai Basin, Tanzania: Part II, expanding the landscape models. *Kaupia, Darmstädter Beiträge zur Naturgeschichte* **6,** 175–221.

Peters, C. R. & O'Brien, E. M. (1981). The early hominid plant-food niche: insights from an analysis of plant exploitation by *Homo, Pan,* and *Papio. Current Anthropology* **22(2),** 127–140.

Peters, C. R. & O'Brien, E. M. (1984). On hominid diet before fire. *Current Anthropology* **22(2),** 127–140.

Peters, C. R. & O'Brien, E. M. (1994). Potential hominid plant foods from woody species in semi-arid versus sub-humid subtropical Africa. In D. J. Chivers & P. Langer, eds., *The Digestive System in Mammals: Food, Form and Function,* pp. 166–192. Cambridge: Cambridge University Press.

Peters, C. R., O'Brien, E. M. & Drummond, R. B. (1992). *Edible Wild Plants of Sub-saharan Africa.* Kew, UK: Royal Botanic Gardens.

Peters, G. (1986). Mixed herds of common and defassa waterbuck, *Kobus ellipsiprymnus* (Artiodactyla: Bovidae), in northern Kenya. *Bonner zoologische Beiträge* **37,** 183–193.

Pettars, S. W. (1991). *Regional Geology of Africa.* New York: Springer-Verlag.

Petter, F. (1973). Addition à la liste des rongeurs myomorphes de la république centrafricaine: *Zelotomys instans* Thomas 1915. *Mammalia* **37,** 683.

Petter, F. & Genest, H. (1970). Liste préliminaire des rongeurs myomorphes de république centrafricaine. Description de deux souris nouvelles: *Mus oubanguii* et *Mus goundae. Mammalia* **34,** 451–458.

Petter, F. (1967). Contribution à la faune du Congo (Brazaville). Mission A. Viliers et A. Descarpentries. *Bulletin de l'Institut Fondamental d'Afrique Noire* XXIX, A, 2, 815–820.

Petter, G. (1987). Small carnivores (Viverridae, Mustelidae, Canidae) from Laetoli. In M. D. Leakey & J. M. Harris, eds., *Laetoli: A Pliocene Site in Northern Tanzania,* pp. 194–234. Oxford: Clarendon Press.

Péwé, T. L. (1959). Sand-wedge polygons (tesselations) in the McMurdo Sound region, Antarctica—a progress report. *American Journal of Science* **257,** 545–552.

Pickford, M. (1983). Sequence and environments of the lower and middle Miocene hominoids of western Kenya. In R. L. Ciochon and R. S. Corruccini, eds., *New Interpretations of Ape and Human Ancestry,* pp. 421–439. New York: Plenum Press.

Pickford, M. (1987). The chronology of the Cercopithecoidea of East Africa. *Human Evolution* **2,** 1–17.

Pickford, M. (1990). Uplift of the roof of Africa and its bearing on the evolution of mankind. *Human Evolution* **5,** 1–20.

Pickford, M. (1993). Climate change, biogeography and *Theropithecus.* In N. G. Jablonski, ed., *Theropithecus—The Rise and Fall of a Primate Genus,* pp. 465–486. Cambridge: Cambridge University Press.

Pickford, M. & Mein, P. (1988). The discovery of fossiliferous Plio-Pleistocene cave fillings in Ngamiland, Botswana. *Comptes Rendus de l'Académie des Sciences de Paris Séries II* **307,** 1681–1686.

Pickford, M., Mein, P. & Senut, B. (1992). Primate bearing Plio-Pleistocene cave deposits of Humpata, Southern Angola. *Human Evolution* **7,** 17–33.

Pickford, M., Senut, B. & Hadoto, D. (1993). *Geology and palaeobiology of the Albertine Rift Valley, Uganda-Zaire.* Vol. 1, Geology. Occasional Paper 1993/24. Orleans, France: International Centre for Training and Exchanges in the Geosciences (CIFEG).

Pilbeam, D. (1984). Reflections on early human ancestors. *Journal of Anthropological Research* **40,** 14–22.

Pilbeam, D., Morgan, M., Barry, J. & Flynn, L. (1996). European MN units and the Siwalik faunal sequence of Pakistan. In R. L. Bernor, V. Fahlbusch & H.-W. Mittmann, eds., *The Evolution of Western Eurasian Neogene Mammal Faunas,* pp. 96–105. New York: Columbia University Press.

Pilbeam, D. R., Rose, M. D., Barry, J. C. & Shah, S. M. I. (1990). New *Sivapithecus* humeri in Pakistan and the relationship of *Sivapithecus* and *Pongo. Nature* **348,** 237–239.

Pilbeam, D. & Walker, A. (1968). Fossil monkeys from the Miocene of Napak, northeast Uganda. *Nature* **229,** 408–409.

Pimm, S. L., Rosenzweig, M. L. & Mitchell, W. (1985). Competition and food selection: field tests of a theory. *Ecology* **66,** 798–807.

Pisias, N. G. & Moore, T. C. Jr. (1981). The evolution of Pleistocene climate: a time series approach. *Earth and Planetary Science Letters* **52,** 450–458.

Platnick, N. I. & Nelson, G. (1978). A method of analysis for historical biogeography. *Systematic Zoology* **27,** 1–16.

Plummer, T. W. & Bishop, L. C. (1994). Hominid paleoecology at Olduvai Gorge, Tanzania as indicated by antelope remains. In J. S. Oliver, N. E. Sikes & K. M. Stewart, eds., *Early Hominid Behavioural Ecology Journal of Human Evolution* **27,** 47–75.

Plummer, T., Bishop, L. C., Ditchfield, P. & Hicks, J.

(1998). Research on late Pliocene Oldowan sites at Kanjera South, Kenya. *Journal of Human Evolution* **34,** in press.

Pocock, T. N. (1987). Plio-Pleistocene fossil mammalian microfauna of southern Africa. A preliminary report including description of two new fossil muroid genera (Mammalia, Rodentia). *Paleontologia Africana* **26(7),** 69–91.

Poincare, H. (1913). L'invention mathematique. *Bulletin Institute General Psychology* **8,** 175–187.

Pomel, A. (1897). Homme, singe, carnassiers, Equides, suilliens, ovides. Les Equides. *Carte geologie, Algerie, Paléontologie Monographies,* pp. 1–44.

Popp, J. (1978). Male baboons and evolutionary principles (Ph.D. diss.). Cambridge, MA: Harvard University.

Popper, K. (1978). Natural selection and the emergence of mind. *Dialectica* **32,** 339–355.

Post, D. (1978). Feeding and ranging behavior of the yellow baboon (*Papio cynocephalus*) (Ph.D. dissertation). New Haven, CT: Yale University.

Potts, R. (1988). *Early Hominid Activities at Olduvai Gorge.* New York: Aldine de Gruyter.

Potts, R. (1989). Olorgesailie: new excavations and findings in early and middle Pleistocene contexts, southern Kenya rift valley. *Journal of Human Evolution* **18,** 477–484.

Potts, R. (1994). Variables versus models of early Pleistocene hominid land use. *Journal of Human Evolution* **27,** 7–24.

Potts, R., Ditchfield, P., Hicks, J. & Deino, A. (1997). Paleoenvironments of late Miocene and early Pliocene strata of Kanam, western Kenya. *American Journal of Physical Anthropology, Supplement* **24,** 188–189.

Pratt, D. J., Greenway, P. J. & Gwynne, M. D. (1966). A classification of East African Rangeland with an appendix on terminology. *Journal of Applied Ecology* **3,** 369–382.

Pratt, D. J. & Gwynne, M. D., eds. (1977). *Rangeland Management and Ecology in East Africa.* Huntington, NY: Krieger.

Prell, W. L. (1985). Pliocene stable isotope and carbonate stratigraphy (Holes 572C and 573A): paleoceanographic data bearing on the question of Pliocene glaciation. In L. Mayer & F. Theyer, eds., *Initial Reports of the Deep Sea Drilling Project* no. 85, pp. 723–734. Washington, DC: U.S. Government Printing Office.

Prell, W. L. & Kutzbach, J. E. (1992). Sensitivity of the Indian monsoon to forcing parameters and implications for its evolution. *Nature* **360,** 647–652.

Prentice, M. L. & Denton, G. H. (1988). The deep-sea oxygen isotope record, the global ice sheet system and hominid evolution. In F. E. Grine, ed., *Evolutionary History of the "Robust" Australopithecines,* pp. 383–403. New York: Aldine de Gruyter.

Prentice, M. L. & Matthews, R. K. (1991). Tertiary ice sheet dynamics. The snow gun hypothesis. *Journal of Geophysical Research* **26,** 6811–6827.

Pringle, J. A. (1974). The distribution of mammals in Natal. Part 1. Primates, Hyracoidea, Lagomorpha (except *Lepus*), Pholidota and Tubulidentata. *Annals of the Natal Museum* **22,** 173–186.

Pringle, J. A. (1977). The distribution of mammals in Natal. Part 2. Carnivora. *Annals of the Natal Museum* **23,** 93–115.

Pritchard, J. M. (1979). *Landform and Landscape in Africa.* London: Edward Arnold.

Prost, J. H. (1980). Origin of bipedalism. *American Journal of Physical Anthropology* **52,** 175–189.

Provine, W. B. (1983). The development of Wright's theory of evolution: Systematics, adaptation, and drift. In M. Greene, ed., *Dimensions of Darwinism,* pp. 43–70. New York: Cambridge University Press.

Provine, W. B. (1986). *Sewall Wright and Evolutionary Biology.* Chicago: University of Chicago Press.

Provine, W. B. (1989). Founder effects and genetic revolutions in microevolution and speciation: an historical perspective. In V. L. Giddins, K. Y. Kaneshiro & W. W. Anderson, eds., *Genetics, Speciation, and the Founder Principle,* pp. 43–76. New York: Oxford University Press.

Provine, W. B. (1991). Mechanisms of speciation: a review. In S. Osawa & T. Honjo, eds., *Evolution of Life,* pp. 201–214. Tokyo: Springer-Verlag.

Pulliam, H. R. (1988). Sources, sinks, and population regulation. *The American Naturalist* **135(5),** 652–661.

Pyke, G. H. (1982). Local geographic distributions of bumblebees near Crested Butte, Colorado: competition and community structure. *Ecology* **63,** 555–573.

Qiu, Z., Huang, W. & Guo, Z. (1987). Chinese hipparionines from the Yushe Basin. *Palaeontologica Sinica Series C* **175(25),** 1–250.

Quade, J., Cater, J. M. L., Ojha, T. P., Adam, J. & Harrison, T. J. (1995). Late Miocene environmental change in Nepal and the northern Indian subcontinent: Stable isotopic evidence from paleosols. *Geological Society of America Bulletin* **107**(12), 1381–1397.

Quade, J., Cerlin, T. E. & Bowman, J. R. (1989). Development of Asian monsoon revealed by marked ecological shift during the latest Miocene in northern Pakistan. *Nature* **342,** 161–166.

Rahmstorf, S. (1994). Rapid climate transitions in a coupled ocean-atmosphere model. *Nature* **372,** 82–85.

Ramirez Rozzi, F. V. (1992). *Le développement dentaire des hominidés du Plio-Pléistocène de l'Omo, Ethiopie.* Paris: Thèse du Muséum National d'Histoire Naturelle.

Ramirez Rozzi, F. V. (1993a). Aspects de la chronologie du développement dentaire des hominidés plio-pléistocènes de l'Omo, Ethiopie. *Comptes Rendus de l'Académie des Sciences Paris* **316,** 1155–1162.

Ramirez Rozzi, F. V. (1993b). Le développement dentaire des hominidés plio-pléistocènes. *Bulletin et Mémoires de la Societé d'Antrhopologie de Paris,* n.s. **5,** 131–142.

Ramirez Rozzi, F. V. (1993c). Aspects du développement dentaire et leur contribution à la connaissance de la croissance chez les hominidés plio-pléistocènes. *Annales de la Fondation Fyssen* **8,** 59–67.

Ramirez Rozzi, F. V. (1993d). Teeth development in east African *Paranthropus. Journal of Human Evolution* **24,** 429–454.

Ramirez Rozzi, F. V. (1995). Time of formation in Plio-Pleistocene hominid teeth. In J. Moggi-Cecchi, ed., *Aspects of Dental Biology: Palaeontology, Anthropology, and Evolution,* pp. 217–238. Florence: International Institute for the Study of Man.

Ramirez Rozzi, F. V. (1996). Comment on the causes of thin enamel in Neandertals. *American Journal of Physical Anthropology* **99,** 625–626.

Ramirez Rozzi, F. V. (1997). Les Hominidés du Plio-Pléistocène de l'Omo, Ethiopie. Charactérisation et Modification au Cours du Temps de Leur Développement Dentaire¿ Partir de l'Étude de la Microstructure de L'émail. Paris: CNRS.

Ramirez Rozzi, F. V. (1998). Can enamel microstructure be used to establish the presence of different species of Plio-Pleistocene hominids from Omo, Ethiopia? *Journal of Human Evolution* **35,** 543–576.

Raunkaier, C. (1934). *The Life Form of Plants and Statistical Plant Geography.* Oxford: Clarendon Press.

Raup, D. M. (1966). Geometric analysis of shell coiling: general problems. *J. Paleontology* **40,** 1178–1190.

Raup, D. M. & Sepkoski, J. J. (1986). Periodic extinction of families and genera. *Science* **231,** 833–836.

Rautenbach, I. L. (1978). A numerical re-appraisal of the southern African biotic zones. *Bulletin of Carnegie Museum of Natural History* **6,** 175–187.

Rautenbach, I. L. (1982). *Mammals of the Transvaal.* Pretoria: Ecoplan Monograph No. 1.

Raymo, M. E. (1991). Geochemical evidence supporting T. C. Chamberlain's theory of glaciation. *Geology* **19,** 344–347.

Raymo, M. E. (1994). The Himalaya, organic carbon burial, and climate in the Miocene. *Paleoceanography* **9,** 339–404.

Raymo, M. E., Hodell, D. & Jansen, E. (1992). Response of deep-ocean circulation to initiation of Northern Hemisphere glaciation (3-2MA). *Paleoceanography* **7,** 645–672.

Raymo, M. E., Grant, B., Horowitz, M. & Rau, G. H. (1996). Mid-Pliocene warmth: stronger greenhouse and stronger conveyor. *Marine Micropaleontology* **27,** 313–326.

Raymo, M. E. & Ruddiman, W. F. (1992). Tectonic forcing of Late Cenozoic climate. *Nature* **359,** 117–122.

Raymo, M. E., Ruddiman, W. F., Backman, J., Clement, B. M. & Martinson, D. G. (1989). Late Pliocene variation in Northern Hemisphere ice sheets and North Atlantic deep water circulation. *Paleoceanography* **4,** 413–446.

Raymo, M. E., Ruddiman, E. F. & Froelich, P. N. (1988). Influence of late Cenozoic mountain building on ocean geochemical cycles. *Geology* **16,** 649–653.

Raymo, M. E., Ruddiman, W. F., Shackleton, N. J. & Oppo, D. W. (1990). Evolution of Atlantic-Pacific $\delta^{13}C$ gradients over the last 2.5 m.y: *Earth and Planetary Science Letters* **97,** 353–368.

Rayner, R. G. & Masters, C. J. (1995). A good loser is still a loser: competition and the South African fossil record. *South African Journal of Science* **91,** 184–189.

Rayner, R. J., Moon, B. P. & Masters, J. C. (1993). The Makapansgat Australopithecine environment. *Journal of Human Evolution* **24,** 219–231.

Reader, J. (1988). *Missing Links,* 2nd ed. London: Pelican.

Reed, K. E. (1996). The paleoecology of Makapansgat and other African Pliocene hominid sites (Ph.D. dissertation). Stony Brook: State University of New York at Stony Brook.

Reed, K. E. (1997). Early hominid evolution and ecological change through the African Plio-Pleistocene. *Journal of Human Evolution* **32,** 289–322.

Reid, E. M. & Chandler, M. E. J. (1933). *The London Clay Flora.* London: British Museum.

Rensch, B. (1947). *Neuere Probleme der Abstammungslehre. Die transspezifische Evolution.* Stuttgart: Ferdinand Enke.

Rensch, B. (1959). *Evolution Above the Species Level.* New York: Columbia University Press.

Retallack, G. J. (1992). Middle Miocene fossil plants from Fort Ternan (Kenya) and evolution of African grasslands. *Paleobiology* **18,** 383–400.

Retallack, G. J., Dugas, D. P. & Bestland, E. A. (1990). Fossil soils and grasses of a middle

Miocene East African grassland. *Science* **247,** 1325–1328.
Reumer, J. W. F. (1995). The effect of paleoclimate on the evolution of the Soricidae (Mammalia, Insectivora). In E. S. Vrba, G. H. Denton, T. C. Partridge & L. H. Burckle, eds., *Paleoclimate and Evolution,* pp. 133–147. New Haven, CT: Yale University Press.
Revel, M., Sinko, J. A., Grousset, F. E. & Biscage, P. (1996). Sr and Nd isotopes as tracers of North Atlantic lithic particles: Paleoclimatic implications. *Paleoceanography* **11,** 95–113.
Richards, P. W. (1952). *The Tropical Rain Forest.* Cambridge: Cambridge University Press.
Rieppel, O. (1994). Species and history. In R. W. Scotland, D. J. Siebert & D. M. Williams, eds., *Models in Phylogeny Reconstruction,* pp. 31–50. Oxford: Clarendon Press.
Rightmire, G. P. (1993). Did climatic change influence human evolution? *Evolutionary Anthropology* **2,** 43–45.
Rind, D. & Chandler (1991). Increased ocean heat transports and warmer climate. *Journal Geophysical Research* **96,** 7437–7461.
Risnes, S. (1985). A scanning electron microscopy study of the three dimensional extent of Retzius lines in human dental enamel. *Scandinavian Journal of Dental Research* **93,** 145–152.
Risnes, S. (1986). Enamel apposition rate and prism periodicity in human teeth. *Scandinavian Journal of Dental Research* **94,** 394–404.
Ritchie, J. C. & Haynes, C. V. (1987). Holocene vegetation zonation in the eastern Sahara. *Nature* **330,** 645–647.
Roberts, D. F. (1978). *Climate and Human Variability.* Menlo Park, CA: Cumming.
Roberts, N., Taieb, M., Barker, P., Damnati, B., Icole, M. & Williamson, D. (1993). Timing of the Younger Dryas event in East Africa from lake-level changes. *Nature* **366,** 146–148.
Rodgers, W. A., Owen, C. F. & Homewood, K. M. (1982). Biogeography of East African forest mammals. *Journal of Biogeography* **9,** 41–54.
Rodman, P. S. & McHenry, H. M. (1980). Bioenergetics and origins of bipedalism. *American Journal of Physical Anthropology* **52,** 103–106.
Rogers, M. E., Maisels, F., Williamson, E. A., Tutin, C. E. G. & Fernandez, M. (1992). Nutritional aspects of gorilla food choice in the Lopé Reserve, Gabon. In N. Itoigawa, Y. Sugiyama, G. P. Sackett & R. K. R. Thompson, eds., *Topics in Primatology,* vol. 2. *Behavior, Ecology, and Conservation,* pp. 255–266. Tokyo: University of Tokyo Press.
Rogers, M. J., Harris, J. W. K. & Feibel, C. S. (1994). Changing patterns of land use by Plio-Pleistocene hominids in the Lake Turkana Basin. *Journal of Human Evolution* **27,** 139–158.
Rogl, F. & Daxner-Hock, G. (1996). Late Miocene Paratethys correlations. In R. L. Bernor, V. Fahlbusch & H.-W. Mittmann, eds., *The Evolution of Western Eurasian Neogene Mammal Faunas.* New York: Columbia University Press.
Rogl, F., Zapfe, H., Bernor, R. L., Brzobohaty, R., Daxner-Hock, G., Draxler, I., Fejfar, O., Gaudant, J., Herrmann, P., Rabeder, G., Schultz, O. & Zetter, R. (1993). Die Primatenfundstelle Gotzendorf an der Leitha (Obermiozan des Wiener Beckens, Niederosterreich). *Jahrbuch Geologie* **136,** 503–526.
Roosevelt, T. & Heller, E. (1915). *Life Histories of African Game Animals.* London: John Murray.
Root, R. (1967). The niche exploitation pattern of the blue-grey gnatcatcher. *Ecological Monographs* **37,** 317–350.
Rose, J. C. (1979). Morphological variations of enamel prisms within abnormal striae of Retzius. *Human Biology* **51,** 139–151.
Rose, M. D. (1983). Miocene hominoid postcranial morphology: monkey-like, ape-like, or both? In R. L. Ciochon and R. S. Corruccini, eds., *New Interpretations of Ape and Human Ancestry,* pp. 405–417. New York: Plenum.
Rose, M. D. (1984). Food acquisition and the evolution of positional behaviour: the case of bipedalism. In D. J. Chivers, B. A. Wood & A. Bilsborough, eds., *Food Acquisition and Processing in Primates,* pp. 509–524. New York: Plenum.
Rose, M. D. (1993). Locomotor anatomy of Miocene hominoids. In D. L. Gebo, ed., *Postcranial Adaptation in Nonhuman Primates,* pp. 252–272. DeKalb: Northern Illinois University Press.
Rosen, B. R. (1988). Biogeographic patterns: a perceptual overview. In A. A. Myers & P. S. Giller, eds., *Analytical Biogeography,* pp. 23–55. London: Chapman & Hall.
Rosen, D. E. (1978). Vicariant patterns and historical explanations in biogeography. *Systematic Zoology* **27,** 159–188.
Rosen, D. E. (1979). Fishes from the uplands and intermontane basins of Guatemala: revisionary studies and comparative geography. *Bulletin of the American Museum of Natural History* **162,** 267–376.
Rosenzweig, M. L. (1966). Community structure in sympatric Carnivora. *Journal of Mammalogy* **47,** 602–612.
Rosenzweig, M. L. (1973). Evolution of the predator isocline. *Evolution* **27,** 89–94.
Rosenzweig, M. L. (1987a). Community organization from the point of view of habitat selectors. In J. H. R. Gee & P. S. Giller, eds., *Organization of Communities Past and Present,* pp. 469–490. Oxford: Blackwell Scientific.
Rosenzweig, M. L. (1987b). Habitat selection as a

source of biological diversity. *Evolutionary Ecology* **1,** 315–330.

Rosenzweig, M. L. (1989). Habitat selection, community organization, and small mammal studies. In D. W. Morris, B. J. Fox, Z. Abramsky & M. Willig, eds., *Patterns in the Structure of Mammalian Communities,* pp. 5–21. Lubbock, TX: Texas Tech University Press.

Rosenzweig, M. L. (1991). Habitat selection and population interactions: the search for mechanism. *American Naturalist* **137,** 51–55.

Rosenzweig, M. L. (1995). *Species Diversity in Time and Space.* Cambridge: Cambridge University Press.

Rosenzweig, M. L., Brown, J. S. & Vincent, T. L. (1987). Red Queen and ESS: the coevolution of evolutionary rates. *Evolutionary Ecology* **1,** 59–94.

Rosenzweig, M. L. & McCord, R. D. (1991). Incumbent replacement: evidence for long-term evolutionary progress. *Paleobiology* **17,** 23–27.

Rosevear, D. (1969). *The Rodents of West Africa.* London: British Museum National History Publications.

Ross, C. F. & Henneberg, M. (1995). Basicranial flexion, relative brain size, and facial kyphosis in *Homo sapiens* and some fossil hominids. *American Journal of Physical Anthropology* **98,** 575–594.

Ross, C. F. & Ravosa, M. J. (1993). Basicranial flexion, relative brain size, and facial kyphosis in nonhuman primates. *American Journal of Physical Anthropology* **91,** 305–324.

Rowe-Rowe, D. T. (1978). The small carnivores of Natal. *The Lammergeyer* **25,** 1–48.

Rowe-Rowe, D. T. (1991). *The Ungulates of Natal.* Pietermaritzburg: Natal Parks Board.

Ruddiman, W. F. & Kutzbach, J. E. (1989). Forcing of late Cenozoic Northern Hemisphere climate by plateau uplift in Southern Asia and the American West. *Journal of Geophysical Research* **94,** 18, 409–18, 427.

Ruddiman, W. F. & Kutzbach, J. E. (1991). Plateau uplift and climatic change. *Scientific American* **264,** 66–75.

Ruddiman, W. F. & Raymo, M. E. (1988). Northern hemisphere climate régimes during the past 3 Ma: possible tectonic connections. *Philosophical Transactions of the Royal Society of London* B **318,** 411–430.

Ruddiman, W. F., Raymo, M. E., Martinson, D. G., Clement, B. M. & Backman, J. (1989). Pleistocene evolution: Northern Hemisphere ice sheets and North Atlantic Ocean. *Paleoceanography* **4,** 353–412.

Ruddiman, W. F., Raymo, M. E. & McIntyre, A. (1986a). Matuyama 41,000-year cycles: North Atlantic Ocean and Northern Hemisphere ice sheets. *Earth and Planetary Science Letters* **80,** 117–129.

Ruddiman, W. F., Shackleton, N. J. & McIntyre, A. (1986b). North Atlantic sea-surface temperatures for the past 61 million years. In C. P. Summerhayes & N. J. Shackleton, eds., *North Atlantic Paleoceanography.* Special Publication of the Geological Society, 21, pp. 155–173. London: Geological Society of London.

Ruff, C. B. & Walker, A. (1993). Body size and body shape. In A. Walker & R. E. Leakey, eds., *The Nariokotome H. ergaster Skeleton,* pp. 234–265. Berlin: Springer Verlag.

Ruff, C. B. (1991). Climate and body shape in hominid evolution. *Journal of Human Evolution* **21,** 81–105.

Ruff, C. B. (1993). Climatic adaptation and hominid evolution: the thermoregularity imperative. *Evolutionary Anthropology* **2(2),** 53–60.

Ruff, C. B. (1994). Morphological adaptation to climate in modern and fossil hominids. *Yearbook of Physical Anthropology* **37,** 65–107.

Rusciano, D. & Burger, M. M. (1992). Why do cancer cells metastasize into particular organs? *BioEssays* **14,** 185–194.

Ruse, M. (1989). Is the theory of punctuated equilibria a new paradigm? *Journal of Social Biological Structure* **12,** 195–212.

Rutherford, M. C. & Westfall, R. H. (1986). *Biomes of Southern Africa—An Objective Categorization.* Memoirs of the Botanical Survey of South Africa no. 54. Pretoria: Botanical Research Institute.

Ruxton, A. E. & Schwarz, E. (1929). On hybrid hartebeests and on the distribution of the *Alcelaphus buselaphus* group. *Proceedings of the Zoological Society of London for 1929,* pp. 567–583.

Ryan, A. S. & Johanson, D. C. (1989). Anterior dental microwear in Australopithecus afarensis: comparisons with human and nonhuman primates. *Journal of Human Evolution* **18,** 235–268.

Sabatier, M. (1978). *Les rongeurs des sites à Hominidés de Hadar et Melka Kunturé (Ethiopie). Thès 3è cycle U.S.T.L.* Thesis, University of Montpellier.

Sabatier, M. (1982). Les rongeurs du site Pliocène à hominidés de Hadar (Ethiopie). *Paleovertebrata* **12(1),** 1–56.

Sakai, A., Paton, D. M. & Wardle, P. (1981). Freezing resistance of trees of the south temperate zone, especially subalpine species of Australia. *Ecology* **62,** 563–570.

Salthe, S. N. (1985). *Evolving Hierarchical Systems.* New York: Columbia University Press.

Salthe, S. N. (1989). [Commentary]. In M. K. Hecht, ed., *Evolutionary Biology at the Crossroads,* pp. 175–176. Flushing, NY: Queens College Press.

Salthe, S. N. (1993). *Development and Evolution. Complexity and Change in Biology.* Cambridge, MA: MIT Press.

Saltzman, B. & Maasch, K. A. (1990). A first-order global model of late Cenozoic climatic change. *Transactions of the Royal Society of Edinburgh, Earth Sciences* **81,** 315–325.

Sarich, V. M. (1971). A molecular approach to the question of human origins. In P. Dolhinow & V. M. Sarich, eds., *Background for Man,* pp. 60–81. Boston: Little, Brown.

Sarich, V. M. & Wilson, A. C. (1967). Immunological time scale for hominid evolution. *Science* **158,** 1200–1203.

Sarna-Wojcicki, A. M., Meyers, C. E., Roth, P. H. & Brown, F. H. (1985). Ages of tuff beds at East African early hominid sites and sediments in the Gulf of Aden. *Nature* **313,** 306–308.

Sarnthein, M., Winn, K., Jung, S. J. A., Duplessy, J.-C., Labeyrie, L., Erienkeuser, H. & Ganssen, G. (1994). Changes in east Atlantic deepwater circulation over the late 30,000 years: eight time slice reconstructions. *Paleoceanography* **9,** 209–267.

SAS Institute, Inc. (1985). *SAS User's Guide: Statistics, Version 5 Edition.* Cary, NC: SAS Institute, Inc.

Savage, D. E. & Russell, D. E. (1983). *Mammalian Paleofaunas of the World.* London: Addison-Wesley.

Savage, R. J. G. (1978). Carnivora. In V. J. Maglio & H. B. S. Cooke, eds., *Evolution of African Mammals,* pp. 249–267. Cambridge, MA: Harvard University Press.

Savage, R. J. G. & Hamilton, R. (1973). Introduction to the Miocene mammal faunas of Gebel Zelten, Libya. *Bulletin of the British Museum of Natural History, Geology Series* **22,** 483–511.

Savin, S. M., Douglas, R. G. & Stehli, F. G. (1975). Tertiary marine paleotemperatures. *Geol. Soc. Am. Bull.* **86,** 1499–1510.

Schaffer, W. M. & Rosenzweig, M. L. (1978). Homage to the Red Queen I, Coevolution of predators and their victims. *Theoretical Population Biology* **9,** 1395–157.

Schliewen, U. K., Tautz, D. & Paabo, S. (1994). Sympatric speciation suggested by monophyly of crater lake cichlids. *Nature* **368,** 629–632.

Schlosser, P., Bonisch, G., Rheim, M. & Bayer, R. (1991). Reduction of deepwater formation in the Greenland Sea during the 1980's: evidence from tracer data. *Science* **251,** 1054–1056.

Schlüchter, C. (1988). The deglaciation of the Swiss Alps: a paleoclimatic event with chronological problems. *Bulletin de l'Association francaise pour l'etude du Quaternaire* **2/3,** 141–145.

Schmid, P. (1991). The trunk of the australopithecines. In Y. Coppens & B. Senut, eds., *Origine(s) de la bipedie chez les Hominidés,* pp. 225–234. Paris: Editions du CNRS.

Schmidt-Nielsen, K. (1984). *Scaling—Why is Animal Size So Important?* Cambridge: Cambridge University Press.

Schmitt, R. J. (1982). Consequences of dissimilar defenses against predation in a subtidal marine community. *Ecology* **63,** 1588–1601.

Schneider, S. H., Thompson, S. L. & Barron, E. J. (1985). Mid-Cretaceous continental surface temperatures: are high CO_2 concentrations needed to simulate above freezing winter conditions? In T. Sundquist & W. S. Broecker, eds., *The Carbon Cycle and Atmospheric CO_2. Natural Variations Archean to Present.* American Geophysical Union Monograph 32, pp. 554–560. Washington, DC: American Geophysical Union.

Schnitker, D. (1980). North Atlantic oceanography as possible cause of Antarctic glaciation and eutrophication. *Nature* **284,** 615–616.

Scholes, R. J. & Walker, B. H. (1993). *An African Savanna. Synthesis of the Nylsvley Study.* Cambridge: Cambridge University Press.

Schopf, T. J. M. (1981a). Evidence from findings of molecular biology with regard to the rapidity of genomic change: implications for species durations. In K. J. Niklas, ed., *Paleobotany, Paleoecology and Evolution,* pp. 135–192. New York: Praeger.

Schopf, T. J. M. (1981b). Punctuated equilibrium and evolutionary stasis. *Paleobiology* **7,** 156–166.

Schopf, T. J. M. (1982a). A critical assessment of punctuated equilibria. I. Duration of taxa. *Evolution* **36,** 1144–1157.

Schopf, T. J. M. (1982b). Historical approaches versus equilibrium approaches to evolutionary data. In M. H. Nitecki, ed., *Biochemical Aspects of Evolutionary Biology,* pp. 1–8. Chicago: University of Chicago Press.

Schopf, T. J. M. (1984). Rates of evolution and the notion of "living fossils". *Annual Reviews of Earth and Planetary Sciences* **12,** 245–292.

Schour, I. & Poncher, H. C. (1937). The rate of apposition of human enamel and dentin as measured by the effects of acute fluorosis. *American Association for Diseases of Children* **54,** 757–776.

Schouteden, H. (1943). Catalogue des mammifères du Congo belge et du Ruanda-Urundi. *Revue de Zoologie et de Botanique Africaine* XXXVII, **1,** 104–125.

Schouteden, H. (1944–1946). De Zoogdieren van Belgisch-Congo en van Ruanda-Urundi. *Annales du Musée du Congo Belge, C. - Zoologie, Serie II* **3(1-3),** 1–176.

Schrenk, F., Bromage, T. G., Betzler, C. G., Ring, U. & Juwayeyi, Y. M. (1993). Oldest Homo and Pliocene biogeography of the Malawi Rift. *Nature* **365,** 833–836.

Schrenk, F., Bromage, T. G., Gorthner, A. & Sandrock, O. (1995). Paleoecology of the Malawi Rift. Vertebrate and invertebrate faunal contexts of the Chiwondo beds, northern

Malawi. *Journal of Human Evolution* **28,** 59–70.

Schwarz, E. (1929). On the local races and distribution of the black and white colobus monkeys. *Proceedings of the Zoological Society of London for 1929,* pp. 585–598.

Schwarzacher, W. (1993). *Cyclostratigraphy and the Milankovitch Theory.* Developments in Sedimentology 52. New York: Elsevier.

Scott, K. (1985). Allometric trends and locomotor adaptations in the Bovidae. *Bulletin of the American Museum of Natural History* **179,** 197–288.

Scott, K. M. (1987). Allometry and habitat-related adaptations in the postcranial skeleton of Cervidae. In C. M. Wemmer, ed., *Biology and Management of the Cervidae,* pp. 65–80. Washington, DC: Smithsonian Institution Press.

Scott, L. (1995). Pollen evidence for vegetational and climatic change in southern Africa during the Neogene and Pleistocene. In E. S. Vrba, G. H. Denton, T. C. Partridge & L. H. Burckle, eds., *Paleoclimate and Evolution with Emphasis on Human Origins,* pp. 65–76. New Haven, CT: Yale University Press.

Sefve, I. (1927). Die Hipparioninen Nord-Chinas. *Palaeontologic Sinica, Series C* **4(2),** 1–54.

Sen, S. (1996). Present state of magnetostratigraphic studies in the continental Neogene of Europe and Anatolia. In R. L. Bernor, V. Fahlbusch & H.-W. Mittmann, eds., *The Evolution of Western Eurasian Neogene Mammal Faunas,* pp. 56–63. New York: Columbia University Press.

Sen, S., Sondaar, P. Y. & Staesche, U. (1978). The biostratigraphical applications of the genus *Hipparion* with special references to the Turkish representatives. *Proceedings, Koninklijke Nederlandse Akademie van Wetenschappen Series B* **96,** 151–216.

Senut, B. (1996). Pliocene hominid systematics and phylogeny. *South African Journal of Science* **92,** 165–166.

Senut, B., Pickford, M., Mein, P., Conroy, G. & Van Couvering, J. (1992). Discovery of 12 new Late Cenozoic fossiliferous sites in palaeokarsts of the Otavi Mountains, Namibia. *Comptes Rendus de l'Académie des Sciences de Paris Séries II* **314,** 727–733.

Senut, B. & Tardieu, C. (1985). Functional aspects of Plio-Pleistocene hominid limb bones: implications for taxonomy and phylogeny. In E. Delson, ed., *Ancestors: The Hard Evidence,* pp. 193–201. New York: Alan R. Liss.

Setzer, H. W. (1956). Mammals of the Anglo-Egyptian Sudan. *Proceedings of the United States National Museum* **106,** 3377, 447–586.

Shackleton, N. J. (1987). Oxygen isotopes, ice volume and sea level. *Quaternary Science Review* **6,** 183–190.

Shackleton, N. J. (1995). New data on the evolution of Pliocene climatic variability. In E. Vrba, ed., *Palaeoclimate and Evolution with Emphasis on Human Origins,* pp. 242–248. New Haven, CT: Yale University Press.

Shackleton, N. J., Backman, J., Zimmerman, H., Kent, D. V., Hall, M. A., Roberts, D. G., Schnitker, D., Baldauf, J. G., Desprairies, A., Homrighausen, R., Huddlestun, P., Keene, J. B., Kaltenback, A. J., Krumsiek, K. A. O., Morton, A. C., Murray, J. W. & Westberg-Smith, J. (1984). Oxygen-isotope calibration of the onset of ice-rafting and history of glaciation in the North Atlantic region. *Nature* **307,** 620–623.

Shackleton, N. J., Berger, A. & Peltier, W. R. (1990). An alternative astronomical calibration of the lower Pleistocene timescale based on ODP site 677. *Transactions of the Royal Society of Edinburgh, Earth Sciences* **81,** 251–261.

Shackleton, N. J. & Boersma, A. (1981). The climate of the Eocene ocean. *Journal of the Geological Society of London* **138,** 153–157.

Shackleton, N. J., Crowhurst, S., Hagelberg, T., Pisias, N. G. & Schneider, D. A. (1995b). A new late Neogene time scale: application to Leg 138 sites. *Proceeding of the Ocean Drilling Program, Scientific Results* **138,** 73–101.

Shackleton, N. J., Duplessy, J.-C., Arnold, M., Maurice, P., Hall, M. A. & Cartlidge, J. (1988). Radiocarbon age of last glacial Pacific deep water. *Nature* **335,** 708–711.

Shackleton, N. J. & Hall, M. A. (1989). Stable isotope history of the Pleistocene at ODP site 677. *Proceedings of the Ocean Drilling Program Scientific Results* **111,** 295–316.

Shackleton, N. J., Hall, M. A. & Pate, D. (1995a). Pliocene stable isotope stratigraphy of site 846. *Paleoceanography* **138,** 337–355.

Shackleton, N. J. & Kennett, J. P. (1975a). Paleotemperature history of Cenozoic and initiation of Antarctic glaciation: oxygen and carbon isotopic analyses in DSDP sites 277, 279, and 281. *Initial Reports Deep Sea Drilling Project* **29,** 743–755.

Shackleton, N. J. & Kennett, J. P. (1975b). Late Cenozoic oxygen and carbon isotope changes at DSDP site 284: implications for glacial history of the Northern Hemisphere and Antarctica. In J. P. Kennett, R. E. Houtz et al., eds., *Initial Reports Deep Sea Drilling Project* **29,** 801–807.

Shackleton, N. J. & Opdyke, N. D. (1973). Oxygen isotope and paleomagnetic stratigraphy of equatorial Pacific core V28-238: Oxygen isotope temperatures and ice volumes on a 10^5 and 10^6 year scale. *Quaternary Research* **3,** 39–55.

Shackleton, N. J. & Opdyke, N. D. (1976). Oxygen isotope and paleomagnetic stratigraphy of Pacific core V28-239, Late Pliocene to latest Pleistocene. *Geological Society of America Memoir* **145,** 449–464.

Sharman, M. (1980). Feeding, ranging and social organization of the Guinea baboon, *Papio papio* (Ph.D. dissertation). St. Andrews: University of St. Andrews.

Shellis, R. P. (1984). Variations in growth of the enamel crown in human teeth and a possible relationship between growth and enamel structure. *Archives of Oral Biology* **29,** 697–705.

Shipman, P. (1986). Paleoecology of Fort Ternan reconsidered. *Journal of Human Evolution* **15,** 193–204.

Short, D. A. & Mengel, J. G. (1986). Tropical climatic phase lags and Earth's precession cycle. *Nature* **323,** 48–50.

Short, R. V. (1975). The evolution of the horse. *Journal of Reproduction and Fertility (Supplement)* **23,** 1–6.

Shortridge, G. C. (1934). *The Mammals of South West Africa. A Biological Account of the Forms Occurring in that Region.* London: William Heinemann Limited.

Sikes, N. E. (1995a). Early Hominid habitat preferences in East Africa: paleosol carbon isotopic evidence. In J. S. Oliver, N. E. Sikes & K. M. Stewart, eds., *Early Hominid Behavioral Ecology Journal of Human Evolution* **27,** 25–45.

Sikes, N. E. (1995). Early hominid habitat preferences in East Africa: stable isotopic evidence from paleosols (Ph.D. dissertation). Urbana: University of Illinois-Urbana.

Sikes, N. E. (1996). Hominid habitat preferences in lower Bed II: Stable isotope evidence from paleosols. "Four Million Years of Hominid Evolution in Africa: in Honour of Dr. Mary D. Leakey's Outstanding Contribution in Palaeoanthropology," *Kaupia* 16, 231–238.

Sikes, N. E., Potts, R. & Behrensmeyer, A. K. (1997). Isotopic study of Pleistocene paleosols from the Olorgesailie Formation, southern Kenya rift. *Journal of Human Evolution* **32,** A20–A21.

Sillen, A., Hall, G. & Armstrong, R. (1995). Strontium calcium ratios (Sr/Ca) and strontium isotopic ratios (^{87}Sr/^{86}Sr) of *Australopithecus robustus* and *Homo sp.* from Swartkrans. *Journal of Human Evolution* **28,** 277–286.

Simon, H. A. (1962). The architecture of complexity. *Proceedings of the American Philosophical Society* **106,** 467–482.

Simons, E. L. (1994). New monkey (Prohylobates) and ape humerus from the Miocene Moghara Formation of northern Egypt. *Proceedings of the XIV International Primatology Congress,* pp. 247–253.

Simpson, G. G. (1944). *Tempo and Mode in Evolution.* New York: Columbia University Press.

Simpson, G. G. (1951). The species concept. *Evolution* **5,** 285–298.

Simpson, G. G. (1953). *The Major Features of Evolution.* New York: Columbia University Press.

Simpson, G. G. (1960). The history of life. In S. Tax, ed., *The Evolution of Life: Its Origin, History, and Future,* pp. 117–180. Chicago: University of Chicago Press.

Simpson, G. G. (1961). *Principles of Animal Taxonomy.* New York: Columbia University Press.

Sinclair, A. R. E. (1979). The eruption of the ruminants. In A. R. E. Sinclair & M. Norton-Griffiths, eds., *Serengeti. Dynamics of an Ecosystem,* pp. 82–103. Chicago: University of Chicago Press.

Sinclair, A. R. E. (1985). Does interspecific competition or predation shape the African ungulate community? *Journal of Animal Ecology* **54,** 899–918.

Sinclair, A. R. E. (1995). Serengeti past and present. In A. R. E. Sinclair & P. Arcese, eds., *Serengeti II: Dynamics, Management and Conservation of an Ecosystem,* pp. 3–30. Chicago: University of Chicago Press.

Skelton, R. R. & McHenry, H. M. (1992). Evolutionary relationships among early hominids. *Journal of Human Evolution* **23,** 309–349.

Skelton, R. R., McHenry, H. M. & Drawhorn, G. M. (1986). Phylogenetic analysis of early hominids. *Current Anthropology* **27,** 21–43.

Skinner, J. D. & Smithers, R. H. N. (1990). *The Mammals of the Southern African Subregion.* Pretoria: University of Pretoria.

Sloan, L. C., Crowley, T. J. & Pollard, D. (1996). Modeling of middle Pliocene climate with NCAR GENESIS general circulation model. *Marine Micropaleontology* **27,** 51–61.

Slobodkin, L. B., Dunn, K. & Bossert, P. (1987). Evolutionary constraints and symbiosis in hydra. In P. Calow, ed., *Evolutionary Physiological Ecology,* pp. 151–167. London: Cambridge University Press.

Smart, C. (1976). The Lothagam 1 fauna: its phylogenetic, ecological, and biogeographic significance. In Y. Coppens, F. C. Howell, G. L. Isaac & R. E. F. Leakey, eds., *Early Man and Environments in the Lake Rudolph Basin,* pp. 361–369. Chicago: University of Chicago Press.

Smith, B. N. (1972). Natural abundance of the stable isotopes of carbon in biological systems. *BioScience* **22,** 226–231.

Smith, R. J. (1996). Biology and body size in human evolution: statistical inference misapplied. *Current Anthropology* **37,** 451–482.

Smith, B. N. & Epstein, S. (1971). Two categories of ^{13}C/^{12}C ratios for higher plants. *Plant Physiology* **47,** 380–384.

Smith, F. & Spencer, F. (1982). *The History of*

American Physical Anthropology. New York: Academic Press.

Smith, F. D. M., May, R. M. & Harvey, P. H. (1994). Geographical range of Australian mammals. *Journal of Animal Ecology* **63,** 441–450.

Smithers, R. H. N. (1971). The mammals of Botswana. *Museum Memoir, The Trustees of the National Museums and Monuments of Rhodesia* **4,** 1–340.

Smithers, R. H. N. & Lobao Tello, J. L. P. (1976). Check list and atlas of the mammals of Mocambique. *Museum Memoir, The Trustees of the National Museums and Monuments of Rhodesia* **8,** 1–184.

Smithers, R. H. N. & Wilson, V. J. (1979). Check list and atlas of the mammals of Zimbabwe Rhodesia. *Museum Memoir, The Trustees of the National Museums and Monuments, Zimbabwe* **9,** 1–193.

Snow, D. W., ed. (1978). *An Atlas of Speciation in African Non-Passerine Birds.* London: Trustees of the British Museum (Natural History).

Sober, E. (1984). *The Nature of Selection. Evolutionary Theory in Philosophical Focus.* Cambridge, MA: MIT Press.

Sober, E. (1985). Darwin on natural selection: a philosophical perspective. In D. Kohn, ed., *The Darwinian Heritage,* pp. 867–899. Princeton, NJ: Princeton University Press.

Sober, E. (1993). *Philosophy of Biology.* Oxford: Oxford University Press.

Solounias, N. & Dawson-Saunders, B. (1988). Dietary adaptations and paleoecology of the late Miocene ruminants from Pikermi and Samos in Greece. *Palaeogeography, Palaeoclimatology, Palaeoecology* **65,** 149–172.

Solounias, N., Walker, A. & Teaford, M. (1988). Interpreting the diet of extinct ruminants: the case of a non-browsing giraffid. *Paleobiology* **14,** 287–300.

Sondaar, P. Y. (1968). The osteology of the manus of fossil and recent Equidae with special reference to phylogeny and function. *Proceedings, Koninklijke Nederlandse Akademie van Wetenschappen Series B* **25,** 1–76.

Sondaar, P. Y. & Eisenmann, V. (1995). The vertebrate locality Maramena (Macedonia, Greece) at the Turolian-Ruscinian boundary (Neogene). 12. The hipparions (Equidae, Perissodactyla, Mammalia). *Munchner Geowissenshaften Abhanlungen* **28,** 137–142.

Spencer, L. M. (1994). Dietary adaptations of Plio-Pleistocene Bovidae: implications for habitat reconstruction [abstract]. *American Journal of Physical Anthropology Supplement* **18,** 184.

Spoor, F., Wood, B. A. & Zonnefeld, F. (1994). Implications of early hominid labyrinthine morphology for evolution of human bipedal locomotion. *Nature* **369,** 645–648.

Spoor, F., Wood, B. A. & Zonnefeld, F. (1996). Evidence for a link between human semicircular canal size and bipedal behaviour. *Journal of Human Evolution* **30,** 183–187.

Stanley, S. M. (1979). *Macroevolution: Pattern and Process.* San Francisco: W. H. Freeman.

Stanley, S. M. (1988). Paleozoic mass extinctions: shared patterns suggest global cooling as a common cause. *American Journal of Science* **288,** 334–352.

Stebbins, L. (1950). *Variation and Evolution in Plants.* New York: Columbia University Press.

Stein, R. & Robert, C. (1985). Siliciclastic sediments as Sites 588, 590, and 591: Neogene and Paleogene evolution in the southwest Pacific and Australian climate. In J. P. Kennett & C. C. von der Borch, eds., *Initial Reports of the Deep Sea Drilling Project* no. 90, pp. 1437–1454. Washington, DC: U.S. Government Printing Office.

Steininger, F. F., Berggren, W. A., Kent, D. V., Bernor, R. L., Sen, S. & Agusti, J. (1996). Circum-Mediterranean Neogene (Miocene and Pliocene) marine-continental chronologic correlations of European mammal units. In R. L. Bernor, V. Fahlbusch & H.-W. Mittmann, eds., *The Evolution of Western Eurasian Neogene Mammal Faunas,* pp. 7–46. New York: Columbia University Press.

Steininger, F. F., Bernor, R. L. & Fahlbusch, V. (1989). European Neogene marine/continental chronologic correlations. In E. H. Lindsay, V. Fahlbusch & P. Mein, eds., *European Neogene Mammal Chronology,* pp. 15–46. New York: Plenum.

Stern, J. T. & Susman, R. L. (1983). The locomotor anatomy of *Australopithecus afarensis. American Journal of Physical Anthropology* **60,** 279–317.

Stevens, G. C. (1989). The latitudinal gradient in geographical range: how so many species coexist in the tropics. *American Naturalist* **133(2),** 240–256.

Stewart, D. R. M. & Stewart, J. (1963). The distribution of some large mammals in Kenya. *Journal of the East African Natural History Society* **24(3),** 1–52.

Stocker, T. F. & Wright, D. G. (1991). Rapid transitions of the deep ocean circulation induced by changes in surface water fluxes. *Nature* **351,** 729–732.

Stone, P. H. (1978). Constraints on dynamical transports of energy on a spherical planet. *Dynamics of Oceans and Atmosphere* **2,** 123–139.

Stott, K. (1959). Giraffe intergradation in Kenya. *Journal of Mammalogy* **40,** 251.

Stroeven, A. P., Prentice, M. L. & Kleman, J. (1996). On marine microfossil transport and pathways in Antarctica during the late Neogene: evidence from the Sirius Group at Mount Fleming. *Geology* **24,** 727–730.

Stuart, C. & Stuart, T. (1988). *Field Guide to the Mammals of Southern Africa.* London: New Holland.

Stuart, C. T. (1981). Notes on the mammalian carnivores of the Cape Province, South Africa. *Bontebok* **1,** 1–58.

Stute, M., Forster, M., Frischkorn, H., Serejo, A., Clarak, J. F., Schlosser, P., Broecker, W. S. & Bonani, G. (1995). Cooling of tropical Brazil (5°C) during the last glacial maximum. *Science* **269,** 379–383.

Suc, J. P. & Zagwijn, W. H. (1983). Plio-Pleistocene correlations between the northwestern Mediterranean region and northwestern Europe according to recent biostratigraphic and palaeoclimatic data. *Boreas* **12,** 153–166.

Sugden, D. E., Marchant, D. R., Potter, N., Jr., Souchez, R. A., Denton, G. H., Swisher, C. C., III & Tison, J.-L. (1995). Preservation of Miocene glacier ice in East Antarctica. *Nature* **376,** 412–414.

Susman, R. L. (1988). Hand of *Paranthropus robustus* from Member 1, Swartkrans: fossil evidence for tool behaviour. *Science* **240,** 781–784.

Susman, R. L. (1993). Hominid postcranial remains from Swartkrans. In C. K. Brain, ed., *Swartkrans. A Cave's Chronicle of Early Man,* pp. 117–136. Pretoria: Transvaal Museum Monograph no. 8.

Susman, R. L. & Stern, J. T. (1979). Telemetered electromyography of flexor digitorum profundis and flexor digitorum superficialis in *Pan troglodytes* and implications for interpretation of the OH 7 hand. *American Journal of Physical Anthropology* **50,** 565–574.

Susman, R. L. & Stern, J. T. (1982). Functional morphology of *Homo habilis. Science* **217,** 931–934.

Susman, R. L., Stern, J. T. & Jungers, W. L. (1984). Arboreality and bipedality in the Hadar hominids. *Folia Primatologia* **43,** 113–156.

Sutcliffe, A. J. (1985). *On the Track of Ice Age Mammals.* Cambridge: Harvard University Press.

Suwa, G. (1990). *A comparative analysis of hominid dental remains from the Shungura and Usno Formations, Omo Valley, Ethiopia* (Ph.D. thesis). Berkeley: University of California.

Suwa, G., White, T. & Howell, F. C. (1996). Mandibular postcanine dentition from the Shungura formation, Ethiopia: crown morphology, taxonomic allocations, and plio-pleistocene hominid evolution. *American Journal of Physical Anthropology* **101,** 247–282.

Svendsen, J. I., Elverhøi, A. & Mangerud, J. (1996). The retreat of the Barents Sea Ice Sheet on the western Svalbard margin. *Boreas* **25,** 244–256.

Swisher, C. C. III. (1996). New 40Ar/39Ar dates and their contribution toward a revised chronology for the late Miocene nonmarine of Europe and West Asia. In R. L. Bernor, V. Fahlbusch & H.-W. Mittmann, eds.), *The Evolution of Western Eurasian Neogene Mammal Faunas,* pp. 64–77. New York: Columbia University Press.

Swisher, C. C., III, Curtis, G. H., Jacob, T., Getty, A. G., Suprijo, A. & Widiasmoro (1994). Age of the earliest known Hominids in Java, Indonesia. *Science* **263,** 1118–1121.

Swisher, C. C., III, Rink, J., Anton, J. C., Schwarcz, H. P., Curtis, G. H., Suprijo, A. & Widiasmoro (1996). Latest Homo erectus of Java. Potential Contemporaneity with Homo sapiens in Southeast Asia. *Science* **274,** 1870–1874.

Swynnerton, G. H. & Hayman, R. W. (1951). A checklist of the land mammals of the Tanganyika Territory and the Zanzibar Protectorate. *Journal of the East African Natural History Society* **20(6,7),** 274–392.

Szalay, F. S. (1991). The unresolved world between taxonomy and population biology: what is, and what is not, macroevolution? *Journal of Human Evolution* **20,** 271–280.

Szalay, F. S. (1993). Species concepts. The tested, the untestable, and the redundant. In W. H. Kimbel & L. B. Martin, eds., *Species, Species Concepts, and Primate Evolution,* pp. 21–41. New York: Plenum Press.

Szalay, F. S. (1994). *Evolutionary History of the Marsupials and an Analysis of Osteological Characters.* New York: Cambridge University Press.

Szalay, F. S. & Bock, W. J. (1991). Evolutionary theory and systematics: relationships between process and patterns. *Zeitschrift fur zoologische Systematik und Evolutionsforschung* **29,** 1–39.

Szalay, F. S. & Bock, W. J. (1991). Evolutionary theory and systematics: Relationships between process and patterns. *Zeitschrift Zoologische Systematik und Evolutionsforschung* **29,** 1–39.

Szalay, F. S. & Delson, E. (1979). *Evolutionary History of the Primates.* New York: Academic Press.

Szalay, F. S. & Gould, S. J. (1966). Asiatic Mesonychidae (Mammalia, Condylarthra). *Bulletin of the American Museum of Natural History* **132,** 127–174.

Tague, R. G. & Lovejoy, C. O. (1986). The obstetrics of AL 288-1 (Lucy). *Journal of Human Evolution* **15,** 237–255.

Tainton, N. M. & Walker, B. H. (1993). Grasslands of Southern Africa. In R. T. Coupland, ed., *Natural Grasslands: Eastern Hemisphere & Resume,* pp. 265–290. London: Elsevier Science Publishers.

Talbot, M. R. & Johannessen, T. (1992). A high resolution palaeoclimatic record for the last 27,500 years in tropical West Africa from the carbon and nitrogen isotopic composition of lacustrine organic matter. *Earth and Planetary Science Letters* **110,** 23–37.

Taper, M. L. & Case, T. J. (1992). Coevolution among competitors. *Oxford Surveys in Evolutionary Biology* **8,** 63–109.

Taylor, D. M. (1992). Pollen evidence from Muchoya Swamp, Rukiga Highlands (Uganda), for abrupt changes in vegetation during the last ca. 21,000 years. *Bulletin Societe du Géologie France* **1,** 77–82.

Taylor, P. J., Contrafatto, G. & Willan, K. (1994). Climatic correlations of chromosomal variation in the African vlei rat, *Otomys irroratus. Mammalia* **58,** 623–634.

Taylor, P. J. & Meester, J. (1993). Morphometric variation in the yellow mongoose, *Cynictis penicillata* (Cuvier, 1829) (Carnivora: Viverridae), in Southern Africa. *Durban Museum Novitates* **18,** 37–71.

Taylor, P., Rautenbach, I. L., Gordon, D., Sink, K. & Lotter, P. (1995). Diagnostic morphometrics and southern African distribution of two sibling species of tree rat, *Thallomys paedulcus* and *Thallomys nigricauda* (Rodentia: Muridae). *Durban Museum Novitates* **20,** 49–62.

Tchernov, E. (1992). Dispersal: a suggestion for a common usage of this term. *Courier Forschungsinstitut Senckenberg* **153,** 103–123.

Teaford, M. F. (1984). Quantitative differences in dental microwear between primate species with different diets and a comment on the presumed diet of *Sivapithecus. American Journal of Physical Anthropology* **64,** 191–200.

Teaford, M. F. (1993). Dental microwear and diet in extant and extinct *Theropithecus:* preliminary analyses. In N. G. Jablonski, ed., *Theropithecus—The Rise and Fall of a Primate Genus,* pp. 331–349. Cambridge: Cambridge University Press.

Teaford, M. F. (1995). Dental microwear and dental function. *Evolutionary Anthropology* **3,** 17–30.

Teleki, G. (1981). The omnivorous diet and eclectic feeding habits of chimpanzees in Gombe National Park, Tanzania. In R. S. Harding & G. Teleki, eds., *Omnivorous Primates: Gathering and Hunting in Human Evolution,* pp. 303–343. New York: Columbia University Press.

Templeton, A. R. (1981). Mechanisms of speciation—a population genetic approach. *Annual Review of Ecology and Systematics* **12,** 23–48.

Thackeray, J. F. (1995). Do strontium/calcium ratios in early Pleistocene hominids from Swartkrans reflect physiological differences in males and females? *Journal of Human Evolution* **29,** 401–404.

Thomas, H. (1979). Les Bovidae miocènes des rifts est-africains: implications biogéographiques. *Bulletin de la Société geologique de France* XXI, **3,** 395–299.

Thomas, H. (1984). Les Bovidae (Artiodactyla: Mammalia) du Miocene du Sous-Continent Indien, de la Peninsule Arabique et det l'Afrique, biostratigraphie, biogeographie et ecologie. *Palaeogeography, Palaeoclimatology, Palaeoecology* **45,** 251–299.

Thomas, M. F. & Goudie, A. S., eds. (1985). *Dambos: small channelless valleys in the tropics—Characteristics, formation, utilisation. Zeitschrift fur Geomorphologie, Supplementband* 52. Berlin: Gebruder Borntraeger.

Thompson, D. W. (1942). *On Growth and Form.* Cambridge: Cambridge University Press.

Thornthwaite, C. W. (1933). The climates of the earth. *Geographical Review* **23,** 433–440.

Thornthwaite, C. W. (1948). An approach toward a rational classification of climate. *Geographical Review* **38,** 55–94.

Tidwell, W. D. & Nambudiri, E. M. V. (1989). *Tomlinsonia thomasonii,* get. et sp. nov., a permineralized grass from the upper Miocene Ricardo Formation, California. *Reviews in Palaeobotany and Palynology* **60,** 165–177.

Tiedemann, R., Sarnthein, M. & Shackleton, N. J. (1994). Astronomical timescale for the Pliocene Atlantic $\delta^{18}O$ and dust flux records of Ocean Drilling Program site 659. *Paleoceanography* **9,** 619–638.

Tieszen, L. L., Senyimba, M. M., Imbamba, S. K. & Troughton, J. H. (1979). The distribution of C_3 and C_4 grasses along an altitudinal and moisture gradient in Kenya. *Oecologia* **37,** 337–350.

Tinley, K. L. (1982). The influence of soil moisture balance on ecosystem patterns in southern Africa. In B. J. Huntley & B. H. Walker, eds., *The Ecology of Tropical Savannas,* pp. 175–192. Berlin: Springer-Verlag.

Tobias, P. V. (1971). *The Brain in Hominid Evolution.* New York: Columbia University Press.

Tobias, P. V. (1981). *The Evolution of the Human Brain, Intellect, and Spirit.* Adelaide: University of Adelaide.

Tobias, P. V. (1985). Ten climateric events in hominid evolution. *South African Journal of Science* **81,** 271–272.

Tobias, P. V. (1987). The brain of *Homo habilis:* a new level or organisation in cerebral evolution. *Journal of Human Evolution* **16,** 741–761.

Tobias, P. V. (1991a). *Olduvai Gorge Volume 4. The Skull, Endocasts and Teeth of* Homo habilis. Cambridge: Cambridge University Press.

Tobias, P. V. (1991b). The environmental background of hominid emergence and the appearance of the genus *Homo. Human Evolution* **6,** 129–142.

Tobias, P. V. (1994). The craniocerebral interface in early hominids: cerebral impressions, cranial thickening, paleoneurobiology, and a new hypothesis on encephalization. In R. S. Corruccini & R. L. Ciochon, eds., *Integrative Paths to the Past: Paleoanthropological Advances in Honor of F. Clark Howell,* pp. 185–203. Englewood Cliffs, NJ: Prentice Hall.

Tobien, H. (1952). Uber die funktion der seitenzehen tridactyler Equiden. *Neues Jahrbuch für Geologie und Palontologie Abhandlungen* **96,** 137–172.

Tobien, H. (1986). Die jungtertiare Fossilgrabungsstatte Howenegg im Hegau (Sudwestdeutschland) ein Statusbericht. *Carolinea* **44,** 9–34.

Toggweiler, J. R. & Samuels, B. (1992). Is the magnitude of the deep outflow from the Atlantic Ocean actually governed by southern hemisphere winds? In M. Heimann, eds., *The Global Carbon Cycle.* New York: Springer-Verlag.

Toggweiler, J. R. & Samuels, B. (1995). Effect of Drake Passage on the global thermohaline circulation. *Deep-Sea Research I,* **42,** 477–500.

Truswell, E. M. (1991). Antarctica: a history of terrestrial vegetation. In R. J. Tingey, ed., *The Geology of Antarctica,* pp. 499–537. Oxford: Clarendon Press.

Turner, A. (1984). Hominids and fellow travellers: human migration into high latitudes as part of a large mammal community. In R. Foley, ed., *Hominid Evolution and Community Ecology,* pp. 193–217. London: Academic Press.

Turner, A. (1987). New fossil carnivore remains from the Sterkfontein hominid site (Mammalia: Carnivora). *Annals of the Transvaal Museum* **34,** 319–347.

Turner, A. (1990). The evolution of the guild of larger terrestrial carnivores during the Plio-Pleistocene in Africa. *Geobios* **23,** 349–368.

Turner, A. (1992a). Villafranchian-Galerian larger carnivores of Europe: dispersions and extinctions. *Courier Forschungsinstitut Senckenberg* **153,** 153–160.

Turner, A. (1992b). Large carnivores and earliest European hominids: changing determinants of resource availability during the Lower and Middle Pleistocene. *Journal of Human Evolution* **22,** 109–126.

Turner, A. (1993). Species and speciation. Evolution and the fossil record. *Quaternary International* **19,** 5–8.

Turner, A. (1994). The species in paleontology. In D. M. Lambert & H. G. Spencer, eds., *Speciation and the Recognition Concept,* pp. 57–70. Baltimore, MD; The Johns Hopkins University Press.

Turner, A. (1995a). Plio-Pleistocene correlations between climatic change and evolution in terrestrial mammals: the 2.5 Ma event in Africa and Europe. *Acta Zoologica Cracoviensia* **38,** 45–58.

Turner, A. (1995b). The species in palaeontology. In D. M. Lambert & H. G. Spencer, eds., *Speciation and the Recognition Concept: Theory and Application,* pp. 57–70. Baltimore, MD: Johns Hopkins University Press.

Turner, A. (1995c). The Villafranchian large carnivore guild: geographic distribution and structural evolution. *Il Quaternario* **8,** 349–356.

Turner, A. (1997). Further remains of Carnivora (Mammalia) from the Sterkfontein hominid site. *Palaeontologia Africana* **34,** 115–126.

Turner, A. & Antón, M. (1997). *The Big Cats and their Fossil Relatives: An Illustrated Guide to their Evolution and Natural History.* New York: Columbia University Press.

Turner, A. & Antón, M. (in press). Climate and evolution: Implications of some extinction patterns in African and European machairodontine cats of the Plio-Pleistocene. *Estudios Geologicos.*

Turner, A. & Chamberlain, A. T. (1989). Speciation, morphological change and the status of African *Homo erectus. Journal of Human Evolution* **18,** 115–130.

Turner, A. & Paterson, H. E. H. (1991). Species and speciation: evolutionary tempo and mode in the fossil record reconsidered. *Geobios* **24,** 761–769.

Turner, A. & Wood, B. A. (1993a). Comparative palaeontological context for the evolution of the early hominid masticatory system. *Journal of Human Evolution* **24,** 301–318.

Turner, A. & Wood, B. A. (1993b). Taxonomic and geographic diversity in robust australopithecines and other African Plio-Pleistocene mammals. *Journal of Human Evolution* **24,** 147–168.

Turner, J. R. G. (1984). Darwin's coffin and Dr. Pangloss—do adaptationist models explain mimicry? In B. Shorrocks, ed., *Evolutionary Ecology,* pp. 313–361. Oxford: Blackwell.

Turner, J. R. G. (1986). The genetics of adaptive radiation: a neo-Darwinian theory of punctuated equilibrium. In D. M. Raup & D. Jablonski, eds., *Patterns and Processes in the History of Life,* pp. 183–207. Heidelberg: Springer-Verlag.

Turpie, J. K. & Crowe, T. M. (1994). Patterns of distribution, diversity and endemism of larger African mammals. *South African Journal of Zoology* **29(1),** 19–32.

Tutin, C. E. G. & Fernandez, M. (1985). Foods consumed by sympatric populations of *Gorilla g. gorilla* and *Pan t. troglodytes* in Gabon. *International Journal of Primatology* **6,** 27–43.

Tutin, C. E., Frenandez, M., Rogers, M. E. & Williamson, E. A. (1992). A preliminary analysis of the social structure of lowland gorillas in

the Lopé Reserve, Gabon. In N. Itoigawa, Y. Sugiyama, G. P. Sackett & R. K. R. Thompson, eds., *Topics in Primatology,* vol. 2. *Behavior, Ecology, and Conservation,* pp. 245–253. Tokyo: University of Tokyo Press.
Tyson, P. D. (1986). *Climatic Change and Variability in Southern Africa.* Oxford: Oxford University Press.
Uehara, S. (1986). Sex and group difference in feeding on animals by wild chimpanzees in the Mahale Mountains, Tanzania. *Primates* **27,** 1–13.
Ungar, P. S. (1992). Dental evidence for diet in primates. *Anthropologiai Közlemények* **34,** 141–155.
Ungar, P. S. (1998). Dental allometry, morphology and wear as evidence for diet in fossil primates. *Evolutionary Anthropology* **6,** 205–217.
Ungar, P. S. & Grine, F. E. (1991). Incisor size and wear in *Australopithecus africanus* and *Paranthropus robustus. Journal of Human Evolution* **20,** 313–340.
Unger. P. S. & Teaford, M. F. (1994). Non-occlusal surface molar microwear in primates [abstract]. *American Journal of Physical Anthropology Supplement* **18,** 200.
Van Cakenberghe, V. & De Vree, F. (1993). Systematics of African *Nyeteris* (Mammalia: Chiroptera). Part II. The *Nyeteris hispida* group. *Bonner zoologische Beiträge* **44,** 299–332.
Van Couvering, J. A. H. & Van Couvering, J. A. (1975). Early Miocene mammal fossils from East Africa: aspects of geology, faunistics and paleo-ecology. In G. L. Isaac & E. R. McCow, eds., *Human Origins, Louis Leakey and the East African Evidence,* pp. 155–207. Menlo Park, CA: W. A. Benjamin.
Van Couvering, J. H. (1980). Community evolution in East Africa during the late Cenozoic. In A. K. Behrensmeyer & A. Hill, eds., *Fossils in the Making,* pp. 272–298. Chicago: University of Chicago Press.
Van der Klaauw, C. J. (1948). Ecological studies and reviews. IV. Ecological morphology. *Bibliographica Biotheoretica* **4,** 27–111.
Van der Made, J. (1992). Migrations and climate. *Courier Forschungsinstitut Senckenberg* **153,** 27–37.
Van der Wateren, F. M. & Hindmarsh, R. (1995). East Antarctic ice sheet: Stabilists strike again. *Nature* **376,** 389–391.
Van Hoepen, E. C. N. (1930). Fossiele Pferde van Cornelia, O.V. S. *Paleontologie Navorsing Nasionale Museum Bloemfontein* **2,** 13–24.
Van Hoepen, E. C. N. (1932). Die stamlyn van die Sebras. *Paleontologie Navorsing Nasionale Museum Bloemfontein* **2,** 25–37.
Van Neer, W. (1989). Contribution to the Archaeozoology of Central Africa. *Annales Sciences Zoologiques. Tervuren: Musée Royal de l'Afrique Centrale* **259.**
Van Valen, L. (1960). A functional index of hyposodonty. *Evolution* **14,** 531–532.
Van Valen, L. (1973). A new evolutionary law. *Evolutionary Theory* **1,** 1–30.
Van Valen, L. M. (1988). Species, sets, and the derivature nature of philosophy. *Biology & Philosophy* **3,** 49–66.
Van Valkenburg, B. (1987). Skeletal indicators of locomotor behavior in living and extant carnivores. *Journal of Vertebrate Paleontology* **7,** 162–182.
Van Zinderen Bakker, E. M. & Mercer, J. H. (1986). Major late Cainozoic climatic events and palaeoenvironmental changes in Africa viewed in a world wide context. *Palaeogeography, Palaeoclimatology, Palaeoecology* **56,** 217–235.
Verdcourt, B. (1963). The Miocene nonmarine mollusca of Rusinga Island, Lake Victoria, and other localities in Kenya. *Palaeontographica* **121**(Abt. A), 1–37.
Vermeij, G. J. (1973a). Biological versatility and earth history. *Proceedings of the National Academy of Science USA* **70,** 1936–1938.
Vermeij, G. J. (1973b). Adaptation, versatility, and evolution. *Systematic Zoology* **22,** 466–477.
Vermeij, G. J. (1987). *Evolution and Escalation: An Ecological History of Life.* Princeton, NJ: Princeton University Press.
Vermeij, G. J. (1991). When biotas meet: understanding biotic interchange. *Science* **253,** 1099–1104.
Vesey-Fitzgerald, D. F. (1960). Grazing succession among East African game mammals. *Journal of Mammalogy* **41,** 161–172.
Viljoen, S. (1977). Feeding habits of the bush squirrel. *Zoologica Africana* **12,** 459–468.
Viljoen, S. (1983). Feeding habits and comparative feeding rates of three southern African arboreal squirrels. *South African Journal of Zoology* **18,** 378–887.
Vincents, A. (1987). Environments botanique et climatique des hominides de l'Est Turkana, Kenya, entre 2,0 et 1,4 millions d'annees: Apport de la palynologie. In J. A. Coetzee, ed., *Palaeoecology of Africa and the Surrounding Islands* **18,** 257–269.
Vincent, E. & Berger, W. H. (1985). Carbon dioxide and polar cooling in the Miocene: the Monterrey hypothesis. In E. T. Sundquist & W. S. Broecker, eds., *The Carbon Cycle and Atmospheric* CO_2*: Natural Variations Archean to Present,* pp. 455–468. Washington, DC: American Geophysical Union.
Vogel, J. C. (1978). Recycling of carbon in a forest environment. *Oecologia Plantarum* **13,** 89–94.
Vogt, P. R. (1971). Asthenosphere motion recorded by the ocean floor south of Iceland. *Earth and Planetary Science Letters* **13,** 153–160.
Vogt, P. R. (1972). The Faeroe-Iceland-Greenland

aseismic ridge and the Western Boundary Under-current. *Nature* **239,** 79–81.
Vogt, P. R. (1983). The Iceland mantle plume: status of the hypothesis after a decade of new work. In M. H. P. Bott, S. Saxov, M. Talwani & J. Thiede, eds., *Structure and Development of the Greenland-Scotland Ridge,* pp. 191–213. New York: Plenum.
von Höhnel (1891). *Discovery by Count Teleki of Lakes Rudolf and Stefani,* vol. 2. London: Frank Cass and Company.
von Koenigswald, G. H. R. (1969). Miocene Cercopithecoidea and Oreopithecoidea from the Miocene of East Africa. In L. S. B. Leakey, ed., *Fossil Vertebrates of Africa,* vol. 1, pp. 39–51. London: Academic Press.
von Koenigswald, W. (1991). Exoten in der Grossäuger-Fauna des Letzten Interglazials von Mitteleuropa. *Eiszeitalt und Gegenweldt.* **41,** 70–79.
von Koenigswald, W. (1992). Various aspects of migrations in terrestrial mammals in relation to Pleistocene faunas of central Europe. *Courier Forschungsinstitut Senckenberg* **153,** 39–47.
Vrba, E. S. (1974). Chronological and ecological implications of the fossil Bovidae at the Sterkfontein australopithecine site. *Nature* **250,** 19–23.
Vrba, E. S. (1975). Some evidence of chronology and palaeoecology of Sterkfontein, Swartkrans and Kromdraai from the fossil Bovidae. *Nature* **254,** 301–304.
Vrba, E. S. (1976). The fossil bovidae of Sterkfontein, Swartkrans and Kromdraai. *Transvaal Museum Memoir* **21.**
Vrba, E. S. (1979). A new study of the scapula of *Australopithecus africanus* from Sterkfontein. *American Journal of Physical Anthropology* **51,** 117–129.
Vrba, E. S. (1980a). Evolution, species and fossils: how does life evolve? *South African Journal of Science* **76,** 61–84.
Vrba, E. S. (1980b). The significance of bovid remains as an indicator of environment and predation patterns. In A. K. Behrensmeyer & A. P. Hill, eds., *Fossils in the Making,* pp. 247–242. Chicago: University of Chicago Press.
Vrba, E. S. (1981). The Kromdraai australopithecine site revisited in 1980: recent investigations and results. *Annals of the Transvaal Museum* **33(3),** 17–60.
Vrba, E. S. (1982). Biostratigraphy and chronology, based particularly on Bovidae, of southern African hominid-associated assemblages: Makapansgat, Sterkfontein, Taung, Kromdraai, Swartkrans; also Elandsfontein (Saldanha), Broken Hill (now Kabwe) and Cave of Hearths. In H. deLumley & M. A. deLumley, eds., Proceedings of *Congrès International de Paléontologie Humaine,* vol. 2, pp. 707–752. Nice: Union Internationale des Sciences Préhistoriques et Protohistorique.
Vrba, E. S. (1983). Macroevolutionary trends: new perspectives on the roles of adaptation and incidental effect. *Science* **221,** 387–389.
Vrba, E. S. (1984). Evolutionary pattern and process in the sister group Alcelaphini-Aepycerotini (Mammalia: Bovidae). In N. Eldredge & S. M. Stanley, eds., *Living Fossils,* pp. 62–79. New York: Springer.
Vrba, E. S. (1985a). African Bovidae: evolutionary events since the Miocene. *South African Journal of Science* **81,** 263–266.
Vrba, E. S. (1985b). Early hominids in southern Africa: Updated observations on chronological and ecological background. In P. V. Tobias, ed., *Hominid Evolution. Past, Present, and Future,* pp. 195–200. New York: Alan R. Liss.
Vrba, E. S. (1985c). Ecological and adaptative changes associated with early Hominid evolution. In E. Delson, ed., *Ancestors: The Hard Evidence,* pp. 63–71. New York: Alan R. Liss.
Vrba, E. S. (1985d). Environment and evolution: alternative causes of the temporal distribution of evolutionary events. *South African Journal of Science* **81,** 229–236.
Vrba, E. S. (1985e). Palaeoecology of early Hominidae, with special reference to Sterkfontein, Swartkrans and Kromdraai. In Y. Coppens, ed., *L'Environnement des Hominidés au Plio-Pléistocène,* pp. 345–369. Paris: Masson.
Vrba, E. S. (1987a). A revision of the Bovini (Bovidae) and a preliminary revised checklist of Bovidae from Makapansgat. *Palaeontologica Africa* **26,** 33–46.
Vrba, E. S. (1987b). Ecology in relation to speciation rates: some recent case histories of Miocene recent mammal clades. *Evolutionary Ecology* **1,** 283–300.
Vrba, E. (1988). Late Pliocene climatic events and human evolution. In F. Grine, ed., *Evolutionary History of the "Robust" Australopithecines,* pp. 405–426. New York: Aldine de Gruyter.
Vrba, E. S. (1989). Levels of selection and sorting with special reference to the species level. *Oxford Surveys in Evolutionary Biology* **6,** 111–163.
Vrba, E. S. (1992). Mammals as a key to evolutionary theory. *Journal of Mammalogy* **73,** 1–28.
Vrba, E. S. (1993a). Mammal evolution in the Africa Neogene and a new look at the Great American Interchange. In P. Goldblatt, ed., *Biological Relationships between Africa and South America,* pp. 393–432. New Haven, CT: Yale University Press.
Vrba, E. S. (1993b). The pulse that produced us. *Natural History* **5,** 47–71.
Vrba, E. S. (1993c). Turnover-pulses, the Red

Queen, and related topics. *American Journal of Science* **293-a,** 418–452.
Vrba, E. S. (1994). An hypothesis of heterochrony in response to climatic cooling and its relevance to early hominid evolution. In R. Ciochon & R. Corruccini, eds., *Integrative Paths to the Past: Palaeoanthropological Advances in Honour of F. Clark Howell,* pp. 345–376. New York: Prentice Hall.
Vrba, E. S. (1995a). Species as habitat-specific, complex systems. In D. M. Lambert & H. G. Spencer, eds., *Speciation and the Recognition Concept: Theory and Application,* pp. 3–44. Baltimore, MD: Johns Hopkins University Press.
Vrba, E. S. (1995b). On the connections between paleoclimate and evolution. In E. S. Vrba, G. H. Denton, T. C. Partridge & L. H. Burckle, eds., *Paleoclimate and Evolution with Emphasis on Human Origins,* pp. 24–45. New Haven, CT: Yale University Press.
Vrba, E. S. (1995c). The fossil record of African antelopes (Mammalia, Bovidae) in relation to human evolution and paleoclimate. In E. S. Vrba, G. H. Denton, T. C. Partridge & L. H. Burckle, eds., *Paleoclimate and Evolution with Emphasis on Human Origins,* pp. 385–424. New Haven, CT: Yale University Press.
Vrba, E. S. (1996). Climate, heterochrony, and human evolution. *Journal of Anthropological Research* **52,** 1–29.
Vrba, E. S. (1997). New alcelaphine fossils (Bovidae, Mammalia) from Middle Awash, Ethiopia, and phylogenetic analysis of Alcelaphini. *Palaeontologia Africana* **34,** 199–217.
Vrba, E. S. (1998). Multiphasic growth models and the evolution of prolonged growth exemplified by human brain evolution. *Journal of Theoretical Biology* **190,** 227–239.
Vrba, E. S. & Eldredge, N. (1984). Individuals, hierarchies and processes: Towards a more complete evolutionary theory. *Paleobiology* **10,** 146–171.
Vrba, E. S., Denton, G. H., Partridge, T. C. & Burckle, L. H., eds. (1995). *Paleoclimate and Evolution with Emphasis on Human Origins.* New Haven, CT: Yale University Press.
Vrba, E. S., Denton, G. H. & Prentice, M. L. (1989). Climatic influences on early hominid behaviour. *OSSA* **14,** 127–156.
Vrba, E. S. & J. E. Gatesy (1995). New fossils of hippotragine antelopes from the Middle Awash deposits, Ethiopia, in the context of a phylogenetic analysis of Hippotragini (Bovidae, Mammalia). *Palaeontologia Africana* **31,** 1–18.
Vrba, E. S. & Gould, S. J. (1986). The hierarchical expansion of sorting and selection: sorting and selection cannot be equated. *Paleobiology* **12,** 217–228.
Wade, M. J. (1992). Sewall Wright: gene interaction and the shifting balance theory. *Oxford Surveys in Evolutionary Biology* **8,** 35–62.
Wake, D. B., Roth, G. & Wake, M. H. (1983). On the problem of stasis in organismal evolution. *Journal of Theoretical Biology* **101,** 211–224.
Walker, A. C. (1978). Prosimian Primates. In V. J. Maglio & H. B. S. Cooke, eds., *Evolution of African Mammals,* pp. 90–99. Cambridge, MA: Harvard University Press.
Walker, A. C. (1987). Fossil Galaginae from Laetoli. In M. D. Leakey & J. M. Harris, eds., *Laetoli: A Pliocene Site in Northern Tanzania,* pp. 88–90. Oxford: Clarendon Press.
Walker, A. & Leakey, R. (1993). *The Nariokotome Homo erectus Skeleton.* Berlin: Springer-Verlag.
Walker, A. C., Leakey, R. E. F., Harris, J. M. & Brown, F. H. (1986). 2.5 Myr *Australopithecus boisei* from west of Lake Turkana, Kenya. *Nature* **322,** 517–522.
Walker, A. L. (1981). Diet and teeth: dietary hypotheses and human evolution. *Philosophical Transactions of the Royal Society of London* **292,** 57–64.
Walker, A. L. (1993). Perspectives on the Nariokotome discovery. In A. C. Walker & R. E. Leakey, eds., *The Nariokotome Homo erectus Skeleton,* pp. 411–430. Berlin: Springer-Verlag.
Walker, A., Teaford, M., Martin, L. & Andrews, P. (1993). A new species of *Proconsul* from the early Miocene of Rusinga-Mfwangano Islands, Kenya. *Journal of Human Evolution* **25,** 43–56.
Walter, G. H. (1988). Competitive exclusion, coexistence and community structure. *Acta Biotheoretica* **37,** 281–313.
Walter, G. H. (1995). Species concepts and the nature of ecological generalizations about diversity. In D. M. Lambert & H. G. Spencer, eds., *Speciation and the Recognition Concept: Theory and Application,* pp. 191–224. Baltimore, MD: Johns Hopkins University Press.
Walter, G. H. & Paterson, H. E. H. (1994). The implications of paleontological evidence for theories of ecological communities and species richness. *Australian Journal of Ecology* **19,** 241–250.
Walter, H., Lieth, H. & Rehder, G. (1960–67). *Klimadiagramm-Weltatlas.* Jena: Fischer.
Walter, H. (1971). *Ecology of Tropical and Subtropical Vegetation,* trans. D. Mueller-Dombois, ed. J. H. Burnett. New York: Van Nostrand Reinhold.
Walter, H. (1984). *Vegetation of the Earth and Ecological Systems of the Geobiosphere.* Berlin: Springer-Verlag.
Walter, H. & Breckle, S. W. (1985). *Ecological Systems of the Geobiosphere, 2, Tropical and*

Subtropical Zonobiomes, trans. S. Gruber. Berlin: Springer-Verlag.

Walter, R. C., Manega, P. C., Hay, R. L., Drake, R. E. & Curtis, G. H. (1991). Laser-fusion $^{40}Ar/^{39}Ar$ dating of Bed I, Olduvai Gorge, Tanzania. *Nature* **354,** 145–149.

Walter, R. H. & Aronson, J. L. (1993). Age and source of the Sidi Hakoma Tuff, Hadar Formation, Ethiopia. *Journal of Human Evolution* **25(3),** 229–240.

Ward, S. C. & Kimbel, W. H. (1983). Subnasal alveolar morphology and the systematic position of Sivapithecus. *American Journal of Physical Anthropology* **61,** 157–171.

Ward, S. C. & Molnar, S. (1980). Experimental stress analysis of topographic diversity in early hominid gnathic morphology. *American Journal of Physical Anthropology* **53,** 383–395.

Warren, B. A. (1990). Suppression of deep oxygen concentrations by Drake Passage. *Deep-Sea Research* **37,** 1899–1907.

Waser, P. M. (1987). Interactions among primate species. In B. B. Smuts, D. L. Cheney, R. M. Seyfarth, R. W. Wrangham & T. T. Struhsaker, eds., *Primate Societies,* pp. 210–226. Chicago: University of Chicago Press.

Wateren van der, D. & Hindmarsh, R. (1995). Stabilists strike again. *Nature* **376,** 389–391.

Watson, V. & Plug, I. (1995). *Oreotragus major* (Wells, 1952) and *Oreotragus oreotragus* (Zimmerman, 1783) (Mammalia: Bovidae): Two species? Annals of the Transvaal Museum **36,** 183–191.

Webb, P. N. & Harwood, D. M. (1993). Pliocene fossil *Nothofagus* (southern beech) from Antarctica: phytogeography, dispersal strategies, and survival in high latitude glacial-deglacial environments. In J. Alden, ed., *Forest Development in Cold Climates,* pp. 135–165. New York: Plenum Press.

Webb, P. N., Harwood, D. M., McKelvey, B. C., Mercer, J. H. & Stott, L. D. (1984). Cenozoic marine sedimentation and ice-volume variation on the East Antarctic craton. *Geology* **12,** 287–291.

Webb, S. D. (1983). The rise and fall of the late Miocene ungulate fauna in North America. In M. H. Nitecki, ed., *Coevolution,* pp. 267–306. Chicago: University of Chicago Press.

Weber, B. H. & Depew, D. J. (1996). Natural selection and self-organization. *Biology and Philosophy* **11,** 33–65.

Werdelin, L. & Solounias, N. (1991). The Hyaenidae: taxonomy systematics and evolution. *Fossils and Strata* **30,** 1–104.

Werdelin, L. & Solounias, N. (1996). The evolutionary history of hyaenas in Europe and western Asia during the Miocene. In R. L. Bernor, V. Fahlbusch & S. Rietschel, eds., *The Evolution of Western Eurasian Neogene Mammal Faunas,* pp. 290–306. New York: Columbia University Press.

Werdelin, L. & Turner, A. (1996). The fossil and living Hyaenidae of Africa: present status. In K. Stewart & K. Seymour, eds., *The Palaeoecology and Palaeoenvironments of Late Cenozoic Mammals: Tributes to the Career of C. S. Church,* pp. 637–659. Toronto: Toronto University Press.

Werdelin, L., Turner, A. & Solounias, N. (1994). Studies of fossil hyaenids: the genera *Hyaenictis* Gaudry and *Chasmaporthetes* Hay, with a reconsideration of the Hyaenidae of Langebaanweg, South Africa. *Zoological Journal of the Linnean Society* **111,** 197–217.

Werger, M. & Coetzee, B. J. (1978). The Sudano-Zambexi region. In M. Werger, ed., *Biogeography and Ecology of Southern Africa,* pp. 301–462. The Hague: Junk.

Werger, M. (1983). Tropical grasslands, savannas, woodlands: natural and manmade. In W. Holzner, M. Werger & I. Ikusima, eds., *Man's Impact on Vegetation,* pp. 107–137. The Hague: Junk.

Wesselman, H. B. (1982). *Pliocene Micromammals from the Lower Omo Valley, Ethiopia: Systematics and Paleoecology.* PhD thesis, University of California, Berkeley.

Wesselman, H. B. (1984). *The Omo Micromammals, Systematics and Paleoecology of Early Man Sites from Ethiopia.* New York: Karger.

Wesselman, H. B. (1985). Fossil micromammals as indicators of climatic change about 2.4 Myr ago in the Omo Valley, Ethiopia. *South African Journal of Science* **81,** 260–261.

Wesselman, H. (1995). Of mice and almost men: regional paleoecology and human evolution in the Turkana basin. In E. S. Vrba, G. H. Denton, T. C. Partridge & L. H. Burckle, eds., *Paleoclimate and Evolution with Emphasis on Hominid Origins,* pp. 356–368. New Haven, CT: Yale University Press.

West-Eberhard, M. J. (1986). Alternative adaptations, speciation, and phylogeny (a review). *Proceedings of the national Academy of Sciences USA* **83,** 1388–1392.

Western, D. (1975). Water availability and its influence on the structure and dynamics of a Savannah large mammal community. *East African Wildlife Journal* **13,** 265–286.

Wheeler, P. E. (1991a). The influence of bipedalism on the energy and water budgets of early hominids. *Journal of Human Evolution* **21,** 117–136.

Wheeler, P. E. (1991b). The thermoregulatory advantages of hominid bipedalism in open equatorial environments: the contribution of increased convective heat loss and cutaneous evaporative cooling. *Journal of Human Evolution* **21,** 107–116.

Wheeler, P. (1992). The thermoregulatory advantages of large body size for hominid foraging in savannah environments. *Journal of Human Evolution* **23,** 351–362.

Wheeler, P. (1993). The influence of stature and body form on hominid energy and water budgets: a comparison of *Australopithecus* and early *Homo* physiques. *Journal of Human Evolution* **24,** 13–28.

Wheeler, P. E. (1994). The thermoregulatory advantages of heat storage and shade-seeking behaviour to hominids foraging in equatorial savannah environments. *Journal of Human Evolution* **26,** 339–350.

Whistler, D. P. & Burbank, D. W. (1992). Miocene biostratigraphy and biochronology of the Dove Spring Formation, Mojave Desert, California, and characterization of the Clarendonian mammal age (late Miocene) in California. *Geological Society of America Bulletin* **104,** 644–658.

White, A. F. & Blum, A. E. (1995). Effects of climate on chemical weathering in watersheds. *Geochimica et Cosmochimica Acta* **59.**

White, F. (1981). The history of the Afromontane archipelago and the scientific need for its conservation. *African Journal of Ecology* **19,** 33–54.

White, F. (1981). *UNESCO/AETFAT/UNSO Vegetation Map of Africa.* Paris: UNESCO/AETFAT/UNSO.

White, F. (1983). *The Vegetation of Africa. A Descriptive Memoire to Accompany the UNESCO/AETFAT/UNSO Vegetation Map of Africa. Natural Resources Research,* vol. 20. Paris: UNESCO.

White, F. (1984). *Vegetation Map of Africa. Africa South of the Sahara 1984–1985,* 14th ed. Europa Publications Ltd.

White, F. (1986). *La végétation de l'Afrique.* Paris: UNESCO.

White, F. (1993). Refuge theory, ice-age aridity and the history of tropical biotas: an essay in plant geography. *Fragmenta Floristica et Geobotanica Supplementum* **2,** 385–409.

White, M. J. D. (1978). *Modes of speciation.* San Francisco: Freeman.

White, T. D. (1985). African suid evolution: the last six million years. *South African Journal of Science* **81,** 271.

White, T. D. (1988). The comparative biology of "robust" *Australopithecus:* clues from context. In F. E. Grine, ed., *Evolutionary History of the "Robust" Australopithecines,* pp. 449–483. New York: Aldine de Gruyter.

White, T. D. (1995). African omnivores: global climatic change and Plio-Pleistocene hominids and suids. In E. S. Vrba, G. H. Denton, T. C. Partridge & L. H. Burckle, eds., *Paleoclimate and Evolution with Emphasis on Human Origins,* pp. 369–384. New Haven, CT: Yale University Press.

White, T. D., Suwa, G. & Asfaw, B. (1994). *Australopithecus ramidus,* a new species of early hominid from Aramis, Ethiopia. *Nature* **371,** 306–312.

White, T. D., Suwa, G. & Asfaw, B. (1995). Corrigendum: *Australopithecus ramidus,* a new species of early hominid from Aramis, Ethiopia. *Nature* **375,** 88.

White, T. D., Suwa, G., Hart, W. K., Walters, R. C., WoldeGabriel, G., de Heinzelin, J., Clark, J. D., Asfaw, B. & Vrba, E. (1993). New discoveries of *Australopithecus* at Maka in Ethiopia. *Nature* **366,** 261–265.

Wiley, E. O. (1978). The evolutionary species concept reconsidered. *Systematic Zoology* **27,** 17–26.

Wiley, E. O. (1981). *Phylogenetics. The Theory and Practice of Phylogenetic Systematics.* New York: John Wiley and Sons.

Wilkinson, L. (1990). *SYGRAPH: The System for Graphics.* Evanston, IL: SYSTAT, Inc.

Williams, G. C. (1966). *Adaptation and Natural Selection.* Princeton: Princeton University Press.

Williams, G. C. (1985). A defense of reductionism in evolutionary biology. *Oxford Surveys in Evolutionary Biology* **2,** 1–26.

Williams, G. C. (1992). *Natural Selection: Domains, Levels, and Challenges.* New York: Oxford University Press.

Williamson, P. G. (1985). Evidence for an early Plio-Pleistocene rainforest expansion in East Africa. *Nature* **315,** 487–489.

Willis, J. C. (1973). *A Dictionary of the Flowering Plants and Ferns,* 8th ed., revised by H. K. Airy Shaw. Cambridge: Cambridge University Press.

Wilson, D. E. & Reeder, D. M. (1993). *Mammal Species of the World. A Taxonomic and Geographic Reference,* 2nd ed. Washington, DC: Smithsonian Institution Press.

Wilson, D. S. (1975). The adequacy of body size as a niche difference. The *American Naturalist* **109,** 769–784.

Wilson, E. O. (1983). Sociobiology and the Darwinian approach to mind and culture. In D. S. Bendall, ed., *Evolution from Molecules to Men,* pp. 545–553. Cambridge: Cambridge University Press.

Winkler, A. J. (1992). Systematics and biogeography of middle Miocene rodents from the Muruyur beds, Baringo district, Kenya. *Journal of Vertebrate Palaeontology* **12(2),** 236–249.

Winkler, A. J. (1994). Middle Miocene rodents from Maboko Island, western Kenya: contributions to understanding small mammal evolution during the Neogene [abstract]. *Journal of Vertebrate Paleontology* **14,** 53A.

Winkler, A. J. (1997). Systematics, paleobiogeography and paleoenvironmental significance of rodents from the Ibole Member, Manonga Valley, Tanzania. In T. Harrison, ed., *Neogene Paleontology of the Manonga Valley, Tanzania: A Window into the Evolutionary History of East Africa,* pp. 311–332. New York: Plenum.

Winterbottom, J. M. (1978). Birds. In M. J. A. Werger & A. C. Van Bruggen, eds., *Biogeography and Ecology of Southern Africa,* pp. 949–979. The Hague: W. Junk.

WoldeGabriel, G., White, T. D., Suwa, G., Renne, P., de Heinzelin, J., Hart, W. K. & Heiken, G. (1994). Ecological and temporal placement of early Pliocene hominids at Aramis, Ethiopia. *Nature* **371,** 330–333.

Wolfe, J. A. (1978). A paleobotanical interpretation of Tertiary climates in the Northern Hemisphere. *American Scientist* **66,** 694–703.

Wolfe, J. A. (1979). Temperature parameters of humid to mesic forests of eastern Asia and relation to forests of other regions of the Northern Hemisphere and Australasia. *Professional Papers of the U.S. Geological Survey 1* **106,** 1–37.

Wolfe, J. A. (1980). Tertiary climates and floristic relationships at high latitudes in the Northern Hemisphere. *Palaeogeography, Palaeoclimatology, Palaeoecology* **30,** 313–323.

Wood, B. A. (1991). *Koobi Fora Research Project,* vol. 4. *Hominid Cranial Remains.* Oxford: Clarendon Press.

Wood, B. (1992a). Origin and evolution of the genus *Homo. Nature* **355,** 783–790.

Wood, B. A. (1992b). Early hominid species and speciation. *Journal of Human Evolution* **22,** 351–365.

Wood, B. A. (1992c). Evolution of Australopithecines. In S. Jones, R. Martin, D. Pilbeam & S. Bunney, eds., *The Cambridge Encyclopedia of Human Evolution,* pp. 231–240. Cambridge: Cambridge University Press.

Wood, B. A. (1993). Rift on the record. *Nature* **365,** 789–790.

Wood, B. (1994). The oldest hominid yet. *Nature* **371,** 280–281.

Wood, B. (1995). Evolution of the early hominin masticatory system: mechanisms, events and triggers. In E. S. Vrba, G. H. Denton, T. C. Partridge & L. H. Burckle, eds., *Paleoclimate and Evolution with emphasis on Human Origins,* pp. 348–448. New Haven, CT: Yale University Press.

Wood, B. A. (in press). *Homo habilis, Homo rudolfensis* and *Homo ergaster.* In E. Delson, I. Tattersall, J. A. Van Couvering & A. S. Brooks, eds., *Encyclopedia of Human Evolution and Prehistory.* New York: Garland Publishing.

Wood, B. A. & Abbott, S. A. (1983). Analysis of the dental morphology of Plio-Pleistocene hominids: I. Mandibular molars: crown area measurements and morphological traits. *Journal of Anatomy* **136,** 197–219.

Wood, B. A., Abbott, S. A. & Graham, S. H. (1983). Analysis of the dental morphology of Plio-Pleistocene hominids: II. Mandibular molars—study of cusp areas, fissure pattern and cross sectional shape of the crown. *Journal of Anatomy* **137,** 287–314.

Wood, B. A. & Chamberlain, A. T. (1986). *Australopithecus:* grade or clade? In B. A. Wood, L. Martin & P. Andrews, eds., *Major Topics in Primate and Human Evolution,* pp. 248–270. Cambridge: Cambridge University Press.

Wood, B. A., Abbott, S. A. & Uytterschaut, H. (1988). Analysis of the dental morphology of Plio-Pleistocene hominids: IV. Mandibular postcanine root morphology. *Journal of Anatomy* **156,** 107–139.

Wood, B. A., Yu, L. & Willoughby, C. (1991). Intraspecific variation and sexual dimorphism in cranial and dental variables among higher primates and their bearing on the hominid fossil record. *Journal of Anatomy* **174,** 185–205.

Woodburne, M. O. & Bernor, R. L. (1980). On superspecific groups of some Old World hipparionine horses. *Journal of Paleontology* **54,** 1319–1348.

Woodburne, M. O., Bernor, R. L. & Swisher, C. S. III (1996). An appraisal of the stratigraphic and phylogenetic bases for the "Hipparion Datum" in the Old World. In R. L. Bernor, V. Fahlbusch & H.-W. Mittmann, eds., *The Evolution of Western Eurasian Neogene Faunas,* pp. 124–136. New York: Columbia University Press.

Woodruff, F., Savin, S. M. & Douglas, R. G. (1981). Miocene stable isotope record: a detailed deep Pacific Ocean study and its paleoclimatic implications. *Science* **212,** 665–668.

Woodward, F. I. (1987). *Climate and Plant Distribution.* Cambridge: Cambridge University Press.

Wooller, R. D. (1984). Bill shape and size in honeyeaters and other small insectivorous birds in Western Australia. *Australian Journal of Zoology* **32,** 657–661.

Wrangham, R. W., Gittleman, J. L. & Chapman, C. A. (1993). Constraints on group size in primates and carnivores: population density and day-range as assays of exploitation competition. *Behavioral Ecology & Sociobiology* **32,** 199–209.

Wrangham, R. W. & Van Zinnicq Bergmann Riss, E. (1990). Rates of predation on mammals by Gombe chimpanzees 1972–1975. *Primates* **31,** 157–170.

Wright, J. D. & Miller, K. G. (1996). Control of North Atlantic deep water circulation by the

Greenland-Scotland Ridge. *Paleoceanography* **11,** 157–170.

Wright, S. (1932). The roles of mutation, inbreeding, crossbreeding, and a selection in evolution. *Proceedings of the VIth International Congress on Genetics* **1,** 356–366.

Wright, S. (1982). Character change, speciation, and the higher taxa. *Evolution* **3,** 427–443.

Wright, S. (1988). Surfaces of selective value revisited. *American Naturalist* **131,** 115–123.

Wynne-Edwards, V. C. (1962). *Animal Dispersion in Relation to Social Behavior.* London: Oliver & Boyd.

Yadava, P. S. (1991). Savannas of north-east India. In P. A. Werner, ed., *Savanna Ecology and Management,* pp. 41–50. Oxford: Blackwell.

Yalden, D. W. & Largen, M. J. (1992). The endemic mammals of Ethiopia. *Mammal Review* **22,** 115–150.

Yalden, D. W., Largen, M. J. & Kock, D. (1976–1986). Catalogue of the mammals of Ethiopia. *Monitore Zoologico Italiano, N.S. Supplemento* **8,** 1–118, **9,** 1–52, **13,** 169–272, **19,** 67–221, **21,** 31–103.

Yamagiwa, J., Mwanza, N., Yumoto, T. & Maruhashi, T. (1992). Travel distances and food habits of eastern lowland gorillas: a comparative analysis. In N. Itoigawa, Y. Sugiyama, G. P. Sackett & R. K. R. Thompson, eds., *Topics in Primatology,* vol. 2. *Behavior, Ecology, and Conservation,* pp. 267–281. Tokyo: University of Tokyo Press.

Ye, Q., Matthews, R. K., Frohlich, C. & Gam, S. (1993). High-frequency glacioeustatic cyclicity in the Early Miocene and its influence on coastal and shelf depositional systems, NW Gulf of Mexico Basin. In J. M. Armentrout, R. Bloch, H. C. Olson & B. F. Perkins, eds., *Rates of Geologic Processes: Tectonics, Sedimentation, Eustay and Climate: Implications for Hydrocarbon and Exploration,* pp. 287–298. Program with Papers, 14th Annual Research Conference, Gulf Coast Section, Society of Economic Paleontologists and Mineralogists Foundations.

Ye, Q., Galloway, W. E. & Matthews, R. A. (1995). High-frequency glacier eustatic cyclicity of the Miocene in central Texas and western Louisiana and its application in hydrocarbon exploration. *Gulf Coast Association of Geological Societies Transactions* **45,** 587–594.

Yemane, K., Bonnefille, R. & Faure, H. (1985). Paleoclimatic and tectonic implications of Neogene microflora from the Northwestern Ethiopian Highlands. *Nature* **318,** 653–656.

Yemane, K., Robert, C. & Bonnefille, R. (1987). Pollen and clay mineral assemblages of a Late Miocene lacustrine sequence from the Northwestern Ethiopian Highlands. *Palaeogeography, Palaeoclimatology, Palaeoecology* **60,** 123–141.

Yom-Tov, Y., Green, W. O. & Coleman, J. D. (1986). Morphological trends in the common brushtail possum, *Trichosusrus vulpecula,* in New Zealand. *Journal of Zoology* **208,** 583–593.

Zachos, J. C., Bergren, W. A., Aubry, M. P. & Mackensen, A. (1992). Isotope and trace element geochemistry of Eocene and Oligocene foraminifers from Site 748, Kerguelen Plateau. *Proceedings of the ODP* **120(B),** 839–854.

Zachos, J. C., Quinn, T. M. & Salamy, K. A. (1996). High-resolution (10^4 years) deep-sea foraminiferal stable isotope records of the Eocene-Oligocene climate transition. *Paleoceanography* **11,** 251–266.

Zagwijn, W. H. (1960). Aspects of the Pliocene and Early Pleistocene vegetation in the Netherlands. *Mededelingen van de Geologische Stichting, Series C* **III** 1, 5:1–78.

Zagwijn, W. H. (1989). Vegetation and climate during the warmer intervals in the Late Pleistocene of western and central Europe. *Quaternary International* **3-4,** 57–66.

Zagwijn, W. H. (1992). Migration of vegetation during the Quaternary in Europe. *Courier Forschungsinstitut Senckenberg* **153,** 9–20.

Zagwijn, W. H. & Doppert, J. W. C. (1978). Upper Cenozoic of the southern North Sea Basin: Palaeoclimatic and palaeogeographic evolution. *Geologie En Mijnbouw* **57,** 577–588.

Zagwijn, W. H. & Suc, J. P. (1984). Palynostratigraphie du Plio-Pléistocène d'Europe et de mediterranée nord-occidentales: corrélations chronostratigraphiques, histoire de la vegetation et du climat. *Paleobiological Contributions* **14,** 475–483.

Zhegallo, V. I. (1978). The hipparions of Central Asia. *Transactions of the Joint Soviet-Mongolian Paleontological Expedition* **7,** 1–156.

Zuckerkandl, E. (1983). Molecular basis for directional evolution. In J. Chaline, ed., *Medalités Rhythmes et Mécanismes de l'Evolution Biologique,* pp. 337–350. Paris: Colloques Internationaux de Centre national de la Recherche Scientifique.

Taxon Index

Subject Index